# ANISOTROPY AND LOCALIZATION OF PLASTIC DEFORMATION

## Proceedings of PLASTICITY '91: The Third International Symposium on Plasticity and Its Current Applications

Proceedings of the Third International Symposium on Plasticity and its Current Applications (PLASTICITY '91), held at Grenoble, France, 12–16 August 1991.

# ANISOTROPY AND LOCALIZATION OF PLASTIC DEFORMATION

## Proceedings of PLASTICITY '91: The Third International Symposium on Plasticity and Its Current Applications

*Edited by*

JEAN-PAUL BOEHLER

and

AKHTAR S. KHAN

ELSEVIER APPLIED SCIENCE
LONDON and NEW YORK

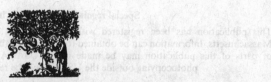

ELSEVIER SCIENCE PUBLISHERS LTD
Crown House, Linton Road, Barking, Essex IG11 8JU, England

*Sole Distributor in the USA and Canada*
ELSEVIER SCIENCE PUBLISHING CO., INC.
655 Avenue of the Americas, New York, NY 10010, USA

WITH 18 TABLES AND 352 ILLUSTRATIONS



**British Library Cataloguing in Publication Data**

Anisotropy and localization of plastic deformation.
I. Boehler, Jean-Paul II. Khan, Akhtar S.
620.112

ISBN 1 85166 688 5

**Library of Congress CIP data applied for**

# PREFACE

Present developments in materials science, mechanics and engineering, as well as the demands of modern technology, result in a new and growing interest in plasticity and in bordering domains of the mechanical behavior of materials. This growing interest is attested to by the success of both *The International Journal of Plasticity*, which after its inception rapidly became the leading journal for plasticity research, and the series of International Symposia on Plasticity and Its Current Applications, which is now the premier international forum for plasticity research dissemination.

The First International Symposium on Plasticity and Its Current Applications was conceived and organized by Professor Akhtar S. Khan, and was held at the University of Oklahoma (Norman, Oklahoma, USA) from July 30 to August 3, 1984. It was attended by over one hundred scientists from fifteen countries. "Plasticity '89: the Second International Symposium on Plasticity and Its Current Applications" was held at Mie University (Tsu, Japan) from July 31 to August 4, 1989; this symposium was co-chaired by Professors Khan and Tokuda. The main emphasis of this meeting was on dynamic plasticity and micromechanics, although it included other aspects of plasticity as well. It was attended by over two hundred researchers from twenty-three nations.

"Plasticity '91: the Third International Symposium on Plasticity and Its Current Applications" will be held in Grenoble (France) from August 12 to 16, 1991. The choice of France as the site for this symposium is to recognize the contribution of the "French School" in plasticity research and to facilitate the participation of the French scientific community in particular and the European community in general in this prominent international meeting. Also, this will provide an opportunity to the international plasticity community to visit beautiful France, taste her cuisine and experience her culture. It is expected to be attended by over two hundred scientists from twenty-six countries around the world. The main emphasis of the symposium will be on Anisotropic Plasticity and Deformation Localization; however, it will also include many other multi-faceted aspects of the Plastic and Viscoplastic Behavior of materials, such as single-crystalline and polycrystalline metals and superalloys, shape memory alloys, ceramics, polymers, composite materials, ice, clays and reinforced soils, sand and granular materials, porous materials, rocks, and jointed rocks.

The symposium will treat different plastic aspects of anisotropy, hardening, localization, damage, failure, fracture, microscopic and macroscopic behavior, relations between microstructure and overall properties in materials science and engineering, as well as applications in mechanical and civil engineering. Besides general sessions, the Symposium will include topical sessions dealing with the following specific topics:

- Thermodynamical Aspects of Constitutive Laws
- Mechanics of Plastic Deformation and Overall Behavior
- Self-organization of Microstructure and Strain Localization
- Simplified Analysis in Elastoplasticity
- Instability Phenomena in Finite Plastic Deformation Processes
- Constitutive Modeling in Dynamic Plasticity
- The Response of Bars and Plates to Impact Loadings
- Elastic-Plastic Materials as Simple Materials
- Kinematical and Constitutive States of Elastic-Plastic Material
- Anisotropic Inelasticity of Polymers and Polymer Composites
- Finite Plasticity

This proceedings volume contains 161 compact versions of papers to be presented at the "Plasticity '91" symposium. Other papers were received too late to be included in this volume. Due to the time limitation and desirability to have this proceedings ready at the symposium, it was not possible to correct minor English mistakes in the paper. Also, the variation of the English language from one country to another may interest the readers.

Such an international symposium cannot be organized without financial support. The chairmen are therefore indebted to the French Centre National de la Recherche Scientifique and the French Ministry of Defence (DRET) for the funds they contributed to the symposium. Other organizations which will bring their financial support in the meantime will be acknowledged in the final program. Finally, we would like to gratefully acknowledge Ms Meena Khan, and Ms Wei-Wei He, as well as Elsevier Science Publishers for their excellent job in publishing this volume. We would like to express our thanks to all of the French colleagues who gave their help and support during the preparatory phase of this meeting. A word of appreciation is owed to Dr Monique Piau, Director of the Institut de Mécanique de Grenoble, and to Dr Gérard Biguenet for their valuable assistance in the local organization.

JEAN-PAUL BOEHLER
*University Joseph Fourier, Grenoble*

AKHTAR S. KHAN
*The University of Oklahoma, Norman*

# CONTENTS

## TEXTURE, DISLOCATION AND MICRO-MECHANICAL ANALYSES

## BEHAVIOR OF GRANULAR POROUS MEDIA

## FINITE PLASTICITY

## DYNAMIC PLASTICITY AND VISCO-PLASTICITY

## CYCLIC PLASTICITY

## POLYMERS, POLYMER COMPOSITES, PHASE TRANSITIONS AND SUPERALLOYS

## THERMODYNAMIC CONSIDERATIONS AND THERMAL EFFECTS

## METAL FORMING, STRUCTURAL ANALYSES AND COMPUTATIONAL ASPECTS

# INTERNATIONAL SCIENTIFIC COMMITTEE

**Chairmen**

| | |
|---|---|
| J.P. Boehler | University Joseph Fourier, France |
| A.S. Khan | University of Oklahoma, USA |

**Members**

| | |
|---|---|
| J. Aboudi | Tel-Aviv University, Israel |
| E.C. Aifantis | Michigan Technological University, USA |
| L. Anand | Massachusetts Institute of Technology, USA |
| S.N. Atluri | Georgia Institute of Technology, USA |
| R.C. Batra | University of Missouri-Rolla, USA |
| J.F. Bell | Johns Hopkins University, USA |
| M. Berveiller | Université de Metz, France |
| J.F. Besseling | Technische Universiteit Delft, The Netherlands |
| O.T. Bruhns | Ruhr-Universität Bochum, Germany |
| J. Casey | University of California-Berkeley, USA |
| J.L. Chaboche | ONERA, Chatillon, France |
| N. Cristescu | University of Bucharest, Romania |
| Y.F. Dafalias | University of California-Davis, USA |
| C. Davini | Universita degli Studi di Udine, Italy |
| P. Dawson | Cornell University, USA |
| D.C. Drucker | University of Florida, USA |
| R.N. Dubey | University of Waterloo, Canada |
| G.J. Dvorak | Rensselaer Polytechnic Institute, USA |
| J. de Fouquet | ENSMA, Poitiers, France |
| D. François | Ecole Centrale des Arts et Manuf., France |
| P. Germain | Académie des Sciences, Paris, France |
| K.S. Havner | North Carolina State University, USA |
| T. Inoue | Kyoto University, Japan |
| J. Jonas | McGill University, Canada |
| M. Kleiber | Institute of Fundamental Tech. Research, Poland |
| J. Kratochvil | Czech. Academy of Sciences, Czechoslovakia |
| E. Krempl | Rensselaer Polytechnic Institute, USA |
| L.P. Kubin | ONERA, Chatillon, France |
| A. Lagarde | Université Poitiers, France |

| | |
|---|---|
| Th. Lehmann | Ruhr Universität Bochum, Germany |
| J. Lemaitre | Université Paris 6, France |
| H. Lippmann | Technische Universität München, Germany |
| G.A. Maugin | Université Pierre et Marie Curie, France |
| D. McDowell | Georgia Institute of Technology, USA |
| A. Molinari | Université de Metz, France |
| Z. Mroz | Institute of Fundamental Tech. Research, Poland |
| I. Müller | Technische Universität Berlin, Germany |
| S. Murakami | Nagoya University, Japan |
| K.W. Neale | Université de Sherbrooke, Canada |
| N. Ohno | Nagoya University, Japan |
| E.T. Onát | Yale University, USA |
| D. Owen | Carnegie Mellon University, USA |
| R. Penelle | Université Paris-Sud, France |
| P. Perzyna | Institute of Fundamental Tech. Research, Poland |
| J. Philibert | Université Paris-Sud, France |
| A. Pineau | Ecole Nat.Sup.des Mines de Paris, Evry, France |
| Y. Rajapakse | Office of Naval Research, USA |
| F. Sidoroff | Ecole Centrale de Lyon, France |
| J. Simo | Stanford University, USA |
| P. Suquet | L.M.A., Marseille, France |
| S. Tanimura | University of Osaka Prefecture, Japan |
| C. Teodosiu | Université Paris-Nord, France |
| M. Tokuda | Mie University, Japan |
| G.J. Weng | Rutgers University, USA |
| A. Zaoui | Ecole Polytechnique, Palaiseau, France |
| J. Zarka | Ecole Polytechnique, Palaiseau, France |
| M. Zyczkowski | Politechnika Krakowska, Poland |

# METAL MATRIX COMPOSITES
# AND LAMINATED MATERIALS

# RECENT DEVELOPMENTS IN THE ANALYSIS OF METAL MATRIX COMPOSITES BY THE METHOD OF CELLS

JACOB ABOUDI
Faculty of Engineering, Tel-Aviv University,
Ramat-Aviv 69978, Israel.

## ABSTRACT

The method of cells is a composite model that is capable to predict the overall behavior of composite materials from the knowledge of the properties of the phases. The method was recently employed to generate initial and subsequent yield surfaces of metal matrix composites, and to establish their instantaneous stiffnesses. The overall instantaneous properties of metal matrix composites are employed for the determination of plastic bifurcation buckling loads of metal matrix laminated plates. Applications are given for elastic-viscoplastic matrices reinforced by perfectly elastic fibers.

## INTRODUCTION

The method of cells is a micromechanical model which proved to be capable to predict the overall behavior of composite materials from the knowledge of the fiber and matrix behavior. The capability of the theory in providing the response of elastic, thermoelastic, viscoelastic and metal matrix composites was demonstrated in a recent review (Aboudi (1989)). More recent developments include the prediction of the initial and subsequent yield surfaces of metal matrix composites (Aboudi (1990)), and the determination of their overall instantaneous properties (Paley and Aboudi (1991a)). The prediction of the latter is very important since they can be utilized for the determination of plastic buckling of metal matrix composite structures (Paley and Aboudi (1991b)). These items are the subject of the present article.

### Initial and Subsequent Yield Surfaces

Yielding of a metal matrix composite which consists of elastic fibers is caused by the yielding of the metallic matrix. A micromechanical prediction of initial yield surfaces of metal matrix composites by the method of cells was presented by Pindera and Aboudi (1988). Suppose that the metal matrix composite is loaded in the plastic region. At each stage of plastic deformation, a new yield surface called subsequent yield surface is established. The method of cells can be employed to predict subsequent yield surfaces generated by various types of combined loading at any stage of loading. Fig. 1 exhibits the initial and subsequent yield surfaces of a boron/aluminum composite in the axial $\bar{\sigma}_{11}$ - transverse $\bar{\sigma}_{22}$ space. A combined effect shifting and distortion is clearly observed. Further examples can be found in Aboudi (1990).

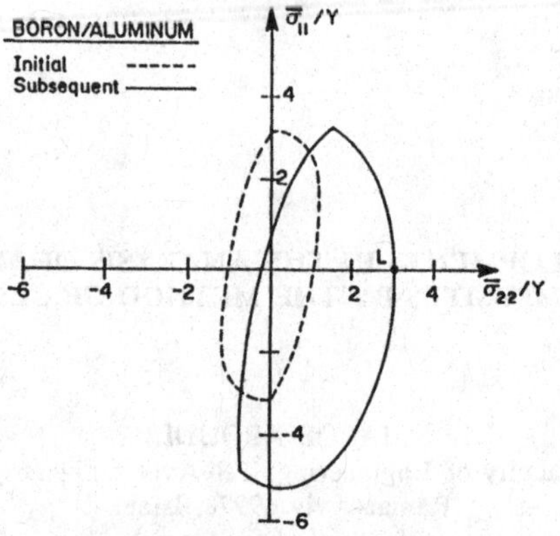


Figure 1. Initial and subsequent (after loading to point L) yield surfaces of a boron/aluminum composite.

**Instantaneous Properties of Metal Matrix Composites**

The consititutive relations of elastic viscoplastic homogeneous materials can be represented in the form

$$\dot{\underset{\sim}{\sigma}} = \underset{\sim}{C}^{VP}\ \dot{\underset{\sim}{\epsilon}}$$

where $\underset{\sim}{\sigma}$ and $\underset{\sim}{\epsilon}$ are the stress and strain tensors, respectively, and dot represents differentiation with respect to time. The fourth order tensor, $\underset{\sim}{C}^{VP}$, represents the instantaneous viscoplastic moduli of the material. Assuming isotropic behavior in the elastic region, it can be shown (Paley and Aboudi (1991a)) that

$$C^{VP}_{ijkl} = \mu(\delta_{ik}\delta_{jl} + \delta_{il}\delta_{jk}) + \lambda\delta_{ij}\delta_{kl} - 2\mu\ \frac{s_{rt}\dot{\epsilon}^{(P)}_{rt}}{s_{pq}\dot{\epsilon}_{pq}} \cdot \frac{s_{ij}s_{kl}}{s_{mn}s_{mn}}$$

where $\lambda$, $\mu$ are the Lamé constants, $\delta_{ij}$ is the Kronecker delta, $s_{ij}$ is the deviatoric stress and $\epsilon^{(P)}_{ij}$ is the plastic strain.

The determination of the instantaneous viscoplastic moduli of metal matrix composites necessitates the use of a micromechanical analysis. This can be achieved by employing the method of cells. This provides the effective composite constitutive law

$$\dot{\bar{\underset{\sim}{\sigma}}} = \underset{\sim}{C}^{*VP}\ \dot{\bar{\underset{\sim}{\epsilon}}}$$

where $\bar{\underset{\sim}{\sigma}}$ and $\bar{\underset{\sim}{\epsilon}}$ are the average stress and strain and $\underset{\sim}{C}^{*VP}$ is the overall fourth order instantaneous tensor. The detailed expressions of the latter can be found in Paley and Aboudi (1991a). In Fig. 2 the overall instantaneous Young's modulus of symmetric angle-ply boron/aluminum laminates, normalized with respect to the effective elastic Young's modulus, is shown. Further examples can be found in Paley and Aboudi (1991a).

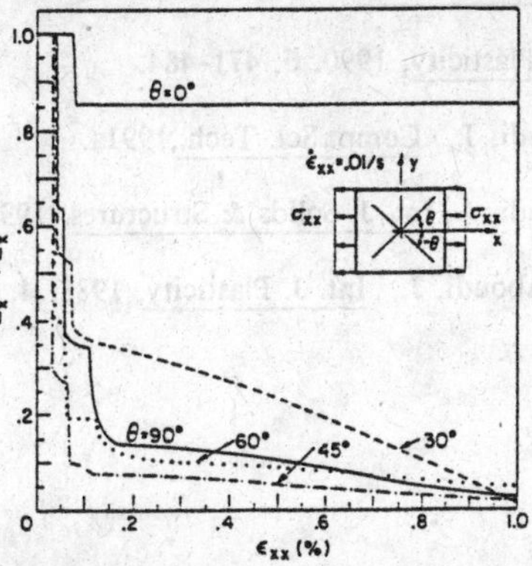


Figure 2. Normalized instantaneous effective Young's modulus in the loading direction of symmetric angle-ply boron/aluminum laminates.

## Plastic Buckling of Metal Matrix Composites

The micromechanically determined instantaneous properties of metal matrix composites can be employed for the determination of their plastic buckling (Paley and Aboudi (1991b)).

Consider for example an anti-symmetric laminated plate and biaxial compressive loading. It can be shown that the buckling condition of the plate is given by

$$N_x \left(\frac{m\pi}{a}\right)^2 + N_y \left(\frac{n\pi}{b}\right)^2 = F(\underset{\sim}{C}^{*VP}) \qquad m,n = 1,2,...$$

where $N_x$, $N_y$ are the loading in the x and y directions, a, b are the dimensions of the plate and F is a specific function. In Fig. 3 the buckling load of an angle-ply boron/aluminum laminate is given for two values of applied strain rate, and compared with the perfectly elastic case. The results are generated by employing a higher order shear deformation theory (HSDT). A detailed discussion and further examples can be found in the latter reference.

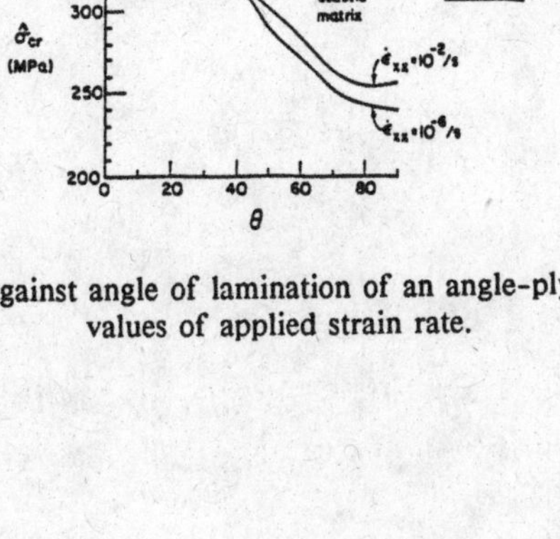

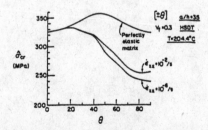

Figure 3. Buckling load against angle of lamination of an angle-ply laminate, for different values of applied strain rate.

## REFERENCES


1. Aboudi, J., Appl. Mech. Rev., 1989, 42, 193-221.

2. Aboudi, J., Int. J. Plasticity, 1990, 6, 471-484.

3. Paley, M. and Aboudi, J., Comp. Sci. Tech.,1991a.

4. Paley, M. and Aboudi, J., Int. J. Solids & Structures, 1991b.

5. Pindera, M.J. and Aboudi, J., Int. J. Plasticity, 1988, 4, 195-214.

# A FINITE STRAIN PLASTICITY THEORY FOR TRANSVERSELY ISOTROPIC MATERIALS

N. ARAVAS
Department of Mechanical Engineering and Applied Mechanics
University of Pennsylvania
Philadelphia, PA 19104, U.S.A

## ABSTRACT

A finite strain theory for anisotropic elastic-plastic materials is presented. The example of a metal-matrix composite reinforced by aligned fibers is analyzed in detail. The plastic spin $\mathbf{W}^p$, which is the average spin of the continuum as seen by an observer spinning with the substructure (fiber), is shown to be $\mathbf{W}^p = \mathbf{nn}\cdot\mathbf{D}^p - \mathbf{D}^p\cdot\mathbf{nn}$, where $\mathbf{D}^p$ is the plastic part of the deformation rate, and $\mathbf{n}$ is the unit vector in the direction of the fiber. The numerical implementation of the developed model is briefly discussed and the example of finite simple shear is presented.

## INTRODUCTION

We consider an elastoplastic material which is characterized by *persistent* transversely isotropic symmetries. A typical example could be a metal-matrix composite reinforced by aligned fibers. The fibers are assumed to follow the deformation of the continuum and to define *locally* the axis of rotational symmetry.

The kinematics of finite elastic plastic deformation is best described by the multiplicative decomposition of the deformation gradient $\mathbf{F} = \mathbf{F}^e \cdot \mathbf{F}^p$, formally introduced in continuum mechanics by LEE [1]. The intermediate unstressed configuration $\mathcal{B}_i$, that is the configuration of the continuum after removal of $\mathbf{F}^e$, is defined in such a way that the orientation of the fibers in $\mathcal{B}_i$ with respect to a global system is the same as the corresponding orientation in the undeformed configuration $\mathcal{B}_0$. The intermediate configuration $\mathcal{B}_i$ defined in such a way is the so-called 'isoclinic configuration' (MANDEL [2]). The velocity gradient $\mathbf{L}$ can be written as

$$\mathbf{L} = \dot{\mathbf{F}} \cdot \mathbf{F}^{-1} = \dot{\mathbf{F}}^e \cdot \mathbf{F}^{e-1} + \mathbf{F}^e \cdot \dot{\mathbf{F}}^p \cdot \mathbf{F}^{p-1} \cdot \mathbf{F}^{e-1} = \mathbf{L}^e + \mathbf{L}^p. \tag{1}$$

The deformation rate $\mathbf{D}$ and the vorticity or spin $\mathbf{W}$, defined as the symmetric and antisymmetric parts of $\mathbf{L}$ respectively, are now written as

$$\mathbf{D} = \mathbf{D}^e + \mathbf{D}^p, \qquad \text{and} \qquad \mathbf{W} = \mathbf{W}^* + \mathbf{W}^p, \tag{2}$$

where $(\mathbf{D}^e, \mathbf{D}^p)$ and $(\mathbf{W}^*, \mathbf{W}^p)$ are the symmetric and antisymmetric parts respectively of $(\mathbf{L}^e, \mathbf{L}^p)$.

## A CONSTITUTIVE EQUATION FOR THE PLASTIC SPIN

We consider the unit vector $\mathbf{N}$ attached to a reinforcing fiber in the direction of axial symmetry in the undeformed configuration $\mathcal{B}_0$. The requirement that $\mathbf{F}^p$ does not rotate $\mathbf{N}$ means that $\mathbf{N}$ is an eigenvector of $\mathbf{F}^p$ and that its rate of change in the isoclinic configuration $\mathcal{B}_i$ vanishes. Using a well-known result of continuum mechanics we can write

$$\dot{\mathbf{N}} = (\mathbf{W}^p_i + \mathbf{D}^p_i \cdot \mathbf{NN} - \mathbf{NN} \cdot \mathbf{D}^p_i) \cdot \mathbf{N} = \mathbf{0}, \tag{3}$$

Where $\mathbf{D}^p_i$ and $\mathbf{W}^p_i$ are the symmetric and antisymmetric parts of $\dot{\mathbf{F}}^p \cdot \mathbf{F}^{p-1}$. The above equation shows that $\mathbf{N}$ is the axial vector of the antisymmetric tensor $\mathbf{W}^p_i + \mathbf{D}^p_i \cdot \mathbf{NN} - \mathbf{NN} \cdot \mathbf{D}^p_i$, and, therefore, we have the representation

$$\mathbf{W}^p_i + \mathbf{D}^p_i \cdot \mathbf{NN} - \mathbf{NN} \cdot \mathbf{D}^p_i = \alpha(\mathbf{N}_2\mathbf{N}_3 - \mathbf{N}_3\mathbf{N}_2), \tag{4}$$

where $\mathbf{N}_2$, $\mathbf{N}_3$, and $\mathbf{N}$ form an orthonormal basis, and $\alpha$ is an arbitrary constant. The right hand side of (4) is an inconsequential spin about $\mathbf{N}$ in the isoclinic configuration which can be set to zero ($\alpha = 0$), so that (4) reduces to

$$\mathbf{W}^p_i = \mathbf{NN} \cdot \mathbf{D}^p_i - \mathbf{D}^p_i \cdot \mathbf{NN}. \tag{5}$$

In the case of small elastic strains, (5) can be written as

$$\mathbf{W}^p = \mathbf{nn} \cdot \mathbf{D}^p - \mathbf{D}^p \cdot \mathbf{nn}, \tag{6}$$

to within elastic strains, where $\mathbf{n} = \mathbf{R}^e \cdot \mathbf{N}$ is the unit vector in the direction of the fibers in the deformed configuration.

The example of HILL's [4] anisotropic yield criterion with associated flow rule is considered next. In the case of small elastic strains, we can define the material symmetries in the deformed configuration. DAFALIAS [3] has shown that Hill's yield criterion can be written, with respect to an arbitrary cartesian coordinate system, as

$$(G+2F)\mathrm{tr}(\sigma'\cdot\sigma') + 2(M-G-2F)\mathrm{tr}(\sigma'\cdot\sigma'\cdot\mathbf{a}) + (5G+F-2M)\mathrm{tr}^2(\sigma'\cdot\mathbf{a}) - 2\sigma_0^2 = 0, \tag{7}$$

where $\sigma$ is the Cauchy stress, a prime denotes the deviatoric part of a tensor, $\mathbf{a} = \mathbf{nn}$, and $F$, $G$, $M$ and $\sigma_0$ are material parameters characteristic of the current state of the material. The above equation can be also used as a plastic potential, in which case

$$\begin{aligned}\mathbf{D}^p = <\dot{\lambda}> \frac{\partial f}{\partial \sigma} &= 2<\dot{\lambda}> \Big[ -\frac{2}{3}(M-G-2F)\,(\sigma':\mathbf{a})\mathbf{I} + (G+2F)\sigma' \\ &\quad + (5G+F-2M)\,(\sigma':\mathbf{a})\mathbf{a}' + (M-G-2F)(\mathbf{a}\cdot\sigma' + \sigma'\cdot\mathbf{a})\Big],\end{aligned} \tag{8}$$

where $\mathbf{I}$ is the second order identity tensor, $\dot{\lambda}$ is a loading parameter, and $<>$ are the Macauley brackets. Using equation (6) we find that the corresponding equation for the plastic spin becomes

$$\mathbf{W}^p = 2<\dot{\lambda}> M(\mathbf{a}\cdot\sigma - \sigma\cdot\mathbf{a}). \tag{9}$$

## FINITE ELEMENT FORMULATION

The model is completed by writting the elastic part of the constitutive equations in terms of a transversely isotropic elastic potential in the isoclinic configuration. The details for the formulation can be found in ARAVAS [5]. The constitutive equations are implemented in a finite element program and the computations are carried out using the tangent stiffness approach described in [6] with 'equilibrium correction' at the end of each increment. The numerical scheme used for the integration of the elastoplastic equations is described in detail in [5].

The example of simple shear is discussed in the following. Referring to Fig. 1, we can readily show that $\tan\theta = \tan\theta_0/(1+\gamma\tan\theta_0)$, where $\theta_0$ is the corresponding angle when $\gamma = 0$. The material is assumed to be rigid-perfectly-plastic obeying Hill's yield criterion with associated flow rule. It should be noted that, because of incompressibility, the stresses can only be determined to within an arbitrary pressure. Using the yield condition together with the flow rule we find the solution to be [5]

$$\frac{\hat{\sigma}'_{11}}{c_{11}} = \frac{\hat{\sigma}'_{22}}{c_{22}} = \frac{\hat{\sigma}_{12}}{c_{12}} = \frac{\sigma_0\sqrt{2}}{\left[F(c_{11}+2c_{22})^2 + G[(c_{11}-c_{22})^2 + (2c_{11}+c_{22})^2] + 2Mc_{12}^2\right]^{1/2}}, \quad (10)$$

where a caret indicates components with respect to the $\hat{x}_1$-$\hat{x}_2$ coordinate system as shown in Fig. 1, $c_{11} = \sin 2\theta/(12G)$, $c_{22} = -c_{11}(F+2G)/(G+2F)$, and $c_{12} = \cos 2\theta/(4M)$. In Fig. 2 the deviatoric stress components with respect to the $x_1$-$x_2$ coordinate system are plotted versus $\gamma$ for $F = 2.5$, $G = 1$, $M = 3$, and $\theta_0 = 90°$. The open symbols in that figure indicate the results of the elastic-plastic finite element calculations. The elastic moduli used in the finite element calculations are three orders of magnitude larger than $\sigma_0$ so that the role of elasticity becomes secondary. The results of the finite element calculations agree well with the exact solution (10).

## REFERENCES


1. Lee, E. H., Elastic-plastic deformations at finite strains. *J. Appl. Mech.*, 1969, **36**, 1–6.

2. Mandel, J., *Plasticité classique et viscoplasticité*, Courses and Lectures, No. 97, International Center for Mechanical Sciences, Udine, Springer-Verlag, 1971.

3. Dafalias, Y. F., Anisotropy, reference configuration and residual stresses. In *Constitutive Laws for Engineering Materials, Theory and Applications*, eds. Desai, C. S., Krempl, E., Kiousis, P. D., and Kundu, T., Vol. 1., Elsevier, 1987, pp. 69–80.

4. Hill, R., A theory of yielding and plastic flow of anisotropic metals. *Proc. Roy. Soc. A*, 1948, **193**, 281–97.

5. Aravas, N., Finite elastoplastic transformations of anisotropic metals. Manuscript in preparation, 1991.

6. McMeeking, R. M. and Rice, J. R., Finite-element formulations for problems of large elastic-plastic deformation, *Int. J. Solids Structures*, 1975, **11**, 601–16.

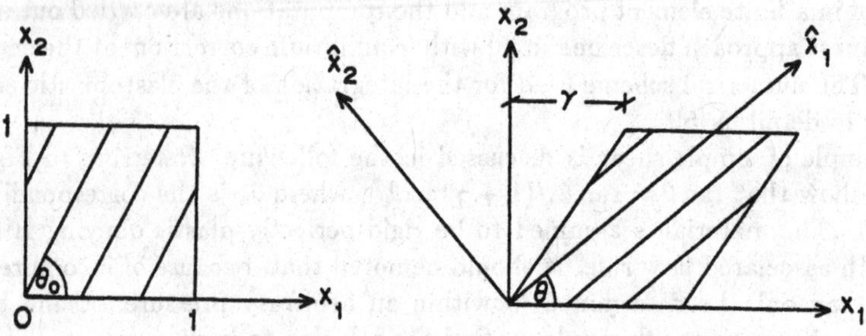


Figure 1: Plane strain simple shear

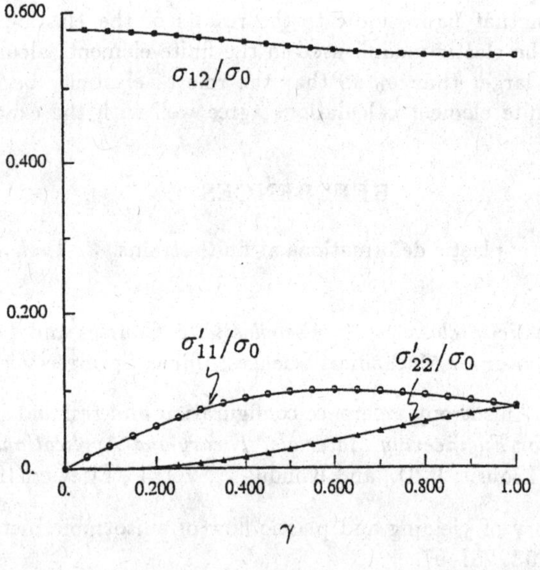


Figure 2: Comparison of finite element and analytical solutions

# THE EFFECT OF LAMINATE THICKNESS ON FIBER/MATRIX DEBONDING IN METAL MATRIX COMPOSITES

JAMES G. BOYD, RODNEY H. JONES, and DAVID H. ALLEN

Center for Mechanics of Composites
Texas A&M University
College Station, TX 77843-3141

## ABSTRACT

In this paper an analysis is performed on laminated metal matrix composites to determine the effects of the edge ply free surface on both effective composite properties and interface stresses between the matrix and fiber when the matrix is elastoplastic and the fiber diameter is almost equal to the ply thickness. The matrix is assumed to be characterized by classical rate independent strain hardening plasticity theory, and the analysis is performed using the finite element method. Results indicate that there is a significant variation in the interfacial stresses for one-, two-, and three-ply laminates. Therefore, it is concluded that composite structures with the characteristics described above may need to be modelled by nonlocal constitutive models in order to predict the evolution of both fiber and matrix cracking.

## INTRODUCTION

Laminated metal matrix composites are typically manufactured from monotapes containing one fiber per ply in the through-thickness direction. Thermomechanical constitutive tests are often performed on four ply boron/aluminum laminates. Under these circumstances, a free-surface boundary layer effect may significantly influence the evolution of damage and plastic deformation. Prior analytic research has suggested that the effect of the free edge in one-, two-, and three-ply unidirectional laminates is not significant [1,2]. However, this research did not account for plasticity in the matrix. In fact, experimental evidence demonstrates that there is significant fiber fracture in four ply laminates prior to failure, whereas none occurs in two ply laminates [3]. Perhaps more significantly, the two ply laminates undergo unstable fracture of the component, whereas the four ply coupons do not [3]. Therefore, there is a significant discrepancy between the elasticity based theory and experimental observations. The current research has thus been undertaken to predict the effect of laminate thickness on the initiation and growth of circumferential debonding cracks in unidirectional laminates subject to isothermal monotonic loading in transverse direction.

The research herein makes the assumption that the inelastic constitution of the matrix is the principal cause of the discrepancy between experimental evidence and previous analyses. Classical rate-independent elasto-plastic constitutive theory is used to model the matrix deformation. Whereas an analytic solution is possible when the fiber and matrix are assumed to both be linear elastic, the same cannot be said when the matrix is elastic-plastic. To date, the most advanced inelastic micromechanics solutions that we are aware of require axisymmetric geometry [4-7]. In the case considered herein, the free edge requires that a fully two-dimensional analysis be performed. Therefore, the micromechanics solution obtained herein is obtained computationally by the finite element method. Results are obtained for both effective properties and fiber-matrix interfacial stresses as functions of number of plies. The details of this analysis are described below.

## ANALYSIS

The computational model utilizes the finite element code NONPLAS developed at Texas A&M University for analysis of elastic-plastic media [8]. The code contains constitutive packages for both elastic and rate independent elastic-plastic media. The plasticity model assumes isotropic material behavior and employs the Prandtl-Reuss equations in conjunction with the Von Mises yield criterion and a combined hardening rule to account for yield surface expansion and translation [8]. Since only monotonic loading is considered in the current paper, hardening is assumed to be isotropic, which is suggested from experimental data on monolithic metals [9]. The computational algorithm uses a load incrementation scheme in conjunction with a Newton-Raphson iterative process on each load increment to account for material nonlinearity [10]. The element used in the code is a constant strain triangle. Computational analysis has been performed form Boron/6061-T6 aluminum on the three geometries shown in Fig. 1, using in-plane extensional loads as shown in the same figure. Results were obtained assuming plane strain conditions for a fiber volume fraction of 50 percent. The elastic moduli and Poisson's ratios were 393 GPa, 67 GPa and 0.20, 0.33 for boron and aluminum, respectively.

## COMPUTATIONAL RESULTS

Converged computational results are shown in Figs. 2 and 3. In Fig. 2 the boundary averaged stresses are plotted against boundary averaged strains for the three laminates subjected to in-plane loading. As can be seen from these results, there is a significant variation in stiffness between the three stacking sequences, thus indicating a substantial nonlocality in effective material properties.

Similarly, normal interface stresses are plotted in Fig. 3. Once again, it is found that the interface stresses are profoundly affected by the free surface. This lends credence to the experimental observations reported in reference 3.

## CONCLUSION

It has been observed experimentally that microstructural damage and instability are dependent on the number of plies in unidirectional metal matrix laminates [3]. Although previous elastic stress analysis [1] has failed to account for this discrepancy, the current research predicts substantial differences in the interfacial stresses between fiber and matrix in one-, two-, and three-ply laminates, thus supporting the contention the inelasticity in the matrix is a significant contributor to the variation in stresses near a free surface. This result in turn suggests that a nonlocal constitutive material model may be warranted for metal matrix laminates, especially when microstructural damage is considered.

## REFERENCES


1. Hulbert, L.E. and Rybicki, E.F., "Boundary Point Least Squares Analysis of the Free Edge Effects in Some Unidirectional Fiber Composites," J. Com. Materials, Vol. 5, p. 164, 1971.

2. Pagano, N.J., "The Role of Effective Moduli in the Elastic Analysis of Composite Laminates," Composite Materials Volume 2 Mechanics of Composite Materials, Sendeckyj, G.P., Ed., Academic Press, New York, pp. 1-22, 1974.

3. Klein, M.J., "Effect of the Filament-Matrix Interface on Off-Axis Tensile Strength," Composite Materials Volume 1 Interfaces in Metal Matrix Composites, Metcalfe, A.G., Ed., Academic Press, New York, pp. 169-210, 1974.

4. Hill, R., "Theory of Mechanical Properties of Fibre-Strengthened Materials: II. Inelastic Behavior," J. Mech. Phys. Solids, Vol. 12, pp. 213-218, 1964.

5. Mulhern, J.F., Rogers, T.G., and Spencer, A.J.M., "Cyclic Extension of an Elastic Fiber with an Elastic-Plastic Coating," J. Inst. Maths Applics, Vol. 3, pp. 21-40, 1967.

6. Ebert, L.J. and Gadd, J.D., "A Mathematical Model for Mechanical Behavior of Interfaces in Composite Materials," Fiber Composite Materials, ASTM, pp. 89-113, 1964.

7. Lee, J.W. and Allen, D.H., "An Analytical Solution for the Elastoplastic Response of a Continuous Fiber Composite under Uniaxial Loading," Research in Structures, Structural Dynamics and Materials 1990, NASA Conference Publication 3064, pp. 55-66, 1990.

8. Allen, D.H., Introduction to the Mechanics of Elastoplastic and Viscoplastic Media, 1991 (to appear).

9. Hunsaker, B., Jr., "An Evaluation of Four Hardening Rules of the Incremental Theory of Plasticity," Texas A&M University Thesis, December, 1973.

10. Groves, S.E., Allen, D.H., and Haisler, W.E., "An Efficient and Accurate Alternative to Subincrementation," Computers & Structures, Vol. 20, No. 6, pp. 1021-1031, 1985.

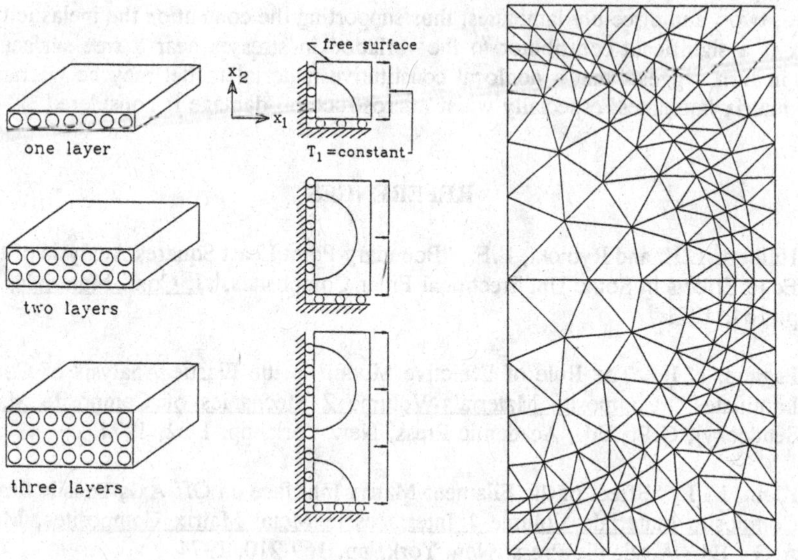


**Fig. 1. Geometry, Boundary Conditions, and a Representative Mesh**

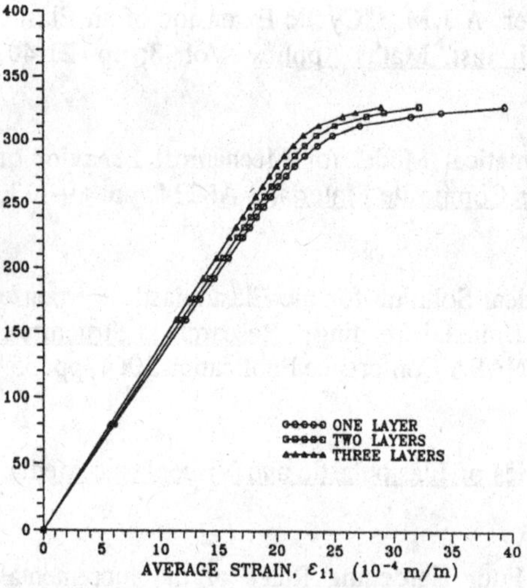


**Fig. 2. Composite Stress-Strain Curves for One-, Two- and Three-Ply Laminates**

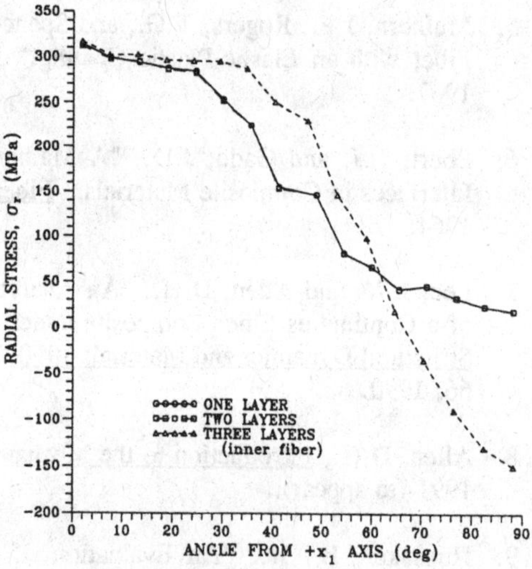


**Fig. 3. Normal Interface Stresses as a Function of Angle for Laminates Subjected to In-Plane Loading**

# EFFECTIVE ANISOTROPIC PROPERTIES OF CREEPING COMPOSITES

PEDRO PONTE CASTAÑEDA
Mechanical Engineering and Applied Mechanics
University of Pennsylvania
Philadelphia, PA 19104, U.S.A.

## ABSTRACT

A recently developed variational principle for estimating the effective properties of nonlinear composites in terms of the corresponding properties of linear composites with the same microstructural distributions of phases is applied to two model anisotropic composite materials. The model materials considered are laminated materials and fibre-reinforced materials, and they correspond in the dilute limit of the inclusion phase to materials reinforced by aligned platelets and fibers, respectively. For simplicity, the power exponent of one of the phases will be taken to be unity corresponding to linear behavior, while the other phase will assumed to satisfy a pure-power creeping law. Both phases will be assumed to be isotropic and incompressible.

## INTRODUCTION

A new procedure for estimating the effective properties of composite materials with phases exhibiting nonlinear constitutive behavior has been proposed recently by Ponte Castañeda [6]. The straightforward and versatile procedure expresses the effective properties of the nonlinear composite in terms of the effective properties of a family of linear composites with the same distribution of phases as the nonlinear composite. Thus, bounds and estimates for the effective properties of linear composites can be translated directly into bounds and estimates for the corresponding nonlinear composite. Appropriate references for the linear theory of composites are provided by the works of Christensen [1]. The new procedure was applied in the above reference [6] to composite materials containing a nonlinear isotropic matrix either weakened or reinforced by isotropic distributions of voids or rigid particles, respectively. The case of a ductile matrix reinforced by incompressible elastic particles was studied in [7]. The results of these studies were given in the form of estimates and rigorous bounds for the effective properties of such materials. The Hashin-Shtrikman bounds obtained via the new method directly from the Hashin-Shtrikman [3] bounds for the linear comparison material were found to be an improvement over the corresponding bounds obtained in [8] for the same class of nonlinear materials using the nonlinear extension of the Hashin-Shtrikman variational principle [3] proposed by Talbot and Willis [9]. Recently, however, Willis [11] has shown that the bounds obtained via the new method can also be obtained by the Talbot-Willis method with a better choice of their comparison material. More generally, the new procedure can make use of other bounds and estimates for the linear comparison material to yield corresponding bounds and estimates for nonlinear materials. In fact, the new procedure can be shown [2] to yield exact results for nonlinear composites with special microstructures.

In this paper, we apply the new procedure to composite materials containing a creeping phase of the power-law type and a linearly creeping phase. Although the phases will be assumed to be isotropic, the composite itself will be assumed to be *anisotropic* by identifying preferred orientations in the microstructural distribution of the phases. Two examples will be considered, both of which have transversely isotropic overall symmetries: laminated materials and fibre-reinforced materials. The first class of materials can be given the interpretation of materials reinforced by aligned platelets for dilute concentrations of the linear phase; whereas, the second class of materials can be given the interpretation of materials reinforced by short fibers for dilute concentrations of the linear phase. For simplicity, we assume that the phases are incompressible and perfectly bonded to each other. In the next three sections, we will give, respectively, a brief definition of effective properties, an introduction to the new variational principle of [6], and application of the principle to the two classes of materials discussed above. More detailed studies including specific results for the nonlinear laminates and fiber-reinforced materials can be found in [2] and [5], respectively.

## EFFECTIVE PROPERTIES

Consider a two-phase composite occupying a region of unit volume $\Omega$, such that the local stress potential $U(\boldsymbol{\sigma},\mathbf{x})$ is expressed in terms of the homogeneous phase potentials $U^{(r)}(\boldsymbol{\sigma})$ via

$$U(\boldsymbol{\sigma},\mathbf{x})=\sum_{r=1}^{2}\chi^{(r)}(\mathbf{x})U^{(r)}(\boldsymbol{\sigma}), \tag{1}$$

where $\chi^{(r)}$ is the characteristic function of phase $r$. The phases are assumed to be incompressible and isotropic, so that the potentials $U^{(r)}(\boldsymbol{\sigma})$ can be assumed to depend only on the effective stress $\sigma_e=\sqrt{\tfrac{3}{2}\mathbf{S}\cdot\mathbf{S}}$, where $\mathbf{S}$ is the deviator of $\boldsymbol{\sigma}$. Thus, we write $U^{(r)}(\boldsymbol{\sigma})=f^{(r)}(\sigma_e)$, where the $f^{(r)}$ are scalar-valued functions. Then, the local constitutive relation for the creeping material is given by

$$\mathbf{D}=\frac{\partial U}{\partial\boldsymbol{\sigma}}(\boldsymbol{\sigma},\mathbf{x}), \tag{2}$$

where $\mathbf{D}$ is the rate-of-deformation (strain-rate) tensor.

To define the effective properties of the heterogeneous material we introduce, following Hill [4], a uniform constraint boundary condition

$$\boldsymbol{\sigma}\mathbf{n}=\overline{\boldsymbol{\sigma}}\mathbf{n},\qquad \mathbf{x}\in\partial\Omega, \tag{3}$$

where $\partial\Omega$ denotes the boundary of the composite, $\mathbf{n}$ is its unit outward normal, and $\overline{\boldsymbol{\sigma}}$ is a given constant symmetric tensor. It follows that the average stress is precisely $\overline{\boldsymbol{\sigma}}$, and we *define* the average strain-rate $\overline{\mathbf{D}}$ in a similar manner.

Then,the effective behavior of the composite, or the relation between the average stress and the average strain-rate follows from the principle of minimum complementary energy, which can be stated in the form

$$\tilde{U}(\overline{\boldsymbol{\sigma}})=\min_{\boldsymbol{\sigma}\in S(\overline{\boldsymbol{\sigma}})}\int_{\Omega}U(\boldsymbol{\sigma},\mathbf{x})\,dV, \tag{4}$$

where $S(\overline{\boldsymbol{\sigma}})=\{\boldsymbol{\sigma}\,|\,\sigma_{ij,j}=0 \text{ in } \Omega, \text{ and } \sigma_{ij}n_j=\overline{\sigma}_{ij}n_j \text{ on } \partial\Omega\}$ is the set of statically admissible stresses, and where we have assumed convexity of the nonlinear potential $U(\boldsymbol{\sigma},\mathbf{x})$. Thus, assuming that $\tilde{U}(\overline{\boldsymbol{\sigma}})$ is differentiable, we have that

$$\overline{\mathbf{D}}=\frac{\partial\tilde{U}}{\partial\overline{\boldsymbol{\sigma}}}(\overline{\boldsymbol{\sigma}},\mathbf{x}). \tag{5}$$

The task will be to determine bounds and estimates for $\tilde{U}(\overline{\boldsymbol{\sigma}})$, which is known to be convex.

## NONLINEAR VARIATIONAL PRINCIPLES

A new variational principle for determining bounds and estimates for the effective properties of nonlinear composites in terms of the effective properties of linear composites was proposed by Ponte Castañeda [6]. In this section, we specialize this result for the case where both phases are incompressible, and phase 2 is linear so that

$$U^{(2)}(\boldsymbol{\sigma})=\frac{1}{6\mu^{(2)}}\sigma_e^2.$$

The new variational principle is based on a representation of the potential of the nonlinear material in terms of the potentials of a family of linear *comparison* materials. Thus, for a homogeneous nonlinear material with "stronger than quadratic" growth in its potential, $U(\boldsymbol{\sigma})$, and certain additional convexity hypothesis, we have that

$$U(\boldsymbol{\sigma})=\max_{\mu>0}\{U_o(\boldsymbol{\sigma})-V(\mu)\}\ , \tag{6}$$

where

$$V(\mu)=\max_{\boldsymbol{\sigma}}\{U_o(\boldsymbol{\sigma})-U(\boldsymbol{\sigma})\} \tag{7}$$

and where $U_o(\boldsymbol{\sigma})$ is the potential of a linear comparison material with shear modulus $\mu$.

The new variational principle is obtained by making use of relation (6) applied to the nonlinear phase 1 in the complementary energy principle (4) to obtain the following relation for the nonlinear composite

$$\tilde{U}(\overline{\boldsymbol{\sigma}})=\max_{\mu^{(1)}(\mathbf{x})}\left\{\tilde{U}_o(\overline{\boldsymbol{\sigma}})-\int_{\Omega^{(1)}}V^{(1)}(\mu^{(1)})dV\right\}, \tag{8}$$

where

$$\tilde{U}_o(\overline{\boldsymbol{\sigma}})=\min_{\boldsymbol{\sigma}\in S(\overline{\boldsymbol{\sigma}})}\int_{\Omega}U_o(\boldsymbol{\sigma},\mathbf{x})dV, \tag{9}$$

is the effective potential of a linear comparison material with local potential $U_o(\boldsymbol{\sigma},\mathbf{x})$ and shear moduli $\mu^{(1)}$ and $\mu^{(2)}$ in phases 1 and 2, respectively. Further, we note that, in general, the comparison moduli $\mu^{(1)}$ are functions of $\mathbf{x}$.

The variational principle described by (8) roughly corresponds to solving a linear problem for a heterogeneous material with arbitrary moduli variation within the nonlinear phase, and then optimizing with respect to the variations in moduli within the nonlinear phase. Thus, the nonlinear material can be thought of as a "linear" material with variable moduli that are determined by prescription (8) in such a way that its properties agree at each $\mathbf{x}$ with those of the nonlinear material. This suggests that if the fields happen to be constant over the nonlinear phase, then the variable moduli $\mu^{(1)}(\mathbf{x})$ can be replaced by constant moduli $\mu^{(1)}$. More generally, however, we have the following lower bound for $\tilde{U}(\overline{\boldsymbol{\sigma}})$, namely,

$$\tilde{U}_-(\overline{\boldsymbol{\sigma}})=\max_{\mu^{(1)}>0}\{\tilde{U}_o(\overline{\boldsymbol{\sigma}})-c^{(1)}V^{(1)}(\mu^{(1)})\}, \tag{10}$$

where $c^{(1)}$ is the volume fraction of phase 1.

## APPLICATION TO LAMINATES

In this section, we summarize from referencc [2] the results of applying the new variational principle to a laminated material. For this geometry, it is well known that the stress and strain fields take on different constant values within each phase. This suggests that taking $\mu^{(1)}$ to be constant within the nonlinear phase will lead to an exact result as given by (8) or (10) (in this case $\tilde{U}$ and $\tilde{U}_-$ are identical). The key ingredient in this development is the exact result for the effective potential of the linear laminate. This result can be expressed in the form

$$\tilde{U}(\overline{\sigma}) = \frac{1}{6\hat{\mu}}\tau_n^2 + \frac{1}{6\overline{\mu}}\left[\tau_p^2 + (\sigma_n - \sigma_p)^2\right], \tag{11}$$

where $\tau_n$, $\tau_p$, $\sigma_n$ and $\sigma_p$ are the transversely isotropic invariants of the stress tensor corresponding respectively to the out-of-plane tensile, in-plane hydrostatic, out-of-plane shear and in-plane shear stresses. Also, $\overline{\mu}$ and $\hat{\mu}$ stand, respectively, for the Voigt and Reuss estimates of the shear modulus.

Application of this result into the new variational principle can then be shown to lead to the following simple expression for the effective energy of the laminate

$$\tilde{U}(\overline{\sigma}) = \min_{\omega}\left\{c^{(1)} f^{(1)}(s^{(1)}) + c^{(2)} f^{(2)}(s^{(2)})\right\}, \tag{12}$$

where $s^{(1)}$ and $s^{(2)}$ are functions of $\omega$ given by the relations $s^{(1)} = \sqrt{(1+c^{(2)}\omega)^2(\sigma_e^2 - \tau_n^2) + \tau_n^2}$ and $s^{(2)} = \sqrt{(1-c^{(1)}\omega)^2(\sigma_e^2 - \tau_n^2) + \tau_n^2}$.

Analogous forms can be derived for the fibre-reinforced material, except that for this geometry, the effective energy cannot be given explicitly by (8), but only characterized in the form of bounds by means of (10). Detailed results of these calculations are given in [5], where they are also compared to the results of Talbot and Willis [10] for the same microstructure using the Talbot-Willis variational principle. Finally, both types of results can be averaged over all possible orientations to yield results depicting approximately the effect of inclusion shape on the effective properties of isotropic composites.

## ACKNOWLEDGEMENTS

This work was supported by the Air Force Office of Scientific Research under grant 91-0161.

## REFERENCES

[1] CHRISTENSEN, R.M. (1979) *Mechanics of Composite Materials*, Wiley, New York, 137

[2] DE BOTTON, G. and PONTE CASTAÑEDA, P. (1991) To appear.

[3] HASHIN, Z. and SHTRIKMAN, S. (1962) *J. Mech. Phys. Solids* **10**, 335.

[4] HILL, R. (1963) *J. Mech. Phys. Solids* **11**, 357.

[5] PONTE CASTAÑEDA, P. (1991) To appear.

[6] PONTE CASTAÑEDA, P. (1991) *J. Mech. Phys. Solids* **39**, 45.

[7] PONTE CASTAÑEDA, P. (1991) In *Inelastic Deformation of Composite Materials* (ed. G.J. DVORAK), Springer-Verlag, New York, 216.

[8] PONTE CASTAÑEDA, P. and WILLIS, J. R. (1988) *Proc. R. Soc. Lond. A* **416**, 217.

[9] TALBOT, D. R. S. and WILLIS, J. R. (1985) *IMA J. Appl. Math.* **35**, 39.

[10] TALBOT, D. R. S. and WILLIS, J. R. (1991) In *Inelastic Deformation of Composite Materials* (ed. G.J. DVORAK), Springer-Verlag, New York, 527.

[11] WILLIS, J. R. (1991) *J. Mech. Phys. Solids* **39**, 73.

# ON THERMAL HARDENING AND UNIFORM FIELDS IN TWO–PHASE COMPOSITE MATERIALS

GEORGE J. DVORAK
Institute Center for Composite Materials and Structures
Rensselaer Polytechnic Institute
Troy, New York 12180, U.S.A.

## ABSTRACT

The effect of uniform thermal changes on the position of yield surfaces in the overall stress space is described for two–phase fibrous and particulate composite materials with an elastic–plastic matrix. An auxiliary uniform strain field is used to evaluate this effect in an exact manner. The thermal change is shown to cause a rigid–body translation of all local branches of the yield surface in the overall stress space, in the direction of the overall stress associated with the uniform strain field. In a similar way, the contribution of a thermal change to the overall strain is evaluated from the solution of a mechanical loading problem, in terms of an equivalent overall stress increment, and a uniform isotropic strain change. Such connections between mechanical and thermal loads in inelastic analysis of composites have been described in [1–4]; their applications in processing of composite laminates appear in [5].

## INTRODUCTION

Hardening of homogeneous elastic–plastic solids typically describes the change in shape and position of the yield or relaxation surface, due to mechanical loading along certain stress or strain path. Uniform changes in temperature may influence the magnitude of the yield stress or strain, but under homogeneous boundary conditions, they do not produce stress or strain increments that would change the shape and position of the respective surfaces. In contrast, heterogeneous materials subjected to such thermal changes may develop internal stress and strain fields which have an effect on the overall surfaces, and also on the magnitude of overall plastic strain or relaxation stress. Since the microstructural geometry is usually quite complex, the local thermal fields and their role in overall hardening are not easily evaluated. However, in certain two–phase composites, problems of this kind can be easily solved by a procedure that relies on the existence of uniform strain fields. In principle, the thermal field is superimposed with an auxiliary overall stress that makes the total local strain field exactly uniform, and isotropic in the matrix. In a plastically incompressible matrix, this step causes no plastic strain. The auxiliary overall stress must then be removed, or subtracted from a prescribed mechanical load increment. This implies that the effect of thermal change can be accounted for by superposition of the auxiliary and actual stress increments.

## UNIFORM STRAIN FIELDS IN TWO–PHASE MEDIA

Consider a representative volume of a statistically homogeneous, two–phase composite material with perfectly bonded phases $r = \alpha, \beta$, of arbitrary geometry and material symmetry. Any applied surface tractions are in equilibrium with a uniform overall stress $\sigma$, and the corresponding surface displacements are compatible with a uniform overall strain $\epsilon$, such that $\epsilon = \mathbf{M}\,\sigma$, and $\sigma = \mathbf{L}\,\epsilon$, where $\mathbf{L}$ and $\mathbf{M}$ are the overall elastic stiffness and compliance tensors. In addition, a uniform thermal change $\theta$ from some reference temperature of a stress–free state may be specified. Under such circumstances, the local strain and stress fields become

$$\epsilon_r(\mathbf{x}) = \mathbf{M}_r\,\sigma_r(\mathbf{x}) + \mathbf{m}_r\,\theta \qquad \sigma_r(\mathbf{x}) = \mathbf{L}_r\,\epsilon_r(\mathbf{x}) + \boldsymbol{\ell}_r\,\theta\,, \quad r = \alpha, \beta, \tag{1}$$

where the $\mathbf{m}_r$ and $\mathbf{l}_r$ are the phase thermal strain and stress tensors.

Note that the thermal contributions are uniform in each phase. If each of the above equations is written for $r = \alpha, \beta$, then each such pair may be solved for the single stress or strain [4]

$$\hat{\sigma} = (\mathbf{M}_\alpha - \mathbf{M}_\beta)^{-1}(\mathbf{m}_\beta - \mathbf{m}_\alpha)\,\theta \qquad \hat{\epsilon} = (\mathbf{L}_\alpha - \mathbf{L}_\beta)^{-1}(\boldsymbol{\ell}_\beta - \boldsymbol{\ell}_\alpha)\,\theta\,, \tag{2}$$

which are both uniform in the entire representative volume. It follows that the uniform stress and strain fields (2) can be created in the composite if the respective fields are applied as overall stress and strain, together with the uniform thermal change $\theta$. Some algebra shows that the two fields correspond to a single deformation state, and that they are both isotropic if both phases are elastically isotropic.

In composites reinforced by aligned continuous fibers, uniform strain fields may be created not only as above, but also by purely mechanical loading [2]. For example, if both phases are transversely isotropic about the $x_1$ axis paralel to the fibers, then the axisymmetric overall stress state $\sigma_1 = S_A$, $\sigma_2 = \sigma_3 = S_T$ causes the stresses

$$\hat{\sigma}_1^\alpha = q(\ell_\alpha\Delta\ell - n_\alpha\Delta k)\,S_T + qk_\alpha E_\alpha(\ell_\beta\Delta\alpha - 2k_\beta\Delta\beta)\,\theta$$

$$\hat{\sigma}_1^\beta = q(\ell_\beta\Delta\ell - n_\beta\Delta k)\,S_T + qk_\beta E_\beta(\ell_\alpha\Delta\alpha - 2k_\alpha\Delta\beta)\,\theta \tag{3}$$

$$\hat{\sigma}_2^\alpha = \hat{\sigma}_2^\beta = \hat{\sigma}_3^\alpha = \hat{\sigma}_3^\beta = S_T\,, \qquad \hat{\sigma} = [S_A, S_T, S_T, 0, 0, 0]^T;$$

$k_r$, $l_r$, $n_r$ are Hill's moduli, $q^{-1} = (\ell_\alpha k_\beta - k_\alpha \ell_\beta) \neq 0$, $\mathbf{m}_r = [\alpha_r, \alpha_r, \beta_r, 0, 0, 0]^T$, and $\Delta k = k_\alpha - k_\beta$.

If one selects $S_T$ as

$$S_T = \{qk_\beta E_\beta(\ell_\alpha\Delta\alpha + 2k_\alpha\Delta\beta)\,[1 - q(\ell_\beta\Delta\ell - n_\beta\Delta k)]^{-1}\}\,\theta\,, \tag{4}$$

then it is possible to show that the local fields in the matrix phase $r = \beta$ become uniform and isotropic. Hence we conclude that the application of a uniform thermal change $\theta$, together with the auxiliary overall stresses implied by (2) or (3), cause only uniform or piecewise uniform fields in the composite. In certain cases, the matrix fields are also isotropic.

## THERMAL HARDENING AND PLASTIC FLOW

We now limit our attention to those systems which admit isotropic stress fields in the matrix, i. e., to composites with isotropic phases of any geometry, and to fibrous materials where one or both phases may be transversely isotropic or isotropic. The matrix is assumed to be elastic–plastic and plastically incompressible. These limitations are not particularly serious, as the group would include most actual metal matrix composite materials. Suppose that a micromechanical model is available for incremental inelastic analysis of the representative volume of the composite under purely mechanical loading. Such model may discretize each phase in the representative volume into many local volumes or subelemets k = 1, 2, ... N. Our objective is to extend the model to applications which involve combined thermomechanical loading.

Each local volume in the matrix has a certain yield function in the local stress space, to which there corresponds a branch of the overall yield surface in the overall stress space. The same is true for the respective relaxation surfaces in the local and overall strain space. In the elastic range, the average local stress or strain in each local volume k is related to the overall strain or stress by the mechanical concentration factors. Hence one can find the following stress increments in the local yield surfaces due to the indicated stress increments within the overall surface,

$$(\mathrm{d}\sigma_k - \mathrm{d}\alpha_k) = \mathrm{B}_k(\mathrm{d}\sigma - \mathrm{d}\overline{\alpha}_k) \qquad (\mathrm{d}\epsilon_k - \mathrm{d}\beta_k) = \mathrm{A}_k(\mathrm{d}\epsilon - \mathrm{d}\overline{\beta}_k), \tag{5}$$

where the translations of the respective centers are denoted by $\mathrm{d}\alpha_k$, $\mathrm{d}\overline{\alpha}_k$, $\mathrm{d}\beta_k$, and $\mathrm{d}\overline{\beta}_k$.

Examine now the effect of a pure thermal change $\mathrm{d}\theta \neq 0$ at $\mathrm{d}\sigma = 0$, or $\mathrm{d}\epsilon = 0$, on the position of the centers of the two surfaces in overall stress or strain space. To that end, apply the $\mathrm{d}\theta$ together with an auxiliary overall stress or strain that creates uniform, isotropic increments in all local volumes of the matrix. By definition, only the deviatoric parts affect the local surface, hence both the local stress and strain terms and the corresponding local translation vectors in (5), are taken as equal to zero. This gives the translation of all matrix k–branches of the overall yield surface as $\mathrm{d}\overline{\alpha}_k = \mathrm{d}\hat{\sigma}$, $\mathrm{d}\overline{\beta}_k = \mathrm{d}\hat{\epsilon}$, where the $\mathrm{d}\hat{\sigma}$ and $\mathrm{d}\hat{\epsilon}$ follow from (2) or (3). The figures illustrate the various increments involved, for k = m, a single matrix volume.

A similar procedure is followed in evaluation of the overall strains or stresses under incremental thermomechanical loading. If a $\mathrm{d}\theta \neq 0$ is applied together with a mechanical increment, the thermal load is accounted for in an elastic loading step with the auxiliary mechanical load, which is then removed in the plastic loading step by subtraction from the actual mechanical load increments. The total overall strain or stress increments follow in terms of the overall instantaneous stiffness or compliance tensors $\mathscr{L}$ and $\mathscr{M}$, and the local fields from the instantaneous mechanical influence functions $\mathscr{A}_r(\mathbf{x})$, $\mathscr{B}_r(\mathbf{x})$:

$$\mathrm{d}\sigma_r(\mathbf{x}) = \mathrm{d}\hat{\sigma}_r + \mathscr{B}_r(\mathbf{x})(\mathrm{d}\sigma - \mathrm{d}\hat{\sigma}) \qquad \mathrm{d}\epsilon_r(\mathbf{x}) = \mathrm{d}\hat{\epsilon}_r + \mathscr{A}_r(\mathbf{x})(\mathrm{d}\epsilon - \mathrm{d}\hat{\epsilon}), \qquad (r = \alpha, \beta)$$

$$\mathrm{d}\epsilon = \mathrm{d}\hat{\epsilon} + \mathscr{M}(\mathrm{d}\sigma - \mathrm{d}\hat{\sigma}) \qquad \mathrm{d}\sigma = \mathrm{d}\hat{\sigma} + \mathscr{L}(\mathrm{d}\epsilon - \mathrm{d}\hat{\epsilon}). \tag{6}$$

Similar relations can be derived for multiphase media, albeit in less simple form.

## REFERENCES


1. Dvorak, G.J., "Thermal Expansion of Elastic–Plastic Composite Materials," Journal of Applied Mechanics, 1986, **53**, pp. 737–743.

2. Dvorak, G.J., "On Uniform Fields in Heterogeneous Media," Proc. R. Soc. Lond. A, 1990, **431**, 89–110.

3. Dvorak, G.J., "Plasticity Theories for Fibrous Composite Materials," Metal Matrix Composites: Mechanisms and Properties, eds. R. K. Everett and R. J. Arsenault, Academic Press, Boston, 1991, 1–77.

4. Benveniste, Y. and Dvorak, G.J., "On a Correspondence Between Mechanical and Thermal Effects in Two–Phase Composites," Micromechanics and Inhomogeneity, eds. G.J. Weng, M. Tāya and H. Abe, Springer–Verlag, New York, 1990, 65–81.

5. Bahei–El–Din, Y. A., "Yielding and Thermal Hardening in Fibrous Composite Laminates," to be published.


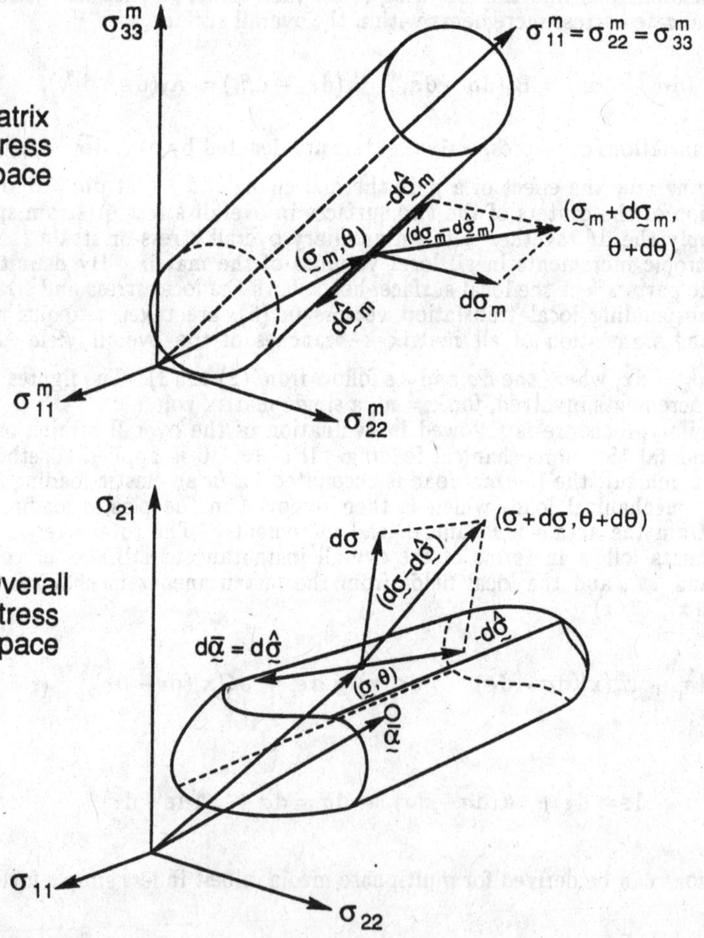

# FRACTURE ANALYSIS OF MICROLAMINATE COMPOSITES WITH PLASTICALLY DEFORMING MATRICES

G.A. KARDOMATEAS
School of Aerospace Engineering, Georgia Institute of Technology
Atlanta, Georgia 30332-0150

## ABSTRACT

Composite systems consisting of sheet reinforcement and polymeric or metallic matrix which can undergo plastic deformation present unique and unexplored issues with respect to the mechanics of their deformation, especially in the presence of cracks. For example, microlaminates consisting of layers of metal sheet reinforced with sheets of nonmetallic composite material are a family of new structural composite material systems.

In this class of composites, which consist of very strong and stiff sheets in a plastic matrix, the elastic and plastic distributions of stress and strain around a crack can be determined by assuming that the transverse displacements around the crack are negligible, by analogy from the corresponding case for longitudinal shear.

An infinite yield strength in the reinforcing sheets then gives a corresponding plastic zone and on a microscopic scale, the plastic flow within the matrix between the last cracked and the first uncracked reinforcing sheet is discussed and used to determine the stress concentration in the uncracked sheet. Other issues associated with the quantitative understanding of crack growth are discussed.

## INTRODUCTION

Microlaminates consisting of layers of metal sheet reinforced with sheets of nonmetallic composite material are a family of new structural composite material systems. An example is the aramid aluminum laminates which consist of layers of thin aluminum alloy sheet bonded by adhesive impregnated with high strength unidirectional Aramid fibers.

The issues addressed in the present paper pertain to any material system with sheet reinforcement and plastically deforming matrix (metal or polymetric) and are therefore not restricted to the particular aramid/aluminum system. In general, the toughness of high strength materials can be enhanced in at least three ways. First, introducing free surfaces normal to the crack direction allows geometrical readjustments and transfer of the load to neighboring elements. The free surfaces may be manufactured, as in a fibrous structure, or may be developed during fracture, as with secondary cracks that blunt the main crack. Second, the toughness is enhanced by a certain amount of viscosity, so that cracks become blunt under low loads and can then withstand greater occasional overloads than if they had remained sharp. A third method of toughening is by using a matrix that deforms plastically, so that its yield strength will limit the stress applied to the high strength elements. Such plastic yielding not only occurs in metallic crystals but is also an approximation to the non-linear viscoelastic flow of polymers subjected to elastic strains of more than a few percent.

Consider a composite with elastic reinforcing elements in sheet form and a plastic matrix. Since the reinforcing elements are in sheet form, plane strain may be assumed.

In this work, we seek the microscopic distribution of stress and strain at the tip of a crack normal to the reinforcing sheet.

## ANALYSIS

**The Classical Anisotropic Elasticity of a Sheet-Reinforced Composite.** The configuration under study, shown in Fig. 1, consists of relatively stiff and strong reinforcing sheets of thickness $t_r$, bonded in a relatively compliant, plastically deforming matrix of thickness $t_m$. The question of concern is the stress distribution in the unbroken platelet next to the tip of a crack. Intuitively one expects the reinforcing sheets to slide over each other with the plastically deforming material acting as a kind of solid lubricant. If the sheets are stiff enough, almost all the deformation will be in the vertical direction. If so, the analysis is much simplified by the analogy with longitudinal shear (Mode III) [1]. We therefore first consider the anisotropic elasticity of the material to see the extent to which the displacement can be assumed unidirectional.

In regions of moderate stress and strain gradients, each layer is subjected to the same strain $\epsilon_{22}$ in the $x_2$ direction and the same stress $\sigma_{11}$ in the $x_1$ direction, as well as being under plane strain so that $\epsilon_{33} = 0$. When these conditions are imposed on the three-dimensional stress-strain relations for each layer, regarded as isotropic with a common Poisson's ratio $\nu$ and moduli of elasticity $E_m$ and $E_r$ for the matrix and reinforcing layers, respectively, an averaging process gives

$$\epsilon_{22} = -\sigma_{11}\frac{\nu(1+\nu)(t_m+t_r)}{E_m t_m + E_r t_r} + \sigma_{22}\frac{(1-\nu^2)(t_m+t_r)}{E_m t_m + E_r t_r}\,, \tag{1}$$

$$\epsilon_{11} = \sigma_{11}\left[\left(\frac{t_m}{E_m}+\frac{t_r}{E_r}\right)\frac{(1-2\nu)(1+\nu)}{(t_m+t_r)(1-\nu)} + \frac{\nu(1+\nu)(t_m+t_r)}{(1-\nu)(t_m E_m + t_r E_r)}\right] -$$

$$-\sigma_{22}\frac{\nu(1+\nu)(t_m+t_r)}{E_m t_m + E_r t_r}\,, \quad \gamma_{12} = \sigma_{12}\left(\frac{t_m}{G_m}+\frac{t_r}{G_r}\right)/(t_m+t_r)\,. \tag{2}$$

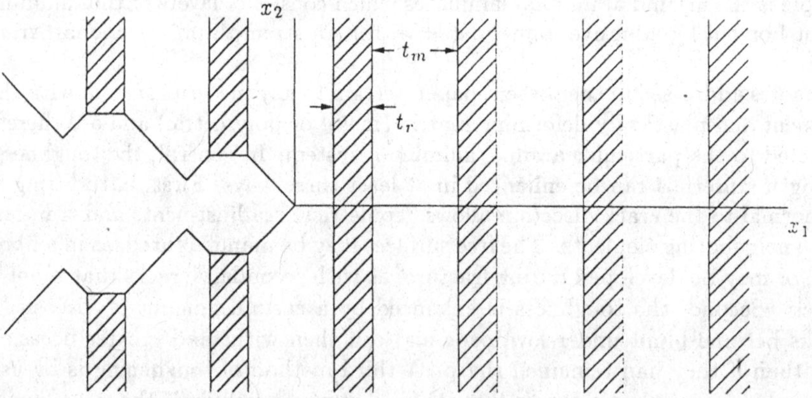


Figure 1. Crack in a laminated composite with ductile matrix.

As an example, consider the following properties: $E_r/E_m = 10$, $\nu_r = \nu_m = 0.3$, $t_m/(t_m + t_r) = 0.7$. In terms of the modulus,of elasticity of the reinforcing sheet, Eqs. (1,2) give: $E_r\epsilon_{11} = (6.93)\sigma_{11} - (1.054)\sigma_{22}$, $E_r\epsilon_{22} = -(1.054)\sigma_{11} + (2.46)\sigma_{22}$, $E_r\gamma_{12} = (18.98)\sigma_{12}$. Note that the transverse strain $\epsilon_{11}$ due to a longitudinal stress $\sigma_{22}$ is indeed small. This means that the effect of transverse strains on the longitudinal stress is also small and suggests that the assumption of an analysis based on displacements in the $x_2$ direction could be appropriate.

**Anisotropic Elastic Stress and Strain Distributions around a Crack Tip.** The stress and strain fields around the tip of a crack in an anisotropic medium under plane strain conditions for cracks at arbitrary angles to the axes of anisotropy have been presented in Refs 2 and 3. In our case the stress-strain relation reduce to

$$\epsilon_{11} = a_{11}\sigma_{11} + a_{12}\sigma_{22}\ , \quad \epsilon_{22} = a_{12}\sigma_{11} + a_{22}\sigma_{22}\ , \quad \gamma_{12} = a_{66}\sigma_{12}\ . \tag{3}$$

The crack tip stress distributions are given in terms of the roots $\mu_i$ of an equation formed from the compliance coefficients of Eq. (3):

$$a_{11}\mu^4 + (2a_{12} + a_{66})\mu^2 + a_{22} = 0\ . \tag{4}$$

Solving for $\mu_i$,

$$\mu_1^2, \mu_2^2 \simeq -\frac{2a_{12} + a_{66}}{a_{11}}\ , \quad -\frac{a_{22}}{2a_{12} + a_{66}}\ . \tag{5}$$

Note that the coefficients of compliance $a_{ij}$ in comparison to their magnitude, indicate that the values of $\mu$ will be imaginary, and that $\mu_1$ is large compared with $\mu_2$. Neglecting $\mu_2$ with respect to $\mu_1$, the equations given in Ref 2 for the stress distribution in terms of a stress intensity factor $k_1$ (defined below) are:

$$\sigma_{11} \simeq \frac{k_1}{\sqrt{2r}}\mathrm{Re}\left[\frac{\sqrt{a_{22}/a_{11}}}{\sqrt{\cos\theta + i\sqrt{a_{66}/a_{11}}\sin\theta}}\right] = \frac{k_1}{\sqrt{2}}\mathrm{Re}\left[\frac{\sqrt{a_{22}/a_{11}}}{\sqrt{x_1 + ix_2\sqrt{a_{66}/a_{11}}}}\right]\ , \tag{6a}$$

$$\sigma_{22} \simeq \frac{k_1}{\sqrt{2}}\mathrm{Re}\left[\frac{1}{\sqrt{x_1 + ix_2\sqrt{a_{22}/a_{66}}}}\right]\ , \tag{6b}$$

$$\sigma_{12} \simeq \frac{k_1}{\sqrt{2}}\mathrm{Re}\left[\frac{i\sqrt{a_{22}/a_{66}}}{\sqrt{x_1 + ix_2\sqrt{a_{66}/a_{11}}}} - \frac{i\sqrt{a_{22}/a_{66}}}{\sqrt{x_1 + ix_2\sqrt{a_{22}/a_{66}}}}\right]\ . \tag{6c}$$

The stress intensity factor $k_1$ differs from one frequently used in that the factor $\pi^{1/2}$ disappears from a number of equations. In particular, for a crack of half length, $c$, in a body subjected to a tensile stress $\sigma_\infty$,

$$k_1 = \sigma_\infty\sqrt{c} = K_I/\sqrt{\pi}\ . \tag{7}$$

The corresponding equations for longitudinal shear (Mode III) in an isotropic material are

$$\sigma_{1'3'} = -\frac{k_{3'}}{\sqrt{2}}\mathrm{Re}\left[\frac{i}{(x_{1'}+ix_{2'})^{1/2}}\right] , \quad \sigma_{2'3'} = \frac{k_{3'}}{\sqrt{2}}\mathrm{Re}\left[\frac{1}{(x_{1'}+ix_{2'})^{1/2}}\right] . \tag{8}$$

To complete the analogy, turn to the equations for displacement. For plane strain with large values of $\mu_1$, the equations given in Ref 2 become

$$u_1 = k_1\sqrt{2}\,\mathrm{Re}\left[a_{12}(x_1+\mu_2 x_2)^{1/2} - \sqrt{a_{11}a_{22}}(x_1+\mu_1 x_2)^{1/2}\right] , \tag{9a}$$

$$u_2 = k_1\sqrt{2}\,\mathrm{Re}\left[-i\sqrt{a_{22}a_{66}}(x_1+\mu_2 x_2)^{1/2}\right] . \tag{9b}$$

For Mode III deformation in anisotropic material

$$u_{3'} = k_3\sqrt{2}\,\mathrm{Re}\left[-i(1/G)(x_{1'}+ix_{2'})^{1/2}\right] . \tag{10}$$

Comparison of Eqs. (9b) and (10) along with (6b,c) and (8) gives the following analogy between variables in isotropic longitudinal shear and anisotropic plane strain tension:

$$u_{3'} \to u_2 \ ; \ G \to 1/\sqrt{a_{22}a_{66}} \ ; \ x_{2'} \to x_2\sqrt{a_{22}/a_{66}} , \tag{11}$$

$$x_{1'} \to x_1 \ ; \ \sigma_{1'3'} \to \sigma_{12}\sqrt{a_{66}/a_{22}} \ ; \ \sigma_{2'3'} \to \sigma_{22} \ ; \ k_{3'} = \sigma_{2'3'}\sqrt{c} \to k_1 = \sigma_{22\infty}\sqrt{c} . \tag{12}$$

<u>**Yielding at the Crack Tip.**</u> For predicting fracture we need the stress in front of the crack. Applying the procedure by Rice [4, 5] to the present problem, we find that well away from the plastic zone, the solution approaches the elastic stress distribution:

$$\sigma_{12} = \frac{-k\sin\theta'/2}{\sqrt{2r'}} , \qquad \sigma_{22} = k\sqrt{\frac{a_{66}}{a_{22}}}\frac{\cos\theta'/2}{\sqrt{2r'}} , \tag{13}$$

where

$$\theta' = \tan^{-1}\left(x_2\sqrt{a_{22}/a_{66}}/x_1\right) , \qquad r' = \frac{\sqrt{x_1^2 + x_2^2 a_{22}/a_{66}}}{(k_1/k)^2(a_{22}/a_{66})} .$$

Close to, but ahead of the crack tip ($|\,\bar{z}'\,| \ll$ and $x_1 > 0$ ), the solution becomes

$$\frac{\sigma_{12}}{k} = -\frac{\theta'}{\pi/2} , \qquad \frac{\sigma_{22}}{k} = \frac{2}{\pi}\sqrt{\frac{a_{66}}{a_{22}}}\ln\frac{\pi^2}{4r'} . \tag{14}$$

<u>**Fully Plastic Flow within a Layer of Matrix.**</u> The stress and strain distribution within the matrix layer at the tip of the crack are needed to predict whether or not the layer will delaminate and thus blunt the crack. The displacements of the reinforcing sheets at the crack tip set boundary conditions on the stress and strain distribution in the matrix layer. Further boundary conditions are set by the transverse tension $\sigma_{11}$, which tends to suck matrix material away from the crack tip, increasing the thickness of the laminar layer. This increased thickness will in turn decrease $\sigma_{11}$. So far, no

such analysis is available. Therefore we neglect this thickening action and inquire what transverse stress develops from two nonunique, fully plastic flow fields.

The simplest flow field consists of pure sliding along one interface, as shown in Fig. 2a. The slip line field shown by the dashed lines demonstrates that plastic flow need not spread out from the corner if the rest of the matrix remains rigid. The stress distribution is indeterminate, except for a yield strength in shear $k$ along the active interface. Any strain-hardening would tend to thicken the zone of plastic flow; so we seek a field with flow more widely distributed through the matrix layer.

A field with plastic flow in the matrix is shown in Fig. 2b. It was derived assuming geometrical similarity as the crack opens. A gradually thickening layer of the matrix is drawn upwards at half the displacement rate of the cracked reinforcing sheet.

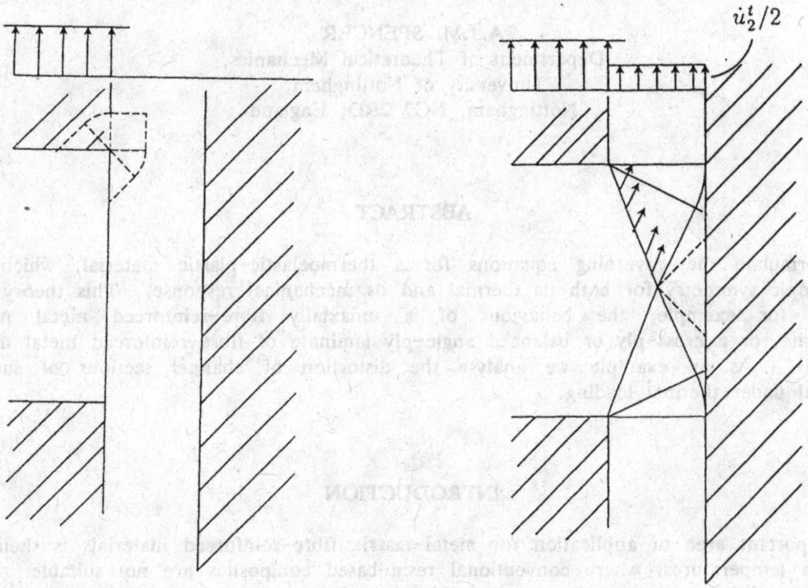


a) Stationary matrix b) Sliding matrix

Figure 2. Plastic flow in matrix.

## REFERENCES


1. McClintock F.A., On the Plasticity of Growth of Fatigue Cracks, in Fracture of Solids, D.C. Drucker and J.J. Gilman eds., Interscience, New York, 1963, pp. 65-102.
2. Sih G.C. and Liebowitz H., Mathematical Theories of Brittle Fracture, in Fracture, vol. 2, H. Liebowitz ed., Academic Press, New York, 1968, pp. 68-190.
3. Paris P.C. and Sih G.C., Stress Analysis of Cracks, ASTM STP 381, Am. Soc. Testing Mat., 1965, pp. 30-81.
4. Rice J.R., Mechanics of Crack Tip Deformation and Extension by Fatigue, ASTM STP 415, Am. Soc. Testing Mat., 1967, pp. 247-309.
5. Rice J.R., On the Theory of Perfectly Plastic Anti-Plane Straining, Report NSF GT 286/2, Div. of Eng., Brown University, Providence, R.I., March 1966.

# THERMOELASTIC-PLASTIC DEFORMATIONS OF ORTHOTROPIC MATERIALS

A.J.M. SPENCER
Department of Theoretical Mechanics,
University of Nottingham,
Nottingham, NG7 2RD, England

## ABSTRACT

We formulate the governing equations for a thermoelastic-plastic material, which has orthotropic symmetry for both its thermal and its mechanical response. This theory may model, for example, the behaviour of a uniaxially fibre-reinforced metal matrix composite, or a cross-ply or balanced angle-ply laminate of fibre-reinforced metal matrix material. As an example we analyse the distortion of channel sections of such a material under thermal loading.

## INTRODUCTION

An important area of application for metal-matrix fibre-reinforced materials is their use at high temperatures, where conventional resin-based composites are not suitable. It is therefore of interest to study the thermal stress behaviour of metal-matrix composites. In the theory presented in this paper we have in mind application to metal-matrix materials reinforced with aligned fibres, constructed so as to have orthotropic symmetry on the macroscopic scale. The case of a transversely isotropic material, with a single preferred direction, may be treated as a special case. Such symmetry may be obtained by uniaxial reinforcement, by cross-ply lamination, or by balanced angle-ply lamination.

Dvorak and Rao [1] have investigated thermal stress in metal-matrix composites on the micromechanics scale, and have concluded that localized thermo-plastic behaviour is significant. Here we deal with the macroscopic behaviour, and treat the composite material as an elastic-plastic continuum, taking into account the effect of thermal expansion. We formulate constitutive equations for such a material, along the lines of classical plasticity theory, using a yield condition which is a generalization of von Mises condition for isotropic materials.

As an application we analyse the thermal distortion of channel sections. The corresponding thermoelastic problem was studied by O'Neill, Rogers and Spencer [2]. We show that, in the case of a thermoelastic-plastic section, even for a body supporting residual stress as a result of previous plastic deformation, the thermal response is identical to that of the thermoelastic material, and that the distortion due to heating is reversible and does not lead to permanent deformation.

## ORTHOTROPIC THERMOELASTIC–PLASTIC SOLID

We consider a thermoelastic–plastic material with, locally, orthotropic symmetry and introduce an orthogonal curvilinear coordinate system $Ox_i$ (i=1,2,3) such that the coordinate axes coincide with the orthotropic axes. The components of the stress tensor $\underset{\sim}{\sigma}$ in this system are denoted by $\sigma_{ij}$. If yielding is independent of hydrostatic stress it can be shown by standard invariance arguments [3,4] that the most general quadratic yield function (that is, the natural generalization of von Mises' isotropic yield function to orthotropic symmetry) can be expressed in the form

$$\frac{(\sigma_{11}-\sigma_{22})(\sigma_{11}-\sigma_{33})}{Y_1^2} + \frac{(\sigma_{22}-\sigma_{33})(\sigma_{22}-\sigma_{11})}{Y_2^2} + \frac{(\sigma_{33}-\sigma_{11})(\sigma_{33}-\sigma_{22})}{Y_3^2}$$

$$+ \frac{\sigma_{23}^2}{k_1^2} + \frac{\sigma_{31}^2}{k_2^2} + \frac{\sigma_{12}^2}{k_3^2} \leq 1. \tag{1}$$

Here $Y_1$, $Y_2$ and $Y_3$ may be interpreted as yield stresses for uniaxial tension or compression in the $x_1$, $x_2$ and $x_3$ directions respectively; $k_1$ is the yield stress for shear in the $x_3$ direction on the plane $x_2$ = constant, and $k_2$ and $k_3$ have analogous interpretations.

It is assumed that the strain-rate tensor $\underset{\sim}{d}$, with components $d_{ij}$ in the specified system, may be decomposed into the sum of three parts

$$\underset{\sim}{d} = \underset{\sim}{d}_e + \underset{\sim}{d}_t + \underset{\sim}{d}_p , \quad d_{ij} = d_{ij}^e + d_{ij}^t + d_{ij}^p , \tag{2}$$

representing elastic, thermal and plastic strain-rate respectively.

The elastic part is assumed to be related to the stress-rate $\dot{\underset{\sim}{\sigma}}$ (suitably defined) according to the orthotropic elastic stress-strain relations

$$\begin{aligned} d_{11}^e &= s_{11}\dot{\sigma}_{11} + s_{12}\dot{\sigma}_{22} + s_{13}\dot{\sigma}_{33} , \\ d_{22}^e &= s_{12}\dot{\sigma}_{11} + s_{22}\dot{\sigma}_{22} + s_{23}\dot{\sigma}_{33} , \\ d_{33}^e &= s_{13}\dot{\sigma}_{11} + s_{23}\dot{\sigma}_{22} + s_{33}\dot{\sigma}_{33} , \\ d_{23}^e &= s_{44}\dot{\sigma}_{23} , \quad d_{31}^e = s_{55}\dot{\sigma}_{31} , \quad d_{12}^e = s_{66}\dot{\sigma}_{12} , \end{aligned} \tag{3}$$

where $s_{ij}$ are the compliance coefficients for the material.

The thermal part of the strain-rate is given by

$$d_{11}^t = \alpha_1\dot{T} , \quad d_{22}^t = \alpha_2\dot{T} , \quad d_{33}^t = \alpha_3\dot{T} , \tag{4}$$

where $\dot{T}$ denotes the rate of change of temperature, and $\alpha_1$, $\alpha_2$ and $\alpha_3$ are coefficients of thermal expansion in the respective coordinate directions.

For the plastic part of the strain-rate we adopt the flow rule associated with the yield function (1) as plastic potential, which gives

$$\begin{aligned} d_{11}^p &= \dot{\lambda}\left\{Y_1^{-2}(2\sigma_{11}-\sigma_{22}-\sigma_{33}) + (Y_2^{-2}-Y_3^{-2})(\sigma_{22}-\sigma_{33})\right\} , \\ d_{22}^p &= \dot{\lambda}\left\{Y_2^{-2}(2\sigma_{22}-\sigma_{33}-\sigma_{11}) + (Y_3^{-2}-Y_1^{-2})(\sigma_{33}-\sigma_{11})\right\} , \\ d_{33}^p &= \dot{\lambda}\left\{Y_3^{-2}(2\sigma_{33}-\sigma_{11}-\sigma_{22}) + (Y_1^{-2}-Y_2^{-2})(\sigma_{11}-\sigma_{22})\right\} , \end{aligned} \tag{5}$$

$$d^p_{23} = \dot{\lambda} k_1^{-2} \sigma_{23}\ , \qquad d^p_{31} = \dot{\lambda} k_2^{-2}\ \sigma_{31}\ , \qquad d^p_{12} = \dot{\lambda} k_3^{-2} \sigma_{12}\ ,$$

where $\dot{\lambda}$ is a loading parameter or, in the case of a perfectly plastic material, a factor of proportionality. We note that the material is plastically incompressible in the sense that $d^p_{kk} = 0$.

## THERMAL DISTORTION OF CHANNEL SECTIONS

Suppose an orthotropic elastic solid to be formed into the shape of a sector of a circular cylindrical tube, in such a manner that, in terms of cylindrical polar coordinates $r,\theta,z$, the coordinate surfaces are surfaces of material symmetry. In general, the plastic deformation that leads to this configuration will lead to residual stress; we allow any equilibrium residual stress field that does not violate the yield condition.

The body is now raised to a temperature T above the reference temperature (or, in the case of a body formed at high temperature, reduced to −T below the forming temperature). We identify the coordinates $x_1$, $x_2$, $x_3$ with $r,\theta,z$ respectively, and seek solutions of the form

$$u_r = u_r(r,\theta)\ , \qquad u_\theta = u_\theta(r,\theta)\ , \qquad u_z = u_z(z)\ ,$$
$$\dot{\sigma}_{rz} = \dot{\sigma}_{\theta z} = 0\ , \tag{6}$$

where vector and tensor components in $(r,\theta,z)$ coordinates are denoted in the usual way. Then

$$d_{rr} = \frac{\partial v_r}{\partial r}\ , \qquad d_{\theta\theta} = \frac{1}{r}\left[v_r + \frac{\partial v_\theta}{\partial \theta}\right], \qquad d_{zz} = \frac{\partial v_z}{\partial z}\ ,$$
$$d_{r\theta} = \frac{1}{2}\left[\frac{1}{r}\frac{\partial v_r}{\partial \theta} + \frac{\partial v_\theta}{\partial r} - \frac{v_\theta}{r}\right], \qquad d_{rz} = d_{\theta z} = 0\ . \tag{7}$$

Now the governing equations reduce to

$$s_{11}\dot{\sigma}_{rr}+s_{12}\dot{\sigma}_{\theta\theta}+s_{13}\dot{\sigma}_{zz}+\alpha_1\dot{T}+\dot{\lambda}\left\{Y_1^{-2}(2\sigma_{rr}-\sigma_{\theta\theta}-\sigma_{zz})+(Y_2^{-2}-Y_3^{-2})(\sigma_{\theta\theta}-\sigma_{zz})\right\} = \frac{\partial v_r}{\partial r}\ ,$$

$$s_{12}\dot{\sigma}_{rr}+s_{12}\dot{\sigma}_{\theta\theta}+s_{23}\dot{\sigma}_{zz}+\alpha_2\dot{T}+\dot{\lambda}\left\{Y_2^{-2}(2\sigma_{\theta\theta}-\sigma_{zz}-\sigma_{rr})+(Y_3^{-2}-Y_1^{-2})(\sigma_{zz}-\sigma_{rr})\right\}$$
$$= \frac{1}{r}\left[v_r + \frac{\partial v_\theta}{\partial \theta}\right]\ ,$$

$$s_{13}\dot{\sigma}_{rr}+s_{23}\dot{\sigma}_{\theta\theta}+s_{33}\dot{\sigma}_{zz}+\alpha_3\dot{T}+\dot{\lambda}\left\{Y_3^{-2}(2\sigma_{rr}-\sigma_{rr}-\sigma_{\theta\theta})+(Y_1^{-2}-Y_2^{-2})(\sigma_{rr}-\sigma_{\theta\theta})\right\} = \frac{\partial v_z}{\partial z}\ ,$$

$$s_{66}\dot{\sigma}_{r\theta} + \dot{\lambda} k_3^{-2}\sigma_{r\theta} = \frac{1}{2}\left[\frac{1}{r}\frac{\partial v_r}{\partial \theta} + \frac{\partial v_\theta}{\partial r} - \frac{v_\theta}{r}\right]\ , \tag{8}$$

$$s_{44}\dot{\sigma}_{\theta z} + \dot{\lambda} k_1^{-2}\sigma_{\theta z} = 0\ , \qquad s_{55}\dot{\sigma}_{rz} + \dot{\lambda} k_2^{-2}\sigma_{rz} = 0\ ,$$

where the stress components satisfy the inequality (1), together with the equilibrium equations.

This system of equations has the solution

$$\dot{\underset{\sim}{\sigma}} = 0\ , \qquad \dot{\lambda} = 0\ , \tag{9}$$

provided that

$$\frac{\partial v_r}{\partial r} = \alpha_1 \dot{T} , \quad \frac{1}{r}\left[v_r + \frac{\partial v_\theta}{\partial \theta}\right] = \alpha_2 \dot{T} ,$$
$$\frac{\partial v_z}{\partial z} = \alpha_3 \dot{T} , \quad \frac{1}{r}\frac{\partial v_r}{\partial \theta} + \frac{\partial v_\theta}{\partial r} - \frac{v_\theta}{r} = 0 , \tag{10}$$

which are essentially the equations that arose in the analogous thermoelastic problem [2]. They have the solution

$$v_r = \alpha_1 r\dot{T} , \quad v_\theta = (\alpha_2-\alpha_1)r\theta\dot{T} , \quad v_z = \alpha_3 z\dot{T} , \tag{11}$$

and the corresponding displacements, for small deformations, are

$$u_r = \alpha_1 r(T-T_0), \quad u_\theta = (\alpha_2-\alpha_1)r\theta(T-T_0), \quad u_z = \alpha_3 z(T-T_0) , \tag{12}$$

where $T_0$ is the reference temperature. The angular displacement of a plane $\theta$ = constant is

$$\Delta\theta = u_\theta/r = (\alpha_2-\alpha_1)\theta(T-T_0). \tag{13}$$

Thus for a circular section with sector angle $\beta$, the increase $\Delta\beta$ in sector angle in a temperature increase $T-T_0$ is

$$\Delta\beta = (\alpha_2-\alpha_1)\beta(T-T_0) . \tag{14}$$

This analysis shows that on uniform heating, the pre–stressed section considered will undergo a distortion of shape as well as volume, given by (12), but no change of stress. Thus, if it is in equilibrium and below yield at the reference temperature, it will remain so during the thermal distortion. The deformation is reversible, and the body reverts to its reference shape on returning to the reference temperature.

The analysis is easily extended in the manner described in [2], to the case of a chananel section of uniform thickness but arbitrary cross–section. The main conclusion is that (14) remains valid if $\theta$ is interpreted as the angle that the normal to the curved surface of the channel makes with the plane on which $u_\theta = 0$.

In conclusion, we note that the simplicity of the solution presented here is a consequence of the essentially two–dimensional nature of the problem. In three–dimensional problems for anisotropic solids, thermal expansion will in general cause a change in internal stress, which may result in the occurrence of plastic deformation.

## REFERENCES


1. Dvorak, G.J. and Rao, M.S.M., Thermal stresses in heat–treated fibrous composites. J. Appl. Mech., 1976, **43**, 619–624.

2. O'Neill, J.M., Rogers, T.G. and Spencer A.J.M., Thermally induced distortions in the moulding of laminated channel sections. Math. Engng. Ind., 1988, **2**, 65–72.

3. Spencer, A.J.M., The formulation of constitutive equations for anisotropic solids. In Mechanical Behaviour of Anisotropic Solids, ed. J.P. Boehler, Editions du CNRS, Paris, 1982, pp.3–26.

4. Spencer, A.J.M., Constitutive theory for strongly anisotropic solids. In Continuum Theory of the Mechanics of Fibre–reinforced Composites, ed. A.J.M. Spencer, CISM Courses and Lectures No. 282, Springer–Verlag, Wien, 1984, pp.1–32.

# ADIABATIC SHEAR BANDS

# AND OTHER LOCALIZED DEFORMATION

# A SHEAR BAND ANALYSIS IN ELASTOPLASTIC GRANULAR MATERIAL

J.P. BARDET and J.PROUBET
*Civil Engineering Department, KAP 210*
*University of Southern California, Los Angeles, CA 90089-2531, USA*

## ABSTRACT

Based on the numerical simulations of an assembly of two-dimensional particles, the paper assesses the assumptions and limitations of the linear stability analyses proposed for shear bands in granular media. It shows that the linear stability analyses that are based on a micropolar generalization of flow and deformation theories of plasticity are capable of accounting for the thickness of shear bands, the rotations and tangential displacements of the particles within shear bands in idealized granular soils.

## INTRODUCTION

The deformations within granular soils are commonly observed to concentrate in shear bands, the thickness of which is 8 to 10 times the mean grain diameter [1]. Mühlhaus and Vardoulakis [1] and Vardoulakis [2] introduced couple stresses into the constitutive equations of the deformation and flow theories of plasticity, and successfully predicted the emergence, inclination and thickness of the shear bands in sands. They determined the thickness of shear bands by using the linear stability analysis in the post-bifurcation range. The linear stability analysis is appropriate to investigate the emergence and initial inclination of strain localization within homogeneously strained and stressed solids. However, its application in the post-bifurcation range requires additional assumptions about the uniformity of stress and strain, which need to be verified. At the present, the structure of shear bands in soils has not been examined as in metals [3], due to the lack of experimental data on the localized deformations in soils.

The main objective of the present work is to assess the assumptions and limitations of the theories [1, 2] based on the results of the numerical simulations on idealized granular media. Following the introduction, the second section reviews the results of the numerical simulations. The third section summarizes the results of shear band analyses based on the deformation theory of plasticity.

## NUMERICAL SIMULATIONS

We summarize only the results of the numerical simulations presented in [4]. The principles of the numerical simulations used hereafter are slightly different from those of Cundall [5]. The contacts between cylindrical particles have an elastic-perfectly plastic behavior that is characterized by the normal and tangential stiffnesses, $k_n$ and $k_s$, and the intergranular friction angle, $\phi_\mu$. The equilibrium equations of each particle are solved by using an Adaptative Dynamic Relaxation (ADR) technique [6]. The ADR algorithm controls the dynamic transition between two static states, restrains the amplitude of the transient motions and avoids the "shaking" of particles between two equilibrium states. It prevents the spurious oscillations of the particles, which artificially decreases the porosity and increases the shear strength of granular materials. The ADR method has been implemented into the computer program JP2 [7].

Fig.1a shows the undeformed sample after a random generation and an isotropic consolidation. It is made of a even distribution of 2000 disks with radius 1.25 R and 0.75R where R is the average particle radius. All stresses are normalized by the pressure $p_o$ applied on the flexible lateral boundary, a stress-controlled boundary similar to the rubber membrane of triaxial tests. The upper and lower platens are rigid and have the same friction coefficients and spring stiffnesses as the particles ($\phi_\mu$=25°, $k_n$=$k_s$=100$p_o$). No gravity force is applied to the particles. The initial state of the sample is isotropic. During the biaxial test, $\sigma_{11}$ is kept constant on the flexible boundary while the upper and lower platens are moved closer in small steps.

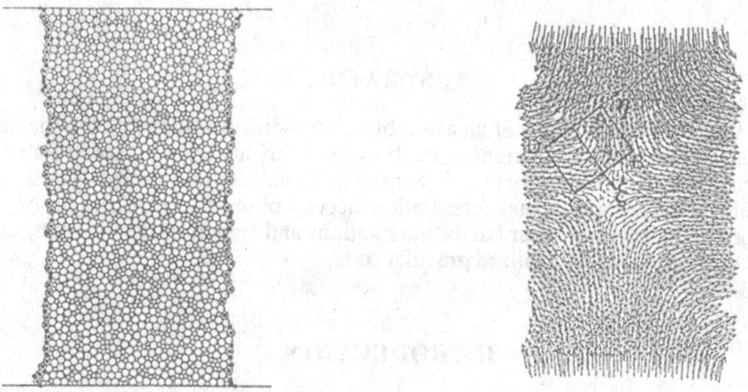

Figure 1. Initial state of the sample and displacements of particles at $\gamma$ = 44%.

Fig.2 shows the circular areas $V_o$ that are selected to calculate the average stress and strain from contact forces and particle displacements. Area No.1 is centered inside the sample whereas areas No.2 and 3 are located inside and outside the shear band, respectively. Fig.2 shows the stress-strain responses in terms of the stress ratio $\tau/|p| = \sin\phi$ versus the shear strain $\gamma$. $\phi$ is the mobilized friction angle, $\tau$ is the second invariant of deviator stress and p is the mean pressure. The shear strain $\gamma$ is the second invariant of the deviator strain. All the stress-strain responses coincide and are almost linear when $\gamma$ is smaller than 4%. When $\gamma$ is about 4%, small spikes are detected in the stress-strain response curves before the main peak. They correspond to the beginning of rapid structural changes in the granular assemblies which ultimately form a failure mechanism with shear bands. As shown in Fig.2, samples No.1 and No.2 have similar stress-strain response curves. It is concluded that the stresses and strains inside and outside the shear bands coincide approximately, a conclusion which supports the assumptions of uniform stress and strain made in [1, 2].

Fig.1b shows the displacement of the particles for $\gamma$=44%. Two shear bands form a "X", the branches of which make an angle of 52°±1° and 38°±1° with the horizontal direction. These inclinations are predicted by the Mohr-Coulomb theory, i.e., $\theta = \pi/4 \pm \phi/2$ when $\phi = \phi_r$ =14°. The upper part of the shear band inclined at 52° does not interact with any rigid boundary. It finds no kinematic constraint as it intersects the flexible boundary; it is assumed to be representative of the shear bands within the granular masses of infinite dimension. The selected fragment of the persistent shear band is covered by the window ABCD shown in Fig.1b. The shear band is parallel to the ξ-axis, and normal to the η-axis. There are about 12 average particles across the ξ and η directions of the window ABCD.

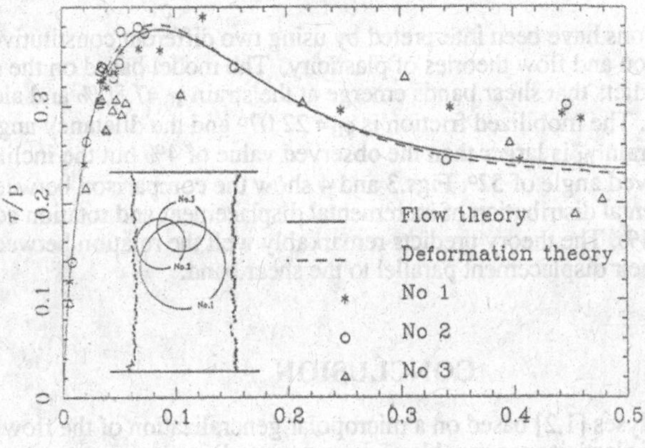


Figure 2. Measured and predicted variations of stress ratio τ/|p| versus shear strain γ.

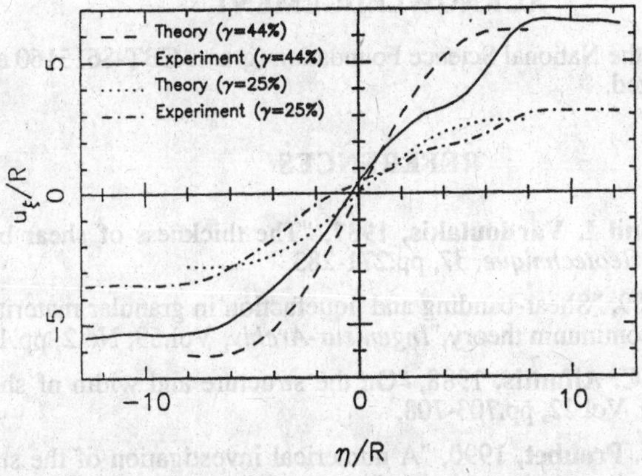


Figure 3. Measured and predicted average tangential displacement across the shear band at $\gamma$ = 25% and 44%.

Fig.3 shows the variation of the average tangential displacement versus η for $\gamma$ = 25% and 44%. $u_\xi$ is constant outside the shear band and varies linearly inside the shear band. $u_\xi$ keeps a similar shape but its amplitude is amplified for $\gamma$= 44%. The thickness of the shear band is about 15 times the mean grain radius R at $\gamma$=25% and 44%. The average particle rotation $\theta^c$ across the shear band were also obtained at $\gamma$ = 25% and 44% .The rotations outside the shear bands were smaller than 10°, and opposite to the rotations within the shear band. The distance between the two points across the shear band where $\theta^c=0$ was exactly 15 times R. This width coincides with the one found from the displacement field.

## SHEAR BAND ANALYSIS WITH ELASTOPLASTICITY

The numerical simulations have been interpreted by using two different constitutive equations based on the deformation and flow theories of plasticity. The model based on the deformation theory of plasticity predicts that shear bands emerge at the strain $\gamma_B$=7.80% and along the inclination $\theta_B$ = 51.83°. The mobilized friction is $\phi_B$= 22.07° and the dilatancy angle is $\psi_B$= 4.70°. The predicted strain $\gamma_B$ is larger than the observed value of 4% but the inclination $\theta_B$ is very close to the observed angle of 52°. Figs.3 and 4 show the comparison between the predicted and experimental distribution of incremental displacement and rotation across the band at $\gamma$ = 25% and 44%. The theory predicts remarkably well the relation between the particles rotation and their displacement parallel to the shear band.

## CONCLUSION

The linear stability analyses [1,2] based on a micropolar generalization of the flow and deformation theories of plasticity are capable of reproducing the emergence, inclination and thickness of shear bands in idealized granular materials. The stress and strain were found to be sufficiently uniform to extend the linear stability analysis in the post-bifurcation range. The theory describes well the tangential displacements and rotations of the particles inside the shear bands. The importance of particle rotations within shear bands strongly supports the micropolar description.

## ACKNOWLEDGEMENT

The financial support of the National Science Foundation (grants CBT-8615160 and MSM 8657999) is acknowledged.

## REFERENCES

1. **Mühlhaus, H.B., and I. Vardoulakis,** 1987, "The thickness of shear bands in granular materials," *Géotechnique, 37,* pp.271-283.
2. **Vardoulakis, I.,** 1989, "Shear-banding and liquefaction in granular materials on the basis of a Cosserat continuum theory,"*Ingenieur-Archiv,* Vol.59, No.2, pp.106-114.
3. **Zbib, H.M. and E.C. Aifantis,** 1988, "On the structure and width of shear bands," *Scripta Metallurgica,* Vol.22, pp.703-708.
4. **Bardet, J.P., and J. Proubet,** 1990, "A numerical investigation of the structure of persistent shear bands in granular materials," accepted for publication in *Géotechnique.*
5. **Cundall, P.A.,** 1989, "Numerical experiments on localization in frictional materials," *Ingenieur-Archiv,* Vol.59, No.2, pp.148-159.
6. **Bardet, J.P., and J. Proubet,** 1989, "Adaptative dynamic relaxation for statics of granular materials," *Computers & Structures, in press.*
7. **Bardet, J.P., and J. Proubet,** 1989, "JP2, a program to simulate the behavior of two-dimensional granular materials," *Univ. of Southern California, L.A.*

# HIERARCHY OF LOCALIZATION PHENOMENA INSIDE, AT THE BOUNDARIES AND INTERFACES OF SOLIDS

AHMED BENALLAL and RENE BILLARDON
Laboratoire de Mécanique et Technologie
E.N.S. de Cachan/C.N.R.S./Université Paris 6
61, Avenue du Président Wilson, 94235-CACHAN, France

## ABSTRACT

The conditions for the localization of the deformation in rate-independent materials, understood as a first order bifurcation, are obtained by the analysis of the rate problem. Three types of localization phenomena are exhibited from this analysis : the deformation may localize inside the body, at its boundaries or at its possible interfaces. The corresponding conditions are given mechanical interpretations and used as indicators of local rupture of the material.

## INTRODUCTION

Here, we give a general view of various bifurcation and localization phenomena for possibly heterogeneous solids made of rate-independent (including elastic-plastic, damageable) materials. Under the small strain assumption, the behaviour of these materials is described by the following piece-wise linear rate constitutive laws :

$$\text{(1)} \qquad \dot{\sigma} = \mathbb{L} : \varepsilon(\mathbf{v}) \quad \text{with} \quad \begin{cases} \mathbb{L} = \mathbb{E} \ \text{when } f < 0, \ \text{or } f = 0 \text{ and } \mathbf{b} : \mathbb{E} : \varepsilon(\mathbf{v}) < 0 \\ \mathbb{L} = \mathbb{H} \ \text{when } f = 0 \text{ and } \mathbf{b} : \mathbb{E} : \varepsilon(\mathbf{v}) \geq 0 \end{cases}$$

where $\dot{\sigma}$ and $\varepsilon(\mathbf{v}) = \dot{\varepsilon}$ respectively denote the stress and strain rates, $\mathbf{v}$ the velocity, and $f$ the yield function.This paper constitutes a unified and augmented presentation of results given in [5-9] where it was shown that in general different types of localization phenomena may occur, depending on the failure of one of three conditions which are described in the following .

## LOCALIZATION CONDITIONS

Let us consider a possibly heteregeneous body (see figure 1); it may represent an inclusion within a matrix, a fibre in a composite, .... Qualitative results can be exhibited from the analysis of the rate problem for general incrementally linear solids with tangent modulus $\mathbb{H}$. The corresponding linear boundary value problem *is well-posed if and only if* the following conditions are met :

**The ellipticity condition :**
The rate equilibrium equations must be elliptic in the closure of the body $\Omega$, i.e.

$$\det(\mathbf{n} \cdot \mathbb{H} \cdot \mathbf{n}) \neq 0 \qquad \text{for any vector } \mathbf{n} \neq \mathbf{0}, \ \text{and any point } M \in \overline{\Omega}.$$

*this ellipticity condition* is very classical. It has been related to *stationary acceleration waves* [1,2,3]. Its failure is also the condition for localization given by Rice [4] and Rudnicki and Rice [12] and linked to the appearance of *deformation modes involving discontinuities of the velocity gradient* and used to describe shear banding phenomena. Necessary conditions for the occurence of such modes in the case of the bi-linear constitutive relations (1) were also given by Rudnicki and Rice [13]. These conditions where shown to be in fact necessary and sufficient by Borré and Maier [11]. In neither of these studies, were the interactions of shear bands with boundaries and interfaces taken into account. We have included these effects in [9] and [14]. In [9], we have given the *necessary and sufficient conditions* for which a discontinuity surface for the velocity gradient appears at, or reaches the boundary of the solid. These conditions are given below for the constitutive laws (1) with

$$\mathbb{H} = \mathbb{E} - \frac{(\mathbb{E} : \mathbb{a}) \otimes (\mathbb{b} : \mathbb{E})}{h}$$

where it is assumed that $h > 0$, and $\mathbb{E}$ is strictly positive definite.

At a point P of the boundary $\Gamma$ where only surface traction rates $\dot{F}$ are applied, *the necessary and sufficient conditions for continuous localization* [i.e. the material is in loading ($\mathbb{L} = \mathbb{H}$) on each side of the singular surface] are

(2a)
i) there exists $\dot{\varepsilon}_0$ such that $m \cdot \mathbb{H} : \dot{\varepsilon}_0 = \dot{F}$
ii) $\det (n \cdot \mathbb{H} \cdot n) = 0$
iii) $(m \cdot \mathbb{E} \cdot n) \cdot (n \cdot \mathbb{E} \cdot n)^{-1} \cdot (n \cdot \mathbb{E} : \mathbb{a}) = m \cdot \mathbb{E} : \mathbb{a}$

These criteria have been applied to the determination of the orientation of cracks appearing on the free surface of specimen subjected to various types of loadings in [15].

At a point P of the boundary $\Gamma$ where only surface traction rates $\dot{F}$ are applied, *the necessary and sufficient conditions for discontinuous localization* [i.e. the material is in loading ($\mathbb{L} = \mathbb{H}$) on one side and in unloading ($\mathbb{L} = \mathbb{E}$) on the other side of the singular surface] are

(2b)
i) there exists $\dot{\varepsilon}_0$ such that $m \cdot \mathbb{H} : \dot{\varepsilon}_0 = \dot{F}$ and $\mathbb{b} : \mathbb{E} : \dot{\varepsilon}_0 > 0$
ii) $\det (n \cdot \mathbb{H} \cdot n) < 0$
iii) $(m \cdot \mathbb{E} \cdot n) \cdot (n \cdot \mathbb{E} \cdot n)^{-1} \cdot (n \cdot \mathbb{E} : \mathbb{a}) = m \cdot \mathbb{E} : \mathbb{a}$

In these conditions, **m** denotes the unit outward normal to the boundary of the solid in P, whereas **n** denotes the unit normal to the singular surface in P.

In [14], we have looked for the same problem for an interface and we have seeked conditions for which a discontinuity surface of the velocity gradient appears at, or reaches the interface. At a point Q of an interface I, two types of singular surface may occur. The singular surface *can* either stop at, or cross the interface (see figure 2). In the latter case, the singular surface can meet the interface at different angles on each side. In this case, analogous conditions to conditions (2) can be exhibited and for simplicity, an example is given here for continuous localization :

At a point Q of the interface I, *the necessary conditions for continuous localization* [i.e. the material is in loading ($\mathbb{L} = \mathbb{H}$) on each side of the singular surface and in both materials], when the acoustic tensor $(m \cdot \mathbb{H}^a \cdot m)$ is not singular, are

(3)
i) $\det (n^a \cdot \mathbb{H}^a \cdot n^a) = 0$
ii) $\det (n^b \cdot \mathbb{H}^b \cdot n^b) = 0$
iii) $(m \cdot (\mathbb{H}^a - \mathbb{H}^b) \cdot n^a) \cdot (n^a \cdot \mathbb{E}^a \cdot n^a)^{-1} \cdot (n^a \cdot \mathbb{E}^a : \mathbb{a}^a) =$
$(m \cdot \mathbb{H}^b \cdot m) \cdot (m \cdot \mathbb{H}^a \cdot m)^{-1} \cdot (m \cdot (\mathbb{H}^a - \mathbb{H}^b) \cdot n^b) \cdot (n^b \cdot \mathbb{E}^b \cdot n^b)^{-1} \cdot (n^b \cdot \mathbb{E}^b : \mathbb{a}^b)$

the indices a and b are related to each of the materials constituting the heteregeneous body, **m** is the unit normal to the interface in Q , $n^a$ and $n^b$ are the unit normals to the singular surfaces in the two materials in Q.

**The boundary complementing condition :**
This relation between the coefficients of the field and boundary operators must be satisfied at every point P belonging to the boundary $\Gamma$ where the boundary conditions are formally written as $\mathbb{B}(v) = g$. This condition is easily phrased in terms of an associated problem on a half space defined by $z > 0$. It requires for every vector

$\mathbf{k} = (k_1, k_2, 0) \neq \mathbf{0}$, that the only solution to the rate equilibrium equations with constant coefficients (equal to those of the operator at point P), in the form

$$v(x, y, z) = w(z) \exp[i (k_1 x + k_2 y)]$$

with bounded $\mathbf{w}$ and satisfying the homogeneous boundary conditions $\mathbb{B}(v) = \mathbf{0}$, is the zero solution $\mathbf{v} \equiv \mathbf{0}$.

*The boundary complementing condition* governs instabilities at the boundary of the solid. Its failure leads to *deformation modes localized at the boundary* and is related to *stationary surface waves* (for instance Rayleigh waves).

**The interfacial complementing condition :**
This relation between the coefficients of the field operators in $\Omega_1$ and $\Omega_2$ must be satisfied at every point Q of the interface I between $\Omega_1$ and $\Omega_2$. This condition is again easily phrased in terms of an associated problem on the whole space divided by the plane interface z = 0. It requires for every vector $\mathbf{k} = (k_1, k_2, 0) \neq \mathbf{0}$, that the only solution to the rate equilibrium equations with constant coefficients (equal to those of the operators at point Q, in $\Omega_1$ for z < 0 and in $\Omega_2$ for z > 0), in the form

$$(v_1(x, y, z), v_2(x, y, z)) = (w_1(z), w_2(z)) \exp[i (k_1 x + k_2 y)]$$

with bounded $(w_1, w_2)$ and satisfying the continuity requirements (continuity of the velocity and the traction rates) across the interface z = 0, is the identically zero solution $(v_1(z), v_2(z)) \equiv (0, 0)$ ; (where $v_1$ and $v_2$ are the solutions, respectively for z < 0 and for z > 0).

*The interfacial complementing condition* governs instabilities at interfaces. Its failure leads to *deformation modes localized at each side of the interface* and is related to *stationary interfacial waves* (Stonely waves).

**Remarks :**

- when these three conditions are fulfilled, the rate boundary problem admits *a finite number of linearly independent solutions which depend continuously on the data and which constitute diffuse modes of deformation ;*
- *these three conditions are local,* and this is particulally important when considering their numerical implementation ;
- the above-given results remain valid for an arbitrary number of non-intersecting interfaces, an interfacial condition being written for each interface ;
- the failure of these conditions can be interpreted as localization criteria . *These localization criteria can also be used as indicators of the local failure of the material* (Rice[4], Billardon & Doghri[10]);
- *both boundary and interfacial complementing conditions fail in the elliptic regime* of the equilibrium equations, or at the latest, when the ellipticity condition fails. Thus, localized modes of deformation at the boundary or at the interface generally occur before the onset of so-called shear banding modes ;
- conditions (2) are to be compared with the corresponding *conditions inside the solid* given by Borré & Maier and stated as

(4) $$\det (\mathbf{n} \cdot \mathbb{H} \cdot \mathbf{n}) \leq 0$$

with *equality corresponding to continuous,* and *inequality corresponding to discontinuous localization ;*
- given conditions (2) and (3), singular surfaces of the type discussed here *generally* appear *first inside the body ;*
- similar conditions to conditions (2) can be exhibited for general boundary conditions. (Note that *for displacement boundary conditions, relation (3) applies both inside the body and at the boundary.*) ;
- conditions (2) and (3) are a priori unrelated to the boundary and interfacial complementing conditions.

## APPLICATIONS

As a simple application allowing closed form calculations, we consider here the case of isotropic incrementally linear materials. The tangent modulus H is defined by two constitutive parameters L and G which are dependent on the history of the deformation so that

(5) $$H_{ijkl} = L\, \delta_{ij}\, \delta_{kl} + G(\delta_{ik}\, \delta_{jl} + \delta_{il}\, \delta_{jk})$$

where δ is the Kronecker delta.The results can be applied to linear elasticity but may be of less practical interest. The loss of ellipticity (shear bands) occurs when G = 0 or L + 2G = 0. and the normal to the plane of localization is arbitrary.
For surface modes, one has to consider two cases :

- two-dimensional situations where the boundary complementing condition fails either when $L + 3G = 0$ for displacement boundary conditions or when $L + G = 0$ for traction boundary conditions.

- three-dimensional situations where the boundary complementing condition fails only for traction boundary conditions when $L + G = 0$.

Finally, for interfacial modes, conditions are much more complex and we give them here in the case of traction boundary conditions and in two dimensional situations ; in this case, the interfacial boundary complementing condition fails when

$$\left[\frac{2G_1(L_1+2G_1)}{L_1+3G_1}+\frac{2G_2(L_2+2G_2)}{L_2+3G_2}\right]^2$$

$$+\left[\frac{2G_1(L_1+2G_1)}{L_1+3G_1}-\frac{2G_2(L_2+2G_2)}{L_2+3G_2}\right].\left[L_1+\frac{(L_1+G_1)(L_1+2G_1)}{L_1+3G_1}-L_2-\frac{(L_2+G_2)(L_2+2G_2)}{L_2+3G_2}\right]=0$$

## REFERENCES


1. J. HADAMARD, Leçons sur la propagation des ondes et les équations de l'hydrodynamique (Paris, 1903).
2. R. HILL, J. Mech. Phys. Solids, 10, 1-16 (1962).
3. J. MANDEL, J. de Mécanique, 1,.3-30 (1962).
4. J.R. RICE in Theoretical and Applied Mechanics, 207-220, edited by W.T. Koiter (North-Holland, 1976).
5. A. BENALLAL, R. BILLARDON & G. GEYMONAT in Cracking and Damage, 247-258, edited by J. Mazars & Z.P. Bazant (Elsevier, 1989).
6. A. BENALLAL , R. BILLARDON & G. GEYMONAT, C.R. Acad. Sci. Paris, 308 (II), 893-898 (1989).
7. A. BENALLAL , R. BILLARDON & G. GEYMONAT in Actes Congrès Français de Mécanique, 1, 242-243 (AUM, Metz 1989).
8. A. BENALLAL , R. BILLARDON & G. GEYMONAT, C.R. Acad. Sci. Paris, 310 (II), 679-684 (1990).
9. A. BENALLAL , R. BILLARDON & G. GEYMONAT, Constitutive Laws for Engineering Materials, 387-390, edited by C.S. Desai & G. Frantziskonis (Elsevier,1991).
10. R. BILLARDON & I. DOGHRI, C. R. Acad. Sci. Paris, 308 (II),.347-352 (1989).
11. G. BORRE & G. MAIER, Meccanica, 24, 36-41 (1989).
12. J. RUDNICKI & J.R. RICE, J. Mech. Phys. Solids, 23, 371-394 (1975).
13. J. RUDNICKI & J.R. RICE, Int. J. Solids Struct., 16, 597-605 (1980).
14. A. BENALLAL & R. BILLARDON, Internal report, LMT-Cachan (1991).
15. A. BENALLAL ,D.BOUDON& E. LAJOIX, Internal report, LMT-Cachan (1991).


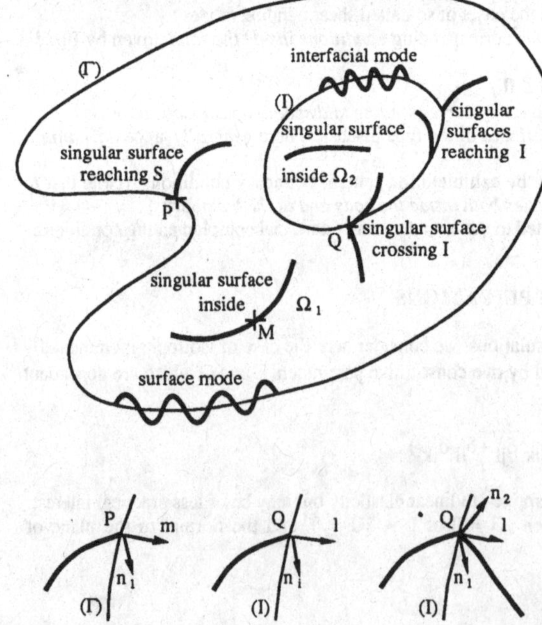


Figures 1. & 2. Different possible types of localization modes

# ON NECK FORMATION IN THE COLD DRAWING OF VISCOELASTIC FIBERS

BERNARD D. COLEMAN & DANIEL C. NEWMAN
College of Engineering, Rutgers University
Piscataway, New Jersey 08855-0909

## ABSTRACT

An asymptotic expression has been obtained for the axial tension in a slender rod of a non-linear viscoelastic material undergoing inhomogeneous stretch. This expression, which is valid to within an error of fourth order in diameter, gives the tension per unit cross-sectional area as a functional of the history of the local stretch and its first two spatial derivatives. The theory that results from treating the expression as exact is consistent with thermodynamical principles, and it yields a functional-differential equation for the stretch field that is amenable to numerical analysis. Computed solutions for creep under static loads show that for an appropriate class of materials with slowly fading memory there is a range of applied loads for which an initially homogeneous deformation evolves into a well defined neck whose edges, after a period of relatively quiescent incubation, advance rapidly along the fiber and in so doing transform moderately stretched material into highly stretched, i.e., drawn, material. The calculated fiber profiles and the predicted dynamics of neck formation and growth are in good accord with familiar observations.

## CONSTITUTIVE RELATIONS

We here discuss a recently obtained extension to viscoelastic materials [1] of results obtained previously for elastic materials susceptible to cold drawing [2, 3]. To describe a motion of a three-dimensional viscoelastic body, one may write $\chi_{[t]}(x, \zeta)$ for the place in space at time $\zeta$ of the material point that is at the place $x$ at a time $t$, which we interpret as the present time. If one puts $F_{[t]}(\zeta) = \nabla_x \chi_{[t]}(x, \zeta)$, then the symmetric, positive-definite tensor,[1] $F_{[t]}(\zeta)^T F_{[t]}(\zeta)$, is the right Cauchy-Green tensor at time $\zeta$ computed using the configuration at time $t$ as the reference. Now, let $G(\zeta) = [F_{[t]}(\zeta)^T F_{[t]}(\zeta)]^{-1}$, and define the *history up to time t of* $G$ to be the function $G^t$ on $[0, \infty)$ for which

$$G^t(s) = G(t-s) = F_{[t]}(t-s)^{-1}[F_{[t]}(t-s)^{-1}]^T \qquad (s \geq 0). \tag{1}$$

We are concerned with incompressible materials for which the stress tensor $S$ of Cauchy obeys a constitutive relation of the form[2]

$$S(t) = -p\mathbf{1} + 2\sum_{j=1}^{n} \int_0^\infty K_j(s) G^t(s) \nabla H_j(G^t(s))\, ds\,; \tag{2}$$

here $\nabla H_j$ is the tensor-valued gradient of a scalar-valued isotropic function $H_j$; because $H_j$ is isotropic, $H_j(G) = h_j(\xi_1, \xi_2, \xi_3)$, where $h_j$ is a completely symmetric function of the proper numbers $\xi_1, \xi_2, \xi_3$ of $G$; each of the functions $K_j$ is positive on $[0, \infty)$ with $dK_j(s)/ds \geq 0$.

When, as in calculations discussed below, one is treating semicrystalline polymers, the functions $H_j$ and $K_j$ are chosen such that, although the material described by (2) is a fluid with gradually fading memory, the memory fades so slowly that the material does not have a finite steady-state viscosity: it cannot undergo either a motion of steady[3] extension [4] or a steady[3] shearing flow at a finite level of stress.

Consider now a fiber, i.e., a body that in an undistorted reference configuration $\mathcal{R}$ is a long, thin, cylindrical rod of circular cross section and diameter $D$. We suppose the motion such that, in a cylindrical coordinate system with the $Z$-axis along the axis of the rod

$$\theta = \Theta, \qquad r = \nu(Z, t)R, \qquad z = z(Z, t) \qquad (0 \leq R \leq D/2) \tag{3}$$

where $\theta$, $r$, $z$ are the coordinates at time $t$ of the material point with coordinates $\Theta$, $R$, $Z$ in $\mathcal{R}$. The condition that the motion be isochoric is $\nu = \lambda^{-1/2}$ with $\lambda$ the (axial) *stretch* defined by $\lambda = \lambda(Z, t) = \partial z(Z, t)/\partial Z$. After much work, one can show that it follows from (2) and (3) that, under neglect of terms $O(D^4)$, at each time $t$ the axial tension (per unit of area in $\mathcal{R}$) is given by

---

[1] $F_{[t]}(\zeta)^T$ is the transpose of the tensor $F_{[t]}(\zeta)$.

[2] Here, $-p\mathbf{1}$, with $\mathbf{1}$ the unit tensor, is an unspecified hydrostatic pressure.

[3] "Steady" here implies at *constant rate for all time.*

$$T(t) = \int_0^\infty \frac{\partial}{\partial \lambda(t)} f\left(s, \frac{\lambda(t)}{\lambda(t\text{-}s)}\right) ds - \frac{1}{2}\int_0^\infty \frac{\partial}{\partial \lambda(t)} \left[ g\big(s, \lambda(t), \lambda(t\text{-}s)\big)\, \omega_{(t)}(s)^2 \right] ds$$

$$+ \frac{\partial}{\partial Z} \int_0^\infty \frac{g\big(s, \lambda(t), \lambda(t\text{-}s)\big)}{\lambda(t\text{-}s)}\, \omega_{(t)}(s) ds, \tag{4}$$

where

$$\omega_{(t)}(s) = \frac{\partial}{\partial Z}\left[\frac{\lambda(Z, t)}{\lambda(Z, t-s)}\right]. \tag{5}$$

The functions $f$ and $g$ are determined by the material functions $K_j$ and $h_j$ through the relations

$$f(s, \xi) = \sum_{j=1}^{n} K_j(s) h_j(\xi^2, \xi^{-1}, \xi^{-1}), \tag{6}$$

$$g(s, \alpha, \beta) = \frac{-D^2}{32\alpha\beta}\left[(\alpha/\beta)^3 - 1\right]^{-1} \frac{\partial}{\partial \alpha} f(s, \alpha/\beta). \tag{7}$$

Here $\xi$, $\alpha$, $\beta$ play the roles of $\lambda(t)/\lambda(t\text{-}s)$, $\lambda(t)$, $\lambda(t\text{-}s)$, respectively. It will be noticed that the function $g$ is proportional to $D^2$ and is determined by the function $f$ which is independent of $D$ and governs the fiber's response to homogeneous deformations.

## NONHOMOGENEOUS CREEP

In the accompanying figure one sees fiber profiles calculated using equations (4)–(7) with $n = 3$ and choices of the functions $K_j$ and $h_j$ appropriate to typical semicrystalline polymeric materials, such as the high-density polyethylene studied by Crissman and Zapas [5]. The numerical problem treated here is that of solving equation (4) to obtain $\lambda$ as a function of $z$ and $t$ under the assumption that $T(t) = 0$ for $t < 0$ and $T(t) = T°$ for $t > 0$, with $T°$ a positive constant in a range for which earlier calculations indicated that homogeneous creep should become an unstable motion at a time between 54 and 55 seconds. An area-reduction technique was employed to localize the place where neck-formation might begin; i.e., the fiber was weakened by a slight reduction of initial cross-sectional area in the region whose boundaries are indicated in the figure with vertical marks. In the case presented here, the time $t_B$ required for the neck to attain a high rate of elongation is about 82 seconds. Our numerical study indicates that $t_B$ is, within limits, independent of the amount of area-reduction. There is a broad range of the area-reduction parameter in which the more a fiber is weakened, the sooner a neck starts to form, but the longer the neck must "incubate" before its edges rapidly advance into homogeneously deformed material.

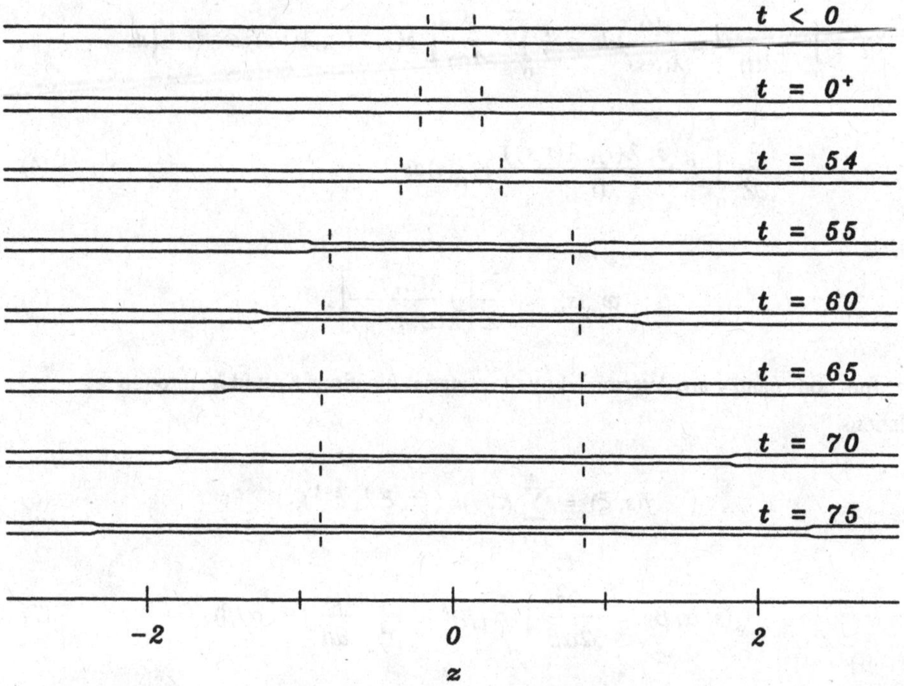


## REFERENCES

1. Coleman, B. D., & Newman, D. C., On the rheology of cold drawing: II. Viscoelastic materials, (submitted for publication).

2. Coleman, B. D., Necking and drawing in polymeric fibers under tension, *Arch. Rational Mech. Anal.*, 1983, **83**, 115-137.

3. Coleman, B. D., & Newman, D. C., On the rheology of cold drawing: I. Elastic materials, *J. Polym. Sci. B, Polymer Physics,* 1988, **26**, 1801-1822.

4. Coleman, B. D., & Noll, W., Steady extension of incompressible simple fluids, *Phys. Fluids*, 1962, **5**, 840-843.

5. Crissman, J. M., & Zapas, L. J., Creep failure and fracture of polyethylene in uniaxial extension, *Polym. Eng. & Sci.* 1979, **19**, 99-103.

Research supported by the U. S. National Science Foundation and by the Donors of the Petroleum Research Fund, administered by the American Chemical Society.

# MICROPOLAR ELASTO-PLASTICITY AND ITS ROLE IN LOCALIZATION ANALYSIS

ANDREAS DIETSCHE, PAUL STEINMANN, KASPAR WILLAM
Institute of Mechanics, University of Karlsruhe, D-7500 Karlsruhe 1
CEAE-Department, University of Colorado at Boulder, CO 80309-0428

**Abstract**

In this paper we focus on micropolar elasto-plasticity and the conditions for discontinuous solutions of IBVP's. To start with, we briefly review the kinematic relations, the balance conditions and introduce linear-elastic and elasto-plastic constitutive models for micropolar continua. An 'augmented' localization tensor is then developed which reveals discontinous bifurcation at the constitutive level. This localization condition is constrained however by the balance of angular momentum which restricts the bifurcated stress and couple-stress fields, see also ref. [5].

## 1. PRELIMINARIES

In Cosserat theory each material particle is endowed with three translational and three rotational degrees of freedom [2]. The translatory motion is described by the displacement field $\mathbf{u}$, whereas the rotatory motion is described by the rotation $\boldsymbol{\omega}$ of the directors (hidden triad). For infinitesimal deformations the non-symmetric strain tensor is comprised of the displacement gradient and the rotation tensor such that $\boldsymbol{\epsilon} = \nabla \mathbf{u} + \mathbf{e} \cdot \boldsymbol{\omega}$. In addition, the curvature tensor is defined by the gradient of the rotation field, i.e. $\boldsymbol{\kappa} = \nabla \boldsymbol{\omega}$. Omitting body forces and inertia effects the local balance of linear momentum results in the traditional divergence statement $\operatorname{div}(\boldsymbol{\sigma}^t) = \mathbf{0}$. However, there are additional couple-stresses $\mathbf{m}$ present, which are work-conjugate to the curvatures. Hence, the local balance of angular momentum results in the statement $\operatorname{div} \mathbf{m} + \mathbf{e} : \boldsymbol{\sigma} = \mathbf{0}$ when body couples and rotatory inertia effects are omitted. Therefore, the stress tensor is in general non-symmetric in contrast to the Boltzmann axiom.

In the case of linear elasticity the stress and strain tensors are related by the linear mapping $\boldsymbol{\sigma} = \mathbf{E} : \boldsymbol{\epsilon}$. For isotropic behavior the elasticity tensor $\mathbf{E}$ involves three rather than two independent material moduli,

$$\mathbf{E} = 2\mu \mathbf{1}_4^{sym} + 2\mu_c \mathbf{1}_4^{skw} + \lambda \mathbf{I} \otimes \mathbf{I} \tag{1}$$

because of the skew-symmetric contribution of the micropolar shear modulus $\mu_c \neq 0$. Similarly, the couple-stresses are related to the curvatures by the linear mapping $\mathbf{m} = \mathbf{C} : \boldsymbol{\kappa}$. The material tensor $\mathbf{C}$ involves three additional material moduli for isotropic conditions,

$$\mathbf{C} = 2\alpha \mathbf{1}_4^{sym} + 2\alpha_c \mathbf{1}_4^{skw} + \gamma \mathbf{I} \otimes \mathbf{I} \tag{2}$$

In the case of plastic flow the yield-condition $F = F(\boldsymbol{\sigma}, \mathbf{m}) = 0$ delimits the elastic range of the micropolar-continuum by a single scalar-valued function of stress and couple-stress, see e.g. refs. [1,2]. In analogy to classical plasticity the strain rate and the curvature rate are decomposed into an elastic and a plastic contribution. Assuming normality the magnitudes of the plastic strain rate and the plastic curvature rate are controlled by the same plastic multiplier $\dot{\lambda}$ if a single yield condition governs the plastic flow.

$$\dot{\epsilon}_p = \dot{\lambda}\frac{\partial F}{\partial \sigma} = \dot{\lambda}\mathbf{n}^\sigma; \qquad \dot{\kappa}_p = \dot{\lambda}\frac{\partial F}{\partial \mathbf{m}} = \dot{\lambda}\mathbf{n}^m \tag{3}$$

The resulting tangential elasto-plastic constitutive relations compacts into

$$\begin{pmatrix} \dot{\sigma} \\ \dot{\mathbf{m}} \end{pmatrix} = \begin{pmatrix} \mathbf{E}_{ep} & \mathbf{D}_{ep} \\ \mathbf{D}_{ep}^t & \mathbf{C}_{ep} \end{pmatrix} : \begin{pmatrix} \dot{\epsilon} \\ \dot{\kappa} \end{pmatrix} \tag{4}$$

where the suboperators are defined below

$$\mathbf{E}_{ep} = \mathbf{E} - \frac{\mathbf{E} : \mathbf{n}^\sigma \otimes \mathbf{n}^\sigma : \mathbf{E}}{\mathbf{n}^\sigma : \mathbf{E} : \mathbf{n}^\sigma + \mathbf{n}^m : \mathbf{C} : \mathbf{n}^m}; \quad \mathbf{D}_{ep} = -\frac{\mathbf{E} : \mathbf{n}^\sigma \otimes \mathbf{n}^m : \mathbf{C}}{\mathbf{n}^\sigma : \mathbf{E} : \mathbf{n}^\sigma + \mathbf{n}^m : \mathbf{C} : \mathbf{n}^m} \tag{5}$$

$$\mathbf{C}_{ep} = \mathbf{C} - \frac{\mathbf{C} : \mathbf{n}^m \otimes \mathbf{n}^m : \mathbf{C}}{\mathbf{n}^\sigma : \mathbf{E} : \mathbf{n}^\sigma + \mathbf{n}^m : \mathbf{C} : \mathbf{n}^m}$$

## 2. LOCALIZATION AS A BIFURCATION PROBLEM

As point of departure, we assume that the homogeneously deformed solid is subjected to quasi-static increments of deformation. While the displacement- and rotation fields remain $C^0$-continuous, i.e.

$$[[\mathbf{u}]] = \mathbf{u}^+ - \mathbf{u}^- = 0; \qquad [[\omega]] = \omega^+ - \omega^- = \mathbf{0} \tag{6}$$

the fields of displacement- and rotation gradients may exhibit a jump across a discontinuity surface whose orientation is defined by the normal $\mathbf{N}$.

$$[[\nabla\mathbf{u}]] = \nabla\mathbf{u}^+ - \nabla\mathbf{u}^- \neq \mathbf{0}; \qquad [[\nabla\omega]] = \nabla\omega^+ - \nabla\omega^- \neq \mathbf{0} \tag{7}$$

Maxwells compatibility conditions require that these two jump conditions must form rank-one tensors, thus

$$[[\nabla\mathbf{u}]] = \gamma^u \mathbf{M}^u \otimes \mathbf{N}; \qquad [[\nabla\omega]] = \gamma^\omega \mathbf{M}^\omega \otimes \mathbf{N} \tag{8}$$

Analogous to the argument in classical continuum theory, the vector $\mathbf{M}$ denotes the polarization direction and $\gamma$ the localization amplitude. The corresponding discontinuities in the strain rate and curvature rate are defined by the tensor product of the unit vectors $\mathbf{N}$ and $\mathbf{M}$

$$[[\dot{\epsilon}]] = [[\nabla\dot{\mathbf{u}}]] + \mathbf{e} \cdot [[\dot{\omega}]] = \dot{\gamma}^u \mathbf{M}^u \otimes \mathbf{N}; \quad [[\dot{\kappa}]] = [[\nabla\dot{\omega}]] = \dot{\gamma}^\omega \mathbf{M}^\omega \otimes \mathbf{N} \tag{9}$$

Introducing these jump conditions of the stress and couple-stress rates into the elasto-plastic material operator, the balance of linear and angular momentum across the plane of discontinuity leads to the critical localization condition for micropolar continua:

$$\begin{pmatrix} \mathbf{Q}_{ee} & \mathbf{Q}_{ec} \\ \mathbf{Q}_{ce} & \mathbf{Q}_{cc} \end{pmatrix} \cdot \begin{pmatrix} \dot{\gamma}^u \mathbf{M}^u \\ \dot{\gamma}^\omega \mathbf{M}^\omega \end{pmatrix} = \begin{pmatrix} \mathbf{0} \\ \mathbf{0} \end{pmatrix} \tag{10}$$

Localization initiates when this singularity condition is satisfied by the combination of critical directions $\mathbf{N}$ and $\mathbf{M}$. Thereby the suboperators of the localization tensor are defined as

$$\mathbf{Q}_{ee} = \mathbf{N} \cdot \mathbf{E}_{ep} \cdot \mathbf{N}; \quad \mathbf{Q}_{ec} = \mathbf{N} \cdot \mathbf{D}_{ep} \cdot \mathbf{N}; \quad \mathbf{Q}_{cc} = \mathbf{N} \cdot \mathbf{C}_{ep} \cdot \mathbf{N} \tag{11}$$

## 3. CONFLICTING EQUILIBRIUM CONDITIONS

In the case of discontinuous bifurcation the bifurcated stress and couple-stress fields exhibit the following discontinuities

$$\boldsymbol{\sigma}^{+} = \boldsymbol{\sigma} + \varsigma_{\sigma}[[\dot{\boldsymbol{\sigma}}]]; \quad \boldsymbol{\sigma}^{-} = \boldsymbol{\sigma} + (\varsigma_{\sigma} - 1)[[\dot{\boldsymbol{\sigma}}]] \tag{12}$$

$$\mathbf{m}^{+} = \mathbf{m} + \varsigma_{m}[[\dot{\mathbf{m}}]]; \quad \mathbf{m}^{-} = \mathbf{m} + (\varsigma_{m} - 1)[[\dot{\mathbf{m}}]] \tag{13}$$

Here $\varsigma_\sigma, \varsigma_m$ denote indeterminate scalar factors which distribute the jumps across the discontinuity surface. Requiring balance of linear and angular momentum of the bifurcated fields renders two constraint conditions

$$\operatorname{div}(\boldsymbol{\sigma}^{+,t}) = \operatorname{div}(\boldsymbol{\sigma}^{t}) + \varsigma_{\sigma}\operatorname{div}[[\dot{\boldsymbol{\sigma}}]] \doteq \mathbf{0} \tag{14}$$

$$\operatorname{div}(\mathbf{m}^{+}) + \mathbf{e} : \boldsymbol{\sigma}^{+} = \operatorname{div}(\mathbf{m}) + \mathbf{e} : \boldsymbol{\sigma} + \varsigma_{m}\operatorname{div}[[\dot{\mathbf{m}}]] + \varsigma_{m}\mathbf{e} : [[\dot{\boldsymbol{\sigma}}]] \doteq \mathbf{0} \tag{15}$$

Since the amplitude $\varsigma_m$ of the discontinuity can not vanish, the second constraint condition requires that

$$\mathbf{e} : [[\dot{\boldsymbol{\sigma}}]] = \mathbf{e} : (\dot{\gamma}^{u}\mathbf{E}_{ep} : (\mathbf{M}^{u} \otimes \mathbf{N}) + \dot{\gamma}^{\omega}\mathbf{D}_{ep} : (\mathbf{M}^{\omega} \otimes \mathbf{N})) \doteq \mathbf{0} \tag{16}$$

This implies that discontinuous bifurcation may only take place if the localization tensor is singular, and additionally, if the bifurcated stress field is self-equilibrated. This condition holds either when the jumps in the strain- and curvature fields are symmetric, i.e. when $\mathbf{M} = \mathbf{N}$, or when the two material operators are symmetric, i.e. $\mathbf{E}_{ep} = \mathbf{E}^{t}_{ep}$ and $\mathbf{D}_{ep} = \mathbf{D}^{t}_{ep}$. This is the case in the degenerate couple-stress theory of symmetric stresses, when the micropolar shear modulus approaches zero.

## 4. EXAMPLES and CONCLUSIONS

For illustration a rectangular specimen of the size 120 × 60 mm has been subjected to uniform compression under plane strain. The limit load /bifurcation problem is studied using a consistent refinement of four node bilinear and biquadratic Cosserat elements. The loading is applied via arc-length control using equality constraints of the vertical motion at the top surface to enforce uniform compression. The material formulation is based on the extension of the $J_2$-model to micropolar elasto-plasticity [1,2], where the yield condition is $F = F(\boldsymbol{\sigma}, \mathbf{m}) = \sqrt{3J_2} - Y = 0$, and its generalization to pressure-sensitive behavior follows the Drucker-Prager model $F = F(\boldsymbol{\sigma}, \mathbf{m}) = \sqrt{3}\alpha_F I_1 + \sqrt{3J_2} - Y = 0$. Here $I_1 = \boldsymbol{\sigma} : \mathbf{I}$ and $J_2$ denotes the generalized second invariant of deviatoric stress $J_2 = \frac{1}{4}(\mathbf{s} : \mathbf{s} + \mathbf{s} : \mathbf{s}^t) + \frac{1}{2}\mathbf{m} : \mathbf{m}$. The elastic material moduli are $E = 4000\ MPa$, $\nu = 0.25$, $\ell_c = 1.0\ mm$, $\mu_c = 2000\ MPa$,. The strain softening plastic properties are defined by $Y_o = 100\ MPa$, $h = -400\ MPa$, $\alpha_F = 0.2$. Fig. (1) depicts the response predictions of the von Mises and the Drucker-Prager models at the peak response when the 12 × 24-mesh is utilized. Only the upper right hand quadrant was analyzed because of the underlying symmetry assumption. The grey-tone indicates the intensity of micropolar rotation which measures the extent of localized failure at peak. In both cases localization was induced by a 10% reduction of the yield strength in the lower left hand corner of the quarter specimen. No discontinous bifurcation took place in either of the two model problems. Therefore, the numerical solutions did not exhibit strong mesh-sensitivity and converged with mesh-refinement as observed in ref. [1]. Considering however the two very different diffuse failure modes, their relationship to the actual failure mechanism has to be questioned?

Clearly, the introduction of a higher order continuum theory with additional field variables, which in turn introduces a length scale into the constitutive description, delays and often prohibits the loss of ellipticity when compared to classical continuum theory. However, there is no assurance that the Cosserat formulation will suppress discontinuous localization altogether, particularly when lack of normality and strain softening strongly destabilize the governing tangential operators.

## REFERENCES

[1] R. de Borst, "A Generalisation of $J_2$-Theory for Polar Continua", TNO-IBBC Report BI-89-195, Delft University of Technology, Delft (1989).

[2] E. Cosserat and F. Cosserat, "Théorie des Corps Déformables", Herman et fils, Paris, (1909).

[3] H.-B. Mühlhaus, "Application of Cosserat Theory in Numerical Solutions of Limit Load Problems" Ingenieur Archiv, Vol. 59, No.2, pp. 124-137 (1989).

[4] P. Steinmann and K. Willam, "Localization in Micropolar Elasto-Plasticity", Proc. 3rd Int. Conf. 'Constitutive Laws for Engineering Materials', C.S.Desai et al. eds., ASME-Press New York, pp. 461-465 (1991).

[5] P. Steinmann and K. Willam, "Localization within the Framework of Micropolar Elasto-Plasticity", V. Mannl ed., 60th Anniv.Volume Prof. Lippmann, Springer Verlag Berlin, (1991).

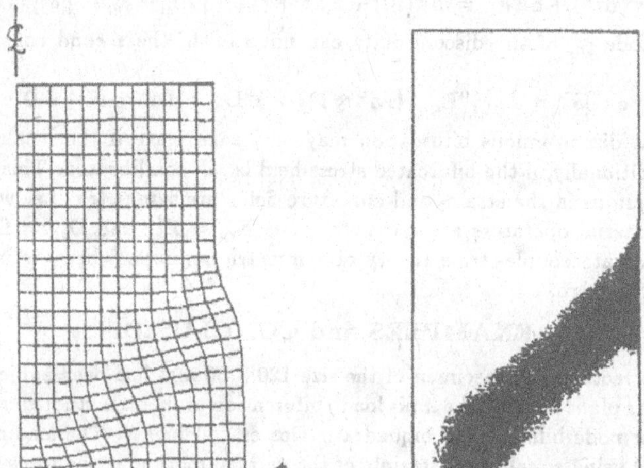

Figure 1 von Mises Model: Deformed Mesh and Intensity of Micro-Polar Rotations

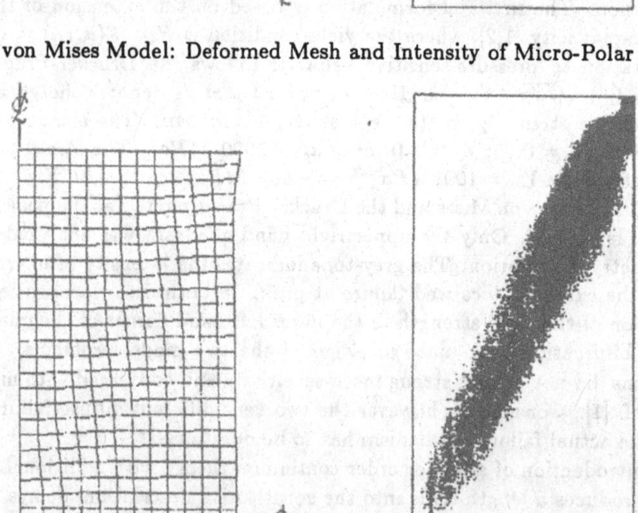

Figure 2 Drucker-Prager Model: Deformed Mesh and Intensity of Micro-Polar Rotations

# ADIABATIC SHEAR BAND LOCALIZATION IN ELASTIC-PLASTIC SINGLE CRYSTALS

MARIA K. DUSZEK-PERZYNA and PIOTR PERZYNA

Institute of Fundamental Technological Research

Polish Academy of Sciences, Warsaw, Poland

## ABSTRACT

The main objective of the paper is the investigation of shear band localization criteria for finite elastic-plastic deformations of single crystal subjected to adiabatic process. The next objective is to focus attention on temperature dependent plastic behaviour of single crystal considered. A constitutive model is developed within the thermodynamic framework of the rate type covariance constitutive structure, i.e., it is invariant with respect to diffeomorphism.

## INTRODUCTION

To described experimentally observed facts, namely that the shear band localization in single crystals may occur for small positive value of the plastic hardening rate $h_{crit}$, and that the formed shear bands are misaligned with the active slip systems in the crystal's matrix we intend to consider the synergetic effects generated by simultaneous incorporation of the spatial covariance effects and the thermomechanical couplings. In chapter 2 the experimental observations of localizations of plastic deformations in single crystals are discussed. Heuristic considerations of the different localization modes and physical motivation for the present research are presented. Chapter 3 is devoted to the development of a constitutive model within the thermodynamic framework of the rate type covariance constitutive structure. The Lie derivative is used to define objective rates. Thermomechanical couplings are investigated. In Chapter 4 the macroscopic shear band formation during adiabatic process for single slip is studied. For some simplified cases the criteria for adiabatic shear band localization are obtained in exact analytical form. The discussion of the results obtained is presented in chapter 5.

## EXPERIMENTAL AND PHYSICAL MOTIVATION

From the analysis of experimental investigations of localized shearing in single crystals performed by CHANG and ASARO [2] and SPITZIG [5] during the uniaxial tests it follows that in the first stage of the process a crystal specimen undergos uniform extension in single slip. At the moment when the load-engineering strain trajectory reaches its maximum, a crystal specimen exhibits slight amount of very diffuse necking. At this stage of the process the thermomechanical coupling effects begin to play crucial role. With continued extension macroscopic, adiabatic shear bands have soon developed within the diffusely necked region. It has been observed that at the inception of the macroscopic, adiabatic shear band, the critical value of the strainhardening rate $h_{crit} = \frac{\partial \tau}{\partial \gamma}$ is small but positive and that the macroscopic shear bands are not aligned with the active slip systems in the crystal's matrix but are misaligned by angle $\delta$.

To describe these experimentally observed facts, we intend to consider synergetic effects resulting from taking into account spatial covariance effects and thermomechanical couplings.

## CONSTITUTIVE STRUCTURE FOR CRYSTAL

### Constitutive postulates

For rate independent behaviour of single crystals we adopt the Schmid law. Namely, a slip initiates on system $\alpha$ when the Schmid resolved shear stress $\tau^{(\alpha)}$ reaches a critical value $\tau_c^{(\alpha)}$,

$$\tau^{(\alpha)} = \tau_c^{(\alpha)}. \tag{1}$$

The work hardening-softening law is postulated in the from

$$\tau_c^{(\alpha)} = \tau_c^{(\alpha)}(\gamma, \vartheta), \tag{2}$$

where $\gamma = \sum_{\alpha=1}^{n} \gamma^{(\alpha)}$ is shear strain and $\vartheta$ is temperature.

### Thermodynamic restrictions

Introducing the following assumptions: (i)conservation of mass, (ii) balance of momentum, (iii) balance of moment of momentum, (iv) balance of energy, (v) entropy production inequality, as well as two fundamental postulates: (i) existence of the free energy function $\hat{\psi}$ and (ii) the axiom of entropy production, we get the results as follows

$$\boldsymbol{\tau} = \rho_{Ref} \frac{\partial \hat{\psi}}{\partial \mathbf{e}}, \quad \eta = -\frac{\partial \hat{\psi}}{\partial \vartheta}, \tag{3}$$

$$\vartheta \hat{i} - \frac{1}{\rho \vartheta} \mathbf{q} \cdot \operatorname{grad} \vartheta \geq 0, \tag{4}$$

where $\mathbf{e}$ is the Eulerian strain tensor and

$$\vartheta \hat{i} = -\sum_{\alpha=1}^{n} \frac{\partial \hat{\psi}}{\partial \gamma^{\alpha}} \dot{\gamma}^{(\alpha)} \tag{5}$$

denotes the rate of the internal dissipation.

Introducing the denotation

$$\chi\tau^{(\alpha)} = -\rho\frac{\partial\hat{\psi}}{\partial\gamma^{(\alpha)}} \tag{6}$$

we have

$$\vartheta\hat{i} = \chi\sum_{\alpha=1}^{n}\frac{1}{\rho}\tau^{(\alpha)}\dot{\gamma}^{(\alpha)}. \tag{7}$$

The coefficient $\chi$ is determined from the equation

$$\chi\rho_{Ref}\frac{\partial\hat{\psi}}{\partial\mathbf{e}} : \mathbf{N}^{(\alpha)} = -\rho\frac{\partial\hat{\psi}}{\partial\gamma^{(\alpha)}}, \tag{8}$$

where $\mathbf{N}^{(\alpha)}$ is the symmetric part of the tensor $\mathbf{s}^{(\alpha)}\mathbf{m}^{(\alpha)}$, and $\mathbf{s}^{(\alpha)}$, $\mathbf{m}^{(\alpha)}$ denote the slip direction vector and the slip plane normal vector, respectively.

## Rate type constitutive relation

Assuming fixed prior history and operating on the stress relation $(3)_1$ with the Lie derivative we obtain

$$(\overset{el}{L_{\boldsymbol{v}}\boldsymbol{\tau}}) = \mathcal{L}^e : \mathbf{d}^e - \theta\mathbf{Z}\dot{\vartheta}, \tag{9}$$

where

$$(\overset{el}{L_{\boldsymbol{v}}\boldsymbol{\tau}})^{ij} = \dot{\tau}^{ij} - \tau^{ik}(d^e_{lk} + w^e_{lk})g^{lj} - \tau^{kj}(d^e_{lk} + w^e_{lk})g^{il} \tag{10}$$

$$\mathcal{L}^e = \rho_{Ref}\frac{\partial^2\hat{\psi}}{\partial\mathbf{e}^2}, \quad \theta\mathbf{Z} = -\rho_{Ref}\frac{\partial^2\hat{\psi}}{\partial\mathbf{e}\partial\vartheta}. \tag{11}$$

Denoting by

$$\mathbf{b}^{(\alpha)} = (\mathbf{N}^{(\alpha)} + \mathbf{W}^{(\alpha)})\cdot\boldsymbol{\tau} + \boldsymbol{\tau}\cdot(\mathbf{N}^{(\alpha)} - \mathbf{W}^{(\alpha)}) \tag{12}$$

we finally have the resulting constitutive law of the rate type in the form

$$L_{\boldsymbol{v}}\boldsymbol{\tau} = \mathcal{L}^e : \mathbf{d} - \sum_{\alpha=1}^{n}(\mathcal{L}^e : \mathbf{N}^{(\alpha)} + \mathbf{b}^{(\alpha)})\dot{\gamma}^{(\alpha)} - \theta\mathbf{Z}\dot{\vartheta}. \tag{13}$$

## Adiabatic process

It is proved that for the adiabatic process the evolution equation for temperature can be simplified to the form

$$\rho c_p\dot{\vartheta} = \chi\sum_{\alpha=1}^{n}\tau^{(\alpha)}\dot{\gamma}^{(\alpha)}, \tag{14}$$

where the coefficient $\chi$ is determined by Eq.(8).

## Adiabatic process for single slip

Differentiation of (1) written for a single slip with respect to time gives

$$\dot{\gamma} = \frac{1}{h}\dot{\tau} + \pi\dot{\vartheta}, \tag{15}$$

where

$$h = \frac{\partial \tau_c}{\partial \gamma}, \quad \pi = -\frac{1}{h}\frac{\partial \tau_c}{\partial \vartheta}. \tag{16}$$

Operating on the relation defінding Schmid resolved shear stress for single slip with the Lie derivative, we obtain

$$\dot{\tau} = (\overset{el}{L_{\mathbf{v}}}\boldsymbol{\tau}) : \mathbf{N} + 2\boldsymbol{\tau} \cdot (\mathbf{N} + \mathbf{W}) : \mathbf{g} \cdot \mathbf{d}^e. \tag{17}$$

Relations (15), (17) and (13) yields

$$\dot{\gamma} = \frac{1}{h}(\mathcal{L}^e : \mathbf{N} + \mathbf{b}) : (\mathbf{d} - \mathbf{N}\dot{\gamma}) - \frac{1}{h}\theta\mathbf{Z} : \mathbf{N}\dot{\vartheta} + \pi\dot{\vartheta}. \tag{18}$$

Eliminating $\dot{\gamma}$ and $\dot{\vartheta}$ from the set of equations (13), (14), (18) we obtain the fundamental rate equation in the form

$$L_{\mathbf{v}}\boldsymbol{\tau} = \mathbb{L} : \mathbf{d} \tag{19}$$

where

$$\mathbb{L} = \left[\mathcal{L}^e - \frac{(\mathbf{Q}^* + \frac{\Theta}{\mu}\mathbf{Z}\boldsymbol{\tau} : \mathbf{N})\mathbf{Q}^*}{h - \Pi\boldsymbol{\tau} : \mathbf{N} + (\mathbf{Q}^* + \frac{\Theta}{\mu}\mathbf{Z}\boldsymbol{\tau} : \mathbf{N}) : \mathbf{N}}\right], \tag{20}$$

$$\mathbf{Q}^* = \mathcal{L}^e : \mathbf{N} + \mathbf{b}, \tag{21}$$

$$\Theta = \frac{\chi\theta}{\rho c_p}\mu, \quad \Pi = \frac{\chi}{\rho c_p}h\pi. \tag{22}$$

The scalar coefficients $\Theta$ and $\Pi$ describe the thermal expansion and thermal plastic softening effects respectively ($\mu$ denotes the elastic shear modulus).

## Shear band localization criteria

Taking advantage of the constitutive rate equations (19)-(21) and applying standard bifurcation method (cf.RICE [3] and RUDNICKI and RICE [4]), as well as perturbation procedure developed by ASARO and RICE [1] we obtain the critical hardening rate at the onset of localization and the orientation shear band as follows:

$$h_{crit} = \Pi\tau + \frac{1}{4}\tau^2\left(\frac{\Theta}{\mu}\mathbf{Z} : \mathcal{N} + 2\mathbf{s}\right) \cdot (\mathbf{s} \cdot \mathcal{M} \cdot \mathbf{s})^{-1} \cdot \left(\frac{\Theta}{\mu}\mathbf{Z} : \mathcal{N} + 2\mathbf{s}\right), \tag{23}$$

$$\mathbf{n} = \mathbf{m} + \frac{1}{2}\tau(\mathbf{s} \cdot \mathcal{M} \cdot \mathbf{s})^{-1} \cdot \left(\frac{\Theta}{\mu}\mathbf{Z} : \mathcal{N} + 2\mathbf{s}\right), \tag{24}$$

where

$$\begin{aligned} \mathcal{M} &= \mathcal{L}^e - (\mathcal{L}^e \cdot \mathbf{m}) \cdot (\mathbf{m} \cdot \mathcal{L}^e \cdot \mathbf{m})^{-1} \cdot (\mathbf{m} \cdot \mathcal{L}^e), \\ \mathcal{N} &= \mathbf{s1} - \mathbf{m}(\mathbf{m} \cdot \mathcal{L}^e \cdot \mathbf{m})^{-1} \cdot (\mathbf{m} \cdot \mathcal{L}^e \cdot \mathbf{m}). \end{aligned} \tag{25}$$

## DISCUSSION OF THE RESULTS

### Thermal effects

There are two thermal effects which influence the criteria for adiabatic shear band localization, namely thermal expansion and thermal plastic softening.

Thermal expansion affects the critical hardening rate $(h/\tau)_{crit}$ cf.Eq.(23), as well as the direction of the shear band **n**, cf.Eq.(24), while thermal plastic softening influences only the critical hardening rate $(h/\tau)_{crit}$.

The contribution to the critical hardening rate generated by thermal expansion is small and consists of two terms: $2(\tau/\mu)_{crit}\nu\Theta$ and $2(\tau/\mu)_{crit}\Theta$. The first term is 5 times smaller than the second which has synergetic nature. The second term represent the cooperative phenomena of thermal expansion and spatial covariance effects.

The main contribution to the critical hardening rate is implied by thermal plastic softening represented by $\Pi$. This term dominates the result and is of the order of $10^{-2}$ while the thermal expansion term is of the order of $10^{-4}$.

### Misalignment of the shear bands

The misalignment of the direction of the macroscopic shear bands from the active slip systems in the crystal's matrix is affected by two phenomena, namely by thermal expansion and spatial covariance effects.

The spatial covariance term $\frac{1}{4\nu}(\tau/\mu)_{crit}$ has dominated influence on the direction of the shear band and is 2.5 times larger than the thermal expansion term $\frac{1}{2}\Theta(\tau/\mu)_{crit}$.

The theoretical results for the angle $\delta$ at the inception of the shear band localization have been estimated for aluminum-cooper single crystals tested at 298K and 77K and for nitrogenated Fe-Ti-Mn single crystals tested at 295K. It has been found that obtained values of the angle $\delta$ are small but sufficiently distinct to explain experimentally observed misalignment.

## REFERENCES


1. Asaro, R.J. and Rice, J.R., Strain localizationin ductile single crystals. J.Mech.Phys.Solids, 1977, **25**, 309-338.

2. Chang, Y.W. and Asaro, R.J., An experimental study of shear localization in aluminium-cooper single crystals. Acta Metall., 1981, **29**, 241-257.

3. Rice, J.R., The localization of plastic deformation. In Theoretical and Applied Mechanics, ed. W.T.Koiter, N.-Holland, Amsterdam, 1977, pp.207-220.

4. Rudnicki, J.W and Rice, J.R., Conditions for the localization of deformation in pressure-sensitive dilatant materials. J.Mech.Phys.Solids, 1975, **23**, 371-394.

5. Spitzig, W.A., Deformation behaviour of nitrogenated Fe-Ti-Mn and Fe-Ti single crystals. Acta Metall., 1981, **29**, 1359-1377.

# THE GENESIS OF S-LIKE FRACTURE SHAPES DURING TRANSVERSE SHEARING OF METALS

EDWARD STANISLAW DZIDOWSKI
Technical University of Wrocław, I-24
Wybrzeże Wyspiańskiego 27, Poland

## ABSTRACT

A new approach to the genesis and the possibility of changing the S-like shape of the fracture's that forms during the transverse shearing of metals is presented. It is shown that the shape depends on the development of the location of strains in microscopic shear bands and particularly on the macroscopic range of the location and the interaction between the shear bands and the grain boundaries.

## INTRODUCTION

Controlled decohesion is the essence of a large group of plastic working processes connected with the shaping or separation (shearing) of materials by the overcoming of their cohesion.

At present, one can state that an insufficient knowledge of the mechanism of shear fracture is the main barrier for the further development of the technology of shearing of metals.

The model of shear fracture (resulting from the nucleation and the coalescence of oval voids) does not explain a lot of the phenomena that occur at large plastic strains.

The aim of this paper is to prove that the genesis of the S-like shape of the fracture surface can be explained by the changes in the structure that accompany the deformation of the sheared metal.

## MATERIALS AND METHODS

Tests were conducted on 0.04% C armco iron. Specimens of 100x25x15 (2.4) mm were sheared in a device powered by a hydraulic press. The temperature of shearing was 293 K and the rate of shearing was about 0.02 mm/s. The gap between the cutting-edges was about 0.05 mm. Gradual slitting of the specimens was adopted. The side surface of the specimens were polished and etched before each slitting operation. It made possible later investigation of the development of shear bands and the course of cracking depending on the shape of grains and the state of their boundaries. An optical microscope and a scanning electron microscope were used for the studies. The course of shearing force $P_c$ was recorded as a function of displacement h of the cutting-edge of the moving cutting tool.

## RESULTS

The macroscopic course of cracking: Fig. 1 shows the development of the macrolocation of strains depending on displacement h of the cutting-edge and the value of shearing force $P_c$. The macrolocation of strains starts directly after the loss of stability by the shearing force (Fig. 1b). The macroscopic zone of the location of strains assumes initially the shape of a biconvex lens. The shape changes as the fractures increase (Fig. 1d, e, f). At the same time, the heigt of the macroscopic zone of strains location and its width increase. As it follows from the figure, the macroscopic development of cracking in closely bound up with the boundaries of the location zone. An increase in the width of the location zone is accompanied by increased divergence of both the opposite fractures.

The microscopic course of cracking: The electronography shown in Fig. 2 illustrates the development of shearing bands SBs within the strains location zone and the way in which the shear bands interact with grain boundary $GB_1$.

The cracking propagates along the shear bands situated near the boundary of the strains location zone where the greatest separations of the material along shear bands SB (displace-

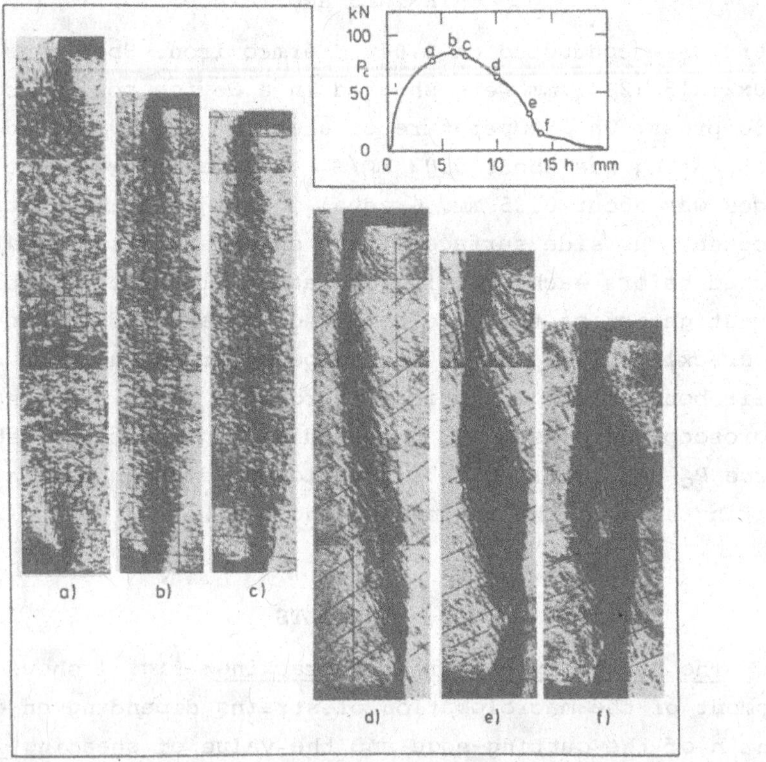


Figure 1. The evolution of the shape of zone MLS depending on displacement h of the blade: a, b, c, d, e - displacement increment $\Delta h$ = 1 mm, f - $\Delta h$ = 2 mm. Steel 0.04% C. Optical microscop.

ment $\Delta l$) are observed - Fig. 2. The shear bands that develop within the location zone usually do not cross over into the neighbouring grains. As a result, wedge-shaped microcracks appear along boundary $GB_1$.

## CONCLUSIONS

1. The course of cracking during the classic transverse shearing of metals depends on the development of the location of strains in the shear bands and the effects of the interactions between the bands and the grain boundaries.
2. The macroscopic shape of the fracture's surface is determined mainly by the presence and the intensity of displacements

Figure 2. Electronography (SEM) of the border area of zone MLS. Visible shear bands SBsI and boundary wedge microcracks and transverse displacements of the material along boundary $GB_1$: Δl - displacement associated with the total superaction of the material along one of shear bands SBsI. Steel 0.04% C.

of the material along the defective grain boundaries.

3. An intervention into the course of the location of strains and the change of the shape of grains and the state of their boundaries may be the key to the solution of the fundamental problems of controlled decohesion, i.e. to the elimination of cracking or the control of the course of cracking aimed at obtaining the assumed shape of the fracture.

# A NOTE ON SHEAR BANDING AND MATERIAL INSTABILITY IN FINITELY DEFORMED COMPRESSIBLE MATERIALS: AN EXAMPLE IN THIN SHEETS

Y. K. LEE
George W. Woodruff School of Mechanical Engineering
Georgia Institute of Technology
Atlanta, GA 30332-0405

## ABSTRACT

A finite elastoplasticity model for compressible materials is reviewed in order to emphasize some special features of the model. The criterion for material instability has an invariant form and is embodied exclusively in the constitutive relation. Shear band analysis accounts for both noncoaxial boundary loading and noncoaxial strain path. The influence of both the plastic shear modulus and the stress state over the shear band behavior is identified. As an illustrative example, the case of proportional loading is discussed. The results are in good agreement with experimental data from the open literature and thus demonstrate the efficacy of the model.

## INTRODUCTION

This paper continues a study made in a previous paper[1] in which necessary and sufficient conditions ıor the onset of shear banding were studied for plane strain and axisymmetric deformations. In the analysis, shear banding was regarded as a geometrical instability and the onset of the instability was defined at the instant when the complete governing field equations cease to be elliptic. The analysis results showed that, under a normality condition, shear bands might occur before the material reached its peak load. In addition, the results also showed that shear bands might form at angles other than 45 degrees inclined with respect to the principal axes. The outcome of the analysis provided plausible explanations for many experimental observations. These findings were attributed to the fact that the analysis considered the rate change of the equilibrium equations and accounted for both noncoaxial boundary loading and noncoaxial strain path, which are characteristic of finite strain deformations. As this new development has evolved, we now extend the analysis to deformations in plane stress. The objective is two fold: (1) to identify key analysis features that are distinctive from traditional bifurcation studies, and (2) to validate the elastoplasticity model by comparing the analysis results with the experimental data obtained from the open literature.

As in the previous study[1], the present shear banding analysis is posted as a boundary value problem. The analysis properly treats the noncoaxialities caused by large rotation and enforces the equilibrium requirements throughout the deformation. Without the complete representation of the equilibrium state, the analysis may not capture the noncoaxial characteristics of the deformation. As a result, the analysis may unknowningly retain a locally small-strain assumption.

The material instability on the basis of physical grounds can be defined at an instant when the material shows no further load-carrying capacity, i.e., when the constitutive relation loses its invertibility between the stress rates and the rates of deformation[2]. Under this definition, material instability has no direct bearing on shear band localization and is governed by the constitutive relation alone.

## FORMULATION OF THE RATE PROBLEM

We now elaborate previous development[2] for the general formulation of a finite elastoplastic flow theory for compressible materials. In particular, we focus on two key elements: 1. the establishment of a consistent reference frame for analysis and 2. the enforcement of equilibrium requirements during deformation. Treating the porous material as a continuum, we describe the total deformation of $B_o$ into

$B$ in a fixed reference frame by the mapping

$$x^i = x^i(X^J, t), \qquad |x^i{}_{,J}| \neq 0. \tag{1}$$

In (1), $x^i$ are spatial coordinates of material particles comprising domain $B$ at time $t > 0$, and $X^J$ are the Lagrangian positions. Both $x^i$ and $X^J$ refer to the fixed coordinates $y^i$. The instantaneous motion of the body is described by the velocity field $v^i = v^i(x^i, t)$, from which the velocity gradient can be computed as $v_{i;j} = d_{ij} + w_{ij}$ with $d_{ij} = (v_{i;j} + v_{j;i})/2$ and $w_{ij} = (v_{i;j} - v_{j;i})/2$. The contravariant components of the Eulerian or Cauchy stress tensor are denoted $\sigma^{ij}$ with their rate change given by $\dot{\sigma}^{ij} = \partial\sigma^{ij}/\partial t + \sigma^{ij}{}_{;k}v^k$. The Cauchy stress tensor is a Eulerian field variable and provides a complete description of the instantaneous loading state. At the surface of the deforming body, the loading state is described via surface tractions $t^i$ that follow the Cauchy formula $t^i = \sigma^{ij}n_j$, where $n_j$ are the covariant components of unit outer normal vector to the surface. At any instant of time, the instantaneous rate of change of the tractions can be found by differentiating the Cauchy fomula directly. The result is

$$\dot{t}^i = (\dot{\sigma}^{ij} + \sigma^{ij}n^l n_k v^j{}_{;l} - \sigma^{ik}v^j{}_{;k})n_j \tag{2}$$

The traction rate in (2) accounts for the change of both area and direction of the deforming surface; thus it characterizes any deformation sensitive loading. Terms associated with the velocity gradients are deformation sensitive and usually do not vanish. They provide a means to describe any noncoaxial boundary loading that is caused by an instantaneous surface deformation.

### Equilibrium requirements

Since deformation is considered as a rate problem, requirements of mechanical equilibrium must be applied not only to the instantaneous state of the stresses but also to their rate change. To maintain the deforming body in equilibrium during time varying loading, it requires that the time rate of the net applied force be zero. The result is the stress rate equations of equilibrium

$$\dot{\sigma}^{ij}{}_{;j} - \sigma^{ki}{}_{;j}v^j{}_{;k} = 0 \qquad \text{in} \quad B. \tag{3}$$

This equation gives a complete description for the equilibrium state of material particles both in the interior and at the boundary of the deforming body. Equation (3) in fact can be taken as an axiom describing the law of quasistatic motion for a deforming body. The equation is in a rate form, and thus it governs the stress field irrespective of deformation magnitude and material structure. Satisfaction of (3) not only implies total stress equilibrium but also assures that, given an equilibrated stress field, equilibrium is maintained in the presence of time varying loading. The equations include terms accounting for the deformation sensitive loading and the convective effect of the nonhomogeneous field. These two elements usually are coupled and need to be addressed simultaneously. Without the load sensitive and the convective terms, the equation would reduce to that for small strain deformation. Equations (2) and (3) fully describe the equilibrium requirements for the rate deformation problems and can be adopted to accommodate a broad variety of constitutive models.

### Constitutive relation

The derivation of elastoplastic flow equations employs the Jaumann rate of Cauchy stress. In deriving the constitutive relation, three primary assumptions were adopted:

A1) Additive splitting of the deformation rate into elastic ($e$) and plastic ($p$) parts, i.e., $d_{ij} = d^{(e)}_{ij} + d^{(p)}_{ij}$;

A2) Existence of a load function, $F = \Phi(I_1, J_2) - \kappa(W^{(p)}(\rho)) \leq 0$;

A3) Drucker's hypothesis $\overset{\circ}{\sigma}{}^{ij}d^{(p)}_{ij} \geq 0$, where the Jaumann rate $\overset{\circ}{\sigma}{}^{ij} = \dot{\sigma}^{ij} + \sigma^i{}_m w^{mj} - \sigma_m{}^j w^{im}$.

In (A1) the elastic part of the deformation rate follows the generalized Hooke's law while the plastic part obeys the flow rule associated with the yield function $\Phi$ in (A2), where $\Phi$ is the apparent yield function of the aggregate which depends upon the first invariant of the stress tensor $I_1$ and the second invariant of the stress deviator $J_2$. The inclusion of $I_1$ gives the provision for treating aggregate dilation and compressibility. Note that in (A2), $\Phi$ is a scalar measure of the stress vector. The apparent strain hardening function $\kappa$ is also a scalar which measures the degree of hardening experienced by the aggregate and depends solely on the prior plastic strain energy density $W^{(p)}$. Here $W^{(p)}$ is an implicit function of the relative density $\rho$ of the aggregate, which is defined as the ratio of the aggregate density to the matrix density. The inequality sign holds for the elastic states of stress; the equality sign holds for the plastic states and represents a yield surface. Statement (A3) implies the convexity of the yield surface, normality of $d_{ij}^{(p)}$ to that surface, and the existence of a linear relation between $\hat{\sigma}_{ij}$ and $d_{ij}^{(p)}$[3].

With the above bases, a general flow rule for compressible materials can be derived in the form

$$2\,\mu d_{ij} = [(g_{ik}g_{jl} + g_{il}g_{jk})/2 - \nu g_{ij}g_{kl}/(1+\nu) + \beta(\partial\Phi/\partial\sigma^{ij})(\partial\Phi/\partial\sigma^{kl})]\hat{\sigma}^{kl} \tag{4}$$

$$\beta = \{1/[(1/3)(2\mu_{oc}^{(p)}/\tau_{oc})(\dot{\tau}_{oc}/\tau_{oc} - d_{nn})/(\dot{\tau}_{oc}/\tau_{oc})]\}\{2\mu/[(\partial\Phi/\partial\sigma^{rs})\sigma^{rs}]\} \tag{5}$$

and the inverse is

$$\hat{\sigma}^{ij} = D^{ijkl}d_{kl} \tag{6}$$

$$D^{ijkl} = \mu\,\{(g^{ik}g^{jl} + g^{il}g^{jk}) + [2\,\nu/(1-2\,\nu)]g^{ij}g^{kl}\} - \frac{2\,\mu(\phi_{ab} + [\nu/(1-2\nu)]\phi_{pp}g_{ab})(\phi_{cd} + [\nu/(1-2\nu)]\phi_{qq}g_{cd})g^{ai}g^{bj}g^{ck}g^{dl}}{1/\beta + \phi_{mn}(\phi_{rs} + [\nu/(1-2\nu)]\phi_{tt}g_{rs})g^{mr}g^{ns}} \tag{7}$$

where $\mu$ is the modulus of the aggregate and can be related to that of the matrix material, $\mu_m$, by $\mu = \rho\mu_m$; $\nu$ is the Poisson's ratio of the aggregate, the same as that of the matrix material; $\phi_{ij} = \partial\Phi/\partial\sigma^{ij}$ and $\phi_{ii}$ measures the degree of compressibility of the plastic strain. With the subscripts $oc$ referring to the octahedral planes of stress and strain, the octahedral plastic shear modulus of the matrix material $\mu_{oc}^{(p)}$ is defined as $2\mu_{oc}^{(p)} = \dot{\tau}_{oc}/\dot{\gamma}_{oc}^{(p)}$, in which $\gamma_{oc}^{(p)}$ is the octahedral plastic shear strain and can be found by integrating the rate of octahedral plastic shear strain $d_{oc}^{(p)}$ with respect to time, i.e., $\gamma_{oc}^{(p)} = \int_t d_{oc}^{(p)}dt$. This modulus can be measured from a single experiment on the constituent matrix material at full density. Furthermore the apparent strain hardening rate represented by the quantity inside the first bracket in (5) can be completely determined from that of the matrix material (i.e., $2\mu_{oc}^{(p)}/\tau_{oc}$) because the total dilation rate $d_{nn}$ which is defined as $-\dot{\rho}/\rho$ can be found from the flow data for a specific deformation.

Equation (6) is a general elastoplastic continuum model for compressible materials. Some special features of the constitutive model are:

1. The apparent strain hardening rate of the aggregate is a scalar function of stress and strain invariants.
2. Since load carrying capacity of the aggregate diminishes when the constitutive relation looses its invertibility, the term $(\dot{\tau}_{oc}/\tau_{oc} - d_{nn})$ in (5) actually is an invariant that defines a criterion for material instability during deformation.
3. The material instability may be regarded as a material property because of its invariant character.

The above results suggest that the detection of material failure for deforming solids can be fully incorporated in the constitutive model and becomes a natural outcome of the analysis. This is different from an uncoupled study, where a stress analysis is carried out and then a failure criterion is imposed.

To include compressibility and dilatancy, we require that the first stress invariant be included in the yield function and take $\Phi = [2J_2/3 + (I_1/f)^2]^{1/2}$ with $f = -(2/\rho)\ln(1-\rho)$. The yield function gives an exclusive dependence of the aggregate yield on the properties of the constituent matrix material[2].

So far, the complete governing field equations are written from a spatial viewpoint, in terms of the tensor of Cauchy stress rate $\dot{\sigma}^{ij}$ and velocity vector $v^i$. All the field variables refer to the fixed coordinate system and are derived based on an instantaneous configuration of the continuum. Because the reference coordinate system is fixed in the space, the coordinates of a particle always represent a Lagrangian position. Therefore, the description of the deformable body requires only a direct update of the Cauchy stress $\sigma^{ij}$ and the coordinate $x^i$ at the instantaneous position, by adding the corresponding rates $\dot{\sigma}^{ij}$ and $v^i$ that are to be solved from the field equations. Based upon this formulation, the solution procedure virtually becomes routine because it requires no transformation for the stress rates between two reference states. Hence it incurs no transformation error in the analysis. This is especially important when the rate problem is regarded as incremental and is solved by numerical means[4].

## SHEAR BANDING IN THIN SHEETS

Representing the Jaumann rate in terms of the Cauchy stress rate and the rate of deformation in terms of velocity gradient in (6) and then combining the result with (3), one would have the Navier's equations for the compressible elastoplastic flow. For planar cases, the characteristic equation of the Navier's equations is given by a canonical form[1]

$$a(1+\bar{N})\Delta^4 + 2b\Delta^2 + c(1-\bar{N}) = 0 \tag{8}$$

where $\bar{N}$ is the normalized principal tangential stress and is defined as $\bar{N} = (\sigma_{11} - \sigma_{33})/2\mu$ with the principal stresses $\sigma_{11}$ and $\sigma_{33}$. Coefficients $a, b$, and $c$ are functions deduced from $D^{ijkl}$. For $\bar{N} < 1$, the necessary and sufficient condition for the onset of shear banding (i.e., $\Delta$ to be real) is that[1]

$$\text{(i)}\quad |\bar{N}| = [(ac-b^2)/ac]^{1/2} \qquad \text{and} \qquad \text{(ii)}\quad b < 0 \tag{9}$$

hold simultaneously. The corresponding characteristic directions can be found by solving (8).

Equation (9)-(i) gives the critical value of $\beta$ and hence the critical value of $2\mu_{oc}^{(p)}$ for shear banding. By neglecting the terms involving the plastic compressibility in (7), it can be shown for plane stress deformation that

$$2\mu_{oc}^{(p)}\Big|_{cr} = \frac{3(1+\nu)}{2\mu} \cdot \frac{\dot{\tau}_{oc}}{(\dot{\tau}_{oc}/\tau_{oc} - d_{nn})} \cdot \frac{\phi_{11}^2 \phi_{33}^2 \sigma_{oc}^2}{\phi_{11}\sigma_{11} + \phi_{33}\sigma_{33}}, \tag{10}$$

below which shear bands may occur. By defining a stress ratio $R \equiv \sigma_{33}/\sigma_{11}$, we rewrite the above equation as

$$2\mu_{oc}^{(p)}\Big|_{cr} = \frac{27(1+\nu)\sigma_{oc}\dot{\tau}_{oc}}{4\mu(\dot{\tau}_{oc}/\tau_{oc} - d_{nn})} \cdot \frac{[(1+R)^2/f^4 + (1+R)^2/9f^2 + (2-R)(2R-1)/81]^2}{[(1+R)^2/f^2 + 2(2R^2-R+1)/9](R^2-R+1)} \tag{11}$$

At the onset of material instability, $\mu_{oc}^{(p)}\big|_{cr}$ becomes $\infty$ because $(\dot{\tau}_{oc}/\tau_{oc} - d_{nn})$ vanishes. This indicates that dilation, thus the increase of porosity, would promote shear band initiation. However, whether or not shear bands will occur depends on the inequality condition in (9)-(ii) as well. This inequality condition gives stress states which are favorable for shear banding. In terms of the stress ratio $R$, the inequality relation can be written as

$$\left(\frac{1}{f^4} + \frac{1}{9f^2} - \frac{2}{81}\right)R^2 + \left(\frac{2}{f^4} + \frac{2}{9f^2} + \frac{5}{81}\right)R + \left(\frac{1}{f^4} + \frac{1}{9f^2} - \frac{2}{81}\right) + O\left(\frac{\sigma_{oc}}{\mu}\right)^2 < 0. \tag{12}$$

The characteristic directions can be found from (8). At the onset of material instability, it can be shown that

$$\theta = \tan^{-1}\Delta \approx \tan^{-1}\left\{\pm\left[\frac{-[(1+R)/f^2 + (2-R)/9]}{(1+R)/f^2 + (2R-1)/9}\right]^{1/2}\right\}. \tag{13}$$

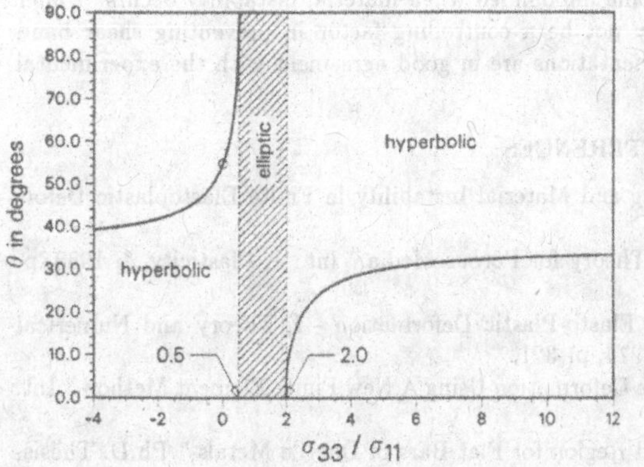


Fig. 1 $\theta$ as a function of stress ratio.

Equation (13) is shown graphically in Fig. 1 for the limit case at $\rho = 1$. Shear bands may occur in biaxial tension but not when $1/2 < R < 2$ if $\rho = 1$. As indicated by the circle, shear bands would occur along lines which form angles of $54^{\circ}44'$ with the tension axis $x_1$ during a homogeneous deformation. This is consistent with what has been shown in most textbooks. In general, however, the angle may vary because of the biaxial stress effect caused by necking. In fact, the density decrease evolved by biaxial tension is not favorable for shear banding, a converse concept from that deduced in (11).

For plane stress deformation, immediate integration of (4) is feasible under proportional loading. In this case, all stress components grow at a constant rate throughout the material and $\hat{\sigma}_{ij} = \dot{\sigma}_{ij}$. By expressing the coefficients in terms of their principal values, we then integrate (4) with respect to time. Consequently, the strain ratio $\alpha$ ($\equiv \int_t d_{33}dt / \int_t d_{11}dt$) can be found as

$$\alpha = \frac{(1+R)^3/f^4 + (1+R)(1+2R^2)/9f^2 + (2R-1)(2R^2-2R+2)/81}{(1+R)^3/f^4 + (1+R)(2+R^2)/9f^2 + (2-R)(2R^2-2R+2)/81} \tag{14}$$

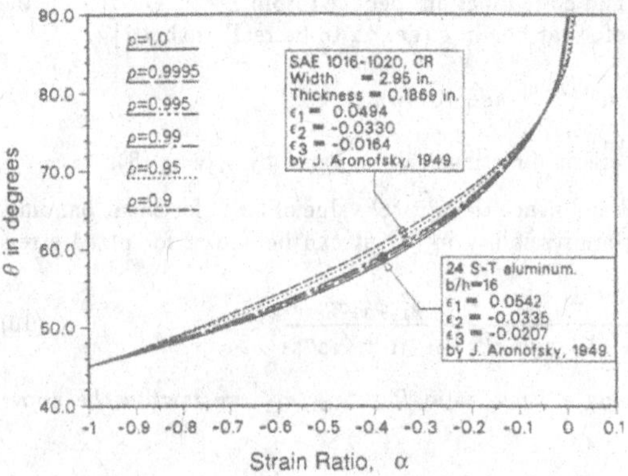


Fig. 2 $\theta$ as a function of strain ratio.

for $1/\beta = 0$. Figure 2 shows $\alpha$ vs $\theta$ for different values of $\rho$. Also shown in Fig. 2 are experimental data obtained by Aronofsky[5]. For both materials shown, the prediction of shear band orientations at $\rho = 1$ show merely a fraction of a degree of difference from the measurements. In viewing the narrow range of the variation of the predicted data, we may conclude that the theoretical model indeed gives a reliable prediction for the orientations of shear bands. Also noticeable from Fig. 2 is that, for a pure shear deformation (i.e., $\alpha = R = -1$), the orientation of shear bands is independent of the relative density and is always $45^o$ inclined from the principal axes. However, the angles would be different if diffuse necking occurs prior to shear band initiation.

## CONCLUSIONS

In this study, we have a clear pathway to the field equations for finite elastoplastic flow. The formulation accounts for both noncoaxial boundary loading and noncoaxial strain path. The field equations are realized in a straightforward manner, and initial and boundary conditions are explicit. There is no need to employ any transformation caused by the changing spatial frames during deformation because the field equations are derived strictly based on the fixed reference frame.

By virture of physical interpretation of the constitutive model, material instability can be defined at the instant when the constitutive relation loses its invertibility. The criterion for the instability is an invariant. Unlike the material instability, geometrical instability or shear banding does not result as a matter of necessity but that, depending on the stress state, it does sometimes arise. In the analysis, the influence of both the plastic shear modulus and the stress state over the shear band behavior is clearly identified. Under a condition of plane stress, the critical plastic shear modulus for the onset of shear banding is always positive and may become unbounded when material instability occurs. Under this circumstance, the material character may not be a controlling factor in preventing shear band localization. The predictions on shear band orientations are in good agreement with the experimental data.

## REFERENCES


1. Lee, Y. K., "Conditions for Shear Banding and Material Instability in Finite Elastoplastic Deformation," Int. J. Plasticity, 5, 1989, p. 197.
2. Lee, Y. K., "A Finite Elastoplastic Flow Theory for Porous Media," Int. J. Plasticity, 4, 1988, p. 301.
3. Osias, J. R. and Swedlow, J. L., "Finite Elasto-Plastic Deformation - I, Theory and Numerical Examples, " Int. J. Solids Structure, 10, 1974, p. 321.
4. Lee, Y. K., "Analysis of Finite Elastoplastic Deformation Using A New Finite Element Method," Int. J. Plasticity, 6, 1990, p. 521.
5. Aronofsky, J., "Investigation of the Necked Region for Flat Bars of Ductile Metals," Ph.D. Thesis, University of Pittsburgh, 1949, PA.

# SHEAR LOCALIZATION IN CRYSTALLOGRAPHIC BANDS IN ROLLED COPPER AND BRASS

T. LEFFERS AND N. HANSEN
Materials Department
Risø National Laboratory
DK-4000 Roskilde, Denmark

## ABSTRACT

Microbands in copper and bundles in brass both represent shear localization in bands parallel to {111}. The two types of bands are compared, and the possible relation to shear bands is discussed.

## INTRODUCTION

Two different types of shear localization have been observed in fcc materials rolled to moderate reductions: microbands (in our terminology second-generation microbands) in copper [1,2], aluminium-magnesium [3] and nickel [4] and bundles in brass and other fcc alloys with low stacking fault energy [5,6]. Both microbands and bundles are thin (0.5 μm or less) plates or bands closely parallel to {111} in which the strain (slip parallel to the plane of the band) is far greater than in the surrounding material.

Together with these obvious similarities between second-generation microbands and bundles there are a number of differences - rather great differences which raise the question whether the above similarities reflect a real similarity in the underlying physical processes or just a formal similarity. In the present work we investigate and discuss these similarities and differences. We also discuss the effects of microbands and bundles on the overall deformation pattern. Finally we discuss the possible relation between the microscopic shear localization as represented by the microbands and the bundles and the more macroscopic, non-crystallographic shear localization in shear bands.

## DESCRIPTION OF MICROBANDS

In copper rolled to moderate reductions there are two types of grains: grains with an equiaxed cell structure, LWD (low wall density) grains, and grains with a parallelogram-shaped pattern of dislocation walls, HWD (high wall density) grains [7]. Second-generation microbands form preferentially in the former type of grains, but they may also be observed in the latter. Leffers and Ananthan [8] have investigated the possible relation between the type of structure in a grain and its crystallographic orientation. No convincing overall relation was found. The only clear relation is that grains with orientations close to {110}<001> always have an LDW structure - with many second-generation microbands. Thus, the preliminary result is that there is no obvious general relation between the crystallographic orientation and the frequency of microbands. More detailed investigations of the possible orientation dependence will be presented in the lecture.

The investigation of the second-generation microbands is hampered by their structural instability as described in detail by Ananthan et al. [2,7]. It is a very common observation in TEM investigations that parts of microbands have vanished. Ananathan et al. [2] suggest that the microbands are eroded by subsequent slip processes, while Hatherly [9] suggests that the microbands disappear during thin-foil preparation. The reason for this instablity seems quite clear: when formed the walls of the microbands do not consist of geometrically necessary dislocations [10] since they do not separate areas with different lattice orientations (the concentrated slip on slip planes parallel to the walls does not per se create any orientation change in the material in the microbands, cf. [2]). According to the density of second-generation microbands observed by TEM the microbands only account for about 10% of the strain in the grains with microbands (the LWD grains) [2], but because of the instability this is obviously an underestimate. Surface observations [9] have been quoted to indicate that the microbands account for the majority of the strain in a range of moderate rolling reductions.

There is no doubt that the second-generation microbands represent shear localization and hence a type of plastic instability. The well defined morphology and crystallography of the microbands indicate that the localization/instability is connected to the local dislocation configurations - as opposed to the more macroscopic process of shear-band formation. At low strains one normally only observes one system of parallel second-generation microbands in a given grain [2]. If the microbands carry a substantial fraction of the strain in the LWD grains, the situation in these grains at low strain may be similar to that in the grains with bundles in brass (see next section) with predominant slip on one single slip plane (a situation very different from that considered in the Taylor model [11]). At somewhat higher strains there are typically two or three intersecting systems of second-generation microbands in each LWD grain.

## DESCRIPTION OF BUNDLES

The bundles in brass are plate-shaped zones, consisting of a composite of thin twin lamellae and untwinned matrix, which carry the great majority of strain (as slip parallel to the plane of the bundles) in the grains with bundles. Thus the bundles represent a plastic instability as the second-generation microbands do in copper. Normally there is only one system of parallel bundles in a grain. This means that in the grains with bundles slip is largely restricted to one single slip plane. This predominance of one single slip plane is confirmed by investigations of the crystallographic orientation of the twin lamellae [6,12]: slip on a single slip plane will mean that the twin lamellae remain parallel to the {111} plane on which they formed originally, whereas multiple slip according to the Taylor model [11] will lead to a substantial deviation from the original orientation parallel to a {111} plane. The former turns out to be by far closest to the experimental observations.

At moderate rolling reductions there are bundles in ~50% of the grains [6,8,12]. The formation of bundles is governed by the crystallographic orientation of the grains [6,8]. For instance, bundles are never observed in grains with orientations close to {110}<001>. It is worth noticing that this is the one and only orientation which in copper has a well defined pattern of behaviour: grains with this orientation always develop an LWD structure, i.e. a structure with a high frequency of second-generation microbands. At moderate reduction the deformation pattern is as follows: the grains with bundles, deforming with slip on one single slip plane, cannot follow the macroscopic strain. Intergranular strain accommodation is accomplished by the grains without bundles which deform by multiple slip - necessarily also without following the macroscopic strain. Thus, the deformation pattern is clearly different from the Taylor pattern [11] which is based on the suggestion that strain continuity is maintained by all grains following the macroscopic strain.

## RELATION TO SHEAR BANDS

Shear bands are non-crystallographic platelets with concentrated shear, i.e. they also represent plastic instability. Typical thicknesses of shear bands are in the range 0.01-1 μm both in copper and in brass (in copper clusters of shear bands may be up to 3 μm thick [13]).

Korbel and Martin [3] have suggested that a (second-generation) microband (on a {111} plane) in one grain may continue as a shear band (with no specified crystallographic habit plane) in a neighbouring grain, i.e. that shear bands nucleate from microbands. However, shear bands are common only at high strains where the frequency of microband formation is very low, e.g. [13]. This clearly goes against the relation between microbands and shear bands suggested by Korbel and Martin. There is absolutely no indication that shear bands in brass nucleate from bundles. Thus, one must conclude that there is no direct relation between the crystallographic plastic instabilities and shear bands.

## CONCLUSIONS

There is shear localization/shear instability in bands parallel to {111} planes in copper (and a number of other fcc materials with high to intermediate stacking fault energy), namely second-generation microbands, and in brass (and other fcc alloys with low stacking fault energy), namely bundles. However, there are clear differences in structure and behaviour between microbands and bundles. In brass the bundles lead to slip on one single slip plane in the grains containing bundles. Second-generation microbands may have a similar effect in the LWD grains in copper at low strain.

There is no direct connection between the crystallographic bands and shear bands.

## REFERENCES


1. Malin, A.S. and Hatherly, M., Microstructure of cold-rolled copper. Metal Sci., 1979, **13**, 463-472.

2. Ananthan, V.S., Leffers, T. and Hansen, N., Characteristics of second-generation microbands in cold-rolled copper. Scripta metall. mater., 1991, **25**, 137-142.

3. Korbel, A. and Martin, P., Microscopic versus macroscopic aspect of shear bands deformation. Acta metall., 1986, **34**, 1905-1909.

4. Bay, B., Hansen, N., Hughes, D.A. and Kuhlmann-Wilsdorf, D., Evolution of fcc deformation structures in polyslip. Acta metall. mater., (in press).

5. Duggan, B.J., Hatherly, M., Hutchinson, W.B. and Wakefield, P.T., Deformation structures in cold-rolled 70:30 brass. Metal Sci., 1978, **12**, 343-351.

6. Leffers, T. and Bilde-Sørensen, J.B., Intra- and intergranular heterogeneities in the plastic deformation of brass during rolling. Acta metall. mater., 1990, **38**, 1917-1926.

7. Ananthan, V.S., Leffers, T. and Hansen, N., Cell and band structures in cold-rolled polycrystalline copper. Mater. Sci. Techn. (in press).

8. Leffers, T. and Ananthan, V.S., Plastic instability in copper and brass and its relation to microstructure and texture. In Proceedings ICOTOM 9, Gordon and Breach Publishers (in press).

9. Hatherly, M., Deformation at high strains. In Proceedings ICSMA6, ed. R.C. Gifkins, Pergamon Press, Oxford, 1983, pp. 1181-1195.

10. Ashby, M.F., The deformation of plastically non-homogeneous materials. Phil. Mag., 1970, **21**, 399-424.

11. Taylor, G.I., Plastic strain in metals. J. Inst. Metals, 1938, **62**, 307-324.

12. Leffers, T. and Juul Jensen, D., The relation between texture and microstructure in rolled fcc materials. In Proceedings ICOTOM 9, Gordon and Breach Publishers (in press).

13. Ananthan, V.S., Leffers, T. and Hansen, N., Deformation structures in rolled copper (to be published).

# ON THE STABILITY OF STEADY SHEAR FLOWS OF THERMO-VISCOPLASTIC MATERIALS

ALAIN MOLINARI* and YVES M. LEROY**

* *Laboratoire de Physique et Mécanique des Matériaux, U.R.A. C.N.R.S. no. 1215, Université de Metz, Ile du Saulcy, 57045 Metz, Cedex 1, France,*

** *Shell Research (Koninklijke/Shell Exploratie en Produktie Laboratorium), Postbus 60, 2280 AB, Rijswijk Z.H., The Netherlands.*

## ABSTRACT

We study the stability of steady plane-Couette flows of thermo-viscoplastic materials. Of interest are the role of the boundary conditions and the spatial variation of the dominant mode of instability. In a Neumann problem, it is found that the rheological parameters determine whether one- or two-dimensional mode of instability prevails whereas, in a Dirichlet problem, all stationary solutions are stable.

## INTRODUCTION

We study the structure of shear zones that are formed in materials that are strain-rate and heat sensitive. Examples of these materials include structural steels that are submitted to high rates of strain [1] and geomaterials sustaining a ductile mode of deformation in the earth's crust.

We discuss the influence of the boundary conditions on the stability of the stationary laminar solutions of the plane-Couette flow. The boundary conditions are introduced as a linear combination of the shear force and the boundary velocity. Such linear combination are pertinent to geological situations [2] and can be shown to prevail during the main part of dynamic tests in a torsional Kolsky bar [3]. Exact analytical results are presented for the stability of the laminar stationary solutions when perturbed by one-dimensional modes.

Several observations have been reported that show the limitations of one-dimensional models. Multi-dimensional stability analysis are required to explain, for example, the variation in the shear band thickness reported in [1] and the existence of structured shear zones in the earth's lower crust, that was inferred from the interpretation of deep seismic reflection surveys [4]. The results of a two-dimensional stability analysis of the steady laminar flows, proposed by Molinari and Leroy [5], are presented. For Neumann boundary conditions, it is shown that a one-dimensional or a two-dimensional regime prevails depending on the exact value of the rheological parameters.

## CONSTITUTIVE MODEL AND FUNDAMENTAL SOLUTIONS

The model problem considered is one of a layer of unitary half-width and of infinite extent, containing a thermo-viscoplastic solid and sustaining a simple shear mode of deformation overall, Fig.1. The constitutive equations are:

$$\sigma_{ij} - p\delta_{ij} = 2\exp[-\beta(\theta - 1)]\dot{\gamma}^{m-1} D_{ij}\,, \tag{1}$$

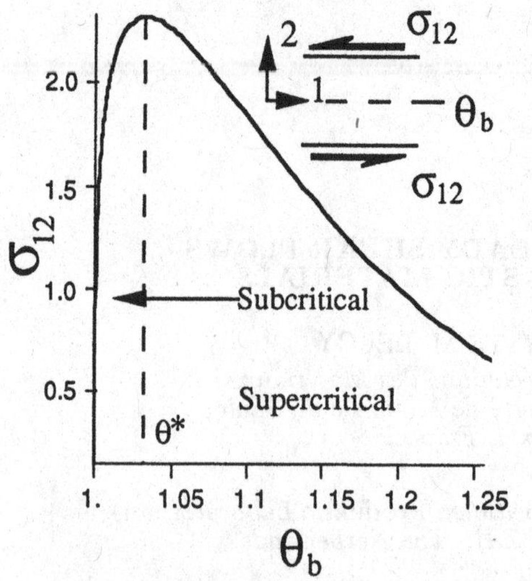


**Figure 1.** Admissible values of the stress $\sigma_{12}$ and the band temperature $\theta_b$ in the steady state.

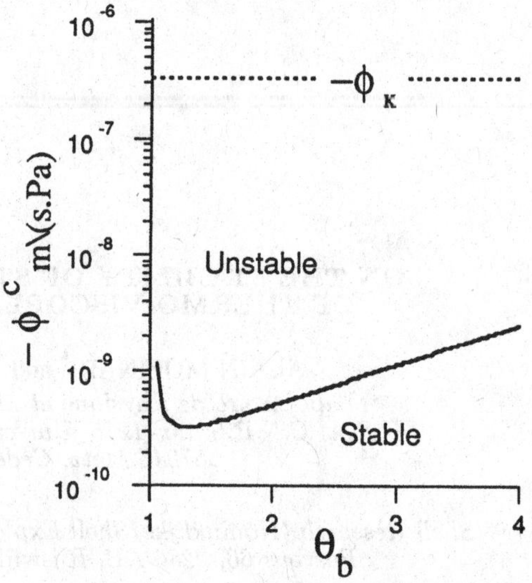


**Figure 2.** Evolution of the critical boundary parameter $\phi^c$ for neutral stability for a range of central temperatures relevant to the Kolsky bar test. The actual parameter $\phi_K$ is found to be three orders of magnitude larger then the minimum necessary to have stability. Wave number: $\xi = 0$.

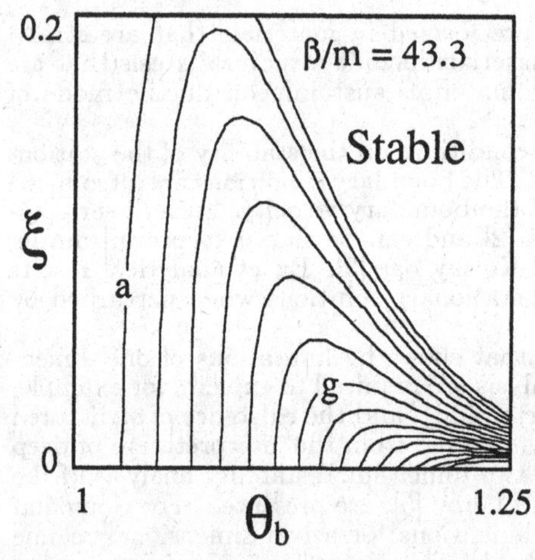


**Figure 3.** Rate of growth of the perturbation as a function of its wave number and the central temperature. 15 iso-contours ranging from 0 to +28. The first iso-contour (a) limits the region of instability.
Boundary parameter: $\phi = -\infty$.

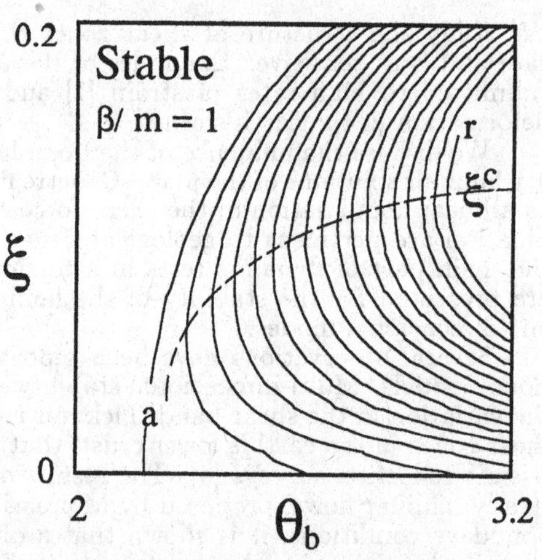


**Figure 4.** Rate of growth of the perturbation as a function of its wave number and the central temperature. 18 iso-contours ranging from 0 to +17. The first iso-contour (a) limits the region of instability.
Boundary parameter: $\phi = -\infty$.

in which $\sigma_{ij}, D_{ij}$ are the components of the stress tensor, the rate-of-deformation tensor and $\theta, p, \dot{\gamma}$, the temperature, the pressure and the effective strain rate, respectively. The parameters $m$ and $\beta$ are the strain-rate sensitivity and the thermal softening exponents, respectively.

At the boundaries, we assume that the temperature is the external reference one and that the second component of the velocity is null. A linear boundary condition between the stress an velocity is considered:

$$(u_1 \pm \overline{u}_1) - \phi(\sigma_{12} - \overline{\sigma}_{12})\big|_{x_2=\pm1} = 0\big|_{x_2=\pm1}, \forall x_1 \in \mathbb{R} \tag{2}$$

in which $\overline{u}_1$ and $\overline{\sigma}_{12}$ are the prescribed velocity in a Dirichlet problem ($\phi = 0$) and applied shear stress in a Neumann problem ($\phi = -\infty$), respectively. In a typical geological setting, the boundary conditions are often among the unknowns of the problem. Melosh and Ebel [2] were the first to consider explicitly a linear combination between velocity and stress as a possible boundary condition. In the case of the torsional Kolsky bar tests [1], it can be proved that the boundary condition that prevails during the main part of the test is of the form (2). For the Kolsky bar test, an expression for the coefficient $\phi_K$ has been given by Leroy and Molinari [3].

Closed-form expressions for the fundamental laminar solutions of this plane-Couette problem has been presented by various authors (e.g. [3,6,7]). The corresponding variation of the shear stress with the central temperature $\theta_b$ is plotted in Fig. 1. If the applied stress is larger than the maximum observed on this curve, no stationary solutions are attainable, whereas for a stress lower than this maximum, two steady states are possible. The steady states on the right of the maximum are called supercritical while the branch of the curve on the left of the maximum is named the subcritical branch. Note that the temperature associated with the maximum shear stress in Fig. 1 is approximated by the linear relation : $\theta^* \simeq 1 + 5/4(m/\beta)$. For geological applications and metal plasticity problems, typical values for the ratio $m/\beta$ are in a range of $10^{-2}$ to $5.10^{-2}$ [3]. It can thus be concluded that a small temperature increase with respect to the external temperature is sufficient for the stationary solutions of physical interest to be in the supercritical range.

## LINEAR STABILITY ANALYSIS

### 1-D analysis

It has been recently proved that the supercritical branch is unstable [6,7] and that the subcritical solutions are stable when perturbed by a one-dimensional mode, in a Neumann problem. With velocity boundary conditions, all steady states are found to be stable [3,6]. If mixed boundary conditions are considered, the subcritical solutions are stable but the stability of the supercritical branch depends on the exact value of the boundary parameter $\phi$. For the case of the Kolsky bar test, we have plotted in Fig. 2 the variation of the critical parameter $\phi^c$, that is required to have neutral stability, with the central temperature. In this particular instance, it can be seen that the steady states are unstable. For a hypothetical value of the Kolsky bar parameter $\phi_K$ of $-10^{-9} m/(sPa)$, the stationary solutions are stable at low and high temperatures on the supercritical branch (Fig. 2).

### 2-D analysis

We now study the dimension of the dominant mode of instability and present some of the results of Molinari and Leroy concerning the two-dimensional stability of the laminar solutions [5]. In this work, a normal mode decomposition of the perturbation in the longitudinal direction of the band is introduced, as well as a weak form of the linearized problem and a finite-element approximation. The stability problem is then phrased as an algebraic generalized eigenvalue problem with matrices having complex coefficients. The eigenvalue having the largest real part, which is found to be always real, governs the stability of the flow.

For Neumann boundary conditions, we have plotted in Fig. 3 the variation of the largest real part of the eigenvalues for a range of central temperatures and wave numbers $\xi$. It is seen that the stability transition at the lowest temperature is one-dimensional. In the supercritical regime, perturbations of various wave number have a positive rate of growth. However, for a given temperature, the perturbation having the fastest growth has a zero wave number. Instability in the supercritical range is thus dominated by a one-dimensional mode.

We now consider a ratio $\beta/m$ of 1 and, in contrast with Fig. 3, we observe now in Fig. 4 that in the supercritical regime the perturbation having the fastest growth has a non-zero wave number. The evolution of this critical wave number $\xi^c$ with $\theta_b$ is represented by the dashed line in this figure. We can thus conclude that, for materials that are weakly heat-sensitive and very rate-sensitive ($\beta \to 0^+, m \to 1$), two-dimensional modes dominate. Fig. 5 depicts the real part of the perturbation in temperature for $\theta_b = 2.5$ and the corresponding critical wavenumber of 0.09 (Fig. 4).

For the same range of values as considered in Figs. 3 and 4 and for Dirichlet boundary conditions, the eigenvalue having the largest real part is found to be real but its value remains always negative.

**Acknowledgements**
This paper is published by the permission of Shell Internationale Research Maatschappij.

## REFERENCES

1 Marchand, A. and Duffy, J. 1988, "An experimental study of the formation process of adiabatic shear bands in a structural steel", J. Mech. Phys. Solids, **36**, 251-283.
2 Melosh, H.J. and Ebel, J. 1979, "A simple model for thermal instability in the asthenosphere", Geophys. J.R. Astr. Soc., **59**, 419-436.
3 Leroy, Y.M. and Molinari, A. 1991, "Stability of steady states in shear zones", J. Mech. Phys. Solids, to appear.
4 Blundell, D.J., Reston T.J.and Stein, A.M. 1989, "Deep crustal structural controls on sedimentary basin geometry", in Geophysical Monograph 48, IUGG Volume 3
5 Molinari, A. and Leroy, Y.M. , " Structures in shear zones due to thermal effects", submitted for publication.
6 Molinari, A. and Leroy, Y.M. 1990, " Existence and stability of stationary shear bands with mixed-boundary conditions",C.R. Acad. Sci. Paris, **310**, 1017-1023.
7 Chen, H.T., Douglas, A.S. and Malek-Madani, R. 1989, "An asymptotic stability condition for inhomogeneous simple shear", Quart. of Appl. Math., **47**, 247-262.

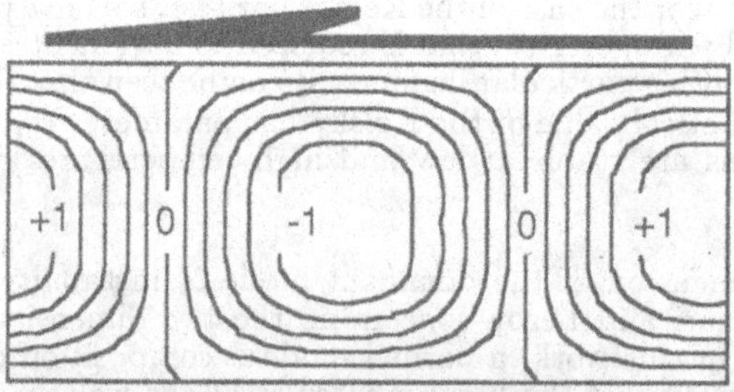


**Figure 5.** Iso-contours for the distribution of the perturbation in temperature, on a grid of thickness 2 and length equal to the perturbation wavelength. ( A different length scale is chosen in each direction.) The amplitude of the perturbations is arbitrary. Boundary parameter: $\phi = -\infty$.

# THE EFFECT OF HEAT RELEASE ON THE CHARACTER OF PLASTIC FLOW: LINEAR STABILITY ANALYSIS

A. MOLINARI*, Y. ESTRIN**, and D. DUDZINSKI*
*LPMM, UA CNRS 1215, Faculté des Sciences, Université de Metz
Ile du Saulcy, 57045 Metz Cedex 1, France
**Technical University Hamburg–Harburg,
P.O.Box 901052, 2100 Hamburg 90, F.R.G.

## ABSTRACT

Thermomechanical instability associated with heat generation during thermally activated plastic flow is considered. The analysis is focussing on deformation of metallic materials at extremely low temperatures in the cryogenics range (4.2 K to 10 K). The instability criterion obtained by linear stability analysis is shown to depend on the imposed deformation conditions. With regard to the plastic strain rate, a lower and an upper bound exist under constant imposed strain rate condition, while for constant imposed load condition there is no pronounced upper bound. A continual weakening of unstable behavior is found instead. The available experimental data are discussed on the basis of the instability criteria obtained.

## INTRODUCTION

One of the most plausible reasons for discontinuous yielding often observed at very low temperatures is thermomechanical coupling [1]. Due to heat release accompanying plastic deformation the plastic strain rate is enhanced leading, in turn, to increased heat generation. Deformation is prone to instabilities unless the stabilizing effect of heat removal prevails. Instability conditions could be derived by assuming uniform [1] or extremely localized [2] heat generation.

A spectacular consequence of the instability condition [1] is the prediction of a lower and an upper bound with regard to strain rate. In a semi–quantitative test [3], the existence of a lower bound to the instability range was validated. The situtation with the upper bound was less clear: while violent stress serrations disappeared above a certain applied strain rate, the deformation curves remained somewhat irregular.

In the present communication we revise the instability criteria by looking into the role of the imposed deformation conditions. In addition, temperature dependences of the relevant thermal parameters, which were partially neglected in the previous analysis, as well as gradient terms are included. A detailed presentation of all aspects of the analysis will be given elsewhere [4].

## GOVERNING EQUATIONS OF THE MODEL

The model is specified by the following set of equations:

(i) Heat conduction equation

$$\rho C_v \dot{T} = \beta \sigma \dot{\epsilon}_p - \frac{1}{S}\frac{\partial(qS)}{\partial x} - \frac{2h}{R}(T - T_0) \; ; \; q = -\kappa \partial T/\partial x \tag{1}$$

where $T(x,t)$ is the absolute temperature in the cross–section $x$ at time $t$, $\dot{T}$ is the partial time derivative, $\rho$ is the density, $C_v$ is the heat capacity (at constant volume), $q(x,t)$ is the heat flux, $\kappa$ is the heat conductivity, $T_0$ is the exterior temperature, and $h$ is the heat exchange coefficient. The source term $\beta\sigma\dot{\epsilon}_p$ represents the fraction $\beta$ of the plastic strain rate and $\sigma$ the axial stress. $R$ and $S$ denote the specimen radius and the cross–sectional area, respectively.

(ii) Equation expressing conservation of momentum

$$\partial(\sigma S)/\partial x = 0 \tag{2}$$

(iii) Superposition rule The axial strain rate $\dot{\epsilon}$ in each cross–section is taken as the sum of the plastic strain rate $\dot{\epsilon}_p$ and the elastic strain rate $\dot{\epsilon}_e$ :

$$\dot{\epsilon} = \dot{\epsilon}_p + \dot{\epsilon}_e \, . \tag{3}$$

(iv) Hooke's law (with $E$ = Young's modulus):

$$\epsilon_e = \sigma/E \tag{4}$$

(v) Plastic flow law:

$$\dot{\epsilon}_p = \varphi(T, \epsilon_p, \sigma) \quad \text{or} \quad \sigma = \psi(T, \epsilon_p, \dot{\epsilon}_p) \, . \tag{5}$$

Equations (5) are specified by using the Arrhenius form

$$\dot{\epsilon}_p = \dot{\epsilon}_0 \exp\left[-\Delta G/k_B T\right] , \tag{6}$$

where $k_B$ is the Boltzmann constant and the preexponential factor $\dot{\epsilon}_0$ can be considered constant. The Gibbs free energy of activation $\Delta G$ is linearized with respect to stress; introducing $V$, the activation volume:

$$\Delta G = \Delta G_0 - V\sigma \tag{7}$$

(vi) Plastic incompressibility condition yields

$$S = [1 + (1-2\nu)\epsilon_e] \, S(x,0) \exp\left[-\epsilon(x,t)\right] . \tag{8}$$

## BOUNDARY CONDITIONS

Two types of mechanical boundary conditions are considered: (A) the specimen is subjected to a constant load (as in constant load creep test); (B) the imposed overall strain rate $\dot{\epsilon}_a$ is kept constant. The thermal boundary conditions are specified as follows. The lateral heat transfer has been explicitly included in the heat conduction equation, eqn. (1), via the term $-(2h/R)(T - T_0)$. With $Bi = Rh/2\kappa << 1$, the temperature is nearly uniform within a cross–section. Adiabatic boundary conditions at the specimen ends $(\partial T/\partial x = 0)$ have been chosen.

## LINEAR STABILITY ANALYSIS

With the above set of governing equations, the uniform quasi–steady state solution is easily obtained. Its stability is checked by "probing" the system with small perturbations at various stages of deformation history. For example, the perturbation in temperature is taken in the form

$$\delta T = (\delta T)^{(0)} \exp\,[i\xi x + \eta t] \tag{9}$$

where $(\delta T)^{(0)}$ is an amplitude factor of the perturbation, $\eta$ is generally a complex number, and $\xi$ is the wave number. A similar form of perturbations is considered for all other variables ($\sigma$, $\epsilon_p$, etc.) as well. When $\mathrm{Re}\,\eta > 0$, the perturbation grows with time. The uniform steady state solution is thus unstable for $\mathrm{Re}\,\eta > 0$ and stable otherwise. Generally, the instability condition depends on the wave number $\xi$. A full analysis taking into account the temperature dependence of $C_v$, $\kappa$, and $h$ has been carried out. Both thermomechanical coupling and the geometrical effect due to variation of cross–sectional area have been taken into account. The values of $\eta$ are found as the roots of the following quadratic equations which result from equating to zero the determinant of the governing equations linearized in the perturbations. (The most unstable mode with $\xi = 0$ is considered.)

<u>For constant load boundary condition (A):</u>

$$\left[\rho C_v^{(0)} \frac{1}{\sigma} \frac{\partial \psi^{(0)}}{\partial \dot{\epsilon}_p}\right] \cdot \eta^2 + \left\{ \beta \frac{\partial \psi^{(0)}}{\partial T} + \frac{1}{\sigma} \cdot \frac{\partial \psi^{(0)}}{\partial \dot{\epsilon}} \left[ \frac{2}{R^{(0)}} \left[\frac{\partial h}{\partial T}\right]^{(0)} (T^{(0)} - T_0) \right.\right.$$
$$\left.\left. + 2\frac{h^{(0)}}{R^{(0)}} \right] - \rho C_v^{(0)} \left(e^{\epsilon^{(0)}} - 2\nu \frac{\sigma^{(0)}}{E}\right)^{-1} \right\} \cdot \eta + \left[ \frac{2}{R^{(0)}} \left[\frac{\partial h}{\partial T}\right]^{(0)} (T^{(0)} - T_0) \right.$$
$$\left. + 2\frac{h^{(0)}}{R^{(0)}} - \beta \dot{\epsilon}_\rho^{(0)} \frac{\partial \psi^{(0)}}{\partial T} + \frac{h^{(0)}}{R^{(0)}} \cdot \frac{\partial \ell n \psi^{(0)}}{\partial T} (T^{(0)} - T_0) \right] \left[2\nu \frac{\sigma^{(0)}}{E} - e^{\epsilon^{(0)}}\right]^{-1} = 0 \tag{10}$$

<u>For constant imposed strain rate condition (B):</u>

$$\left[\rho C_v^{(0)} \frac{1}{E} \frac{\partial \psi^{(0)}}{\partial \dot{\epsilon}_p}\right] \cdot \eta^2 + \left[ \rho C_v^{(0)} + \frac{2}{R^{(0)}} \frac{1}{E} \frac{\partial \psi^{(0)}}{\partial \dot{\epsilon}_p} \left[\frac{\partial h}{\partial T}\right]^{(0)} (T^{(0)} - T_0) \right.$$
$$\left. + \frac{2h^{(0)}}{R^{(0)}} \frac{1}{E} \frac{\partial \psi^{(0)}}{\partial \dot{\epsilon}_p} + \beta \frac{\sigma^{(0)}}{E} \frac{\partial \psi^{(0)}}{\partial T} \right] \cdot \eta + \left[ \frac{2}{R^{(0)}} \left[\frac{\partial h}{\partial T}\right]^{(0)} (T^{(0)} - T_0) \right.$$
$$\left. + \frac{2h^{(0)}}{R^{(0)}} - \beta \dot{\epsilon}_p^{(0)} \frac{\partial \psi^{(0)}}{\partial T} - \frac{1-2\nu}{E} \cdot \frac{h^{(0)}}{R^{(0)}} \frac{\partial \psi^{(0)}}{\partial T} e^{-\epsilon^{(0)}} (T^{(0)} - T_0) \right] = 0\,. \tag{11}$$

The subscript (o) refers to the uniform quasi–steady state.

The results of the computation of the normalized instability growth parameter, Re $\eta/\dot{\epsilon}^{(0)}$ , which is most relevant as a measure of linear stability, can be summarized as follows: a comparison shows a higher degree of stability under constant imposed strain rate than at constant imposed load. The lower bound of the instability range is shifted in case B to higher values of strain rate. Besides, the level of Re $\eta/\dot{\epsilon}^{(0)}$ is lower in case B than in case A . Finally, for the Al–2%Mg alloys tested at 4.2 K in gaseous He [3], a true upper bound was theoretically found at constant imposed strain rate, while at constant imposed load, the normalized growth parameter Re $\eta/\dot{\epsilon}^{(0)}$ decreases gradually, asymptotically tending to zero but remaining positive, however.

These results seem to be in conflict with the experimental observations [3] indicating that there is no pronounced upper bound for the case of *constant imposed strain rate*. The contradiction is resolved by realizing that in experiment, "mixed" imposed conditions are operative. Indeed, in eq. (11) referring nominally to constant imposed strain rate, the parameter E should be regarded as a combined elastic modulus of specimen and testing machine, rather than Young's modulus. Thus, in Ref. 3, E was one order of magnitude smaller than Young's modulus. Computations [4] show that for sufficiently low values of E , such as in experiment [3], the calculated Re $\eta/\dot{\epsilon}^{(0)}$ profile is nearly the same as for the case of constant imposed load.

## CONCLUSIONS

Linear stability analysis of uniform deformation taking into account the effect of thermomechanical coupling and the geometrical effect has been carried out. The role of the imposed deformation conditions has been studied. The results of the general analysis have been applied to rationalize unstable plastic flow of Al alloys observed at cryogenic temperatures.

## REFERENCES

1. Estrin, Y., and Kubin, L.P., Criterion for Thermomechanical Instability of Low Temperature Plastic Deformation, Scripta metall., 1980, 14, pp. 1359–1364.

2. Estrin, Y., and Kubin, L.P., Thermomechanical Instabilities of Low Temperature Plastic Flow, Continuum Models of Discrete Systems 4, eds. O. Brulin and R.K.T. Hsieh, North–Holland, 1981, pp.13–20.

3. Estrin, Y., Tangri, K., Thermal Mechanism of the Anomalous Temperature Dependence of the Flow Stress, Scripta metall., 1981, 15, pp. 1323–1328.

4. Molinari, A., Estrin, Y., and Dudzinski, D., Thermomechanical Instability Associated with Heat Release During Plastic Flow, Intl.J.Plasticity (submitted).

# COLLECTIVE DISLOCATION BEHAVIOUR AND PLASTIC INSTABILITIES - MICRO AND MACRO ASPECTS

HARTMUT NEUHÄUSER
Institut für Metallphysik und Nukleare Festkörperphysik
Technische Universität Braunschweig
Mendelssohnstr.3, D-3300 Braunschweig, F.R.Germany

## ABSTRACT

The occurrence of instabilities in plastic deformation under various conditions is reviewed and their intimate connection with localization of shear is emphasized. This is a consequence of the collective behaviour of, e.g. dislocations in crystals, and is enhanced by macroscopic changes of specimen geometry. Various physical mechanisms of instabilities are compiled and experimental techniques to observe them are compared. A few examples, i.e. the evolution of slip bands and shear bands, and the propagation of Portevin-LeChatelier bands and Lüders bands are considered in some detail.

## INTRODUCTION

A plastic instability may be defined as a catastrophic event of excessive plasticity, e.g., a sudden elongation of a specimen in tensile deformation. Sometimes it leads to early rupture of the specimen. Often, however, due to the elastic spring constant of the loading machine ($F_m$) and of the specimen itself ($F_s = E \cdot A(t)/l(t)$ with Young's modulus E, cross section A and length l), a strong rapid plastic elongation rate $\dot{l}_{pl}(t) > \dot{l}_m$ (= cross-head speed of the machine) according to

$$\dot{l}_{pl}(t) + (\dot{P}(t) \cdot (1/F_m + 1/F_s(t)) = \dot{l}_m \qquad (1)$$

causes a rapid unloading ($-\dot{P}(t)$) of the specimen which stops further plastic shear and is followed by elastic reloading, resulting in a serrated or oscillating load trace.

In practically all cases the mechanisms causing the plastic relaxation tend to localize the deformation in one or several narrow regions of the specimen, and we will examine how and to which extent this localization enhances or even causes the instability.

In a phenomenological approach Estrin & Kubin (1) gave a useful classification of the different kinds of plastic instabilites Taking the flow stress $\tau$ to be determined by the average strain $\epsilon$ and strain rate $\dot{\epsilon}$

$$d\tau = (\partial\tau/\partial\epsilon)d\epsilon + (\partial\tau/\partial \ln\dot{\epsilon})d\ln\dot{\epsilon} = h\cdot d\epsilon + S\cdot d\ln\dot{\epsilon} \qquad (2)$$

(with work hardening rate h and strain rate sensitivity S) the linear stability analysis ( $\delta\epsilon(t) = \delta\epsilon_0 \exp(\lambda t)$ ) shows that the sign of

$$\lambda = \epsilon\cdot(\tau - h)/S \qquad (3)$$

determines whether ($\lambda > 0$) or not ($\lambda < 0$) an instability occurs. This will be considered below for examples in a variety of conditions (low and high temperatures, pure and defected crystals, polymers and glasses) and with different physical origin. An important common feature is the tendency to localize plastic shear. This produces micro stresses which affect the microscopic units carrying the deformation (i.e., dislocations in crystalline, local shear transformations in amorphous materials). The resulting collective effects of these units may enhance the instability further, and have to be known if the observed macro stress are to be interpreted, in particular as it in turn controls the micro behaviour.

**MECHANISMS**

The linear stability analysis (1) can be used to classify the following types of plastic instabilities (1), cf. refs in (2):

**a)** The **type M instability** is connected with the change of slip geometry by rotation of the lattice due to deformation. Examples are texture softening in certain polycrystals, or grip stresses in single crystals.

**b)** The **type h instability** ($h < \tau$, $S > 0$) occurs when the hardening rate h is insufficient. Well-known examples are
(i) necking in tensile deformation, according to Considère if $h < \tau$, according to Hart if $h < \tau - S$. Because of the reduction of the cross section of the specimen, catastrophic failure occurs in tension.
(ii) structural softening by obstacle destruction in the deformed volume, e.g. by cutting precipitates or irradiation-produced defect clusters, or by destroying "alien" dislocation structures (formed previously in another strain path), favouring further slip in (or adjacent to) the already sheared parts of the specimen. Usually rapid (local) work hardening by the extensive local deformation stops the instability when a certain amount of shear is reached.
(iii) dislocation multiplication in the early stages of deformation, when the grown-in dislocation density is low. This corresponds to a local effective **negative** work hardening rate as, e.g. in slip lines at the beginning of plastic deformation of crystals. Again the rapid increase of local work hardening by the local deformation soon stops this instability.
(iv) dynamic recovery and recrystallization gives an example for an oscillating instability in dislocation populations of

high density occurring at elevated temperatures due to the non-linear relations for dislocation production and annihilation.
(v) dislocation inertial effects leading to serrated instabilities at very low temperatures may occur when the phonon drag is small enough for the dislocations to overshoot their equilibrium positions at the obstacles and to overcome them more easily. The resulting drop in load with accompanied decrease of dislocation velocity will again soon stop this instability.

c) The **type S instability** ($h > \tau$, $S < 0$) with a negative strain rate sensitivity $S = d\tau/d\ln\dot{\varepsilon} < 0$ has been observed in (i) the Portevin-LeChatelier (PLC) effect, or yielding by repeated serrations at elevated temperature in crystalline alloys. It is commonly explained by the repeated breakaway of dislocations from their solute cloud, probably supported by pile-up stresses. The stress drops to a value where the aging time (= waiting time at obstacles) for (nearly) fresh dislocations becomes long enough to reestablish the cloud, followed by elastic reloading up to the breakaway stress.
(ii) twinning of the crystal structure which occurs in some crystal systems at high stresses (often low T) and may lead to instabilites, if the nucleation stress exceeds the propagation stress of the twin lamellae (for not too low stacking fault energy). As a diffusionless structural transformation twinning produces extremely rapid load drops down to values where the propagation of the twin boundary stops.

d) The **type T instability** is due to the fact that > 85% of the work expended by the deformation machine during deformation is dissipated as heat in the specimen. If the deformation is localized, if thermal conductivity is small, and if the specific heat is low, the resulting local temperature rise facilitates further local shear. Examples are low T deformation of bcc metals, shear band development in metallic glasses, and deformation of polymers, as well as at ambient temperature the formation of "adiabatic" shear bands in high speed deformation.

As indicated above, the various mechanisms for instabilities favour localization of shear, e.g. in crystals by the rapid multiplication of dislocations forming groups which sometimes (b)(ii), c)(i),(v), d)) move more easily than single dislocatio lateral propagation of the plastic front from the deforming region into adjacent fresh regions of the crystal (sometimes limited regions distributed at random along the crystal, sometimes propagating orderly along the specimen length), either by activating potential sources via micro stress fields of the moving groups and macro stresses around the deformed constricted region, or by creating new sources via mechanisms like double cross slip of dislocations. Often these slip transfer processes seem to govern the overall strain rate of the specimen and as they are affected by local stresses, the externally measured stress must be interpreted with caution.

## EXPERIMENTAL TECHNIQUES

The most simple and direct method to observe plastic instabilities is based on eq.(1) which is applicable, if the plastic relaxation occurs during a time >> l/c (l = specimen length, c = speed of sound). Otherwise the complicated inertial behaviour of specimen and straining system with their acoustic coupling has to be accounted for, cf. indications in (3).

A much more sensitive but more indirect method to monitor plastic instabilities is the acoustic emission produced by plastic events although there are problems of calibrating the signal and of selecting the relevant frequencies. The method is claimed to even resolve the jumplike motion of single dislocations segments, which is well-known from direct TEM (in situ straining) and etch pit work (cf. refs. in (2)).

The most direct method, which is able to detect spatial and temporal correlations in a field of about 0.3 mm diameter and to record the growth of shear steps (local strain and strain rate) with quite high resolution (6 nm in step height, 6 µs in time) is the light microscopic observation of slip bands during deformation. It has been successfully applied in particular in situations with repeated propagation of a plastic front like Lüders or PLC bands (2, 3).

## SPECIAL EXAMPLES

In the following we will discuss a few selected examples combining macro and micro techniques:

(i) The propagation of a Lüders band in irradiated and alloyed Cu (destroyable obstacles) in stage I of single glide oriented crystals and the formation of single slip bands ahead of the front and within the front indicate the decisive role of the slip transfer process to neighbouring planes.

(ii) The propagating mode of the PLC effect in AlMg and in CuAl as well as the random PLC mode in CuAl combining measurements of stress serrations and slip band recording, indicate the importance of the breakaway of aged and successive multiplication of (nearly) fresh dislocations.

(iii) The shear band formation in heavily predeformed crystals receives recently increasing interest as well in macro and micro experiments as in theoretical modeling.

(iv) The unstable formation of shear bands in amorphous materials such as metallic glasses indicates the production of free volume in the dilated region around a local shear transformation, possibly enhanced by local heating. Analogous processes are expected to occur in shear banding of polymers.

## ACKNOWLEDGEMENTS

Financial support by the Deutsche Forschungsgemeinschaft (SFB 319 - A9) is gratefully acknowledged.

## REFERENCES

1. Estrin, Y. and Kubin L.P., Res Mechanica, 1988, 23, 197-221
2. Neuhäuser, H. in Patterns, Defects and Materials Instabilities, eds. D.Walgraef and N.M.Ghoniem, Kluwer Acad.Publ., Dordrecht, 1990, pp. 241-276
3. Neuhäuser, in Mechanical properties of Solids: Plastic Instabilities, eds. V.Balakrishnan and E.C.Bottani, World Scientific, Singapore, 1986, pp. 209-252

# INFLUENCE OF THERMODYNAMICAL COUPLINGS ON ADIABATIC SHEAR BAND LOCALIZATION FAILURE

PIOTR PERZYNA
Institute of Fundamental Technological Research
Polish Academy of Science, Warsaw, Poland

## ABSTRACT

**The paper aims at the investigation of the influence of thermomechanical couplings on adiabatic shear band localization fracture. In technical dynamical processes fracture can occur as a result of an adiabatic shear band localization generally attributed to a plastic instability generated by thermal softening during plastic deformation.**

**The paper consists of three parts according to three main stages of the thermodynamic plastic flow process considered. The first part is devoted to the investigation of the influence of thermodynamical couplings on criteria for localization of plastic deformation along the shear band. The initial part of the thermodynamic plastic flow process is treated in quasi–static and adiabatic approximation. A thermo–elastic–plastic model of a material with internal micro–damage effects is used. A procedure has been developed which allows us to use the standard bifurcation method in the examination of the adiabatic shear band localization criteria. The influence of two important thermal effects, namely thermal expansion and thermal plastic softening on the criteria of localization of plastic deformation is investigated. Similar influence of spatial covariance terms is also examined. The second part of the paper brings an analysis of the influence of thermomechanical couplings on micro–damage mechanism. Post–critical behaviour within the shear band region is described by a thermo–elastic–viscoplastic model of a material with advanced micro–damage process. Both models, which describe pre– and post–critical responses of a material are developed within the thermodynamic framework of the rate type constitutive structure with internal state variables. Both these models are also consistent with the requirement that the elastic–plastic response of a porous solid can be obtained as a limit case for quasi–static processes of the elastic–viscoplastic behaviour. The third part examines the influence of thermomechanical couplings on final failure mechanism.**

## INTRODUCTION

It is postulated that during the first stage of the deformation process a thermo–elastic–plastic model of a material with internal micro–damage effects is utilized. The model accounts of thermo–mechanical couplings as well as a combination of kinematic and isotropic hardening effects of a porous ductile, rate independent behaviour of a material.

At some instant of the process considered the adiabatic shear band localization may occur. So, a criterion for adiabatic shear band localization has to be investigated.

Along shear band region the localized plastic deformation process is very much influenced by strain rate sensitivity, the intrinsic macro–damage effects and thermomechanical couplings. The nucleation and growth mechanisms of microcracks become very important in this stage of the dynamic deformation process. The micro–damage process is highly localized to the shear band region because the threshold stress for microcrack nucleation is much lower. This preferential nucleation is caused by higher temperature inside the shear band, with an attendant flow stress decrease. On the other hand the large plastic strain along the shear band region causes the intense growth of microcracks.

At some instant of the inelastic process the coalescence of microcracks begins. In This stage the fracture mechanism occurs. The coalescence of microcracks generates a macrocrack which propagates along the damaged shear band region and causes a final failure.

It is assumed that the response of a material along the shear band region is described by a thermo–elastic–viscoplastic model with very pronounced influence of the micro–damage process. The model takes account of thermo–mechanical couplings as well as of a combined kinematic and isotropic hardening of a porous ductile, rate sensitive plastic material.

The final failure of a body is described by a simple micromechanical model of the propagation of a macrocrack in damaged solid. The main idea of the model proposed is based on the experimental observations that the coalescence of microcracks can be treated as a nucleation and growth process on a smaller scale.

## THERMODYNAMIC PLASTIC FLOW PROCESS

### The initial–boundary–value problem

For a thermodynamic plastic flow process the initial–boundary–value problem is as follows (cf. [2]):

Find $\phi$, $\boldsymbol{v}$, $\vartheta$ and $\boldsymbol{\mu}$ as function of $\mathbf{x}$ and $t$ such that

i) the field equation

$$
\begin{aligned}
\varrho\dot{\boldsymbol{v}} &= \operatorname{div}\boldsymbol{\sigma},\\
L_{\boldsymbol{v}}\boldsymbol{\tau} &= \mathcal{L} : \mathbf{d} - \mathcal{Z}\dot{\vartheta},\\
L_{\boldsymbol{v}}\boldsymbol{\mu} &= \mathrm{m}(s)\langle\frac{1}{H}\{\mathbf{P} : [\dot{\boldsymbol{\tau}} - (\mathbf{d}\cdot\boldsymbol{\alpha} + \boldsymbol{\alpha}\cdot\mathbf{d})] + \pi\dot{\vartheta}\}\rangle,\\
\varrho c_p\dot{\vartheta} &= \operatorname{div}(k\operatorname{grad}\vartheta) + \vartheta\frac{\varrho}{\varrho_{Ref}}\frac{\partial\boldsymbol{\tau}}{\partial\vartheta} : \mathbf{d}\\
&\quad + \varrho c_p\chi\langle\frac{1}{H}\{\mathbf{P} : [\dot{\boldsymbol{\tau}} - (\mathbf{d}\cdot\boldsymbol{\alpha} + \boldsymbol{\alpha}\cdot\mathbf{d})] + \pi\dot{\vartheta}\}\rangle;
\end{aligned}
\tag{1}
$$

ii) the boundary conditions
traction $(\boldsymbol{\tau}\cdot\mathbf{n})^a$ are prescribed on $\partial\mathcal{B}$,
temperature $\vartheta$ is prescribed on $\partial\mathcal{B}$;

iii) the initial conditions
$\phi$, $\boldsymbol{v}$, $\boldsymbol{\mu}$ and $\vartheta$ are given at $\mathbf{x}\in\mathcal{B}$ at $t = 0$;

are satisfied.

In the formulation of the initial–boundary–value problem the internal state vector $\boldsymbol{\mu}$ is assumed as follows

$$\boldsymbol{\mu} = (\boldsymbol{\zeta}, \boldsymbol{\alpha}, \xi), \tag{2}$$

where $\boldsymbol{\zeta} \in V_{n-7}$ denotes the new internal state vector which describes the dissipation effects generated by plastic flow phenomena only; $\boldsymbol{\alpha}$ is the residual stress (the back stress) and aims at the description of the kinematic hardening effects and $\xi$ denotes the porosity or the volume fraction parameter brought in to take account of micro–damage effects.

The yield criterion has a particular form

$$f - \kappa = 0, \tag{3}$$

where

$$f = J_2 + [n_1(\vartheta) + n_2(\vartheta)\xi]J_1^2 \tag{4}$$

with

$$J_1 = \tilde{\tau}^{ab} g_{ab}, \quad J_2 = \frac{1}{2}\tilde{\tau}'^{ab}\tilde{\tau}'^{cd} g_{ac} g_{bd}, \quad \tilde{\boldsymbol{\tau}} = \boldsymbol{\tau} - \boldsymbol{\alpha}, \tag{5}$$

$L_v\boldsymbol{\tau}$ denotes the Lie derivative of the Kirchhoff stress tensor $\boldsymbol{\tau}$, $\mathbf{g}$ is the metric tensor in the current configuration, $\kappa$ the isotropic hardening – softening parameter given by the material function

$$\begin{aligned} \kappa &= \hat{\kappa}\left(\in^P, \vartheta, \xi\right) \\ &= \left[\kappa_1 + (\kappa_0 - \kappa_1)e^{-h(\vartheta)\in^P}\right]^2 \left[1 - \frac{\xi}{\xi^F(\vartheta)}\right]\left(1 - b\overline{\vartheta}\right), \end{aligned} \tag{6}$$

$\kappa_0$, $\kappa_1$ and $b$ are constants,

$$\begin{aligned} \overline{\vartheta} &= \frac{\vartheta - \vartheta_0}{\vartheta_0}, \ \in^P = \int_0^t \left(\frac{2}{3}\mathbf{d}^P : \mathbf{d}^P\right)^{\frac{1}{2}} dt, \\ \mathbf{d} &= \mathbf{d}^e + \mathbf{d}^p, \quad H = H^* + H^{**}, \\ H^* &= -\frac{1}{2\sqrt{J_2}}\{n_1 J_1^2 + [\kappa_1 + (\kappa_0 - \kappa_1)e^{-h\in^P}]^2 \frac{1 - b\overline{\vartheta}}{\xi^F}[k_1(\sqrt{J_2} + AJ_1) + 3Ak_2] \\ &+ \frac{h(\kappa_1 - \kappa_0)}{\sqrt{3J_2}}[\kappa_1 + (\kappa_0 - \kappa_1)e^{-h\in^P}]\}[1 - \frac{\xi}{\xi^F}](1 - b\overline{\vartheta})(1 + 6A^2)^{\frac{1}{2}} e^{-h\in^P}, \\ H^{**} &= \tau(\frac{1}{2} + 3A^2), \ \tau = \tau_1 + \tau_2, \ \pi = \frac{1}{2\sqrt{J_2}}\frac{\partial\varphi}{\partial\vartheta}, \\ P &= \frac{1}{2\sqrt{J_2}}\frac{\partial\varphi}{\partial\vartheta}, \ P_{ab} = \frac{1}{2\sqrt{J_2}}\tilde{\tau}'^{cd} g_{ca} g_{db} + A g_{ab}, \\ A &= \frac{1}{\sqrt{J_2}}(n_1 + n_2\xi)J_1, \ k_1 = k_1(s), \ k_2 = k_2(s), \\ \mathbf{m}(s) &= \begin{cases} \mathbf{Z}(s) \\ \tau_1\mathbf{P} + \tau_2\frac{\mathbf{P}:\mathbf{P}}{\tilde{\boldsymbol{\tau}}:\mathbf{P}}\tilde{\boldsymbol{\tau}}, \\ k_1\tilde{\boldsymbol{\tau}} : \mathbf{P} + k_2\mathbf{P} : \mathbf{g}, \end{cases} \end{aligned} \tag{7}$$

and $s$ denotes the intrinsic state which consists of a set of variables as follows

$$s = (\mathbf{e}, \mathbf{F}, \boldsymbol{\vartheta}; \boldsymbol{\mu}), \tag{8}$$

where $\mathbf{e}$ denotes the Eulerian strain tensor and $\mathbf{F}$ is the deformation gradient.

It is postulated that the free energy function exists and has the form

$$\psi = \hat{\psi}(s). \tag{9}$$

Finally the matrix $\mathcal{L}$ and the tensor $\mathcal{Z}$ have the form as follows

$$\begin{aligned} \mathcal{L} &= \left[ I - \frac{\frac{1}{H}\mathcal{L}^e \cdot \mathbf{PP}}{1 + \frac{1}{H}\mathcal{L}^e \cdot \mathbf{P} : \mathbf{P}} \right] \cdot \left[ \mathcal{L}^e - \frac{1}{H}\mathcal{L}^e \cdot \mathbf{P}(\mathbf{P} \cdot \tilde{\boldsymbol{\tau}} + \tilde{\boldsymbol{\tau}} \cdot \mathbf{P}) \right], \\ \mathcal{Z} &= \left[ I - \frac{\frac{1}{H}\mathcal{L}^e \cdot \mathbf{PP}}{1 + \frac{1}{H}\mathcal{L}^e \cdot \mathbf{P} : \mathbf{P}} \right] \cdot \left[ \mathcal{L}^{th} - \frac{1}{H}\pi \mathcal{L}^e \cdot \mathbf{P} \right], \end{aligned} \tag{10}$$

where

$$\mathcal{L}^e = \rho_{Ref} \frac{\partial^2 \hat{\psi}}{\partial \mathbf{e}^2}, \ \mathcal{L}^{th} = -\rho_{Ref} \frac{\partial^2 \hat{\psi}}{\partial \mathbf{e} \partial \vartheta}. \tag{11}$$

the coefficient $\chi$ and the specific heat $c_p$ in the heat equation $(1)_4$ are also determined.

**Quasi–static and adiabatic approximation**

In many practical situation the thermodynamic plastic flow process can be treated as quasi–static and adiabatic.

Then Eqs. $(1)_1$ and $(1)_4$ reduce to:

$$\begin{aligned} \operatorname{div} \boldsymbol{\sigma} &= 0, \\ c_p \dot{\vartheta} &= \frac{\vartheta}{\varrho_{Ref}} \frac{\partial \boldsymbol{\tau}}{\partial \vartheta} : \mathbf{d} + \chi \langle \frac{1}{H} \{ \mathbf{P} : [\dot{\boldsymbol{\tau}} - (\mathbf{d} \cdot \boldsymbol{\alpha} + \boldsymbol{\alpha} \cdot \mathbf{d})] + \pi \dot{\vartheta} \} \rangle. \end{aligned} \tag{12}$$

For adiabatic plastic flow process the evolution equation for temperature $(12)_2$ can be written in the form

$$\dot{\vartheta} = \mathbf{M} : L_{\boldsymbol{v}} \boldsymbol{\tau} + \mathbf{N} : \mathbf{d}\,, \tag{13}$$

where

$$\mathbf{M} = \frac{\chi \mathbf{P}}{H c_p - \chi \pi}, \ \mathbf{N} = \left[ \frac{\vartheta H}{\varrho_{Ref}} \frac{\partial \boldsymbol{\tau}}{\partial \vartheta} + \chi (\mathbf{P} \cdot \tilde{\boldsymbol{\tau}} + \tilde{\boldsymbol{\tau}} \cdot \mathbf{P}) \right] (H c_p - \chi \pi)^{-1}. \tag{14}$$

Substitution (14) into $(1)_2$ gives

$$L_{\boldsymbol{v}} \boldsymbol{\tau} = \mathbb{L} : \mathbf{d} \tag{15}$$

where

$$\mathbb{L} = \left( \mathbf{I} - \frac{\mathcal{Z}\mathbf{M}}{1 + \mathcal{Z} : \mathbf{M}} \right) \cdot (\mathcal{L} - \mathcal{Z}\mathbf{N})\,. \tag{16}$$

The last result allows to use in the investigation of criteria for localization along shear bands the standard bifurcation method.

## THERMODYNAMIC INELASTIC FLOW PROCESS

When the localization of plastic deformation along the shear band takes place the process considered for an entire body is no longer adiabatic.

The response of a material along the shear band region is assumed to be elastic-viscoplastic with the intrinsic micro–damage effects.

So, the initial–boundary–value problem for postcritical behaviour of a body is as follows:

Find $\phi$, $\boldsymbol{v}$, $\vartheta$ and $\boldsymbol{\mu}$ as function of $\mathbf{x}$ and $t$ such that

i) the field equation

$$\begin{aligned} \varrho\dot{\boldsymbol{v}} &= \operatorname{div}\boldsymbol{\sigma} \\ L_{\boldsymbol{v}}\boldsymbol{\tau} &= \mathcal{L}^e : \mathbf{d} - \mathcal{L}^{th}\dot{\vartheta} - \frac{\lambda}{\beta}\langle\Phi(f-\kappa)\rangle\mathcal{L}^e : \mathbf{P}, \\ L_{\boldsymbol{v}}\boldsymbol{\mu} &= \mathbf{m}(s)\frac{\lambda}{\beta}\langle\Phi(f-\kappa)\rangle, \\ \varrho c_p\dot{\vartheta} &= \operatorname{div}(k \operatorname{grad}\vartheta) + \vartheta\frac{\varrho}{\varrho_{Ref}}\frac{\partial\boldsymbol{\tau}}{\partial\vartheta} : \mathbf{d} \\ &\quad + \varrho\chi\frac{\lambda}{\beta}\langle\Phi(f-\kappa)\rangle; \end{aligned} \tag{17}$$

ii) the boundary conditions
tractions $(\boldsymbol{\tau}\cdot\mathbf{n})^a$ are given on $\partial\mathcal{B}$,
temperature $\vartheta$ is prescribed on $\partial\mathcal{B}$;

iii) the initial conditions
$\boldsymbol{\phi}$, $\boldsymbol{v}$, $\vartheta$ and $\boldsymbol{\mu}$ are given at $\mathbf{x}\in\mathcal{B}$ at $t=\tau_c$;

are satisfied.

In (17) the symbol $\langle\Phi(f-\kappa)\rangle$ is defined as follows

$$\langle\Phi(f-\kappa)\rangle = \begin{cases} 0 & \text{if } f-\kappa \le 0 \\ \Phi(f-\kappa) & \text{if } f-\kappa > 0 \end{cases}, \tag{18}$$

$\lambda$ denotes the viscosity coefficient and $\beta$ is the control function. The viscoplastic overstress function $\Phi$ can be determined basing on avilable experimental results for dynamic loading processes.

It is noteworthy that the empirical form for $\Phi$ is suggested by the physical consideration of the thermally–activated mechanism on crystalline slip systems.

## FINAL FAILURE ALONG THE SHEAR BAND REGION

### Failure phenomena along the shear band

As it was observed experimentally (cf. Grebe, Pak and Mayers [3]; Cho, Chi and Duffy [1]; Marchand and Duffy [4]) the onset of fracture along the shear band is directly related to the coalescence of ellipsoidal microcracks. From then on the microcrack coalescence and growth along the shear band result in the elongated macrocracks (cavities). The final failure occurs when one the most developed elongated macrocrack starts to propagate along the damaged shear band region.

### Criterion of fracture

When the micro–damage process within the shear band region is sufficiently advanced the coalescence of microcracks begins.

During the dynamic process it is very difficult to control plastic deformation for different stages. Therefore it seems natural to base a criterion for coalescence and fracture on the control of porosity $\xi$.

Let denote by $\xi^c$ a value of porosity at which the coalescence of microcracks begins and by $\xi^F$ a value of porosity at fracture along the shear band region.

If porosity $\xi$ is treated as a main parameter of the process (which can be controlled at every stage of the process), then the interval $[\xi^c, \xi^F]$ represents the final mechanism of fracture. During this interval (i.c. $\xi \in [\xi^c, \xi^F]$) the coalescence of microcracks plays a dominating role.

It is postulated that the fracture phenomena occurs when

$$\xi = \xi^F \Longrightarrow \in^P = \in_F^P \left(\dot{\in}^P, \vartheta\right), \tag{19}$$

what leads to the condition

$$\kappa = \hat{\kappa}\left(\dot{\in}^P, \xi, \vartheta\right)\Bigg|_{\substack{\xi = \xi^F \\ \in^P = \in_F^P}} = 0. \tag{20}$$

The condition (20) expresses the fact that fracture means a catastrophe or the intrinsic failure when the material along the shear band region loses its stress carrying capacity.

It is noteworthy that in the previous papers of the author (cf. Perzyna [5–7]) the directional character of the micro–damage process has been described by introducing an additional set of the internal state variables.

Taking advantage of this idea a simple micromechanical model of final failure has been proposed. The main conception of the model is based on the experimental observation that the coalescence of microcracks can be treated as nucleation and growth processes on smaller scale. This gives very simple description of the propagation of a macrocrack in the damage region along the shear band.

## REFERENCES


1. Cho, K., Chi, Y.C. and Duffy, J., Microscopic observation of adiabatic shear bands in three different steels. Brown University Report, September 1988.

2. Duszek, M.K., Perzyna, P., The localization of plastic deformation in thermoplastic solids. Int. J. Solids Structures, 1991, **27**, pp. 1419-1443.

3. Grebe, H.A., Pak, H.R. and Meyer, M.A., Adiabatic shear band localization in titanium and Ti-6PctAl-4PctV alloy. Metall. Trans., 1985, **16A**, pp. 761-775.

4. Marchand, A. and Duffy, J., An experimental study of the formation process of adiabatic shear bands in a structural steel. J. Mech. Phys. Solids, 1988, **36**, pp. 251-283.

5. Perzyna, P., Constitutive modelling of dissipative solids for postcritical behaviour and fracture. ASME J. Engng. Mater. Technol., 1984, **106**, pp. 410-419.

6. Perzyna, P., Internal state variable description of dynamic fracture of ductile solids. Int. J. Solids Structures, 1986, **22**, pp. 797-818.

7. Perzyna, P., Influence of anisotropic effects on micro–damage process in dissipative solids. IUTAM/ICM Symposium on Yielding, Damage and Failure of Anisotropic Solids, Villerd-de-Lance, August 1987; Proc., 1990.

# EFFECT OF CRYSTAL ANISOTROPY ON THE INTERACTION BETWEEN TWO HOLES

A.M. PHILIP and J.-H. SCHMITT
Génie Physique et Mécanique des Matériaux, Institut National Polytechnique de Grenoble
Unité Associée au CNRS 793, ENSPG, BP 46, 38402 Saint Martin d'Hères, France

## ABSTRACT

The effect of the coupling between the plastic anisotropy and damage anisotropy is studied through theoretical analyses and experiments on aluminum single crystals containing two identical circular holes. The observations and the theoretical predictions of the plastic domains around the holes and of the slip activities are investigated. The influence of the crystal anisotropy is proved on the amplitude of the hole interaction and on the coalescence process.

## INTRODUCTION

Previous work has shown the effect of the hole distribution on the localization of the deformation in an isotropic material [1]. To study the influence of the plastic anisotropy on the interaction between holes, the present paper focusses on the coupling between the plastic anisotropy and the hole distribution. Tensile experiments are performed on aluminum single crystals containing two identical circular holes. Different crystal orientations simulate various plastic anisotropic behaviors. The hole distribution is characterized by the distance between the holes and by the angle between the line connecting the hole centers and the tensile direction.

A theoretical analysis [2] shows that the influence of the hole distribution on their interaction is strongly dependent on the nature of the plastic anisotropy. To confirm this first result, experiments were performed on two different orientations of crystals having a same hole distribution. The plastic domains and the slip activity are observed on the surface of the sample and are compared with the theoretical predictions. The effects on the coalescence process between these two holes are also investigated.

## EXPERIMENTS

Large grain specimens were obtained by the critical strain technique from a pure aluminum plate. Single crystalline samples were spark-cut, the gage dimensions being 30 x 10 x 1 $mm^3$. The crystal orientations were measured from X ray back-reflection data. Two identical circular holes, with a diameter of 0.6 mm, were spark-cut in a symmetrical position with respect to the center of the sample.

To allow the slip lines to be observed, the sample surface was smoothed through a series of mechanical and electrochemical polishings. The single crystals were continuously pulled on a small tensile device mounted on an optical microscope. Nomarski interferometry was used to characterize the slip lines which develop on the surface during the deformation.

The evolution of the neck between the two holes was observed during the coalescence using a scanning electron microscope.

## RESULTS

Two different crystal orientations ($\mathbb{H}$ : $(421)[6\bar{7}\bar{9}]$, and $\mathbb{G}$ : $(13\bar{2})[023]$) were studied for almost the same hole distribution :

- <u>crystal $\mathbb{H}$</u> : the distance d between the hole centers equals about 1.3 mm ; the angle $\Phi$ between the tensile direction ($[6\bar{7}\bar{9}]$) and the hole alignment is 116° ; the angle $\alpha$ between the trace of the primary slip plane (i.e. the plane corresponding to the maximum Schmid factor in uniaxial tension) and the hole alignment is about 40° (Figure 1b),
- <u>crystal $\mathbb{G}$</u> : d is about 1 mm, $\Phi$ is equal to 118° and the holes are roughly aligned with the trace of the primary slip plane ($\alpha$ = -10°).

The holes induce a heterogeneous stress field in the sample. The plastic flow begins at the hole edge ; the plastic domains are initially located at about 90° from the tensile direction on each side of the holes. The exact location of these zones is a function of the crystal orientation, the distance d and the angle $\Phi$ [2]. The primary slip system is mainly active in the plastic areas.

As the imposed deformation increases, the plastic domains extend around the holes and propagate in the sample from the hole region, the primary slip system being still dominant. In the hole vicinity, activity of secondary slip systems is noted, while a few domains exhibit no evidence of plastic deformation along directions closely parallel to the tensile axis.

Beyond these general features, different plastic behaviors occur in the neck depending on the crystal orientation :

- <u>crystal $\mathbb{G}$</u> : numerous and fine slip lines appear between the holes from the early steps of the deformation (Figure 1a). It characterizes an intense plastic deformation which indicates a strong interaction between the holes.
- <u>crystal $\mathbb{H}$</u> : plastic deformation seems to be more homogeneous in this sample : at a given imposed strain, the extent of the secondary slip systems is less important around the holes than in crystal $\mathbb{G}$ ; the strain is not concentrated in the neck (Figure 1b).

These surface observations allow to conclude that, for a given hole distribution, various crystal orientations, namely different plastic anisotropic behaviors, lead to different intensities of the interaction between the holes.

## DISCUSSION

To quantitatively investigate the influence of the plastic anisotropy on the hole interaction, a model [3] has been developed which predicts, for small strain, the extent of the plastic domains around a hole and the slip activities in these domains. The heterogeneous stress field is calculated in the sample assuming it is a two-dimension infinite elastic medium [4]. A plastic criterion is defined using the generalized Schmid law : the resolved shear stress is calculated for any slip system. A plastic domain is thus determined as the set of points for which the resolved shear stress reaches a critical value on at least one slip system.

For a crystal containing a single hole, the surface observations fairly agree with the prediction of this model [3]. Its extension to the case of two holes seems thus an acceptable approximation. Theoretical results are presented elsewhere [2] for various hole distributions in a given crystal. It was proved that the local behavior between holes is strongly connected with the mutual misorientation $\alpha$ between the hole alignment and the primary slip trace direction : the hole interaction is maximum for $\alpha$ about 0°.

The two cases under investigation correspond to different values of the angle $\alpha$. Figure 2 presents the theoretical evolutions of the plastic domains at three different steps of the imposed deformation. In the crystal $\mathbb{G}$, the plastic domains expand largely around the holes, and, from the smaller strain, the two holes are linked by a continuous plastic zone (Figure 2a). Conversely, the strain concentration is much less important in the crystal $\mathbb{H}$ and the neck is fully plastic only at larger macroscopic strain amount (Figure 2b). This result appears in rather good agreement with the observations of the slip lines. The validity of the prediction is confirmed when the directions of the predicted and observed slip traces are compared [2].

The distance d between the two hole centers is however slightly different in the two studied crystals. In the crystal $\mathbb{G}$, the smaller d value could increase the interaction between the holes. A theoretical simulation [3] has been performed to estimate only the influence of the

Figure 1 . Surface observations of the deformed crystals (average imposed strain about 4%).

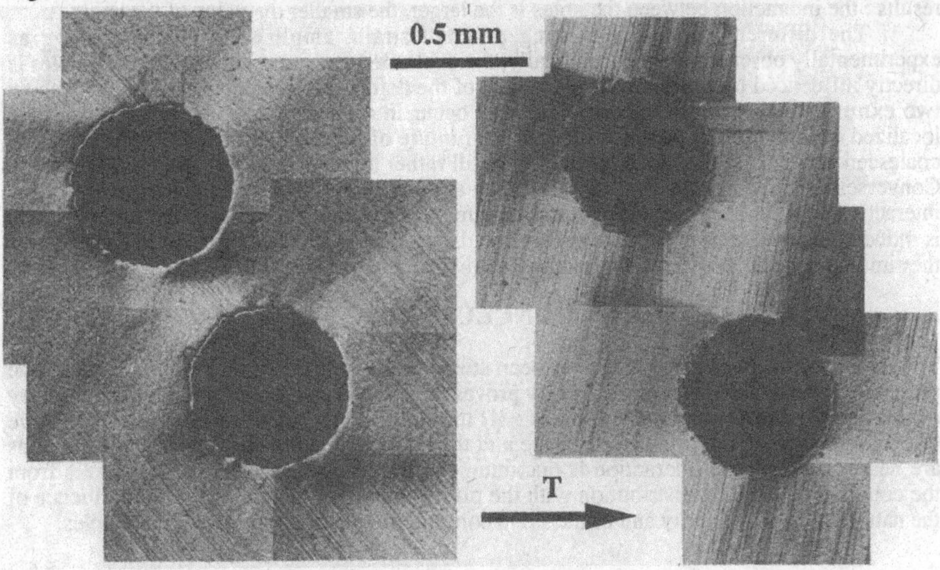


a) crystal G T indicates the tensile direction. b) crystal H

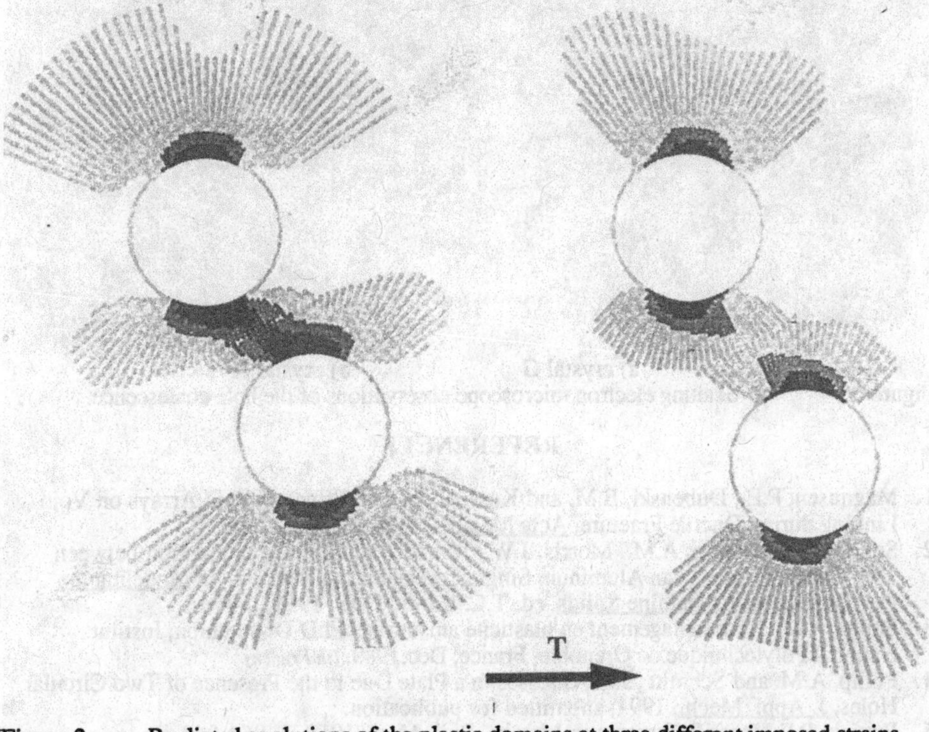


Figure 2 . Predicted evolutions of the plastic domains at three different imposed strains.

crystallographic orientation for a same spacing d. This confirms the main trends of the present results : the interaction between the holes is the larger, the smaller the value of the angle $\alpha$.

The different behaviors existing at small strain, amplify during the loading as experimentally observed. More specifically, the coalescence process between the two holes is directly influenced by the nature and intensity of the deformation in the neck. Figures 3 show two extreme modes of coalescence that may occur in a material [5]. In the crystal $\mathbb{G}$, the localized deformation in the neck induces the rupture of the ligament between the hole ; the coalescence occurs while the deformation is still rather homogeneous all around (Figure 3a). Conversely, for the crystal $\mathbb{H}$, the coalescence appears not to be a direct consequence of the interaction between the holes. The crystal rupture results from the necking of the sample which is induced by the decrease of its actual section due to the holes. This is proved in figure 3b by the simultaneous occurence of the fracture between and outside the holes.

## CONCLUSIONS

The interaction between two holes has been studied in two different aluminum crystals. The present experimental and theoretical study proves the influence of the crystal plastic anisotropy on the local behavior around the holes : **(i)** the hole interaction depends on the relative orientation between the primary slip trace and the hole alignment ; **(ii)** when these directions are nearly parallel, the interaction is maximum ; **(iii)** the coalescence process results from the coupling of the hole distribution with the plastic anisotropy as shown by the influence of the nature of the slip activity and of the strain concentration in the neck between the holes.

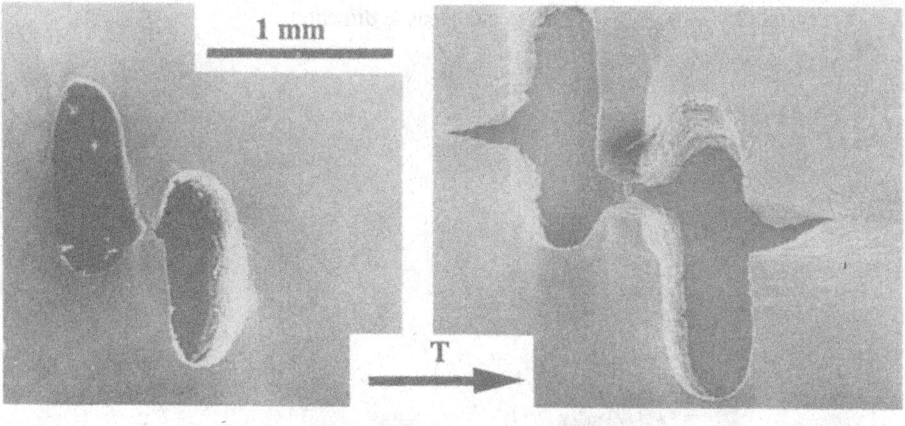


a) crystal $\mathbb{G}$ b) crystal $\mathbb{H}$

Figure 3 . Scaning electron microscope observations of the hole coalescence.

## REFERENCES

1. Magnusen, P.E., Dubenski, E.M. and Koss, D.A., The Effect of Void Arrays on Void Linking during Ductile Fracture, Acta Metall., 1988, **36**, 1503-1509.
2. Schmitt, J.H., Philip, A.M., Morris, J.W., Jr. and Baudelet, B., Interaction between Two Circular Holes in an Aluminum Single Crystal during Tension. In Modeling the Deformation of Crystalline Solids, ed. T.C. Lowe, TMS, 1991, in press.
3. Philip, A.M., Endommagement en plasticité anisotrope. PhD Dissertation, Institut National Polytechnique de Grenoble, France, Dec.1989. *(in French)*
4. Philip, A.M. and Schmitt , J.H., Stresses in a Plate Due to the Presence of Two Circular Holes, J. Appl. Mech., 1991, submitted for publication.
5. Puttick, P.E., Ductile Fracture in Metals, Phil. Mag., 1959, **4**, 964-969.

# EFFECT OF TRIAXIAL TENSION ON FLOW LOCALIZATION FOR DIFFERENT PLASTIC SPINS

VIGGO TVERGAARD[‡] and ERIK VAN DER GIESSEN[†]
[‡]The Technical University of Denmark, Dep. of Solid Mechanics, DK-2800 LYNGBY, Denmark
[†]Delft University of Technology, P.O. Box 5033, NL-2600 GA DELFT, The Netherlands

## ABSTRACT

Corotational stress rates involving different constitutive laws for the plastic spin are introduced into a material model for combined isotropic–kinematic hardening of a porous ductile solid. The model is used to study the localization and post-localization behaviour in shear bands for various levels of the stress triaxiality.

## INTRODUCTION

Some recent constitutive theories for anisotropic large strain plasticity have introduced a corotational stress rate determined by the plastic spin, which accounts for the spin of directions of anisotropy relative to the material. Using the simple constitutive relationship for the plastic spin suggested by Dafalias [1] and Loret [2], we have in a previous work incorporated such stress rates into Tvergaard's [3] version of a Gurson-type constitutive model for porous solids with combined isotropic–kinematic hardening and void nucleation [4]. The resulting material models for different plastic spin parameters were used to analyse strain localization in shear bands under conditions of uniaxial (plane strain) tension. It was found that the localization behaviour is quite sensitive to the amount of plastic spin. The differences found in the post-localization regime, where large shear strains occur, were expected based on well-known results for solids subject to simple shear, but the onset of localization occurs prior to large shear strains, where differences of the predictions cannot be explained the same way; instead, these differences must be interpreted as a strong sensitivity to small differences in the constitutive law, as emphasized by Rice [5].

In this paper, we extend these analyses by considering cases characterized by a considerably higher triaxiality of the stress state, such as occurs typically in notched tensile bars or in the neck of a tensile specimen. For contrast, we also consider one case of a reduced triaxiality, which is relevant for thick-walled tubes under internal pressure. Employing different plastic spin parameters, we study the onset of localization as well as the post-localization behaviour leading to ductile fracture in a void-sheet.

## PROBLEM FORMULATION

The constitutive model used in this study is a modification of the Gurson-type model for a porous ductile material described in [3], featuring combined kinematic/isotropic hardening and strain controlled void

TABLE 1
Basic constitutive equations for porous ductile material

| | |
|---|---|
| • yield condition: | $\Phi = \frac{\bar{\sigma}_e^2}{\sigma_F^2} + 2q_1 f^* \cosh\left(\frac{\mathrm{tr}\,\bar{\sigma}}{2\sigma_F}\right) - 1 - (q_1 f^*)^2 = 0\,,\quad f^* = \begin{cases} f & f \le f_C \\ f_C + \frac{f_U^* - f_C}{f_F - f_C}(f - f_C) & f \ge f_C \end{cases}$ |
| | $\bar{\sigma} = \sigma - \alpha\,,\ \bar{\sigma}_e^2 = \frac{3}{2}\mathrm{tr}\,\bar{s}^2\,,\ \bar{s} = \bar{\sigma} - (\frac{1}{3}\mathrm{tr}\,\bar{\sigma})I\,;\quad \alpha_e^2 = \frac{3}{2}\mathrm{tr}\,a^2\,,\ a = \alpha - (\frac{1}{3}\mathrm{tr}\,\alpha)I$ |
| • failure condition: | $f = f_F\,,\ f^* = f_U^* = 1/q_1$ |
| • hardening: | $\sigma_F = (1-b)\,\sigma_y + b\,\sigma_M\,;\ \varepsilon_M = \frac{\sigma_y}{E}\left(\frac{\sigma_M}{\sigma_y}\right)^n$ for $\sigma_M \ge \sigma_y$ |
| • evolution relations: | $\dot{f} = (\dot{f})_{\mathrm{growth}} + (\dot{f})_{\mathrm{nucleation}}\,;\ (\dot{f})_{\mathrm{growth}} = (1-f)\,\mathrm{tr}\,d^P\,,\ (\dot{f})_{\mathrm{nucleation}} = \mathcal{A} h_\alpha \dot{\varepsilon}_M^P$ |
| | $\mathcal{A} = \frac{1}{h_\alpha}\frac{f_N}{s\sqrt{2\pi}}\exp\left[-\frac{1}{2}\left\{\frac{\varepsilon_M^P - \varepsilon_N}{s}\right\}^2\right]\,;\ h_\alpha = \frac{EE_t}{E - E_t}\,,\ E_t = \dot{\sigma}_M/\dot{\varepsilon}_M$ |

nucleation. For future reference, some of the central governing equations are listed in Table 1. The modification involved (see [4]) relates to the corotational rate of change of the Cauchy stress $\boldsymbol{\sigma}$ and the back stress $\boldsymbol{\alpha}$ entering through the hypoelastic law

$$\overset{\circ}{\sigma} = R : d^E \tag{1}$$

and the finite strain generalization

$$\overset{\circ}{\alpha} = \dot{\mu}\,(\sigma - \alpha)\,,\ \dot{\mu} \ge 0, \tag{2}$$

of the Ziegler kinematic hardening rule ($\boldsymbol{R}$ is the usual tensor of elastic moduli and $d^E$ is the elastic part of the strain rate, $d^E = d - d^P$). On the basis of the work of Dafalias [1] and Loret [2], the corotational rate in (1) is defined by

$$\overset{\circ}{\sigma} = \dot{\sigma} - \omega^E \cdot \sigma + \sigma \cdot \omega^E \tag{3}$$

and similarly for $\overset{\circ}{\alpha}$, in terms of the "elastic" part $\omega^E$ of the continuum spin, $\omega^E = \omega - \omega^P$. The plastic spin $\omega^P$ represents the spin of the material relative to the directions of anisotropy induced by the back stresses $\boldsymbol{\alpha}$ and is here taken to be governed by the following simple constitutive relation:

$$\omega^P = \frac{1}{2}\varrho\,(\alpha \cdot d^P - d^P \cdot \alpha)\,, \tag{4}$$

with $\varrho$ an additional material parameter. Following [1, 2], we shall consider constant values of $\varrho$ for simplicity; but we will also consider $\varrho$ to be specified by

$$\varrho = \sqrt{\frac{3}{2}}\frac{12\alpha_e}{h_\alpha^2 + 3\alpha_e^2} \tag{5}$$

as suggested in [6]. Notice that for isotropic hardening ($b = 1$) or for combined hardening ($0 \le b < 1$) with $\varrho = 0$, the corotational rate (°) reduces to the Jaumann rate and our model coincides with that in [3].

The effect of different plastic spins on strain localization in shear bands is studied by means of simple model analyses (see e.g. [3, 4, 7]). These analyses consider the development of a shear band in an otherwise uniformly strained solid out of a plane slice of material containing an initial material inhomogeneity, oriented at an initial angle $\psi_I$ from the major principal stress direction (see inset Fig. 1). With

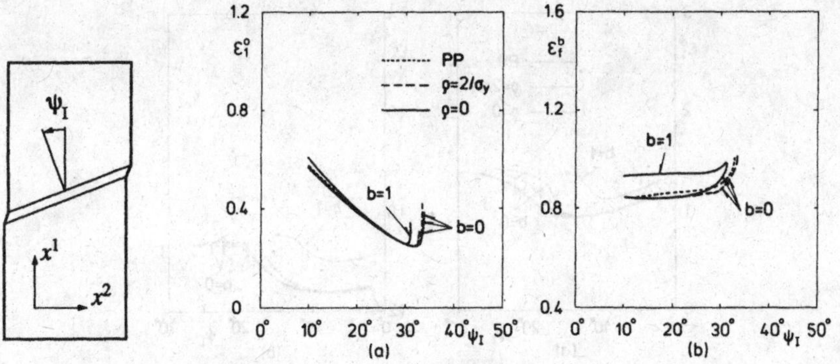


Figure 1. Plane strain tension, $\xi = 0$ and $\varkappa = 0.25$. (a) Localization strain vs. initial band orientation; (b) Fracture strain.

superscripts $b$ and 0 denoting quantities inside and outside the band, respectively, the uniform state outside the band is specified by the ratios $\xi = \varepsilon_3^0/\varepsilon_2^0$ and $\varkappa = \sigma_2^0/\sigma_1^0$ of the principal logarithmic strains $\varepsilon_i$ and principal stresses $\sigma_i$. The initial inhomogeneity is here taken as an additional volume fraction $\Delta f_N = f_N^b - f_N^0$ of void nucleating particles.

## RESULTS

The previous study [4] of the effect of plastic spin on shear localization and on ductile fracture by void-sheet failure inside shear bands focussed on uniaxial plane strain tension ($\xi = \varkappa = 0$) and on uniaxial axisymmetric tension ($\xi = 1$, $\varkappa = 0$). Here, the effect on localization is studied further, by considering the influence of stress triaxiality.

The material considered is power hardening, with $\sigma_y/E = 0.0033$, $n = 10$, $\nu = 0.3$, $q_1 = 1.5$, $f_C = 0.15$ and $f_F = 0.25$. Plastic strain controlled nucleation is assumed, with $\varepsilon_N = 0.3$, $s = 0.1$, $f_N^0 = 0.01$ and $\Delta f_N = 0.001$, and there are no voids initially. Figures 1 and 2 show results for plane strain tension ($\xi = 0$), with $\varkappa = 0.25$ and $\varkappa = -0.25$, respectively, and Fig. 3 shows results for axisymmetric tension ($\xi = 1$) and $\varkappa = 0.25$. In all three figures, $\varepsilon_1^0$ is the strain outside the band at the onset of localization, and

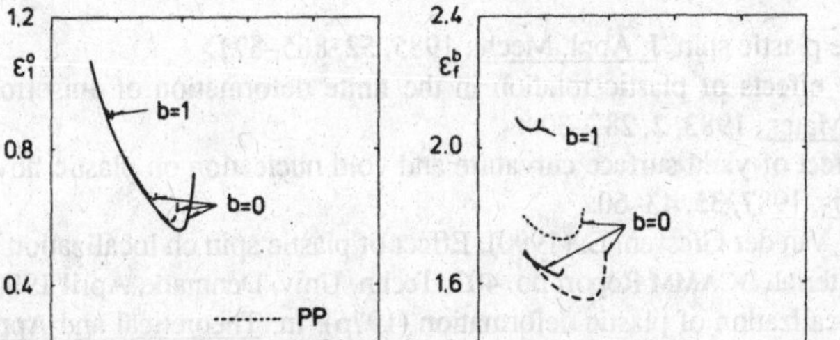


Figure 2. Plane strain tension, $\xi = 0$ and $\varkappa = -0.25$. (a) Localization strain vs. initial band orientation; (b) Fracture strain.

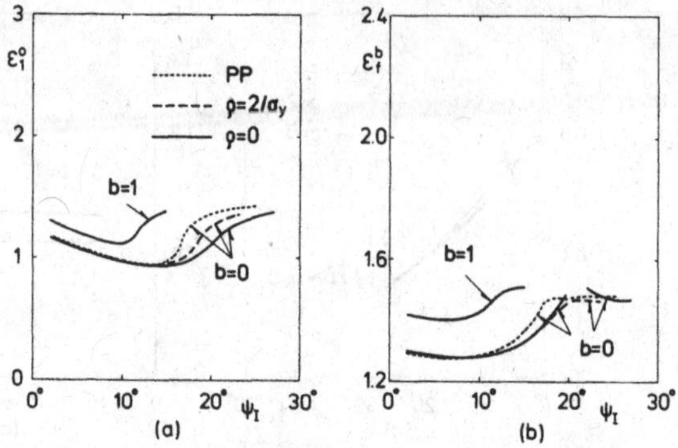


Figure 3. Axisymmetric tension, $\xi = 1$ and $\varkappa = 0.25$. (a) Localization strain vs. initial band orientation; (b) Fracture strain.

$\varepsilon_f^b$ is the maximum principal logarithmic strain inside the band when the void volume fraction reaches the value 0.20 (just before failure). Compared to previous results [4], Figs. 1 and 3 show that an increased hydrostatic tension reduces the critical strain for the onset of localization, as was known [3, 7], but also significantly reduces the differences between the results for different plastic spins (PP indicates $\varrho$ according to (5)). In agreement with this trend, Fig. 2 for lower hydrostatic tension shows increased localization strains and a stronger effect of corotational stress-rate differences.

It is noted that increased triaxiality (Figs. 1 and 3) occurs at a notch in a tensile specimen, while reduced triaxiality (Fig. 2) occurs near the inner surface of a thick-walled pressurised tube [8]. In Fig. 3, all curves show a somewhat surprising behaviour at $\psi_I$ values above that corresponding to the minimum localization strain. This is a consequence of $f$ exceeding the value $f_C$, so that the approximate representation of progressive void coalescence has become active in the material model.

**Acknowledgement**
The work of EvdG was made possible by a fellowship of the Royal Netherlands Academy of Arts and Sciences.

## REFERENCES

1. Dafalias, Y.F., The plastic spin. J. Appl. Mech., 1985, **52**, 865–871.
2. Loret, B., On the effects of plastic rotation in the finite deformation of anisotropic elastoplastic materials. Mech. Mater., 1983, **2**, 287–304.
3. Tvergaard, V., Effect of yield surface curvature and void nucleation on plastic flow localization. J. Mech. Phys. Solids, 1987, **35**, 43–60.
4. Tvergaard, V. and Van der Giessen, E. (1990). Effect of plastic spin on localization predictions for a porous ductile material. DCAMM Report no. 402, Techn. Univ. Denmark, April 1990.
5. Rice, J.R., The localization of plastic deformation (1976). In: Theoretical and Applied Mechanics, ed. W.T. Koiter, North-Holland, Amsterdam, 1976, pp. 207–220.
6. Paulun, J.E. and Pecherski, R.B., Remarks on the description of anisotropic hardening in finite deformation plasticity. Arch. Mech. (to appear).
7. Mear, M.E. and Hutchinson, J.W., Influence of yield surface curvature on flow localization in dilatant plasticity. Mech. Mater., 1985, **4**, 395–407.
8. Larsson, M., Needleman, A., Tvergaard, V., Storåkers, B., Instability and failure of internally pressurized ductile metal cylinders. J. Mech. Phys. Solids, 1982, **30**, 121–154.

# SUSCEPTIBILITY TO SHEAR BAND FORMATION IN WORK HARDENING MATERIALS

T. W. WRIGHT
Ballistic Research Laboratory
Aberdeen Proving Ground
MD 21005, U.S.A.

## ABSTRACT

The linearized equations that describe the evolution of small perturbations to a state of homogeneous shearing are described for a work hardening, thermo/viscoplastic material. Simple analysis of special cases shows that arbitrary initial conditions are followed by a temporal boundary layer, which reduces the initial degrees of freedom, and that scaling laws, which govern the timing of localization and the sensitivity to imperfections, can be found from the linearized equations alone.

## EQUATIONS

In nondimensional form the equations that describe quasi-static simple shearing of a narrow strip of rigid/plastic material may be written as follows

$$\text{Momentum:} \quad s_y = 0 \tag{1}$$

$$\text{Energy:} \quad \vartheta_t = k\vartheta_{yy} + sv_y$$

$$\text{Flow Rule:} \quad s = F(\kappa, \vartheta, v_y)$$

$$\text{Work Hardening:} \quad \kappa_t = M(\kappa, \vartheta)sv_y$$

with $-1 \le y \le +1$ and $0 \le t \le \infty$. Boundary conditions are assigned as $v(\pm 1, t) = \pm 1$ and $\vartheta_y(\pm 1, t) = 0$. In (1) the dependent variables are shear stress $s$, temperature $\vartheta$, work hardening parameter $\kappa$, and strain rate $v_y$.

Solutions to these equations that depend only on t and not on y are called homogeneous solutions

$$s = S(t)\ ,\quad \vartheta = \Theta(t)\ ,\quad \kappa = K(t)\ ,\quad v_y = 1 \tag{2}$$

The functions $K(\cdot)$ and $\Theta(\cdot)$ are monotonically increasing, but $S(\cdot)$ increases at first with the initial work hardening and then reaches a maximum at a finite value of t. This qualitative behavior of homogeneous solutions is the key to understanding the behavior of perturbations, together with the fact that the strain rate sensitivity, defined as $m \equiv F_{\dot\gamma}/F$ when evaluated on the homogeneous solution, is a small number (for metals at least), $m \ll 1$.

Small perturbations of the homogeneous solution lead to the linearized equations

$$\tilde{s} = 0 \tag{3}$$

$$\tilde{\vartheta}_t = k\tilde{\vartheta}_{yy} - (S_\vartheta/m)\tilde{\vartheta} - (F_\kappa/m)\tilde{\lambda}$$

$$\tilde{\lambda}_t = M_\kappa S\tilde{\lambda} - kM\tilde{\vartheta}_{yy}\ ,\qquad \text{where } \tilde{\lambda} \equiv \tilde{\kappa} - M(t)\tilde{\vartheta}$$

$$\tilde{v}_y = -(F_\kappa/mS)\tilde{\lambda} - (S_\vartheta/mS)\tilde{\vartheta}$$

It is important to note that with the exception of k in (3.2) and (3.3), the coefficients of all dependent variables in these equations are functions of time since they must be evaluated on the homogeneous solution. Equation (3.1) is obtained after applying the boundary conditions to the basic perturbation equations. Equations (3.2) and (3.3) must be solved simultaneously subject to boundary conditions $\tilde{\vartheta}_y(\pm1,t) = 0$ and $\tilde{v}(\pm1,t) = 0$, and initial conditions $\tilde{\vartheta}(y,0) = \tilde{\vartheta}_0(y)$ and $\tilde{\kappa}(y,0) = \tilde{\kappa}_0(y)$, where $\tilde{\vartheta}_0$ and $\tilde{\kappa}_0$ are arbitrary functions.

## A SPECIAL SOLUTION

An important special case occurs when $k = 0$, M depends only on $\kappa$ and m is constant. In that case the solution may be written explicitly as

$$\tilde{\lambda} = \tilde{\lambda}_0(y)\,\frac{M(t)}{M_0} \tag{4}$$

$$\tilde{\vartheta} = \tilde{\vartheta}_0(y)\,S^{-1/m}(t) - \frac{\tilde{\lambda}_0(y)}{mM_0}\,S^{-1/m}(t)\int_0^t F_\kappa(t')M(t')S^{1/m}(t')\,dt'$$

Equation (4.1) shows that $\tilde{\lambda}$ varies only slowly, since the plastic modulus M varies only slowly, but that $\tilde{\vartheta}$ varies rapidly at first, since m has been assumed to be small. For times less than the time at which S reaches its maximum, an asymptotic representation for (4.2) is

$$\tilde{\vartheta} \sim \tilde{\vartheta}_0 S^{-1/m} - \frac{\tilde{\lambda}_0}{M_0}\frac{F_\kappa M}{S_\vartheta}\left[1 - S^{-1/m}\right] + O(m\tilde{\lambda}_0) \tag{5}$$

In the boundary layer $S^{-1/m}(t) \to 0$, and then the remainder may be written as

$$F_\vartheta(t)\tilde{\vartheta}(y,t) + F_\kappa(t)\tilde{\kappa}(y,t) = \tilde{\lambda}_0(y)S_\vartheta(t)O(m) \tag{6}$$

In effect equation (6) states that to first order in m the perturbations in temperature and work hardening are not independent of each other.

Near the maximum in S the second term in (4.2) peaks up sharply and may be shown to have the asymptotic representation

$$\tilde{\vartheta} \sim -\tilde{\lambda}_0(y)\left\{\frac{F_\kappa \mathcal{M} S}{F_{\dot\gamma}}\right\}\left\{\frac{-2}{SS_{\vartheta\vartheta}}\right\}^{1/2}\frac{\sqrt{\pi \bar{m}}}{2}\left[1 + \mathrm{erf}(z/\bar{m}^{1/2})\right]e^{z^2/\bar{m}} \tag{7}$$

where the terms in braces are to be evaluated at the time when S reaches a maximum and $z^2/\bar{m}$ is defined by

$$z^2/\bar{m} = -\int_{t_m}^{t}\frac{S_\vartheta(t')S(t')}{F_{\dot\gamma}(t')}\,dt' \tag{8}$$

An estimate of the time when intense localization ensues may be found from (3.4) when $\tilde{v}_y = 1$ for then the perturbed strain rate is of the same order as the applied rate, and the perturbation solution may be presumed to have broken down. This line of reasoning leads to a relationship between the initial functions and the time past peak homogeneous stress, expressed through the homogeneous temperature

$$\xi e^{\frac{1}{2}\xi^2}\frac{1}{2}\left[1 + \mathrm{erf}(\xi/\sqrt{2})\right] = \frac{1-\beta}{\sqrt{2\pi}\,\beta} \tag{9}$$

where the critical variable $\xi$ is defined by

$$\frac{1}{2}\xi^2 \equiv -\frac{1}{2}\left(\frac{S_{\vartheta\vartheta}}{F_{\dot\gamma}}\right)_{t_m}(\vartheta - \vartheta_m)^2 \tag{10}$$

and the imperfection $\beta$ is defined by

$$\beta = -\left(\frac{F_\kappa \mathcal{M}}{F_{\dot\gamma}}\right)_{t_m}\tilde{\lambda}_0(0) \tag{11}$$

As an example consider the case where the flow rule is given by $F = \kappa e^{-a\vartheta}v_y^m$, and the work hardening modulus is given by $M = \frac{n}{\psi}\kappa^{-1/n}$ where a, m, n, and $\psi$ are all constants. Since

$S_\vartheta = 0$ at peak homogeneous stress, it is easy to work out that

$$\sqrt{(-S_{\vartheta\vartheta}/F_{\dot\gamma})} = \sqrt{1+n}\,\frac{a}{\sqrt{mn}} \;, \quad a\vartheta = \frac{n-a\psi}{1+n} \;, \quad F_\kappa M/F_{\dot\gamma} = \frac{a\psi}{mn} \qquad (12)$$

where the left hand sides of (12) are all evaluated at peak homogeneous stress.

Complete details of the above calculations, as well as some estimates of the effect of heat conduction on the early evolutionary stages of small perturbations will be given in a forthcoming paper.

# ON THE STRUCTURE AND WIDTH OF SHEAR BANDS IN FINITE ELASTOPLASTIC DEFORMATIONS

H.M. ZBIB* and E.C. AIFANTIS**
*Department of Mechanical and Materials Engineering
Washington State University, Pullman, WA 99164, USA
**Department of Mechanical Engineering and Engineering Mechanics
MM Program, MTU, Houghton, MI 49931, USA
Physical and Mathematical Sciences, AUT, Thessaloniki 54006, GREECE

## ABSTRACT

This paper addresses the problem of shear banding by considering a gradient-dependent theory of finite elastoplasticity including kinematic hardening and the plastic spin. The numerical analysis reveals the significant effect of higher order gradients, anisotropy and spin on the evolution of the shear band in the post localization regime.

## INTRODUCTION

In previous work [1] the authors have taken up an earlier suggestion [2] of including higher order gradient, of micro (dislocation densities) or macro (strain) variables into plasticity theory in order to capture the fascinating phenomenon of deformation patterning. In fact, a gradient-dependent plasticity theory was obtained by simply including second-order gradients of the equivalent plastic strain into the yield condition. This simple modification has equipped the structure of classical theory with an internal length and provided a possibility for describing pattern-forming instability in plasticity similar to those analyzed in standard fluid mechanics approaches.

The simplest form of the gradient modification of plasticity theory involves one extra coefficient c entering into the yield condition in the form

$$\tau = \kappa\,(\gamma) - c\,\nabla^2\,\gamma, \tag{1}$$

where $\tau$ is the equivalent stress, $\gamma$ is the equivalent plastic strain, $\kappa\,(\gamma)$ is the usual homogeneous flow stress, and $c = c\,(\gamma)$ is a gradient coefficient. The implications of

(1) have been found within both the rate-independent and rate-dependent versions of plasticity theory, and for both pressure-insensitive and pressure-sensitive materials. In particular, an expression for the width of a stationary shear band and its variation with the strain was derived. The appropriate values of c were then obtained by inferring to shear band width experimental data.

The next level of complexity of a gradient-dependent plasticity model that has been assumed in order to analyze the structure of a shear band is reflected in a slightly more general yield condition of the form

$$\tau=\kappa(\gamma,\dot{\gamma})-c_1\nabla^2\gamma-c_2\nabla\gamma\cdot\nabla\gamma, \tag{2}$$

where now the flow stress $\tau$ depends on the equivalent rate $\dot{\gamma}$ and a second gradient coefficient $c_2$ is introduced to account for the effect of a term quadratic in the first gradient of plastic strains, in a manner similar to that adopted in the theory of fluid interfaces [2a]. The gradient coefficients $c_1$ and $c_2$ were determined in relation to both shear band width and the true stress-strain relation of the material inside the band [3]. Depending on the rolling direction of the tested material, different values of $c_1$ and $c_2$ were obtained. This dependence of the gradient coefficients $c_1$ and $c_2$ on the rolling direction, indicates that a strong relation exists between texture or deformation-induced anisotropy and shear band width. To reveal this relation a finite theory of plastic deformation is required.

## FINITE ELASTOPLASTICITY

In connection with the question of examining the width and structure of the shear band in the post localization regime within a large deformation plasticity theory, it is pointed out that a scale-invariance argument and a maximization procedure have been used recently by Aifantis and co-workers to obtain the structure of phenomenological models of plasticity based on the process of crystal slip and dislocation glide. Such a "scale invariance approach" may be viewed as a compromise between "multislip" crystal plasticity averaging models and purely continuum plasticity theories. It was shown, in particular, that small-deformation plasticity models have a direct extension to large deformations (mathematically analyzed on the basis of the decomposition $\mathbf{F} = \mathbf{R}\,\mathbf{F}^e\,\mathbf{F}^p$, with $\mathbf{F}$ denoting the total deformation gradient R the material rotation, $\mathbf{F}^e$ the elastic deformation gradient, and $\mathbf{F}^p$ the plastic deformation gradient) by properly measuring the evolution of deformation-induced anisotropy. For kinematic hardening plasticity models, in particular, the plastic anisotropy is usually represented by the back stress $\boldsymbol{\alpha}$ whose evolution is determined, in turn, by the plastic spin $\mathbf{W}^p$. In other words, we have

$$\begin{aligned}&\hat{\boldsymbol{\alpha}}=\mathbf{f}(\boldsymbol{\alpha},\boldsymbol{D}^p),\\&\hat{\boldsymbol{\alpha}}=\dot{\boldsymbol{\alpha}}-\boldsymbol{\omega}\boldsymbol{\alpha}+\boldsymbol{\alpha}\boldsymbol{\omega}\quad;\quad\boldsymbol{\omega}=\boldsymbol{W}-\boldsymbol{W}^p,\\&\boldsymbol{W}^p=\boldsymbol{W}(\boldsymbol{\alpha},\boldsymbol{D}^p)\end{aligned} \tag{5}$$

where $\mathbf{f}$ and $\boldsymbol{\omega}$ are specific material functions, $\mathbf{D}^p$ is the plastic strain rate, $\boldsymbol{\omega}$ is the material spin, $\mathbf{W}$ in the vorticity or continuum spin, and $\mathbf{W}^p$ is the plastic spin.

The particular form of **f** and **w** assumed in this work read as

$$\mathbf{f} = h\boldsymbol{D}^P - C\dot{\gamma}\boldsymbol{\alpha}, \quad \boldsymbol{W}^P = \zeta(\boldsymbol{\alpha}\boldsymbol{D}^P - \boldsymbol{D}^P\boldsymbol{\alpha}), \tag{6}$$

where h, C and $\zeta$ are material parameters. Based on equation (3) and (4) supplemented by the remaining standard equations of elastoplasticity theory we perform a linear stability analysis and determine the critical nature of stress, strain, orientation etc. at the onset of localization. In order to investigate the material response and the structure of the band in the post localization regime, we solve the resulting nonlinear problem numerically. It is shown that the plastic spin does not affect the onset of instability of plastic flow (also as suggested by Tvergaard and Giessen [4]). However, it has a considerable influence on the development of severe localization, an increase in the plastic spin (resulting to a decrease in softening and texture development) causes a delay in the formation of a severe shear band. This is actually attributed to the fact that in the absence of the plastic spin, the internal (back) stress inside the band exhibits oscillatory behavior producing apparent softening. The plastic spin, however, tends to eliminate such an effect and thus delaying localization. The numerical analysis, in addition, reveals that the band width is also influenced by the strain rate sensitivity and back stress. This suggests that internal variables play, in general, an important role in determining the structure of the band. This can be deduced from the numerical result shown in Figure 1 where the deformed mesh and strain contours for a tensile specimen are shown. It can be seen that the width of the band w spans over a few elements.

## ACKNOWLEDGMENT

The support of the US National Science Foundation under grants CES-8800459 (ECA), MSS-8920700 (ECA) and MSS-807748 (HMZ) and that of the US Army Research Office under contract DAAL03-90-G-0151 (ECA & HMZ) is gratefully acknowledged.

## REFERENCES

1. a) Zbib, H.M. and Aifantis, E.C., On the Localization and Postlocalization Behavior of Plastic Deformation, I, II, III, *Res. Mechancia*, **23**, 261-305 (1988)
b) Zbib, H.M. and Aifantis, E.C., On the Structure and Width of Shear Bands, *Scripta Metall.*, **22**, 703-708 (1988)
c) Zbib, H.M. and Aifantis, E.C., A Gradient-Dependent Theory of Plasticity, Application to Metal and Solid Instabilities, *Appl Mech Review*, **42** 295-304 (1989).

2. a) Aifantis, E.C., On the Microstructural Origin of Certain Inelastic Models, Trans, ASME, *J. Eng. Mat. Techn.*, **106**, 326-330 (1984)
b) Aifantis, E.C., The Physics of Plastic Deformation, *Int. J. Plasticity*, 3, 211-247 (1987).

3. Zbib, H.M. and Aifantis, E.C., On the Gradient-Dependent Theory of Plasticity and Shear Banding, *Acta Mech* in press.

4. Tvergaard, V. and van der Giessen, E., Effect of Plastic Spin on Localization. Predictions for a Porous Ductile Material, DCAM, Report No. 402, 1990, The Technical University of Denmark.

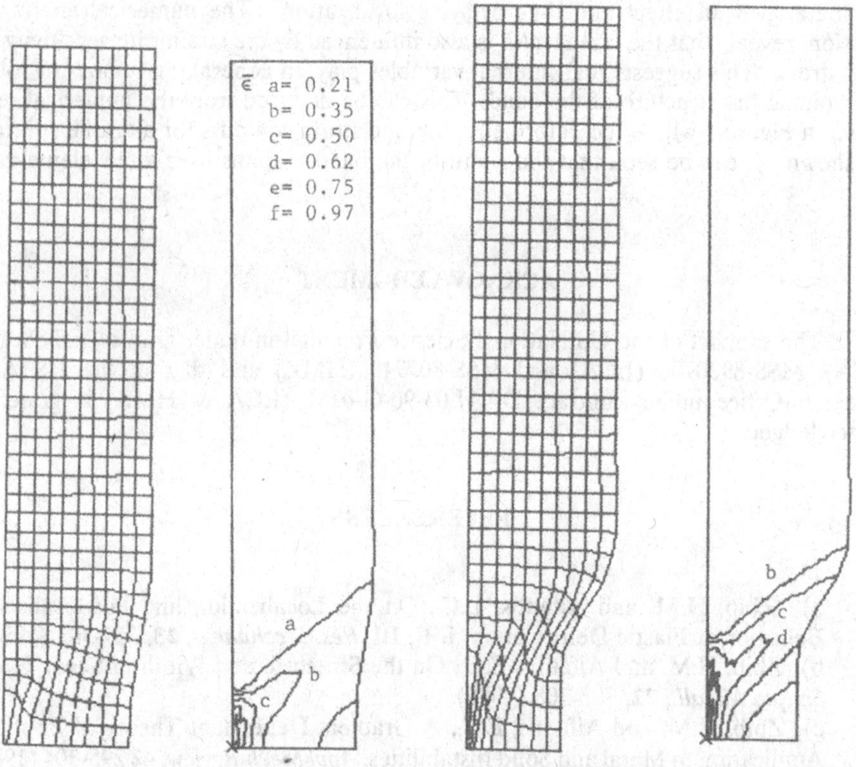


Figure 1. Effect of anisotropy on shear band formation: Numerial Simulation

# SHEAR BAND DEVELOPMENT IN A VISCOPLASTIC CYLINDER

X.-T. ZHANG and R. C. BATRA
Department of Mechanical and Aerospace Engineering
and Engineering Mechanics
University of Missouri - Rolla
Rolla, MO 65401-0249, USA

## ABSTRACT

We study the development of shear bands in a viscoplastic hollow circular cylinder containing two elliptical voids on a radial line and located symmetrically about the center. The material of the cylinder is assumed to exhibit strain and strain-rate hardening and thermal softening. The impact loading on the inner surface of the cylinder is simulated by applying a known radial velocity to particles on the inner surface; the outer surface of the cylinder is taken to be traction free. Both of these surfaces are taken to be perfectly insulated. It is found that shear bands initiate from void tips closer to the inner surface sooner than those from the other void tips. The effect of strain-hardening is to delay the initiation of shear bands.

## INTRODUCTION AND FORMULATION OF THE PROBLEM

Shear bands are narrow regions of intense plastic deformation that form in such high strain-rate processes as machining, forging, ballistic penetration, and ore crushing. They have been observed in both ferrous and non-ferrous alloys, and usually precede material fracture. Much interest in the subject was generated by the work of Zener and Hollomon [1], who not only observed 32 $\mu$m shear bands during the punching of a hole in a low carbon alloy steel, but added that heating of the material due to plastic work done softened the material, and the material became unstable when this thermal softening equalled or exceeded the combined effects of strain and strain-rate hardening. The reader is referred to the recent articles by Shawki and Clifton [2], and Batra and Zhang [3] for additional references on the subject. Here we extend our previous work [3] to include the effects of strain-hardening.

We describe the deformations of the body with respect to a set of rectangular Cartesian coordinate axes with $x_1$-axis passing through the void centroids. The deformations of the cylinder are assumed to be symmetrical about the $x_1$- and $x_2$-axes. Thus, deformations of the material in the first quadrant are analyzed. It is assumed that a plane strain state of deformation prevails. In order to conserve space, we refer the reader to the paper by Batra and Zhang [6] for equations expressing the balance of mass, linear momentum and internal energy, the pertinent boundary conditions, a sketch of the cross-section of the prismatic body, locations of the voids, and the finite element mesh used to analyze the problem. Even though equations there are written in the spatial description, we use the updated Lagrangian description and the Galerkin approximation to obtain a set of coupled nonlinear ordinary differential equations which are integrated by the Gear stiff method. We employ the following constitutive relation that accounts for work-hardening of the material through the internal variable $\psi$.

$$\sigma_{ij} = -B\,((\rho/\rho_0) - 1)\,\delta_{ij} + 2\,\mu\,D_{ij} - \nu K \theta \delta_{ij},$$

$$2\mu = \sigma_0\,(1 + bI)^m\,(1 - \alpha\theta)\,(1 + \psi/\psi_0)^n/(\sqrt{3}\,I),$$

$$2\,I^2 = D_{ij}\,D_{ij},$$

$$\dot{\psi} = 4\,\mu\,I^2\,/\,(\sigma_0\,(1 + \psi/\psi_0)^n), \qquad D_{ij} = D_{ij} - \frac{1}{3}\,D_{kk}\,\delta_{ij}.$$

Here $\sigma_{ij}$ is the Cauchy stress tensor, $D_{ij}$ the strain-rate tensor, K the bulk modulus, $\nu$ the coefficient of thermal expansion, $\sigma_0$ the yield stress in a quasistatic simple compression test, $\theta$ the temperature rise, $\rho$ the present mass density, $\rho_0$ the mass density in the reference configuration, and $\psi$ an internal variable used to describe the work hardening of the material. The various material and geometric parameters are assigned the same values as in [3], except that now $\alpha = 0.667 \times 10^{-3}/°C$, $\psi_0 = 0.017$, $n = 0.1$, $\nu = 1.08 \times 10^{-5}/°C$, K = 168 GPa.

## NUMERICAL RESULTS

Figure 1 depicts the evolution of the maximum principal logarithmic strain, defined as the natural logarithm of the maximum principal stretch, at points Q and R adjacent to the left and right void tip, respectively. These reveal that the consideration of strain hardening effects delays the initiation of shear bands and reduces the slope of the strain vs. the elapsed time t curve when shear bands do initiate. The peak strain reached at t = 0.04 at point Q is more for the work hardening case, because the higher value of the temperature (cf. Fig. 2) softens the material more. This is further evidenced by the plots included in Fig. 3 of the effective stress, defined as the square-root of the second invariant of the deviatoric part of the Cauchy stress tensor. Also included in Fig. 3 is the evolution of the effective stress at points P and S on the inner and the outer surface, respectively. The work hardening effects increase the effective stress initially at points Q and R till the thermal softening caused by the heating of material due to plastic working overcomes the combined effects of strain and strain-rate hardening. The rate of stress drop is sharper for the work hardening case.

The contours of the temperature rise $\theta$ when n = 0.10 are shown in Fig. 4; those for n = 0.0 look similar.

## CONCLUSIONS

The consideration of work hardening effects delays the initiation of shear bands, results in higher temperature rise within the band, reduces the initial rate of increase of strain at the time of the initiation of the shear band, and gives rise to steeper drop in the effective stress within the band.

Acknowledgement: This work was partially supported by the U. S. Army Research Office Grant DAAL03-91-G-0084 to the University of Missouri - Rolla.

## REFERENCES


1. Zener, C. and Holloman, J. H., J. Appl. Phys., 1944, **14**, 22-32.
2. Shawki, T. G. and Clifton, R. J., Mechs. Materials, 1989, **8**, 13-43.
3. Batra, R. C. and Zhang, X.-T., Acta Mechanica, 1990, **85**, 221-234.


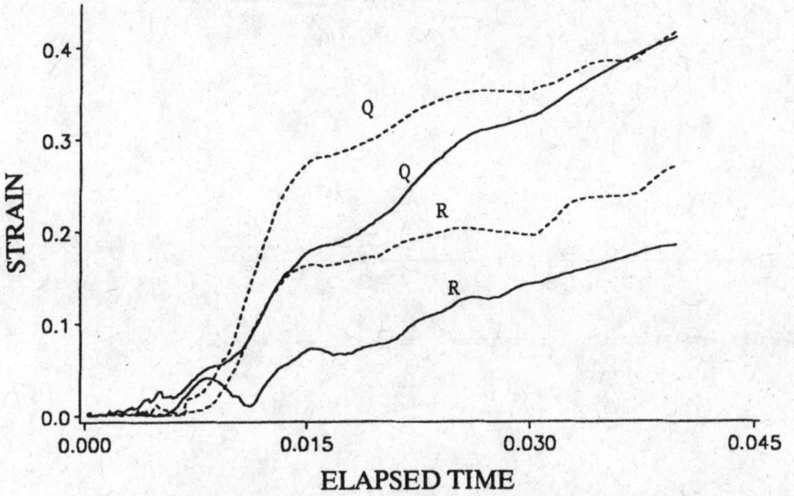


Fig. 1. Evolution of the maximum principal logarithmic strain at points Q and R adjacent to the left and right void tips, respectively. ——— with work-hardening, --------- without work-hardening.

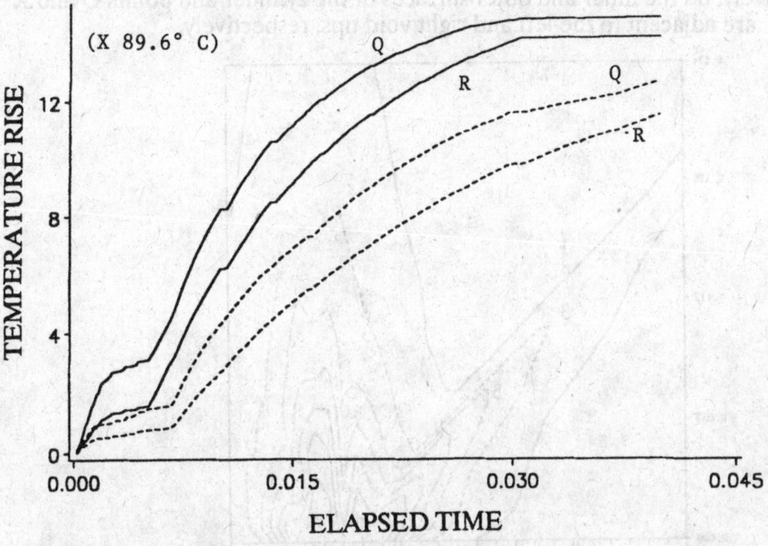


Fig. 2. Evolution of the temperature rise at points Q and R adjacent to the left and right void tips, respectively. ——— with work-hardening, --------- without work-hardening.

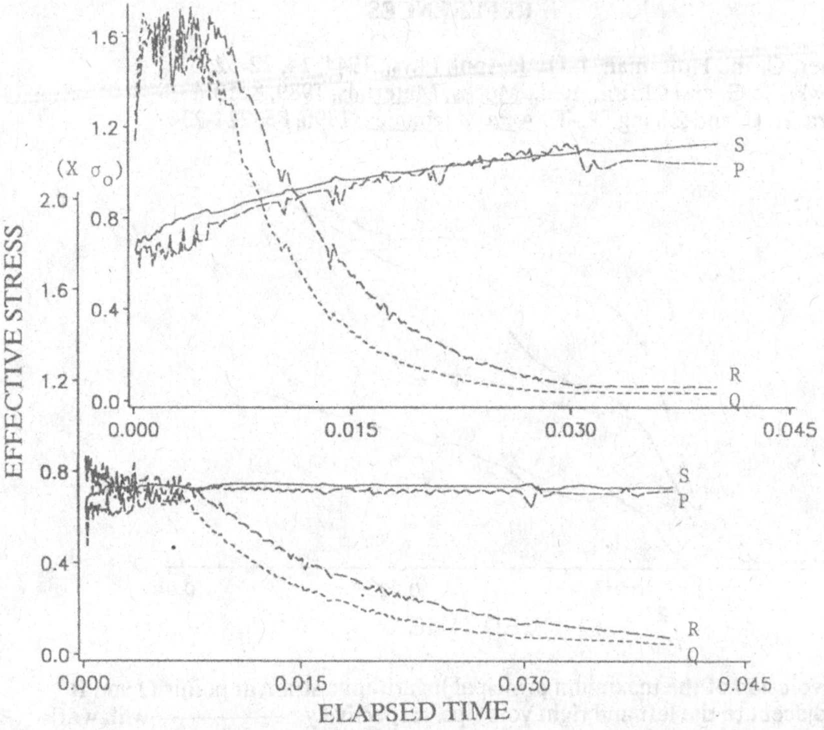


Fig. 3. Evolution of the effective stress at points P, Q, R, and S. Points P and S are, respectively, on the inner and outer surfaces of the cylinder and points Q and R are adjacent to the left and right void tips, respectively.

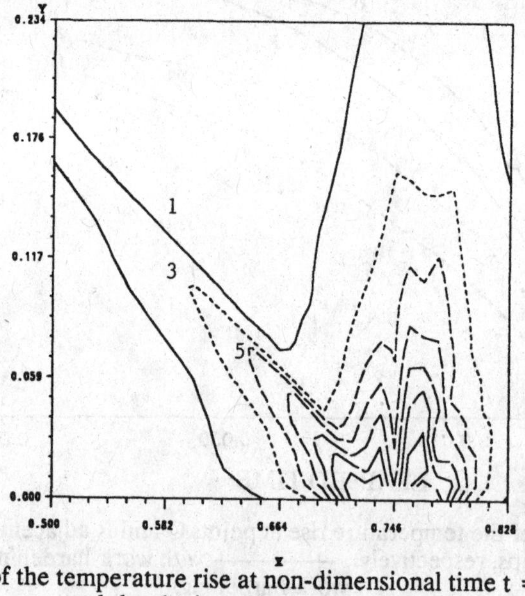


Fig. 4. Contours of the temperature rise at non-dimensional time t = 0.052 when work-hardening effects are considered.

# TEXTURE, DISLOCATION

# AND MICRO-MECHANICAL ANALYSES

# REPRESENTATION OF MICROSTRUCTURE AND MECHANICAL BEHAVIOR OF POLYCRYSTALS

B. L. ADAMS, J. P. BOEHLER, M. GUIDI and E. T. ONAT
Yale University, New Haven, CT, USA
and
Institut de Mécanique de Grenoble, France

## ABSTRACT

We review the group theoretical basis of the result that the observed mechanical behavior of a material can be represented by constitutive (differential) equations which govern the evolution of state variables and that these variables are even rank irreducible tensors. On the other hand microscopic observations of internal structure of a polycrystal produce functions that are defined on "curved" objects such as the unit sphere of directions or the set of distinct orientations of a cube, etc. We show, in terms of an example (the crystallite orientation distribution function for a macroscopically homogeneous polycrystal composed of grains of a cubic crystalline solid) that representations of such functions give rise to Fourier coefficients that are also irreducible tensors. The tensorial state variables will be related to these tensorial Fourier coefficients. A major problem of mechanics of materials is to develop methods that enable one, for a given material and for a given purpose, to extract tensorial state variables and the laws for their evolution from the knowledge obtained from the studies of the microstructure and behavior of the material.

## A SUMMARY OF THE PAPER

A fundamental task of the engineer is to predict the behavior of a structure composed of a given material under extreme operating conditions. The structure may be a piece of material undergoing a forming operation or a turbine disk or a part of a space vehicle. With this task in mind, phenomenological and microscopic studies are conducted on the mechanical behavior of the material of interest. The material will, as a rule, exhibit heterogeneities at various scales of magnification. For the sake of simplicity consider here the case where the material can be taken to be a heterogeneous continuum. Let this continuum be in a state of equilibrium at temperature $\theta_o$.

To conduct an isothermal phenomenological (thought) experiment we may extract a spherical element of radius *R* from the material and apply to its *boundary* a time dependent homogeneous deformation F($\tau$) while keeping the boundary temperature at $\theta_o$. Here $\tau \geq 0$ denotes the time and F is a linear invertible map on $\mathbb{R}^3$ with F(0) = I. The deformation applied to the boundary will cause the development of (Cauchy) stresses within the element and these stresses will vary from point to point because of the assumed heterogeneity of the material. Let $\sigma(\tau)$ denote the *average* of this field over the element at time $\tau$ (Note that $\sigma(0) = 0$).

The relationship $\Phi$ that exists, for a given element of the material, between the histories of *applied deformation* F($\tau$) and the *observed average stress* $\sigma(\tau)$ is of great interest to the engineer and material scientist. The applied deformation will be accompanied by changes that take place in the internal structure of the element. Indeed, it is these changes that impose the relationship $\Phi$ on the mechanical behavior. Hence, the need for microscopic studies of internal structure at various stages of a phenomenological experiment. It must be emphasized that the relationship $\Phi$ will depend, because of the assumed heterogeneity of the material, not only on the constituents and internal structure of the material but also on the radius $R$ and the location $x_C$ of the center of the element that was extracted from the material:

$$\Phi = \Phi(x_C, R)$$

Suppose that the relationship $\Phi(x_C, R)$ ceases to depend on $x_C$ as $R$ increases and tends to the limit $\Phi_0$. If there exists a minimal radius $R_\varepsilon$ such that $R>R_\varepsilon$ implies that $\Phi_0$ approximates $\Phi(x_C, R)$, to within an error $\varepsilon$, then we say that the material is macroscopically homogeneous from the point of view of mechanical behavior.

If for a desirable $\varepsilon$ the minimal radius $R_\varepsilon$ is much smaller than some of the characteristic lengths associated with the structure, loading, etc., one can, for purposes of structural analysis, replace the actual heterogeneous material with one which is homogeneous and exhibits the constitutive relationship $\Phi_o$.

The next task is to develop constitutive equations that reproduce, *with various levels of accuracy,* the "homogenized" mechanical behavior $\Phi_o$. The engineer will then choose from this set of constitutive equations those that are appropriate to the problem at hand and will use these equations, together with the conservation laws of mechanics and appropriate numerical techniques to obtain the fields of average stress, strain, temperature, etc. over the structure.

*Representation of Mechanical Behavior*

It has become clear in recent years that (homogenized) isothermal mechanical behavior of non-aging inelastic solids can be represented, in a coordinate free manner, by differential equations that govern the evolution of the tensorial variables that measure the state and orientation of the material of interest. The representation has the form (cf. Geary and Onat [1]):

$$\sigma(t) = f(S(t)) \qquad f: \Sigma \rightarrow T_2^s, \Sigma \subset \mathbb{R}^n$$

$$\frac{dS(t)}{dt} = g(S,D) + T_\Omega S(t) \qquad g: \Sigma \times T_2^s \rightarrow \mathbb{R}^n$$

where, for simplicity we omit the initial conditions and the invariance and smoothness requirements on f and g.

Here $\sigma(t)$ denotes the *average* Cauchy stress carried by the material element at time $t$ and $T_2^s$ is the space of symmetric second-rank tensors. D and $\Omega$ are, respectively, the tensors of rate of deformation and rotation at that time defined as follows:

$$\dot{F}(t)F^{-1}(t) = D + \Omega$$

where F is the homogeneous deformation applied to the boundary of the element and the dot denotes derivative with respect to time. Also $S(t) \in \Sigma \subset \mathbb{R}^n$ stands for the n parameters that measure those aspects of the present microstructure within a sufficiently large material element ($R \geq R_\varepsilon$) that are relevant to an (approximate) description of the future mechanical behavior of the element. We refer to S as the *internal state and orientation of the material;* although, we recognize that S cannot describe all the details of the microstructure if n is not extremely large.

It is known that S will be composed of a number of *even* rank irreducible tensors:

$$S=(s_1, \ldots, s_m)$$

Here by irreducible tensors we mean scalars, vectors and all higher rank tensors which are completely symmetric with respect to their indices and are traceless. These tensors serve as measures of symmetry, and the tensorial state variables with rank higher than zero make their appearance when the material is initially anisotropic or when it develops anisotropy during the course of deformation.

A major problem of mechanics of materials is to develop methods that enable one, for a given material and for a given purpose, to extract tensorial state variables and the laws for their evolution from the knowledge obtained from the studies of the microstructure and behavior of the material. The present paper may be considered as a contribution to the study of this problem.

*The crystallite orientation distribution function.*

As a material undergoes deformation and heating, its internal structure will change. The accompanying evolutions of observable aspects of microstructure, such as the crystallite orientation distribution function (codf), constitute measurable signatures of these internal changes. It is known from recent observations and computations (cf. Molinari et al, [2]; Harren and Asaro, [3]; Adams and Field, [4]) that the current distributions of crystal size, shape and orientation influence in many important ways the future elastic -plastic behavior of a polycrystalline solid.

The codf, introduced by Bunge ([5], [6]) and Roe ([7]) contains, in particular, first-order statistical information about the distribution of interatomic bonding within the solid. It is not surprising, therefore, that the codf would correlate well with a number of important physical and mechanical properties.

Thus the state variables ($s_1, ..., s_m$) needed for a more faithful representation of the mechanical behavior of a polycrystal at large deformations will have to represent *relevant* aspects of its evolving codf and of other statistical measures of microstructure such as the distribution of dislocations etc. A deep understanding of micromechanics of the solid, however, is needed to decide which aspects of what measures of microstructure are *relevant* for a description of the mechanical behavior of the solid. Nevertheless, in choosing and defining the tensors ($s_1, ..., s_m$) that comprise S in the representation of mechanical behavior as functions of the statistical measures of interest, the first step would be to describe these measures in terms of tensors also.

Here we reconsider the well studied problem of representation of the codf for polycrystals composed of grains of a crystalline solid(cf. Bunge, [6]). For definiteness we assume that the crystalline solid is cubic. The main purpose of the paper is to construct, with the help of the theory of group representations (Onat, [8]), a simple coordinate free tensorial representation of the codf for cubic materials .

We show that the codf $f : SO(3) \rightarrow \mathbb{R}^+$ where SO(3) denotes the group of rigid body rotations accepts the following representation

$$f(g) = V_o + V_4.f_4(g) + V_6.f_6(g) + V_8.f_8(g) + V_9.f_9(g) + V_{10}.f_{10}(g) +$$

$$V_{12}^{(1)}.f_{12}^{(1)}(g) + V_{12}^{(2)}.f_{12}^{(2)}(g) + \cdots$$

where the coefficents $V_o$, $V_4$ ... and the basis functions $f_4(g)$, $f_6(g)$, ... are tensor valued and the dot denotes the classical inner product appropriate to the rank indicated by the subscripts.

This representation confirms Bunge's result and it is equivalent to it except that the basis functions here are generated (without recourse to generalized spherical harmonics and to lengthy calculations) from irreducible tensors that have the symmetries of the cube. An advantage of the above representation is that its coefficients $V_l$ transform as tensors appropriate to the rank $l$ under the rigid body rotations of the sample.

Thus if the codf f(g) for a material element has the the Fourier coefficients $V_o$, $V_4$, $V_6$, ... then the codf $p_k f(g)$ of the k-rotated element will have the coefficients $p_k V_o$, $p_k V_4$, $p_k V_6$, ... , where $p_k$ denotes the tensor transformation appropriate to the rank $l$ of $V_l$ . Since tensors can be defined without the benefit of a coordinate frame the representation is coordinate free. Moreover any (processing) symmetries that the material element may possess are expressed in these tensorial coefficients.

## ACKNOWLEDGEMENT

This work was supported by a Materials Research Group Award of the National Science Foundation (DMR-9001378).


## REFERENCES


1. Geary, J. E., and Onat, E. T. "Representation of Nonlinear Hereditary Mechanical Behavior", *Oak Ridge National Laboratory Report, ORNL-TM-4525*. (1974)

2. Molinari, A., Canova. G. R. and Ahzi, S. "A Self Consistent Approach of the Large Deformation Polycrystal Viscoplasticity"*Acta Metall.*, **35**, 2983. (1987)

3. Harren, S. V., and Asaro, R. J."Non-uniform Deformations in Polycrystals and Aspects of the Validity of the Taylor Model", *J. Mech. Phys. Solids*, **37**, 191. (1989)

4. Adams, B. L., and Field, D. P."Statistical Theory for Creep of Polycrystals"to appear in *Acta Metall. Mater.*(1991)

5. Bunge, H. J." Zur Darstellung allgemeiner Texturen", *Z. Metallkunde*, **56**, 872. (1965)

6. Bunge, H. J.*Texture Analysis in Materials Science*,Butterworths, London. (1982)

7. Roe, R. J. " Description of Crystallite Orientation in Polycrystalline Materials. III.General Solution to Pole Figure Inversion", *J. Appl. Phys.*, **36**, 2024. (1965)

8. Onat, E. T. " Group Theory and Representation of Microstructure and Mechanical Behavior of Materials", to appear in Proc. Symposium on *Modeling the deformation of Crystalline Solids*,TMS Meeting, New Orleans,February (1991).

# MECHANICAL PROPERTIES, DEFORMATION TEXTURES, ELASTOPLASTIC BEHAVIOUR OF BCC POLYCRYSTALLINE MATERIALS.

**DOMINIQUE CECCALDI, FATIMA YALA, THIERRY BAUDIN, RICHARD PENELLE**
**Laboratoire de Métallurgie Structurale, URA CNRS 1107, bat. 413, 91405 ORSAY.**
**FRANCOIS ROYER**
**Université de Metz, IPEM/LPMC, boulevard Arago, 57070 METZ.**

## ABSTRACT

**A certain number of uniaxial tensile tests have been performed as well along the rolling and transverse directions as out of the principal axes of anisotropy in a low carbon rimmed steel sheet with equiaxed grains. The deformation textures and the Lankford's ratio are calculated for these loading pathes in the framework of different Taylor's type models generalized to all the classes of symmetry allowed for the texture even though the previous works only be focused on the orthotropic symmetry. Different glide assumptions ( {110}<111> crystallographic slip, {110}<111> + {112}<111> mixed conditions, pencil glide ) are explored. Furthermore some models are applied to the calculation of the deformation textures obtained during the rolling process.**

## INTRODUCTION

**One of the main purpose of industrial manufacturers is to be able to predict the general behaviour of a material from a minimal set of experimental data. A way to answer such a problem can be given by the connections between microscopic and macroscopic phenomena during plastic flow.**
**The anisotropy of a polycrystal is as well described by the results of mechanical tests as by the crystallographic texture both being however strongly linked. The orientation distribution function ( O.D.F. ) also called texture function F( g ) defined in the Euler space G ( $g \in G$ ) has been calculated from the incompleted pole figures and developped over a basis set of generalized spherical harmonics. The knowledge of the active slip systems deduced from the assumptions of the models based on a plastic behaviour of the polycrystal will led to the calculations of the deformation textures.**

## EXPERIMENTAL DATA

**The initial sample is a recrystallized sheet with equiaxed grains with a mean diameter of 20 µm and a small amount of lamellar perlite. From a mechanical point of view different samples cut out every 15° in the plane of the sheet have been deformed with fixed grips under a constant strain rate of $2.10^{-4}$ $s^{-1}$ until deformations of 10%, 15% and 20%. The studied sample presents a yield point and so on a lower and upper yield stress. The experimental**

textures measured by x-rays diffraction have been completed by an original iterative method [1]. The measured textures are now included in several models in order to describe the plastic behaviour of the sample.

## MODELLING AND RESULTS

A particular attention is put in this paper first on the study of the variations of the Laankford's coefficient in the plane of the sheet and secondly on a simulation of the deformation textures in the case of a triclinic symmetry of the texture.

For the B.C.C. materials if it is commonly assumed that the glide directions belong to <111> the slip planes experimentaly observed can be {110}, {112} and {123}. To get a good description of the deformation mechanism by multislip the two slip systems {110}<111> and {112}<111> with a variable ratio of the critical resolved shear stresses for the two {110} and {112} glide planes are taken into account in the framework of the Taylor's assumptions (full constraint model justified by equiaxed grains ). Instead of using the principles of linear programmation [2] an original formulation is developped [3] using the critical polyhedron determined by Orlans-Joliet [4] in the case of mixed slip. The vertices of this polyhedron are then classified in 15 irreducible groups each of which is associated with combinaisons of 5, 6 or 8 slip systems. Furthermore this model is modified by a perturbation based on the introduction of an inhomogeneous microscopic behaviour of elastoplastic polycrystals at large strains proposed by Arminjon [5].

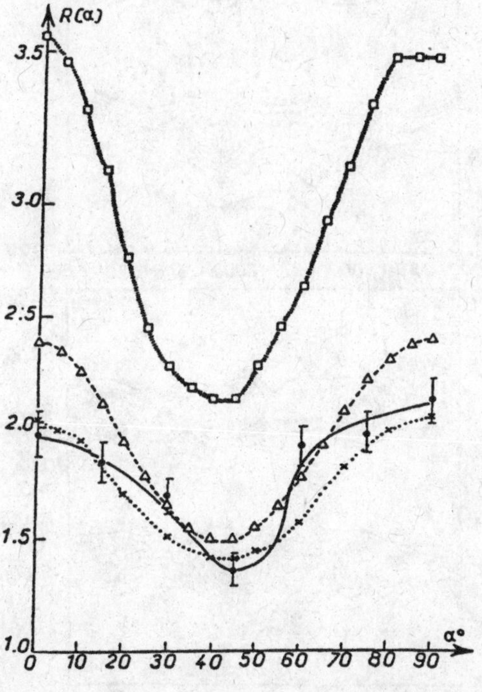


FIGURE 1 : Comparison of the experimental and calculated values of the Lankford's coefficient : (□) {110}<111> crystallographic slip, (Δ) pencil glide, (x) mixed conditions, (●) experiment.

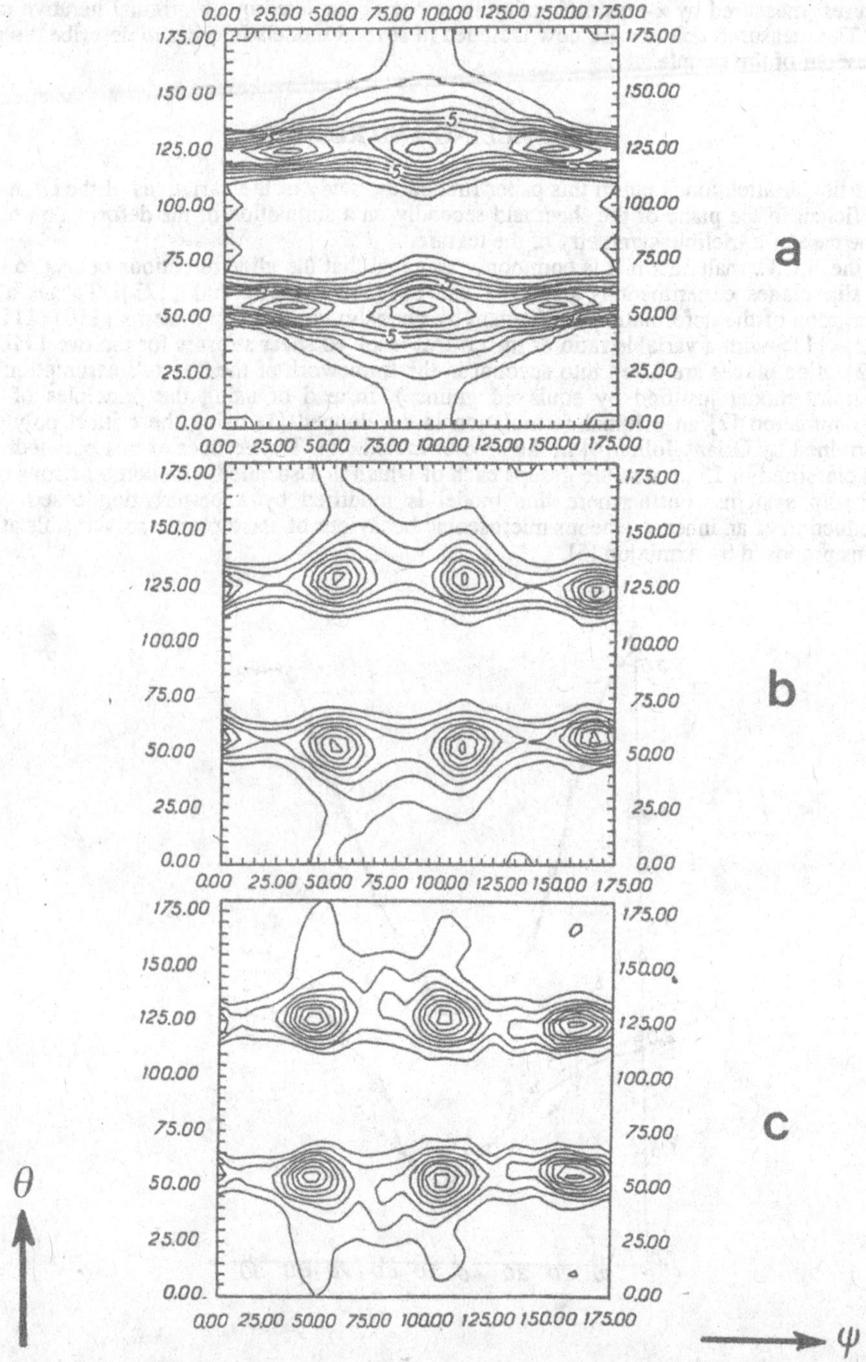


FIGURE 2 : cut at φ = 45° (a) initial texture [ F(g)=18 ] (b) sample 20% deformed at 45° of RD [ experiment F(g)=13 ] (c) simulation by {110}<111> + {112}<111> glide [$F(g)_{HM}$=13 ].

To complete the glide hypothesis, the "pencil glide" model [6] ( 3 or 4 active systems ) is also used in some comparison. In this case the slip plane is referred to the maximal resolved shear stress and can be of a non-crystallographic type.
FIGURE 1 summarize some results obtained for the Lankford's coefficient the ratio of the critical shear stresses being $\zeta = 0.93$ . There is a large agreement between the experimental data and the calculated values with mixed slip however the pencil glide assumption must not be excluded for this material. Crystallographic glide only on the {110}<111> systems has to be clearly rejected for B.C.C. materials.
One can noticed that in most of the previous works the Taylor's factor ( sum of the absolute values of the amount of glide on the active slip systems ) is developped on the same basis of harmonics as the texture. In our mixed slip conditions the mean value of this factor used in the calculation of $R(\alpha)$ is obtained by a three dimensional integration (trapezium method) in the Euler space the texture function appears then as a weight parameter.
The influence of a prestrain can also be pointed out by including the experimental deformation textures in the calculations of $R(\alpha)$. Such a study shows that there exists a strong correlation between $R(\alpha)$ and the O.D.F. . With equal scattering every variation of the maximum of the O.D.F. induces variations of $R(\alpha)$ which increases with the maximum of the texture function.
The results of FIGURE 1 are associated to uniaxial tension but the formalism can easily be extended to any kind of strain tensors. Concerning the determination of the deformation textures the situation of a tensile test at 45° of the rolling direction is very interesting in the sense that modelling leads to components not present on the experimental pole figure. The calculated texture function F(g) is then developped on the basis of spherical harmonics and $F(g)_{HM}$ determined. The subscript HM specifies that the function is generated by the harmonic method. Such an approach shows that the harmonic method smooth down the sharpness of the calculated texture and permits to reproduce with a good approximation the experimental textures. FIGURE 2 illustrates these remarks.
For the calculations of rolling textures the inhomogeneous model proposed by Arminjon is used. The microscopic deformation amount of a grain is submitted to fluctuate in the the neighborhoud of the macroscopic strain ratio and the mixed conditions are always taken into account.
In conclusion is can be put in evidence that the mixed slip describes correctly the homogeneous multislip in the B.C.C. polycrystals and the so obtained results are in good agreement with experimental O.D.F.

## REFERENCES


[1] Baudin, T., Penelle, R., Ceccaldi, D. and Royer, F., Mém. Et. Sci. Rev. Métallurgie, 1990, 10, 611-617.
[2] van Houtte, P., Mater. Sci. Engng., 1982, 55, 69.
[3] Ceccaldi, D., Sedrati-Yala, F., Baudin, T., Penelle, R., and Royer, F., to be published.
[4] Orlans-Joliet, B., Bacroix, B., Montheillet, F., Driver, J.H., Jonas, J.J., Acta Metall., 1988, 36, 1365-1380.
[5] Arminjon, M., Acta Metall., 1987, 35, 615-630.
[6] Royer, F., Tavard, C., and Penning, P., J. Appl. Cryst., 1979, 12, 436-441.

# VARIATIONAL PROBLEMS IN DEFECTIVE CRYSTALS

CESARE DAVINI, IRENE FONSECA, GARETH PARRY

*Istituto di meccanica teorica ed applicata, Universita' di Udine, Udine, Italy*
*Department of mathematics, Carnegie Mellon University, Pittsburgh, U.S.A.*
*School of mathematical sciences, Univesity of Bath, Bath, U.K.*

## ABSTRACT

We discuss general notions, and boundary value problems, associated with a simple continuum model of defective crystals.

## GENERAL NOTIONS

Here we present just some of the ideas introduced in [2], developed in [3],[4],[6],[7]. Let $\mathbf{d}_a(\cdot)$ ,$a$ = 1,2,3 denote three linearly independent fields of (average) lattice vectors defined over a region $\mathcal{B}$. We call $\Sigma = \{ \mathbf{d}_a(\cdot), \mathcal{B}\}$ a *state* of a solid crystal, and agree that in the *elastic deformation* corresponding to an invertible, differentiable mapping $\mathbf{u} : \mathcal{B} \longrightarrow \mathbf{u}(\mathcal{B})$, lattice vectors $\mathbf{d}_a(\mathbf{x})$ map to $(\nabla\mathbf{u})(\mathbf{x})\,\mathbf{d}_a(\mathbf{x}) = \mathbf{d}_a^*(\mathbf{u}(\mathbf{x}))$, say,so the fields $\mathbf{d}_a^*(\cdot)$ are defined over $\mathbf{u}(\mathcal{B})$.

The Burgers' integrals of the classic theory of continuous distributions of dislocations are the circuit integrals

$$\oint_c \mathbf{d}^a \cdot d\mathbf{x}$$

where $\{\mathbf{d}^a\}$ are the dual (average) lattice vectors, so $\mathbf{d}^1 \det\{\mathbf{d}_a\} = \mathbf{d}_2 \times \mathbf{d}_3$, for example. It is easy to check that the Burgers' integrals are *elastic invariants* , in the sense that if $c^* = \mathbf{u}(c)$, then

$$\oint_c \mathbf{d}^a \cdot d\mathbf{x} = \oint_{c^*} \mathbf{d}^{a*} \cdot d\mathbf{x}^* .$$

One can sensibly assert , then , that Burgers' integrals measure some aspect of the "defectiveness" of the crystal, since the conventional wisdom has it that elastic deformations do not change (whatever is meant by the) defectiveness. By Stokes' theorem, if $c$ is the boundary of a surface $\mathcal{S}$,

$$\oint_c \mathbf{d}^a \cdot d\mathbf{x} = \int_{\mathcal{S}} \nabla\times\mathbf{d}^a \cdot \mathbf{n}\, da ,$$

and one refers to $\nabla\times\mathbf{d}^a$ as the Burgers' vector, or the density corresponding to the Burgers' integral. To capture rather more aspects of the defectiveness one might try to find other integrals (not necessarily circuit integrals) which are elastic invariants, in the obvious sense. We find, in [4], that there is an infinite list of such integrals, an infinite number of corresponding densities, but that there is a functional basis for this list consisting just of the densities

$$\mathbf{b}^a = \nabla \times \mathbf{d}^a , \; s^{ab} = \nabla \times \mathbf{d}^a \cdot \mathbf{d}^b , \; n = \mathbf{d}^1 \cdot \mathbf{d}^2 \times \mathbf{d}^3 , \; \nabla(s^{ab}/n) \times \mathbf{d}^c .$$

Said differently, if these four densities match pointwise in possibly different states $\{ \mathbf{d}_a(\cdot), \mathcal{B}\}$, $\{ \mathbf{d}_a'(\cdot), \mathcal{B}\}$, then *all* densities of appropriate form also match.

Now if all densities match in the two states $\{ \mathbf{d}_a(\cdot), \mathcal{B}\}$, $\{ \mathbf{d}_a'(\cdot), \mathcal{B}\}$ one would like to believe that the defectiveness of the two states has to be the same, in some sense. The natural question, then, is "what is the connection between such fields $\mathbf{d}_a(\cdot)$, $\mathbf{d}_a'(\cdot)$?". It is the answer to this question that makes us believe the model is useful, for such fields must be connected by combinations of elastic deformations and *slip* (see[3],[4] for details). By *slip* we understand a translation of material points in lines or in surfaces where the lattice vectors are constant, with lattice vectors carried unchanged by the translation. So slip amounts to the classic glide mechanism of plasticity theory, and it is a notion which emerges here from abstract ideas formulated without prejudice regarding the kinematical nature of defects. Moreover the result provides convincing evidence that defining defectiveness by means of the elastic invariants is a sufficiently exhaustive procedure.

## SLIPS IN A PERFECT CRYSTAL

By a perfect crystal let us understand $\Sigma = \{ \mathbf{e}_a, \mathcal{B}\}$ with $\{ \mathbf{e}_a\}$ the canonical basis of $\mathbb{R}^3$. It is easy to calculate the class of states which have the same defectiveness as this perfect crystal - if one restricts attention to states defined over the same region $\mathcal{B}$, they have the form $\{ (\nabla\mathbf{v})^{-1}\mathbf{e}_a, \mathcal{B}\}$ with $\mathbf{v} : \mathcal{B} \longrightarrow \mathbf{v}(\mathcal{B})$ an invertible, differentiable, unimodular mapping, $\det(\nabla\mathbf{v}) = 1$. Since elastic deformations do not change defectiveness, by hypothesis, it follows that this class also includes

$$\{ \mathbf{l}_a(\mathbf{y}), \mathbf{u}(\mathcal{B}) \} , \tag{1}$$

where $\mathbf{l}_a(\mathbf{u}(\mathbf{x})) = \nabla\mathbf{u}\,\{(\nabla\mathbf{v})^{-1}\mathbf{e}_a\}$, with $\mathbf{u}$, $\mathbf{v}$ as before, where $\mathbf{u}$ represents the elastic deformation, and $\mathbf{v}$ represents the slip which we might call the plastic part of the change of state. In fact (1) represents all such states and so provides an explicit factorisation of an important class of changes of state via purely kinematical ideas. One might contrast the procedure with the common "thought experiment" of allowing sufficiently small pieces of crystal to relax to a supposed natural state and calling the residual strain the plastic part of the deformation. In this last procedure there is an ambiguity which comes from the requirement of objectivity - the plastic part of the deformation is only defined to within a rotation. Likewise there is a nonuniqueness in the representation (1), though it is a kinematical feature and independent of any concept of stress. We find, in [6], that if displacement conditions are given on $\partial\mathcal{B}$, then

$$\nabla \mathbf{u}\,(\nabla \mathbf{v})^{-1} = \nabla \tilde{\mathbf{u}}\,(\nabla \tilde{\mathbf{v}})^{-1} \quad \text{in } \mathcal{B} \tag{2}$$

if and only if $\mathbf{u} = \tilde{\mathbf{u}} \circ \mathbf{f}$, $\mathbf{v} = \tilde{\mathbf{v}} \circ \mathbf{f}$ with $\mathbf{f}$ such that $\mathbf{f}(\mathbf{x}) = \mathbf{x}$ on $\partial\mathcal{B}$ and $\det(\nabla \mathbf{f}) = 1$ in $\mathcal{B}$.

## VARIATIONAL PROBLEMS

Commonly, the study of variational problems in elasticity theory is regarded as a reasonable use of time, for elastic changes are supposed to be reversible. Contrastingly the irreversibility of plastic change, attributed to friction, to the tangling of dislocation loops, to interstitials lost in forests, and the like, leaves scant hope that variational problems are realistic in a general context (see [8]). However they may be useful in the grey area where changes of state do not change the invariant integrals described above, where according to section 2 changes are elastic or represent "easy glide" in the surfaces where lattice vector fields are constant (the reader may refer to [3],[6] for more extensive justification ). So we suppose that an energy functional $\mathcal{E}$ is defined over states of the form (1),

$$\mathcal{E} = \int_{\mathcal{B}} W(\nabla \mathbf{u}\,(\nabla \mathbf{v})^{-1})\,dx \quad , \tag{3}$$

where W is a non-negative energy density function with $W(\mathbb{I})=0$, with symmetry properties appropriate to a perfect crystal (see [1],[5],[6] for relevant details), and where we assume displacement boundary conditions $\mathbf{u}=A\mathbf{x}$ on $\partial\mathcal{B}$ (A is a constant matrix). If we try to minimise $\mathcal{E}$ by choice of the two functions $\mathbf{u}$, $\mathbf{v}$ we find that, generally, $\mathbf{u}$, $\mathbf{v}$ are the limits of piecewise differentiable functions $\mathbf{u}_\kappa$, $\mathbf{v}_\kappa$ with ever increasing numbers of surfaces of discontinuity. The limit functions themselves are nowhere differentiable, in general, but it turns out that significant volume averages are well-defined, e.g. the numbers

$$\lim_{\kappa \to \infty} \frac{1}{\mathrm{vol}(\mathcal{V})} \int_{\mathcal{V}} \mathbf{u}_\kappa \, dx, \qquad \lim_{\kappa \to \infty} \frac{1}{\mathrm{vol}(\mathcal{V})} \int_{\mathcal{V}} (\nabla \mathbf{u}_\kappa)(\nabla \mathbf{v}_\kappa)^{-1}\, dx,$$

all exist for arbitrarily chosen $\mathcal{V} \subset \mathcal{B}$. So we can speak of the average limiting elastic and plastic parts of the average limiting lattice vectors. Also, one can make sense of the average Cauchy stress tensor, and find that the average limiting Cauchy stress is *isotropic* . Though the analysis is technical, this rather startling result has some appeal if we judge it by statements in the literature that crystals offer almost no resistance to glide in appropriate lattice surfaces, the implication being that shear stresses relax by the development of sufficiently finely distributed, properly oriented shear bands (with this observed phenomenon mirrored by the differentiability properties of $\mathbf{v}$).

## INVARIANT ENERGY FUNCTIONALS

The nonuniqueness of decomposition of lattice vectors given by (2) has no bearing on functionals of the form (3), for $\mathcal{E}$ is well-defined whatever representation of lattice vectors is chosen. On the other hand, one might choose to consider functionals which also penalise slip, perhaps to model the work done by friction. We give a couple of examples;

(i) if $$\mathcal{E} = \int_{\mathcal{B}} \{ W(\nabla u \, (\nabla v)^{-1}) + g\,(\nabla v)\}\, dx \;,$$

then g must be such that

$$\int_{\mathcal{B}} g\,(\nabla v)\, dx = \int_{\mathcal{B}} g\,(\nabla(v \circ f))\, dx$$

since the slip is not known uniquely. The class of all such g is determined in [7].

(ii) the functional $$\mathcal{E} = \int_{\mathcal{B}} W(\nabla u\, (\nabla v)^{-1})\, dx + \int_{\partial v(\mathcal{B})} \Gamma(\mathbf{n})\, d\sigma \;,$$

where $\mathbf{n}$ is the normal to $\partial v(\mathcal{B})$ and $\Gamma(\mathbf{n})$ is a given function, is invariant in the appropriate sense since $\partial v(\mathcal{B}) = \partial v \circ f(\mathcal{B})$, from (2). It turns out that the limiting elastic and plastic parts (in this latter example) *decouple* in the sense that $\partial v(\mathcal{B})$ is determined by the Wulff costruction which provides minimisers of the "surface energy" type term in $\mathcal{E}$.

These two last examples are ,perhaps, not particularly well set by the physics, but they illustrate that the nonuniqueness of decomposition can be perfectly accounted for in formulating the corresponding boundary value problems. In summary, the thrust of the work described briefly here is twofold;

(a) mechanisms associated with plasticity arise naturally in this simple model, long-standing problems in the classic theory have analogues with solutions,

(b) the relevant variational problems are novel from the point of view of the calculus of variations.

**REFERENCES**

[1] Chipot, M. and D. Kinderlehrer (1988) Equilibrium configurations of crystals, *Arch. Rat. Mech. Anal.* 103, 237-277.

[2] Davini, C. (1986) A proposal for a continuum theory of defective crystals, *Arch. Rational Mech. Anal.* 96, 295 -317.

[3] Davini, C. and G. P. Parry (1989) On defect-preserving deformations in crystals, *Int. J. Plasticity* 5, 337-369.

[4] Davini,C.and G.P.Parry (1991) A complete list of invariants for defective crystals, to appear in *Proc. Roy. Soc. London. A.*

[5] Fonseca,I. (1987) Variational methods for elastic crystals, *Arch. Rat. Mech. Anal.* 97, 189-220.

[6] Fonseca, I. and G.P. Parry (1991) Equilibrium configurations of defective crystals, to appear.

[7] Fonseca, I.and G.P. Parry (1991) A class of invariant integrals, to appear.

[8] Ericksen,J,L. (1983) Thermoelastic considerations for continuously dislocated crystals, *Proc. Int. Symp.* on *Mechanics of Dislocations,* Houghton, MI.

# SIMULATION OF DEFORMATION-INDUCED ANISOTROPY IN METAL FORMING APPLICATIONS

PAUL R. DAWSON, YVAN B. CHASTEL, AND ANTOINETTE M. MANIATTY
Sibley School of Mechanical and Aerospace Engineering
Cornell University, Ithaca, NY 14853, U.S.A.

## ABSTRACT

A micromechanical based polycrystalline analysis for modeling strain-induced anisotropy in metal forming applications is presented. Several examples involving rolling are presented in order to demonstrate the ability of the algorithm to predict textures and residual stresses.

## INTRODUCTION

When modeling industrial forming processes where the workpiece undergoes large deformations, it is important to be able to accurately characterized the evolving state of the material as it deforms. For metals, in particular, the texture, hardness, and elastic strains can be used to describe the anisotropic state. A method for modeling the evolution of these characteristics based on micromechanics for steady-state industrial applications is presented herein. A description of the problem formulation and the method used to solve it are described. The formulation is employed to simulate two rolling problems. In the first problem, the focus is on computing the texture through the thickness for a B.C.C. material and comparing the results with experimentally obtained textures. The second problem examines the effect of the strain-induced anisotropy on the residual stress distribution in an F.C.C. material by comparing the residual stresses obtained in this analysis with those obtained in an isotropic analysis.

## DEFINITION OF THE PROBLEM

The usual boundary value problem consisting of equilibrium and boundary conditions needs to be satisfied on the two-dimensional domain of interest B giving

$$\operatorname{div} \mathbf{T_g} + \mathbf{b} = \mathbf{0} \qquad \text{on } B \tag{1}$$

$$\mathbf{v} \cdot \mathbf{e}_i = \hat{v}_i \quad \text{on } \partial B_{1i}\,, \qquad \mathbf{T_g n} \cdot \mathbf{e}_i = \hat{t}_i(v_i) \quad \text{on } \partial B_{2i} \tag{2}$$

where $\mathbf{T_g}$ is the macroscopic Cauchy stress, $\mathbf{b}$ is the body force, $\hat{v}_i$ is the velocity boundary condition, $\hat{t}_i$ is the traction boundary condition which may be a function of the velocity $v_i$, (e.g. in the case of friction), $\mathbf{n}$ is a unit outward normal, and $\mathbf{e}_i$ is an orthonormal basis for the two-dimensional domain, i=1,2. Also $\partial B_{1i} \cup \partial B_{2i} = \partial B$ and $\partial B_{1i} \cap \partial B_{2i} \equiv \emptyset$, i=1,2, where $\partial B$ is the boundary of B.

In addition, information regarding the material response in the form of constitutive relations must be specified. First, focusing on single crystals, the structure of kinematics follows that given by Peirce et al. [1], where the deformation gradient is assumed to have a multiplicative decomposition into an elastic (lattice) component and a plastic component. This gives

$$\mathbf{F} = \mathbf{R}^L\mathbf{U}^e\mathbf{F}^p \tag{3}$$

where $\mathbf{F}$ is the deformation gradient, $\mathbf{R}^L$ is the lattice rotation, $\mathbf{U}^e$ is the right elastic (lattice) stretch, and $\mathbf{F}^p$ is the plastic deformation gradient. The plastic part of the deformation is assumed to be isochoric and to be due only to slipping along crystallographic slip systems. Therefore, the plastic velocity gradient on the intermediate relaxed configuration $\tilde{\mathbf{L}}^p$ can be expressed as

$$\tilde{\mathbf{L}}^p = \dot{\mathbf{F}}^p\mathbf{F}^{p^{-1}} = \sum_{\alpha=1}^{N_s} \dot{\gamma}^\alpha\,(\mathbf{s}^\alpha \otimes \mathbf{m}^\alpha) \tag{4}$$

where $\dot{\gamma}^\alpha$ is the rate of shearing on the $\alpha$ slip system, $\mathbf{s}^\alpha$ is the slip direction, $\mathbf{m}^\alpha$ is the slip plane normal, and $N_s$ is the number of slip systems. The rate of shearing on an individual slip system is related to the resolved shear stress on that system through a power law relationship of the form

$$\dot{\gamma}^\alpha = \dot{a}\left(\frac{\tau^\alpha}{g}\right)\left|\frac{\tau^\alpha}{g}\right|^{1/m - 1} \tag{5}$$

where $\dot{a}$ and m are material parameters and g is the grain hardness which is taken to be the same on all the slip systems. Furthermore, an evolution equation for the hardness g is required. This is assumed to take the following form

$$\dot{g} = h(\gamma, g)\,\dot{\gamma} \qquad \dot{\gamma} \equiv \sum_{\alpha=1}^{N_s} |\dot{\gamma}| \tag{6}$$

where h is a prescribed function. Finally, a constitutive equation for the elasticity is also needed, and for cubic crystals, this can be written as

$$\overline{\mathbf{T}} = 2\mu\,\overline{\mathbf{E}}^e + \lambda\,\mathrm{tr}(\overline{\mathbf{E}}^e) - 2\beta\,\mathcal{C} : \overline{\mathbf{E}}^e \tag{7}$$

where $\mu$, $\lambda$, and $\beta$ are the elastic moduli, and $\mathcal{C}$ is a fourth order tensor determined by the symmetries of a cubic crystal. In the cases presented here, $\overline{\mathbf{E}}^e$ is the logarithmic elastic strain and $\overline{\mathbf{T}}$ is the work conjugate stress.

The macroscopic response of the material is then determined by making an assumption about the grain interaction. A Taylor model is assumed initially here which states that the global deformation is prescribed on each grain $\mathbf{D}_g = \mathbf{D}$, and the macroscopic stress is then taken to be the average of the stresses in all the grains $\mathbf{T}_g = \langle\mathbf{T}\rangle$. This has been found to be a good approximation for cubic crystals with fairly equiaxed grains. When the grains become distorted, a weaker constraint on the kinematics is necessary and a "relaxed" constraints approach may be used for a certain volume fraction of the grains. In this case, certain components of the global deformation and certain components of the global stress are prescribed on each grain. For details see Ref. [2]. In addition, pencil glide, and for relaxed grains, restricted glide is also included in the formulation [3] for B.C.C. crystals.

## SOLUTION PROCEDURE

The problem then is to solve the boundary value problem given in equations (1), (2) with the material response prescribed by equations (3) - (7). The solution is divided into two parts, the solution of the boundary value problem on B and the integration of the constitutive equations along particle pathlines through the domain B. The boundary value problem is solved by expressing the equilibrium equation (1) in the weak form, and then solving for the velocity and pressure fields using the finite element method. In order to accomplish this, it is necessary to write a relationship for the deviatoric macroscopic stress $\mathbf{T}_g^{'}$ in terms of $\mathbf{D}_g$. Using equations (3) - (6), a relation of the form

$$\mathbf{T}_g^{'} = \mathcal{K} : \mathbf{D}_g - \mathbf{Z}^e \tag{8}$$

can be obtained where $\mathcal{K}$ is a fourth order stiffness tensor and $\mathbf{Z}^e$ is a stress due to the elastic component. A volumetric constraint is also needed and can be obtained from the elastic constitutive equation (7). For details of this development see Ref. [4]. In order to update the state of the material through the domain, the constitutive equations for aggregates of grains at representative locations in the domain are integrated along particle pathlines which can be found using the velocity field. The integration procedure for the case neglecting elasticity can be found in Ref. [2] and for the case with elasticity in Ref. [4]. Then the updated state is used in equation (8) to give the material response, and the boundary value problem is resolved. The alternate solution of the boundary value problem and integration of the constitutive equations continues until convergence.

## NUMERICAL EXAMPLES

Two examples involving rolling were analyzed. The Voce-Kocks model described in Ref. [2] is used to describe the hardness evolution in equation (7) for both cases. As a validation of the anisotropic metal forming model described above, the texture development during a cold rolling process of silicon steel (a B.C.C. material) was modeled. Since the focus was only on modeling the texture evolution, the elastic straining part of the deformation was neglected because the elastic strains are generally small compared to the plastic strains and have an insignificant effect on the texture development. Fifteen rolling passes were simulated so as to reproduce, as closely as possible, those of experiments performed at Allegheny Ludlum [5]. PopLA [6] was used to discretize the initial experimental pole figures into "weights files" which provided the starting texture for the rolling simulations at four positions through the sheet thickness. The evaluation of the material parameters, the choice of the friction model and of the rule for linking the microscopic and macroscopic length scales are described in detail in Ref. [3]. A comparison of the <200> and <112> pole figures is shown in Figure (1).

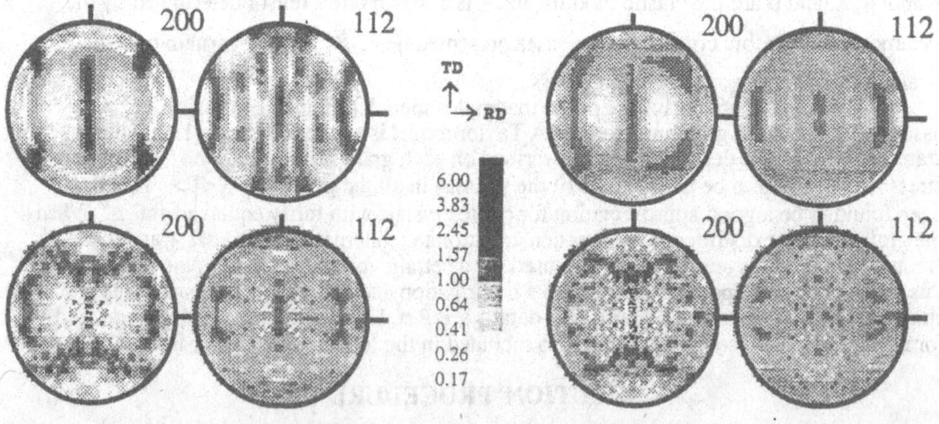


Figure 1. Textures after 70% reduction. Experiments (top) and predictions (bottom). Along centerline (left) and at the subsurface s = .8 (right).

As can be seen in Figure (1), the formulation can account for through thickness texture variations caused by the inhomogeneity of the deformation. The subsurface textures are at a distance of s = .8 from the centerline where s is such that s = 1 at the surface.

In the second example, the elastic part of the deformation is included in order to predict the residual stresses for 1100-aluminum (an F.C.C. material) and to observe the effect of strain-induced anisotropy on the stresses as compared to an isotropic analysis. For the isotropic case, Hart's model is used for the viscoplastic part of the deformation. The parameters used for both the anisotropic and the isotropic cases are given in Ref. [4]. In this case, only a simple Taylor model is used. The rolling geometry with the finite element mesh used is shown Figure (2) where a = 4 cm, b = .5a, the roll radius R=2.5a, and the roll velocity at the surface is $v_o$ = 1m/s. Sticking friction was assumed. The $T_{xx}$ component of the residual stresses normalized with respect to the maximum effective residual stress $\sigma_{ef}$ is plotted for each case in Figure (3). The curves are not significantly different from each other, although the anisotropic case gives consistently lower normalized residual stresses. The actual residual stresses (non-normalized) for the anisotropic case were higher than for the isotropic case, but the effective stress $\sigma_{ef}$ was also higher for the anisotropic case. For this particular case, the anisotropy seems to have little effect on the residual stress pattern. For other textures, this may not be the case [4].

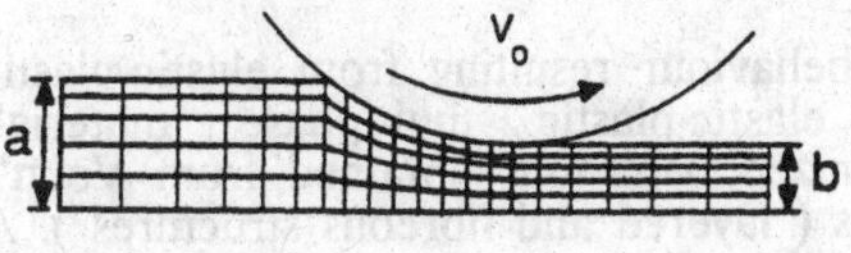


Figure 2

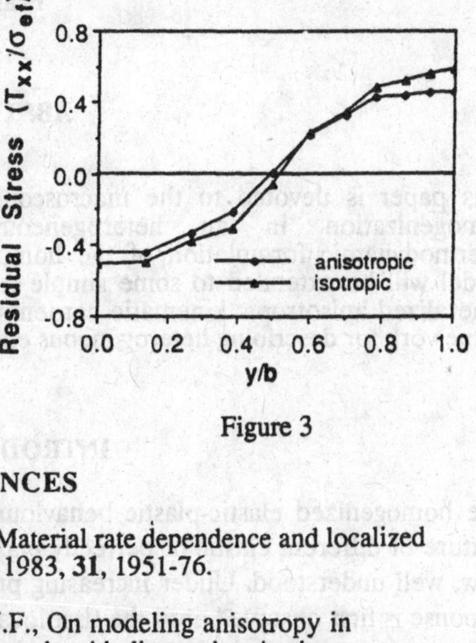


Figure 3

## REFERENCES


1. Peirce, D., Asaro, R.J., and Needleman, A., Material rate dependence and localized deformation in crystalline solids. Acta Metall., 1983, **31**, 1951-76.

2. Mathur, K.K., Dawson, P.R., and Kocks, U.F., On modeling anisotropy in deformation processes involving textured polycrystals with distorted grain shape. Mechanics of Materials, 1990, **10**, 183-202.

3. Chastel, Y.B. and Dawson, P.R., Comparison of polycrystalline rolling simulations with experiments. In press in the proceedings of TMS Conference, 1991 in New Orleans.

4. Maniatty, A. M., Eulerian Elasto-Viscoplastic Formulation for modeling steady-state deformations with strain induced anisotropy, Ph.D. Thesis, Cornell University, 1991.

5. Salsgiver, J., private communication, 1989.

6. Kallend, J.S., Kocks, U.F., Rollett, A.D., and Wenk, H.R., Operation texture analysis, Mater. Sci. Eng., in press.

# ELASTIC-PLASTIC ANISOTROPIC BEHAVIOUR RESULTING FROM MICROHETEROGENEITY

ABDERRAHIM ELOMRI and FRANCOIS SIDOROFF

Ecole Centrale de Lyon,
Département de Mécanique des Solides
BP 163, 69131 Ecully Cedex,
France.

## ABSTRACT

This paper is devoted to the macroscopic behaviour resulting from elastic-plastic homogenization in an heterogeneous elastic-plastic multiphase material. Thermodynamic formulation of the homogenized behaviour obtained from Voigt's model will be extended to some simple cases ( layered and fibreous structures ). A generalized anisotropic kinematic hardening will be shown to provide an appropriate framework for describing heterogeneous elastic-plastic multiphase material.

## INTRODUCTION

The homogenized elastic-plastic behaviour of a multiphase material consisting in a mixture of different elastic or perfectly plastic materials is, from a qualitative point of view, well understood. Under increasing proportional loading, for instance, an elastic response is first observed until the first local plastification occurs. This is followed by a macroscopic hardening resulting from the extension of the plastic zone, eventually ending up with a global yielding for complete plastification. In cyclic loading, similarly, a typical hysteretic behaviour is obtained which can be approached by the Masing model resulting from Voigt's homogenization.

From a quantitative point of view, however very little is known. The Voigt's approximation mentioned earlier cannot be considered as satisfactory. In particular it ignores the anisotropy resulting from the geometrical structure of the heterogeneity. Numerical simulations of course, can be performed but they must be repeated for each loading path and if they may give some physical insight, they will not be of much help for constructing explicit homogenized constitutive equation.

The purpose of the present work is to lay the basis for the formulation of such models for the simplest geometrical structures. This will be based on a thermodynamic formulation within the framework of generalized standard materials.

An heterogeneous elastic perfectly plastic material is considered. The local (respectively global) quantities will be denoted by small letters: $\boldsymbol{\sigma}(x)$ (stress), $\boldsymbol{\epsilon}(x)$ (strain) and $\mathbf{a}(x)$ (elastic stiffness) (respectively $\boldsymbol{\Sigma}$, $\in$ and $\mathbf{A}$). The averaging operator will be denoted by brackets:

$$\boldsymbol{\Sigma} = \langle \boldsymbol{\sigma}(x) \rangle \qquad \in = \langle \boldsymbol{\epsilon}(x) \rangle \; . \tag{1}$$

The material at each point will be assumed to be elastic perfectly plastic with a local plastic criterium:

$$f(x;\boldsymbol{\sigma}(x)) = f_0(\boldsymbol{\sigma}(x)) - Y(x).$$

## VOIGT'S HOMOGENIZATION

We shall begin with the thermodynamic formulation of the homogenized model resulting from Voigt's approximation. This will allow us to fix the structure to be generalized later.

Voigt model assumes uniform strain ( $\boldsymbol{\epsilon}(x) = \in$ ). The local behaviour is then

$$\boldsymbol{\sigma}(x) = \mathbf{a}(x)\,(\in - \boldsymbol{\epsilon}^p(x)) \; . \tag{2}$$

The homogenized behaviour relating macroscopical variables is obtained by averaging

$$\boldsymbol{\Sigma} = \mathbf{A}\,(\in - \in^p) \; ; \tag{3}$$
$$\mathbf{A} = \langle \mathbf{a}(x) \rangle \; , \qquad \in^p = \mathbf{A}^{-1} \langle \mathbf{a}(x)\; \boldsymbol{\epsilon}^p(x) \rangle \; .$$

Hence the local stress is given by:

$$\boldsymbol{\sigma}(x) = \mathbf{a}(x)\,\mathbf{A}^{-1}\,\boldsymbol{\Sigma} - \mathbf{a}(x)\mathbf{i}(x) \; ; \quad \mathbf{i}(x) = \boldsymbol{\epsilon}^p(x) - \in^p \; , \tag{4}$$

where $\mathbf{i}(x)$ defines the plastic incompatibility tensor.

The elastic energy $W$ of such medium is given classically by :

$$W = \langle w(x) \rangle = 1/2 \, \langle (\in - \boldsymbol{\epsilon}^p(x))\mathbf{a}(x)(\in - \boldsymbol{\epsilon}^p(x)) \rangle \tag{5}$$
$$= 1/2 \, (\in - \in^p)\,\mathbf{A}\,(\in - \in^p) + \hat{W}(\boldsymbol{\epsilon}^p(x))$$

with

$$2\hat{W}(\boldsymbol{\epsilon}^p(x)) = \langle \boldsymbol{\epsilon}^p(x)\;\mathbf{a}(x)\;\boldsymbol{\epsilon}^p(x) \rangle - \in^p \mathbf{A} \in^p = \langle \mathbf{i}(x)\;\mathbf{a}(x)\;\mathbf{i}(x) \rangle$$

where $\hat{W}(\boldsymbol{\epsilon}^p(x))$ is the hardening energy which remains after unloading and results from the residual stresses $\boldsymbol{\sigma}^r(x) = \mathbf{a}(x)\,\mathbf{i}(x)$.

The thermodynamic forces and their correspondent fluxes will be obtained from the differential :

$$dW = \epsilon^e \, \mathbf{A} \, d\epsilon^e + < \mathbf{i}(x) \; \mathbf{a}(x) \; d\epsilon^p(x) > . \tag{6}$$

The dissipation energy $d\varphi$ is then :

$$d\varphi = \mathbf{\Sigma} \, d\epsilon - dW = (\mathbf{\Sigma} - \mathbf{A}\epsilon^e) \, d\epsilon^e + \mathbf{\Sigma} \, d\epsilon^p + <\sigma^r(x) \; d\epsilon^p(x)> . \tag{7}$$

Classically we obtain, from this expression, the elastic constitutive equation

$$\mathbf{\Sigma} = \mathbf{A} \, \epsilon^e = dW/d\epsilon^e \tag{8}$$

and $d\varphi$ is :

$$d\varphi = \mathbf{\Sigma} \, d\epsilon^p + < \sigma^r(x) \; d\epsilon^p(x) > , \tag{9}$$

showing that $\mathbf{\Sigma}$ and $\sigma^r(x)$ respectively are the thermodynamic forces associated to the macroscopic and the local plastic strain $\epsilon^p$ and $\epsilon^p(x)$. The generalized standard evolution equations are :

$$d\epsilon^p = <d\lambda(x) \frac{\partial F}{\partial \mathbf{\Sigma}}(x)> , \qquad d\epsilon^p(x) = d\lambda(x)\frac{\partial F}{\partial \sigma^r}(x) , \tag{10}$$

$$F(x;\mathbf{\Sigma},\sigma^r(x)) = f(x;\sigma(x)) = f( \mathbf{a}(x)\mathbf{A}^{-1}\mathbf{\Sigma} - \sigma^r(x) ) \leq 0$$

which are easily proved to be the same as the local plastic evolution equation.

Therefore the Voigt homogenized material can be considered as a generalized standard material with $\epsilon^p$ and $\mathbf{i}(x)$ as internal variables, and $W$ given by (5) and the plastic condition by (10).

Several special cases can be discussed as :

- Elastic homogeneity ( $\mathbf{a}(x) = \mathbf{A}$ ) greatly simplifies the model.
- A two phase material with one elastic phase can be shown to result in the usual Prager model, etc..

## LAYERED STRUCTURE

We shall now show that these results can be extended to other homogenization frameworks and in particular to the case of a layered structure normal to the $x_3$ axis.

Introducing the decomposition of any symmetric tensor $\mathbf{t}$ into its plane $\mathbf{t}_H$ and antiplane $\mathbf{t}_V$ components :

$$\mathbf{t} = \mathbf{t}_H + \mathbf{t}_V \; ; \qquad \mathbf{t}_H = \begin{bmatrix} t_{11} & t_{12} & 0 \\ t_{12} & t_{22} & 0 \\ 0 & 0 & 0 \end{bmatrix} \; ; \qquad \mathbf{t}_H = \begin{bmatrix} 0 & 0 & t_{13} \\ 0 & 0 & t_{23} \\ t_{13} & t_{23} & t_{33} \end{bmatrix}$$

The localisation conditions can be written as :

$$\sigma_V(x) = \mathbf{\Sigma}_V \qquad \epsilon_H(x) = \epsilon_H \tag{11}$$

The elastic law is given by :

$$\begin{bmatrix} \epsilon_H - \epsilon_H^p(x) \\ \epsilon_V(x) - \epsilon_V^p(x) \end{bmatrix} = \begin{bmatrix} \lambda_{HH}(x) & \lambda_{HV}(x) \\ \lambda_{VH}(x) & \lambda_{VV}(x) \end{bmatrix} \begin{bmatrix} \sigma_H(x) \\ \Sigma_V \end{bmatrix} \qquad (12)$$

where $\lambda(x)$ denotes the elastic stiffness and $\epsilon^p(x)$ the plastic strain tensor.
A partial inversion gives

$$\begin{bmatrix} \sigma_H \\ \epsilon_V(x) - \epsilon_V^p(x) \end{bmatrix} = \begin{bmatrix} \chi_{HH}(x) & \chi_{HV}(x) \\ \chi_{VH}(x) & \chi_{VV}(x) \end{bmatrix} \begin{bmatrix} \epsilon_H - \epsilon_H^p(x) \\ \Sigma_V \end{bmatrix} \qquad (13)$$

$$\chi_{HH}(x) = \lambda_{HH}(x)^{-1}; \quad \chi_{VV}(x) = \lambda_{VV}(x) - \lambda_{VH}(x)\ \lambda_{HH}^{-1}(x)\ \lambda_{HV}(x);$$
$$\chi_{HV}(x) = -\lambda_{HH}^{-1}(x)\ \lambda_{HV}(x) = -\chi_{VH}(x)^T$$

The homogenized elastic is then easily obtained :

$$\begin{bmatrix} \Sigma_H \\ \epsilon_V \end{bmatrix} = \begin{bmatrix} \langle\chi_{HH}(x)\rangle & \langle\chi_{HV}(x)\rangle \\ \langle\chi_{VH}(x)\rangle & \langle\chi_{VV}(x)\rangle \end{bmatrix} \begin{bmatrix} \epsilon_H \\ \Sigma_V \end{bmatrix} + \begin{bmatrix} -\langle\chi_{HV}(x)\,\epsilon_H^p(x)\rangle \\ \langle\epsilon_V^p(x)\rangle - \langle\chi_{HV}(x)\,\epsilon_H^p(x)\rangle \end{bmatrix} \qquad (14)$$

or in a more classical for

$$\epsilon_H^p = \langle\chi_{HH}(x)\rangle^{-1}\ \langle\chi_{HH}(x)\ \epsilon_H^p(x)\rangle$$
$$\epsilon_V^p = \langle\epsilon_V^p(x)\rangle - \langle\chi_{VH}(x)\,\epsilon_H^p(x)\rangle + \langle\chi_{VH}(x)\rangle\epsilon_H^p;$$

$$\begin{bmatrix} \Sigma_H \\ \Sigma_V \end{bmatrix} = \begin{bmatrix} A_{HH}(x) & A_{HV}(x) \\ A_{VH}(x) & A_{VV}(x) \end{bmatrix} \begin{bmatrix} \epsilon_H - \epsilon_H^p \\ \epsilon_H - \epsilon_H^p \end{bmatrix}; \qquad (15)$$

$$A_{HH} = \langle\chi_{HH}(x)\rangle - \langle\chi_{HV}(x)\rangle\ \langle\chi_{VV}(x)\rangle^{-1}\ \langle\chi_{VH}(x)\rangle; \quad A_{VV} = \langle\chi_{VV}(x)\rangle^{-1};$$
$$A_{HV} = \langle\chi_{HV}(x)\rangle\ \langle\chi_{VV}(x)\rangle^{-1} = -A_{VH}^T.$$

The microstress and macrostress are related by :

$$\sigma_H = C_H\ \Sigma_H + C_V\ \Sigma_V - \chi_{HH}(x)\ i_H(X) \qquad (16)$$
$$i_H(X) = \epsilon_H^p - \epsilon_H^p(x)$$
$$C_H = \chi_{HH}(x)\langle\chi_{HH}(x)\rangle^{-1}\ ; \quad C_V = \chi_{HV}(x) - \chi_{HH}(x)\ \langle\chi_{HH}(x)\rangle^{-1}\ \langle\chi_{HV}(x)\rangle$$

The elastic energy is calculated as in the previous case and will have the following form:

$$W = \langle w(x)\rangle = 1/2\langle\epsilon_H^e(x)\chi_{HH}(x)\,\epsilon_H^e(x)\rangle + \Sigma_V\langle\chi_{VV}(x)\rangle\Sigma_V \qquad (17)$$

with $\quad \epsilon_H^e(x) = \epsilon_H - \epsilon_H^p(x) \qquad \epsilon_H^e = \epsilon_H - \epsilon_H^p$

It can be proved by using (13) that the strain energy can be written in the same form as in the previous case (5)

$$W = 1/2\ \epsilon_H^e\ \mathbf{A}\ \epsilon_H^e + \hat{W}(\epsilon_H^P(x)) \tag{18}$$

with
$$\hat{W}(\epsilon_H^P(x)) = 1/2 < \mathbf{i}_H(x)\ \chi_{HH}(x)\ \mathbf{i}_H(x) >$$

The differential dW is then

$$dW = \epsilon^e\ \mathbf{A}\ d\epsilon^e + < \mathbf{i}_H(x) \chi_{HH}(x) d\epsilon_H^P(x) >. \tag{19}$$

The dissipation function reads :

$$d\varphi = \mathbf{\Sigma}_H\ d\epsilon_H^P + \mathbf{\Sigma}_V\ d\epsilon_V^P - < \mathbf{i}_H(x)\ \chi_{HH}(x)\ d\epsilon_H^P(x) > \tag{20}$$

Relations (16), (18) and (20) obviously extends (4), (5) and (9) in the case of layered structure. They provide the basis for a thermodynamic formulation of the elastic plastic composite within the frameworks of generalized standard materials.

## FIBREOUS STRUCTURE

A fibreous structure along $x_3$ can be analyzed in a similar way with the approached localization conditions :

$$\sigma_{33}(x) = \Sigma_{33} \quad ; \qquad \epsilon_Q(x) = \epsilon_Q \tag{21}$$

$$\mathbf{t} = t_{33}\ \vec{e}_3 \otimes \vec{e}_3 + \mathbf{t}_Q(x) \quad ; \quad \mathbf{t}_Q = \begin{bmatrix} t_{11} & t_{12} & t_{13} \\ t_{12} & t_{22} & t_{23} \\ t_{13} & t_{23} & 0 \end{bmatrix} \tag{22}$$

leading to similar results.

## CONCLUSION

More generally, as soon as the elastic and residual homogenization are available, the elastic-plastic behaviour can be described through a generalized standard formulation with :

$$W = 1/2\ \epsilon^e\ \mathbf{A}\ \epsilon^e + < \mathbf{i}(x)\ \mathbf{B}(x)\ \mathbf{i}(x) >$$
$$d\varphi = \mathbf{\Sigma}\ d\epsilon^P - < \boldsymbol{\sigma}^r(x)\ d\boldsymbol{\epsilon}^P(x) >$$
$$F(x;\mathbf{\Sigma},\boldsymbol{\sigma}^r(x)) = f(x;\boldsymbol{\sigma}(x))$$

The anisotropy of this response essentially depends on the material anisotropy and the geometrical anisotropy. This general formulation however is of a little help unless some special assumption are made leading to a discrete model, for instance the Prager model is the simplest case. In such a case the anisotropic structure of the tensors **A** and **B** and of the function F can be obtained from the preceding analysis.

# MICROSCOPIC AND CRYSTALLOGRAPHIC ASPECTS OF FLOW STRESS ANISOTROPY

N. HANSEN and D. JUUL JENSEN
Materials Department, Risø National Laboratory, DK-4000 Roskilde, Denmark

## ABSTRACT

The plastic anisotropy of cold-rolled aluminium has been investigated at plastic strains from 0.1 to 3.0. Flow stresses (0.2% offset) have been measured at room temperature as a function of the angle between the tensile axis and the rolling direction and textures have been determined by neutron diffraction. It has been found that for most experimental conditions, texture effects alone cannot explain the observed anisotropy and the microstructural anisotropy has to be taken into account. Causes for the microstrutural anisotropy are discussed especially the effect of the spatial arrangement of dislocation boundaries.

## INTRODUCTION

The flow stress of deformed metals can be highly anisotropic which have implications both for design and for metals forming. The anisotropy may have many causes related to the crystallographic texture and the microstructure. Consequently it has been observed that anisotropy depends on materials and process parameters. The effect of crystallographic texture on flow stress anisotropy has been studied fairly extensively and it has been found that texture has a significant effect. With good accuracy this effect can be accounted for on the basis of experimental texture determinations in combination with modelling of the deformation behaviour (1). The identification of the contribution of texture has however demonstrated that in many cases texture alone cannot account for the anisotropy observed. Other causes (often of similar strength) must therefore be sought. A cause can be microstructural anisotropy as shown in recent work (2, 3). In specific terms the anisotropy may be related to the presence of dislocation boundaries (4). The microstructural evolution as a function of strain must therefore be studied in parallel with the development of the deformation texture. This has been the objective of a recent study (4) on cold-rolled high-purity aluminium (99.996%) and commercial purity aluminium (99.5%). Typical results from this work will be summarized below together with new results.

## FLOW STRESS

The flow stress after cold-rolling to two reductions in the range 0.05-3.0 (5 to 95% reduction in thickness) are shown in Fig. 1 for commercial purity aluminium having a grain size of 80 μm. The specimens for tensile testing were cut out at fixed angles to the rolling direction, and the flow stress is plotted as a function of this angle, $\alpha$.

## TEXTURE

The crystallographic texture was measured by neutron diffraction and the texture development with strain followed the typical pattern for aluminium. At low strains the texture consists of remaining initial texture components and weak rolling components. At higher strains the typical rolling texture is observed.

The effect of texture on the tensile flow stress at a given strain ($\varepsilon$) is normally accounted for by the equation

$$\sigma\,(0.2\%) = M(\varepsilon)\cdot\tau$$

where $\tau$ is the shear stress and $M(\varepsilon)$ is the relative strength which takes the texture into account. $M(\varepsilon)$ can be determined using various deformation models, e.g. the full constraint (FC) Taylor model or various relaxed-constraints (RC) models. $M(\varepsilon)$ values based on the FC-approach have been calculated as a function of $\alpha$ (4). At low strains the variation of $M(\varepsilon)$ with $\alpha$ is small whereas at high strain the texture effect on $M(\varepsilon)$ can amount to about 10%.

The texture contribution to the flow is taken into account by calculating $\sigma(0.2\%)/M(\varepsilon)$ as a function of $\alpha$, see Fig. 2. At low strains it is observed that texture alone cannot account for the observed anisotropy, which must be ascribed to other causes, e.g. microstructural anisotropy. At large strains texture can be the major cause of the anisotropy (see Fig. 2), however, also high strains other causes for anisotropy may be of importance (4). These conclusions are not affected significantly by choosing an RC-approach instead of the FC-approach (4).

## MICROSTRUCTURE

Various factors related to the microstructure can cause anisotropy for example directionality of the grain shape or of the distribution of second phase particles. Other causes are latent hardening which is most pronounced at low strains and macroscopic shear bands which may form at large strains. These different effects have been discussed in Ref. 4 and it has been concluded that they can only account insignificantly to the observed flow stress anisotropy. The microstructural evolution with increasing strain is therefore considered.

Deformation microstructure of aluminium and other cell forming metals have been studied by TEM (5). The experimental observations have been explained through the governing principle

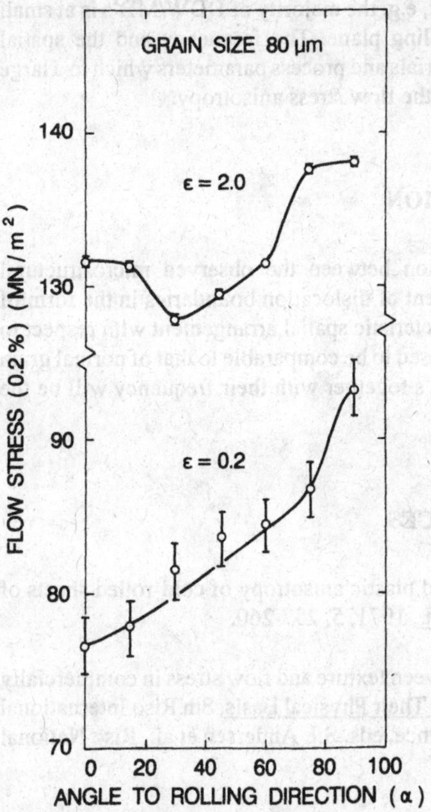


Figure 1. Angular variation of the flow stress (0.2% offset) in the rolling plane of commercial purity aluminium deformed to ε = 0.2 and ε = 2.0

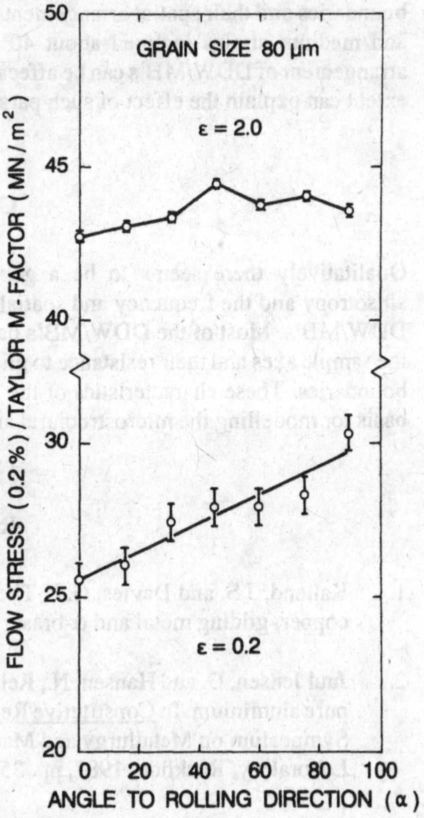


Figure 2. Angular variation of the flow stress from Figure 1 divided by the corresponding M(ε) factor (see text).

that in the course of polyslip, grains break up into volume elements in which fewer slip systems operate than requested by the Taylor model. This grain break-up begins with relatively few domains which in turn break-up into an increasing number of cell blocks on a smaller and smaller scale. The boundaries between the different volume elements accommodate the lattice misorientation arising from different glide systems combinations in neighbouring volumes and are therefore geometrically necessary boundaries (GNB's). In TEM they appear variously as domain walls, dense dislocation walls (DDW's), different types of microbands (MB's) and combinations thereof (DDW/MB's). The misorientation across the GNB's increases with the strain to about 10-15°, i.e. an order of magnitude larger than the lattice rotation across ordinary cell walls (5). The spatial arrangement of the GNB's especially the DDW/MB's with respect to the sample axes may affect the flow stress anisotropy in analogy with the effect of grain boundaries. However, the spacing of the DDW/MB's in rolled specimens is much smaller than that of original grain boundaries and their spatial arrangement is different, e.g. the majority of DDW/MB's is at small and medium strains inclined about 40° to the rolling plane. The frequency and the spatial arrangement of DDW/MB's can be affected by materials and process parameters which to a large extent can explain the effect of such parameters on the flow stress anisotropy.

## CONCLUSION

Qualitatively there seems to be a good correlation between the observed microstructural anisotropy and the frequency and spatial arrangement of dislocation boundaries in the form of DDW/MB's. Most of the DDW/MB's have a characteristic spatial arrangement with respect to the sample axes and their resistance to glide is supposed to be comparable to that of normal grain boundaries. These characteristics of the DDW/MB's together with their frequency will be the basis for modelling the microstructural anisotropy.

## REFERENCES


1. Kallend, J.S. and Davies, G.J., The elastic and plastic anisotropy of cold-rolled sheets of copper, gilding metal and α-brass. J. Inst. Met., 1971, 5, 257-260.

2. Juul Jensen, D. and Hansen, N., Relations between texture and flow stress in commercially pure aluminium. In Constitutive Relations and Their Physical Basis, 8th Risø International Symposium on Metallurgy and Materials Science, eds. S.I. Andersen et al., Risø National Laboratory, Roskilde, 1987, pp. 353-360.

3. Raphanel, J.L., Schmitt, J.-H. and Baudelet, B., Plastic behaviour of prestrained materials: Experiments and analysis through a simple model. In Constitutive Relations and Their Physical Basis, 8th Risø International Symposium on Metallurgy and Materials Science, eds. S.I. Andersen et al., Risø National Laboratory, Roskilde, 1987, pp. 491-496.

4. Juul Jensen, D. and Hansen, N., Flow stress anisotropy in aluminium, Acta metall. mater., 1990, 38, 1369-1380.

5. Bay, B., Hansen, N., Hughes, D.A. and Kuhlmann-Wilsdorf, D., Evolution of fcc deformation structures in polyslip, Acta metall. mater., 1991. Accepted for publication.

# COMPUTATIONAL STUDY OF POLYCRYSTALLINE BEHAVIOUR UNDER COMPLEX LOADING CONDITIONS

FRANTIŠEK HAVLÍČEK and MASATAKA TOKUDA
Department of Mechanical Engineering
Mie University
Kamihama–cho 1515, Tsu 514, Japan

## ABSTRACT

The finite element method is used to simulate macroscopic inelastic responses of a polycrystalline structure. An explicit approach is used in which polycrystalline aggregate consists of the specified number of single crystals (grains) discretized into the finite elements. The global behaviour of the structure is based on the simplified constitutive equations of its constituent grains which have the random three-dimensional crystallographic orientations. The macroscopic behaviour of this model under complex loading histories is studied, specifically, its responses to the corner-type loading paths with the different corner angles. At each loading point the microscopic grain state represented by slip system development is also evaluated.

## INTRODUCTION

Any macroscopic inelastic behaviour (including the tensile test) of polycrystalline materials is a consequence of microscopic inelastic events and their interactions on the single crystal component level. An crystallographic slip which is controlled by nonuniform stress within the grain is supposed to be a basic mechanism of plastic deformation. Theoretically guided and experimentally supported by single crystal tests these inelastic mechanisms can be summarized in the generally accepted following simplifications:

- the slip occurs along certain crystal directions on certain crystal planes;
- the slip depends on the resolved shear stress and is independent of the normal stress on the sliding plane;
- the resolved shear stress varies with the amount of the slip.

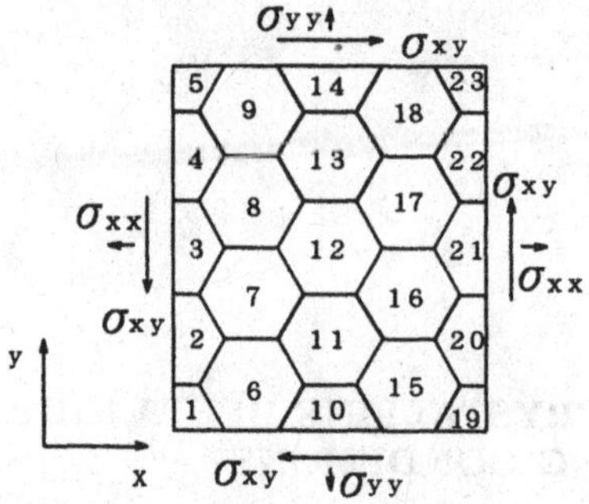

Figure 1: Model polycrystal

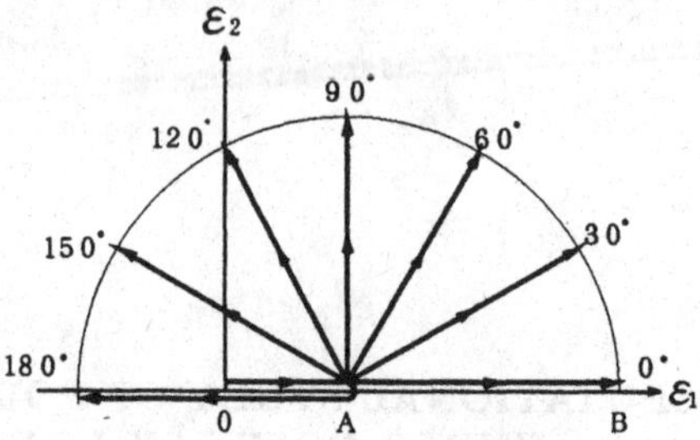

Figure 2 : Strain paths in Ilyushin space

In the implementation of these assumptions into the finite element algorithm described in [1], the mathematical expression for the resolved shear stress $\tau^{(n)}$ on the n-th slip system can be formulated as

$$\tau^{(n)} = m^{(n)} \cdot \sigma \cdot s^{(n)} \tag{1}$$

where $m^{(n)}$ is the unit vector along the n-th slip direction, $s^{(n)}$ is the unit normal to the n-th slip plane and $\sigma$ is the stress tensor. The constitutive description of the plasticity on each slip system is expressed in terms of its resolved shear stress and the slip rate on that system as

$$\dot{\gamma}^{(n)} = \dot{\gamma}_0\{exp[\delta(\tau^{(n)} - \tau_y^{(n)})/\tau_y^{(n)}] - 1\} \tag{2}$$

where $\dot{\gamma}_0$ and $\delta$ are the material constants and $\tau_y^{(n)}$ is the critical resolved shear stress on the n-th slip system which initiates or causes the continuation of slip and is assumed to be controlled by the linearized Taylor hardening law as

$$\tau^{(n)} = \tau_0 + \sum_{m=1}^{N} H_{nm}\gamma^{(m)} \tag{3}$$

where $\tau_0$ is the initial resolved shear stress, $H_{nm}$ is the hardening matrix, $\gamma^{(m)}$ is the accumulated plastic strain for the m-th slip system and $N$ is the number of the slip systems in the single crystal.

## COMPUTATIONAL ANALYSES

The two-dimensional model polycrystalline aggregate in Fig. 1 is analysed. It consists of 23 grains as it is shown in the figure. The initial three-dimensional crystallographic orientation, with respect to the global coordinate system $(x, y, z)$, of each grain in Fig. 1 is randomly selected to represent initial isotropy. The elastic properties for each single grain are determined by considering the pure aluminium. The other material constants from Eqs. 2 and 3 are given as

$$\dot{\gamma}_0 = 100.0\text{s}^{-1},\ \delta = 0.1,\ \tau_0 = 36.5\text{MPa},\ H_{nm} = 0.0\text{Mpa}\ (n, m = 1, 2, ..., 12) \tag{4}$$

The numerical experiments were done by applying complex strain histories to the model polycrystal, which were represented by the vector $\varepsilon = (\varepsilon_1, \varepsilon_2)$ in the two-dimensional

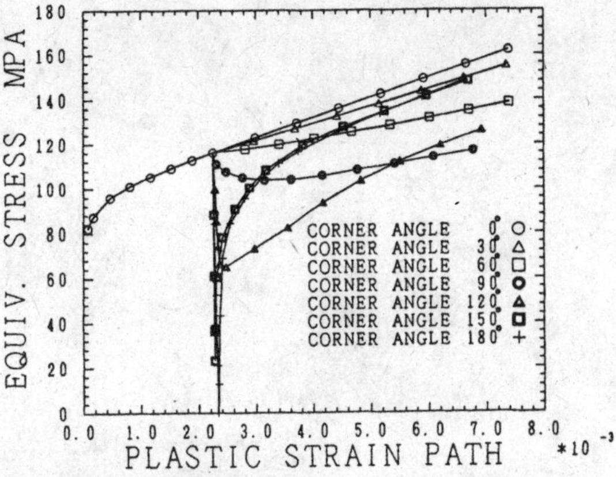


Figure 3: Relation $\sigma_{eq} - L$ for strain paths with several corner angles

Ilyushin space in Fig. 2 with its vector components given as

$$\varepsilon_1 = 2/\sqrt{3}\varepsilon_{xx},\ \varepsilon_2 = 2/\sqrt{3}\varepsilon_{xy} \tag{5}$$

where $\varepsilon_{xx}$ and $\varepsilon_{xy}$ are the components of the strain tensor related to the coordinate system $(x, y, z)$. In Fig. 2, the pre-strain length OA is 0.37%; the length of the strain path AB is 0.80%. Fig. 3 shows the macroscopic responses for the various corner angles expressed through the equivalent stress $\sigma_{eq}$ and the length of the total plastic strain path. The development of the internal state within the grain no.12 in Fig. 1 represented by slip system activities during loading is shown in Fig. 4.

## CONCLUSION

The numerical experiments based on the finite element method for the model polycrystal under complex loading conditions in an inelastic range were carried out. The results can be outlined as follows:

- the model polycrystal used here can well reproduce experimentally observed characteristic features of polycrystal macroscopic responses;
- it seems that the Bauschinger effect is not so strongly related with interactions among grains as simply modeled here, but is affected by another mechanisms associated with the slip, for example, proposed by Orowan;
- the simulated slip system activities are nonhomogeneously distributed within the grain and are closely related with loading mode.

## REFERENCE

[1] Havlíček, F., Kratochvíl, J., Tokuda, M. and Lev, V., Finite element model of plastically deformed multicrystal. Int. J. Plast., 1990, **6**, 281 – 292.

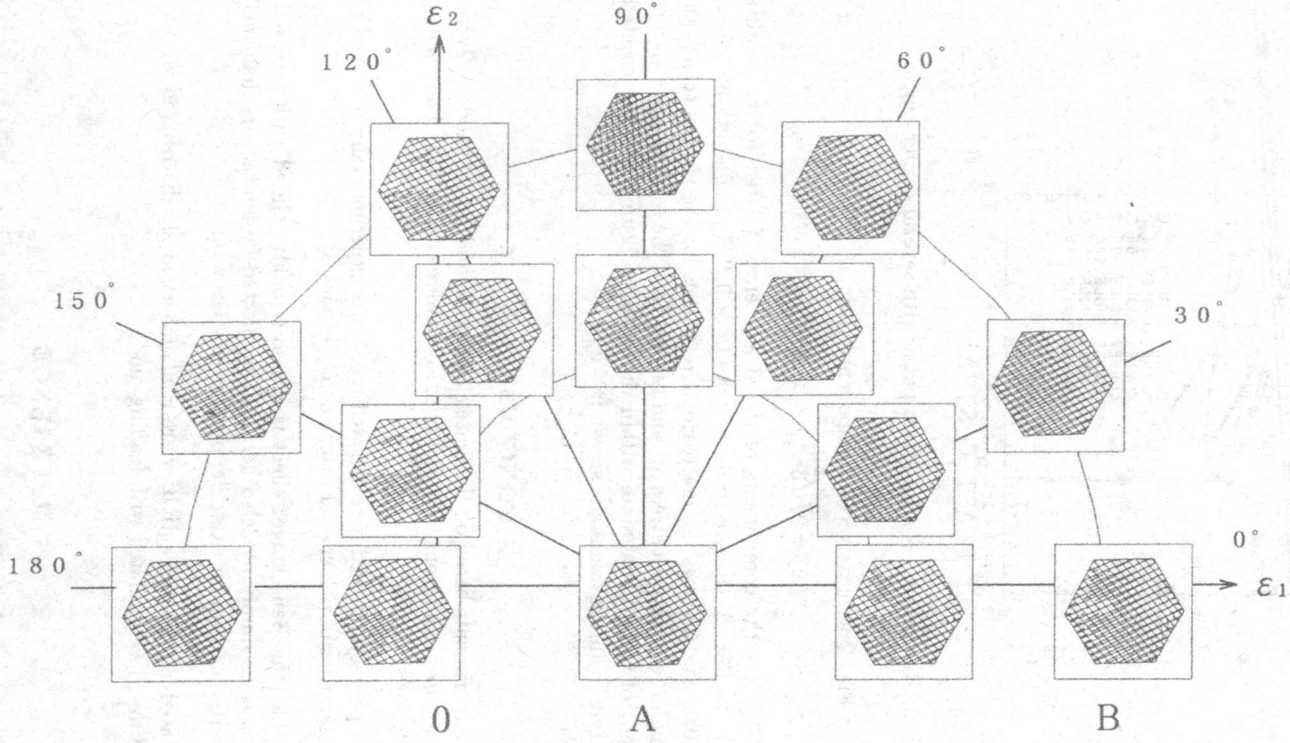


Figure 4: Slip line development in grain no.12 for strain paths with several corner angles

# ON THE ACCURACY OF THE TAYLOR ASSUMPTION IN POLYCRYSTALLINE PLASTICITY

S. R. KALIDINDI, C. A. BRONKHORST and L. ANAND
Department of Mechanical Engineering,
Massachusetts Institute of Technology,
Cambridge, MA 02139, USA.

## ABSTRACT

The accuracy of a Taylor-type polycrystal constitutive model, in which compatibility is satisfied but equilibrium between grains is violated, is evaluated by comparing the predictions from this model for the overall stress-strain response and crystallographic texture evolution during simple shear of fcc metals against predictions from a full finite element model in which both compatibility and equilibrium in the polycrystalline aggregate are satisfied. Our calculations show that the Taylor-type model is in reasonable first-order agreement with the results of the finite element calculations and also with the results from corresponding physical experiments on single-phase copper.

## INTRODUCTION

Asaro and Needleman [1] have developed an elastic-plastic, rate-dependent polycrystalline model for low homologous temperatures in which deformation within the individual crystals is taken to be by crystallographic slip alone. To predict the global response of the polycrystal, the transition from the micro-response of the individual grains to the macro-response of the polycrystalline aggregate, they follow the pioneering work of Taylor [2] and assume that all grains have equal volume, and that the deformation gradient is uniform throughout the aggregate. In this approximate model, compatibility is satisfied and equilibrium holds in each grain, but equilibrium is usually violated between grains. This simple averaging procedure gives that the macroscopic volume average Cauchy stress in the polycrystal is simply the number average of the Cauchy stress in each crystal. In this model, the deformation producing mechanisms of twinning, diffusion and grain boundary sliding are not considered, and other sources of anisotropy due to the morphological effects of grain shape, size and arrangement are not accounted for. In a recent paper [3] we have presented a fully implicit time integration procedure for a slightly modified form of Asaro and Needleman's Taylor-type model [1], and implemented it in a finite element code to simulate non-steady, non-homogeneous deformations in polycrystalline fcc metals. In this brief note, we evaluate the accuracy of Taylor assumption by comparing the predictions from this model for the evolution of crystallographic texture and stress-strain curves in simple shear of initially isotropic OFHC copper against predictions from a finite element polycrystalline model which satisfies equilibrium and compatibility in the entire polycrystalline aggregate. A detailed report of our investigation, which includes other modes of deformation, has been recently submitted for publication [4].

## TAYLOR-TYPE POLYCRYSTAL MODEL

The constitutive equation for the stress in a single crystal is

$$\mathbf{T}^* = \mathcal{L}[\mathbf{E}^*], \tag{1}$$

$$\mathbf{E}^* \equiv (1/2)\left\{\mathbf{F}^{*T}\mathbf{F}^* - \mathbf{1}\right\}, \qquad \mathbf{T}^* \equiv \mathbf{F}^{*-1}\{(\det\mathbf{F}^*)\mathbf{T}\}\,\mathbf{F}^{*-T}, \qquad \mathbf{F}^* \equiv \mathbf{F}\mathbf{F}^{p-1}, \tag{2}$$

where $\mathbf{T}$ is the symmetric Cauchy stress tensor, $\mathbf{F}$ is the deformation gradient, $\mathcal{L}$ is the fourth order elasticity tensor, and $\mathbf{F}^p$ with $\det \mathbf{F}^p = 1$, is the plastic deformation gradient. The flow rule for the single crystal is

$$\dot{\mathbf{F}}^p \equiv \mathbf{L}^p\mathbf{F}^p, \qquad \mathbf{L}^p = \sum_\alpha \dot{\gamma}^\alpha \mathbf{S}_o^\alpha, \qquad \mathbf{S}_o^\alpha \equiv \mathbf{m}_o^\alpha \otimes \mathbf{n}_o^\alpha, \tag{3}$$

$$\dot{\gamma}^\alpha = \dot{\gamma}_o\left|\frac{\tau^\alpha}{s^\alpha}\right|^{1/m}\text{sign}(\tau^\alpha), \qquad \tau^\alpha \approx \mathbf{T}^*\cdot\mathbf{S}_o^\alpha, \tag{4}$$

where $\mathbf{m}_o^\alpha$ and $\mathbf{n}_o^\alpha$ are (time-independent) orthonormal unit vectors which define the slip direction and slip plane normal of the slip system $\alpha$ in a fixed reference configuration; $\dot{\gamma}^\alpha$, $\tau^\alpha$ and $s^\alpha$ are respectively the plastic shearing rate, resolved shear stress and slip resistance; and $m$ is the slip rate sensitivity parameter. The slip system hardening rule is taken as

$$\dot{s}^\alpha = \sum_\beta h^{\alpha\beta}\left|\dot{\gamma}^\beta\right|, \qquad h^{\alpha\beta} = q^{\alpha\beta}h^{(\beta)}, \qquad h^{(\beta)} = h_o\left\{1 - \frac{s^\beta}{s_s}\right\}^a, \tag{5}$$

where $h^{\alpha\beta}$ is the rate of strain hardening on slip system $\alpha$ due to a shearing on the slip system $\beta$. The parameters $h_o$, $s_s$ and $a$ are single slip hardening parameters, and $q^{\alpha\beta}$ are latent hardening coefficients.
For the polycrystal, Taylor-type assumptions [1] lead to:

$$\bar{\mathbf{T}} = \frac{1}{N}\sum_{k=1}^{N}\mathbf{T}^{(k)}, \tag{6}$$

where $\mathbf{T}^{(k)}$ denotes the Cauchy stress in the $k^{th}$ crystal, $\bar{\mathbf{T}}$ is the volume averaged macroscopic Cauchy stress, and $N$ is the total number of grains comprising a material point.

We examine the applicability of this model to polycrystalline fcc copper. Annealed OFHC copper is assumed to be initially "isotropic" with an initial texture adequately representable by a suitable set of randomly oriented 400 crystals (represented by Euler angles). Initial values for $s^\alpha$ were assumed to be the same for all slip systems. The latent hardening coefficients are assumed equal to 1.4 when the slip systems $\alpha$ and $\beta$ are noncoplanar and 1.0 otherwise [1]. The single slip hardening parameters were estimated by curve-fitting simulated stress-strain curves using the Taylor-type model to an experimental stress-strain curve from a simple compression test. The slip system rate sensitivity parameter $m$ was assumed to be equal to the macroscopic strain rate sensitivity, and the latter was determined from a strain rate jump experiment in simple compression.

## FINITE ELEMENT MODEL FOR POLYCRYSTAL

We simulate simple shearing deformation in a polycrystalline aggregate using a finite element mesh consisting of 400 two dimensional plane strain elements. Each element of the mesh is assumed to represent a grain and is assigned an orientation from the set of 400 crystal orientations chosen to represent an initially random texture in the polycrystalline aggregate. The results from the finite element model satisfy both equilibrium and compatibility throughout the polycrystalline aggregate.

It should be noted that our calculations differ from those of Harren and Asaro [5] in that we use the actual three-dimensional slip system structure for fcc materials, rather than an idealized two-dimensional slip system structure.

## EVALUATION OF TAYLOR-TYPE MODEL

The stress-strain curves for simple shear predicted using the Taylor-type polycrystalline model and the finite element model of a polycrystal are shown in Fig. 1, and the simulated crystallographic textures are shown in Fig. 2. In these figures we also show the results from corresponding physical experiments. We observe that while both the Taylor model and the FEM model are in good agreement with experiments for the shear stress versus shear strain response, the FEM model predictions for the normal stress is in better agreement with the experiment than the Taylor model. Also, the texture predicted by the Taylor model is sharper than either that obtained from the FEM model or the experiment.

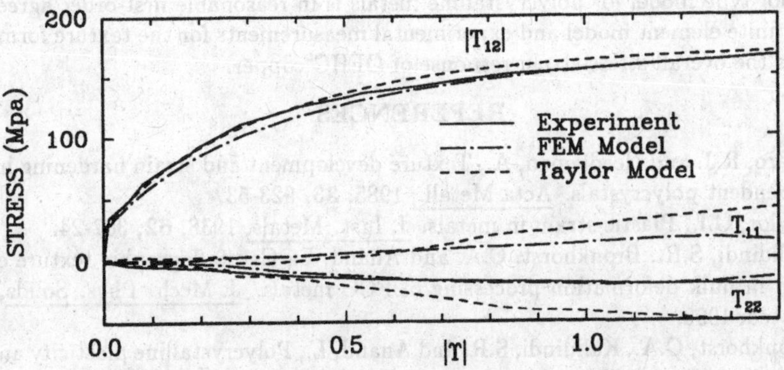


Fig. 1. Stress-strain curves in simple shear deformation.

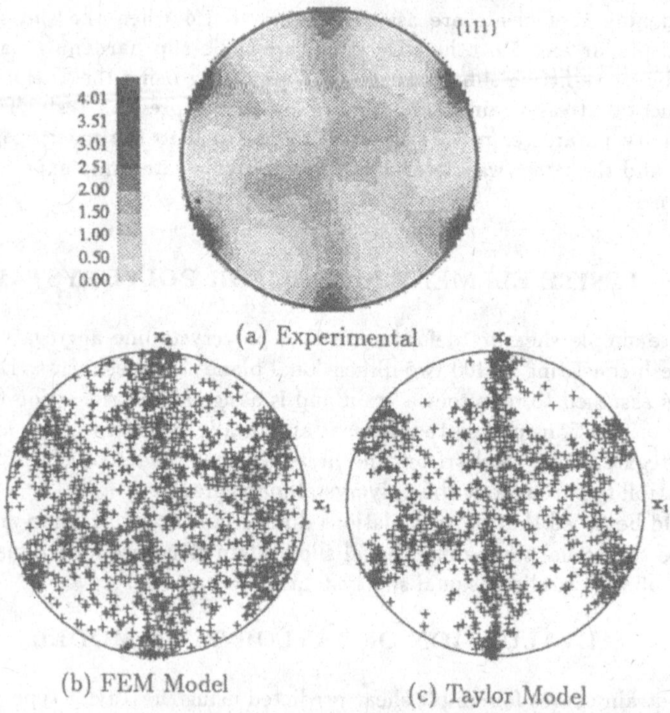


Fig. 2. {111} pole figures of OFHC copper deformed in simple shear to $\gamma = -1.4$.

## CONCLUSION

The Taylor-type model for polycrystalline metals is in reasonable first-order agreement with the finite element model and experimental measurements for the texture formation as well as the overall stress-strain response of OFHC copper.

## REFERENCES

1. Asaro, R.J. and Needleman, A., Texture development and strain hardening in rate dependent polycrystals. Acta Metall., 1985, **33,** 923-53.
2. Taylor, G.I., Plastic strain in metals. J. Inst. Metals, 1938, **62,** 307-24.
3. Kalidindi, S.R., Bronkhorst, C.A. and Anand, L., Crystallographic texture evolution in bulk deformation processing of FCC metals. J. Mech. Phys. Solids, submitted, 1990.
4. Bronkhorst, C.A., Kalidindi, S.R. and Anand, L., Polycrystalline plasticity and the evolution of crystallographic texture in F.C.C. metals. Phil. Trans. R. Soc. Lond. Series A, submitted, 1991.
5. Harren, S.V. and Asaro, R.J., Nonuniform deformations in polycrystals and aspects of the validity of the Taylor Model. J. Mech. Phys. Solids, 1989, **37,** 191-232.

# ULTRASONIC NONDESTRUCTIVE EVALUATION OF MICRO SLIP BAND AND PLASTIC ANISOTROPY GROWTH

MICHIAKI KOBAYASHI
Department of Mechanical Engineering,
Kitami Institute of Technology, Kitami 090,
Hokkaido, JAPAN

## ABSTRACT

The purpose of the present paper is to evaluate microstructural changes of the aluminum alloy induced by plastic deformation by using the proposed ultrasonic nondestructive evaluation method.

## INTRODUCTION

It is well known that plastic deformation is accompanied with micro-structural material property changes, and also these changes affect macroscopic response of the materials. For example, the elastic moduli and Lankford's value vary depending upon both the plastic flow localization and the texture change taking place under the plastic deformation, therefore plastic anisotropy develops due to the anisotropic changes of the elastic properties. Moreover, the inhomogeneous localization of the plastic strain, such as localized slip bands, characterizes the plastic instability and the forming limits of the materials.

The purpose of the present paper is to evaluate onset of micro slip bands and growth of plastic anisotropy by using the ultrasonic non-destructive method.

## OUTLINE OF THEORETICAL ANALYSIS

In this paper, we outline the fundamental equations, and details of the theoretical analysis can be found in the author's earlier paper [ 1 ]. If once we measured the second-, third-order elastic constants and the initial texture effect at undeformed state, we can define the elastic constitutive relation at the finite plastic deformation state, predeformed state, using pull back $\mathbf{S}=\phi^{*}(\boldsymbol{\sigma})$ and push forward $\mathbf{d}^{e}=\mathbf{L}_{v}\phi_{*}(\mathbf{E}^{e})$ in which $\mathbf{S}$ is piola-kirchhoff stress, $\boldsymbol{\sigma}$ is Cauchy stress, $\mathbf{d}^{e}$ is the elastic component of the rate of deformation, $\mathbf{L}_{v}$ is Lie derivative and $\mathbf{E}^{e}$ is the incremental Lagrangian elastic strain.

Considering two types of the hardening behavior, i.e., one is the strain hardening in the long-range stress field and the other is in the short-range stress field, we can obtain the definition of the plastic deformation rate tensor $d_{ij}^{p}$ which consists of the plastic strain rate governed by the associative flow rule, and elastoplastic coupling strain rate $d_{ij}^c$ generated at the yield surface vertex as follows:

$$d_{ij}^p = \delta \cdot \hat{\sigma}'_{ij} d\bar{\lambda} + d_{ij}^c , \qquad d_{ij}^c = \lambda_{ij}^c \overset{\nabla}{\sigma}'_{ij} - \frac{1}{3} \delta_{ij} \lambda_{KK}^c \overset{\nabla}{\sigma}'_{kk} \tag{1}$$

where

$$d\bar{\lambda} = \bar{\Lambda}_{rs} \overset{\nabla}{\sigma}'_{rs}, \qquad \bar{\Lambda}_{rs} = \frac{3\hat{\sigma}'_{rs}}{2H' \hat{\sigma}'_{kl} \hat{\sigma}'_{kl}}, \qquad \lambda_{ij}^c = -\frac{K_{ijklmn}}{(H')^2} \cdot \frac{dC_v}{d\bar{\varepsilon}^p} \cdot \frac{\partial\bar{\sigma}/\partial\sigma'_{kl}}{\partial\bar{\varepsilon}^p/\partial\varepsilon_{mn}^p} \tag{2}$$

and $\delta$ is a loading-unloading flag. In eq.(1), $\hat{\sigma}'_{ij}$ is the deviatoric stress whose direction coincides with the local normal of the yield surface at the loaded point and can be obtained by using the equivalent stress $\bar{\sigma}$ as follows:

$$\hat{\sigma}'_{ij} = \hat{\sigma}_{ij} - \frac{1}{3} \delta_{ij} \hat{\sigma}_{kk}, \qquad \hat{\sigma}_{ij} = \frac{2}{3} \bar{\sigma} \frac{\partial\bar{\sigma}}{\partial s'_{ij}}, \qquad s'_{ij} = \sigma'_{ij} - m_B \varepsilon_{ij}^p . \tag{3}$$

Now, we define the equivalent stress $\bar{\sigma}$ using the unsymmetric yield function and the equivalent plastic strain using the ordinal form as follows:

$$\left(\frac{1}{3}\bar{\sigma}^2\right)^{3/2} = \left(\frac{1}{2} C_{ijkl} s'_{ij} s'_{kl}\right)^{3/2} + p\left(\frac{1}{3} s'_{ij} s'_{jk} s'_{ki}\right),$$

$$C_{ijkl} = \frac{1}{2}\{(\delta_{ik} + A \cdot \varepsilon_{ik}^p)(\delta_{jl} + A \cdot \varepsilon_{jl}^p) + (\delta_{il} + A \cdot \varepsilon_{il}^p)(\delta_{jk} + A \cdot \varepsilon_{jk}^p)\} \tag{4}$$

in which A is the anisotropic parameter, $m_B$ is a Bauschinger modulus, and p is the anisotropic coefficient of the 3rd invariant of stresses. Through the use of eq.(1) and the elastic constitutive relation we can derive the following incremental constitutive relation in the current state through some complicated manipulations as:

$$\begin{aligned}
\overset{\nabla}{\sigma}_{ij} = {} & 2G\left(2 - \frac{G}{G_{ij}} - 2G\lambda_{ij}^c - \varepsilon_{kk} - 2Gx_2\sigma_{kk}\right) d'_{ij} - 4G^2 \delta \hat{\sigma}'_{ij} \bar{\Lambda}_{rs} d'_{rs} - 4G(Gx_3\sigma_{ir} - \varepsilon_{ir}^e - [\varepsilon_{ir}^p]) d'_{jr} \\
& - 4G(Gx_3\sigma_{jr} - \varepsilon_{jr}^e - [\varepsilon_{jr}^p]) d'_{ir} - 2\bar{G}\left\{G\left(x_2 + \frac{2}{3}x_3\right)\sigma_{ij} - \frac{1}{3}\left(1 + 2\frac{G}{\bar{G}}\right)(\varepsilon_{ij}^e + [\varepsilon_{ij}^p])\right\} d_{kk} - \sigma_{ij} d_{kk} + \sigma_{jr} \frac{\partial v_i}{\partial x_r} \\
& + \sigma_{ir} \frac{\partial v_j}{\partial x_r} + \frac{1}{3} \delta_{ij} \bar{G}\left[\left\{1 - \frac{1}{3}\left(1 + 4\frac{G}{\bar{G}}\right)\varepsilon_{mm} - \bar{G}\left(6x_1 + 3x_2 + \frac{2}{3}x_3\right)\sigma_{mm} + 2G\left(x_2 + \frac{2}{3}x_3\right)\sigma_{mm}\right\} d_{kk}\right. \\
& \left. + 2\left\{\left(1 - 2\frac{G}{\bar{G}}\right)(\varepsilon_{rs}^e + [\varepsilon_{rs}^p]) - G\left(3x_2 + 2x_3 - 4\frac{G}{\bar{G}}x_3\right)\sigma_{rs}\right\} d'_{rs} + \frac{G}{\bar{G}}\left(\frac{2G - \bar{G}}{G_{KK}} + 4G\lambda_{KK}^c\right) d'_{kk}\right]
\end{aligned} \tag{5}$$

in which the brackets $[\varepsilon_{ij}^p]$ indicate the following operation:

$$[\varepsilon_{ij}^p] = \begin{cases} \varepsilon_{ij}^p & \text{if } \bar{\varepsilon}^p \text{ is less than or equal to } \bar{\varepsilon}_{th}^p \\ \text{constant} & \text{if } \bar{\varepsilon}^p \text{ is greater than } \bar{\varepsilon}_{th}^p \end{cases}$$

where $\bar{\varepsilon}_{th}^p$ is the threshold plastic strain.

Considering infinitesimal harmonic wave propagation with the wave number $\gamma$ and the angular frequency $\omega$, superposed on the predeformed configuration, and using by the constitutive relation eq.(5) and the equations of motion, a system of equations for the wave amplitude $\tilde{U}_j$, called generalized Christoffel equations, is obtained as:

$$(A_{ijpq} n_j n_q - \rho V^2 \delta_{ip}) \tilde{U}_p = 0, \quad V = \omega/\gamma \tag{6}$$

in which n is a unit wave vector, V is the wave velocity, and $\rho$ is the mass density of the predeformed state. The detailed expression of the generalized Christoffel tensor $A_{ijpq}$ in eq.(6) can be found in the previous paper [1].

Then, the coordinate axes **x** are aligned with the principal directions of stress and strain, and the plane wave propagates in $x_3$-direction : one of the principal directions. Equation (6) leads to the explicit formulars for the longitudinal wave velocity $V_L$ and the transverse wave velocities $V_{T1}$ and $V_{T2}$, respectively, as follows:

$$\rho_0 V_L^2=\lambda+2G+\frac{(\lambda+2G)^2}{2G^2}(g_{11}+g_{22}+g_{33})-2\left(1+\frac{\lambda}{G}\right)(g_{11}+g_{22})-\frac{4}{9}G^2(\lambda_{11}^c+\lambda_{22}^c)-\frac{16}{9}G^2\lambda_{33}^c$$
$$+\left\{5+2\frac{\lambda}{G}-4G(3\lambda+2G)\left(x_2+\frac{2}{3}x_3\right)-\frac{8}{3}G^2x_3\right\}\sigma_3-\frac{1}{3}\left\{\frac{6\lambda}{3\lambda+2G}\left(\frac{\lambda}{G}+2\right)\right.$$
$$\left.+(3\lambda+2G)^2\left(6x_1+3x_2+\frac{2}{3}x_3\right)-4G\lambda(3x_2+2x_3)-\frac{8}{3}G^2x_3\right\}(\sigma_1+\sigma_2+\sigma_3)+4(\lambda+2G)\{\varepsilon_3^p\}, \qquad (7)$$

$$\left.\begin{matrix}\rho_0 V_{T1}^2\\ \rho_0 V_{T2}^2\end{matrix}\right\}=\bar{\Gamma}+\frac{1}{2}\{g_{13}+g_{23}-2G^2(\lambda_{13}^c+\lambda_{23}^c)\}\pm\Big[\Gamma^2+(\Gamma^p)^2+2\Gamma\cdot\Gamma^p\cos 2(\theta-\theta_c)$$
$$+\{g_{13}-g_{23}-2G^2(\lambda_{13}^c-\lambda_{23}^c)\}(\Gamma\cos 2\theta+\Gamma^p\cos 2\theta_c)+\frac{1}{4}\{g_{13}-g_{23}-2G^2(\lambda_{13}^c-\lambda_{23}^c)\}^2\Big]^{1/2} \qquad (8\cdot a, b)$$

in which $g_{ij}$ is inherent anisotropy. In eqs.(8), we emphasize that the elastoplastic coupling strain rate induced by vertexlike yield effects influences the transverse wave velocities in the same manner as the inherent anisotropy $g_{ij}$ which corresponds to the initial texture. Therefore, it is supposed intuitively that the coupling effect corresponds to the plastically induced texture.

## RESULTS AND DISCUSSIONS

The numerical simulation and experiment were performed under uniaxial tension tests for aluminum alloy specimens(A6063-T5) annealed at 250 °C for 4 hours furnace cooling. For the accurate measurement of the changes in ultrasonic wave velocities, the sing-around technique was adopted. Details of the measuring system, experimental procedures and findings can be found in the author's earlier paper [2]. Also details of the numerical simulation procedure can be found in the author's another paper [1].

Figure 1 shows the fractional changes of transverse wave comparing between the simulated and experimental results. In Fig.1, the discrepancy between solid and dot-dash lines indicates the texture change. Therefore, it is predicted that the texture may change to hardening in the loading and the transverse directions together. From Fig.1 the good agreement between the experimental and simulated results with setting the threshold plastic strain $\bar{\varepsilon}_{th}^p$=3.2% suggests that the inhomogeneous deformation such as localized microshear-bands maybe occured in the deformation range over about 3.2% axial tensile strain.

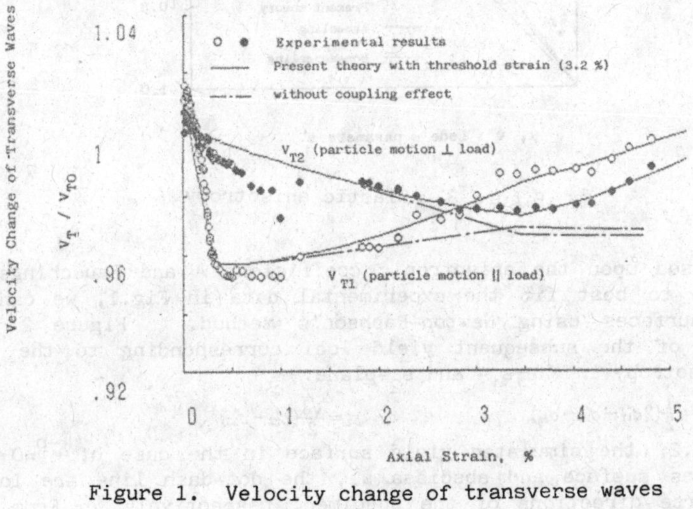


Figure 1. Velocity change of transverse waves

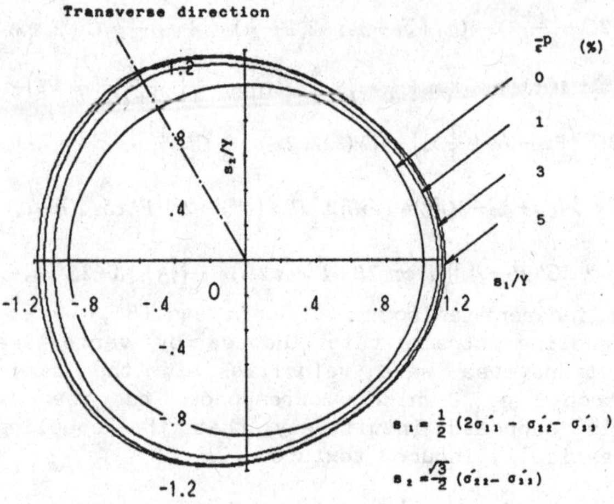


Figure 2. Predicted yield surfaces

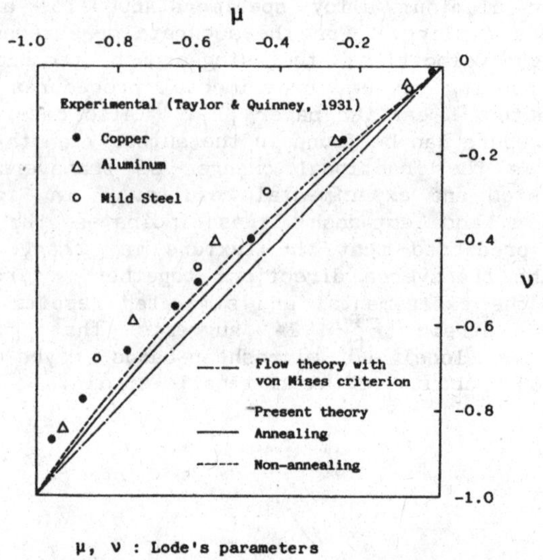


Figure 3. Plastic anisotropy

Now based upon the anisotropic coefficient A and Bauschinger modulus $m_B$ selected to best fit the experimental data in Fig.1, we can estimate the yield-surfaces using Newton-Raphson's method. Figure 2 shows the predictions of the subsequent yield loci corresponding to the growth of plastic anisotropy in the $s_1$- and $s_2$-plane:

$$s_1=\frac{1}{2}(2\sigma'_{11}-\sigma'_{22}-\sigma'_{33}), \qquad s_2=\frac{\sqrt{3}}{2}(\sigma'_{22}-\sigma'_{33}).$$

In Fig.2, the simulated yield surface in the case of $\bar{\varepsilon}^p=0$ indicates initial Mises sueface and abscissa $s_1$, the dot-dash line are loading and the transverse directions of the specimen, respectively. From Fig.2, it is supposed that the plastic anisotropy growth evolves to hardening in the loading and the transverse directions together and has strong correlation with the velocity changes of transverse waves in Fig.1.

Figure 3 shows the plastic anisotropy simulated by the present theory under combined stress states represented by Lode's parameters $\mu$ and $\nu$. As seen from Fig.3, it becomes clear that annealing makes the strain hardening due to the plastic deformation to be isotropic.

## CONCLUDING REMARKS

In the present paper, we applied the proposed ultrasonic nondestructive evaluation method to evaluate microstructural property changes of the aluminum alloy induced by plastic deformation and investigated the plastically deformed state, i.e., yield surface, texture change due to the plastic deformation, and the occurrence of the instability associated with the micro slip band.

Moreover, we predicted the plastic anisotropy growth under combined stress states using the ultrasonic wave velocity changes under uniaxial tensile test.

## REFERENCES


1. Kobayashi, M., Murakami, H., and Kayaba, T., Numerical and experimental studies on plastically induced acoustoelastic effects, Proc. 6th Int. Congr. Experim. Mech., 1988, pp. 416-421.
2. Kobayashi, M., Acoustoelastic theory for plastically deformed solids, JSME Int. Jour., 1990, 33, pp. 310-318.

# MICROASPECTS OF PLASTIC DEFORMATION AND GLOBAL MECHANICAL PERFORMANCE OF METALS

A. KORBEL* and M. BERVEILLER**
*Institute for metals Working and Physical Metallurgy
The Academy of Mining and Metallurgy, Cracow, Poland
**Laboratoire de Physique et Mécanique des Matériaux (UA CNRS)
Faculté des Sciences, Unviversité de Metz, Metz, France

**ABSTRACT**

In the work, the emphasis will be made to the process of heterogeneisation of plastic deformation in crystals and its role on the macroscopic behavior of metals. There will be considered one particular form of heterogeneisation of strain, which is highly cooperative transubstructural glide of dislocations. It may self-induce during monotonic straining as well as it may be induced by the change of loading path.

## INTRODUCTION

Prediction of the mechanical performance of metallic materials in forming operations appears a basic target of the mechanical analyses and modelling. The accuracy of the analitycal predictions depends, however, upon the assumptions made in modelling which in general concern :

a. the mode of deformation,
b. the evolution of the metal structure during deformation and
c. the relationships between the structure, mode of deformation and global properties of a metal.

Although there is an obvious feedback between the structure evolution and evolution of the mechanical characteristic of a metal, variety of intrinsic (metallurgical) and extrinsic (scheme of loading, loading conditions) factors causes that different forms of constitutive law are used to explain a specific aspect of mechanical behaviour in specific materials (eg. Luders deformation. Portevin-LeChatelier Effect, dynamic recovery, dynamic recrystallization, texture formation etc.). The most fundamantal difficulty which sets the limits upon the practical value of the modelling comes, however, from the lack of the equivalence between the local and global properties of a material. Despite of a significant progress in formulation of the constitutive equation for a polycrystalline agregates (1, 2, 3), which resolves itself into allowing for heterogeneity of the stress and strains within the body being deformed, and which shows the way to take these heteregeneities into account, the relationship between the evolution of the structure and properties appears still as crucial as well as the weakest point of the analysis. Hence, instead of making "a priori" choice of the law a better way seems to follow some well established rules which concern the local properties of crystals.

The purpose of this work is to summarized these rules.

## ANALYSIS

The experimental observations of the behaviour of single and polycrystalline metals indicate that :

1. homogeneous deformation requires the slip events to be evenly distributed in the crystals volume (fine slip).

2. The highest strain hardening rate ($10^{-3}$ - $10^{-2}$ $\mu$) correlates with homogeneous (fine) multislip. In terms of the structure evolution this means a continuous built up of the tough obstacles network of successively diminishing meshlength (4).

3. Obstacles netword is unstable with respect to temperature and strain path change (activation of a new slip system) and the latent hardening ratio (when higher than 1.1) determines the structure instability stress (5).

4. The instability of the structure leads to localization of strains in meso-scale (transsubstructural coarse slip) causing that metals behaves like a two phase structure composed of softer matrix and harder slip bands. The global hardening must then reflect a complex effect of homogeneous (matrix) and heterogeneous (bands) deformation.

5. Rotation of crystals during deformation and internal stresses (pile-up stress, comptatibility stress) lead to the change in a set of operating systems (selfinduced change of deformation path) causing instability of structure and in turn localization of strain in bands of coarse slip. The later may convert into shear bands (transgranular localization) giving rise to formation of a geometric defect (neck).

The links between the microaspect of deformation and global mechanical behaviour of metals are summarized in the table 1. In fact, the Table shows the correlation between the type of spatial organization of slips in crystals and a macroscopic response of a bulk sample.

The discussion of the mechanisms reponsible for particular type of slip organization is given elsewhere (6). It is of value, however, to emphazis that the change of the type of organization of slips (from A to E) results from the change in operating slip systems. Figure 1 a,b shows a very dramatic effect of selfinduced change of strain path (entry of critical (a) and sequentialy cross and critical (b) slip systems in Cu-Al single crystals (7), while the effect of the forced change of deformation path in single crystals of copper is shown in Fig. 2 (8).

In polycrystalline metals the change of strain path always leads to a catastrophic flow (types D & E) regardless the kind of material, leading to more dramatic effects with increasing prestrain (Fig. 3 (9)). A selfinduced change in a set of operating systems in individual grains is expected from early stages of deformation because of the strain compatibility argument. Therefore, a coarse slip (type B) superposing fine homogeneous slip (type A) is observed already after small strains and strain hardening rate shows tendency to decrease with the strain.

In the ligth of these observations we incline to assume, that in modelling of the mechanical performance of a metalic material a better approximation may be achieved when the evolution of the internal structure of metals is considered in terms of the criteria of structure destabilization (activation of an alient slip system) and associated changes in spatial organization of slips.

Table 1
Microaspect of deformation and global mechanical behaviour

| TYPE | SLIP DISTRIBUTION and deformation mode | FLOW BEHAVIOUR: Deformation, load Strain hardening rate | EXAMPLES |
|---|---|---|---|
| FINE SLIP (A) | Evenly distributed slips λ↓ when ε↑ multislip | homogen, stable $\approx 10^{-3}$ μ | second stage of defor. in single crystals |
| SLIP BANDS (B) | Clusters of slip lines (parallel) single slip system | meso heterog,. stable low > 0 | easy glide |
| COARSE SLIP (C) | Transsubstructural long distance glide single slip system | meso. heterog. instable zero | Luder's fronts in single crystals : (onset of yielding in alloys, oveshoot instability LH instability) |
| A + C | Fine multislip + + coarse single slip complex | complex stable decreasing with strain or low, constant | Parabolization of $\sigma_{vs}\varepsilon$ (dynamical recovery) IV-stage of devormation (linearization of $\sigma_{vs}\varepsilon$) |
| MICROSHEAR BANDING (D) | Transgranul shear Pseudo single slip | macro-heterog., instable zero | Luder's front in polycrystals Diffusing necks |
| SHEAR BANDING (E) | Clusters of μshear bands pseudo single slip | macro-heterog. <0 | Necking |

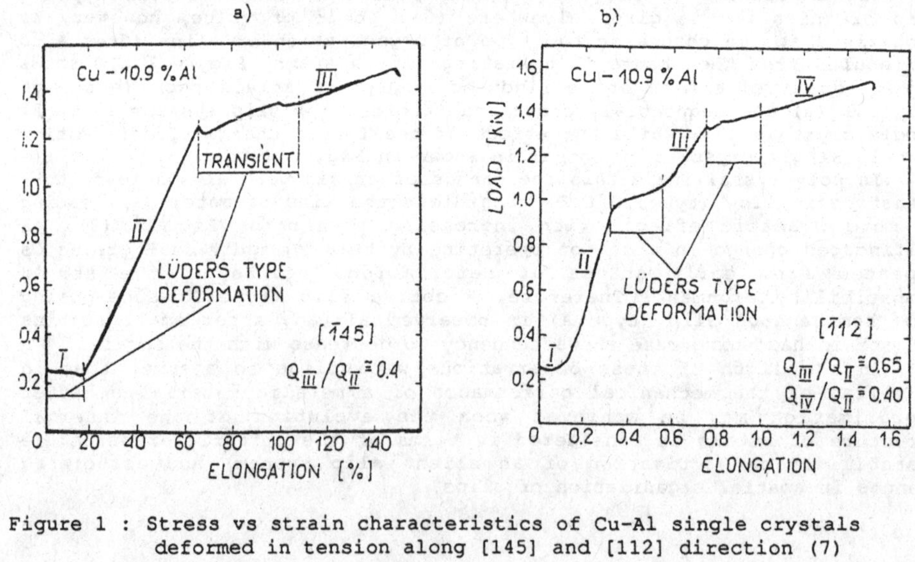


Figure 1 : Stress vs strain characteristics of Cu-Al single crystals deformed in tension along [145] and [112] direction (7)

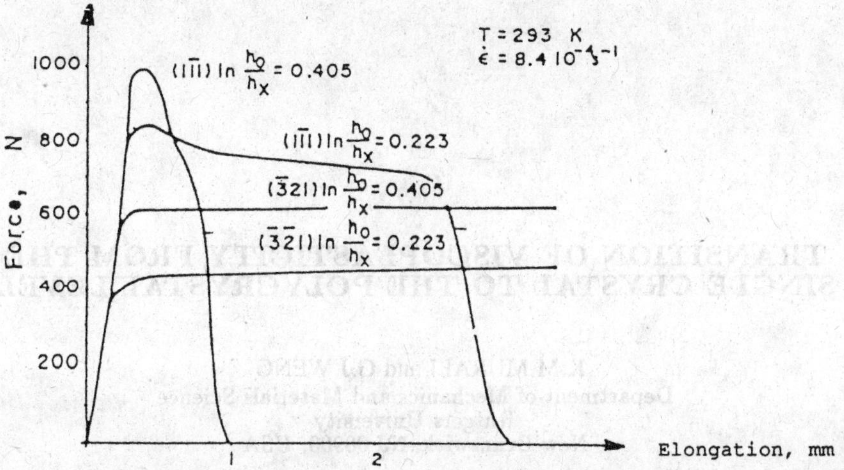


Figure 2 : Effect of the crystal orientation and the amount of rolling deformation on the load-elongation curve in tension (8)

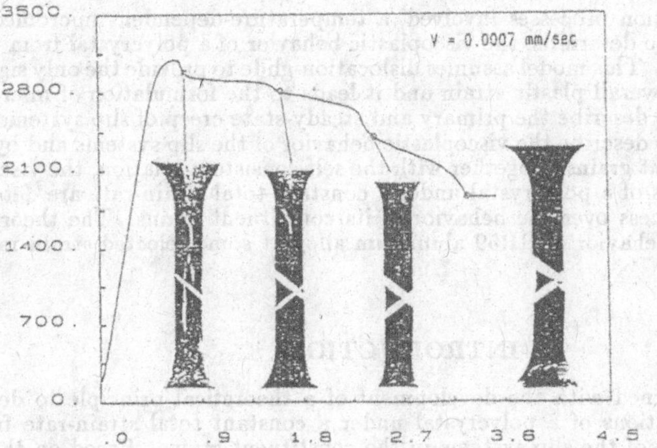


Figure 3 : Mechanical characteristic of Armco Iron in tension after the change of strain path (tension along TD of the 20% primary rolling deformation) (9)

REFERENCES


1. M. Berveiller, A. Zaoui, J. Mech. Phys. Sol., 26, 325 (1979)
2. P. Lipinski, M. Berveiller, Int. Jour. Plasticity, 5, 149 (1989)
3. P. Lipinski, J. Krier, M. Berveiller, Rev. Phys. Appl. 25, 361, 1990.
4. S. Mader, A. Seeger, Acta Metall., 8, 513 (1960)
5. Z.S. Basinski, P.J. Jackson, Phys. Stat. Sol., 9, 805 (1965)
6. A. Korbel, Archives of Metallurgy, 35, 177 (1990)
7. A. Korbel, M. Szczerba, Rev. Phys. Appl., 23, 706, (1988)
8. A. Korbel, M. Szczerba, Acta Metall., 30, 1961 (1982)
9. A. Korbel, P. Martin, Acta Metall, 36, 2575 (1988)

# TRANSITION OF VISCOPLASTICITY FROM THE SINGLE CRYSTAL TO THE POLYCRYSTAL LEVEL

K.M.MURALI and G.J.WENG
Department of Mechanics and Materials Science
Rutgers University
New Brunswick, NJ 08903, USA

## ABSTRACT

Based on the activation processes involved, a temperature-dependent micromechanical theory is developed to determine the viscoplastic behavior of a polycrystal from those of its constituent grains. This model assumes dislocation-glide to provide the only significant contribution to the overall plastic strain and it leads to the formulation of microconstitutive equations that describe the primary and steady-state creep of slip systems. These equations are used to describe the viscoplastic behavior of the slip systems and by which, that of the constituent grains. Together with the self-consistent relation, the viscoplastic stress-strain relations of a polycrystal under a constant total strain-rate are determined by an averaging process over the behavior of its constituent grains. The theory is applied to model the behavior of RR59 aluminum alloy at some selected strain rates and temperatures.

## INTRODUCTION

This paper is concerned with the development of a theoretical principle to determine the stress-strain relations of a polycrystal under a constant total strain-rate from the viscoplastic behavior of the slip systems in the constituent grains. Based on the observation that plastic deformation of metals is inherently a rate-controlling process, a set of temperature-dependent constitutive equations which can account for both the primary and steady-state creep of slip systems will be introduced first. The viscoplastic behavior of a constituent grain - or a single crystal - under a given strain-rate (the sum of the elastic and creep rates) will be derived by considering the simultaneous contribution of creep activities of all slip systems. Then, by means of a self-consistent relation, the transition of the viscoplastic behavior to the polycrystal level will be established by an orientational average under a constant external total strain-rate.

## VISCOPLASTIC BEHAVIOR OF THE SLIP SYSTEMS AND THE CONSTITUENT GRAINS

According to Frost and Ashby's [1] study of deformation mechanisms for polycrystalline metals, 'power-law' creep results when the applied stress and temperature are relatively high. In this regime, the steady-state creep rate depends on the stress by a power-law ($\dot{\epsilon}\,\alpha\,\sigma^n$). It has also been established that the high-temperature creep of polycrystalline pure metals and alloys is a thermally activated process. Creep rate can be represented by the Arrhenius-type law, $\dot{\epsilon}\,\alpha\, exp(\frac{-Q_{app}}{RT})$, where $Q_{app}$ is the apparent activation energy of creep. The apparent activation energy, $Q_{app}$, may depend on the applied stress $\sigma$, the temperature $T$ and the structure $st$ of the material. Based on these factors and the dependence on temperature and shear modulus, the stress dependence of creep rate of a slip-system can be written as

$$\dot{\gamma}^c \sim (\frac{1}{T})\, exp(\frac{-Q_{app}}{RT})\Big(\frac{\tau}{\mu}\Big)^n, \tag{1}$$

where $\mu$ is the elastic shear modulus. Consequently, the steady-state creep rate of, say system k, takes the form

$$\overset{(k)}{\dot{\gamma}^c_s} = \kappa\,(\frac{1}{T})\, exp(\frac{-Q_{app}}{RT})(\frac{\overset{(k)}{\tau}}{\mu})^n. \tag{2}$$

Its primary crep rate, on the other hand, decreases with increasing active and latent hardening; assuming that there is a combined isotropic and kinematic hardening for the slip systems, it may be written as

$$\overset{(k)}{\dot{\gamma}^c_p} = \frac{\eta\,(\frac{1}{T})\, exp(\frac{-Q_{app}}{RT})(\frac{\overset{(k)}{\tau}}{\mu})^n}{\zeta + \sum_l [\alpha + (1-\alpha) cos\ \overset{(k,l)}{\theta}\ cos\ \overset{(k,l)}{\phi}\,](\overset{(l)}{\gamma^c})^m}, \tag{3}$$

where, $\overset{(l)}{\gamma^c}$ is the creep strain of the $l$th system, $\overset{(k,l)}{\theta}$ and $\overset{(k,l)}{\phi}$ are the angles between the slip directions and the slip-plane normals, respectively, of the $k$th and $l$th slip systems, and the sum extends to all systems in the same grain and $\eta$, $\zeta$ and $\alpha$ are material constants like $\kappa$ in equation (2). Parameter $\alpha$ is "the degree of isotropy in work hardening"; when $\alpha = 1$, this corresponds to Taylor's isotropic hardening and when $\alpha = 0$, it reduces to Prager's kinematic hardening. Parameter m is the strain hardening exponent, with m = 1 representing linear hardening.

If the local stress field in the considered grain is denoted by $\sigma_{ij}$, the resolved shear stress is simply given by

$$\overset{(k)}{\tau} = \overset{(k)}{\nu_{ij}}\, \sigma_{ij}, \tag{4}$$

where $\overset{(k)}{\nu_{ij}}$ is the Schmid-factor tensor of the $k$th slip system. In equation (4) and henceforth, the Einstein summation convention for a repeated index is adopted.

Once the creep rate is calculated from the constitutive equations (2) and (3), the creep or plastic strain-rate tensor of the considered grain follows from all contributions of its slip systems

$$\dot{\epsilon}^p_{ij} = \sum_k \overset{(k)}{\nu_{ij}} \overset{(k)}{\dot{\gamma}^p}, \tag{5}$$

where $\dot{\gamma}^p = \dot{\gamma}^p_s + \dot{\gamma}^p_p$, and the summation extends to all slip systems in the grain. The total strain-rate of the grain is then given by

$$\dot{\epsilon}_{ij} = C^{-1}_{ijkl}\dot{\sigma}_{kl} + \dot{\epsilon}^p_{ij}, \tag{6}$$

where $C^{-1}_{ijkl}$ is the elastic compliance tensor of the single crystal. The total strain-rate is seen to depend on the local stress-rate as well as the local stress and plastic strain of the grain.

## TRANSITION TO THE POLYCRYSTAL LEVEL

To caculate the stress-strain relation of a polycrystal under a constant, external total strain-rate $\dot{\bar{\epsilon}}_{ij}$, it is necessary to know the evolution of the local stress in each constituent grain $\sigma_{ij}$ and its viscoplastic or creep strain $\epsilon^p_{ij}$. This can be determined more conveniently in incremental steps, during which the variations of $d\sigma_{ij}$ and $d\epsilon^p_{ij}$ are evaluated for each $dt$. Then, $d\bar{\sigma}_{ij}$ of the polycrystal is calculated for the corresponding time-increment.

We first consider the incremental creep deformation of a polycrystal. By assuming each grain to be elastically isotropic, its plastic strain increment is given by the orientational average

$$d\bar{\epsilon}^p_{ij} = \{d\epsilon^p_{ij}\}, \tag{7}$$

where $d\epsilon^p_{ij} = \dot{\epsilon}^p_{ij} dt$. The corresponding stress variation in the grain can be evaluated approximately by the self-consistent relation [2, 3]

$$d\sigma_{ij} = -2\mu(1-\beta)(d\epsilon^p_{ij} - d\bar{\epsilon}^p_{ij}), \tag{8}$$

where, in terms of Poisson's ratio $\nu$, $\beta = 2(4-5\nu)/15(1-\nu)$.

The stress-strain relation of the polycrystal under an imposed constant total strain-rate $\dot{\bar{\epsilon}}_{ij}$ can be calculated incrementally in the following manner. The total strain-rate is the sum of the elastic and plastic strain-rates. The corresponding elastic strain increment therefore can be calculated from

$$d\bar{\epsilon}^e_{ij} = d\bar{\epsilon}_{ij} - d\bar{\epsilon}^p_{ij}, \tag{9}$$

which in turn provides the macroscopic stress increment $d\bar{\sigma}_{ij}$ from the constitutive equation

$$d\bar{\sigma}_{ij} = C_{ijkl} d\bar{\epsilon}^e_{kl} = 2\mu d\bar{\epsilon}^e_{ij} + \lambda\delta_{ij} d\bar{\epsilon}^e_{kk}. \tag{10}$$

Initially, at $t = 0$, $\bar{\sigma}_{ij} = 0$, $\dot{\epsilon}^p_{ij} = \dot{\bar{\epsilon}}^p_{ij} = 0$ and we know only $\dot{\bar{\epsilon}}_{ij}$. After the first increment $dt$, $d\bar{\epsilon}_{ij} = \dot{\bar{\epsilon}}_{ij} \cdot dt$ and using the forward incremental scheme, the stress increment is given by $d\bar{\sigma}_{ij} = C_{ijkl} \cdot d\bar{\epsilon}_{kl}$. At the end of $dt$, the local stress increment of the grain is given by $d\sigma = d\bar{\sigma}_{ij}$, giving rise to a new stress state $\sigma_{ij}(t+dt) = \sigma_{ij}(t) + d\sigma_{ij}(t)$, where $\sigma_{ij}(t=0) = 0$. This $\sigma_{ij}$ is then substituted into (4), to calculate the slip rates by (2) and (3) and the plastic strain rate $\dot{\epsilon}^p_{ij}$ of the grain by (5). The plastic strain increment of the polycrystal $d\bar{\epsilon}^p_{ij}$ then follows from (7) and the corresponding elastic strain increment $d\bar{\epsilon}^e_{ij}$ and stress increment $d\bar{\sigma}_{ij}$ of the aggregate follow from (9) and (10) at the end of the second time increment. For the calculation of the third (and the subsequent) incremental step, there are two sources of $d\sigma_{ij}$ from the previous step, one from $d\bar{\sigma}_{ij}(d\sigma_{ij} = d\bar{\sigma}_{ij})$, and the other from the heterogeneity of plastic deformation in the aggregate. Thus, at the beginning of the n-th step

$$d\sigma_{ij}(n) = d\bar{\sigma}_{ij}(n-1) - 2\mu(1-\beta)(d\epsilon^p_{ij}(n-1) - d\bar{\epsilon}^p_{ij}(n-1)), \tag{11}$$

so that, $\sigma_{ij}(n) = \sigma_{ij}(n-1) + d\sigma_{ij}(n)$. This procedure can be repeated to generate the entire stress-strain curve of the polycrystal under a constant total strain-rate $\dot{\bar{\epsilon}}_{ij}$.

## NUMERICAL RESULTS

The micromechanical theory is applied from the level of slip systems in each grain of a polycrystal. The three dimensional polycrystal model is selected to consist of 216 different grain rotations; they are obtained by rotating a basic crystal, aligned along the material axes, about the three orthogonal axes. The polycrystal considered is RR59 aluminum alloy. It has an fcc structure with four (111) slip planes and three [110] slip directions on each plane.

Based on a simulation of the creep data of RR59 [4] as shown in Fig.1, the material parameters are determined to be $\kappa = 5300$, $\eta = 1.6 \times 10^7$, $\zeta = 10$, $Q_{app} = 50,000$, $n = 2$ and $m = 3$ where stress, strain, time and energy are in the units of MPa, $10^{-5} m/m$, hr and J/mole, in turn. Using these parameters, the $\bar{\sigma}_{11}$ versus $\bar{\epsilon}_{11}$ curve at T = 523 $^o$K is predicted at strain rates of 1, 0.5 and 0.15 $\times 10^{-5} m/m/hr$ as shown in Fig.2. To demonstrate the temperature dependence of the viscoplastic behavior of the polycrystal, the corresponding stress-strain curves at T = 473 $^o$K and 423$^o$K are also plotted in Figs.3 and 4. As the strain-rate increases, the curves shift upward. Thus, one of the important characteristics of stress versus strain curves at a constant strain-rate is brought forward by this prediction. Also, as the strain increases, the curves flatten out, implying that they are approaching a saturation stress which will be attained at steady-state. That, such a saturation stress should be attained, is shown by Kocks [5].

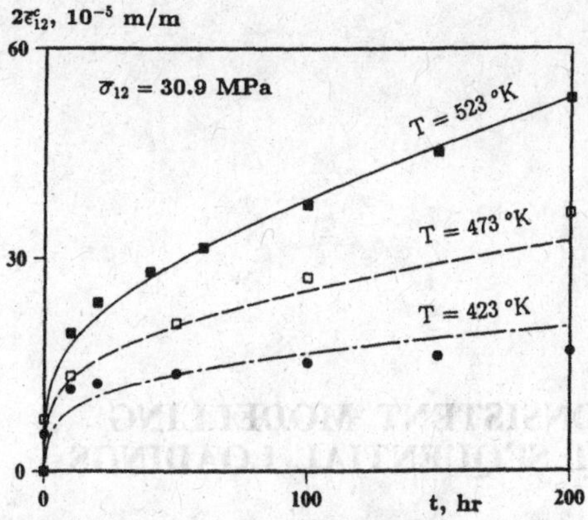


Fig.1 Simulation of creep data.

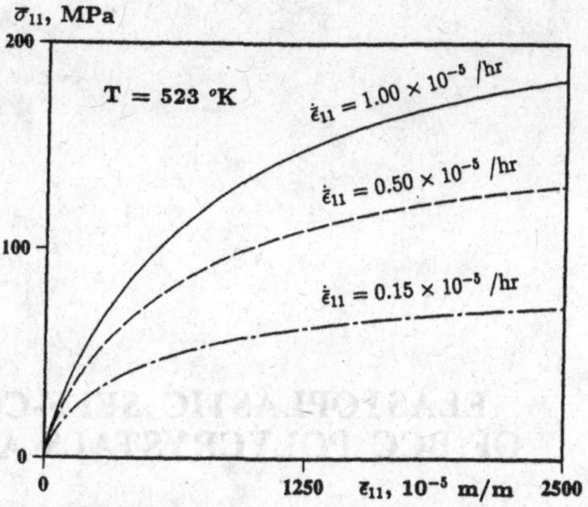


Fig.2 Prediction of $\bar{\sigma}_{11}$ vs. $\bar{\epsilon}_{11}$.

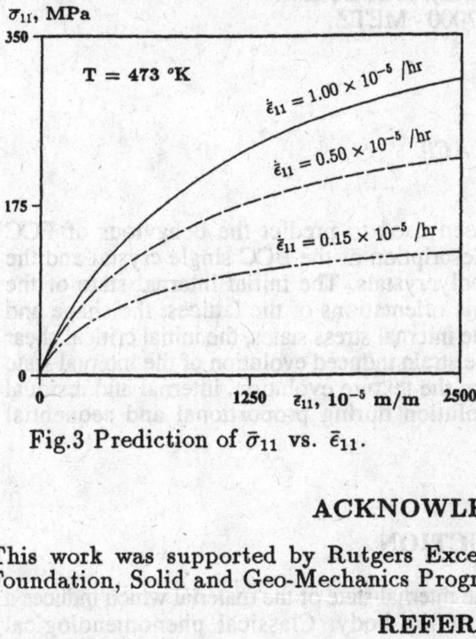


Fig.3 Prediction of $\bar{\sigma}_{11}$ vs. $\bar{\epsilon}_{11}$.

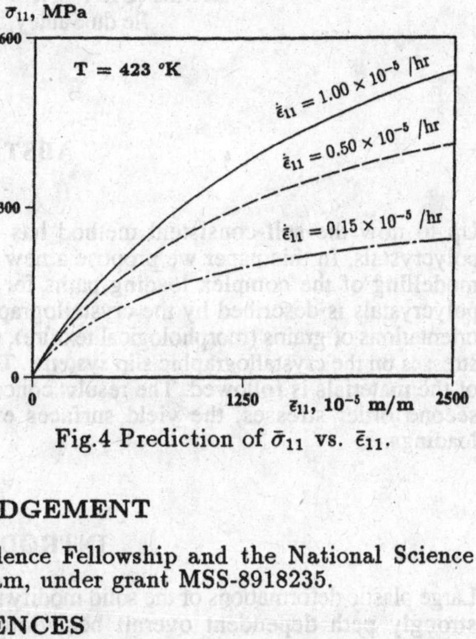


Fig.4 Prediction of $\bar{\sigma}_{11}$ vs. $\bar{\epsilon}_{11}$.

## ACKNOWLEDGEMENT

This work was supported by Rutgers Excellence Fellowship and the National Science Foundation, Solid and Geo-Mechanics Program, under grant MSS-8918235.

## REFERENCES

[1] H. J. Frost and M. F. Ashby. *Deformation-Mechanism Maps*. Pergamon Press, Oxford, 1982.

[2] E. Kröner. Zur plastichen verformung des vielkrystalls. *Acta Metallurgica*, 9:155–161, 1961.

[3] G. J. Weng. Self-consistent determination of time-dependent behavior of metals. *Journal of Applied Mechanics*, 48:41–46, 1981.

[4] A. E. Johnson and N. E. Frost. The temperature dependence of transient and secondary creep of an aluminum alloy to british standard 2142 at temperatures between 20 and 250 °C at constant stress. *Journal of the Institute of Metals*, 81:93–107, 1952.

[5] U. F. Kocks. Laws for work-hardening and low-temperature creep. *Journal of Engineering Materials and Technology*, 98:76–85, 1976.

# ELASTOPLASTIC SELF-CONSISTENT MODELLING OF BCC POLYCRYSTALS AT SEQUENTIAL LOADINGS

A. NADDARI, P. LIPINSKI, M. BERVEILLER

LPMM (URA CNRS 1215), ISGMP, ENIM
Ile du Saulcy, 57000 - METZ

## ABSTRACT

Up to now the self-consistent method has been used to predict the behaviour of FCC polycrystals. In this paper we propose a new description of the BCC single crystal and the modelling of the complex loading paths for polycrystals. The initial internal state of the polycrystals is described by the crystallographic orientations of the lattices, the shape and orientations of grains (morphological texture), the internal stress states, the initial critical shear stresses on the crystallographic slip systems. The strain induced evolution of the internal state of the materials is followed. The results concern the texture evolution, internal and residual second order stresses, the yield surfaces evolution during proportional and sequential loadings.

## INTRODUCTION

Large plastic deformations of the solid modify the internal state of the material which induces a strongly path dependent overall behaviour of the body. Classical phenomenological description of such complex responses needs numerous parameters, rather difficult to identify. It is the reason why micromecanical models have been developped in the last two decades. Their objectif is to relate the actual responses of the material to its internal state.

The general formal analysis of such a kind of problem consists in the formulation of a complicated integral equation, whose exact solution is rather difficult to obtain. For sufficiently disordered polycrystals the self-constistent approach [1], may be considerd as an admissible approximation.

In what follows, we suppose the crystallographic slip to be a unique mechanism of plastic deformation. This mechanism is well described in the case of FCC metals. The results of the numerical simulations for FCC metals may be found in [2], [3]. For BCC materials the ambiguity arises concerning the definition of the slip planes. In order to reproduce the so-called "pencil-glide", we propose to consider two families of slip systems, namely : {110} <111> and {112} <111>.

In such a case, the behaviour of the single crystal of BCC is defined by critical shear stresses of these two families and the appropriate hardening matrix describing the interactions between dislocations on all 48 slip systems.

Results presented below concern the simulations of induced crystallographic texture, transformation of textures, induced yield surfaces and anisotropy of the yield point.

## RESULTS

The polycrystal is represented by 100 initially spherical grains. The elastic behaviour of these grains is supposed to be isotropic and defined by the Lamé's constants. In such a case, the overall elasticity is also isotropic and described by the same constantes. The plastic properties are defined by two initially different critical shear stresses and the hardening matrix H, whose anisotropy is given by A parameter being the ratio between the strong and weak interactions between the systems. The initial crystallocraphic texture is randomely generated. Figure 1 shows the influence of the preloading direction and the definition of the plastic offset on the subsequent yield surfaces. The induced crystallographic texture is also presented. A very important influence of the preloading direction (up to 60 % of equivalent plastic strain) is visible. A complicated evolution of the yield surfaces manifesting the kinematic hardening and local deformation of the surface is obtained. A completely different evolution of the crystallographic texture for various preloading explains the different evolution of the yield surfaces. The second type of results concerns the texture transformation and its influence on the anisotropy of the yield point.

Now, the initial texture of the material is very pronounced as drown on the figure 2a and corresponds to the texture of rolling of BCC material. We simulate the tension test in the direction oriented at 45° with respect to the rolling direction. Figures 2b, 2c and 2d illustrate the transformation of the crystallographic texture during this test. The next figure shows the influence of these phenomenan on the yield point defined as stress corresponding to 0,2 % of plastic strain. An important induced anisotropy of the material can be observed. The comparison with the experimental results by Boehler and al [4] shows a good qualitative and quantitative prediction of the real phenomena.

## REFERENCES

[1] P. Lipinski, M. Berveiller, Elastoplasticity of micro-inhomogeneous metals at large strains, Int. J. Plasticity, **5**, pp 149-172, 1989.

[2] P. Lipinski, J. Krier, M. Berveiller, Elastoplasticité des métaux en grandes déformations: Comportement global et évolution de la stucture interne, Revue Phys. Appl. **25**,361-388, 1990.

[3] P. Lipinski, M. Berveiller, F. Corvasce, Statistical approach to elastoplastic behaviour of polycrystals at finite deformations, Arch. Mech., **40**, 5-6, pp. 725-740,Warszawa 1988.

[4] J-P. Boehler and al, Rapport du contrat DRET n° 85-039,Institut de mécanique de Grenoble, (1990).

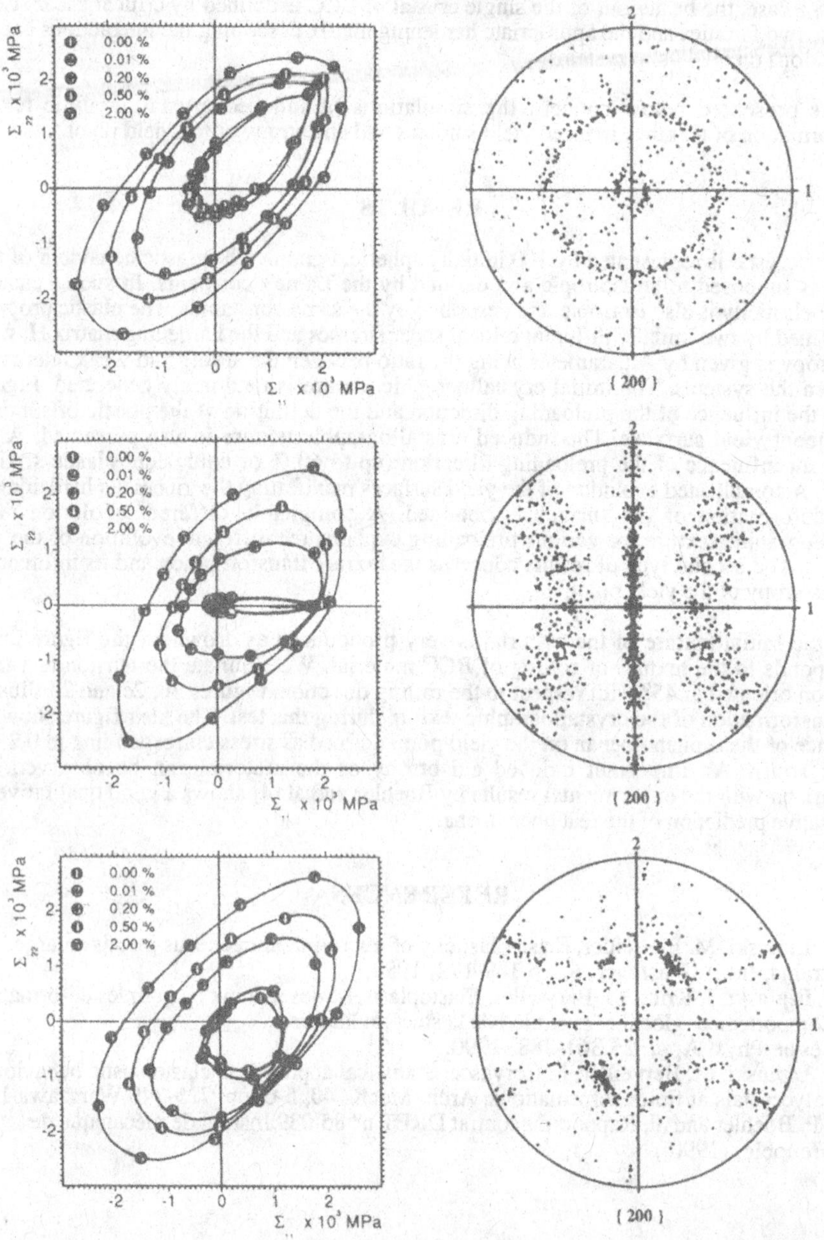


Fig. 1 : The effect of the preloading direction in the stress space on the induced textures ( { 200 } pole figures ) and their consequences on the subsequent yield surfaces for various plastic offset definitions.

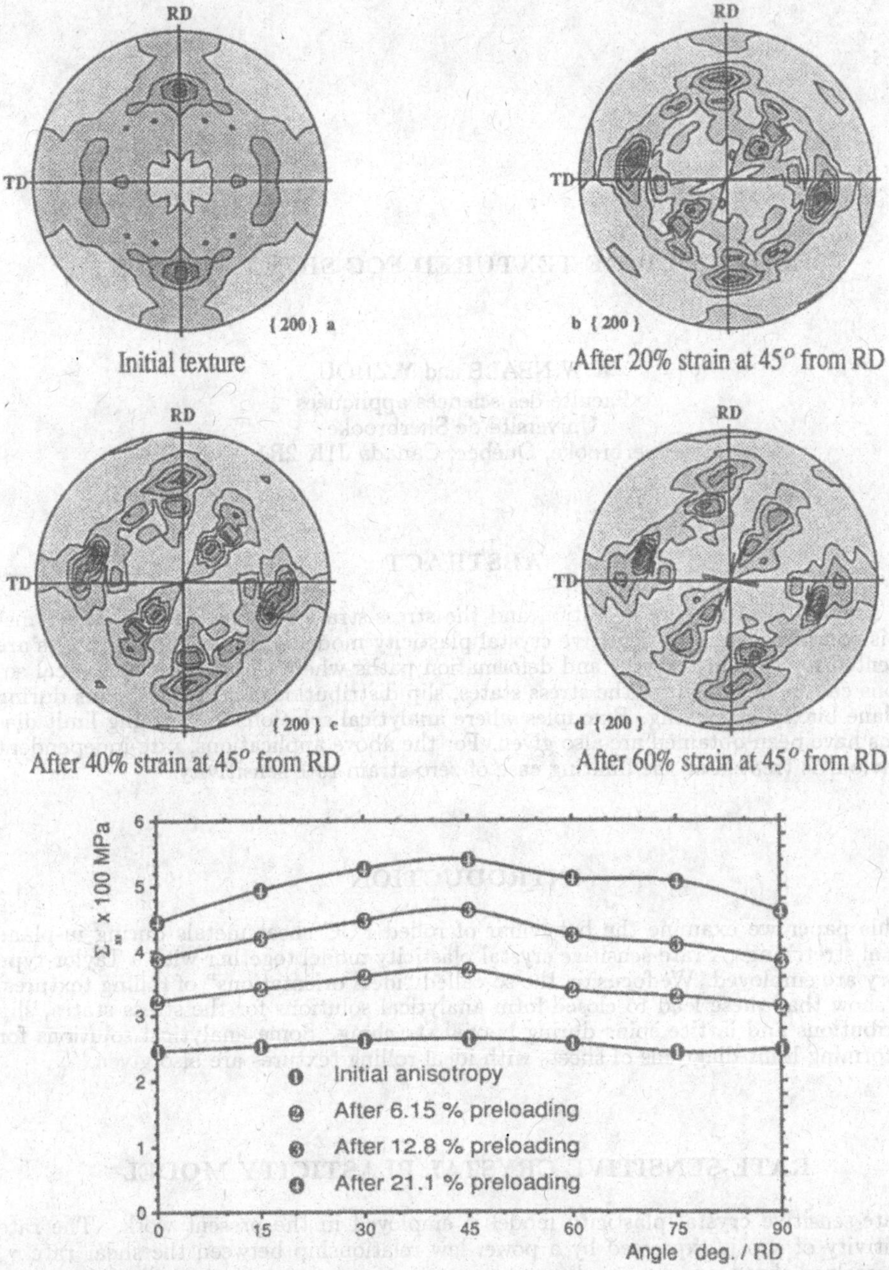


Fig. 2 : Induced texture by a tensile test at 45° of the rolling direction for an initially anisotropic BCC sheet and evolution of the 0.2% yield stress in fonction of the tensile preloding and the angle between tensile test an RD.

# BEHAVIOUR OF TEXTURED FCC SHEET METALS

K.W.NEALE and Y.ZHOU
Faculté des sciences appliquées
Université de Sherbrooke
Sherbrooke, Québec, Canada J1K 2R1

## ABSTRACT

The simulation of texture evolution and the stress-strain response of FCC sheet metals is considered. A rate-sensitive crystal plasticity model is employed and results are presented for certain textures and deformation paths where closed-form analytical solutions can be obtained for the stress states, slip distributions and lattice spins during in-plane biaxial stretching. Examples where analytical solutions for forming limit diagrams have been obtained are also given. For the above applications, rate-independent behaviour is treated as the limiting case of zero strain-rate sensitivity.

## INTRODUCTION

In this paper we examine the behaviour of rolled FCC sheet metals during in-plane biaxial stretching. A rate-sensitive crystal plasticity model together with a Taylor-type theory are employed. We focus on the so-called "ideal orientations" of rolling textures, and show that these lead to closed-form analytical solutions for the stress states, slip distributions and lattice spins during biaxial streching. Some analytical solutions for the forming limit diagrams of sheets with ideal rolling textures are also given.

## RATE-SENSITIVE CRYSTAL PLASTICITY MODEL

A rate-sensitive crystal plasticity model is employed in the present work. The rate sensitivity of slip is expressed by a power-law relationship between the shear rate $\dot{\gamma}_s$ and resolved shear stress $\tau_s$ on slip system $s$

$$\tau_s = \tau_0 \,\mathrm{sgn}\,(\dot{\gamma}_s)|\frac{\dot{\gamma}_s}{\dot{\gamma}_0}|^m = \tau_0 \frac{\dot{\gamma}_s}{\dot{\gamma}_0}|\frac{\dot{\gamma}_s}{\dot{\gamma}_0}|^{m-1} \qquad (1)$$

Here $m$ is the strain-rate sensitivity index, $\tau_0$ and $\dot{\gamma}_0$ are the reference shear stress and shear rate, respectively, assumed to be equal for all slip systems in this analysis. Our analysis is based on crystallographic slip in the 12 {111} < 110 > slip systems of a FCC crystal (listed in [1]).

The relation between the strain-rate tensor **D** and the shear rates $\dot{\gamma}_s$ is

$$D_{ij} = \sum_s \frac{1}{2}(m^s_{ij} + m^s_{ji})\,\dot{\gamma}_s \tag{2}$$

where $\mathbf{m}^s = \mathbf{b}^s \times \mathbf{n}^s$ is the Schmid tensor for the slip system $s$, and $\mathbf{b}^s$ and $\mathbf{n}^s$ denote the unit slip direction and unit slip plane normal of the slip system $s$ in the current configuration, respectively. The Cauchy stress tensor $\boldsymbol{\sigma}$ and the resolved shear stress $\tau_s$ are related as follows:

$$\tau_s = m^s_{ij}\,\sigma_{ij} \tag{3}$$

From the above relations, we obtain the following expression for **D** in terms of $\boldsymbol{\sigma}$:

$$D_{ij} = \frac{\dot{\gamma}_0}{2\,\tau_0^{1/m}} \sum_s (m^s_{ij} + m^s_{ji})\, m^s_{kl}\,\sigma_{kl} |m^s_{mn}\,\sigma_{mn}|^{1/m-1} \tag{4}$$

The crystal orientation changes, i.e. the lattice spins $\dot{\boldsymbol{\Omega}}$, are due to the difference between the velocity gradient **L** and the velocity gradient produced by plastic slip [2]

$$\dot{\Omega}_{ij} = L_{ij} - \sum_s m^s_{ij}\,\dot{\gamma}_s \tag{5}$$

All geometrically possible slip systems are activated with this rate-sensitive model. Furthermore, the deviatoric stresses, shear rates and lattice spins are uniquely determined for a specified strain rate. The rate-insensitive response can be inferred from this model as the limiting case of zero strain-rate sensitivity ($m \to 0$).

The sample coordinate system employed is such that $X_1, X_2$ and $X_3$ correspond to the rolling, transverse and normal directions, respectively. For the case of sheet stretching treated here, it can be assumed that the rolling and normal directions remain constant during deformation. This implies that $L_{21} = L_{31} = L_{32} = 0$ in Equation (5).

## SOLUTIONS FOR BIAXIAL SHEET STRETCHING

We consider thin sheets having ideal rolling textures and subjected to in-plane proportional straining paths. Accordingly, the strain states are such that $D_{22} = \rho D_{11}$, with the strain ratio $\rho = const.$, and $D_{33} = -(D_{11} + D_{22})$. The full constraint Taylor theory is adopted in this analysis.

Closed-form analytical solutions have been obtained using the rate-sensitive crystal plasticity model. For a prescribed strain ratio $\rho$ and given rate-sensitivity index $m$, we have determined the stress states (normalized by $\tau_0$), the slip system shear rates and lattice spins as a function of the applied strain rates. Rate-independent behaviour is obtained by asymptotically setting $m = 0$. Typical solutions, for the case $\rho = 1$ (equi-biaxial stretching), are summarized in Table 1.

**Table 1**
Limiting analytical solutions ($m \to 0$) for biaxial stretching ($\rho = 1$)

| Orien-tation | Stress in Sample Axes (Other $\sigma_{ij}=0$) | Shear Rates in Active Slip Systems | Lat-tice Spin |
|---|---|---|---|
| Cube | $\sigma_{11}=\sigma_{22}=\sqrt{6}\tau_0$ | $\dot\gamma_1=-\dot\gamma_4=-\dot\gamma_7=\dot\gamma_{10}=$ $\dot\gamma_2=\dot\gamma_5=\dot\gamma_8=\dot\gamma_{11}=\frac{1}{4}\sqrt{6}D_{11}$ | 0 |
| Goss | $\sigma_{11}=\frac{1}{2}\sigma_{22}=\sqrt{6}\tau_0$ | $\dot\gamma_3=-\dot\gamma_1=\dot\gamma_4=\dot\gamma_6=\frac{1}{4}\sqrt{6}D_{11}$ $\dot\gamma_7=\dot\gamma_9=-\dot\gamma_{10}=\dot\gamma_{12}=-\frac{1}{2}\sqrt{6}D_{11}$ | 0 |
| Brass | $\sigma_{11}=\frac{4}{5}\sigma_{22}=\frac{4}{3}\sqrt{6}\tau_0$ $\sigma_{12}=-\frac{2}{3}\sqrt{3}\tau_0$ | $\dot\gamma_3=-\dot\gamma_1=\dot\gamma_4=\dot\gamma_6=\frac{1}{4}\sqrt{6}D_{11}$ $\dot\gamma_7=\dot\gamma_9=-\dot\gamma_{10}=\dot\gamma_{12}=-\frac{1}{2}\sqrt{6}D_{11}$ | 0 |
| Copper | $\sigma_{11}=\frac{8}{7}\sigma_{22}=\frac{4}{3}\sqrt{6}\tau_0$ $\sigma_{13}=\frac{4}{3}\sqrt{3}\tau_0$ | $\dot\gamma_5=-\dot\gamma_7=\frac{1}{2}\sqrt{6}D_{11}$ $\dot\gamma_6=-\dot\gamma_9=\frac{3}{4}\sqrt{6}D_{11}$ | $\sqrt{2}D_{11}$ |

The rate-sensitive model has been used in conjunction with the so-called "M-K" analysis [3] to determine, analytically, forming limit diagrams for certain ideal textures. Such diagrams depict, as a function of $\rho$, the strain states $\epsilon_{11}^*, \epsilon_{22}^* = \rho\epsilon_{11}^*$ which lead to localized necking failures during biaxial stretching. The analysis assumes an initial thickness inhomogeneity in the form of a narrow band, in the transverse direction, which is slightly thinner than the nominal sheet thickness. By tracing the evolution of the band thickness with applied deformation, the limiting strains in the nominal sheet region $\epsilon_{11}^*, \epsilon_{22}^*$ can be obtained.

A simple expression for the forming limit curves, corresponding to the rate-independent ($m \to 0$) limiting behaviour of the cube and Goss orientations, has been obtained. This relation takes the implicit form

$$\epsilon_{11}^* = \left[\gamma^{b*}/\sqrt{6} - \ell n\{f\,\tau_0^b\,(\gamma^{b*})/\tau_0\,(\epsilon_1^*)\}\right]/(1+\alpha\,\rho_c) \tag{6}$$

where $\gamma^{b*}$ is determined from the condition $d\tau_0/d\gamma = \tau_0/\sqrt{6}$. (Isotropic slip system hardening is assumed here.) In this expression $f$ is the initial inhomogeneity (the ratio of the band thickness to the nominal thickness), and $\alpha = 1$ and 2 for the cube and Goss orientations, respectively. The parameter $\rho_c$ takes on the following values:

$$\rho_c = \begin{cases} 0 & \text{for} \quad -0.5 \le \rho \le 0 \\ \rho & \text{for} \quad 0 \le \rho \le 1 \end{cases} \tag{7}$$

When $m \neq 0$, only semi-analytical solutions are possible and the resulting equations, although simplified, must be solved incrementally. Typical results of such calculations are shown in the figure below for $f = 0.99$ and various $m$-values. Here the slip hardening law employed corresponds to that given in [4] for $Al$ single crystals. Also shown in this figure are the rate-independent forming limit curves described by Equation (6). It can be seen that the rate-sensitive ($m \neq 0$) results tend to the rate-independent limit (Equation 6) as $m$ approaches zero.

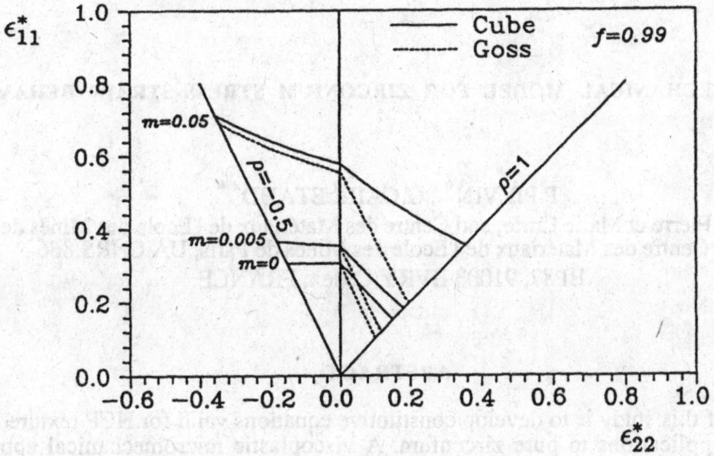


**Figure 1.** Forming limit diagrams for *Al* crystals with cube and Goss orientations.


## ACKNOWLEDGEMENTS

This work was supported by the Natural Sciences and Engineering Research Council of Canada (NSERC) and the Government of the Province of Québec (Programme FCAR).


## REFERENCES


1. Tóth, L. S., Jonas, J. J. and Neale, K. W., Comparison of the minimum plastic spin and rate sensitive slip theories for loading of symmetric crystal orientations. *Proc. R. Soc. Lond.*, 1990, **A427**, 201-219.

2. Kocks, U. F. and Chandra, H., Silp geometry in partially constrained deformation. *Acta Metall.*, 1982, **30**, 695-709.

3. Marciniak, Z. and Kuczyński, K., Limit strains in the processes of stretch-forming sheet metal. *Int. J. Mech. Sci.*, 1967, **9**, 609-620.

4. Zhou, Y., Neale K. W. and Tóth, L. S., (to be published).

# A MICROMECHANICAL MODEL FOR ZIRCONIUM STRESS-STRAIN BEHAVIOUR

P.PILVIN* , G.CAILLETAUD**
*Université Pierre et Marie Curie, and Centre des Matériaux de l'Ecole des Mines de Paris
**Centre des Matériaux de l'Ecole des Mines de Paris, UA CNRS 866
BP87, 91003 EVRY Cedex, FRANCE.

## ABSTRACT

The purpose of this study is to develop constitutive equations valid for HCP textured metals, with special applications to pure zirconium. A viscoplastic micromechanical approach is proposed, with the constitutive equations written on slip systems.The first results, shown in this paper, concern qualitative simulations of loading surfaces and of 1D tension test, performed with various assumptions on the active slip systems.

## DESCRIPTION OF THE MODEL

The model is written in the framework of polycrystalline viscoplasticity ([1]-[5]). It takes into account a physical description of the microstructure, and explicitely defines local stresses, local viscoplastic strains in the grains, resolved shear stresses and viscoplastic shear strains on the slip systems. The constitutive equations are given on a microscale, they define the viscoplastic shear strain rate as a function of the actual state of the material (microscopic internal variables), and of the applied stresses. A first version of the model was previously studied for FCC materials [6],[7]. More recent developments allow us to take into account the plastic accomodation at intergranular level [8]. On the other hand, a flexible numerical implementation [9] makes the model evolution very easy.

The equations are based on the experimental fact that kinematic hardening (ie translation of the elastic domain in stress space) is related with internal stress development, so with heterogeneities. Two heterogeneity levels are retained for metal : grain and system. Due to strain incompatibilities from one grain to the other, or to stress redistribution inside of the grains (precipitates, cell structures, ...), the local stress tensor in each grain and the effective resolved stress on each system should not be directly computed from the macroscopic stress tensor. Two sources are then registered for kinematic hardening. On the other hand, isotropic hardening is related to low range interactions; it is then simulated only at the system level. The equations coming from this very simple approach are summarized in Table 1.
One can find :
- the expression of the local stress in each grain, $\sigma^g$, and of the shear stress on each system, $\tau^s$,
- the expression of the hardening variables for the grain g, $X^g$, and for the system s, $x^s$ and $r^s$,
- the evolution laws of the state variables $\beta^g$, $\alpha^s$ and $v^s$, corresponding respectively to intergranular kinematic hardening, intragranular kinematic hardening, and intragranular isotropic hardening. These constitutive equations are an adaptation of non linear kinematic and isotropic hardening for macroscopic approaches [10] : in particular, the non linear evolution of

TABLE 1 : Summary of the equations of the polycrystalline model

$$\sigma^g = \Sigma - X^g \quad ; \quad \tau^s = \sigma^g : m^s$$

$$X^g = C\,(\beta^g - \sum_{h \in G} f_h\, \beta^h) \quad ; \quad x^s = c_I\, \alpha^s \quad ; \quad r^s = r_I^o + \sum_{r \in g} Q_I\,[1 - h_{rs} \exp(-b_I\, v^r)]$$

$$\dot{\beta}^g = \dot{\varepsilon}^g - D\,\beta^g \sum_{s \in g} \dot{v}^s \quad ; \quad \dot{\alpha}^s = \dot{\gamma}^s_v - d_I\, \alpha^s\, \dot{v}^s \quad ; \quad \dot{v}^s = |\dot{\gamma}^s_v|$$

$$\dot{\gamma}^s_v = \left\langle \frac{|\tau^s - x^s| - r^s}{k_I} \right\rangle^{n_I} \mathrm{signe}(\tau^s - x^s) \quad ; \quad \dot{\varepsilon}^g = \sum_{s \in g} m^s\, \dot{\gamma}^s_v \quad ; \quad \dot{E}^v = \sum_{h \in G} f_g\, \dot{\varepsilon}^g$$

$C$, $D$ = coefficients for intergranular kinematic hardening
$c_I$, $d_I$ = coefficients for intragranular kinematic hardening (sub I denotes slip family)
$r_I^o$, $Q_I$, $b_I$ = coefficients for intragranular isotropic hardening
$k_I$, $n_I$ = coefficients for viscous effect
$h_{rs}$ = components of the interaction matrix
$m^s$ = orientation tensor of system s
$f_g$ = volume fraction of each grain g

intergranular kinematic hardening naturally produces a plastic accommodation, so that the model is equivalent to a Kröner model at the onset of plastic flow, but the intergranular stresses are then relaxed in each loading branch, for cyclic loadings,

- the flow rules for local shear strain $\dot{\gamma}^s_v$ and for the macroscopic viscoplastic strain $E^v$.

In these expressions, the tensorial variables $\sigma^g$, $X^g$, $\beta^g$, $\varepsilon^g$ are defined for each grain, the scalar variables $\tau^s$, $x^s$, $r^s$, $\alpha^s$, $v^s$, are defined for each slip system. It is assumed that each grain presents the same elasticity that the aggregate, so that elastic behaviour is treated at the macroscopic level, using the relations of the classical elasticity (isotropic or anisotropic). Note that, at the moment, the grain to grain interaction remains initially isotropic, but the elastic law may be orthotropic.

## ADJUSTMENT TO HCP CRYSTAL MODELING

The stress-strain behaviour of HCP metals is strongly influenced by the fact that a low number of slip systems are available, giving rise to anisotropy, and to texture formation. If compared with more classical metals, the stress-strain behaviour of zirconium presents other specificities, for instançe the abnormal viscous effect versus temperature : "creep" is present at room temperature, but creep rate is lower at 200°C than at 100°C, for the same stress level [11]. This last effect seems to be related with a modification of the microstructural mechanisms of plasticity, as the activation of a new slip family. It is why the first aim of this study is trying to understand the effect of the texture on simple mechanical properties as the shape of the yield domain, and checking the relative influence of each possible slip system.

We then first have to choose a microstructure, i.e. the number of grains, their orientation in the aggregate, the geometry of the slip planes and of the slip systems. The simulations made in the following are performed with a 17 grain distribution, adapted from

[12]. It simulates the texture of a rolled Zircaloy-2. A weight is assigned to each individual grain, to account for the experimental fraction of poles : the <10$\underline{1}$0> directions are nearly parallel to the rolling direction, and the range of the angles between the normal direction and the c-axis of the crystal is higher on the transverse direction than on the rolling direction. A number of slip families may be involved in plastic deformation of HCP metals [13] . At the moment, the numerical implementation of the model was only made with slip involving a type <a> Bürgers vector : basal slip on {0001}<$\underline{2}$110> (3 systems), prismatic slip on {01$\underline{1}$0}<$\underline{2}$110> (3 systems), pyramidal slip on {01$\underline{1}$1}<$\underline{2}$110> (6 systems). The representation of twinning is not made in this first version, as twinning is not the dominant mode at small strain : it will be considered in a second step, together with {10$\underline{1}$1}<11$\underline{2}$3> pyramidal slip. The model is implemented in the code SiDoLo [9], so that these further developments will be very quick, and that automatic identifications from experimental data can now be performed.

## SIMULATION RESULTS

The numerical simulations shown here are only qualitative, they are designed to allow a better understanding of the mechanical effect of the various slip families. The material coefficients are taken identical on each slip family ($r_I^o$=100MPa, $Q_I$=-10MPa, $b_I$=20, $h_{rs}$=1 if $s \in I$ and $r \in I$ else $h_{rs}$=0, $c_I$=12000MPa, $d_I$=300), and each grain (C=36000MPa, D=40), the viscous effect is low ($k_I$=50MPa, $n_I$=20). Figure 1 shows the result of predicted yield surfaces in the plane of a laminated sheet : TD and RD are respectively on axis 1 and 2. They are obtained with a plastic offset of $10^{-6}$. The slip families are considered individually (fig.1a), or all together (fig.1b). The response with basal slip is quite characteristic, due to the poor representation of the grain distribution (all the c axis are along RD or TD axes). On the other hand, the pyramidal slip family leads to elongated yield surfaces along the biaxial stress direction. The influence of each family man be found on the resulting curve in fig.1b.

The second type of simulation concerns biaxial tension loadings in the TD-RD plane, with $\sigma_{11}=\sigma_{22}$. The slip families are considered individually (fig.2) : a large difference is again found between basal and prismatic slip, and pyramidal slip, that is consistent with the shape of the yield surface. Such large mechanical effect is not present when all the systems are introduced in the calculation, as in fig.3, which illustrates a possible behaviour in 1D tension, with various angle $\psi$ vs rolling direction.

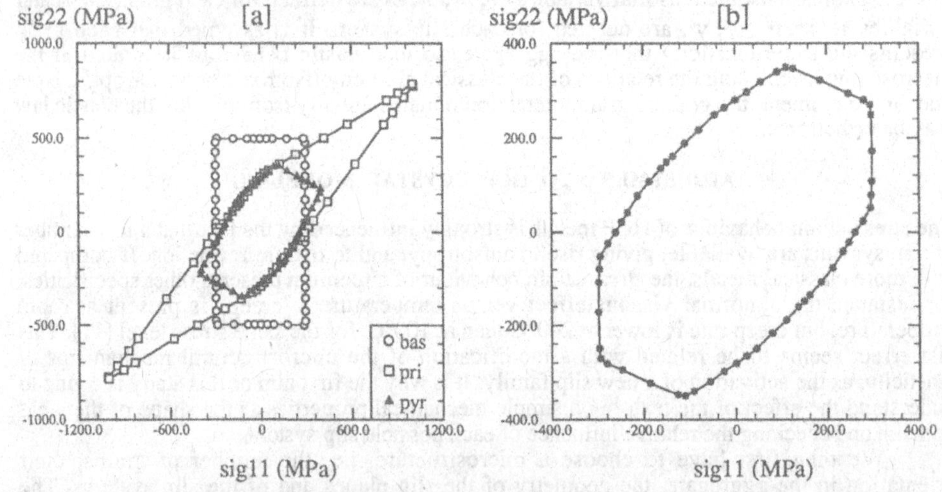


Figure 1. Yield surface on polycrystalline zirconium : [a] slip families taken individually, [b] all the slip families involved.

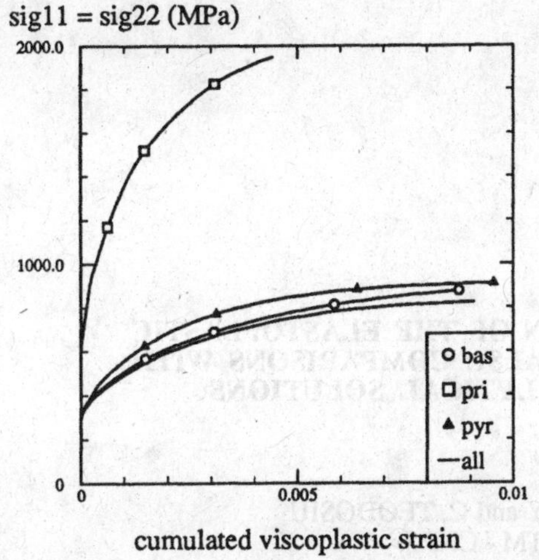


Figure 2. Biaxial tension in TD-RD plane, with each slip family taken individually.

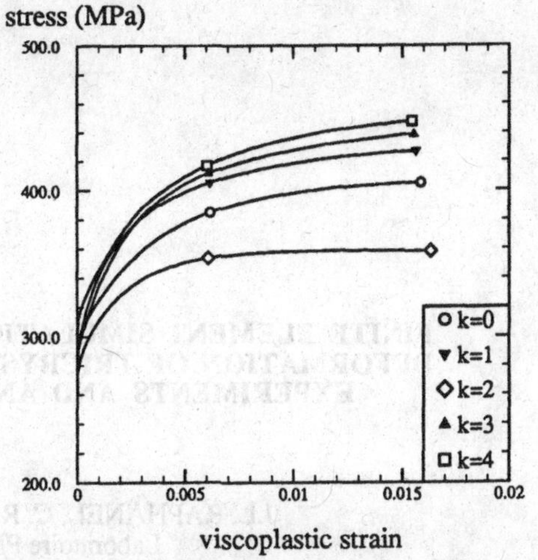


Figure 3. 1D tension in TD-RD plane, with all the slip systems involved ($\psi = k\,\pi/8$).

This type of approach is presently followed by other authors e.g. [14]. The specificity of the present one is to pay attention to the expression of the hardening rules at grain and system levels, so that it is well adapted to complex loading paths.

## REFERENCES


1. KRONER, E., Zur plastischen Verformung des Vielskristalls. Acta Metall., 1961, **9**, 155-161.
2. HILL, R., Continuum Micro-Mechanisms of Elastoplastic Polycrystals. J.Mech.Pys.Solids, 1965, **13**, 89-101.
3. ZAOUI, A., Etude de l'influence propre de la désorientation des grains sur le comportement viscoplastique des métaux polycristallins. Thèse d'état, Univ. Paris, 1970.
4. MANDEL J., Plasticité classique et viscoplasticité. CISM Course 97, Springer Verlag, 1971.
5. HUTCHINSON, J.W., Bounds and Self-consistent Estimates for Creep of Polycrystalline Materials. Proc.Roy.Soc., 1976, **A348**, 101-127.
6. CAILLETAUD, G., Une approche micromécanique du comportement des polycristaux. Rev.Phys. Appl., 1988, **23**, 353.
7. CAILLETAUD, G., Numerical Simulation of Yield Surfaces by a Phenomenological Micromechanical Approach. Proc. Mécamat on "Inelastic Behaviour of Solids", Besançon, France, Aug.30-Sept.1, 1988.
8. PILVIN, P. and CAILLETAUD, G., Intergranular and Transgranular Hardening in Viscoplasticity. In "Creep in Structures IV", Cracow, Poland, Sept.10-14, 1990.
9. PILVIN, P., Approches multi-échelles pour la prévision du comportement anélastique des métaux. Ph.D. Thesis, Univ. Paris VI, Dec.18, 1990.
10. CHABOCHE, J.L., Constitutive Equations for Cyclic Plasticity and Viscoplasticity. Int.J. of Plasticity, 1989, **5**, 247-302.
11. Colloquium : Zirconium, a New Material for Chemical Industry. Société Française de Métallurgie, ENS Lyon, Oct.10-11, 1990.
12. TOME, C. and KOCKS, U.F., The Yield Surface of H.C.P. Crystals. Acta Metall., 1985, **33**, 603-621.
13. TENCKHOFF, E., Deformation, Mechanisms, Texture, and Anisotropy in Zirconium and Zircaloy. ASTM STP 966, 1985.
14. PARKS, D.M. and AHZI, S., Polycrystalline Plastic Deformation and Texture Evolution for Crystals Lacking Five Independent Slip Systems. J.Mech.Phys.Solids, 1990, **38**, 701-724.

# FINITE ELEMENT SIMULATION OF THE ELASTOPLASTIC DEFORMATION OF TRICRYSTALS : COMPARISONS WITH EXPERIMENTS AND ANALYTICAL SOLUTIONS

J.L. RAPHANEL, C. REY and C. TEODOSIU
Laboratoire PMTM - CNRS
Université Paris-Nord
93430 Villetaneuse, France

## INTRODUCTION

The deformation of crystalline aggregates is very often heterogeneous both at an intergranular and intragranular scale. The experimental study of special aggregates made of a very small number of grains limited by grain boundaries of a simple geometrical nature and of well known orientations has allowed a detailed characterization of these heterogeneities. In this particular instance, we consider copper tricrystals with one or two triple junctions which are deformed by tension up to a few percent overall plastic strain. These samples are made of a single layer of grains of centimetric sizes, with grain boundaries orthogonal to a common plane. The overall size of a sample is $40 \times 14 \times 3$ mm$^3$. The heterogeneities of deformation lead to the formation of intragranular domains where different glide systems are active. A finite element model is used in order to simulate the deformation of these aggregates. It allows the computation of the final orientations of the crystallites and the misorientations, the slipmagnitudes on the active systems in different regions, the internal stresses state and the local strains. These values are then compared with results obtained through observations and analytical modeling.

## FINITE ELEMENT MODELING

### Kinematics and constitutive laws

The three dimensional finite element model is intended to describe large deformations of crystalline aggregates.The kinematics rests on the multiplicative decomposition of the deformation gradient into elastic and plastic parts [1,2].

The elastic anisotropy of the crystallites is taken into account via the corresponding hypoelastic law. This allows the evaluation of stresses produced by the incompatibility of the elastic deformation of the grains.

The plastic behavior results from glide on the twelve potential octahedral slip systems of f.c.c. crystals. A viscoplastic power law relates the rate of glide on each system to the ratio between resolved shear stress and "critical" shear on the system. The latter depends on the dislocation densities on all the systems [3] and its evolution is identified by fitting the proposed model [4] on experimental data for copper single crystals [5]. The exponent of the power law is such that for deformation at room temperature of copper, a quasi yield appears in this highly non-linear law and that although all twelve systems are activated, only a few do present a significant amount of glide.

### Finite Element Implementation

The boundary value problem is replaced by the rate form of the principle of virtual power. A forward gradient approximation is used to evaluate the slip increment on each system [6].
Each grain is divided into prismatic pentahedral elements with reduced integration for the volumetric part of the deformation. The mesh has been refined close to the triple junctions. It has also been chosen such that it may allow the development of domains compatible with those observed experimentally.

## EXPERIMENTAL INVESTIGATIONS

The initial geometry and the orientation of each crystallite has been measured. The sample is then deformed and observed in a scanning electron microscope. The nature and the position of the slip planes are recorded. The crystalline misorientations are also determined by Kossel's technique. In some instances, the displacement field is measured by microgrids of 5x5$\mu m^2$.
The formation of intragranular interfaces is one of the most striking feature of the experiments. Two types of such interfaces are observed :
- the ones that are parallel to the grain boundaries;
- those that start at a triple junction and seemingly extend the grain boundary into the adjoining grain.

On Figure 1, a schematic diagram of the observed slip lines is shown with lines that represent the intragranular interfaces [7].

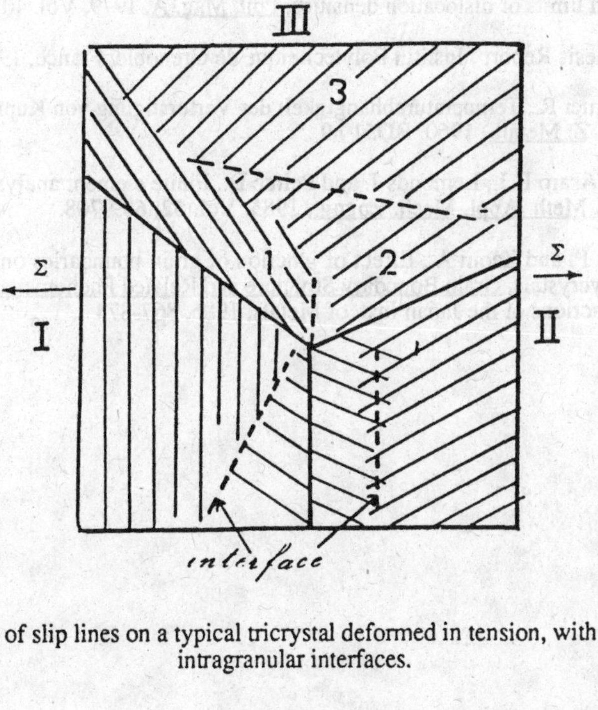


Fig.1 Patterns of slip lines on a typical tricrystal deformed in tension, with formation of intragranular interfaces.

## COMPARISONS

A first analytical treatment of these observations has been proposed [7]. It accounts for the development of internal stresses, based on a continuum theory of dislocations and considering plastic incompatibilities from grain to grain near the triple junction. Despite simplifying assumptions, such as the ones of a semi-infinite medium and an elasto-plastic behavior identified on tensile tests for bicrystals, this model shows that the formation of interfaces starting near a triple junction provides a way of relaxing internal stresses and that the observed pattern is a solution (albeit not the unique one) of the problem.

The finite element simulation takes account of the exact geometry of the specimen and of the end conditions imposed by the grips of the tensile machine, moreover since elastic anisotropy may be included, one may also get an estimate of the stresse produced by the incompatibility of the elastic deformations of the neighboring grains and check how this may be responsible for the occurrence of the subsequent plastic inhomogeneity. The nature of the mostly active slip system is also found and the slip line patterns are compared to the observed ones.

## REFERENCES


1 Teodosiu C., A dynamic theory of dislocations and its applications to the theory of the elastic-plastic continuum. In Fundamental Aspects of Dislocation Theory, eds. J.S. Simmons, R. de Wit and R. Bullough, Natl. Bur. Stand. Spec. Publ., 1970, vol. 2, 837-876 .

2. Mandel J., Plasticité Classique et Viscoplasticité, Springer, Wien-New York, 1971.

3. Essmann U. and Mughrabi H., Anihilation of dislocations during tensile and cyclic deformation and limits of dislocation densities, Phil. Mag. A, 1979, Vol. 40 n°6, 731-756.

4. L. Tabourot, Thesis Report , Institut Polytechnique de Grenoble, France, 1991.

5. Diehl J. and Berner R., Temperaturabhängigkeit der Verferstigung von Kupfer-Einkristallen oberhalb 78 K., Z. Metall., 1960, BD51 h9.

6. Needleman A., Asaro R.J., Lemonds J. and Peirce D., Finite element analysis of crystalline solids, Comput. Meth. Appl. Mech. Engng., 1985, Vol. 52, 689-708.

7. Rey C., Mussot P. and Zaoui A., Effect of junction of grain boundaries on the mechanical behavior of polycrystals, Grain Boundary Structure and Related Phenomena, Proceedings of JIMIS 4, Transactions of the Japan Inst. of Metals, 1986, 867-874

# PLASTIC DEFORMATION OF IONIC CRYSTALS

BERNARD SCHAEFFER
8, Allée Bonaparte
91080 - Courcouronnes, France

## ABSTRACT

A single dislocation pile-up was generated in a transparent crystal. Dislocations stresses were measured. They were simulated numerically by two different methods.

## INTRODUCTION

Although ionic crystals are brittle materials, compared to metals and plastics, they may be deformed plastically. Silver chloride has been used as a model of polycrystalline metals [1]. Plastics also have been used as models because of their transparency, allowing photoelastic measurement of stresses. Beyond the yield stress, plastics are no more linear but in ionic crystals birefringence remains proportional to the stress. By applying very small deformations on specially treated single crystals, it is possible to obtain an excess of only a few thousand edge dislocations in a single glide band.

## MATERIALS AND METHODS

### Experimental

Creating point defects by irradiation is a means to vary at will the mechanical properties of materials. Radiation hardening is particularly effective in lithium fluoride crystals. γ-irradiated LiF crystals (24 x 4 x 3 mm) were obtained by cleavage, then partially annealed in their centre, using a few turns of a resistive electrical wire, in order to localise there the plastic deformation and avoid the effect of stress concentrations at the ends of the specimen. They were deformed by compression while observed through crossed polars to visualise stresses by photoelasticity. The applied force was suppressed immediately after the first glide band had appeared [2,3].

**Numerical**

Two types of calculations were performed:

1) The long range stress distribution was calculated using finite differences, using the dislocation distribution along the glide obtained from photoelastic measurements.

2) The mechanical behaviour of the specimen was simulated with Deform2D [4,5],a dynamical, fully non-linear finite differences program, in the elastic-plastic approximation.

## RESULTS

**Experimental**

A single dislocation pile-up, schematised on fig. 1, was created by careful deformation of an irradiated and partially annealed LiF crystal. The stresses are of opposite signs across (edge dislocations) and along the glide plane (dislocation source) (fig. 2).

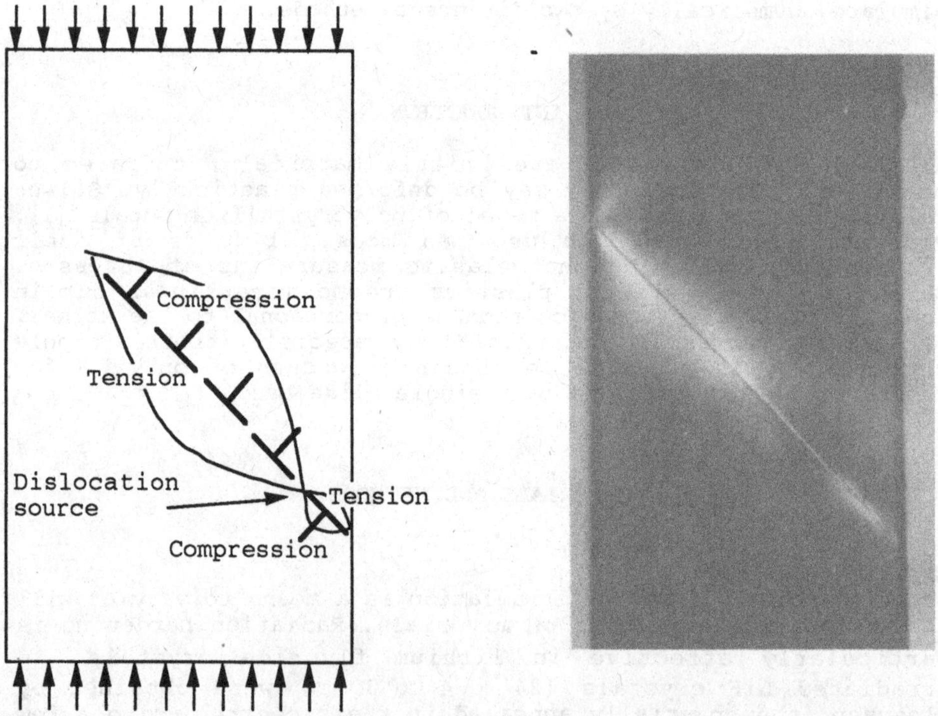


Figure 1. Birefringence around a dislocation pile-up.

The signs of the stresses may be visualised using a sensitive tint plate: one side of the glide plane will be red and the other one blue.The glide plane may be seen on the photograph: it is the fine black line separating two light regions corresponding to the tensile and compressive stresses created by the edge dislocations.

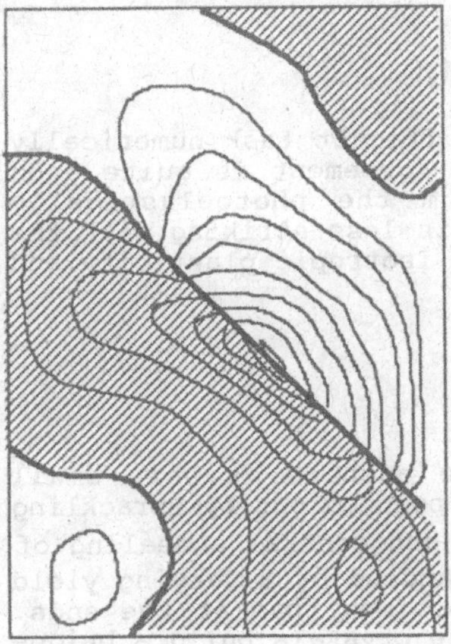

Figure 2. Calculated stress distribution around a dislocation pile-up. - Equidistance of curves is 2 Mpa. From assumed dislocation distribution, obtained from photoelastic measurement along the glide plane and calculated by finite differences, solving the biharmonic equation. Hatched regions correspond to tensile stresses. This is very close to experimental results [3].

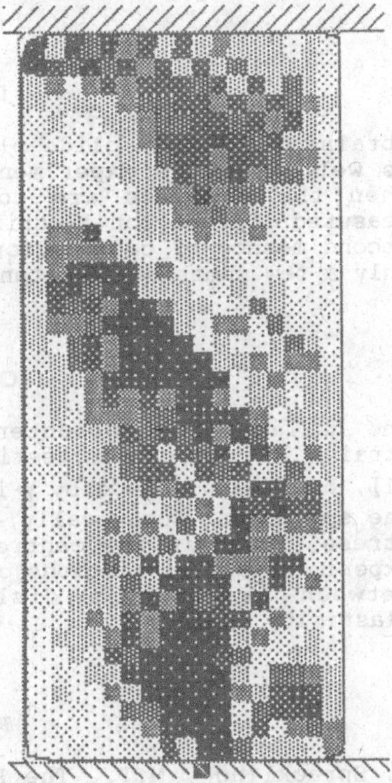

Figure 3. Stress distribution around a glide band, calculated with Deform2D. - The maximum shear in a compression test, being near 45 °, coincids with the glide directions of LiF. At the centre of the specimen the yield stresses are four times smaller than at the ends. Although the stress distribution is not very distinguishable in black and white, the glide band is visible on the picture.

**Numerical**
Figure 2 and 3 show the results of calculations of the stresses produced by plastic deformation in a glide band. On fig. 2, the dislocation distribution is assumed and on fig. 3, a simple shear criterion is used, the material having upper and lower yield stresses.

## DISCUSSION

Obtaining a single glide plane is a difficult task numerically as well as it is experimentally. The agreement is quite good when the stresses are computed from the photoelastically measured dislocation distribution. It is less striking with the second method of calculation, based on isotropic plasticity and only a few numerical constants.

## CONCLUSIONS

The difficulties encountered by some authors at very small strains in compression testing, as was pointed out by Sprackling [1], have been solved by γ-irradiation and central annealing of the specimens. Numerically, this was handled by a varying yield stress, lower in the centre of the specimen than at the ends. Experiment and simulations on LiF single crystals build a bridge between microscopic dislocation theory and macroscopic plasticity theory.

## REFERENCES

1. Sprackling , M.T., The Plastic Deformation of Simple Ionic crystals, Academic Press, London, 1976.

2. Schaeffer, B., Dupuy, C., Saucier, H., Etude photoélastique des dislocations dans LiF coloré, phys. stat. sol., 1965, 9,pp. 753-765.

3. Schaeffer, B., Etude photoélastique d'un empilement de dislocations dans LiF coloré, B.,Bull. Soc. franc. Minér. Crist., 1966, LXXXIX, pp. 297-306.

4. Schaeffer, B., Deform2D: Microcomputer simulation and fracture of solid parts using numerical methods. In Proceedings of the fourth SAS-World Conference, FEMCAD 88, Paris 17-19 October 1988, pp. 133-142.

5. Schaeffer, B., Numerical simulation of Brittle fracture in ductile materials, 8th Biennial European Conference on Fracture, ECF8, Torino, October 1-5, 1990, pp. 885-890.

# SIMULATION OF DISLOCATION PATTERN FORMATION BY CELLULAR AUTOMATA

ELMAR STECK and HANFRIED W. HESSELBARTH
Institut für Allgemeine Mechanik und Festigkeitslehre
Technische Universität Braunschweig, Germany

## ABSTRACT

A computer simulation of the dynamics of dislocations is presented. A two-dimensional Cellular Automata model proposed by Kubin and Lépinoux is extended. Randomly distributed dislocations self-organize under an applied shear stress in a hierarchy of patterns. Simulations of monotonic and cyclic deformation demonstrate the contribution of dislocation arrangement to hardening and the Bauschinger effect.

## INTRODUCTION

Dislocations usually build up characteristic bi- or three- dimensional patterns during monotonic or cyclic plastic deformations. Tangles, walls, veins and more or less complex cell structures are observed by electron microscopy. Several models have been proposed to describe these structures and their relation to macroscopic parameters like yield stress, temperature and strain rate.

Recently Lépinoux and Kubin [1], and Ghonhiem and Amodeo [2] presented simulation models in two dimensions to describe the evolution of dislocation arrangements. While Ghoniem and Amodeo used a technique based upon molecular dynamics methods, Lépinoux and Kubin used the frame-work of Cellular Automata. This second approach is the basis for the present study. Both techniques account for the short range dislocation interactions like annihilation, dipole and multipole formation and dislocation multiplication and to the long range interactions due to the elastic stress fields induced by each dislocation.

## SIMULATION METHOD

The simulation technique used here is based on the concept of Cellular Automata (CA). In CA models space and time are discretized. The possible states of each individual cell is one of a set of discrete values. The system of cells evolves simultaneously in discrete time steps according to a set of predefinded rules. The new state of a cell is determined by its old state and the states of the cells in a given neighbourhood in the previous time step. Although this method is completely deterministic, very simple one- or two-dimensional automata can show very complex behaviour [3]. The short range interactions of dislocations can be treated with a CA model. The long range stress fields which are responsible for the self-organisation must be taken into account by introducing an extended "neighbourhood" to sum up the influence of dislocations up to a given distance.

The system considered here is a configuration of parallel dislocation segments of the same character: screw or edge. The simulation is two-dimensional. The dislocations can glide along their slip plane or normal to it (climb of edges, cross slip of screws). The simulated area is discretized in cells of a fixed size. Every square cell can be occupied by one dislocation with Burgers vector $\pm b$ or it is empty. The cell size should be smaller than the critical annihilation distance to allow two dislocations of opposite sign to annihilate by cross slip or climb, respectively. This critical distance was measured for copper in fatigue at room temperature by Essmann and Mughrabi [4] for screw dislocations $y_c$= 50 nm and for edges $y_c$= 1.6 nm. These data show that the resolution of the simulation should be at least 100 atom distances for screws and only 5 for edge dislocation systems. To simulate effects with a characteristic length of 1 $\mu$m like persistant slip bands (PSB), one needs at least 10 $\mu$m of simulated size, resulting in 6250 cells in one direction.

The dislocation motion is controlled by the local balance of forces. In the slip plane the sum of applied shear force $F_a$ and the interaction forces $F_{ij}$ due to the stress field of other dislocations must exceed a given friction force $F_f$

$$|F_a + F_{ij}| > F_f.$$

For the normal direction the same inequality is used, but the friction force is set higher to account for the more difficult processes of cross slip and climb. The maximum absolute force decides whether the dislocation moves in the glide direction or normal to it. The interaction forces are calculated for each dislocation in each time step according to the Peach - Koehler relation. The components of the stress tensor for straight dislocations are given by linear-elastic theory of dislocations [5]. The short distance events of annihilation are carried out according to a list of possible cases. Dislocation multiplication is introduced in a global manner. The number of dislocations can be fixed. Then, all annihilated dislocations are replaced by pairs of dislocations which are randomly spread into free spaces. The generation rate can be set to a constant value or it can be proportional to the total strain.

## RESULTS

### Monotonic deformation

Material data were taken for copper at room temperature. Fig. 1 shows a simulation of monotonic deformation of an initially random configuration of edge dislocations after 2500 time steps. The automata field consists of 6,250 x 2,500 cells equal to 10 $\mu$m x 2 $\mu$m. The dislocation density was set to $1.5 \cdot 10^{14}$ m$^{-2}$ (6000 dislocations). The cutoff distance is 5 $\mu$m in the glide plane and 0.96 $\mu$m in the climb direction. All annihilated dislocations are replaced by new pairs at random positions. The configuration obtained consists of a wall structure perpendicular to the glide direction. Different evolving regions are in conflict with each other. The average spacing of the density peaks is in the order of 1 $\mu$m in agreement with experiments. If the cutoff distance in the normal direction is chosen close to the field hight, the periodic boundary conditions lead to a stabilizing effect for the wall formation [6]. Gulluoglu et al. [7] found a wall spacing in a molecular dynamic simulation, which was the same as the screening distance in the glide direction. This can be found in our simulations for smaller cutoff distances too. In the examples shown, the cutoff distance is five times higher than the wall spacing observed. The stress-strain curve (Fig. 2) shows hardening. To keep the given strain rate the applied force must rise. This leads to breakage of some weak parts of the walls.

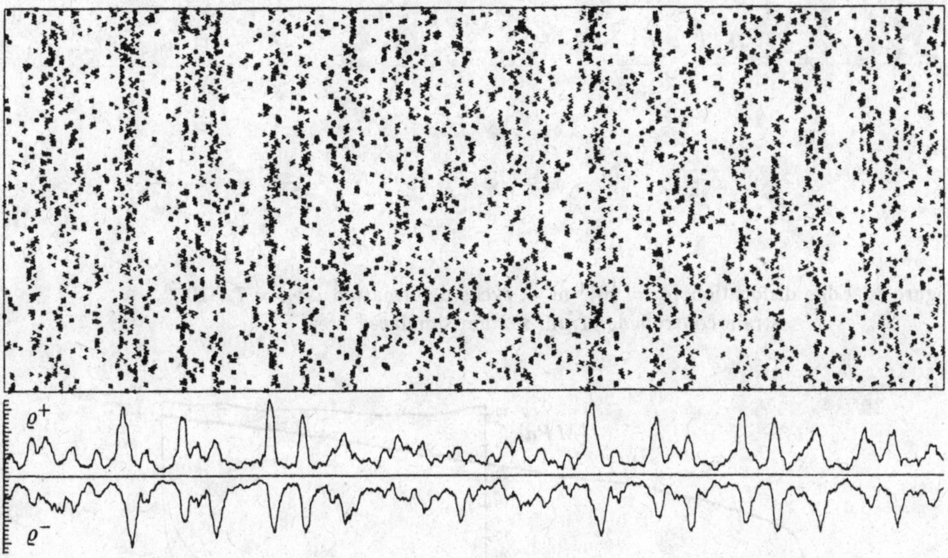


Figure 1. Edge dislocations, $\varrho = 1.5 \cdot 10^{14}\ m^{-2}$, strain controlled, 10 $\mu$m x 4 $\mu$m field. Below: Density per row of positive and negative dislocations.

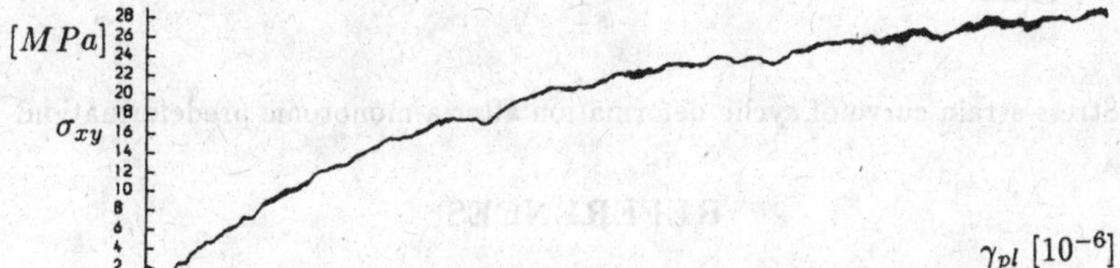


Figure 2. Stress-strain curve, edge dislocations, monotonic deformation.

The dislocations are freed again. The walls are disturbed and a quasi stable state is reached, if only a number of walls corresponding to the cutoff distance remains. Each breakage of a wall leads to a stress drop.

## Cyclic deformation

In cyclic deformation one finds walls of alternating Burgers vector (Fig. 3). The system develops to a limit cycle configuration, which depends on the stress or strain amplitude. The resulting stress-strain curve in Fig. 4 is an example of a cyclic deformation after a monotonic predeformation. The simulated system shows kinematic hardening, the Bauschinger effect and cyclic softening because of the given small strain amplitude in cyclic deformation.

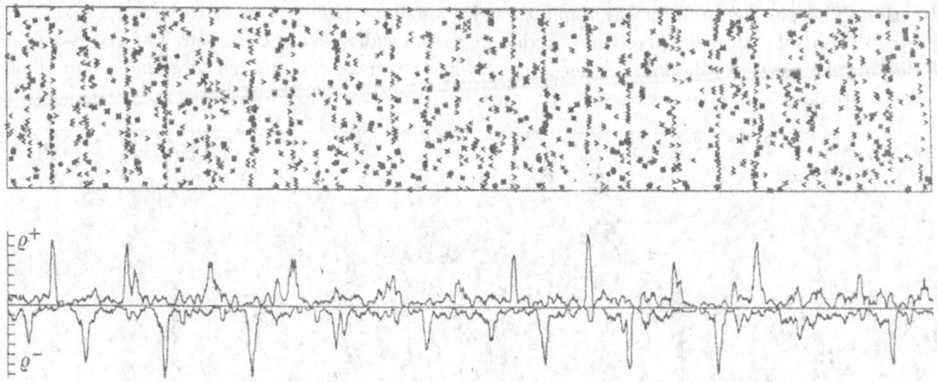


Figure 3. Edge dislocations, $\varrho = 10^{14}$ m$^{-2}$, cyclic deformation $\Delta\gamma_{pl} = 7.7 \cdot 10^{-4}$, strain controlled, 11 $\mu$m x 2 $\mu$m simulated area.

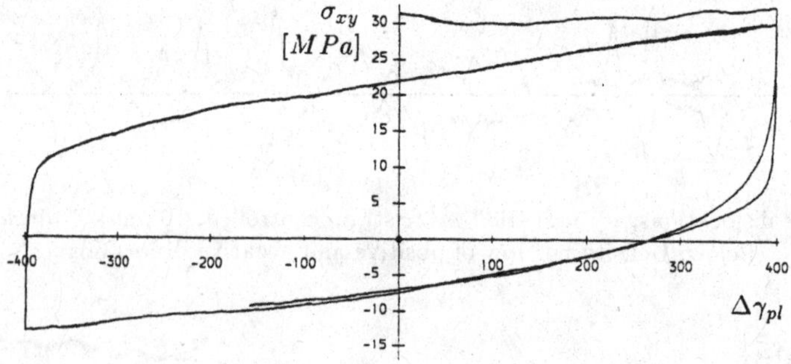


Figure 4. Stress-strain curve of cyclic deformation after a monotonic predeformation.

## REFERENCES

[1] Lépinoux, J. and Kubin, L.P., Scripta Met., 1987, **21**, 833-38.

[2] Ghonhiem, N. M. and Amodeo, R., in Nonlinear Phenomena in Material Science, ed. G. Morton and L. Kubin, Trans. Tech. Publications, Aedermannsdorf, Switzerland, 1988.

[3] Vichniac, G. Y., Physica 10 D, 1984, pp. 96-116.

[4] Essmann, U. and Mughrabi, H., Phil. Mag. A, 1979, **40**, 731-56.

[5] Hirth, J. P. and Lothe, J., Theory of Dislocations, McGraw-Hill, New York, 1968.

[6] Kubin, L. P. and Lépinoux, J., in Strength of Metals and Alloys (ISCMA8), ed. P.O. Kettunen, T.K. Lepistö, M.E. Lehtonen, Pergamon Press, Oxford, 1988, Vol. 1, pp. 35-59.

[7] Gulluoglu, A.N., Srolovitz, D.J., LeSar, R. and Lomdahl, P.S., Scripta Met., 1989, **23**, 1347-52.

# TEXTURE VS. MICROSTRUCTURE IN ANISOTROPIC PLASTICITY

CRISTIAN TEODOSIU
LPMTM-CNRS, Université Paris-Nord
Av. J.-B. Clément, 93430 Villetaneuse, France

## ABSTRACT

The aim of this paper is to review some of the most significant aspects of the texture and microstructural evolution during large plastic deformation of single-phased metallic materials at low temperatures. The implications for the modelling of the anisotropic plastic behaviour are considered within the framework of the internal-variable approach.

## EVOLUTION OF THE INTRAGRANULAR MICROSTRUCTURE

Large plastic, low-temperature deformation of single-phased metallic materials induces significant and specific changes in the microstructure, mainly characterized by dislocation recovery superimposed on strain-hardening and by the formation and/or dissolution of dislocation walls, sheets, and veins (1,2).

The experimental evidence concerning these microscopic events, as revealed by sequences of monotonic tests on single crystals and on polycrystalline materials, is briefly reviewed. Then, a formalism based on an internal variable approach is proposed, which aims at capturing the essential features of the microstructural evolution. The main variables involved characterize the dislocation arrangement in tangled, cell, or banded structures, as well as the influence of these structures on the mean free path of mobile dislocations and on the local stress fields. The equations describing the evolution of these structures take into account the changes produced by monotonic or reversed deformation and, more generally by a varying deformation path.

One of the most striking features of the microstructural organization inside the grains is that dislocation structures evolve towards some steady-state configurations, provided that a sufficient amount of monotonous deformation is allowed for along the same deformation path. Reversed deformation and changes in the deformation mode generally tend to the modification or dissolution of preformed structures and the formation of new ones that correspond to the last deformation mode (3). The transients occurring in the macroscopic bahaviour (e.g. in the stress-strain curves) are in general the phenomenological counterparts of the evolution of some microstructural internal variables towards their steady-state values.

## INTERGRANULAR EFFECTS

The *local texture*, i.e. the crystallographic mismatch across the grain boundaries and the differences in size between the neighbouring grains, may have significant consequences on the heterogeneity of plastic deformation at the grain scale. In particular, it may lead to the partition of the grains in subdomains of different slip activity, as well as to stress concentrations, not only at the triple lines and quadruple points of the grain boundaries, but also over the whole volume of smaller grains, because of the constraints imposed by their neighbours (4,5). Finally, such intergranular effects may be responsible for strain localization and damage. Experimental stress-strain curves are in many cases a rough average representation of the behaviour of a material that may be deforming in a very heterogeneous way.

## THE PLASTIC SPIN

The kinematic description of large deformation plasticity is based on the multiplicative decomposition of the deformation gradient into elastic and plastic parts. On the other hand, the use of this polar decomposition for materials with initial and/or induced plastic anisotropy requires a precise and convenient choice of the director frame that defines the orientation of the intermediate, relaxed configurations.

For a *single crystal* or a grain of a crystalline aggregate the most simple choice is that of isoclinic, relaxed local configurations (6). Namely, these configurations are chosen in such a way that the corresponding lattice orientations be the same for all particles and at any time. Clearly, this choice provides the simplest description of the elastic response, the elastic moduli being almost the same for all particles and at any time. In addition, it leads to a definition of the plastic deformation that is in agreement with the fact that the dislocation glide does not change the average orientation of the lattice, unless the boundary conditions impose a rigid-body and/or a non-uniform rotation of the crystal.

For a *polycrystal*, we adopt after Mandel (7) a kinematic definition of the director frame, its spin being given by the average of the lattice spins of the grains over the macroscopic volume element. This particular choice has some important consequences. First, a maximum simplification of the description of the thermoelastic response is achieved, because the thermoelastic moduli are approximately constant with respect to the director frames of the intermediate configurations. Next, and the most important, the normality structure of the plastic or viscoplastic constitutive laws can be transferred from the grain level to the polycrystalline aggregate. Specifically, this leads to the existence of a plastic or a viscoplastic potential for the plastic strain rate *and* the plastic spin tensors. Finally, the evolution of the texture appears as a consequence of the change in the dispersion of the lattice spin of the crystallites with respect to its mean value (8). The latter, i.e. the plastic spin, incorporates in particular any overall rotation effects, such as the sample spin.

Specific forms of the constitutive laws for the plastic strain rate and plastic spin tensors can be derived by choosing a particular form of the plastic potential (8,9).

## TEXTURE VS. MICROSTRUCTURE

Both the crystallographic and morphologic textures and the intragranular dislocation structure may produce a strong anisotropic plastic behaviour (10). While systematic experiments and simulations attempting to separate and evaluate these effects are scarce, some conclusions can still be drawn, especially from the available evidence concerning sequences of deformations along different deformation paths, performed on cold rolled metal sheets. The use of simple shear as last deformation mode of prestrained materials has proved to be a valuable tool in exploring the mechanisms of the microstructural evolution, because of the quasi-absence of deformation inhomogeneities like necking in tension and barrelling in compression (3).

Polycrystalline f.c.c. metals, heavily cold rolled up to 200% equivalent tensile strain, exhibit a strong plastic anisotropy, which may be almost completely accounted for by the crystallographic texture, probably because the dislocations are arranged in thick-walled, equiaxed cells, providing an almost isotropic hardening (11).

Weakly and medium deformed metal sheets (up to 30% equivalent tensile plastic strain in cold rolling) may also show a strong anisotropy, which is, however, less correlated to texture effects and is apparently due to the directionality of the dislocation structures (sheets of high dislocation densities more or less parallel to the active slip planes) (11). Moreover, sequences of tractions and simple shears on rolled metal sheets provide evidence that after some 15 to 20% of equivalent tensile plastic strain the initial anisotropy of the metal sheet gradually disappears behind the anisotropy induced by the new deformation mode (3). Thus, texture and microstructure can evolve rather independently from each other and eventually reach steady-states, after strain transients that can last some tens of percent.

The most effective way of separating texture from microstructure contributions to anisotropic plasticity remains the use of a micro-macro transition models to calculate the yield surfaces corresponding to experimentally measured crystallographic textures (11,12). Nevertheless, this approach is limited by the assumptions made on the plastic behaviour of each grain and eventually on its interaction with the neighbouring grains. More theoretical and experimental effort is needed in order to select the main features that govern this local behaviour and hence the evolution of plastic anisotropy at large deformations.

## REFERENCES


1. Hansen, N. and Kuhlmann-Wilsdorf, D., Low energy dislocation structures due to unidirectional deformation at low temperatures. *Mater. Sci. Eng.*, 1986, **81**, 141-61.

2. Mughrabi, H., Dislocation clustering and long-range internal stresses in monotonically and cyclically deformed metal crystals. *Revue Phys. Appl.*, 1988, **23**, 367-79.

3. Rauch, E.F. and Schmitt, J.-H., Dislocation substructures in mild steel deformed in simple shear. *Mater Sci. Eng.*, 1989, **A113**, 411-48.

4. Rey, C., Effects of grain boundaries on the mechanical behaviour of grains in polycrystals. *Revue Phys. Appl.*, 1988, **23**, 491-500.

5. Rey, C., Mussot P. and Zaoui, A., Grain boundary structure and related phenomena. *Proc. JIMIS4, Suppl. Japan Inst. Metals*, 1986, 867.

6. Teodosiu, C., A dynamic theory of dislocations and its applications to the theory of the elastic-plastic continuum. In *Fundamental Aspects of Dislocation Theory*, eds. J.A. Simmons, R. deWit and R. Bullough, NBS Spec. Publ. 317, Washington D.C., 1970, pp. 837-76.

7. Mandel, J., Définition d'un repère privilégié pour l'étude des transformations anélastiques du polycrystal. *J. Méc. Théor. Appl.*, 1982, 1, 7-23.

8. Sidoroff, F. and Teodosiu, C., Plastic spin and texture evolution during large deformation of polycrystalline materials. *17th Congr. Theor. Appl. Mech.*, Grenoble, 1988 (unpublished).

9. Teodosiu, C., The plastic spin : microstructural origin and computational significance. In *Computational Plasticity. Models, Software and Applications*, eds. D.R.J. Owen, E. Hinton, E. Onate, Pineridge Press, Swansea, U.K., 1989, pp. 163-75.

10. Hansen, N. and Leffers, T., Microstructures, textures and mechanical properties after large strain. *Revue Phys. Appl.*, 1988, **23**, 519-31.

11. Jensen, D.J. and Hansen, N., Relations between texture and flow stress in commercially pure aluminium. In *Constitutive Relations and their Physical Basis*, eds. S.I.Andersen et al., Riso National Lab., Roskilde, Denmark, 1987, pp. 353-60.

12. Raphanel, J.L., Schmitt, J.-H. and Baudelet, B., Effect of a prestrain on the subsequent yielding of low-carbon steel sheets. *Int. J. Plasticity*, 1986, **2**, 371-78.

# ANALYTICAL REPRESENTATION OF POLYCRYSTAL YIELD SURFACES

LASZLO S. TOTH*, PAUL VAN HOUTTE AND ALBERT VAN BAEL
Department MTM, Katholieke Universiteit Leuven, Belgium
*On leave from Eötvös University, Hungary

## ABSTRACT

A new yield function of the form $f(S_i)=\sum_{s}^{N}|P_i^s\, S_i|^{h+1}/W_s^h$ is proposed to describe the yield surface of textured polycrystals. The parameters of this function can be derived from the rate of plastic work obtained from polycrystal models. f is strictly convex and approximates the yield surface with high precision.

## INTRODUCTION

To modellize forming processes, it is necessary to take into account the deformation texture that evolves during straining. The incorporation of the texture development into a finite element code, however, is rather difficult [1-3]. The finite element technique is much more efficient if the yield surface of the material element is known analytically [2,3].

Several efforts have been made to characterize the anisotropy of materials by phenomenological [4-6] as well as crystallographic based yield functions [7-9]. A general description can be based only on the texture being the main factor affecting the shape of the yield surface. In this work, a new, analytical yield surface is described and the method of obtaining its coefficients from the Taylor factors of polycrystalline materials is analyzed.

## COMMON FEATURES OF POLYCRYSTAL AND SINGLE CRYSTAL YIELD SURFACES

In case of polycrystals, the yield surface can be obtained as the inner envelope of the tangent hyperplanes defined by the strain rate vector $\dot{\varepsilon}_k$ and $\tau_c\bar{M}$, $\tau_c$ being the critical resolved shear stress and $\bar{M}$ is the average Taylor factor of the polycrystal (Fig. 1a). The latter is proportional to the rate of plastic work and $\tau_c\bar{M}$ is the distance of a hyperplane from the origin. The inner envelope is then necessarily a rounded surface which is tangent to each hyperplane. The equation of these planes are [10]:

$$\tau_c\bar{M} = \dot{E}_k S_k \tag{1}$$

where $\dot{E}_k = \dot{\varepsilon}_k/\dot{\varepsilon}_{eqv}$ is the 5 component strain rate vector with normalized length and $\dot{\varepsilon}_{eqv}$ is the equivalent strain rate.

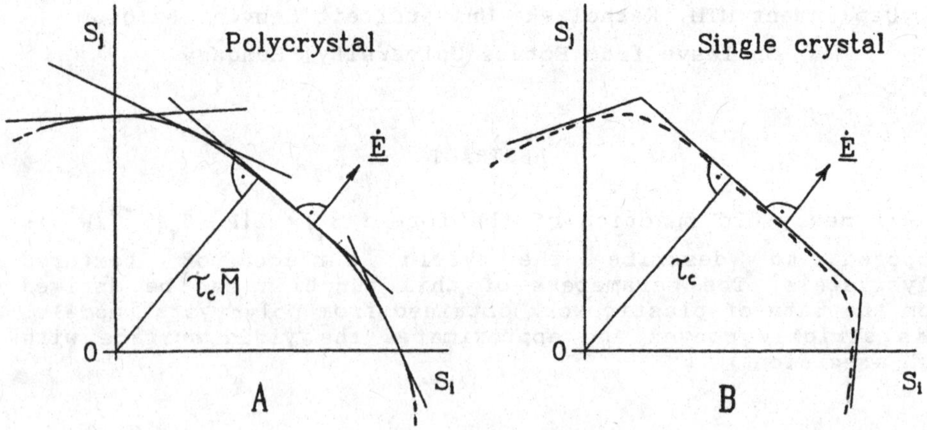


Figure 1. Schematic yield surfaces of polycrystals (a) and single crystals (b).

In case of single crystals, the yield surface is not rounded (Fig. 1b), and consists of the hyperplane-sections of the type

$$\tau_c = m_k^s S_k \tag{2}$$

Here $m_k^s$ is defined by the geometrical position of a slip system. Using the rate sensitive crystallographic slip theory with the constitutive law of

$$\dot{\gamma}_s = \dot{\gamma}_o \operatorname{sgn}(\tau_s/\tau_o^s)\,|\tau_s/\tau_o^s|^{1/m}, \tag{3}$$

it was shown by Tóth et al. [11] that there exists a stress potential function which is rounded, strictly convex, and its limiting m->0 case is identical to the rate insensitive (Bishop and Hill) yield surface. (Here $\dot{\gamma}_o$, $\tau_o^s$ are reference constants, m is the strain rate sensitivity exponent, and $\tau_s$ is the resolved shear stress in slip system s.) Such a stress potential function is shown schematically in Fig. 1b (broken line).

## ANALYTICAL YIELD SURFACE FOR POLYCRYSTALS

There is a strong similarity between the polycrystal yield locus (Fig. 1a) and the single crystal rate sensitive stress potential (Fig. 1b). For low enough m values, the rate sensitive stress potential becomes an inner envelope to the hyperplanes in Fig. 1b. It is therefore suggested here that the rate sensitive stress potential function of the form of single crystals can describe the polycrystal yield surfaces. Such a potential function has the form [11]:

$$f(S)=\sum_s |m_i^s S_i|^{\frac{1}{m}+1} / (\tau_o^s)^{\frac{1}{m}} = \tau_o^{ref}. \qquad (4)$$

Here $\tau_o^{ref}$ is a reference stress factor ($\tau_o^{ref}=1$ belongs to a unit plastic work). It should be noted here that eq. (4) is very similar to the one proposed by Arminjon [12] as a modification to the Schmid law, but is different (in that case: $\sum_s |m_i^s S_i|^{1/m} / (\tau_o^s)^{1/m} = 1$).

To facilitate eq. (4) in the polycrystal case, we consider the tangent planes in eq. (1) as if they were representing slip systems as given in eq. (2). Replacing then the $m_i^s$ values by the $\dot{E}$ vectors prescribed in the polycrystal case, and $\tau_o^s$ by $\tau_c \bar{M}$, eq. (4) becomes

$$\sum_{s=1}^{N} |P_i^s S_i|^{h+1} / W_s^h = \tau_o^{ref}, \qquad (5)$$

where $W_s = \tau_c \bar{M}_s$, $P_i^s = \dot{E}_i^s$ and $h=1/m$. Note that in this case, m or h have nothing to do with strain rate sensitivity. If necessary, strain rate sensitivity could be introduced by making $\tau_c$ as a function of $\dot{\varepsilon}_{eqv}$.

## METHOD TO OBTAIN THE COEFFICIENTS DESCRIBING THE YIELD SURFACE

A very efficient way of obtaining the plastic work for any strain direction is based on the series expansion of the Taylor factors [10]. In this method, the experimental textures can be used as direct input to obtain the $W_s$ quantities. To explore the shape of the yield surface, $W_s$ is calculated for a large number (N) of "sampling directions" for the strain rate vector ($\dot{E}_i^s$). The value of the h parameter is chosen in the range of h>>1.

CONSTRUCTION OF YIELD LOCUS SECTIONS AND PROJECTIONS

A yield locus section is defined by prescribing the direction of the stress only. In this case, the yield point, i.e. the length of the stress vector, can be readily obtained with simple summation procedures. Similarly, the strain rate vector can also be readily calculated for any stress direction.

To obtain a yield locus projection, the strain rate vector has to be specified. In this case, a nonlinear equation system with 5 unknowns for the stress vector has to be solved numerically. This can be done by employing the Newton-Raphson method. In that procedure, the derivatives of the equations are known analytically because of our analytical yield function (eq. 5), which speeds up the calculation time considerably. The mathematical procedures to compute yield locus sections and projections will be described in details in a forthcoming paper.

REFERENCES


1. Mathur. K.K., and Dawson, P.R., Int. J. Plasticity, 1989, 5, 67-94.
2. Van Houtte, P, Mols, K., Van Bael, A., and Aernoudt, E., Textures and Microstructures, 1989, 11, 23-39.
3. Van Bael, A., Van Houtte, P., Aernoudt, E., Pillinger, A., Hartley, P., and Sturgess, C.E.N., this proceedings.
4. Von Mises, R., Gott. Nachricht. Math. Phys., 1913, 582.
5. Hill, R., Proc. Roy. Soc., 1948, A193, 281-297.
6. Hill, R., Math. Proc. Cambr. Phil. Soc., 1979, 85, 179-191.
7. Montheillet, F., Gilormini, P., and Jonas, J.J., Acta Metall., 1985, 33, 705-717.
8. Lequeu, PH., Gilormini, P., Montheillet, F., Bacroix, B., and Jonas, J.J., Acta Metall., 1987, 35, 1159-1174.
9. Arminjon, M., and Bacroix, B., Acta Mechanica, in press.
10. Van Houtte, P., Textures and Microstructures, 1987, 7, 29-72.
11. Tóth, L.S., Gilormini, P., and Jonas, J.J., Acta Metall., 1988, 12, 3077-3091.
12. Arminjon, M., Proc. of 9th Int. Conf. Text. Mat. (ICOTOM9), Avignon, France, Sept. 1990., in press.

# BOUNDS FOR THE OVERALL CREEP BEHAVIOUR OF TEXTURED POLYCRYSTALS

J. R. WILLIS
School of Mathematical Sciences
University of Bath
Bath BA2 7AY, UK.

## ABSTRACT

A method for bounding the overall strain-rate potential of polycrystalline materials will be presented. The formulation contains one or two points of subtlety but the resulting optimization problem is straightforward and can be solved equally easily for any pattern of prescribed loading, and for polycrystals with any texture, giving rise to anisotropic material response. Sample results will be presented for f.c.c. polycrystals, textured so as to display overall transversely isotropic material response, but subjected to general loads that do not respect this symmetry.

## INTRODUCTION

A significant problem associated with the inelastic deformation of real materials is the prediction of their macroscopic behaviour from knowledge of their microstructure. The problem is of interest not only for composites but for any material whose microstructure can be controlled during manufacture. The present work considers polycrystalline metals which may display texture, induced either during initial manufacture or through subsequent operations such as rolling. In most situations, recourse to approximate methods of analysis is unavoidable; schemes of self-consistent type [1, 2, 3] are perhaps the most usual. There are certain classes of problems, however, for which rigorous information can be derived, in the form of bounds. These bounds have some intrinsic interest and, at the more practical level, provide benchmark results against which approximate methods (whose applicability may be wider) can be assessed.

Bounds can be generated when a problem has a formulation as a minimum principle. The present work considers steady-state creep behaviour, during which the integral of a strain-rate potential is minimized over the body, amongst a suitable class of stress fields. The quantity that is bounded directly is the overall strain-rate potential. However, when this is a homogeneous function, of degree $n + 1$ say, bounds on the stress *versus* strain-rate behaviour can be inferred. The methods that are employed are based on a nonlinear generalization [4, 5, 6] of the Hashin-Shtrikman variational principles [7]. Their application to polycrystals requires the numerical solution of an optimization problem in many variables but the problem in fact has a simple structure and its solution is straightforward. The approach has previously been applied to polycrystals without texture by Dendievel, Bonnet and Willis [8].

## THE POLYCRYSTAL

The polycrystalline materials under consideration are composed of a large number of individual crystals, firmly bonded across interfaces (so that grain boundary sliding is not admitted). The steady-state creep deformation of any crystal, when referred to crystallographic axes, is described by a strain-rate potential $F_c$, so that

$$e = F_c'(\sigma) \quad , \tag{1}$$

$\sigma$ and $e$ representing stress and strain-rate tensors, respectively. Relative to a 'laboratory' frame, the orientation of a crystal with label $r$ is defined by a rotation $R_r$, say, and its strain rate potential is

$$F_r(\sigma) = F_c(R_r^{-1}\,\sigma) \quad , \tag{2}$$

where $R_r^{-1}\,\sigma$ represents the operation of the rotation *inverse* to $R_r$, on the stress tensor $\sigma$. In standard tensor notation, this would be written $R_r^T\,\sigma R_r$, the superscript $T$ denoting transpose, but computations are actually performed by representing the stress $\sigma$ as a vector in a space of 6 dimensions.

An individual crystal is taken to deform by slip on a discrete number of slip systems, with system $k$ characterized by a slip plane with normal $n^{(k)}$ and a direction of slip $m^{(k)}$ in the slip plane; $m^{(k)}$ and $n^{(k)}$ are unit vectors. The system is specified by the second-order tensor $\mu^{(k)}$, with components

$$\mu_{ij}^{(k)} = \tfrac{1}{2}(m_i^{(k)}\,n_j^{(k)} + m_j^{(k)}\,n_i^{(k)}) \ . \tag{3}$$

The strain-rate potential $F_c$ is then taken as the sum of individual potentials associated with each slip system:

$$F_c(\sigma) = \sum_k F_c^{(k)}(\tau^{(k)}) \quad , \tag{4}$$

where

$$\tau^{(k)} = \sigma\cdot\mu^{(k)} \tag{5}$$

is the resolved shear stress on the $k$th system.

Although much of the formulation is general, detailed computations are performed for f.c.c. crystals, with four slip planes of the (1, 1, 1) type and three slip directions per plane, of the (1, 1, 0) type, giving 12 slip systems in all. Power-law behaviour is assumed, so that

$$F_c^{(k)}(\tau) = \frac{\alpha\tau_0^{(k)}}{n+1}\left|\frac{\tau}{\tau_0^{(k)}}\right|^{n+1} \tag{6}$$

The normalizing stresses $\tau_0^{(k)}$ could be different for different systems, depending on prior deformation, but they will, in fact, be taken all the same, and equal to $\tau_0$.

## OVERALL BEHAVIOUR

The problem to be addressed concerns polycrystalline material, as described above, which occupies a domain $\Omega$. Units are chosen, for convenience, so that $\Omega$ has unit volume; it is assumed, however, that microstructural dimensions are much smaller than any length scale characterizing $\Omega$ so that, macroscopically, the polycrystalline material appears uniform, with strain-rate potential $\tilde{F}(\sigma)$, say. This overall strain-rate potential can be studied (either theoretically or experimentally) by subjecting $\Omega$ to boundary loads that generate a mean stress $\bar{\sigma}$. The 'complementary energy' principle then yields

$$\tilde{F}(\bar{\sigma}) = \inf_{\sigma \in K} \int_\Omega F(\sigma; x)\, dx \quad , \tag{7}$$

where $F(\sigma; x)$ represents the local strain-rate potential, taking the value $F_r(\sigma)$ in a crystal with label $r$. The infimum in (7) is evaluated over the set $K$ of stress fields that have zero divergence (in the sense of generalized functions) and mean value $\bar{\sigma}$ over $\Omega$:

$$K = \{\, \sigma : \operatorname{div}\sigma = 0\,,\, x \in \Omega \text{ and } \int_\Omega \sigma\, dx = \bar{\sigma} \,\}\,. \tag{8}$$

The definition (7) gives the 'smallest possible' $\tilde{F}(\bar{\sigma})$, since boundary conditions are not specified; it is, in fact, realized when boundary velocities of the form $u_i = \bar{e}_{ij}x_j$ are prescribed over $\partial\Omega$ [9].

The present work is devoted to the construction of bounds for the overall potential $\tilde{F}(\bar{\sigma})$, in the case that the polycrystal displays texture (so that not all rotations $R_r$ are equally likely). The method follows that already applied to the case of untextured polycrystals by Dendievel, Bonnet and Willis [8]. Limitations of space preclude a full description in this outline. However, the use of the variational formulation can be briefly indicated. First, directly from (7), an elementary upper bound can be constructed by substituting into the integrand in (7) the uniform stress $\sigma = \bar{\sigma}$. This yields

$$\tilde{F}(\bar{\sigma}) \le \bar{F}(\bar{\sigma}) \quad , \tag{9}$$

where $\bar{F}(\bar{\sigma})$ represents the mean value of $F(\bar{\sigma}, x)$ over $\Omega$ or, equivalently, the mean value of $F_r(\bar{\sigma})$ over all rotations $R_r$; the bound is crude but evidently sensitive to texture. A family of lower bounds is constructed by introducing a 'comparison' potential $F_0(\sigma)$ and forming a 'dual' potential,

$$(F - F_0)^*(\eta, x) = \sup_\sigma \left[\sigma\cdot\eta - (F - F_0)(\sigma)\right] \quad , \tag{10}$$

so that, for any $\sigma$ and $\eta$,

$$F(\sigma, x) \ge \sigma\cdot\eta + F_0(\sigma) - (F - F_0)^*(\eta, x) \ . \tag{11}$$

It follows from (7) and (11) that

$$\tilde{F}(\bar{\sigma}) \geq \inf_{\sigma \in K} \int_\Omega [\sigma \cdot \eta + F_0(\sigma) - (F - F_0)^* (\eta, x)] \, dx \quad . \tag{12}$$

If the comparison potential $F_0(\sigma)$ is chosen to be quadratic, the minimization required by (12) generates a linear problem that can be solved; the remaining problems then relate to the optimal selection of the parameters defining $F_0$ and the 'polarization' fields $\eta(x)$. The polycrystal is sufficiently complex for these remaining optimization problems to present a challenge, requiring the introduction of further tricks. The result, however, is a formulation which is easy to implement for *any* prescribed $\bar{\sigma}$ and is no more difficult for textured polycrystals than it is for polycrystals without texture; this contrasts, for example, with the self-consistent method of Hutchinson [3], whose implementation is straightforward only for problems displaying axial symmetry. A more complete account, with results, will be presented in the lecture and the full text.

## REFERENCES


1. Hill, R., Continuum micromechanics of elastoplastic polycrystals. *J. Mech. Phys. Solids*, 1965, **13**, 89-101.

2. Hutchinson, J.W., Elastic-plastic behaviour of polycrystalline metals and composites. *Proc. R. Soc. Lond. A*, 1970, **319**, 247-272.

3. Hutchinson, J.W., Bounds and self-consistent estimates for creep of polycrystalline materials. *Proc. R. Soc. Lond. A*, 1976, **348**, 101-127.

4. Willis, J.R., The overall elastic response of composite materials. *J. Appl. Mech.*, 1983, **50**, 1202-1209.

5. Willis, J.R., Variational estimates for the overall response of an inhomogeneous non-linear dielectric. In *Homogenization and Effective Moduli of Materials and Media*, ed. J.L. Ericksen, D. Kinderlehrer, R.V. Kohn and J.-L. Lions, *The IMA Volumes in Mathematics and Its Applications*, Vol.1, Springer-Verlag, New York, 1986, pp.245-263.

6. Talbot, D.R.S. and Willis, J.R., Variational principles for inhomogeneous nonlinear media. *IMA J. Appl. Math.*, 1985, **39**, 215-240.

7. Hashin, Z. and Shtrikman, S., On some variational principles in anisotropic and nonhomogeneous elasticity. *J. Mech. Phys. Solids*, 1962, **10**, 335-342.

8. Dendievel, R., Bonnet, G. and Willis, J.R., Bounds for the creep behaviour of polycrystalline materials. In *Inelastic Deformation of Composite Materials*, ed. G.J. Dvorak, Springer-Verlag, New York, 1991, pp.175-192.

9. Willis, J.R., The structure of overall constitutive relations for a class of nonlinear composites. *IMA J. Appl. Math.*, 1989, **43**, 231-242.

# BEHAVIOR OF GRANULAR AND POROUS MEDIA

# LIMIT LOADS OF REINFORCED STRUCTURES

by

R.ABDI* and J.PASTOR**
* INSTITUT DE MECANIQUE DE GRENOBLE-FRANCE
** LABORATOIRE MATERIAUX COMPOSITES DE CHAMBERY-FRANCE

## ABSTRACT

Various authors such as Le Nizerhy [1], De Buhan [2], Suquet [3] have proposed homogenization techniques for heterogeneous materials. In the present paper a kinematic formulation for limit analysis of structures in reinforced soil is presented. This formulation is based on De Buhan criterion and on a finite element discretization of the homogenized material. The subsequent computer codes are used for determining upper bounds of the limit height of reinforced slopes.

## INTRODUCTION

Direct study through the finite element method of reinforced materials is in fact limited to a finite number of reinforcements. Thus, the use of a homogenization method appears to be an interesting solution. In the present study, it is assumed that the number of the periodically distributed reinforcements is large "enough", so that the chosen criterion plasticityis the De Buhan's one [1], for both thin and very strong reinforcements. After presenting this criterion, it is suggested a finite element formulation for the kinematic method of limit analysis. The so - defined computer codes are then applied to the vertical slope stability problem with Coulomb reinforced material.

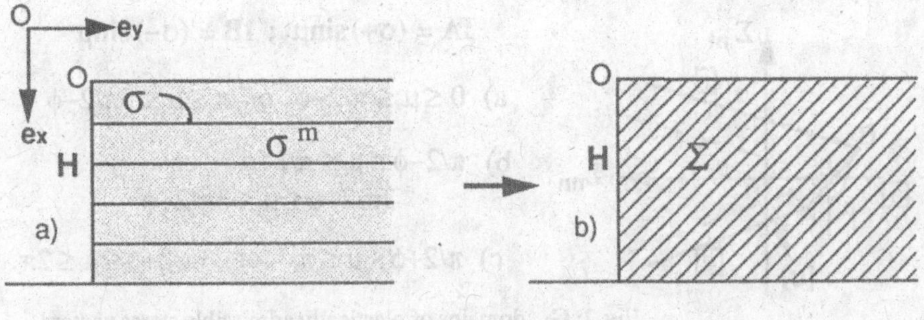


Fig.1: The homogenized material.

## THE PLASTICITY CRITERION

The criterion proposed by De Buhan [2] , in the case of thin and periodically distributed inclusions in plane strain, is the following:

$$\Sigma = \sigma^m + \sigma\, e_y \otimes e_y$$

$$F^{hom}(\Sigma) \le 0 \Leftrightarrow \qquad (1)$$

$$f(\sigma^m) \le 0 \ \text{ et } \ -\sigma^- \le \sigma \le \sigma^+$$

In this relationship $f(\sigma^m)$ is the yield criterion of the unreinforced soil; $\sigma^-$ and $\sigma^+$ are respectively the mean simple tension and compression strength of all the real inclusions(Fig.1).

## THE KINEMATIC APPROACH

Let the $f_i(\sigma^m) \le 0$ set - with i=1 to n, n≥3 - be the convex linear polyhedron approaching from the outside the real Coulomb criterion f. The associated strain rate tensor for the homogenized material is given by :

$$v_{xx} = \sum_{i=1}^{n} \lambda_i \left(\frac{\cos(2\pi i/n)}{\cos\phi} - \tan\phi\right) \quad ; \quad v_{xy} = \sum_{i=1}^{n} \lambda_i \left(\frac{\sin(2\pi i/n)}{\cos\phi}\right)$$

$$v_{yy} = \sum_{i=1}^{n} \lambda_i \left(-\frac{\cos(2\pi i/n)}{\cos\phi} - \tan\phi\right) = \lambda^+ - \lambda^- \quad ; \quad \lambda_i, \lambda^+, \lambda^- \ge 0 \qquad (2)$$

In the case of jump velocity line, the domains of admissible stress vectors in Mohr plane ($\Sigma_{nn}$, $\Sigma_{nt}$ ) is needed. Three cases are found, according to the value of the angle $\theta = (ox,\mathbf{n})$ which is the angle of the **n** normal of the concerned facet to ox (Fig.2).

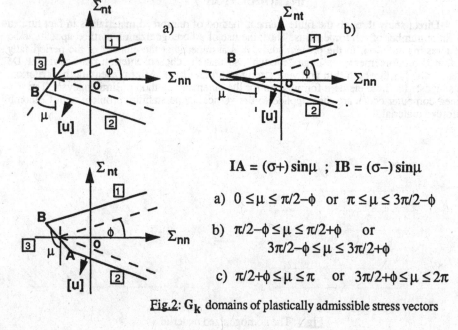

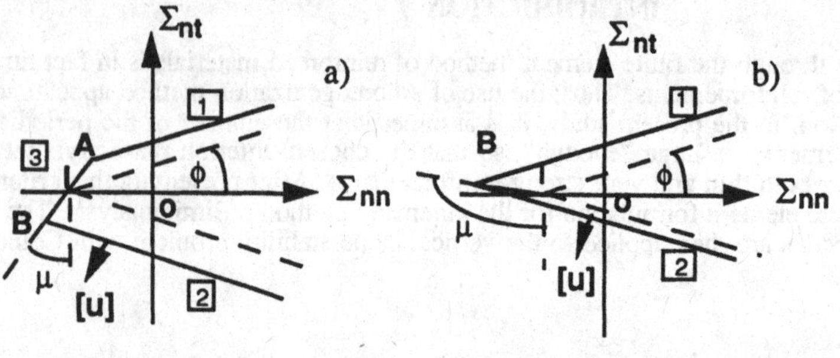


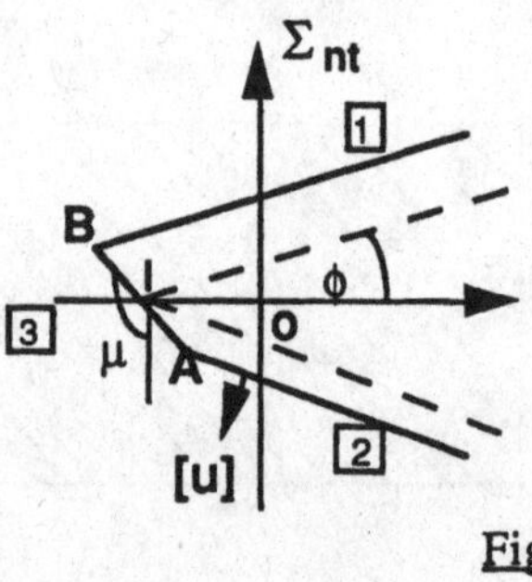


IA = (σ+) sinμ ; IB = (σ–) sinμ

a) $0 \le \mu \le \pi/2-\phi$ or $\pi \le \mu \le 3\pi/2-\phi$

b) $\pi/2-\phi \le \mu \le \pi/2+\phi$ or $3\pi/2-\phi \le \mu \le 3\pi/2+\phi$

c) $\pi/2+\phi \le \mu \le \pi$ or $3\pi/2+\phi \le \mu \le 2\pi$

Fig.2: $G_k$ domains of plastically admissible stress vectors

Using the normality law, the plastically admissibility conditions for the jump $[u] = ([u]_n, [u]_t)$ are determined. **u** is the displacement velocity, and the variables $\mu_k$, k=1 to 2 or 1 to 3, are positive:

$$[u]_n = \sum_k \mu_k \frac{\partial G_K}{\partial \Sigma_{nn}} \quad \text{and} \quad [u]_t = \sum_k \mu_k \frac{\partial G_K}{\partial \Sigma_{nt}} \qquad (3)$$

The final unit dissipated power $\pi^{hom}$ is the sum of the volumic term and that due to the lines of velocity discontinuity, i.e. :

$$\pi^{hom}_{vol} = \text{Min}\,(2\,c\,\Sigma\,\lambda_i + \sigma^- \lambda^- + \sigma^+ \lambda^+)\,;\ \pi^{hom}_{disc} = \text{Min}\,(\Sigma\,A_k\,\mu_k) \quad (4)$$

$$\lambda_i,\ \lambda^+,\ \lambda^- \geq 0 \quad \text{and} \quad \mu_k \geq 0$$

in which $A_k$ is the constant term of the frontier equations of figure 2, too long to be detailed here. The final problem, including discretization and other admissbility conditions, is solved by a linear programming code on Apple MacIIci or Fx.

## TESTS AND COMPARISONS

The application of the two built finite element codes - the static [4] and the kinematic one - was performed, in the case of the vertical slope, with a reinforced Coulomb material. The stability factor $\gamma H/c$ - H, height of the slope - $\gamma$, its volume weight - c, soil cohesion, - is a function of $\phi$, $\sigma^+/c$ and $\sigma^-/c$. In order to show the interest of such approaches, the two following sets of particular values of the above parameters are selected.

**a)** ***$\phi$ variable, $\sigma^+ = 0$, $\sigma^- = \sigma$, $c = 0$***

The stability factor then depends only on $\phi$. Cancelling $\sigma^+$ aims at simulating reinforcement non resistance under compression (caused by local buckling for instance) when side stresses are low.

| $\phi$ | 0 | 5 | 10 | 15 | 20 | 25 | 30 | 35 |
|---|---|---|---|---|---|---|---|---|
| $(\gamma H/\sigma)$ log spiral | 1,33 | 1,70 | 2,16 | 2,74 | 3,46 | 4,37 | 5,50 | 6,89 |
| $(\gamma H/\sigma)$ stat [4] | 1,00 | 1,50 | 2,02 | 2,62 | 3,34 | 4,24 | 5,30 | 6,61 |
| $(\gamma H/\sigma)$ stat [5] | 1,00 | 1,19 | 1,42 | 1,70 | 2,04 | 2,46 | 3,00 | 3,69 |

Table 1 : Tests with pulverulent soil

Here, by using the logarithmic spiral mechanism - see location of [u] in figure 2 - , the kinematic computed values are in agreement with those given by Siad [5]. It can be remarked that the static programme [4] strongly improves the original values obtained by this author.

**b)** *$\phi$ variable, $\sigma^+/c = \sigma^-/c = 2$*

This problem, where $\sigma^+ = \sigma^-$, stands for the case in which the original criterion of the reinforcement is Tresca's, for example.

| $\phi$ | 0 | 10 | 15 | 20 | 25 | 30 | 35 |
|---|---|---|---|---|---|---|---|
| ($\gamma$H/c) log spiral | 7,17 | 9,60 | 11,14 | 12,96 | 15,16 | 17,84 | 21,20 |
| ($\gamma$H/ c) stat [4] | 6,58 | 9,11 | 10,61 | 12,41 | 14,57 | 17,30 | 20,33 |
| ($\gamma$H/ c) kine.(FEM) | 7,14 | 9,97 | 11,92 | 13,99 | 16,53 | 19,26 | 23,81 |

Table 2: Tests with cohesive frictional soil

For $\phi$=0, we could even improve the upper bound given by the logarithmic spirale mechanism. This is a very difficult challenge, well known by all the authors of the kinematic approach. The problem of the slope being by far the most intricate, the two finite element programs prove thus both their robustness and efficiency.

## CONCLUSIONS

The finalized programs are efficient tools as far as reinforced soils are concerned. The linear programming code is now to be improved so that the two programs run fastly on MACIIfx, the CPU time being approximately 30' in case b ( 94 triangles, discontinuity everywhere, 942 lines, 3010 columns ).

One interesting application will be to extend this approach to the field of plane stress, closer to usual composite materials. In particular, this will allow to make an interesting comparison with previous work by Le Nizerhy [1], who is, to our knowledge, one of the first to use Limit Analysis in this area.

## REFERENCES

1 - D. Le Nizerhy, Calcul à la Rupture des Matériaux Composites, Symposium franco-polonais, Cracovie, 1977.

2 - P. De Buhan, Critère de Rupture d'un Matériau Renforcé par Armatures, C.R.A.S.,t 301, série II, n°9, Paris, 1983.

3 - P. Suquet, Analyse Limite et Homogénéisation, C.R.A.S., t 296, série II, Paris, 1983 .

4 - Pastor J., Abdi R., Analyse Limite et Homogénéisation : approche numérique, 9ème Congrès Français de Mécanique, Metz, 1989.

5 - Siad L., Analyse de Stabilité des Ouvrages en Terre Armée par une Méthode d'Homogénéisation, thesis, E.N.P.C., Paris, 1987.

# ELASTOPLASTIC MODELING FOR SOFT ROCKS AND CERAMICS IN THE DUCTILE REGIME

MICHEL AUBERTIN, DENIS E. GILL,
BRANKO LADANYI, ROBERT CHAPUIS
École Polytechnique
C.P. 6079, Succ. A
Montréal, Qc, Canada
H3C 3A7

## ABSTRACT

A viscoplastic model that has been developed for the ductile inelastic behavior of some soft rocks (such as rocksalt and potash) and ceramics (such as alkali halides) is reduced to a basic rate independent or inviscid elastoplastic model. The proposed model includes a von Mises type yield criterion, an associate flow rule (under compressive stresses) and three state variables producing kinematic and isotropic hardening.

## INTRODUCTION

The inelastic behavior of polycrystalline sodium chloride, which is considered to be representative of that of various soft rocks (such as rocksalt and potash) and ceramics (such as alkali halides), is relatively well known. It is usually recognized that, in the ductile regime under compressive stresses, the inelastic strains occurs at an almost constant volume, and that these materials obey the classical normality rule. Under such condition, the inelastic flow can be treated with an associated flow rule and mixed hardening.

A viscoplastic model, recently developed, has been used to describe the inelastic behavior of such materials. This model, named SUVIC (Strain rate history-dependent Unified Viscoplastic model with Internal variables for Crystalline materials), comprises a kinetic law, expressed as a power law function, and three evolution laws associated with three state variables, $B_{ij}$, $R$ and $K$. The first two of these state variables are internal stresses, which generally oppose the applied stress, generating a mixed (isotropic and kinematic) hardening. The third state variable is a scalar used to normalize the active stress [1, 2]. The significance of the various constants and variables included in the SUVIC model equations, as well as the procedure for determining the value of the constants, have been discussed elsewhere [2, 3].

The objective of this paper is to present an elastoplastic version of the SUVIC model, which is believed to be an adequate model for describing the rate independent inelastic behavior of some soft rocks and ceramics in the ductile regime, under compressive stresses.

## ELASTOPLASTIC VERSION

The viscoplastic potential ensuing from the physical and phenomenological considerations that have led to the development of the SUVIC model can be stated as follows [1, 2]:

$$\Omega = \frac{K}{N+1} \left\langle \frac{X_{ae} - R}{K} \right\rangle^{N+1} \tag{1}$$

with

$$X_{ae} = \left[ \frac{3}{2} \ (S_{ij} - B_{ij}) \ (S_{ij} - B_{ij}) \right]^{\frac{1}{2}} \tag{2}$$

where $S_{ij}$ is the deviatoric stress tensor, $N$ is a material constant, and <> are the MacAuley brackets. This potential is almost similar in its mathematical formulation to the empirically derived one used by Mroz [4] and by Lemaitre and Chaboche [5] in their respective viscoplastic models.

In the case of the elastoplastic formulation, the total strain increment $d\epsilon_{ij}$ comprises both the elastic $d\epsilon^e_{ij}$ and plastic $d\epsilon^p_{ij}$ components, so that one may write:

$$d\epsilon_{ij} = d\epsilon^e_{ij} + d\epsilon^p_{ij} \tag{3}$$

The elastic strain is defined here by using a generalized Hooke law in three dimensions, as is the case with SUVIC. The plastic flow, on the other hand, follows the normality rule, so that it can be defined as:

$$d\epsilon^p_{ij} = dH \left( \frac{\delta F}{\delta \sigma_{ij}} \right) \tag{4}$$

where $\sigma_{ij}$ is the stress tensor, $dH$ is a plastic multiplier and $F$ is the yield criterion, which equals the plastic potential for this associated flow rule.

In order to reduce a viscoplastic model with state variables, such as SUVIC, to its elastoplastic counterpart, the viscoplastic potential must be transformed into a yield criterion $F$ [6]. Then, the model can be used to describe the rate-independent inelastic behavior in the context of classical elastoplasticity.

For the SUVIC model, the yield criterion is of the von Mises type, as it is considered that the mean stress does not have any practical influence on the inelastic flow of the material in the ductile regime. By introducing state variables $B_{ij}$ and $R$ in the yield criterion, one obtains:

$$F = X_{ae} - R = 0 \tag{5}$$

It must be mentioned here that, as the initial yield strength is very low for a material in a virgin (completely annealed) state, the initial size of the elastic domain is considered to be negligible.

In this preliminary version of the elastoplastic model, it is admitted that the state variable $K$ takes its saturation value ($K = K' = \sigma_y$, the macroscopic yield strength), so that it becomes a constant, similarly to other classical models of this type. In this case, the plastic strain increment is formulated as:

$$d\epsilon_{ij}^p = \frac{3}{2}\, dH \left[ \frac{(S_{ij} - B_{ij})}{X_{ae}} \right] \tag{6}$$

where $dH$ is obtained from the hardening rules for the condition $F = dF = 0$, associated to plastic flow. One can be reminded that when $F < 0$ or $dF < 0$, there is no plastic flow. Contrary to the viscoplastic version of the model however, the stress state cannot be such that $F > 0$ in the elastoplastic version.

In classical plasticity, many hardening rules have been proposed, such as the linear rules of Prager or Ziegler [4]. With the SUVIC model, these hardening rules which include only the hardening and dynamic recovery functions become:

$$dB_{ij} = \frac{2A_1}{3}\, d\epsilon_{ij}^p - \frac{A_1\, B_{ij}}{B_e'}\, d\epsilon_e^p \tag{7}$$

$$dR = A_3 \left( 1 - \frac{R}{R'} \right) d\epsilon_e^p \tag{8}$$

where $A_1$, $A_2$, $B'_e$ and $R'$ are materials constants, and $\epsilon_e$ is the von Mises equivalent strain. Such equations are similar to the well known hardening functions initialy proposed by Armstrong and Frederick [4].

The plastic multiplier is then equal to the following:

$$dH = \frac{1}{h_e} \left\langle \frac{\delta F}{\delta \sigma_{ij}} : d\sigma_{ij} \right\rangle (HV) \tag{9a}$$

$$= \frac{3}{2h_e} \left\langle \left[ \frac{S_{ij} - B_{ij}}{X_{ae}} \right] : d\sigma_{ij} \right\rangle (HV) \tag{9b}$$

where $HV$ is the Heavyside function (here $HV = 0$ if $F < 0$, $HV = 1$ if $F = 0$), and $h_e$ is the hardening modulus defined as follows:

$$h_e = A_1 - \left[ \frac{3}{2} \left( \frac{A_1 B_{ij}}{B_e'} \right) \left( \frac{S_{ij} - B_{ij}}{X_{ae}} \right) \right] + A_3 \left( 1 - \frac{R}{R'} \right) \tag{10}$$

In an extended version of this article, it will be shown that this elastoplastic version of the SUVIC model can be considered as a particular case of two surface models such as the one proposed by Krieg [7], which are in turn inspired by the multisurface model of Mroz [8]. A more elaborate elastoplastic version of the model will also be discussed. This alternate version, that is closer to its viscoplastic counterpart, includes a contribution of the state variable $K$ to the isotropic hardening of the material. In this case, the strain modulus $h_e$ may be redefined in order to account for such a contribution.

## RÉFÉRENCES


1. Aubertin, M. Développement d'un modèle viscoplastique unifié avec variables internes pour le comportement rhéologique du sel gemme. Ph.D. Thesis, Ecole Polytechnique de Montréal, 1989.

2. Aubertin, M., Gill, D.E., Ladanyi, B. A unified viscoplastic model for the inelastic flow of alkali halides. Mechanics of Materials, 1991, (to be published).

3. Aubertin, M., Gill, D.E., Ladanyi, B. Laboratory validation of a unified viscoplastic model for rocksalt. Proc. 7th Cong. ISRM, 1991, (to be published).

4. Mroz, Z. Phenomenological constitutive models for metals. Modelling Small Deformations of Polycrystals. Elsevier Appl. Sci. Pub., 1986, pp. 293-344.

5. Lemaitre, J., Chaboche, J.L. Mécanique des Matériaux Solides, 2e ed., Dunod-Bordas, 1985.

6. Delobelle, P. Sur les lois de comportement viscoplastique à variables internes. Exemples de deux alliages industriels : inoxydable austénitique 17-12SPH et superalliage INCO718. Revue Phys. Appl., 1988, 23, 1-6.

7. Krieg, R.D. A practical two surface plasticity theory. J. Appl. Mech., ASME, 42, Sept. 1975, 641-646.

8. Mroz, Z. On the description of anisotropic work hardening. J. Mech. Phys. Solids, 1967, 15, 163-175.

# CONSTITUTIVE EQUATIONS FOR ROCK SALT

N. CRISTESCU
Department of Mechanics
University of Bucharest
str.Academiei 14, Bucharest 7olo9,Romania

## ABSTRACT

A procedure to determine an elastic/viscoplastic constitutive equation for rock salt is presented. It is shown how all constitutive coefficients or functions can be determined from the data including the viscoplastic potential. Failure, dilatancy and/or compressibility are all included in the constitutive equation. How this constitutive equation can be simplified, depending on the application desired is also discussed.

## INTRODUCTION

The mechanical properties of rock salt are peculiar as compared with other rocks. We tried to determine the constitutive equation by making as few constitutive assumptions as possible, i.e. the constitutive functions have been determined from the data, without a priori assumptions. Since these may be quite involved, a possible simplification is also thought. Ultimately the comparison with the data is decisive. The experimental data by H. Hunsche are used (1).

## THE CONSTITUTIVE EQUATION

The constitutive equation is of the form (2)

$$\dot{\varepsilon} = \frac{\dot{\sigma}}{2G} + \left(\frac{1}{3K} - \frac{1}{2G}\right)\dot{\sigma}\,\mathbf{1} + k\left\langle 1 - \frac{W^I(t)}{H(\sigma)}\right\rangle \frac{\partial F}{\partial \sigma} \qquad (1)$$

where the elastic constants G and K may be variable: $\langle A\rangle = \frac{1}{2}(A + |A|)$, $W^I(t) = \int_0^t \sigma \cdot \dot{\varepsilon}^I \, dt$ is the stress power used as internal variable (work-hardening parameter), $H(\sigma) = W^I(t)$ is the equation of the stabilization boundary where

$$H(\sigma) := \left[a_1\left(\frac{\sigma - \sigma_1}{\sigma_*}\right)^{10} + \frac{a_2}{\sigma + a_3}\right]\left(\frac{\tau}{\sigma_*}\right)^9 + \left[b_1\left(\frac{\sigma}{\sigma_*}\right)^2 + b_2\frac{\sigma}{\sigma_*} + b_3\right]\left(\frac{\tau}{\sigma_*}\right)^2 + \qquad (2)$$

$$+\left[c_1\left(\frac{\sigma}{\sigma_*}\right)^2 + c_2\frac{\sigma}{\sigma_*} + c_3\right]\frac{\tau}{\sigma_*} + h_2\sigma + h_3 \quad \text{for } \sigma \geq 6\text{MPa}$$

with $a_i, b_i, c_i, h_i, \sigma_1$ constants and $\sigma_* = 1$ MPa. The viscoplastic potential $F(\boldsymbol{\sigma})$ is found in the form

$$kF(\sigma,\tau) = \left[\frac{\psi}{h^2}\left[\left(\frac{\tau}{\sigma_*}\right)^{\frac{r}{s}} f - h\left(\frac{\tau}{\sigma_*}\right)^{\frac{m}{n}}\right]\ln\left[-\left(\frac{\tau}{\sigma_*}\right)^{\frac{r}{s}} + h\sigma\right] +\right.$$

$$\left. + \frac{f\psi}{h}\sigma\right] + g_0\frac{\tau}{\sigma_*} + g_1\left(\frac{\tau}{\sigma_*}\right)^{\frac{3}{2}} + g_2\left(\frac{\tau}{\sigma_*}\right)^5 . \qquad (3)$$

Here $\sigma$ is the mean stress and $\tau$ is the octahedral shear stress related to the second invariant of the stress deviator by $\tau = (\frac{2}{3}\, II_{\sigma'})^{1/2}$, and all other coefficients are constants. Several surfaces H = const. are shown in Fig.1 as dotted lines, the surfaces F = const. as interrupted lines and failure by full lines. In order to match better the data by the simplest possible functions the interval of variation of the mean stress was divided into three parts by $\sigma = 22$ MPa and $\sigma =$ 37 MPa. The compressibility/dilatancy boundary $\partial F/\partial\sigma = 0$ shown as circle-dash line while the boundary $\partial H/\partial\sigma = 0$ as star-dash line; these last two lines are distinct. It is interesting to mention that the compressibility/dilatancy boundary is intersecting the failure line so that at high pressures failure is produced with no dilatancy. Also at high pressures the rock salt has a Mises behaviour and the constitutive equation can be significantly simplified.

## COMPARISON WITH THE DATA

The first check of the prediction of the model is the comparison with the data which have been used just to formulate the model. Such a comparison is shown in Fig.2 for the confining pressure $\sigma = 3o.2$ MPa. The matching of the model prediction (dotted lines) with the data (stars) seems reasonable, including the behaviour of the volume.

## SIMPLIFIED VARIANTS

The two constitutive functions (2) and (3) are certainly quite involved. However, Fig.1 is suggesting that the viscoplastic potential F can be quite well be approximated by a Mises kind of potential, i.e. $F(\boldsymbol{\sigma}) := II_{\sigma'}$. This approximation cannot be done only in the immediate neighbourhood of the failure surface.

What concerns function H if we decompose it in four terms as $H = A\tau^9 + B\tau^2 + C\tau + H_H$ and we estimate the relative magnitude of these terms, we obtain the results shown in Fig.3: a vertical arrow is pointing the compressibility/dilatancy boundary. Thus for relatively small values of $\tau$ the linear and hydrostatic

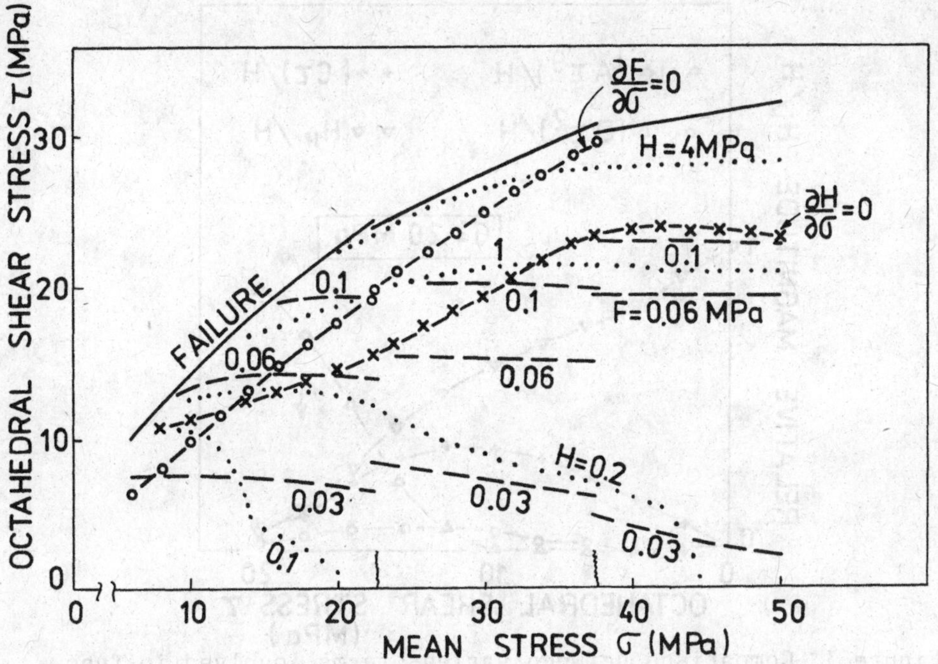


Figure 1. Constitutive domain for rock salt: above the circle-dash line the rock salt is dilatant, under it is compressible.

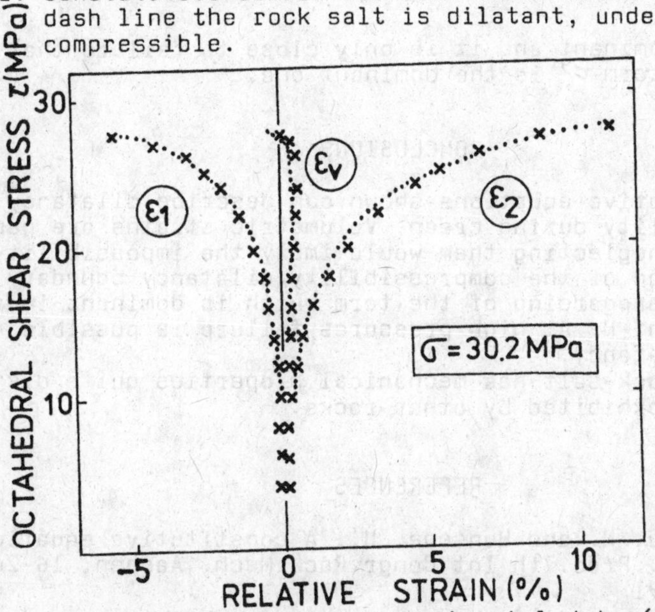


Figure 2. Stress-strain curves: experimental data (stars) compared with model prediction (dots).

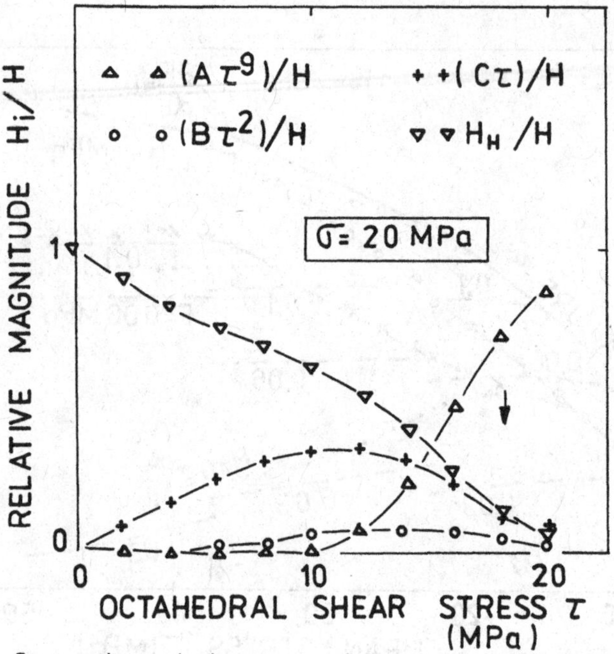


Figure 3. Comparison between various terms involved in function H.

terms are dominant and it is only close to failure that the nonlinear term $\tau^9$ is the dominant one.

## CONCLUSIONS

The constitutive equations shown can describe dilatancy and compressibility during creep. Volumetric strains are generally small, but neglecting them would imply the impossibility of determination of the compressibility/dilatancy boundary and also the disregarding of the term which is dominant in the expression of H. At high pressures failure is possible without dilatancy.

Generally rock-salt has mechanical properties quite distinct from those exhibited by other rocks.

## REFERENCES

1. Cristescu, N. and Hunsche, U., A constitutive equation for salt. Proc.7th Int.Congr.Rock Mech.,Aachen, 16-2o Sept. 1991
2. Cristescu, N. Rock Rheology, Kluwer Academic Publishers, Dordrecht, Holland, 1989, pp.112-15o

# ON FINITE ELASTOPLASTICITY MODELS IN POROUS MEDIA THEORIES

WOLFGANG EHLERS
Universität Essen, FB Bauwesen, Fachgebiet Mechanik
Postfach 103 764, W-4300 Essen 1, F.R.G.

## ABSTRACT

The present note concerns finite elastoplasticity models for liquid-saturated porous solid materials discussed within the framework of porous media theories. In particular, the solid behaviour is outlined by the constitutive example of kinematically hardening soils.

## INTRODUCTION

Within the scope of a macroscopic formulation, liquid-saturated porous solid materials can be described via mixture theories extended by the volume fraction concept ( porous media theories ), thus implying the diverse field functions of the constituents, porous solid matrix and pore liquid, to be represented by average functions of the macroscale. This approach [ 1 - 3 ] is the simplest one to provide an excellent basis for a continuummechanical treatment of saturated porous media where both the thermodynamical properties of the different phases and the mutual interactions between the constituents are taken into account.

In two recent papers [ 4, 5 ], a fundamental discussion on finite elastoplasticity problems in porous media theories was presented, where use was made of the second-grade character of general heterogeneous media. Following this, the kinematical concept of porous media elastoplasticity was described with the aid of a multiplicative split of the first and the second solid deformation gradients, thus taking the plastic parts as internal state variables of the respective deformation histories. Within the constitutive model of an immiscible binary medium consisting of an incompressible solid matrix material and an incompressible pore liquid, the basic set of constitutive equations was directly derived from the dissipation principle for mixtures where, additionally, the yield function was understood as a secondary condition for the thermodynamical process.

Based on these ideas, the present contribution generally covers a wide range of problems. Including convenient constitutive equations for the pore content and the different interactions between the constituents, the model is appropriate to be used for brittle and granular materials ( e.g., rock or soil ) as well as to describe ductile matrices like, e.g., sinter metals. However, the following constitutive model particularly concentrates on saturated soils where the elastic solid deformations are usually small and the plastic range, covering finite deformations, proceeds from kinematical hardening properties.

## CONSTITUTIVE FRAMING

As was shown in [ 4, 5 ], an incompressible liquid-saturated porous solid material is governed by the following basic set of constitutive equations:

$$\begin{aligned} \{\psi^i, \eta^i\} &= f(\Theta, \mathbf{E}_{Se}, \mathbf{E}_{Sp}), \\ \mathbf{q}^S + \mathbf{q}^F &= -\beta_\vartheta \operatorname{grad} \Theta, \\ \mathbf{T}^i &= -n^i p \mathbf{I} + \mathbf{T}^i_E, \\ \mathbf{p}^F &= p \operatorname{grad} n^F + \mathbf{p}^F_E. \end{aligned} \tag{1}$$

Therein, $\psi^i$, $\eta^i$, $\mathbf{q}^i$, $\mathbf{T}^i$ and $n^i$ are the specific free energy and entropy densities, heat influx vectors, partial Cauchy stresses and volume fractions of constituents $\varphi^i$ (i = S: solid phase; i = F: liquid phase), whereas $\mathbf{p}^F = -\mathbf{p}^S$ represents the momentum supply term or, respectively, the interaction force (per unit of bulk volume) between the constituents. Furthermore, $\Theta$ is the common absolute temperature function, $\beta_\vartheta$ is the coefficient of thermal conductivity for the whole system, and the symbol grad means partial differentiation with respect to the spatial position $\mathbf{x}$. The elastic and the plastic Lagrangian strain tensors, $\mathbf{E}_{Se}$ and $\mathbf{E}_{Sp}$, result from an additive decomposition of the total solid strains $\mathbf{E}_S$, compatible with the multiplicative split of the solid deformation gradient: $\mathbf{F}_S = \mathbf{F}_{Se}\mathbf{F}_{Sp}$.

As a direct result of the incompressibility constraint for the present binary model, the solid and the liquid stresses as well as the interaction force, respectively, are expressed by the effective liquid pressure p and so-called "extra quantities" denoted by the additional subscript $(\dots)_E$:

$$\begin{aligned} \mathbf{T}^S_E &= \rho^S \mathbf{F}_S \frac{\partial \psi^S}{\partial \mathbf{E}_{Se}} \mathbf{F}^T_S, \\ \mathbf{T}^F_E &= 2\mu^F \mathbf{D}_F + \nu^F (\mathbf{D}_F \cdot \mathbf{I})\, \mathbf{I}, \\ \mathbf{p}^F_E &= -\alpha_\vartheta \operatorname{grad} \Theta - \mathbf{S}_v (\mathbf{x}'_F - \mathbf{x}'_S). \end{aligned} \tag{2}$$

In these relations, $\rho^S$ is the bulk density function of $\varphi^S$, $\mu^F$ and $\nu^F$ are the macroscopic viscosity parameters of the pore liquid, $\alpha_\vartheta$ is the entropy coupling parameter, $\mathbf{S}_v$ represents the general permeability tensor, and $\mathbf{x}'_F - \mathbf{x}'_S$ is the velocity difference between the constituents. Additionally, note in passing that, in soil mechanics, the solid extra stresses are usually known as the "effective stresses" of the porous soil material. However, for kinematically hardening solid constituents, the dissipation principle furthermore yields a restriction for the back-stress tensor $\mathbf{Y}^S$, viz.:

$$\mathbf{Y}^S = \rho^S \det \mathbf{F}_S\, \mathbf{F}_S \frac{\partial \psi^S}{\partial \mathbf{E}_{Sp}} \mathbf{F}^T_S. \tag{3}$$

Concerning the relations (2) and (3), the above constitutive model proceeds from the "principle of constituent separation" [4] thus implying the liquid free energy function to be independent of the solid deformation state. Following this, it is easily seen that the model also applies, without modifications, to empty porous matrices. In this case, i.e., in the absence of p, the solid extra stresses $\mathbf{T}^S_E$ and the partial stresses $\mathbf{T}^S$ coincide.

## AN ELASTOPLASTICITY MODEL FOR SOILS

To obtain a convenient form for an elastoplasticity model for granular soils, one proceeds from the following free energy split,

$$\psi^S(\Theta, \mathbf{E}_{Se}, \mathbf{E}_{Sp}) = \psi^{Se}(\Theta, \mathbf{E}_{Se}) + \psi^{Sp}(\mathbf{E}_{Sp}),$$
$$\psi^{Se}(\Theta, \mathbf{E}_{Se}) = \bar{\psi}^{Se}(\Theta) + \tilde{\psi}^{Se}(\mathbf{E}_{Se}), \tag{4}$$

where $(4)_2$ additionally corresponds to the assumption of hyperelasticity. It is furthermore known from experiment that the elastic domain of soils is usually very small whereas plastic loading produces considerable deformations.

Thus, the solid extra stresses can be determined by use of an elasticity law of Hookian type, viz.:

$$\hat{\boldsymbol{\tau}}^S_E = 2\,\mu^S\,\hat{\boldsymbol{\Gamma}}_{Se} + \nu^S(\hat{\boldsymbol{\Gamma}}_{Se}\cdot\mathbf{I})\,\mathbf{I},$$
$$\hat{\boldsymbol{\tau}}^S_E = \det \mathbf{F}_S\,\mathbf{F}^{-1}_{Se}\,\mathbf{T}^S_E\,\mathbf{F}^{T-1}_{Se}, \tag{5}$$
$$\hat{\boldsymbol{\Gamma}}_{Se} = \mathbf{F}^{T-1}_{Sp}\,\mathbf{E}_{Se}\,\mathbf{F}^{-1}_{Sp}.$$

Therein, the symbol $(\hat{\dots})$ characterizes quantities defined in terms of the solid plastic intermediate configuration, which must be understood as the most natural kinematical frame for a representation of the elastic and the plastic responses in finite elastoplasticity [5]. With respect to the plastic intermediate configuration, $\hat{\boldsymbol{\tau}}^S_E$ is the Kirchhoff extra stress tensor; the elastic strain $\hat{\boldsymbol{\Gamma}}_{Se}$ is defined as the contravariant push-forward transformation of $\mathbf{E}_{Se}$. Since this transformation is carried out with the plastic part of the solid deformation gradient, $\mathbf{F}_{Sp}$, it is easily seen that the stored energy function must be defined with frozen plastic variables, i.e.:

$$\tilde{\psi}^{Se}(\mathbf{E}_{Se}) = \tilde{\psi}^{Se}(\hat{\boldsymbol{\Gamma}}_{Se}, \mathbf{F}_{Sp})\Big|_{\mathbf{F}_{Sp}=\text{const.}} \tag{6}$$

Thus, $\mu^S$ and $\nu^S$ are the macroscopic Lamé parameters of $\varphi^S$ measured at frozen $\mathbf{F}_{Sp}$.

The plastic range of granular soils is governed by the basic relation for the back-stress tensor, (3), together with convenient forms of a yield condition and a flow rule. Following this, a temperature independent yield function, $\hat{F}$, can be introduced by extending de Boer's yield condition [6] towards the finite kinematically hardening range, viz.:

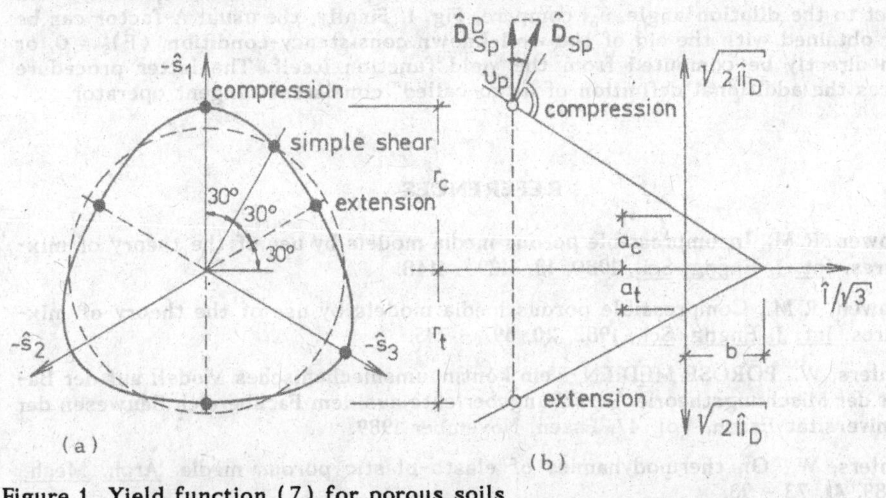


Figure 1. Yield function (7) for porous soils
(a) octahedral plane: $\hat{s}_1/\hat{s}_2/\hat{s}_3$: principal values of $\hat{\boldsymbol{\tau}}^S_E - \hat{\mathbf{Y}}^S$
(b) hydrostatic plane: $a_c = \sqrt{2}\,\varkappa/[1-(2/\sqrt{27})\,\gamma]^{1/3}$
$a_t = \sqrt{2}\,\varkappa/[1+(2/\sqrt{27})\,\gamma]^{1/3}$
$b = \varkappa/(\sqrt{3}\,\beta)$

$$\hat{F} = (\hat{II}_D)^{1/2} \, (1 + \gamma \frac{\hat{III}_D}{(\hat{II}_D)^{3/2}})^{1/3} + \beta \, \hat{I} - \varkappa = 0. \tag{7}$$

Therein, $\hat{I}$, $\hat{II}_D$ and $\hat{III}_D$ are the first invariant and the deviatoric second and third invariants of the difference tensor $\hat{\tau}^S_E - \hat{Y}^S$, whereas the parameters $\beta$, $\gamma$ and $\varkappa$ are material functions depending on the angle of internal friction, $\varphi$, and the cohesion, $c$:

$$\beta = \tfrac{1}{3} \sin \varphi, \qquad \gamma = \tfrac{4}{3} \sin \varphi, \qquad \varkappa = c \cos \varphi. \tag{8}$$

By pull-back of $\mathbf{Y}^S$ from the actual towards the solid intermediate configuration, the following version of the back-stress tensor can be shown to yield:

$$\begin{aligned} \hat{\mathbf{Y}}^S &= c_1 \, (\det \mathbf{F}_{Sp})^{-2/3} \, \hat{\mathbf{K}}^D_{Sp} + c_2 \det \mathbf{F}_{Sp} \, \mathbf{I}, \\ \hat{\mathbf{Y}}^S &= \mathbf{F}^{-1}_{Se} \, \mathbf{Y}^S \, \mathbf{F}^{T-1}_{Se}, \\ \hat{\mathbf{K}}_{Sp} &= \mathbf{R}_{Sp} \, \mathbf{E}_{Sp} \, \mathbf{R}^T_{Sp}, \end{aligned} \tag{9}$$

where the material functions $c_1$ and $c_2$ must be calculated in comparison with experimental data, e.g., from triaxial compression/extension tests. Note in passing that the plastic Karni-Reiner strain, $\hat{\mathbf{K}}_{Sp}$, is defined as the push-forward rotation of $\mathbf{E}_{Sp}$ with the plastic spin $\mathbf{R}_{Sp}$, the topscript $(\dots)^D$ denotes the deviatoric part of $(\dots)$.

To complete the set of equations governing the plastic range, a non-associated flow rule is given by assuming that

$$\begin{aligned} \hat{\mathbf{D}}_{Sp} &= \tfrac{1}{2} \, (\hat{II}_D)^{-1/2} \, \Lambda \, [\, (\hat{\tau}^S_E - \hat{\mathbf{Y}}^S)^D + \zeta \, \hat{I} \, \mathbf{I} \,], \\ \hat{\mathbf{D}}_{Sp} &= \mathbf{F}^{T-1}_{Sp} \, (\mathbf{E}_{Sp})'_S \, \mathbf{F}^{-1}_{Sp}. \end{aligned} \tag{10}$$

Therein, $\hat{\mathbf{D}}_{Sp}$ is the purely plastic rate of deformation tensor given as the contravariant push-forward transformation of the Lagrangian strain rate, $(\mathbf{E}_{Sp})'_S$; the symbol $(\dots)'_S$ defines the material time derivative corresponding to $\varphi^S$. However, the parameter $\zeta$ serves to further fit the flow rule on experimental data, especially, with respect to the dilation angle, $\upsilon_p$, compare, Fig. 1. Finally, the usual $\Lambda$-factor can be either obtained with the aid of the well-known consistency-condition, $(\hat{F})'_S = 0$, or it can directly be computed from the yield function itself. The latter procedure requires the additional definition of a "so-called" consistent tangent operator.

## REFERENCES


1. Bowen, R.M., Incompressible porous media models by use of the theory of mixtures. Int. J. Engng. Sci., 1980, **18**, 1129 - 1148.

2. Bowen, R.M., Compressible porous media models by use of the theory of mixtures. Int. J. Engng. Sci., 1982, **20**, 697 - 735.

3. Ehlers, W., PORÖSE MEDIEN – ein kontinuumsmechanisches Modell auf der Basis der Mischungstheorie. Forschungsberichte aus dem Fachbereich Bauwesen der Universität Essen, Vol. 47, Essen, November 1989.

4. Ehlers, W., On thermodynamics of elasto-plastic porous media. Arch. Mech., 1989, **41**, 73 - 93.

5. Ehlers, W., Towards finite theories of liquid-saturated elasto-plastic porous media. Int. J. Plasticity, 1991, **7**, forthcoming.

6. de Boer, R., On plastic deformation of soils. Int. J. Plasticity, 1988, **4**, 371 - 391.

# DESCRIPTION OF THE BEHAVIOUR OF REINFORCED SOIL USING THE HOMOGENIZATION METHOD

PRUCHNICKI ERICK and SHAHROUR ISAM
Laboratoire de Mécanique de Lille CNRS URA 1441D
Département Sols-Structures
Institut Industriel du Nord
59651 Villeneuve d'Ascq Cédex

## ABSTRACT

The homogenization and the Hill-Mandel processes are used to analyse the macroscopic behaviour of the reinforced soils. An elastoplastic behaviour is assumed for both the soil and the reinforcement material. The bases of the homogenization method are first presented, and then applied to a multi-layered media equivalent to the reinforced soil. A macro-constitutive law involving two internal parameters is finally presented and used to predict the macroscopic behaviour of the reinforced soil.

## INTRODUCTION

Reinforced earthworks can be analysed as composite structures composed of soil (matrix) and reinforcement material (fibre). Finite element calculation of such structures can be carried out using macroscopic constitutive laws which link the macroscopic stress and srain tensors. The macroscopic constitutive laws can be set up using the homogenization process (for periodic structures) or the Hill-Mandel method. Since the behaviour of soil and reinforcement material is strongly non-linear and irreversible, the description of the macroscopic behaviour was carried out assuming an elastoplastic constitutive law for both soil and reinforcement material. In order to simplify the elastoplastic calculation, a multi-layered media has been associated to the reinforcement earthworks. In this case the homogenized macroscopic law can be described by an elastoplastic model involving two internal parameters.

## ANALYSIS OF THE MACROSCOPIC BEHAVIOUR IN THE ELASTOPLASTIC DOMAIN

The macroscopic constitutive law of a reinforced earthwork is given by the relation between the macroscopic stress $\Sigma$ (which represents the mean microscopic stress over the basic cell) and strain E, and eventually some internal parameters. Assuming that both soil and reinforcement material have an elastoplastic behaviour, the macroscopic response ( E(t) , $t \leq T$ ) to a given macroscopic loading path ( $\Sigma(t)$ , $t \leq T$ ) is determined by the resolution of the following variational problem:

$$E^*:<a>:E+E^*:<a:e(v)>=E^*:\Sigma+E^*:<a:e^p>$$

$$<a:e(v^*)>:E+<a:e(v^*):e(v)>=<a:e(v^*):e^p>$$

$e^p$ represents the plastic strain.

v and $v^*$ have to respect the following macro-micro conditions: for the method of Hill Mandel v is zero on $\partial Y$, for the method of homogenization v is Y periodic [1], where Y and $\partial Y$ represent the basic cell and its boundary respectively.

## APPLICATION TO REINFORCED EARTHWORKS

**Association of a multi-layered media**

The resolution of the above elastoplastic problem has been carried out using the finite element program Pecplas [2]. The Lagrange-multiplier was used to introduce the periodic conditions on the displacement field. In order to simplify the elastoplastic calculation, a multi-layered media is associated to the reinforcement earthworks. Since the reinforcement material works only in compression and extension, the associated multi-layered media must conserve the axial stiffness and resistance of the reinforcement material. A Mohr-Coulomb and Tresca laws were assumed for the soil and the reinforcement material respectively. The following table gives the mechanical properties and volumic proportion (w) for both the soil and the reinforcement.

| | E (MPa) | ν | C (MPa) | φ | w |
|---|---|---|---|---|---|
| reinforcement | 10500 | 0.22 | 6 | 0° | 0.00784 |
| soil | 150 | 0.3 | 0 | 30° | 0.99216 |

**Macroscopic behaviour of reinforced earthworks.**

The macroscopic behaviour of reinforced earthwork was investigated along macroscopic compression loading paths ( $\Sigma(t), t \leq T$ ) applied in various directions ($\Sigma 2$) (figure 1).

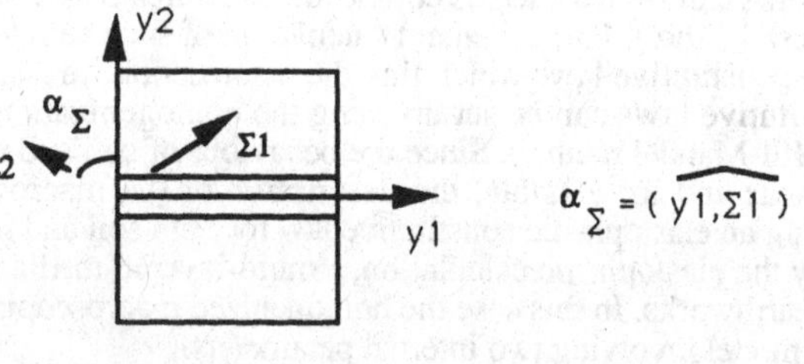


figure 1: direction of loading path

The macroscopic strain was calculated using the finite element method. The macroscopic response of the basic cell to a loading path ($\alpha_\Sigma = 30°$) is illustrated in figures 2 and 3, which shows the evolution of the macroscopic stress invariant Q ( ( $\Sigma 1-\Sigma 2$) ) and PM ( ( $\Sigma 1+\Sigma 2$ ) / 2) ), in terms of the equivalent macroscopic strain:

$$EQ=\sqrt{E_{11}{}^2+E_{22}{}^2+2E_{12}{}^2}$$

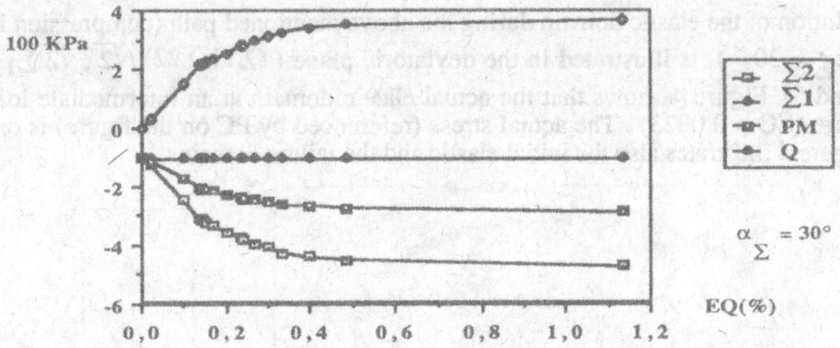


figure 2 : equivalent evolution law given by method of Hill-Mandel

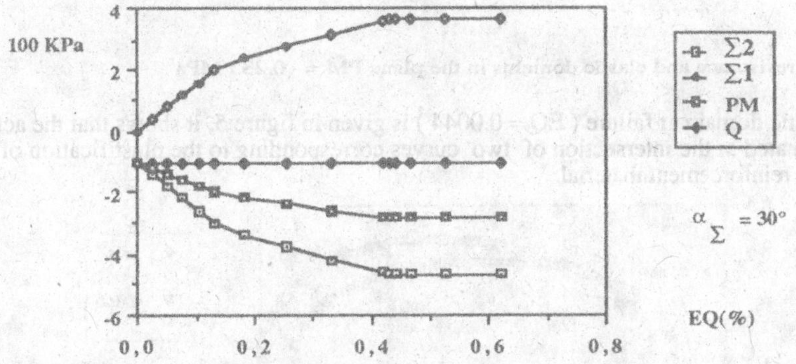


figure 3 : equivalent evolution law given by method of homgenization

Results obtained by the Hill-Mandel method shows that after a linear elastic phase, plastification induces a continuous decrease in the macroscopic stiffness up to failure (figure 2). Whereas the homogenization method shows that the linear-elastic phase is followed by a strong decrease in the macroscopic stiffness which is induced by the plastification of one constituant and that failure occurs as a result of the plastification of the second constituant (figure 3).

## A MACROSCOPIC HOMOGENIZED LAW WITH TWO INTERNAL PARAMETERS

Application of the homogenization method to a bi-layered media shows that the microscopic stress and the elastic and plastic strains are constant in each constituant. This result permits the formulation of a simple macroscopic elastoplastic law for reinforced earthworks [3]. This model is described by its initial elastic domain and its evolution during loading.

The initial elastic domain is given by :

$$DEI = \{ \Sigma , F_s ( \Sigma ) < 0 , F_r ( \Sigma ) < 0 \}$$

$F_s$ and $F_r$ represent the elastic domain of the soil and the reinforcement material respectively. For a given plastic deformation , the actual elastic domain is given by:

$$DE = \{ \Sigma , F_s ( \Sigma + T_s ) < 0 , F_r ( \Sigma + T_r ) < 0 \}$$

The $T_r$ and $T_s$ tensors depend on the microscopic residual stresses and the geometry of the basic cell. The expression of the DE shows that the elastic domain undergoes an anisotropic hardening. In the case of plane strains , we have $T_r = ( (t_{11})_r ,0,0 )$ and $T_s = ( (t_{11})_s,0,0 )$.

The evolution of the elastic domain during the above mentioned path (compression in the direction $\alpha_\Sigma = 30°$ ) is illustrated in the deviatoric plane ( $(\Sigma 11-\Sigma 22)/\sqrt{2}$, $\sqrt{2}\,\Sigma_{12}$ ) ( figures 4 and 5). Figure 4 shows that the actual elastic domain at an intermediate loading before failure (EQ = 0.0023) . The actual stress (referenced by PC on the figure) is on this domain. Figure 4 illustrates also the initial elastic and the failure domains.

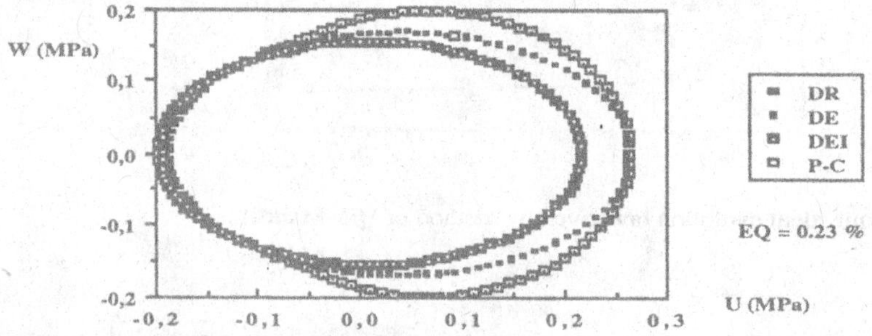


**figure 4 : resistance and elastic domains in the plane PM = - 0.233 MPa**

The elastic domain at failure ( EQ = 0.0044 ) is given in figure 5, it shows that the actual stress is situated at the intersection of two curves corresponding to the plastification of the soil and the reinforcement material.

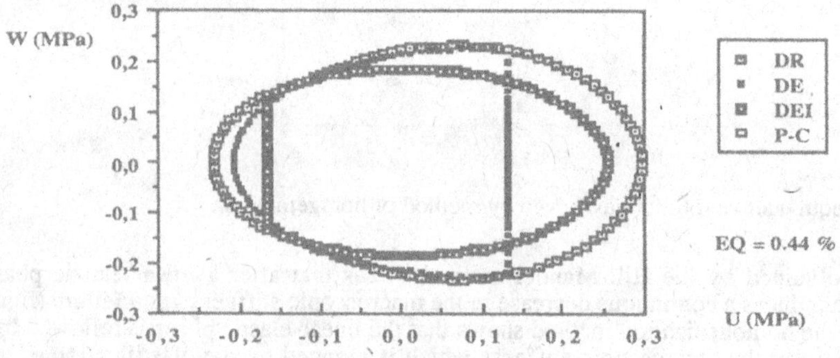


**figure 5 : resistance and elastic domains in the plane PM = - 0.28 MPa**

## CONCLUSION

The macroscopic behaviour of reinforced earthworks in the elastoplastic domain can be described by the homogenization process. By associating a multi-layered media for reinforced earthworks, its macroscopic behaviour can be described by a simplified elastoplastic constitutive law involving two anisotropic internal parameters. This constitutive law can be incorporated into a finite element program and used, after validation, to reinforced earthworks design.

## REFERENCES

1. LENE F. , Contribution à l'étude des matériaux composites et de leur endommagement. Thèse d'état de l'Université de Paris VI, 1984.
2. SHAHROUR I. , Modélisation et validation en génie civil. Habilitation à diriger des recherches de l'Université des Sciences et techniques de Lille, 1988.
3. PRUCHNICKI E. , Contribution à l'homogénéisation en phases linéaire et non linéaire. Application au renforcement des sols. Thèse de Doctorat de l'Université des Sciences et techniques de Lille, 1991.

# SOFT CLAY BEHAVIOR VIA ELASTO-VISCOPLASTICITY

MUSTAFA EMIR

Dept. of Civil Engineering, Concordia University

1455 de Maisonneuve W., Montreal, PQ, Canada, H3G 1M8

## ABSTRACT

The mechanical behavior of very soft clays suggest that an overstress type viscoplastic law is suitable for predictive analyses. The Original Cam Clay yield criterion is used to describe the yielding of clay and excess stresses above yield cause creep deformations. The effects of vibrational loading on the creep of soft clay is established in a dynamic slump test and it is concluded that vibrations influence the rate dependent behavior. The finite element analyses of the problem show that when the vibratory nature of the loading is accounted for, the frequently observed lateral flow of the foundation can be predicted.

## INTRODUCTION

The proliference of land reclamation projects in South East Asia and Japan has recently increased the need for advanced studies on the mechanical behavior of soft clays. One of the most important aspects of soft clays is time dependent deformation characteristics. The subject of this paper is the application of an overstress type elasto-viscoplastic constitutive equation to the analysis of ground response to sustained vibration loading due to vehicle traffic.

This study is based on an actual case history where unexpected and excessive lateral deformations were observed during the construction of a landfill in the Okayama prefecture in Japan. Since the only activity on the fill was the continuous traffic of heavy construction equipment, the engineering report concluded that the cause of the lateral flow might be the triggering of creep deformations by vibration. The case was brought to Kyoto University for further research and analysis.

## THE CONSTITUTIVE EQUATION

The constitutive equations which will be presented here were initially developed by Adachi and Okano [1] at Kyoto University. The underlying feature of the model is the combination of Perzyna's viscoplastic constitutive laws with the Cambridge Critical State Theory yield criteria. The general form of the constitutive equation developed by Perzyna is given as

$$\dot{\varepsilon}_{ij} = \frac{\dot{s}_{ij}}{2G} + \frac{\dot{\sigma}_m}{3K}\delta_{ij} + \gamma\langle\Phi(F)\rangle\frac{\partial F}{\partial\sigma_{kl}} \quad (1)$$

with

$$\Phi(F) = 0 \qquad \text{for } F \le 0$$
$$\Phi(F) = \Phi(F) \quad \text{for } F > 0$$

$$F = \frac{f_d\left(\sigma_{ij},\varepsilon^{vp}_{ij}\right)}{f_s\left(\sigma_{ij},\varepsilon^{vp}_{ij}\right)} - 1 \quad (2)$$

where G, K are elastic shear and bulk moduli respectively, $ds_{ij}$ is the deviatoric stress increment tensor, $d\sigma_m=\sigma_{ii}/3$ is the first invariant of the total stress increment tensor, $\gamma$ is a viscosity constant of the material, $f_d(\sigma_{ij},\varepsilon^{vp}_{ij})=k_d$ and $f_s(\sigma_{ij},\varepsilon^{vp}_{ij})=k$ are the dynamic and static yield criteria respectively, F is the overstress function and $\Phi(F)$ is the dynamic property function. Note that the term dynamic refers to deformation precesses with non-constant strain rates. The functions $f_d$ and $f_s$ are similar in shape so that they coincide as a limiting case.

$$f_d = \frac{\sqrt{2J_{2(d)}}}{M^*\,\sigma'_{m(d)}} + \ln\sigma'_{m(d)} = \ln\sigma'_{my(d)} = k_d \quad (3)$$

The hardening parameter $k_d$ expresses the change of the dynamic yield function $f_d$ due to inelastic strain rate $d\varepsilon^p_{ij}$ and the inelastic volumetric strain rate $dv^p$. The strain rate effects on the hardening rule are taken account by virtue of the overstress function F. The derivation of the complete stress-strain relations can be found in reference [2]. The final form of the equation has been given as

$$\dot{\varepsilon}_{ij} = \frac{\dot{s}_{ij}}{2G} + \frac{\kappa}{1+e}\frac{\dot{\sigma}'_m}{\sigma'_m}\frac{\delta_{ij}}{3} + C\exp\left[m'\left(\frac{\sqrt{2J_{2(d)}}}{\sigma'_{m(d)}} + \ln\frac{\sigma'_{m(d)}}{\sigma'_{me}} - \frac{1+e}{\lambda-\kappa}\varepsilon^p_{kk}\right)\right]\left\{\frac{s_{ij}}{\sqrt{2J_2}} + \left(M^* - \frac{\sqrt{2J_2}}{\sigma'_m}\right)\frac{\delta_{ij}}{3}\right\} \quad (4)$$

In the above, $\lambda$, $\kappa$, $M^*$ are the critical state parameters, e is the initial voids ratio, $\sigma'_{me}$ is the equivalent stress on the virgin consolidation line, C and m' are rate dependent parameters. The parameter m' is the slope of the equi-inelastic volumetric strain line for a given $M = M^*\sqrt{(3/2)}$

in a plot of $(q/\sigma'_m)$ vs. $(\ln d\varepsilon_{11})$ obtained from undrained triaxial compression tests. The value of C is determined from the constitutive equation once the strain rate, the stress ratio, the plastic volumetric strain and m' have been obtained.

The viscoplastic parameter C has been identified as a measure of the fact that the structure of the material is necessarily altered during deformations. This alteration is dependent on the strain rate and time. It is seen that the derivative of C with respect to strain rate gives the actual effect of the alterations: during steady creep it is decreasing and during acceleration creep it is increasing. This derivative gives a physical representation of the state of the structural integrity of the material at any given time during creep and the equivalent physical variable is deduced to be *entropy*. The parameter m' has been identified as the equivalent of the stress factor associated with energy barrier height in physical models. This variable determines the relative movement of particle clusters past each other [3].

## VIBRATIONAL SLUMP TESTS

A new test has been developed at Kyoto University to determine the time dependent material parameters for cylindrical soft clay samples subjected to horizontal vibration in the range 1-50 Hz. The soil tested was remoulded Osaka Clay with high plasticity and preconsolidated to 8 kPa. The objective of the test are: (i) to determine material parameters relevant for vibrational loading conditions; (ii) to determine the effect of vibration on the mechanical and physical behavior of soft clays. The results show that the vibration of the unconfined soft clay sample leads to a softer response. The fact that higher strain rates are maintained for higher frequencies implies that the vibration of the sample prevents the establishment of an equilibrium state where strain rates tend to zero. Thus within the frequency range studied, the vibrations prolong the steady creep stage for a given soil. The effects are defined by a strain rate-frequency-time surface and hence influence the parameter C. Higher frequencies lead to higher strain rates and define softer deformation behavior, thus the unexpected observations referred to earlier.

## PREDICTIONS

The constitutive equation (3) has been implemented in a finite element program in the light of the observed vibration effects. The viscoplastic tangential stiffness matrix is defined implicitly. The parameter C is no longer a constant, but a function of stress state and strain rate (thus frequency of vibrations). The loading is a standard truck with a weight of 24.5 tonnes imposing vibrations of 3 Hz. The truck is assumed to travel on a linear path with a velocity of

15 km/h. Each cyle of loading corresponds to an increment for the algorithm, and the load is moved horizontally with the specified velocity at the beginning of each increment. The loading was continuously applied for an equvalent of 360 days. The soil properties are that of remoulded Osaka Clay and are presented in Table 1.

Table 1: Material properties for viscoplastic foundation analysis

| $e_o$ | G (kPa.) | $\lambda$ | $\kappa$ | M* | C | m' | $p_o$ (kPa.) |
|---|---|---|---|---|---|---|---|
| 4.25 | 8000.0 | 0.74 | 0.08 | 1.05 | $1\,e^{-5}$ | 0.45 | 16.0 |

The results of the analysis are presented in terms of lateral movements. They show that (Fig. 1) when the effects of vibration are accounted by the viscoplastic parameters, the predictive ability of the model is increased. In either case, the description of lateral movenets are better than elaso-plastic models where the parameters are time independent and fail to register inelastic deformations under low stress levels that are produced by a truck.

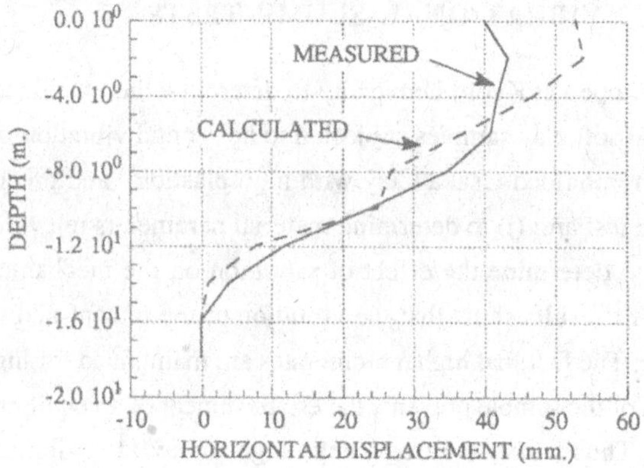


Figure 1. Horizontal displacement of soft clay foundation loaded by a single travelling truck for a duration of 360 days.

## REFERENCES

1. ADACHI, T., OKANO, M., 1974, "A constitutive equation for normally consolidated clay," Soils and Foundations, v.14, no.4, 55.

2. ADACHI, T., OKA, F., MIMURA, M., 1987, "Mathematical structure of an overstress elasto-viscoplastic model for clay," Soils and Foundations, v.27,no.3, 31.

3. EMIR, M., 1991, Behavior of very voft vround under vibration loading from traffic or machinery, Ph. D. Thesis, Cincordia University, Montreal, Canada.

# A NEW CONCEPT OF THE NON-ASSOCIATIVE PLASTIC FLOW

STANISLAW JEMIOŁO, MAREK KWIECIŃSKI
Warsaw University of Technology/Institute of Structural Mechanics
Armii Ludowej 16, 00-637 Warsaw, Poland

## ABSTRACT

The new concept of a "fuzzy" departure from the associative flow rule is formulated following the basic concept of the classical theory of hardening plasticity. General constitutive relations are derived to describe elastic hardening/softening plastic behaviour. The second law of thermodynamics for plastic deformation is used. As an example, constitutive equations are proposed for a plain concrete under short-term monotonic loading. Only one additional material parameter due to dilatancy is involved. The obtained numerical results are compared to existing experimental evidence. Good agreement has been observed.

## GENERAL CONCEPT

The concept of a "fuzzy" departure from the associative flow rule is based on the classical theory of hardening plasticity. However, it differs from the elasto-plastic constitutive descriptions in that it adopts the mathematical tools from fuzzy set theory.

In the paper, the derivation of the constitutive equations is based on the following assumptions:

1) The total strain increment $\dot{\mathbf{E}}$ can be divided into the sum of the elastic $\dot{\mathbf{E}}_e$ and plastic part $\dot{\mathbf{E}}_p$.

2) The elastic strain increments $\dot{\mathbf{E}}_e$ are related to the stress increments $\dot{\mathbf{T}}$ through Hooke's law: $\dot{\mathbf{T}} = \mathbf{C}_e\dot{\mathbf{E}}_e$ ,where $\mathbf{C}_e$ is the fourth order constitutive tensor.

3) For an irreversible process, the second law of thermodynamics requires that the plastic work increment during a plastic flow be non-negative, i.e.

$$\dot{W}_p = trT\dot{E}_p \geq 0. \tag{1}$$

An equivalent stress $\sigma_o$ is defined in connection with the equivalent plastic strain increment $\dot{\varepsilon}_p$:

$$trT\dot{E}_p = \sigma_o \dot{\varepsilon}_p \tag{2}$$

for which the plastic work increment is identical.

4) The yield condition can be described by the yield function with nonlinear hardening rule in the following form

$$f(T,\sigma_o) = 0, \quad \sigma_o = \sigma_o(\varepsilon_p). \tag{3}$$

5) The plastic potential function is defined by the same initial yield condition and hardening rule, but the evolution law also characterizes a stepwise departure from associativity, described by the membership function $\gamma \in \langle 0,1 \rangle$, $\gamma = \gamma(\sigma_o)$,

$$g(T,\sigma_o,\gamma) = 0. \tag{4}$$

The membership function $\gamma$ determines the "degree of associativity" of the flow rule: for $\gamma=0$, $g=f$; for $\gamma=1$, the difference between the function $g$ and $f$ is at a maximum, implying that the greatest departure from the associative flow rule has been reached. The latter case occurs when the equivalent stress $\sigma_o$ is at a maximum.

6) The plastic strain increments are assumed to be obtainable from a plastic potential function $g$, as

$$\dot{E}_p = \lambda \partial g/\partial T . \tag{5}$$

## EXAMPLE, CONSTITUTIVE MODEL FOR PLAIN CONCRETE

The problem of constitutive modelling of concrete by using elastic strain or work-hardening nonassociated plasticity has been investigated by many researchers [1]. A plasticity based constitutive model for compressive softening of concrete was proposed by Ohtani and Chen [2]. In the region of small plastic strains the behaviour of concrete materials can be described by the associated flow rule [1]. However for larger plastic strains, the discrepancy between the predicted values based on the associated flow rule and the experimental results begins to increase [3].

### Yield Condition and Plastic Potential Function

Our considerations are limited herein to the plane stress state problem. The biaxial compressive yielding criterion [4], widely used for concrete materials is formulated as follows (Fig.1):

$$f(\mathbf{T},\sigma_o) = [\beta(3tr\mathbf{S}^2/2) + \alpha tr\mathbf{T}]^{1/2} - \sigma_o, \tag{6}$$

where $\mathbf{S}=\mathbf{T}-(tr\mathbf{T})\mathbf{I}/3$ is the deviatoric stress, $\beta=1.355$, $\alpha=\alpha'_o\sigma_o$, $\alpha'_o= 0.355$.

The plastic-hardening modulus (2.13) derived from the Madrid parabola is given by the formula [4,5]:

$$H = E_o(\sqrt{\varepsilon_{op}/2\varepsilon_p} - 1) . \tag{7}$$

The plastic potential function takes a similar form to the yield condition (Fig.1):

$$g(\mathbf{T},\sigma_o,\gamma) = [\beta(3tr\mathbf{S}^2/2) + \alpha^* tr\mathbf{T}]^{1/2} - \sigma_o, \tag{8}$$

where $\alpha^*=(1-\gamma+\beta_1\gamma)\alpha$ and $\beta_1$ denotes the dilatancy parameter.

The membership function $\gamma$ is defined as

$$\gamma = [(\sigma_o/\alpha'_o f'_c) - 1]\alpha'_o/(1 - \alpha'_o). \tag{9}$$

where $f'_c$ denotes the ultimate uniaxial compressive strength of concrete. In the softening range the function $\gamma$ is constant and equal to unity, the full departure from associativity is assumed.

Both the loading criterion, the consistency condition and the plastic work condition (1) are met by imposing the following restriction:

$$H > -\sigma_o(\partial f/\partial\mathbf{T}).\mathbf{C}_e/(\partial g/\partial\mathbf{T})(\partial f/\partial\sigma_o)tr(\mathbf{T}\partial g/\partial\mathbf{T}). \tag{10}$$

As the value H is negative for softening material, the restriction (10) imposes additional conditions.

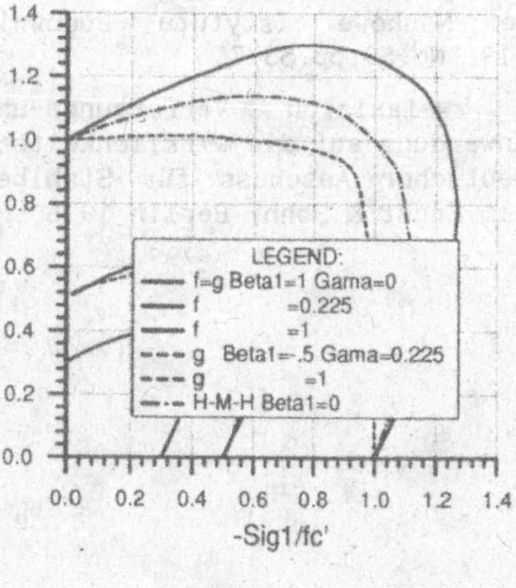


Fig.1

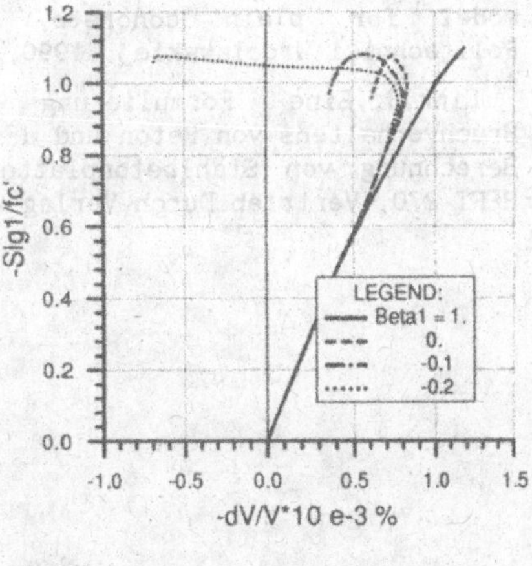


Fig.2

**Comparision with Experimental Results**

The proposed model is compared with experimental data [6]. The uniaxial and biaxial compression of plane concrete is considered. The numerical analysis is performed by direcet step-by-step integration of the constitutive relationship [5]. For the concrete subjected to biaxial compression $\sigma_2/\sigma_1$=0.525, the computed responses are plotted in Fig.2 and the effect of the dilatancy parameter $\beta_1$ is shown. The assumption of dilatancy parameter $\beta_1$=1 is equiponderant with the acceptance of the associated flow rule for the whole loading process. In this case the volumetric response for the stresses greater than about 0.75 $f'_c$ is essentially different than that observed experimentally for plain concrete [6]. For $\beta_1 \neq 1$ we have the nonassociated flow rule, but only the assumption of $\beta_1 \leq 0$ ($\beta_1 \in$ <0,-0.2>) is adequate in describing volumetric response of the plain concrete.

## REFERENCES


1. Chen,W.F., Plasticity in Reinforced Concrete, McGraw-Hill, New York, 1982
2. Ohtani,Y. and Chen, W.F.,A plastic-softening model for concrete materials, Comp. Str.,1989, Vol.33,No.4, pp. 1047-55
3. Hu,H.T. and Schnobrich,W.C., Constitutive modelling of concrete by using nonassociated plasticity, J.Mat.Civ.Engng, 1989,Vol.1,No.4, pp.199-216
4. Owen,D.R.J.,Figueiras, J.A. and Damjanic,F.,Finite element analysis of reinforced and prestressed concrete structures including thermal loading, Comp.Meth.App.Mech.Engng,1983, 41,pp.323-66
5. Jemioło,S. and Lewiński P.M., Elasto-plastic hardening constitutive model for plain concrete, Prace Naukowe Istytutu Budownictwa Politechniki Wrocławskiej, 1990, Vol 19, No.59,pp.55-71
6. Link,J.,Eine Formulierung des zweiaxialen Verformungs-und Bruchverhaltens von Beton und deren Anwendung auf die wirklichkeitsnache Berechnung von Stahlbetonplatten, Deutscher Asschuss für Stahlbeton, HEFT 270, Vertrieb Durch Verlag Wilhelm Ernst & Sohn, Berlin 1976

# A HIGH CAPACITY CUBICAL DEVICE AND MULTI-AXIAL TESTING FOR CONSTITUTIVE MODELING OF CONCRETE

ANANT R. KUKRETI, MUSHARRAF ZAMAN and DHANADA K. MISHRA
School of Civil Engineering and Environmental Science
University of Oklahoma, Norman, OK 73019, USA

## ABSTRACT

A high capacity (30,000 psig) cubical device was designed and fabricated to conduct testing of materials under general three-dimensional loading. The test set-up and procedures are described, and some selected test results and back-prediction using a plasticity-based two surface constitutive model are reported for plain concrete.

## INTRODUCTION

Laboratory testing under three-dimensional (3-D) loading plays an important role in accurately determining the material constants of a constitutive law that describes the load-deformation response of an engineering material. To this end, a high capacity (30,000 psig) cubical device (HCCD) has been designed and fabricated at the University of Oklahoma and is being used to conduct a comprehensive series of tests on pressure sensitive materials such as concrete, rocks and coal. This paper provides an overview of the important components of the HCCD and the procedure developed to conduct tests under general 3-D loading. As an application, load-deformation behavior of plain concrete is investigated using a two surface constitutive model which is based on the theory of plasticity.

## AN OVERVIEW OF THE CUBICAL DEVICE

The HCCD consists of five major components: (i) Cubical space frame and six side walls; (ii) Deformation measuring system; (iii) Hydraulic pressure control system or Servo pumps; (iv) Digital data acquisition system; and (v) Central monitoring unit.

### Cubical Space Frame, Side Walls and Deformation Measurement

A photographic view of the HCCD in assembled form is presented in Fig. 1. The cubical cell consists of a 10" x 10" x 10" rigid steel space frame with 4" x 4" x 4" central cavity and six 10" x 10" square 4" thick steel side walls. The 4" cubical specimen is centered inside the cell and the side walls are bolted on to the space frame. The loads are applied to the specimen using a piston and leather pad arrangement. Each piston has a 1.5" thick and 6" diameter circular base with a 4" square and 2" thick part that transfers the load on to the specimen via a bevelled leather pad. The piston is housed in the side wall and the hydraulic pressure is applied at the back of the circular base which has a double O-ring pressure seal installed. The pressure is thus magnified by a factor of approximately 2.5 on the face of the specimen. The deformation measuring system consists of six Linearly Variable Differential Transducers (LVDT) and SBEL model-250 signal conditioning unit. Each LVDT plunger is directly connected to the piston so as to measure the movement of the piston.

**Hydraulic Pressure System**

The hydraulic system consists of three servo-controlled hydraulic pumps having a capacity of 3,000 psig. Three pressure intensifiers are used to magnify the maximum capacity of the pump by a factor of ten to 30,000 psig. The pumps are connected to dial gages and pressure transducers for both visual inspection and computer storage of the applied loading (pressure). These pumps, driven by the computer software SERVO and the SBEL pressure control system (SBEL model-547) (Fig. 2), are capable of delivering accurate pressure under monotonic and cyclic multi-axial loading conditions.

**Data Acquisition and Central Monitoring Unit**

A 16 channel data acquisition system is used in conjunction with two micro-computers (Fig. 2) for acquisition and storage of test data. Three of the available channels are used for pressure readings and six channels are used for LVDT readings. One of these channels (load or deformation) can be monitored by displaying them continuously during the test. The central monitoring unit consists of a micro-computer, a software called SERVO, the pressure control system, the data conversion cards and other peripheral accessories. The pressure control system has three independent channels to

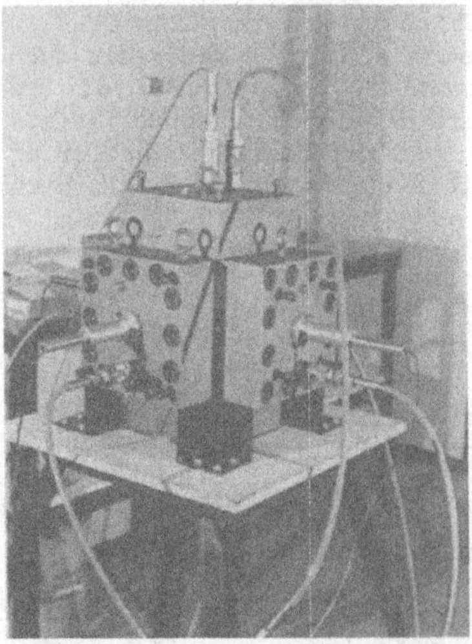

Fig. 1 High Capacity Cubical Device in Assembled Form

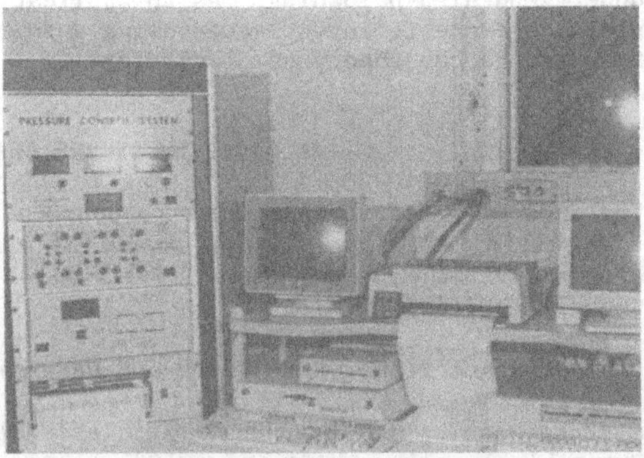

Fig. 2 Pressure Control, Data Acquisition and Monitoring Units

control the three pumps. The software SERVO is used to define independent commands for the three pumps according to the stress path to be followed.

## TYPICAL TEST PROCEDURE

Conducting a typical test using the cubical device consists of the following major steps : (i) Specimen preparation (concrete specimens are used here for illustration); (ii) Calibration; (iii) Assembly and testing; (iv) Processing of test data. A brief description of the test procedure is given below.

(i) The concrete specimens are prepared by casting them in cubical molds of internal dimensions about 1/16" less than 4". Once the specimen is stripped and cured, the surfaces are sand blasted and surface defects are repaired using a resin based compound and finally finished with a sand paper. (ii) Various components of the cubical device such as the LVDTs, the servo control system, the SERVO software and the cubical cell itself must be properly calibrated and the calibration factors stored before conducting a test. (iii) The specimen is placed in the central cavity and the six side walls are mounted and bolted. The bolts are tightened to a pre-specified torque level. The pressure chambers are filled with the hydraulic fluid and the LVDTs are mounted externally. Finally the side walls are connected to the servo pumps such that each pump controls the pressure along a given axis. All the electrical connections are checked and the software SERVO is invoked using DOS. Designated loadings (stress paths), their rates and other pertinent specified data are input and the pumps are switched on. Once the execution command is given, the computer runs the test in an automated mode. At the end of the test duration, the execution stops automatically. (iv) The LVDTs record the combined deformation of the specimen, the leather pad and the cubical cell. To eliminate deformation of the pad and the cell, the test results from calibration tests performed on an aluminum cube and leather pad are used. A computer program has been developed that uses the raw test data and the calibration test data for a given stress path for computing the specimen deformations.

## APPLICATION TO PLAIN CONCRETE

A series of uniaxial, biaxial and triaxial tests are conducted on a 4000 psi design strength plain concrete mix ( cement : sand : gravel = 1.0 : 2.37 : 1.63, cement : water = 1.67 by weight) using the HCCD. A total of 36 cubes were cast in three equal batches along with test cylinders for determining compressive strength. Tests along various stress paths (both compression and extension) were conducted at different confining pressures and the results were compared both qualitatively and quantitatively with similar results in literature [1]. A plasticity-based two surface constitutive model, developed by Faruque and Chang [2] and subsequently modified by Najjar [3], is used to represent the load-deformation response. Ten material constants are needed in this model which consist of two elastic constants, four hardening parameters and four failure parameters. Only selected results are presented here due to space limitation. Figure 3 shows a comparison of the experimental data and model prediction for the triaxial compression (TC) test at a confining pressure of 8 ksi. This test was used in determining the model parameters. The correlation is considered satisfactory from practical considerations. Figure 4 presents a similar comparison for the simple shear (SS) stress path. Although this stress path was not used in the parameter evaluation, the predicted response compares favorably with the experimental data. Experimental data (not presented here) for uniaxial and biaxial loading cases show that the HCCD can be used effectively in simulating these loading conditions also.

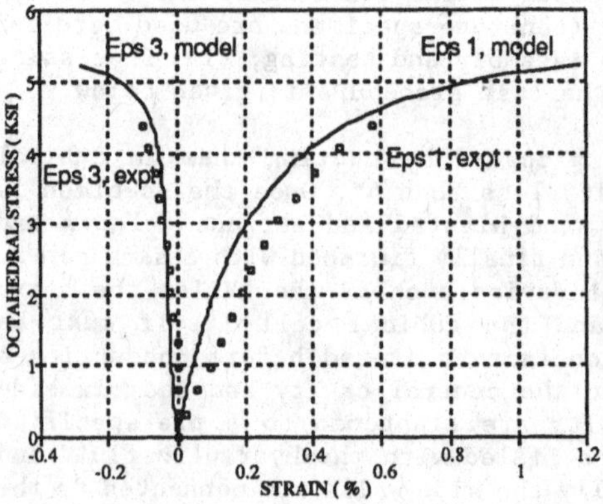


Fig. 3 Test and Predicted Response for TC Stress Path

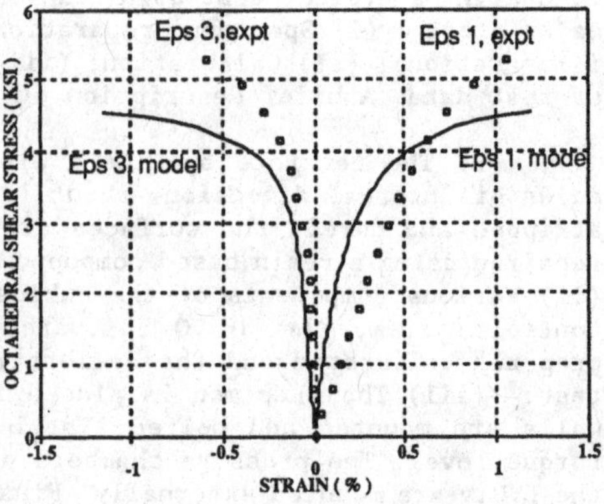


Fig. 4 Test and Predicted Response for SS Stress Path

## CONCLUDING REMARKS

The HCCD is a sophisticated and useful apparatus to study the yielding, fracture and failure behavior of pressure sensitive materials. From the tests conducted so far, the HCCD has proven to be effective in characterizing load-deformation response of materials under uniaxial, biaxial and truly 3-D loading conditions and in developing reliable constitutive laws for various engineering materials.

## ACKNOWLEDGMENTS

The High Capacity Cubical Device was fabricated by a grant from the National Science Foundation (MSM-8705972) and matching funds provided by the Oklahoma Department of Commerce, University of Oklahoma Research Council and Oklahoma Mining and Mineral Resources Research Institute.

## REFERENCES

1. Scavuzzo, R., Stankowski, T., Gerstle, K.H, and Ko, H.Y., Stress-strain curves for concrete under multiaxial load histories. Report, Dept. of Civil, Env. and Arch. Eng., Univ. of Colo., Boulder, CO, USA, 1983.

2. Faruque, M.O. and Chang, C.J., A new cap model for failure and yielding of pressure sensitive materials. ASCE, J. of Eng. Mech. Div., Vol. 112, No. 10, 1986, pp. 1041-1053.

3. Najjar, Y.M., Constitutive modeling and finite element analysis of ground subsidence due to mining. Ph.D. Thesis, Univ. of Okla., Norman, OK, USA, 1990.

# AN ANISOTROPIC HARDENING RULE FOR SATURATED CLAYS

A. S. KUMBHOJKAR
Department of Civil and Architectural Engineering
University of Miami, Coral Gables, FL 33124.

## ABSTRACT

The paper describes a generalized anisotropic hardening rule that uses $d\varepsilon_{ij}^p$ as a hardening parameter. It allows modeling of the change in the degree of anisotropy as a part of updating stress history, as well as inherent and induced anisotropy. The steady state $K_0^{nc}$ expression based on this rule is a function of 4 soil parameters: $\lambda$, $\kappa$, $\phi'$, $\upsilon$, and its predictions compare well with the $K_0^{nc}$ measurements reported in the literature.

## NOTATIONS AND ABBREVIATIONS

d = Prefix showing incremental quantity; $E_{ij}$ = Deviatoric strain tensor; $e_0$ = Initial void ratio; F = Yield function; $J_2$, $J_3$ = Second and third invariants of deviatoric strain tensor; K = Hardening parameter; $K_0^{nc}$ = Coefficient of lateral stress at rest for NC soils; M = Slope of failure line in p-q space; NC, OC = Normally and over consolidated resp; p= Mean effective stress; q = Deviatoric stress; $S_{ij}$, $S_{ij}$ = Deviatoric and reduced deviatoric stress tensor; $\varepsilon_{ij}$, $\sigma_{ij}$ = Principal strain and stress; $\lambda$, $\kappa$ = Slopes of consolidation and swelling lines in e-ln(p) space; $\upsilon$ = Poisson's ratio; $\phi'$ = Effective friction angle; Subscripts : v = Volumetric; d = Deviatoric; Superscripts : 0 = at maximum loading.

## INTRODUCTION

Inherent and induced anisotropy, and changing the degree of anisotropy while updating the stress history are key aspects of anisotropic soil behavior. The success of plasticity based models in simulating these phenomena depends on the capabilities of the yield function and hardening rule or equivalent features adopted in their formulation. The concept of hardening is embedded in the yield function - $F(\sigma_{ij}, K) = 0$ - itself. The classical isotropic and kinematic hardening laws respectively allow a yield surface to expand and translate in principal stress space. Rules which use both these features cause the yield surface to simultaneously expand and translate and, in association with appropriate yield criterion and mechanisms, model effects of anisotropy. For example, the two-surface model of [1] allows the consolidation surface to expand isotropically, and the loading surface shows translation and expansion. Its results however depend upon the assumed initial value of the hardening modulus. A similar approach in [2] uses von Mises yield criterion. The hardening in [3] uses $p^0$ and $S_{ij}{}^0$ as the vehicles for producing anisotropic response. In the majority of anisotropic soil models, $\varepsilon_v{}^p$ plays the role of hardening parameter. This approach is reasonable for isotropic NC soils, but proves unsatisfactory for OC as well as other anisotropic clays. A rational way of modeling the behavior of these soils is using $\varepsilon_d{}^p$ as a hardening parameter. Formulations in which both $\varepsilon_v{}^p$ and $\varepsilon_d{}^p$ act simultaneously as a combined hardening parameter have also been proposed (for a review refer [4]). The significance of such rules depends on the versatility and the extent they predict the behavior realistically. Any anisotropic hardening law should be a function of the $d\varepsilon_{ij}{}^p$. This paper describes a hardening function that meets these criteria.

## THE GENERALIZED ANISOTROPIC HARDENING RULE AND ITS PREDICTIONS

The proposed general expression for the hardening law is :

$$d\sigma^0_{ij} = \frac{1+e_0}{\lambda-\kappa}\frac{\sigma^0_{ii}}{3}[(1-\frac{2}{9}\hat{M})\,d\varepsilon^p_{ii}+\frac{2}{3}\hat{M}d\varepsilon^p_{ij}] = \frac{1+e_0}{\lambda-\kappa}[(d\varepsilon^p_{ii})+\frac{2}{3}\hat{M}(d\varepsilon^0_{ij}-\frac{1}{3}d\varepsilon^p_{ii})] \quad (1)$$

$$\text{Where, } \hat{M} = g(\underline{\theta})-\frac{3\sqrt{3}\,\bar{J}2}{\sigma_{ii}}, \qquad g(\underline{\theta}) = \frac{6\sin\phi'}{3-\sin\phi'\sin 3(\underline{\theta})},$$

$$\underline{\theta} = \frac{1}{3}\sin^{-1}(\frac{3\sqrt{3}}{2}\frac{\bar{J}^3_3}{\bar{J}^3_2}), \qquad \bar{J}^2_2 = \frac{1}{2}\bar{S}_{ij}\bar{S}_{ij}, \qquad \bar{J}^3_3 = \frac{1}{3}\bar{S}_{ij}\bar{S}_{jk}\bar{S}_{ki},$$

$$\bar{S}_{ij} = S_{ij}-\frac{2}{3}\frac{\sigma_{ii}}{\sigma^0_{ii}}S^0_{ij} \qquad S_{ij} = \sigma_{ij}-\frac{1}{3}\sigma_{ii}, \qquad S^0_{ij} = \sigma^0_{ij}-\frac{1}{3}\sigma^0_{ii}$$

The form of this expression is compatible with the anisotropic yield function [3] given in Eqn (2) which represents a series of distorted ellipsoids asymmetrically circumjacent to the line connecting $\sigma_{ij}{}^0$ to the origin of principal stress space.

$$F(\sigma_{ij},\, S^0_{ij},\, p^0) = \frac{3}{2}\frac{1}{g^2(\underline{\theta})}\{\,\bar{S}_{ij}\bar{S}_{ij}-\frac{1}{9}\frac{p}{p^0}s^0_{ij}S^0_{ij}\,\}-pp^0+p^2 = 0 \quad (2)$$

For axisymmetric conditions ($\theta=\pi/6$) Eqn (1) reduces to Eqn (3).

$$dp^0 = \frac{1+e_0}{\lambda-\kappa}p^0 d\varepsilon^p_v, \qquad dq_0 = \frac{1+e_0}{\lambda-\kappa}p^0(M-\frac{q}{p})\,d\varepsilon^p_d \quad (3)$$

Integrating the equation for $dp^0$, we get the well-known 1-D or isotropic consolidation and swelling equations, $de = -\lambda\ln(p)$ and $de = -\kappa\ln(p)$, that have been routinely used as a basis for modeling hardening behavior of saturated clays. The oedometer and isotropic consolidation tests yield measurements of only volumetric compression caused by respectively axial and isotropic stress increments, they provide no basis for modeling deviatoric hardening. It is probably the reason why majority of the hardening rules consider only $d\varepsilon_v^p$ as a hardening parameter, and can not account for the effects of deviatoric strain. In association with the anisotropic yield function, $dp^0$ and $dS_{ij}^0$ allow simultaneous translation, expansion, (kinematic and isotropic hardening) and rotation of the yield surface. It thereby allows simulation of consolidation from initial hydrostatic to an anisotropic state by gradually changing the orientation of the yield surface as shown in Figure 1.

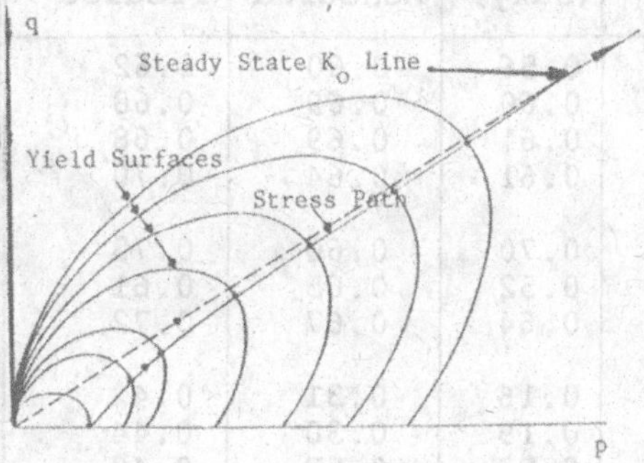


Figure 1: 1-D consolidation of soft clay

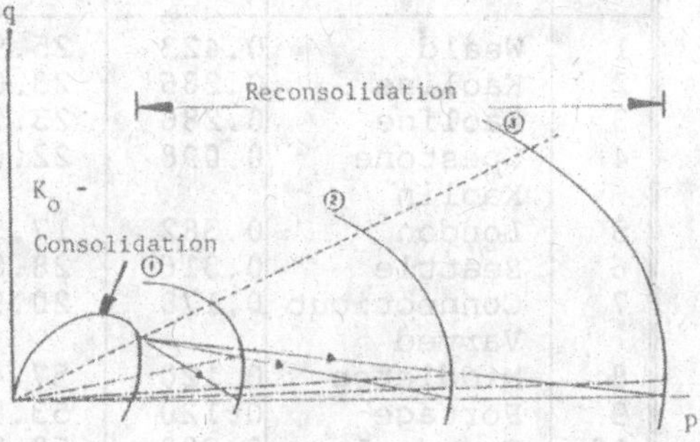


Figure 2: Reconsolidation of $K_0$ clay

When $dq^0/dp^0$ becomes equal to $q^0/p^0$ (Eqn.3), the yield surface no more rotates and the soil reaches a steady state $K_0$ condition. Figure 2 shows an opposite case in which initially anisotropic soil becomes almost isotropic when it is reconsolidated under large isotropic stress increment. This phenomenon is not softening as such, but is simply a reflection of soil response to the applied stress regime. The hardening rule also satisfies the condition at plastic failure, i.e., at failure q/p equals M and the deviatoric stress no more increases. The rule also provides the following expression for steady state $K_0^{nc}$ [4].

$$K_0^{nc} = \frac{6a+b+\sqrt{b^2-4ac}}{6a-2b-2\sqrt{b^2-4ac}} \tag{4}$$

*Where,* $a = \frac{2\kappa}{\lambda}(1+\nu)$, $c = 6(1-2\nu)M$, $b = -[\frac{c}{M}+aM+\frac{3c}{2M}(1-\frac{\kappa}{\lambda})]$

For elastic and perfectly plastic soils, Eqn (4) respectively reduces to $\upsilon/(1-\upsilon)$ and $(5-3\sin\phi')/(5+\sin\phi')$. The first is the well-known elastic relationship, and the second expression, a function of $\phi'$ alone, is similar to Jaky's equation obtained on the assumption of soil as a perfectly plastic material. Table 1 provides a comparison of predictions of Eqn (4) with the $K_0^{nc}$ values reported in the literature for which $\kappa/\lambda$ and $\phi'$ data are also available (references are given in [4]). The Poisson's ratio of 0.3 is assumed in all cases. The reported $K_0^{nc}$ measurements are of course likely to be affected by a large number of factors, and it is difficult to account for their influence. But the good agreement between the predicted and measured values supports the validity of the hardening rule. It also suggests that $K_0^{nc}$, apart from $\phi$, $\kappa/\lambda$ and $\upsilon$ also affect $K_0^{nc}$.

TABLE 1
COMPARISON OF MEASURED AND PREDICTED $K_0^{nc}$ VALUES

| No | Name of Clay/Peat | $\kappa/\lambda$ | $\phi'$ | $K_0^{nc}$ (Jaky) | $K_0^{nc}$ Measured | $K_0^{nc}$ Predict |
|---|---|---|---|---|---|---|
| 1 | Weald | 0.423 | $25.9^0$ | 0.56 | 0.60 | 0.62 |
| 2 | Kaoline | 0.286 | $23.0^0$ | 0.60 | 0.69 | 0.68 |
| 3 | Kaoline | 0.286 | $23.5^0$ | 0.61 | 0.69 | 0.68 |
| 4 | Spestone Kaolin | 0.098 | $22.6^0$ | 0.61 | 0.64 | 0.70 |
| 5 | London | 0.382 | $17.5^0$ | 0.70 | 0.60 | 0.73 |
| 6 | Seattle | 0.316 | $28.8^0$ | 0.52 | 0.65 | 0.61 |
| 7 | Connecticut Varved | 0.178 | $20.9^0$ | 0.64 | 0.67 | 0.72 |
| 8 | Middleton | 0.120 | $57.4^0$ | 0.16 | 0.31 | 0.42 |
| 9 | Portage | 0.120 | $53.8^0$ | 0.19 | 0.30 | 0.44 |
| 10 | Fon du Lac | 0.200 | $50.2^0$ | 0.23 | 0.53 | 0.48 |

## SUMMARY AND CONCLUSIONS

A generalized anisotropic hardening rule is described. It uses $d\varepsilon_{ij}^p$ as a hardening parameter. The rule allows simulation of all key features of anisotropic behavior and provides an explicit expression for steady state $K_0^{nc}$ as a function of $\kappa/\lambda$, $\phi'$ and $\upsilon$. The $K_0^{nc}$ predictions agree with the available data.

## ACKNOWLEDGEMENTS

The research described herein is funded by NSF under grant no. MSS-9009156. The author gratefully acknowledges this support.

## REFERENCES

1. Pietruszczak, S., and Mroz, Z., "On Hardening Anisotropy of $K_0$-Consolidated Clays," Int. J. for Num. and Anal. Methods in Geomech., Vol. 7, No. 9, 1983, pp. 19-36.
2. Prévost, J.H, "Plasticity Theory for Soil Stress-Strain Behavior," Journal of the Soil Mechanics and Foundation Division, ASCE, Vol. 104, EM5, 1978, pp. 1177-1194.
3. Banerjee, P.K., and Yousif, N.B., "A Plasticity Model for Anisotropically Consolidated Clay," Int. J. for Num. Anal. Meth. in Geomech. 10, 1986, pp. 521-541.
4. Kumbhojkar, A.S., "Theoretical and Numerical Analysis of Geotechnical Structures, " dissertation presented to SUNY at Buffalo, NY, 1987, p.205.

# A DOUBLE HARDENING CONSTITUTIVE MODEL FOR BULK SOLIDS

LAURENT LANCELOT and ISAM SHAHROUR

*Laboratoire de Mécanique de Lille*
*CNRS URA 1441D*
*Département Sols-Structures*
*IDN - BP 48*
*59651 Villeneuve d'Ascq Cedex - FRANCE*

## ABSTRACT

In this paper, an elasto-plastic constitutive model for chemical bulk solids at low stresses, developped using experimental results on two powders (organic and mineral), is presented. A first version, including a Mohr-Coulomb failure criterion with an isotropic strain-hardening rule and a non-associated plastic potential, is shown to be insufficient to reproduce experimental data, especially on stress paths involving low deviatoric stress levels like oedometric tests. In an ameliorated version of the model, a second plastic mechanism is added. It includes a purely volumic strain-hardening rule and an associated plastic potential. The model is then validated on various stress paths.

## INTRODUCTION

The increasing production capacity of chemical and food industry, especially in the form of bulk solids, calls for more reliable storage structures. Silo design is usually carried out using Jenike's theory for the determination of stresses in the structure as well as inside the bulk solids (Jenike, 1964). However, this theory does not account for the deformations and other phenomena such as dilatancy.

The recent advances in numerical calculation (especially non-linear) allows the resolution of the problems of silo design, provided an appropriate constitutive model is available for bulk solids.

The study undertaken in the Laboratory of Mechanics of Lille involves two aspects :

- the development of a new specific triaxial equipment adapted for low stresses,
- the development and validation of a constitutive model based on experimental data obtained with the new testing equipment.

The first part of the study has been described elsewhere (Lancelot & Shahrour, 1991). The powders tested – a crystallized organic acid (powder A) and a very fine mineral powder (powder B) – showed a behaviour similar to that of soils. Concepts of soils modelling (critical state, characteristic threshold, strain hardening...), have thus been used to develop the constitutive model presented here.Typical simulations of oedometric and triaxial compression tests on powder A are then given.

## Elasticity

A non-linear elastic behaviour has been assumed for bulk solids. It can be expressed by :

$$E = E_A \cdot \frac{p}{P_A} \tag{1}$$

$$\nu = \nu_o \tag{2}$$

where E and $\nu$ are Young modulus and Poisson's ratio, p is the mean pressure, $E_A$ and $\nu$ are constitutive parameters, and $P_A$ is a reference pressure.

## Yield surface

A Mohr-Coulomb failure criterion has been chosen (Lancelot & Shahrour, 1991). Strain-hardening is governed by a function $R_s$ expressing the evolution of the yield surface expansion in (p,q) plane :

$$f_s(p, q, \kappa) = q - M_f\,(p + C)\,R_s \tag{3}$$

$$R_s = 1 - \exp(-a\,\varepsilon_d{}^p) \tag{4}$$

where q is deviator stress, $\varepsilon_d{}^p$ is plastic shear strain. $M_f$, C and *a* are constitutive parameters.

## Flow rule

Plastic strain increment is supposed to derive from a plastic potential g.

For the powders under investigation, the *characteristic state* concept is satisfied (Lancelot & Shahrour, 1991). Moreover, the ultimate stabilization of $\varepsilon_v{}^p$ observed for powders A and B ("*critical state*") has to be reproduced by the model. The following expression has been assumed for the flow rule :

$$\frac{\partial g}{\partial p} = A_g\left(M_g - \frac{q}{p}\right) \tag{5}$$

where

$$A_g = e^{-\alpha_g \cdot \varepsilon_d^p} \tag{6}$$

$$\frac{\partial g}{\partial q} = 1 \tag{7}$$

These expressions introduce the constitutive parameters $M_g$ and $\alpha_g$.

## Simulation of triaxial and oedometric tests

The model has been used to simulate oedometric (figure 1) and triaxial tests (figure 2) on powder A.

Triaxial tests are well reproduced, though strain-softening is not taken into account by the model. However, for stress paths involving low deviatoric stress levels q/p, such as in oedometric tests, the plastic deformations generated by the model are much too small, resulting in a (non-linear) elastic behaviour.

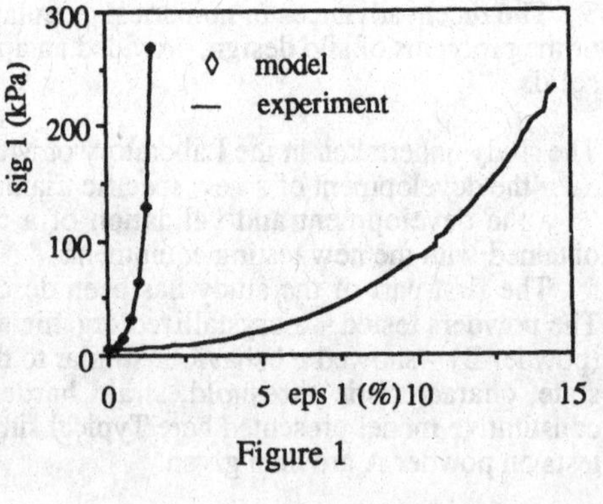


Figure 1

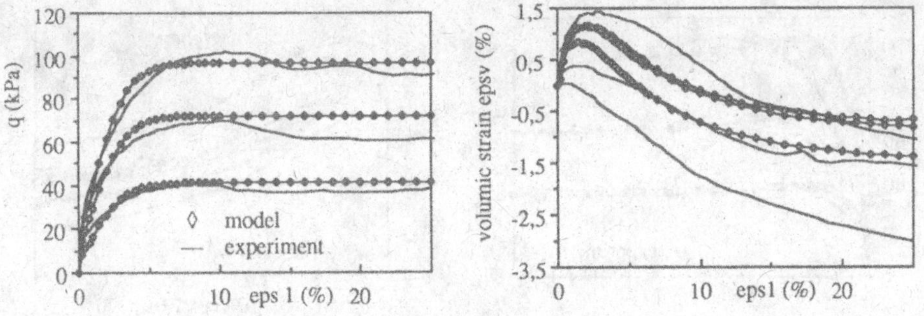


Figure 2

## A DOUBLE-MECHANISM STRAIN-SOFTENING MODEL

### "Cap" yield surface

As a consequence of the simulation for the oedometric test (figure 2), the conical yield surface for the first version of the model has to be completed by a "cap" so that this additional mechanism of plasticity can be mobilized along low deviatoric stress level paths (Lancelot & Shahrour, 1991).

In the present model, a purely volumic yielding mechanism defining a plane yield surface, with an associated flow rule, is introduced :

$$f_c\,(p, \varepsilon_v^p) = \left(\frac{p}{P_{co}}\right)^{1/2} - R_c \qquad (8)$$

$$R_c = \exp\,(\beta\,\varepsilon_v^p) \qquad (9)$$

where $P_{c0}$ and $\beta$ are constitutive parameters.

### Strain softening

As mentioned before, a softening behaviour can be noticed in triaxial test results.

The expression of $R_s$ has then to be modified so that $R_s$ passes through a maximum and then decreases until it stabilizes for large strains. The new expression for $R_s$ is :

$$R_s(\varepsilon_d p\,, \varepsilon_v p\,) = [1 - \exp(-a\,\varepsilon_d p\,)]\,\exp(\mu\,\varepsilon_v p) \qquad (10)$$

where $\mu$ is a new parameter setting the stress peak magnitude.

### Typical simulations

The oedometric test on powder A is now satisfactorily reproduced, although plastic strains are still slightly underestimated (figure 3).

Concerning the simulations of triaxial compression tests on powder A (figure 4), the most representative features of mechanical behaviour are well reproduced: hardening, stress peak, softening and eventual stabilization of deviator stress and volume change for large strains.

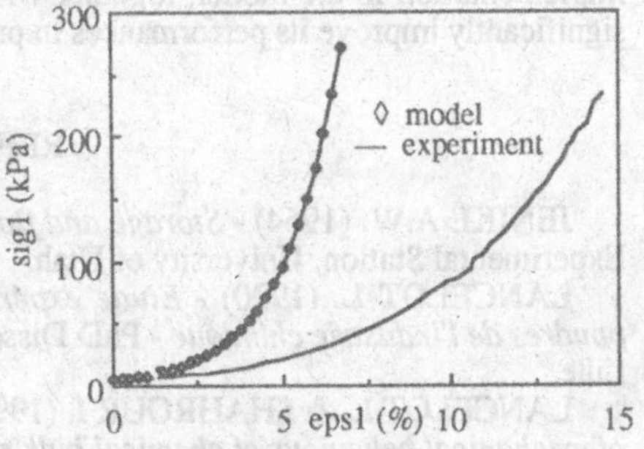


Figure 3

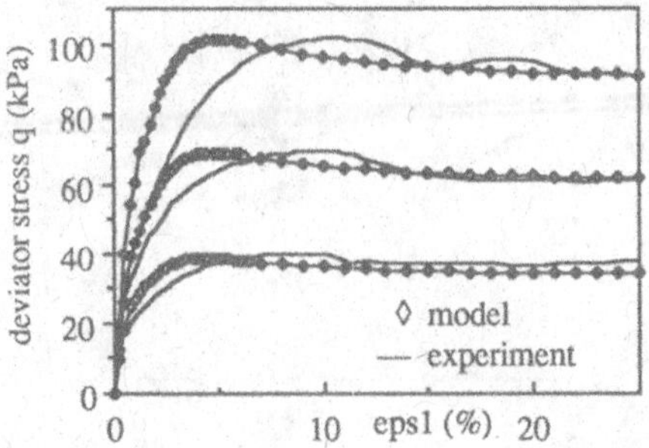


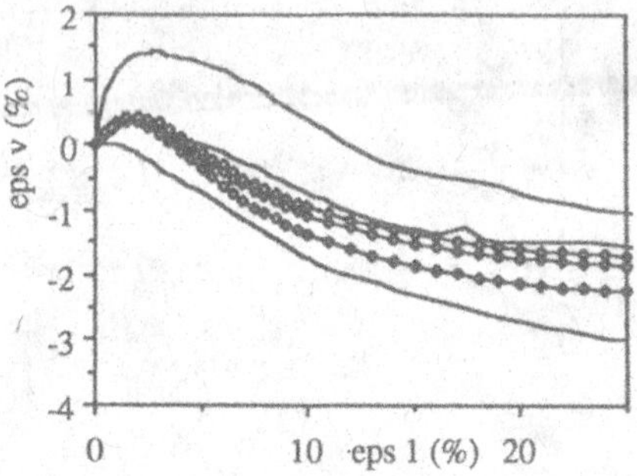


Figure 4

## DISCUSSION

It should be noticed that sample preparation procedure (oedometric loading by compaction in a mould), then sample testing (isotropic loading under given confining pressure, and deviatoric loading until failure) have been taken into account for the simulations (Lancelot, 1990). The elastic radius $R_0$ after oedometric loading due to sample preparation can reach 0.9. Subsequent deviatoric loading in triaxial tests thus remains elastic until $R_s$ reaches $R_0$ (isotropic strain hardening).

This kind of "stress inversion" actually occurs at the transition between the state of stress in a silo after filling, and incipient flow when it is beginning to be emptied. A more progressive and more realistic plastification of the material would be obtained if a kinematic strain hardening rule were chosen for the model.

Moreover, a viscous behaviour should also be taken into consideration for the simulation of flow problems.

## CONCLUSION

The model in its final version is shown to satisfactorily reproduce the response of both crystallized organic powder A and fine mineral powder B (Lancelot, 1990) to homogeneous sollicitations – oedometric, triaxial compression, triaxial extension (Lancelot, 1990). The model is currently used to solve boundary value problems, such as the determination of stress fields in silos.

However, isotropic strain hardening is found to be insufficient to reproduce stress paths involving large stress inversions (such as oedometric loading-unloading followed by triaxial reloading). A kinematic strain hardening rule is required on such stress-paths. Its implementation in the model, together with the implementation of viscosity effects, would significantly improve its performances in problem solving of flow in silos.

## REFERENCES


JENIKE A.W. (1964) - *Storage and flow of bulk solids* - Bulletin 123, Utah Engineering Experimental Station, University of Utah.

LANCELOT L. (1990) - *Étude expérimentale et modélisation du comportement des poudres de l'industrie chimique* - PhD Dissertation, Université des Sciences et Techniques de Lille.

LANCELOT L. & SHAHROUR I. (1991) - *Experimental study and numerical modelling of mechanical behaviour of chemical bulk solids* , to be published in Powder Technology.

# ANALYTICAL STUDY OF THE COALESCENCE OF CAVITIES IN DUCTILE FRACTURE OF METALS

JEAN-BAPTISTE LEBLOND and GILLES PERRIN
*Laboratoire de Modélisation en Mécanique,*
*Université Paris VI, Tour 66, 4 place Jussieu, 75005 Paris, France*
*and Laboratoire de Mécanique des Solides,*
*Ecole Polytechnique, 91128 Palaiseau, France*

## ABSTRACT

*The failure of ductile porous metals generally occurs by coalescence of cavities. This phenomenon has been studied by Tvergaard, Becker* et al. *and Koplik and Needleman by the finite element method, using cell models which represent an elementary volume in a periodic porous medium. An analytical approach to the problem is presented, based on the Rudnicki-Rice-Yamamoto analysis of the localization of deformation for pressure-sensitive materials. It reveals essential to account for the porosity inhomogeneities induced by the deformation; provided this is done, critical porosities for coalescence quite close to those found numerically by Tvergaard, Becker* et al. *and Koplik and Needleman are obtained.*

## INTRODUCTION

Ductile failure of metals generally occurs through nucleation of cavities around inclusions, growth of the former by plasticity and final coalescence. The first two steps are currently described by using the classical macroscopic Gurson [1] model. The third one is generally accounted for by introducing an extra parameter in this model, the *critical porosity for coalescence*. The value of this parameter can be deduced from numerical studies of periodic porous media carried out by various authors [2, 3, 4]. It has been observed to be primarily a function of the *initial* porosity [4].

The aim of this paper is to present, for the first time, an *analytical* approach of the problem allowing for the calculation of the critical porosity. The validity of the theory is assessed through comparison of the critical porosities predicted with those deduced from numerical studies [2, 3, 4]. The success encountered provides motivation for future extensions to more general conditions than those considered in the numerical studies (other loading conditions, random distributions of voids, ...).

Since void coalescence occurs through concentration of the deformation in the ligaments between the cavities, it is natural to try to describe it by using the well-known formalism of strain localization in shear bands. This formalism cannot be applied at the microscopic scale; indeed the material obeys the classical Lévy-Von Mises theory on that scale, and it is known from the work of Rudnicki and Rice [5] that strain localization cannot occur in such a medium for positive values of the hardening modulus. However, on the macroscopic scale, the material exhibits a constitutive softening due to the progressive increase of the porosity; hence strain localization can be envisaged for the "homogenized" model (such as Gurson's) describing the macroscopic behaviour.

In fact, under this simple form, this idea has already been put forward by Tvergaard [2] and observed to yield poor results. The main point of the present paper is to show that an excellent correlation with the results of the numerical studies can however be obtained provided the *porosity inhomogeneities induced by the deformation* are taken into account.

## SIMPLE LOCALIZATION ANALYSIS WITHOUT POROSITY INHOMOGENEITIES

The Gurson yield criterion for a porous ideal-plastic metal reads, with classical notations:

$$\frac{\Sigma_{eq}^2}{\sigma_0^2} + 2qf\cosh\left(\frac{3}{2}\frac{\Sigma_m}{\sigma_0}\right) - 1 - q^2f^2 = 0 .$$

In this expression the Tvergaard q-parameter [2] has been introduced in order to account for interactions between cavities.

The possibility of strain localization for a material obeying such a criterion has been studied by Yamamoto [6], using the formalism previously developed by Rudnicki and Rice [5] for general pressure-sensitive media. If one specializes to an axisymmetric loading with major principal stress directed along the symmetry axis ($\Sigma_{11} = \Sigma_{22} < \Sigma_{33}$), as considered in numerical studies [2, 3, 4], the criterion for strain localization reads

$$\frac{6\sigma_0}{E}q^2(1-f)f\sinh\left(\frac{3TX}{2}\right)\left[\cosh\left(\frac{3TX}{2}\right) - qf\right] = \left[X - qf\sinh\left(\frac{3TX}{2}\right)\right]^2 .$$

In this equation $T = \frac{\Sigma_m}{\Sigma_{eq}} = \frac{(2\Sigma_{11} + \Sigma_{33})/3}{\Sigma_{33} - \Sigma_{11}}$ is the triaxiality and $X = \frac{\Sigma_{eq}}{\sigma_0} = \frac{\Sigma_{33} - \Sigma_{11}}{\sigma_0}$ is connected to f and T through the criterion. This formula results from a calculation of the critical modulus for an arbitrary orientation of the localization plane, followed by a maximization of this modulus over all possible orientations.

Application of this equation to the cases considered in [2, 3, 4] unfortunately yields much too high critical porosities for void coalescence, especially in the case of low triaxialities. In fact, the failure of this first, simple approach is not very surprising, since the above criterion for localization does not include any effect of the *initial* porosity, whereas the critical porosity is known to essentially depend upon this parameter.

## INTRODUCTION OF POROSITY INHOMOGENEITIES

*Figure 1* illustrates the development of porosity inhomogeneities under the effect of deformation. One sees that the cavities progressively concentrate in horizontal bands. It is obvious that strain localization will occur there rather than in the sound material. This suggests to try to apply the localization criterion to the porous bands instead of the whole volume.

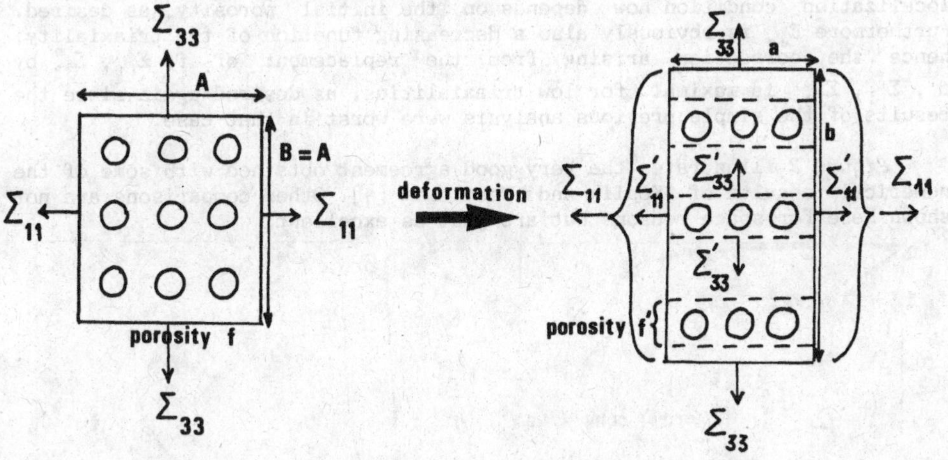


*Figure 1: Strain-induced porosity inhomogeneities*

The first task is to evaluate the porosity f' of the bands and the stresses $\Sigma'_{11}$, $\Sigma'_{33}$ exerted on them. f' is easily calculated as

$$f' = f\frac{b}{a} = f\exp(3E_{eq}/2)$$

where a and b are defined on *Figure 1* and $E_{eq}$ is the Von Mises equivalent strain. Furthermore, equilibrium on the vertical axis implies that $\Sigma'_{33} = \Sigma_{33}$. Finally, $\Sigma'_{11}$ can be deduced from application of the yield criterion to the band (with porosity f'):

$$\frac{(\Sigma_{33} - \Sigma'_{11})^2}{\sigma_0^2} + 2qf'\cosh\left(\frac{\Sigma'_{11} + \Sigma_{33}/2}{\sigma_0}\right) - 1 - q^2f'^2 = 0 .$$

One must now express the localization criterion in terms of the parameters f', $\Sigma'_{11}$, $\Sigma'_{33}$ of the band. When doing so, one must *not* maximize the critical modulus with respect to the orientation of the localization plane as before, since this orientation is no more arbitrary but imposed by that of the porous band. This leads to the condition

$$\frac{3(1-\nu)\sigma_0}{E}q^2(1-f')f'\sinh\left(\frac{3T'X'}{2}\right)\left[\cosh\left(\frac{3T'X'}{2}\right) - qf'\right] = \left[X' - qf'\sinh\left(\frac{3T'X'}{2}\right)\right]^2$$

where $T' = \dfrac{(2\Sigma'_{11} + \Sigma_{33})/3}{\Sigma_{33} - \Sigma'_{11}}$ is the triaxiality in the band and

$$X' = \frac{\Sigma_{33} - \Sigma'_{11}}{\sigma_0}.$$

Since the equivalent strain $E_{eq}$ at localization is obviously a decreasing function of the initial porosity and f' depends on $E_{eq}$, the localization condition now depends on the initial porosity, as desired. Furthermore $E_{eq}$ is obviously also a decreasing function of the triaxiality; hence the correction arising from the replacement of f, $\Sigma_{11}$, $\Sigma_{33}$ by f', $\Sigma'_{11}$, $\Sigma'_{33}$ is maximal for low triaxialities, as desired again since the results of the simple previous analysis were worst in that case.

*Figure 2* illustrates the very good agreement obtained with some of the numerical results of Koplik and Needleman [4]. Other comparisons are not shown here for space reasons but are just as excellent.

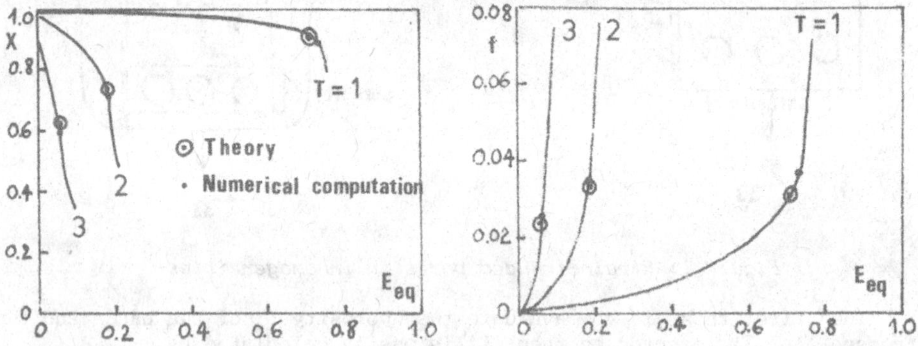


*Figure 2: Comparison with some numerical results [4] ($f_0$ = 0.0013)*

## REFERENCES


1. Gurson, A.L. *Continuum theory of ductile rupture by void nucleation and growth. Part I: Yield criteria and flow rules for porous ductile media.* **J. Engng. Mater. Technol.**, 1977, **99**, 2-15.
2. Tvergaard, V. *On localization in ductile materials containing spherical voids.* **Int. J. Fract.**, 1982, **18**, 237-252.
3. Becker, R., Needleman, A., Richmond, O. and Tvergaard, V. *Void growth and failure in notched bars.* **J. Mech. Phys. Solids**, 1988, **36**, 317-351.
4. Koplik, J. and Needleman, A. *Void growth and coalescence in porous plastic solids.* **Int. J. Solids Structures**, 1988, **24**, 835-853.
5. Rudnicki, J.W. and Rice, J.R. *Conditions for the localization of deformation in pressure-sensitive dilatant materials.* **J. Mech. Phys. Solids**, 1975, **23**, 371-403.
6. Yamamoto, H. *Conditions for shear localization in the ductile fracture of void-containing materials.* **Int. J. Fract.**, 1978, **14**, 347-365.

# PLASTIC FLOW, INSTABILITY AND MODES OF FAILURE IN THE UNIAXIAL COMPRESSION OF POROUS CYLINDERS

J.H.LEE, Y.ZHANG

*Department of Mechanical Engineering*
*University of Alaska Fairbanks, AK 99775, U.S.A.*

## ABSTRACT

Gurson's mixed hardening plasticity model, with strain- and stress-controlled nucleations, was used in a large deformation finite element program to study the plastic flow and damage in the uniaxial compression of cylinders under sticking friction. Modes of deformations, failure and fracture criteria were studied in detail.

## INTRODUCTION

Formability, or workability, is the limit of metal-forming processes that will not result in fracture. Axial compression of cylinders [1] is a relatively simple method for determining the workability of materials. Existing finite element analyses e.g. [2], however, failed to predict fracture, probably because of the following drawbacks. First, rigid-plastic models were used which preclude elastic unloading, which usually accompanies or precedes failure. Second, no fracture criterion was incorporated a priori into the finite element model. Third, the materials of the workpieces were assumed to be homogeneous and void-free; consequently, the progressive damage of the material (due to the presence of voids) cannot be taken into account.

At least for steels, it is clear from metallurgical evidence that final fracture in the upsetting of cylinders is due to the formation of voids. Subsurface voids were found *prior* to final failure. Void formation had been related to plastic instability that was manifested in strain perturbation [3].

The purpose of the present investigation is to study, using the finite element method, the effects of void nucleation, growth and coalescence on the plastic flow, damage, instability and failure in the uniaxial compression of porous cylinders. Effects of strain hardening, nucleation models, yield surface curvature and geometry on the modes of deformation are studied in detail. The possibility of coalescence is also examined. Selected results in localization and failure for various material models and geometries are then presented. Existing fracture criteria and instability conditions are then assessed using the present results.

## FINITE ELEMENT ANALYSIS

The analysis in this paper uses the backward Euler numerical integration method developed in [4, 5] applied to Gurson's mixed hardening model [6].

The yield function in stress space for Gurson's mixed hardening model is [6]

$$\begin{aligned}\Phi &= \Phi(\Sigma_{ij}, A_{ij}, \sigma_F, f)\\ &= \frac{3}{2}\frac{B'_{ij}B'_{ij}}{\sigma_F^2} + 2fq_1 cosh(\frac{3B_m}{2\sigma_F}) - (1+q_2f^2),\end{aligned} \qquad (1)$$

where $\Sigma_{ij}$ are the macroscopic total stresses, $A_{ij}$ are the back stresses and $B_{ij}$ are the shifted stresses.

$$B_{ij} = \Sigma_{ij} - A_{ij}. \tag{2}$$

$B'_{ij}$ is the deviatoric portion and $B_m$ is the spherical portion of the shifted stress. $f$ is the void fraction, $q_1$ and $q_2$ are constants to improve the response of the original Gurson's model.

We assume for mixed hardening,

$$\sigma_F = (1-b)\sigma_y + b\sigma_e, \quad 0 \leq b \leq 1, \tag{3}$$

where $\sigma_y$ is the constant initial yield strength of the matrix, $\sigma_e$ is the yield strength of the matrix.

Evolution of voids is assumed to consist of growth and nucleation portions

$$\dot{f} = \dot{f}_{growth} + \dot{f}_{nucleation},$$

where

$$\dot{f}_{growth} = 3(1-f)\dot{\bar{\epsilon}}^p, \tag{4}$$

and $\dot{\bar{\epsilon}}^p$ is the mean plastic strain rate. When nucleation is dominated by the maximum normal stress, we assume

$$\dot{f}_{nucleation} = \frac{\hat{K}}{\sigma_F}(\dot{\sigma}_e + \frac{\dot{\Sigma}_{kk}}{3}), \tag{5}$$

where $\hat{K}$ is assumed to be statistically distributed:

$$\hat{K} = \frac{k_N}{s_N\sqrt{2\pi}} exp\{-\frac{1}{2}(\frac{\sigma_e + \Sigma_m - \sigma_N}{s_N \sigma_y})^2\}, \tag{6}$$

where $k_N$ is the dimensionless amplitude of the nucleation function, $s_N$ is the dimensionless standard deviation of $\sigma_y$, $\sigma_N/\sigma_y$ is the dimensionless mean stress for nucleation.

When nucleation is dominated by plastic strain, we have

$$\dot{f}_{nucleation} = \hat{F}\dot{\bar{\epsilon}}^p, \tag{7}$$

$$\hat{F} = \frac{\psi}{s_N\sqrt{2\pi}} exp\{-\frac{1}{2}(\frac{\bar{\epsilon}^p - \epsilon_N}{s_N})^2\} \tag{8}$$

where $\bar{\epsilon}^p$ is the accumulated equivalent plastic strain and $\dot{\bar{\epsilon}}^p = (2/3\dot{\gamma}^p_{ij}\dot{\gamma}^p_{ij})^{1/2}$. $\psi$ is the amplitude, $s_N$ is the standard deviation in the equivalent plastic strain rate and $\epsilon_N$ is the mean equivalent plastic strain.

The stress-strain curve of the matrix is expressed as

$$\frac{\sigma_e}{\sigma_y} = (\frac{\sigma_e}{\sigma_y} + \frac{3G}{\sigma_y}\bar{\epsilon}^p)^N.$$

A strain hardening exponent $N = 0.1$ simulates a low hardening material; a strain hardening exponent $N = 0.2$ simulates a high hardening material. The Young's modulus is $207,000MPa$, the Poisson's ratio is 0.3 and the initial yield strength ($\sigma_y$) is $302MPa$. $q_1 = 1.5$, $q_2 = 2.25$.

For plastic strain-controlled nucleation, $\psi = 0.04$. We have chosen two mean strains 0.3 and 0.6 to represent low and high nucleation strains. The standard deviation used is 0.1. For stress-controlled nucleation, $\sigma_N/\sigma_y$ is 2.2, $k_N$ is 0.04 and $s_N = 0.1$. To avoid uncertainty in friction only sticking friction will be used.

For the case $H_0/D_0 = 1$, $\psi = 0.3$, $\Sigma_m/\sigma_e$ versus reduction for four integration points of elements 9 and 10 near the equator, is shown in Fig. 1. The legends for all figures are: $N = 0.1$ (solid circle), $N = 0.2$ (blank circle), von Mises $N = 0.1$ (blank squre). Fig. 2 presents void fraction versus reduction for the same points. The percent of void fraction due to growth (0.02) of the case in Fig. 1 is presented in Fig. 3. The nucleation history for stress-controlled case is given in Fig. 4 and is very much different from that of strain-controlled case. Fig. 5 presents the unloading zone and mesh for the von Mises kinematic hardening model ($b = 0$).

For an integration point undergoing homogeneous deformation, the necessary conditions for incipient strain localizations are analyzed following the works of [7] and [8]. Results of localizations for case of Fig. 1 ($N = 0.1$) at selected stages of deformations are given in Table 1. $0 \le \theta \le 2\pi$, $0 \le \phi \le \pi/2$ are the spherical angles. $\psi$ indicates the mode of localization: $\psi = 90.0$ is pure shear, $\psi = 0.0$ is normal separation. $\Delta$ is the overall strain hardening parameter: $\Delta > 0$ is hardening, $\Delta < 0$ is softening.

## RESULTS AND CONCLUSIONS

Kinematic Gurson and kinematic von Mises models produce a mode of failure (Fig. 5) not in accordance with experimental observations and are not appropriate to the upsetting process studied.

Using Gurson's isotropic hardening model, voids grow at the bulge of the cylinder at latter stages of deformation. The process is nucleation rather than growth dominated. At failure, the maximum void fraction at the bulge is far less than the critical void fraction ($\approx 0.15$) for coalescence. For several cases, points inside the cylinder have higher void fractions than points near the free surface which suggests that subsurface cracking is possible. Different nucleation models can influence significantly the distribution of void fraction. The maximum difference in the mean stress due to the influence of voids at the bulge is approximately 12%.

Localization studies indicate that, in agreement with most experimental results, under sticking friction failure will occur in a plane perpendicular to the $r$-$\theta$ plane and making an angle of approximately 48 degree with the $r$ axis. A longitudinal crack will form and failure is of the "normal" mode. Failure is primarily due to shear band caused by the softening of the material in the presence of voids. No localizations are found for cases using the von Mises criterion. For a few cases, subsurface cracking is suggested by localizations occurring at points inside but close to the free surface. The maximum shear stress at failure $\tau_{max}/\sigma_y$ is approximately constant at failure.

## ACKNOWLEDGEMENT

The results reported here were supported by the National Science Foundation (Solid and Geo- Mechanics Program) under grant MSM-8722786. A grant from The National Center for Supercomputing Applications is also appreciated.

## REFERENCES

1. Kudo, H., and Aoi, K., 1967. Effect of compression test condition upon fracturing of a medium carbon steel. J. Japan Soc. Tech. Plasticity, 8: 17-27.
2. Sowerby, R., O'Reilly, I., Chandrasekaran, N., Dung, N. L., 1984. Materials testing for cold forging. ASME J. Engr. Mater. Tech., 106: 101-106.
3. Lee, P. W., and Kuhn, H. A., 1973. Fracture in cold upset forging - a criterion and model. Metall. Trans., 4: 969-974.
4. Lee, J.H., Zhang, Y., On the numerical integration of a class of pressure-sensitive plasticity models with mixed hardening. Int. J. Numer. Meth. Engr., (in press).
5. Lee, J.H., Final report to NSF.
6. Mear, M. E., and Hutchinson, J. W., 1985. Influence of yield surface curvature on flow localization in dilatant plasticity. Mechanics of Materials, 4: 395-407.

7. Rudnicki, J. W., and Rice, J. R., 1975. Conditions for the localization of deformation in pressure-sensitive dilatant materials. J. Mech. Phys. Solids, 23: 371-394.
8. Ortiz, M., Leroy, Y. and Needleman, A., 1987. A finite element method for localized failure analysis. Computer Methods Appl. Mech. Eng., 61: 89-124.

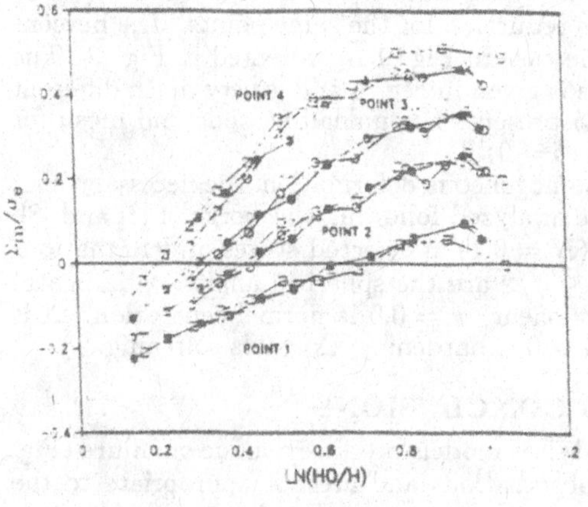


Fig. 1

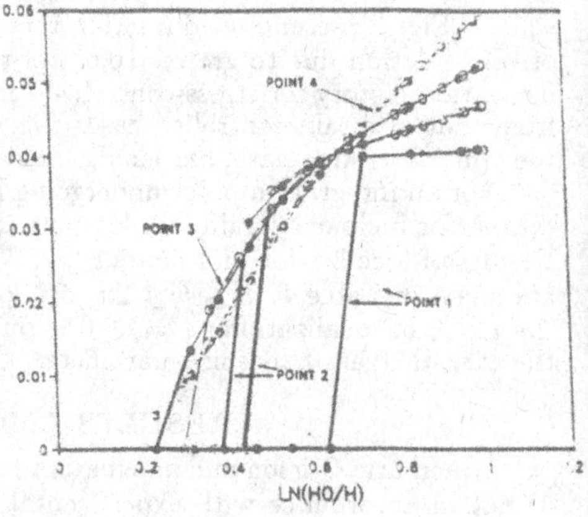


Fig. 2

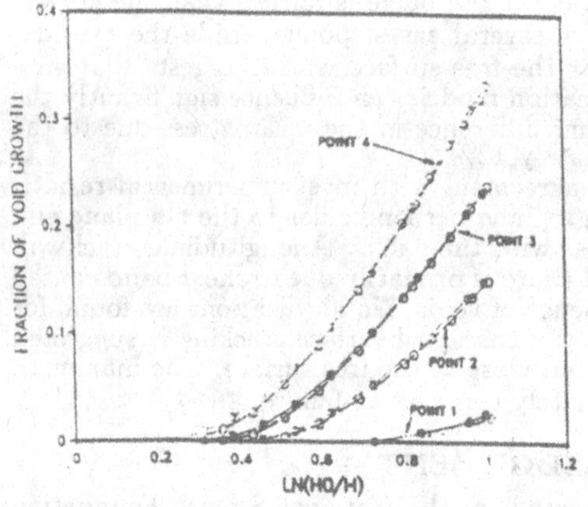


Fig. 3

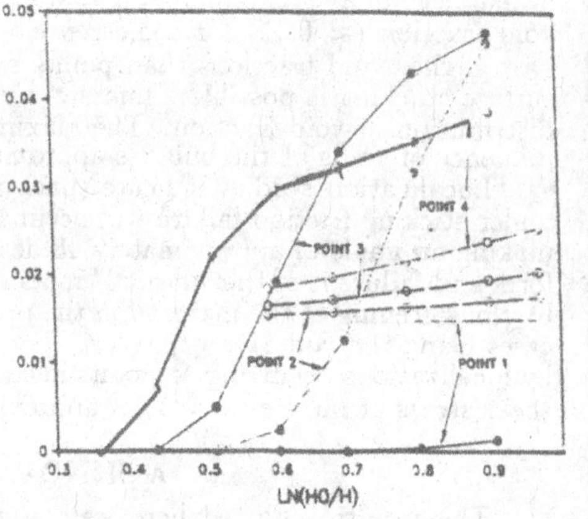


Fig. 4

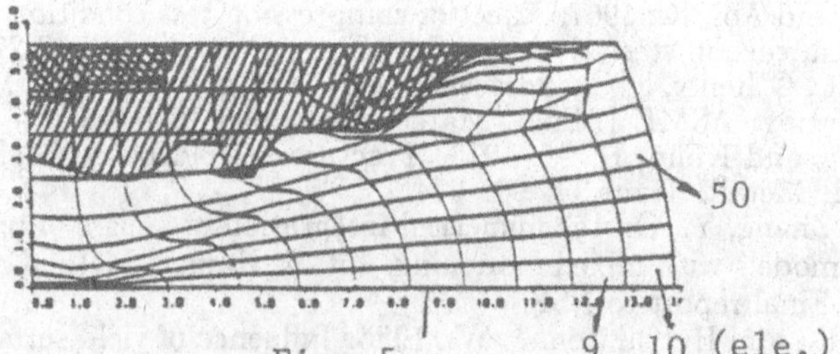


Fig. 5

**Table 1.1 Localization Data**

| $z^*$ (mm) | $ln(H0/H)$ | ELEMENT | GAUSS PT | $\theta$ | $\phi$ | $\psi$ | $\Delta < 0$ |
|---|---|---|---|---|---|---|---|
| 4.0 | 0.916 | 10 | 2 | 47.9 | 0.90 | 84.5 | Yes |
| 4.0 | 0.916 | 10 | 3 | 48.0 | 3.52 | 84.0 | Yes |
| 4.0 | 0.916 | 20 | 2 | 48.1 | 5.32 | 96.0 | Yes |
| 4.0 | 0.916 | 20 | 3 | 48.4 | 8.10 | 95.9 | Yes |
| 4.0 | 0.916 | 30 | 2 | 48.6 | 9.79 | 95.8 | Yes |
| 4.0 | 0.916 | 30 | 3 | 49.2 | 12.15 | 95.8 | Yes |
| 4.0 | 0.916 | 40 | 2 | 49.5 | 14.19 | 95.3 | Yes |
| 4.0 | 0.916 | 40 | 3 | 50.0 | 18.10 | 93.7 | Yes |
| 4.0 | 0.916 | 50 | 2 | 51.4 | 20.62 | 94.3 | Yes |

# CREEP LAW OF ROCK ICE WITH ANISOTROPIC FABRIC INFERRED FROM HOMOGENIZATION

LOUIS LLIBOUTRY
Université Joseph Fourier
B.P. 96, St-Martin d'Hères 38402 cedex, France

ABSTRACT

The steady creep law of rock ice with anisotropic, transversely isotropic fabric, as found in polar ice sheets, has been drawn from homogenization. The crucial point is that local strain is a simple shear on basal planes as with isolated single crystals, but obeying a different law. The strain rate is assumed to be proportional to the third power of the resolved shear stress times an unknown fabric-dependent factor. The local stress is assumed not to differ significantly from the macroscopic one. Seven rheological coefficients appear in the expression of the overall dissipation potential. They are expressed in terms of the four first even moments of the c-axes distribution, times the single unknown factor. With appropriate bases in the 5-D spaces of deviatoric stresses and strain rates, a symmetric creep compliance matrix can be written.

## INTRODUCTION

When calculating the bulk properties of a composite material by homogenization, the known mechanical properties of the microscopic component are assumed to remain the same within the composite. This essential assumption is wrong if we try to infer the creep law of a polycrystal from the creep law of a single crystal. The question is well-documented in case of ice, for effective shear stresses in the range 0.3 bar < $\tau$ < 7 bar.

Mineral ice Ih has a single crystallographic plane of easy glide. It is the basal plane (0001), that contains a large number of screw dislocations (1). When the resolved shear stress on this plane is $\tau_b$, dislocations move at a velocity proportional to $\tau_b$ (about 1 mm $h^{-1}$ $bar^{-1}$ at -20°C). When a single annealed crystal is stressed, the number of dislocations increases with time. In the steady state they appear at a rate more or less proportional to $\tau_b$, and disappear at the crystal boundaries. This explains why in a single crystal strain-softening is observed, and why next the steady strain rate is more or less proportional to $\tau_b^2$.

With a polycrystal there is first strain-hardening, i.e. the strain rate decreases with time. This comes undoubtedly from the progressive piling up of dislocations, that are no longer free to escape. Next a steady (secondary) creep is observed. For macroscopically isotropic rock ice the creep law reads, with standard notations (2):

$$2\,\dot{\varepsilon}_{ij} = \phi\, s_{ij}\,, \qquad \phi = B(T)\,\tau^2 \qquad (2\,\tau^2 = s_{ij}\, s_{ij}) \tag{1}$$

We are interested in the secondary creep of macroscopically anisotropic ice, the most frequent case in polar ice sheets (2, 3). The form of the creep law for any state of stress is unknown. Homogenization can yield it in a closed form because of the third power viscous law.

## IN DEFENCE OF A UNIFORM LOCAL STRESS

When dealing with steady creep, without memory effects, a dissipation potential can be used. With the major assumption that viscous dissipation of energy at the grain boundaries is negligible, the overall dissipation potential is the sum of the dissipation potentials of the individual grains, weighted by their relative volume. This overall potential enters variational theorems. A uniform strain rate and a uniform stress are two complementary trial functions that fix boundaries to the exact solution. Although the approximation of a uniform strain rate has been used so far (4), the approximation of a uniform stress should be much closer to the truth, for two reasons:

(1) Given the power law, to ensure the continuity condition only small departures from the uniform stress approximation are required, whereas very large departures from the uniform strain rate approximation would be required to ensure the equilibrium conditions.

(2) The mismatch of deforming grains can also be overcome by: (a) migration of grain boundaries, as demonstrated by Means and Jessel (5), and observed in the laboratory; (b) micro-slip at the grain boundaries or something similar, linked with the modification of bundles of dislocations that are known to exist at these boundaries.

The dissipation potential of individual crystals within the aggregate, in terms of stress, are assumed to be:

$$\Phi_c = B_b\, \tau_b^{\,4}/4 \tag{2}$$

where $B_b$ is an unknown parameter. It depends on the fabric, that controls the accumulation of dislocations at the grain boundaries.

CALCULATION IN CASE OF A TRANSVERSELY ISOTROPIC FABRIC

A calculation has been done when the statistical distribution of the c-axes has a rotational symmetry about some axis, that is taken as z-axis. We use a modified tensorial notation in 5-D spaces for deviatoric stresses and strain rates:

$$\underline{s} = \begin{bmatrix} s_d \\ s_{ax} \\ s_4 \\ s_5 \\ s_6 \end{bmatrix} = \begin{bmatrix} (s_{xx} - s_{yy})/2 \\ (\sqrt{3}/2)\, s_{zz} \\ s_{yz} \\ s_{zx} \\ s_{xy} \end{bmatrix} , \quad \underline{\dot{\gamma}} = \begin{bmatrix} \dot{\gamma}_d \\ \dot{\gamma}_{ax} \\ \dot{\gamma}_4 \\ \dot{\gamma}_5 \\ \dot{\gamma}_6 \end{bmatrix} = \begin{bmatrix} \dot{\varepsilon}_{xx} - \dot{\varepsilon}_{yy} \\ \sqrt{3}\, \dot{\varepsilon}_{zz} \\ 2\, \dot{\varepsilon}_{yz} \\ 2\, \dot{\varepsilon}_{zx} \\ 2\, \dot{\varepsilon}_{xy} \end{bmatrix} \qquad (3)$$

When the reference frame is rotated about the z-axis, the following quantities are independent invariants of stress:

$$s_{ax} , \quad s_d^2 + s_6^2 = \tau_\perp^2 , \quad s_4^2 + s_5^2 = \tau_\parallel^2 , \quad s_d (s_5^2 - s_4^2) + 2\, s_4 s_5 s_6 = K_3 \qquad (4)$$

There are not other independent invariants. For instance:

$$\tau^2 = s_{ax}^2 + \tau_\perp^2 + \tau_\parallel^2 , \quad \det\left[s_{ij}\right] = K_3 + \frac{2}{\sqrt{3}} s_{ax} (s_{ax}^2/3 - \tau_\perp^2 + \tau_\parallel^2/2) \qquad (5)$$

Let $\gamma$ denote the cosine of the acute angle between the z-axis and the c-axis, and $V(\gamma)$ the relative volume of crystals having this volume smaller than $\gamma$. The overall dissipation function is found to be:

$$\Phi = \frac{1}{4} \Big[ B_{ax} s_{ax}^4 + B_\perp \tau_\perp^4 + B_\parallel \tau_\parallel^4 + 2\, A_{\perp\parallel} \tau_\perp^2 \tau_\parallel^2 + 2\, A_{\parallel ax} \tau_\parallel^2 s_{ax}^2 + 2\, A_{ax\perp} s_{ax}^2 \tau_\perp^2 + 2\, D\, s_{ax} K_3 \Big] \qquad (6)$$

The seven rheological coefficients are given in (2), p. 461 (with $D = -2\sqrt{3}\, C$). Their ratios to $B_b$ are linear expressions of the four even moments $\int_0^1 \gamma^{2m}\, dV$, with m = 1 to 4. (Therefore, only five ones are independent). In the isotropic case, D = 0, and the six others equal $(8/35)B_b$. When the c-axes are clustered in the direction of the z-axis, $B_\parallel = B_b$, and the other coefficients vanish. Creep tests show that $B_b$ is much larger in the latter case.

Since the viscous dissipation of energy per unit volume and unit time is ${}^t\underline{s}\, \underline{\dot{\gamma}}$, the strain rates are the partial derivatives of the dissipation potential:

$$\dot{\gamma} = \frac{\partial \Phi}{\partial s_i} \qquad (i = d,\ ax,\ 4,\ 5,\ 6) \tag{7}$$

With the notation:

$$\begin{bmatrix} \phi_{ax} \\ \phi_{\perp} \\ \phi_{\parallel} \end{bmatrix} = \begin{bmatrix} B_{ax} & A_{ax\perp} & A_{\parallel ax} \\ A_{ax\perp} & B_{\perp} & A_{\perp\parallel} \\ A_{\parallel ax} & A_{\perp\parallel} & B_{\parallel} \end{bmatrix} \begin{bmatrix} s_{ax}^2 \\ \tau_{\perp}^2 \\ \tau_{\parallel}^2 \end{bmatrix} \tag{8}$$

the strain rates are found to be:

$$\dot{\gamma} = \begin{bmatrix} \phi_{\perp} & \frac{D}{2}(s_5^2 - s_4^2) & 0 & 0 & 0 \\ \frac{D}{2}(s_5^2 - s_4^2) & \phi_{ax} & D\, s_5\, s_6 & 0 & 0 \\ 0 & D\, s_5\, s_6 & \phi_{\parallel} - D\, s_d\, s_{ax} & 0 & 0 \\ 0 & 0 & 0 & \phi_{\parallel} + D\, s_d\, s_{ax} & D\, s_{ax}\, s_4 \\ 0 & 0 & 0 & D\, s_{ax}\, s_4 & \phi_{\perp} \end{bmatrix} \underline{s}$$

All the rheological coefficients can be determined by three creep tests in compression and (or) torsion, with a hollowed ice core, when the axis of the fabric coincides with the axis of the core (the usual case). Work is in progress to check the theory and to determine $B_b$ as a function of the fabric.

## REFERENCES


1. Higashi, A. (ed.), Lattice defects in ice crystals, X-ray topographic observations, Hokkaido University Press, Sapporo, 1988.

2. Lliboutry, L., Very slow flows of solids, Kluwer Acad. Publ., Dordrecht, 1987.

3. Lliboutry, L. and Duval, P., Various isotropic and anisotropic ices found in glaciers and polar ice caps and their corresponding rheologies. Annales Geophysicae, 1985, 3, 207-24.

4. Hutchinson, J.W., Creep and plasticity of hexagonal polycrystals as related to single crystal slip. Met. Trans. A, 1977, 8, 1465-9.

5. Means, W.D. and Jessel, M.W., Accomodation migration of grain boundaries. Tectonophysics, 1986, 127, 67-86.

# DYNAMIC STRAIN-LOCALIZATION IN COMPRESSIBLE FLUID-SATURATED POROUS MEDIA

BENJAMIN LORET and OUAHID HARIRECHE
Institute of Mechanics, BP 53X 38041 Grenoble Cedex FRANCE

## ABSTRACT

A stress-point algorithm for coupled inelastic constitutive equations is proposed and analysed. These equations represent the behavior of fluid-saturated elastic-visco-plastic porous media with compressible constituents. The effects of compressibility are further investigated in finite element analyses which display strain-localization.

## INTRODUCTION

In the last decades, the theory of mixtures has been used to describe the linear elastic isotropic properties of fluid saturated porous media. The development of fundamental elastic solutions has been helpful to explain some phenomena observed in geological materials which are dominated by coupled deformation-diffusion effects such as hydraulic fracturing or fault stabilization. Here we address the case where the solid skeleton has an inelastic behaviour.

First, constitutive elastic-plastic equations are proposed. In some sense, they can be considered as a three-dimensional extension of those proposed in [1] where the stress of Nur and Byerlee is used in the elastic equations while the effective Terzaghi's stress enters the yield criterion. Then it turns out that the fluid strain contributes to the change of the effective Terzaghi's stress. This is at variance with the case of incompressible solid consituents. As a consequence, the stress-point algorithm used for water-saturated sands in [2] needs to be modified and one has to devise a stress-point algorithm for truly coupled constitutive equations. For a yield surface and plastic potential of the Drucker-Prager type, the proposed algorithm is unconditionnally stable and explicit. Moreover, it is exact in the large, that is for infinitely large strain-increments. To our knowledge, such a study has not been performed so far even in the simplest context presented here. However, this analysis is thought to be a necessary step before a more realistic material behaviour involving anisotropic hardening can be considered.

In order to describe the development of shear-bands, the viscoplasticity regularization procedure used in [3] for single phase solids and [2] for water-saturated sands is revisited. The regularization procedure retains the well-posedness of the field equations and sets a width to shear-bands where high strain-gradients take place.

## RATE-INDEPENDENT ELASTIC-PLASTIC MIXTURES

### Constitutive equations

The constitutive equations of the mixture are expressed in terms of the partial stresses $\mathbf{t}^\alpha$ acting on the solid ($\alpha = s$) and fluid ($\alpha = w$) phases and consider a compressible fluid whose apparent pressure will be denoted by $p^w$; thus, if $n^\alpha$ denotes the volume fraction of phase $\alpha$, $\mathbf{t}^w = -p^w\delta = -n^w p_w \delta$ with $\delta$ the identity tensor. The infinitesimal strain $\epsilon_\alpha$ in each phase is decomposed into an elastic $\epsilon^e_\alpha$ and a plastic contribution $\epsilon^p_\alpha$ , namely :

$$\epsilon_\alpha = \epsilon^e_\alpha + \epsilon^p_\alpha. \tag{1}$$

The yield criterion and plastic potential depend on the Terzaghi's effective stress $\mathbf{t}'^s$,

$$\mathbf{t}'^s := \mathbf{t}^s - \frac{n^s}{n^w}\mathbf{t}^w, \tag{2}$$

and on a set of internal variables. For the sake of simplicity, only one scalar variable $a$ will be considered. The plastic strain-rates $\dot{\epsilon}_\alpha$ and the rate of the hardening variable $a$ are postulated in the following form :

$$\dot{\epsilon}^p_s = \dot{\lambda}\,\mathbf{P}, \quad \dot{\lambda} \geq 0, \qquad tr\ \dot{\epsilon}^p_w = -\frac{n^s}{n^w}\ tr\ \dot{\epsilon}^p_s, \qquad \dot{a} = \dot{\lambda} \tag{3}$$

where $\mathbf{P}$ is the outward normal to the plastic potential at a smooth point and $tr$ is the trace operator. The above relation between the plastic parts of the strain-rates makes possible to consider mixtures whose constituents are both incompressible.

The resulting rate constitutive equations are obtained by insertion of the above relations into the elastic stress-strain law, namely :

$$\begin{cases} \mathbf{t}^s = \mathbf{E}^s : \epsilon^e_s + [\lambda_{sw}\mathbf{I}] : \epsilon^e_w \\ \mathbf{t}^w = [\lambda_{sw}\mathbf{I}] : \epsilon^e_s + [\lambda_w \mathbf{I}] : \epsilon^e_w \end{cases} \tag{4}$$

where $\mathbf{I}$ denotes the fourth-order tensor, $I_{ijkl} = \delta_{ij}\delta_{kl}$, the symbol : denotes the trace of a scalar product and $\mathbf{E}^s$ is an isotropic elastic tensor with Lamé moduli $\lambda_s$ and $\mu_s$. The elastic coefficients $\lambda_s$, $\mu_s$, $\lambda_{sw}$ and $\lambda_w$ are restricted by the condition of positive-definiteness of the elastic strain-energy function.

The solid skeleton will be assumed to be characterized by a yield surface $f$ and plastic potential $g$ of Drucker-Prager type.

### Stress-point algorithm for porous media with compressible constituents

Given a strain-path $\epsilon_\alpha(t)$ in the time interval $[t_0, t_f]$, and initial values of the stresses and internal variable $a$ at time $t = t_0$, namely $\{\ \mathbf{t}^\alpha(t_0) = \mathbf{t}^\alpha_0, a(t_0) = a_0\ \}$, the stress-point algorithm returns the stresses $\mathbf{t}'^s$ and $\mathbf{t}^w$ and internal variable $a$ at any time

$t \in [t_0, t_f]$. For that purpose, the time interval is divided in subintervals $[t_n, t_{n+1}]$ and the algorithm proceeds step-by-step : typically, for a given set of initial conditions at step n $\{ \mathbf{t}^\alpha\ (t_n) = \mathbf{t}^\alpha_n\ ;\ a(t_n) = a_n \}$ and for a given loading increment $\Delta\epsilon_\alpha$, one seeks for the material response $\{ \mathbf{t}^\alpha\ (t_{n+1}) = \mathbf{t}^\alpha_{n+1}\ ;\ a(t_{n+1}) = a_{n+1} \}$ at step n+1. The restriction that the strain-rate be of fixed direction during the time-step $[t_n, t_{n+1}]$ is adopted for computational simplicity. Here, due to lack of space, only linear hardening is considered. The algorithm proceeds in two stages. First a trial state $\{ \mathbf{t}'^s_{trial}, a_n \}$ is obtained as if the incremental material response during the whole time-step were elastic. If the trial state satisfies the condition $f(\mathbf{t}'^s_{trial}, a_n) \leq 0$, the material response is actually elastic and one sets $\{ \mathbf{t}'^s_{n+1} = \mathbf{t}'^s_{trial};\ a_{n+1} = a_n \}$. Otherwise, plasticity has occurred and a second stage is necessary in order to satisfy the plastic consistency requirement $f(\mathbf{t}'^s_{n+1},\ a_{n+1}) = 0$. For that purpose, the rate equation $(3)_1$ is integrated in implicit form as $\Delta\epsilon^p_s = \Delta\lambda\ \mathbf{P}_{n+1}$. Thus an incremental form of the evolution rules for the hardening variable $a$ and effective stress $\mathbf{t}'^s$ is :

$$\begin{cases} a_{n+1} = a_n \quad + \Delta\lambda \\ \mathbf{t}'^s_{n+1} = \mathbf{t}'^s_{trial} - \Delta\lambda\ \mathbf{E}"^s : \mathbf{P}_{n+1} \end{cases} \tag{5}$$

Here $\mathbf{E}"^s$ is another tensor of elastic moduli with Lamé coefficient $\lambda"_s$:

$$\lambda"_s := \lambda'_s - \frac{n^s}{n^w}\lambda'_{sw}, \quad \lambda'_s := \lambda_s - \frac{n^s}{n^w}\lambda_{sw}, \quad \lambda'_{sw} := \lambda_{sw} - \frac{n^s}{n^w}\lambda_w \tag{6}$$

Introduction of eqs.(5) in the consistency equation $f(\mathbf{t}'^s_{n+1},\ a_{n+1}) = 0$ yields the plastic index $\Delta\lambda \geq 0$, namely $\Delta\lambda = f(\mathbf{t}'^s_{trial},\ a_n)/H$ where the modulus $H$ is assumed to be strictly positive in order to exclude locking materials. The final stress $\mathbf{t}'^s_{n+1}$ can be extracted explicitly from eq. $(5)_2$. The fluid stress $\mathbf{t}^w_{n+1}$ is then easily deduced from the knowledge of $\Delta\lambda$.

## RATE-DEPENDENT ELASTO-VISCO-PLASTIC MIXTURES

Visco-plasticity is introduced as a regularization procedure. Well-posedness of initial value problems requires hyperbolicity of the dynamical equations of motion which in turn requires real and strictly positive wave-speeds. On the other hand, the onset of strain-localization corresponds to a wave-speed becoming equal to zero. Numerical simulations of the development of shear-bands is therefore deemed to face numerical difficulties. Rate-dependency is introduced to unfoil the situation. Actually, the viscoplasticity regularization procedure first proposed in [4] retains hyperbolicity and simultaneously makes possible to mimic the development of shear-bands of finite thickness (see [2,3]). Indeed then the only acceleration waves traveling through the material are elastic waves. The regularization procedure is performed by the introduction of a new parameter $\eta$ called relaxation time. To the above inviscid elastic-plastic mixture referred to as inviscid backbone mixture is associated a one parameter family of rate-dependent mixtures defined by the following rate equations for the visco-plastic strain-rate $\dot{\epsilon}^{vp}_s$ and hardening parameter $\dot{a}$ :

$$\dot{\epsilon}^{vp}_s = \frac{1}{\eta}[\mathbf{E}"^s]^{-1} : (\mathbf{t}'^s - \overline{\mathbf{t}}'^s)\ \mathcal{H}(f(\mathbf{t}'^s, a)), \quad \dot{a} = -\frac{1}{\eta}(a - \overline{a}) \tag{7}$$

where $(\overline{\mathbf{t}}'^{s}, \overline{a})$ denotes the response of the underlying inviscid mixture; in particular, $\overline{\mathbf{t}}'^{s}$ can be viewed as the projection of the stress $\mathbf{t}'^{s}$ on the yield surface. In eq. (7), $\mathcal{H}$ stands for the Heaviside step function. The above equations are supplemented by the following constitutive equations mimicking their inviscid counterparts, namely:

$$tr\ \dot{\epsilon}_w^{vp} = -\ tr\ \frac{n^s}{n^w}\dot{\epsilon}_s^{vp}, \qquad \begin{cases} \dot{\mathbf{t}}'^{s} = & \mathbf{E}'^{s} : (\dot{\epsilon}_s - \dot{\epsilon}_s^{vp}) + [\lambda'_{sw}\mathbf{I}] : (\dot{\epsilon}_w - \dot{\epsilon}_w^{vp}) \\ \dot{\mathbf{t}}^{w} = & [\lambda_{sw}\mathbf{I}] : (\dot{\epsilon}_s - \dot{\epsilon}_s^{vp}) + [\lambda_w\mathbf{I}] : (\dot{\epsilon}_w - \dot{\epsilon}_w^{vp}) \end{cases} \tag{8}$$

Assuming a fixed plastic state $(\overline{\mathbf{t}}'^{s},\ \overline{a})$ [5], the above rate dependent constitutive equations can be integrated analytically over the time-step $[t_n,\ t_{n+1} = t_n + \Delta t]$ .

## NUMERICAL RESULTS

Numerical solutions to dynamic initial and boundary value problems are obtained by using the finite element method. A critical discussion concerning time- and space-integration is reported in [2,6] for both single phase solids and water-saturated porous media with incompressible constituents. The single minor computational difference lies in the fact that now the displacement of the fluid must be kept in memory so that the increment of plastic strain in the fluid phase can be computed. Numerical simulations will be presented at the conference.

## REFERENCES

1. Rice J.R., On the stability of dilatant hardening for saturated rock masses, J. Geophys. Research, 1975, **80**, pp.1531-1536.

2. Loret B. and J.H. Prevost, Dynamic strain-localization in fluid-saturated porous media, J. Engng. Mech., A.S.C.E., to appear.

3. Loret B. and J.H. Prevost, Dynamic strain-localization in elasto-(visco- )plastic solids - Part 1 : General formulation and one-dimensional examples, 1990, Comp. Meth. Appl. Mech. Engng., **83**, pp. 247-273.

4. Duvaut G. and J.L. Lions, Les inéquations en Mécanique et en Physique Dunod, 1972, Paris.

5. Simo J.C., J.G. Kennedy, S. Govindjee and T.J.R. Hughes, Unconditionally convergent algorithms for non-smooth multi-surface plasticity amenable to exact linearization, A.S.M.E. Winter Annual Meeting, Advances in Inelastic Analysis, A.M.D., 1987, **88**, S. Nakazawa, K. Willam and N. Rebello eds., pp. 87-96.

6. Prevost J.H. and B. Loret, Dynamic strain-localization in elasto-(visco-)plastic solids - Part 2 : Plane strain examples , Comp. Meth. Appl. Mech. Engng., 1990, **83**, pp. 275-294.

# CONSTITUTIVE LAW OF POROUS VISCOUS MATERIALS

JEAN-CLAUDE MICHEL and PIERRE SUQUET
CNRS. Laboratoire de Mécanique et d'Acoustique.
31 Chemin Joseph Aiguier. 13402. MARSEILLE. Cedex 09. FRANCE.

**ABSTRACT:**

This paper examines in the light of Micromechanics the overall behavior of nonlinear viscous materials containing voids. A potential depending on two material parameters is proposed. One coefficient is derived from the hollow sphere problem: a closed form solution for a hollow sphere under hydrostatic stress is established for a compressible matrix. The second coefficient is deduced from a variational bounding of the potential.

## 1. CONSTITUTIVE POTENTIALS

A *representative volume element* (r.v.e.) Y of the material, containing a volume V of cavities is submitted to macroscopic stresses $\Sigma$ or strain rates $\dot{E}$ which induce microscopic stresses $\sigma$ and strain rates $\dot{\epsilon}$ at every point of the r.v.e.. The former are the volume averages of the latter, understood in a generalized sense:

$$\Sigma_{ij} = \frac{1}{|Y|}\int_{Y^*} \sigma_{ij}\, dx \ , \quad \dot{E}_{ij} = \frac{1}{|Y|}\int_{\partial Y} \frac{1}{2}\left(\dot{u}_i \nu_j + \dot{u}_j \nu_i\right) ds \tag{1}$$

where $Y^* = Y - V$ , and $\partial Y$ denotes the outer boundary of the r.v.e. with outer normal vector $\nu$. The determination of the local fields $\sigma$ and $\dot{\epsilon}$ in terms of their averages $\Sigma$ and $\dot{E}$ requires the resolution of the local equations at our disposal, namely the equilibrium equations, the constitutive law at the microscopic scale, and the average conditions (1). The boundary conditions are often assumed to take the form of uniform stresses of uniform strains on $\partial Y$ (see HILL [1] or HASHIN [2]) :

$$\sigma_{ij} n_j = \Sigma_{ij} n_j \text{ on } \partial Y, \text{ or } \dot{u}_i = \dot{E}_{ij}\, x_j \text{ on } \partial Y \tag{2}$$

The matrix is supposed to be an incompressible viscous power law material. Its constitutive laws can be defined with the help of potentials:

$$\dot{\epsilon}_{ij} = \frac{\partial\psi}{\partial\sigma_{ij}}(\sigma)\ , \quad \psi(\sigma) = \frac{\sigma_0\dot{\epsilon}_0}{n+1}\left(\frac{\sigma_{eq}}{\sigma_0}\right)^{n+1}\ , \quad \sigma_{eq} = \left(\frac{3}{2}\sigma^D_{ij}\sigma^D_{ij}\right)^{\frac{1}{2}}. \qquad (3)$$

The macroscopic constitutive law can also be derived from a macroscopic potential [3]:

$$\dot{E}_{ij} = \frac{\partial\Psi}{\partial\Sigma_{ij}}(\Sigma,f)\quad ,\quad \Psi(\Sigma,f) = \langle\psi(\sigma)\rangle\ ,\qquad \langle.\rangle = \frac{1}{|Y|}\int_{Y^*} .\ dx \qquad (4)$$

Under the assumption that the voids are spatially distributed at random and exhibit a spheroidal shape with random sizes, the porous material can be reasonably assumed to be isotropic and a scalar variable, namely the volume fraction of the voids f, is sufficient to account for this internal damage. The strain rate potential $\Psi$ is a function of f and of the three invariants of $\Sigma$, but in order to simplify the analysis $\Psi$ is classically assumed to depend only on the two first invariants of $\Sigma$. Using its positive homogeneity of degree (n+1), $\Psi$ can be written as a function of the stress triaxiality ratio X:

$$\Psi(\Sigma_m,\Sigma_{eq},f) = \frac{\sigma_0\dot{\epsilon}_0}{n+1}\left(\frac{\Sigma_{eq}}{\sigma_0}\right)^{n+1} F(X,f)\ ,\ X = \frac{3\Sigma_m}{2\Sigma_{eq}}\ ,\ \Sigma_m = \frac{1}{3}\,Tr(\Sigma). \qquad (5)$$

Several approximate forms of the constitutive potential $\Psi$ or of the function F have been proposed in the literature. Let us mention DUVA and HUTCHINSON [4] for dilute voided materials, DUVA [5] , GUENNOUNI and FRANCOIS [6] basing partly on numerical investigations, PONTE CASTANEDA and WILLIS [7], COCKS [8] deriving variational bounds for $\Psi$. In section 3 we shall adopt the following "quadratic" form for F:

$$F(X,f) = (A(f)X^2+B(f))^{(n+1)/2}, \qquad (6)$$

and our objective is to derive expressions for A and B. The following sections are devoted to the determination of A.

## 2. THE HOLLOW SPHERE PROBLEM FOR A POROUS MATERIAL

The hollow sphere with inner radius a and outer radius b is submitted to an hydrostatic tension S. The material constituting the sphere is viscous and porous with a uniform initial porosity f:

$$\dot{\epsilon}_{ij} = \frac{\partial\psi}{\partial\sigma_{ij}}(\sigma,f)\ ,\ \text{where}\quad \psi(\sigma,f) = \frac{\sigma_0\dot{\epsilon}_0}{n+1}\left(\frac{\sigma_{eq}}{\sigma_0}\right)^{n+1} F(X,f)\ . \qquad (7)$$

By virtue of the spherical symmetry of the problem the velocity field is radial and depends on r only. After due use of the constitutive law, of the compatibility equations and of the equilibrium equations we obtain two differential equations for the triaxiality ratio X, and the equivalent stress $\sigma_{eq}$ (detailed computations are reported in [9]):

$$\frac{\partial X}{\partial r}(r) = \frac{3}{r} + \frac{1}{r} H(X)(X - 1)\ , \quad \frac{\partial \sigma_{eq}}{\partial r}(r) = - \frac{1}{r} \sigma_{eq}(r)\, H(X)\ , \quad H(X) = \frac{\mathcal{N}}{\mathcal{D}}\ , \qquad (8)$$

$$\mathcal{N} = 3((n+1)F(X,f) + (n - X)\, F'(X,f) + (1 - X)F''(X,f)),$$

$$\mathcal{D} = (1 - X)^2 F''(X,f) + 2n(1 - X)F'(X,f) + n(n+1)F(X,f).$$

The two differential equations (8) are decoupled and can be integrated separately with the following boundary conditions $X(a) = 1$, $\sigma_{eq}(b) = \frac{3}{2}\frac{S}{X(b) - 1}$.

### 3. SELF CONSISTENT SCHEME

The above solution for the hollow sphere under hydrostatic tension can be used to derive estimates of the strain potential $\Psi$ in the spirit of the self consistent scheme, at least if the specific "quadratic" form (6) is assumed for F(X). On the one hand the macroscopic dilatation rate is the average of the microscopic dilatation rate, and on the other hand it can be deduced from the macroscopic potential:

$$\mathrm{Tr}(\dot{E}) = \frac{f|\dot{V}|}{|V|} = \frac{f}{|V|}\int_{\partial V} \dot{u}_i \nu_i \, ds, \quad \mathrm{Tr}(\dot{E}) = \frac{\partial \Psi}{\partial \Sigma_m} = \left(\frac{3}{2}\right)^{n+1} \dot{\epsilon}_0\ A^{(n+1)/2} \left(\frac{\Sigma_m}{\sigma_0}\right)^n . \qquad (9)$$

The computation of the first expression of $\mathrm{Tr}(\dot{E})$ in (9) can be performed in the self-consistent spirit by consideration of a single cavity V in an *infinite medium obeying the constitutive law of the unknown homogeneous porous material*. When the macroscopic stress $\Sigma$ is hydrostatic, a solution in an implicit closed form is available from section 2 by letting the outer boundary of the hollow sphere go to infinity ($b \rightarrow +\infty$). With the specific form (6) of F:

$$\mathrm{Tr}(\dot{E}) = \frac{3f\dot{u}(a)}{a} = f\,\frac{3\dot{\epsilon}_0}{2}\left(\frac{\sigma_{eq}(a)}{\sigma_0}\right)^n A^{(n+1)/2}\left(1 + \frac{1}{C}\right)^{(n+1)/2}, \quad C = \frac{A}{B}\ , \qquad (10)$$

and comparison with $(9)_b$ yields: $\displaystyle \frac{1}{C} = \left(f^{-\frac{1}{n}}\,\frac{\Sigma_m}{\sigma_m(a)}\right)^{2n/(n+1)} - 1\ .$

In order to compute $\sigma_m(a)$ we note that:

$$\mathrm{Log}\left(\frac{\Sigma_m}{\sigma_m(a)}\right) = \lim_{b \rightarrow +\infty} \mathrm{Log}\left(\frac{\sigma_m(b)}{\sigma_m(a)}\right) = \lim_{b \rightarrow +\infty} \int_a^b \frac{1}{\sigma_m(r)}\frac{\partial \sigma_m(r)}{\partial r}\, dr\ .$$

With the change of variables $r = r(X)$ and after due use of the differential equations (8) the last integral can be computed [9] and the comparison between the two expressions of $Tr(\dot{E})$ in (9) yields an implicit relation between the material parameter C, the porosity f and the rate sensitivity parameter n:

$$\hat{C}(n^2)\ \mathrm{Log}(f) = -\ \hat{C}(n)\mathrm{Log}(\hat{C}(n)) + \quad + \frac{n-1}{2n^2 C}\ \mathrm{Log}(\hat{C}(1)) + \left(\frac{n-1}{n\sqrt{C}}\right)\left(\mathrm{Arctg}(\sqrt{C}) - \frac{\pi}{2}\right)$$

where $\hat{C}(m) = 1 + \dfrac{1}{mC}$. This implicit equation reduces to the well known linear case when n = 1.

## 4. VARIATIONAL BOUND FOR THE POTENTIAL

Since no explicit solution for the hollow sphere under pur shear is available to the author's knowledge an approximate expression for the coefficient B is derived from a variational estimate of the strain-rate potential Ψ (see PONTE-CASTANEDA (1991), MICHEL and SUQUET (1991)):

$$\Psi(\Sigma) \geqslant \frac{\sigma_0 \dot{\epsilon}_0}{n+1}(1-f)\left(\frac{\left(\frac{9}{4}f\Sigma_m^2 + \left(1+\frac{2}{3}f\right)\Sigma_{eq}^2\right)^{\frac{1}{2}}}{\sigma_0(1-f)}\right)^{n+1}, \text{ thus } B = \frac{1+\frac{2}{3}f}{(1-f)^{2n/(n+1)}}$$

Applications to hot isostatic compaction of ceramic powders will be presented.

## REFERENCES

[1] HILL R.: "Elastic properties of reinforced solids: some theoretical principles". J. Mech. Phys. Solids, 1963, 11, pp 357-372.
[2] HASHIN Z.:"Analysis of composite materials: a survey". J. Appl. Mech., 1983, 50, pp 481-505.
[3] GILORMINI P., LICHT C., SUQUET P.:"Growth of voids in a ductile matrix: a review". Arch. Mech., 1988, 40, p 43.
[4] DUVA J.M., HUTCHINSON J.W.: "Constitutive potentials for dilutely voided nonlinear materials". Mech. Materials, 1984, 3, pp 41-54.
[5] DUVA J.M.: "A constitutive description of nonlinear materials containing voids". Mech. Materials, 1986, 5, pp 137-144.
[6] GUENNOUNI T., FRANCOIS D.:"Constitutive equations for rigid plastic or viscoplastic materials containing voids". Fatigue Fract. Engng Mater. Struct., 1987, 10, p 399.
[7] PONTE CASTANEDA P., WILLIS J.R.:"On the overall properties of nonlinearly viscous composites". Proc. R. Soc. Lond. A, 1988, 416, 217-244.
[8] COCKS A.C.F.: "Inelastic deformation of porous materials". J. Mech. Phys. Solids, 1989, 17, 693-715.
[9] MICHEL J.C., SUQUET P.:"On the constitutive law of nonlinear viscous and porous materials". Submitted.
[10] PONTE-CASTANEDA P.: "The effective mechanical properties of nonlinear isotropic composites".J. Mech. Phys. Solids, 1991, 39, pp 45-71.

# A MODEL TO DESCRIBE THE YIELDING OF AN ELASTIC-PLASTIC POROUS SOLID

S. NAGAKI*, M. GOYA** and R. SOWERBY***
* Okayama University, Okayama, Japan, ** University of the Ryukyus, Okinawa, Japan
*** McMaster University, Hamilton, Ontario, Canada

## ABSTRACT

**In a previous study, the authors proposed an anisotropic yield function for a porous metal. The present investigation extends the previous work by comparing yield loci derived from the yield function and those determined from a FE model. The FE calculations were performed for a flat plate, with a periodic array of cylindrical holes, under biaxial loading; and also for a porous solid with a periodic array of spherical voids under axial load and superimposed hydrostatic pressure.**

## INTRODUCTION

In a previous paper [1], the authors proposed a modification to Gurson's yield function for porous materials [2]. The yield function allowed not only for the influence of the void volume fraction but also for the spatial distribution of the voids. Two loading conditions were analyzed in [1]. The first was a perforated plate (cylindrical holes) under in-plane biaxial stressing, and the second a porous solid (spherical voids) under axial load and superimposed hydrostatic pressure.

## A DAMAGE TENSOR AND ANISOTROPIC YIELD FUNCTION

In 1977, Gurson [2] proposed the following yield function for porous solids, which introduces the void volume fraction f as a scalar parameter

$$F(\sigma, Y, f) = \left(\frac{\bar{\sigma}}{Y}\right)^2 + 2f\cosh\left(\frac{\sigma_{kk}}{2Y}\right) - (1+f^2) = 0 \qquad (1)$$

In the above, $\sigma$ is the "average" macroscopic Cauchy stress for the aggregate, Y the uniaxial flow or yield stress of the matrix material and f the volume fraction of voids. The quantity $\sigma$ is the representative stress defined by

$$\bar{\sigma}^2 = \frac{2}{3}\sigma_{ij}'\sigma_{ij}', \qquad (2)$$

where $\sigma_{ij}' = \sigma_{ij} - 1/3\sigma_{kk}\delta_{ij}$. When f = 0, the yield function described by (1) reduces to that given by von Mises. A number of investigators have modified (1) because they considered it overestimated the yield stress, see Refs. [3-6].

In Ref. 1 the void distribution was assumed to be periodic, and the alignment of the voids to coincide with a set of Cartesian axes. If the hole spacing is different along each of the coordinate axes, see Fig. 1, then so is the projected area of the matrix material in each of the coordinate directions. The difference in projected area of the matrix material can be accounted for by introducing a damage tensor $\Omega$, as was done previously by Murakami and Ohno [7]. The damage tensor was defined as

$$\Omega = \sum \Omega_i \, n_i \, n_i \tag{3}$$

where $\Omega_i$ denotes the void area density over a plane whose normal is the unit vector $\mathbf{n}_i$.

Murakami and Ohno [7] also introduced a net stress tensor S which, because of the reduced area due to the voids, is related to the Cauchy stress $\sigma$ as follows

$$S = \frac{1}{2}(\sigma\,\phi + \phi\,\sigma) \cdot \tag{4}$$

The quantity $\phi$ is a damage effect tensor and related to $\Omega$ in the following way

$$\phi = (1 - c\,\Omega)^{-1} \;, \tag{5}$$

where c is a scalar parameter and was introduced to compensate for effects of stress concentration.

In order to introduce the influence of the spatial distribution of voids on the yielding behaviour it was proposed in [1] to replace the Cauchy stress $\sigma$ in (1) by the net stress S defined in (4). Equation (1) is now modified to

$$F(S, Y, f) = \left(\frac{\bar{S}}{Y}\right)^2 + 2f\cosh\left(\frac{S_{kk}}{2Y}\right) - (1 + f^2) = 0 \;, \tag{6}$$

$$\text{where} \quad \bar{S}^2 = \frac{3}{2} S'_{ij} S'_{ij} \,, \text{ and } \; S'_{ij} = S_{ij} - \frac{1}{3} S_{kk}\,\delta_{ij} \,.$$

For the isotropic case the yield loci predicted by (6) lie inside those derived from the Gurson model. In addition, the yield function (6) predicts anisotropic behaviour when the hole spacing is heterogeneous.

## NUMERICAL ANALYSIS

### Finite Element Analysis

(i) <u>A Perforated Plate Under Biaxial Stress</u>: The variation in hole spacing is shown in Fig. 2(a), while the unit cell with the displacment boundary conditions is shown in Fig. 2(b). The matrix material is assumed to be an elastic-plastic solid which obeys the $J_2$ infinitesimal theory of plasticity. The imposed macroscopic strain components along faces AB and BC, see Fig. 2(b), are

$$E_{11} = \frac{U_1}{L_1} \quad \text{and} \quad E_{22} = \frac{U_2}{L_2} \;, \tag{7}$$

whereas the macroscopic stress, $\sigma_{ij}$, has to be averaged over the specified surface of the unit cell. Different proportional straining paths were examined i.e. maintaining the macroscopic strain ratio, $E_{22}/E_{11}$, constant. Two offset strains, 0.002% and 0.01%, were used to define yielding within a cell.

Calculations were performed for a plate with a void area fraction, f, of either 0.02 or 0.05, and a hole spacing ratio, $L_2/L_1$, of either 1.0 or 2.0. Six loading paths ($E_{22}/E_{11}$ = -1.0, -0.5, -0.25, 0.25, 0.5 and 1.0) were selected in order to determine the shape of the yield locus. The yield loci predicted from the yield function, (6), are nested ellipses, as shown in Fig. 3. For these calculations the value of c in (5) was unity. The yield point states calculated by the FE program are also shown in the figure for comparison. When the hole spacing is modified so that $L_2/L_1$ = 2.0, the yield loci predicted by (6) are again elliptical in shape, but rotated clockwise,see Fig. 4.

(ii) <u>Uniaxial Loading Under Superimposed Pressure</u>: Calculations were performed for a material with a volume fraction of voids of either f = 0.065 or 0.0082, with each of the following aspect ratios: $L_2/L_1 = L_3/L_1$ = 0.75, 1.0 or 2.0. Four straining ratios, $E_{11}/E_{33} = E_{22}/E_{33}$ = –0.5, –0.3, 0.1 and 0.5, were examined and 0.002% proof stress was used to define the yield point. When constructing the yield loci from the yield function (6), the tensile stress (say $\Sigma$) was plotted against the hydrostatic stress $\sigma_H$ ($= \sigma_{kk}/3$); the stress quantities were normalized by the initial yield stress of the material. Loci predicted from eqn. (6) for two volume fractions are shown by the full lines in Fig. 5. As a means of comparison, predicted loci based on the original Gurson model (1), Tvergaard's model [3,4] and Richmond's model [6], are also shown. The finite element results are also illustrated on the figure. Similar FE calculations have been given in ref. 5.

It is possible to make the predictions from (6) correspond with the numerical results shown in Fig. 5. This was achieved, after some trial and error, by selecting $c = 0.5$ in (5) and changing the factor 0.5 in the cosh ( ) term to 0.7 in (6). Calculations were also performed for the anisotropic case i.e. $L_2/L_1 = L_3/L_1 \neq 1$, and compared with the results from the yield model - again using the adjusted numerical factors. Quite good agreement could be obtained between the numerical results and the predicted loci. Adjusting the numerical factors in this way is arbitrary and provides no physical insight. We are currently reviewing the choice of the parameters in Gurson's model.

## CONCLUSIONS

Like the Gurson model, the proposed yield criterion is pressure dependent, and the existence of voids is accounted for by a scalar parameter, termed the void volume fraction. However, the proposed model also allows for the influence of hole or void spacing; and heterogeneous void distribution results in anisotropic behaviour.

Agreement between the proposed yield function and the FE analysis of a biaxially stretched porous plate was quite good, for both homogeneous and heterogeneous hole spacing. The same agreement was not obtained for the yielding of an elastic-plastic porous solid under uniaxial tension and hydrostatic pressure. Predictions from the yield model and the FE calculations could be made to coincide, by changing two scalar parameters in the yield criterion. However, these adjustments are arbitrary and a further examination of the coefficients in Gurson's model is recommended.

## ACKNOWLEDGEMENTS

The authors acknowledge the financial support received from the Natural Sciences and Engineering Council of Canada and the Manufacturing Research Corporation of Ontario.

## REFERENCES

1. Nagaki, S., Sowerby, R. and Goya, M., An Anisotropic Yield Function for Porous Materials, accepted for publication in Mater. Sci. and Eng'g.

2. Gurson, A.L., Continuum Theory of Ductile Rupture by Void Nucleation and Growth: Part I - Yield Criteria and Flow Rules for Porous Ductile Media, J. Eng'g. Mats. and Technology, 1977, 99, 2-15.

3. Tvergaard, V., Influence of Voids on Shear Band Instabilities Under Plane Strain Conditions, Int. J. Fracture, 1981, 17, 389-407.

4. Tvergaard, V., On Localization in Ductile Materials Containing Spherical Voids, Int. J. Fracture, 1982, 18, 237-252.

5. Hom, C.L. and McMeeking, R.M., Void Growth in Elastic Plastic Materials, J. Appl. Mechs., 1989, 56, 309-317.

6. Richmond, O. and Smelser, R.E., 1985, Alcoa Technical Center Report.

7. Murakami, S. and Ohno, N., A Continuum Theory of Creep and Creep Damage in Creep in Structures, eds. A.R.S. Poynter and D.R. Hayhurst, Springer, Berlin, 1981, 422-444.

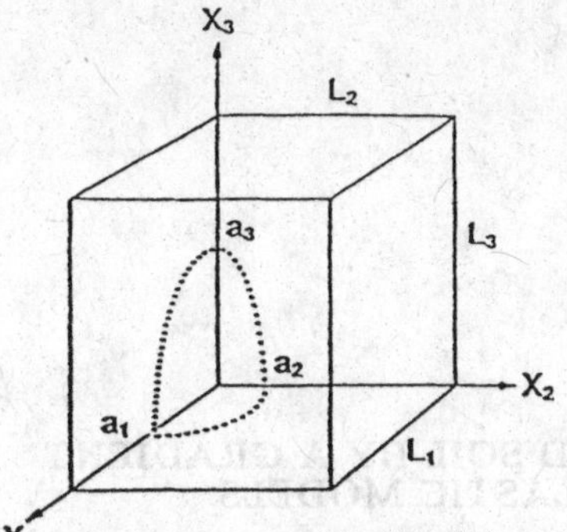


Fig. 1 An ellipsoidal void in a unit cell

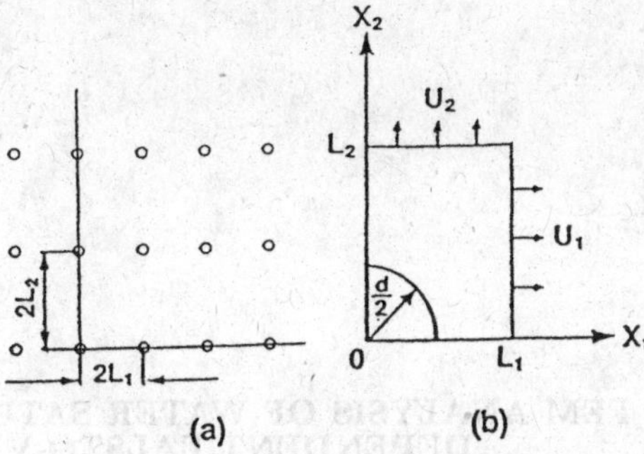


Fig. 2 a) An array of circular holes in a plate
b) The unit cell

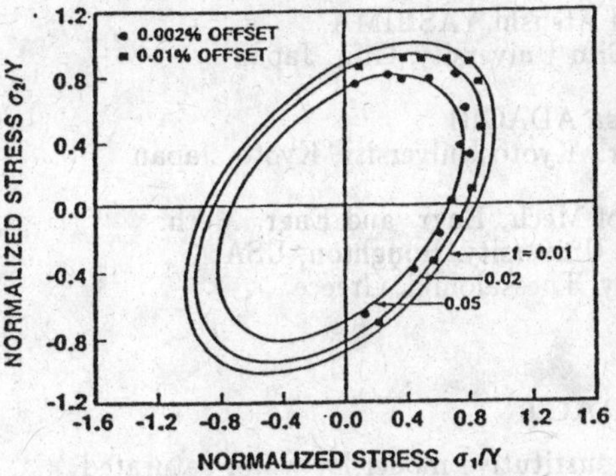


Fig. 3 Yield loci and FE results for the biaxial stretching of a plate - homogeneous hole spacing

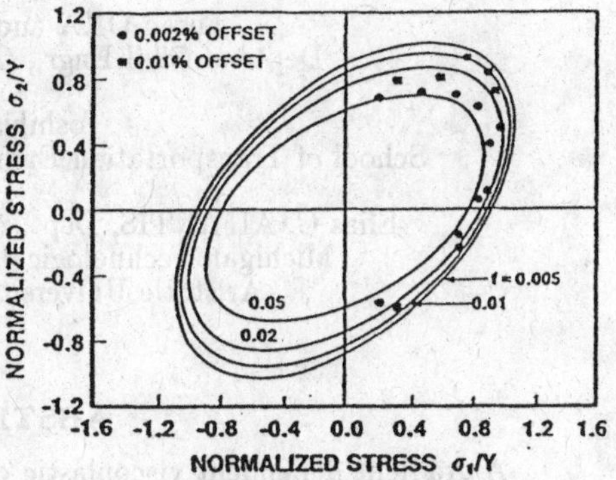


Fig. 4 Yield loci and FE results for the biaxial stretching of a plate - heterogeneous hole spacing

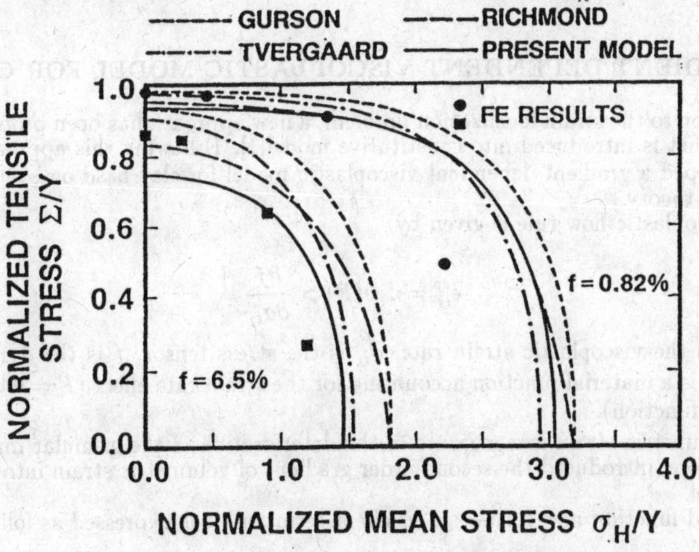


Fig. 5 Yield loci and FE results for uniaxial tension under all round hydrostatic pressure - homogeneous hole spacing

# FEM ANALYSIS OF WATER SATURATED SOIL BY A GRADIENT DEPENDENT EALSTO-VISCOPLASTIC MODELS

Fusao OKA and Atsushi YASHIMA
Dept. of Civil Engr., Gifu University, Gifu, Japan

Toshihisa ADACHI
School of Transportation Engr., Kyoto University, Kyoto, Japan

Elias C. AIFANTIS, Dept. of Mech. Engr. and Engr. Mech.
Michigan Technological University, Houghton, USA
Aristotle University, Thessaloniki, Greece

## ABSTRACT

A gradient dependent viscoplastic constitutive model for water saturated clay is developed and applied to the FEM analysis considering a transport of pore water under plane strain compression.

## GRADIENT DEPENDENT VISCOPLASTIC MODEL FOR CLAY

In relation to the strain localization problem, a new approach has been proposed where strain gradient is introduced into constitutive model[1]. Following this approach, Oka et al.[2] developed a gradient dependent viscoplastic model for clay basd on overstress type viscoplastic theory.

The viscoplastic flow rule is given by

$$\dot{\epsilon}_{ij}^{vp} = < \Phi(F) > \frac{\partial f}{\partial \sigma_{ij}} \tag{1}$$

where $\dot{\epsilon}_{ij}^{vp}$ is the viscoplastic strain rate, $\sigma_{ij}$ is the stress tensor, $f$ is the dynamic yield function, $\Phi$ is a material function accounting for the strain rate effect.($F = 0$ denotes the static yield function).

The volumetric strain is a measure of the deterioration of the granular material. We have, therefore, introduced the second order gradient of volumetric strain into the constitutive model.

The yield function now includes a gradient term and it is expressed as follows:

$$f = \frac{\sqrt{2J_2}}{M^*\sigma'_m} + ln\frac{\sigma'_m}{\sigma'_{my}} - a_3X = 0 \tag{2}$$

where $X$ is the gradient term,i.g. the second order gradient of the viscoplastic volumetric strain[2], $J_2$ is the second invariant of the deviatoric stress tensor $S_{ij}$ , $\sigma'_m$ is the mean effective stress and $\sigma'_{my}$ is the strain hardening parameter. In the following, we will use Terzaghi's effective stress $\sigma'_{ij}$. The material function $\Phi(F)$ is given by

$$\Phi(F) = c\sigma'_m \cdot exp\{m'(\frac{\sqrt{2J_2}}{M^*\sigma'_m} + ln\frac{\sigma'_m}{\sigma'_{me}} - \frac{1+e}{\lambda-\kappa}v^p - a_3X)\} \tag{3}$$

$$X = \nabla^2 v^p \tag{4}$$

We assume that $M^*$, $m'$ ,$c$ and $a_3$ are material constants. In $Eq.$(3), $\sigma'_{me}$ is the initial value of $\sigma'_m$, $\lambda$ is the consolidation index, $\kappa$ is the swelling index, $e$ is the void ratio, and $v^p$ is the volumetric plastic strain. The elastic strain rate is assumed to be obtained by the isotropic Hooke's law.

## FINITE ELEMENT ANALYSIS USING A GRADIENT DEPENDENT VISCOPLASTICITY THEORY

Higher order gradient terms have been introduced into the constitutive equations to simulate localization phenomena in the media of infinite extent. However, there is still a difficult problem to be solved when we apply a constitutive equation with gradient term to solve a boundary value problem defined in a finite domain. Recently, Mühlhaus and Aifantis [3] pointed out the importance of this and proposed a variational principle. In the present study, we assume the weak form of the dynamic yield function that contains the second gradient of viscoplastic strain and the boundary condition in terms of the first gradient of volumetric viscoplastic strain. In the present model, the dynamic yield function corresponds to the volumetric component of the flow rule.

we assume the weak form of the dynamic yield function that contains the second gradient of viscoplastic strain and the boundary condition in terms of the first gradient of volumetric viscoplastic strain. In the present model, the dynamic yield function corresponds to the volumetric component of flow rule.

For the first gradient $v^p$, we consider the flux $\tilde{Q}$ associated with internal structure change. We have

$$\int_V (f - a_1 v^p - a_3\nabla^2 v^p - g(\dot{v}^p))\delta v^p dv + \int_{S_1} (\tilde{Q} - \nabla_n v^p)\delta v^p \tilde{n} ds = 0 \tag{5}$$

where $\tilde{Q}$ is the microstructure flux, $S_1$ is the boundary where $\tilde{Q}$ is defined, and $\delta v^p$ is the virtual volumetric viscoplastic strain associated with virtual displacement $\delta u$ during the deformation process. In the psresent study, viscoplastic strain rate is assumed to always occur. From $Eq.$(5), we obtain the relation between the following equation for the microstructure change flux $\tilde{Q}$.

$$\tilde{Q} = \nabla_n v^p \tag{6}$$

where $\nabla_n$ is the normal gradient to the boundary.

For the total equilibirum equation, we use the well known virtual work theorem. When we implement the proposed model into the FEM code, we need the boundary value of $\tilde{Q}$. In this respect, we postulate that the flux $\tilde{Q}$ is zero at the boundary between rigid or elastic material and viscoplastic one. From a physical point of view, we could say that for a material that never deforms plastically, the flux $\tilde{Q}$ never flows into the material, and also the flux $\tilde{Q}$ never outflows from the viscoplastic material into rigid or elastic materials.

At the boundary where the stresses are specified, like a free surface of the ground, the flux $\tilde{Q}$ is also assumed to be zero. For the deformation analysis with consolidation, we could use the same method as used by Oka, Adachi and Okano[4]. They combined the finite element method with the finite difference method. The equation of motion of fluid phase is discretized by a backward finite difference scheme. The forward finite difference scheme is also used to explicitly calculate the second gradient of viscoplastic strain. In Fig.1, the finite difference grid for the rate of microstructural change flow is shown.

## NUMERICAL EXAMPLE

To examine the effect of gradient term, we have numerically analyzed the strain rate constant compression test under undrained plane strain condition. A weak element is introduced to enhance the growth of inhomogeneity. In the weak element, the material parameter C is three times larger than that in the perfect element. Fig.2 shows the finite element mesh and the boundary conditions. The material parameters used for calculation are as follows: $M^*$=0.865 ,$C = 1.0 \times 10^{-12}$ 1/sec, initial mean effective stress $\sigma'_{me} = 1.5kgf/cm^2$, initial stress ratio = 1.0, $m' = 25$, $\lambda = 0.231$ ,$\kappa = 0.05$ , void ratio = 1.5, elastic shear modulus $G = 20kgf/cm^2$ , permeability coefficient = $1.16 \times 10^{-10} m/sec$, $a_3 = 50.0$, $\dot{\epsilon}_{22}$: average axial strain rate is 0.2 $\%/min$. Figs. 3 and 4 show the distribution of the second invariant of the deviatoric viscoplastic strain rate tensor $\dot{\gamma}^p = (\sqrt{\dot{e}^p_{ij}\dot{e}^p_{ij}})$ for two cases( with and without strain gradient) at the axial strain $\epsilon_{22}$ of 2 %. It is seen that the inhomogeneity grows with the increase of average strain and the localization intensity in the case that the gradient term is introduced is larger than that in the case without the gradient term. In Figure 5, we can observe the checker board pattern at larger average strain of 2.8%.

## REFERENCES


1.Aifantis E.C.(1987), *Int. J. Plasticity* 3, 211-247.

2.Oka,F., Yashima, A.,Adachi,T. & Aifantis, E.C.(1991), *Proc. 3rd Int. Conf. on Constitutive law for Engineering Materials*, ed. by C. S. Desai 1991 Tucson.

3.Mülhaus,H.B. and E.C.Aifantis(1991), Int J. Solid and Structures to appear

4.Oka,F., Adachi, T. and Okano, Y.(1986), *Int. J. Num. Anal. Methods in Geomechanics*, 10, 1-16.


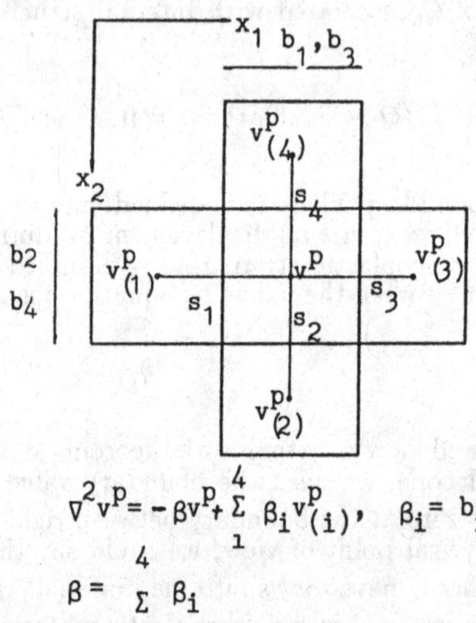


$$\nabla^2 v^p = -\beta v^p + \sum_1^4 \beta_i v^p_{(i)}, \quad \beta_i = b_i/s_i$$

$$\beta = \sum_1^4 \beta_i$$

Fig.1 Evaluation of gradient term $\nabla^2 v^p$

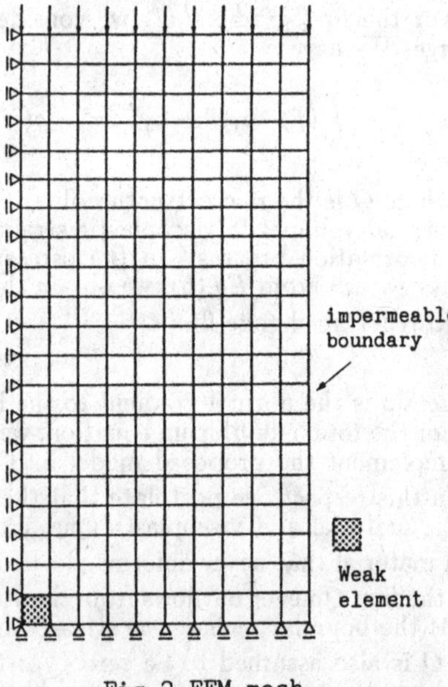


Fig.2 FEM mesh

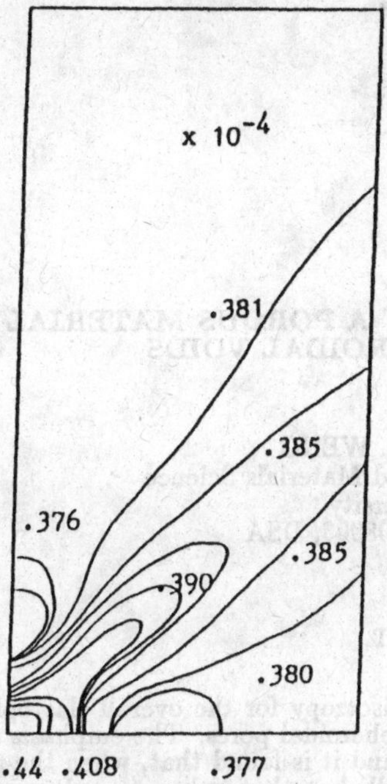


Fig. 3 Contours of $\dot{\gamma}^p$ for $a_3$=0, average strain 2 % (without gradient)

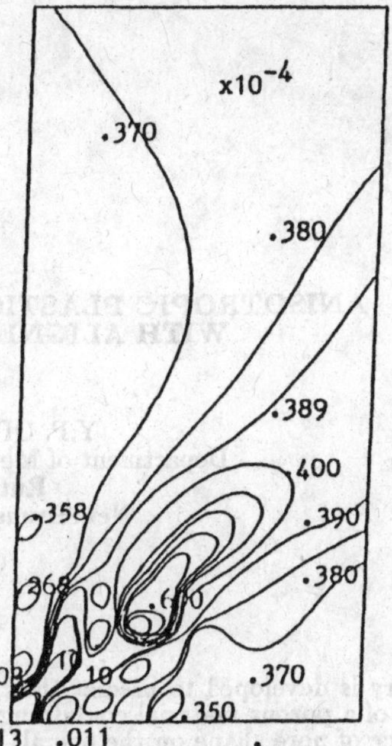


Fig. 4 Contours of $\dot{\gamma}^p$ for $a_3$= 50 , average strain 2 % (with gradient)

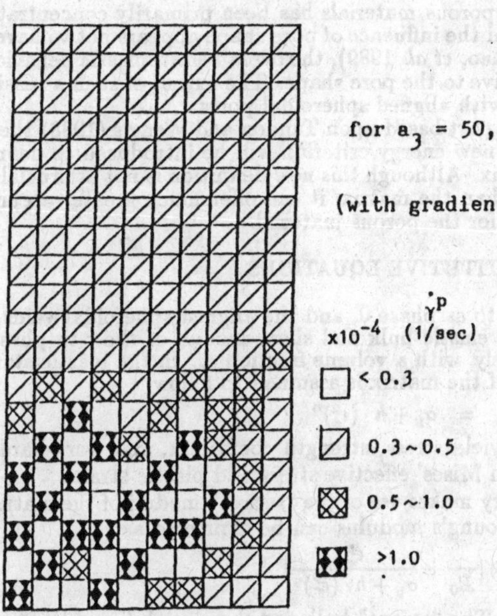


for $a_3$ = 50, average strain 2.8 %

(with gradient term )

Fig. 5 Distribution of $\dot{\gamma}^p$

# ANISOTROPIC PLASTICITY OF A POROUS MATERIAL WITH ALIGNED SPHEROIDAL VOIDS

Y.P. QIU and G.J. WENG
Department of Mechanics and Materials Science
Rutgers University
New Brunswick, NJ 08903, USA

## ABSTRACT

A theory is developed to predict the transverse isotropy for the overall elastoplastic response of a porous material containing aligned spheroidal pores. The emphasis is on the influence of pore shape on the overall behavior, and it is found that, when the pores are in a prolate shape the shape-dependence is small, but that, when they become oblate the pore-shape sensitivity is rather pronounced. The disc or penny-shaped pores are seen to cause the severest damage under the axial tension and the axial shear, but along the transverse tension, transverse shear, and biaxial plane-strain tension, penny-shaped pores provide the least weakening effect.

## INTRODUCTION

Previous study on the plasticity of porous materials has been primarily concentrated on the spherical or cylindrical pores and the influence of pore shape appears not to have been addressed. As in the elastic case (Zhao, *et al.* 1989), the overall elastoplastic behavior of a porous solid is believed to be sensitive to the pore shape. The extent of such a sensitivity is the subject of this investigation with aligned spheroidal pores.

The theory to be developed is in part based upon Tandon and Weng's (1988) theory of particle-reinforced plasticity, but a new energy criterion will be introduced to define the "overall" effective stress of the matrix. Although this new definition is not as crucial when the particles are elastically stiffer than the matrix, it can offer much greater accuracy if the inclusions are softer, especially for the porous material.

## CONSTITUTIVE EQUATIONS

The ductile matrix will be referred to as phase 0, and the aligned spheroidal voids, with an aspect ratio $\alpha$, as phase 1. The elastic bulk and shear moduli of the r-th phase will be denoted by $\kappa_r$ and $\mu_r$, respectively, with a volume fraction $c_r$. In the plastic state the effective stress and strain relation of the matrix is assumed to follow

$$\sigma_e = \sigma_y + h \cdot (\epsilon_e^p)^n , \tag{1}$$

where $\sigma_y$, $h$ and $n$ are the tensile yield stress, strength coefficient, and work-hardening exponent, and $\sigma_e$ and $\epsilon_e^p$ are the von Mises' effective stress and plastic strain.

Tandon and Weng's (1988) theory makes use of the "secant" moduli of the matrix. At a given plastic state the "secant" Young's modulus can be expressed as

$$E_0^s = [\frac{1}{E_0} + \frac{\epsilon_e^p}{\sigma_y + h \cdot (\epsilon_e^p)^n}]^{-1} , \tag{2}$$

in terms of the Young's modulus $E_0$. The "secant" bulk and shear moduli and the "secant" Poisson's ratio, then follow as

$$\kappa_0^s = \kappa_0 = \frac{E_0^s}{3(1-2\nu_0^s)}\,, \quad \mu_0^s = \frac{E_0^s}{2(1+\nu_0^s)}\,, \quad \nu_0^s = \frac{1}{2} - (\frac{1}{2} - \nu_0)\frac{E_0^s}{E_0}\,, \tag{3}$$

from the isotropic relation and plastic incompressibility. The plastic state of the matrix is seen to be characterized by $\epsilon_e^p$, or any of $E_0^s$, $\mu_0^s$ and $\nu_0^s$.

## PORE-SHAPE DEPENDENCE OF THE OVERALL SECANT MODULI

Following Tandon and Weng (1988) and Zhao and Weng (1990), the overall elastoplastic response of the transversely isotropic porous medium can be conveniently determined by introducing a linear homogeneous comparison material. Let the average secant moduli tensor of the matrix be denoted as $L_0^s$, with the components $L_0^s = (3\kappa_0,\ 2\mu_0^s)$, and we also let the comparison material to possess the elastic moduli $L_0^s$ at this instant. The analysis then can be carried out using Mori and Tanaka's (1973) method, by introducing Eshelby's (1957) equivalent transformation strain in the voided regions. From Zhao and Weng's (1990), the five independent overall secant moduli of the transversely isotropic porous material can be deduced to

$$\begin{aligned}
\frac{E_{11}^s}{E_0^s} &= \frac{1}{1+c_1(1-\nu_0^s)\{2(\alpha^2-1)[2\nu_0^s(g-1)-1]+(2-3g)\}/(2c_0p)}\,,\\
\frac{\nu_{12}^s}{\nu_0^s} &= \frac{1-c_1(1-\nu_0^s)[\alpha^2(2-3g)+(1+2\nu_0^s)(\alpha^2-1)g]/(2c_0\nu_0^s p)}{1+c_1(1-\nu_0^s)\{2(\alpha^2-1)[2\nu_0^s(g-1)-1]+(2-3g)\}/(2c_0p)}\,,\\
\frac{\kappa_{23}^s}{\bar{\kappa}_0^s} &= \frac{(1+\nu_0^s)(1-2\nu_0^s)}{1-\nu_0^s-2\nu_0^s\nu_{12}^s+c_1(1-\nu_0^s)[\alpha^2(1+\nu_{12}^s)(2-3g)-(\alpha^2-1)(1-\nu_{12}^s-2\nu_0^s\nu_{12}^s)]/(c_0p)}\,,\\
\frac{\mu_{12}^s}{\mu_0^s} &= 1+c_1\{-1+\frac{c_0}{2(1-\nu_0^s)}[1-2\nu_0^s-\frac{\alpha^2+1}{\alpha^2-1}-\frac{1}{2}(1-2\nu_0^s-\frac{3(\alpha^2+1)}{\alpha^2-1})g]\}^{-1}\,,\\
\frac{\mu_{23}^s}{\mu_0^s} &= 1+c_1\{-1+\frac{c_0}{2(1-\nu_0^s)}[\frac{\alpha^2}{2(\alpha^2-1)}+(1-2\nu_0^s-\frac{3}{4(\alpha^2-1)})g]\}^{-1}\,,
\end{aligned} \tag{4}$$

where

$$p = 2\alpha^2-(4\alpha^2-1)g+(1+\nu_0^s)(\alpha^2-1)g^2\,, \tag{5}$$

and the parameter $g$ depends solely on the aspect ratio $\alpha$ of the voids,

$$\begin{aligned}
g &= \tfrac{\alpha}{(\alpha^2-1)^{3/2}}[\alpha(\alpha^2-1)^{1/2}-cosh^{-1}\alpha]\,, && \text{for prolate shape}\\
&= \tfrac{\alpha}{(1-\alpha^2)^{3/2}}[cos^{-1}\alpha-\alpha(1-\alpha^2)^{1/2}]\,, && \text{for oblate shape.}
\end{aligned} \tag{6}$$

In addition, it takes the values $g = 1,\ 2/3,\ 0$ when $\alpha \to \infty,\ 1,\ 0$, respectively.

The transverse "secant" Young's modulus, if needed, is

$$\frac{1}{E_{22}^s}(=\frac{1}{E_{33}^s}) = \frac{{\nu_{12}^s}^2}{E_{11}^s}+\frac{1}{4\kappa_{23}^s}+\frac{1}{4\mu_{23}^s}\,. \tag{7}$$

## AVERAGE EFFECTIVE STRESS OF THE MATRIX

We now subject the porous medium to a proportionally increasing combined stress,

$$\bar{\sigma}_{ij} = \alpha_{ij}\bar{\sigma}(t)\,, \tag{8}$$

where $\alpha_{ij}$ are the desired proportional constants, and $\bar{\sigma}(t)$ is the current stress magnitude ($t$ being a time-like parameter). The initial response of the porous material at low $\bar{\sigma}(t)$ is elastic, with an overall behavior by setting $\nu_0^s = \nu_0$ in (4) (Tandon and Weng, 1984).

After plastic deformation eq. (1) must be satisfied between $\sigma_e$ and $\epsilon_e^p$ of the matrix. The former in Tandon and Weng's (1988) was calculated by $\sigma_e^{(0)} = (\frac{3}{2}\bar{\sigma}_{ij}'^{(0)}\bar{\sigma}_{ij}'^{(0)})^{1/2}$, in terms of the mean deviatoric stress of the matrix. In a porous material

$$\bar{\sigma}_{ij}^{(0)} = \frac{1}{c_0}\bar{\sigma}_{ij}\,, \tag{9}$$

and thus, when $\bar{\sigma}_{ij}$ is in a state of purely hydrostatic tension, $\sigma_e^{(0)}$ vanishes, and the porous composite, according to this approach, will undergo no plastic deformation. While it is known that, when the inclusions are elastically stiffer than the matrix the plastic strain of the composite containing spherical inclusions under a hydrostatic tension is negligible, it is not so with spherical voids.

To remove this drawback, we recall that von Mises's effective stress essentially represents the elastic distortional energy of the solid, and therefore the overall effective stress of the heterogeneously deformed matrix is perhaps more suitably derived from the energy approach. Then, keeping in mind that the comparison material possessing the equivalent transformation strain is a linear solid, we may write its elastic energy $U_s$ as

$$U_s \;=\; \frac{1}{2}\bar{\sigma} M_s \bar{\sigma} = c_0 U_0^s \,, \tag{10}$$

where $M_s$ is the overall elastoplastic compliances tensor of the porous material. The

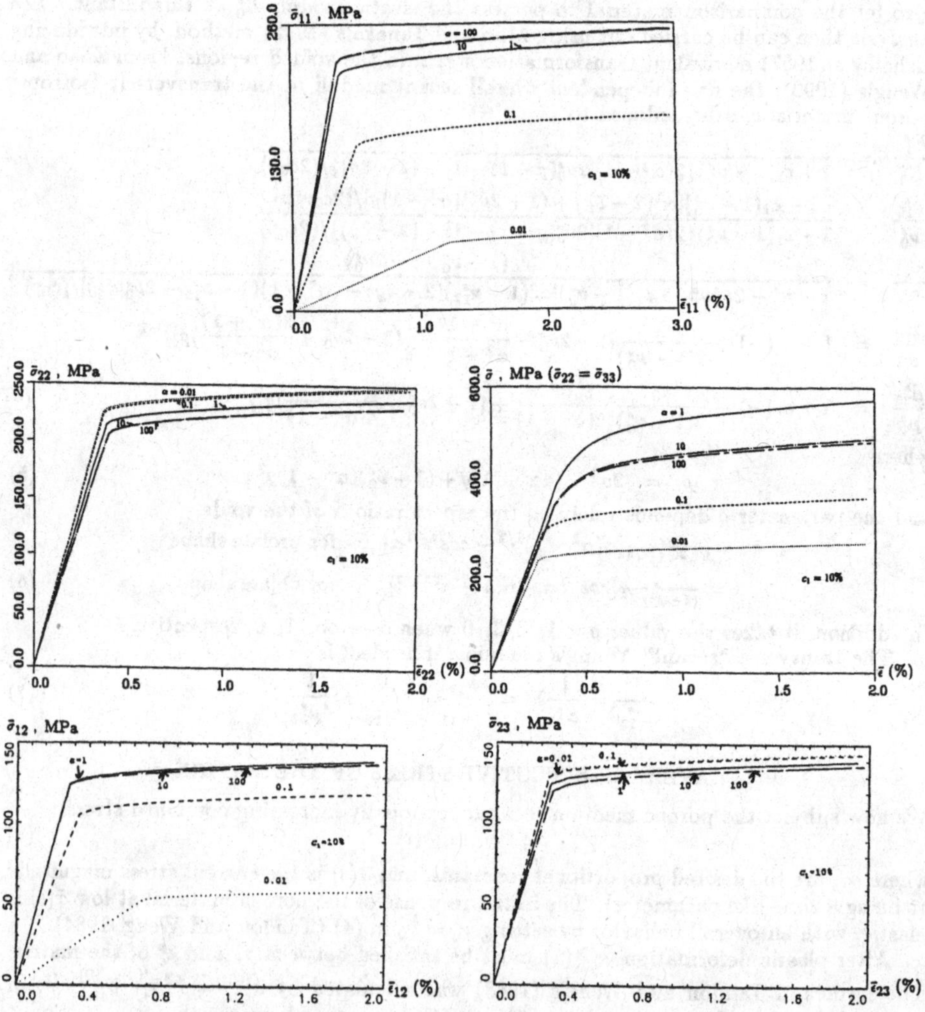


Fig.1 The influence of pore shape on the stress-strain relation of the porous material under five respective loadings.

elastic energy of the matrix $U_0^s$ consists of the distortional energy and the dilatational energy, which can be written as $\frac{1}{6\mu_0^s}\sigma_e^{(0)2}$ and, as an approximation, $\frac{1}{2\kappa_0}\bar{\sigma}_m^{(0)2}$, respectively. The effective stress of the matrix then follows from (10), as

$$\sigma_e^{(0)2} = \frac{3\mu_0^s}{c_0}[\frac{1}{E_{11}^s}\bar{\sigma}_{11}^2 - \frac{2\nu_{12}^s}{E_{11}^s}\bar{\sigma}_{11}(\bar{\sigma}_{22}+\bar{\sigma}_{33}) + \frac{1}{E_{22}^s}(\bar{\sigma}_{22}+\bar{\sigma}_{33})^2 - \frac{1}{\mu_{23}^s}\bar{\sigma}_{22}\bar{\sigma}_{33} + \frac{1}{\mu_{12}^s}(\bar{\sigma}_{12}^2+\bar{\sigma}_{13}^2) + \frac{1}{\mu_{23}^s}\bar{\sigma}_{23}^2] - \frac{3\mu_0^s}{\kappa_0}\frac{1}{c_0^2}\bar{\sigma}_m^2 , \quad (11)$$

after incorporating the components of $M_s$ and the relation $\bar{\sigma}_m = c_0\bar{\sigma}_m^{(0)}$.

When the voids are spherical, the overall response is isotropic and (11) becomes

$$\sigma_e^{(0)2} = \frac{6\mu_0^S}{c_0}[\frac{1}{6\mu_s}\bar{\sigma}_e^2 + \frac{1}{2\kappa_s}\bar{\sigma}_m^2] - \frac{3\mu_0^s}{\kappa_0}\frac{1}{c_0^2}\bar{\sigma}_m^2 , \quad (12)$$

which is seen to possess the desired property that it is non-vanishing even under a pure $\bar{\sigma}_m$.

The stress-strain curves of the porous material under (8) now can be generated in the following way. One may start with a given $\epsilon_e^p$ for the matrix, which provides $E_0^s$, $\mu_0^s$, and $\nu_0^s$ from (3). The corresponding overall secant moduli can be calculated from (4). On the other hand this assigned $\epsilon_e^p$ also corresponds to a $\sigma_e^{(0)}$ from (1), and then, by (11) and (8), the corresponding $\bar{\sigma}(t)$ can be calculated, for any chosen $\alpha_{ij}$, which in turn provides the $\bar{\sigma}_{ij}$ components. By increasing the value of $\epsilon_e^p$, this process may be repeated to generate the entire stress-strain curves under a monotonically increasing combined stress.

## NUMERICAL RESULTS

For numerical illustrations we use the properties of aluminum for the matrix with $E_0 =$ 68.3 GPa, $\nu_0 = 0.33$, $\sigma_y = 250$ MPa, $h = 173$ MPa, and $n = 0.45$. The influence of pore shape on the stress-strain curves of the transversely isotropic solid is illustrated respectively by loading along $\bar{\sigma}_{11}$, $\bar{\sigma}_{22}$, biaxial $\bar{\sigma}_{22} = \bar{\sigma}_{33}$ ($\bar{\epsilon}_{11} = 0$), $\bar{\sigma}_{12}$, and $\bar{\sigma}_{23}$. The corresponding results are shown in Figs. 1(a) to (e) at $c_1 = 10\%$ and five pore shapes $\alpha = 100$, 10, 1, 0.1, and 0.01. The influence of pore shape on the overall stress-strain curves appears to be not strong when it is prolate, but its sensitivity grows when it becomes oblate. The disc or penny-shaped pores are seen to cause the severest damage along $\bar{\sigma}_{11}$, $\bar{\sigma}_{12}$, and $\bar{\sigma}_{22} = \bar{\sigma}_{33}$ ($\bar{\epsilon}_{11} = 0$), but along $\bar{\sigma}_{22}$ and $\bar{\sigma}_{23}$, penny-shaped pores provide the least weakening effect.

## ACKNOWLEDGEMENT

This work was supported by the National Science Foundation, Solid and Geo-Mechanics Program, through MSS-8918235.

## REFERENCES

Eshelby, J.D. (1957). The determination of the elastic field of an ellipsoidal inclusion, and related problems. *Proc. Roy. Soci, London, A241*. 376-396.

Mori, T. and K. Tanaka (1973). Average stress in the matrix and average elastic energy of materials with misfitting inclusions. *Acta Metall. 21*. 571-574.

Tandon, G.P. and G.J. Weng (1984). The effect of aspect ratio of inclusions on the elastic properties of unidirectionally aligned composites. *Polymer Composites, 5*. 327-333.

Tandon, G.P. and G.J. Weng (1988). A theory of particle-reinforced plasticity. *J. Appl. Mech. 55*, 126-135.

Zhao, Y.H., G.P. Tendon and G.J. Weng (1989). Elastic moduli for a class of porous materials. *Acta Mech. 76*, 105-130.

Zhao, Y.H. and G.J. Weng (1990). Theory of plasticity for a class of inclusion and fiber-reinforced composites. in *Micromechanics and Inhomogeneity, The Toshio Mura Anniversary Volume*, ed. G.J. Weng, M. Taya and H. Abe, Spring-Verlag, New York. 599-622.

# ELASTO-VISCOPLASTIC MODELLING OF POROUS ROCK UNDER HIGH CONFINING PRESSURE

J.F.SHAO, M.BEDERIAT, J.P.HENRY

**Laboratory of mechanics of Lille**
**EUDIL 59 655 - Villeneuve d'Ascq Cédex France**

## ABSTRACT
In this study, an elasto-viscoplastic model for porous rock is presented. This model based on the Perzyna's viscoplasticity theory contains two viscoplastic flow mechanisms, a hydrostatic one and a deviatoric one. Simulations of isotropic and triaxial creep tests under different confining pressures are presented.

## INTRODUCTION
Rock behaviour is generally time dependent. Some constitutive models have been proposed to describe the creep behaviour of brittle rock under low confining pressure, for instance Cristescu (1987), Toshihisa et al (1987), Gioda (1981). Viscoplastic modelling of porous rock under high pressure has great interest in petroleum engineering and underground works at great depth. But there are so far few studies in this domain.
In this study, a white porous chalk is used. Its average porosity is 40 %. A new experimental device has been designed to realize creep tests under high confining pressure Bederiat (1991). It is a self pressure compensated and self axial load controlled triaxial cell. It is possible to realize independently (without hydraulic machine) all types of triaxial creep tests. For the above chalk have been performed hydrostatic creep tests under different stresses and triaxial creep tests at different stress deviators under several confining pressures. The obtained results show that this rock presents two viscoplastic flow mechanism, a hydrostatic one related to plastic pore collapse and a deviatoric one to distorsion. Only primary creep has been observed under hydrostatic stress and under relatively low stress deviator. But second and even tertiary creep have been observed under high stress deviator. The details of the experimental investigation will be presented in the full length version.
From the experimental observations, an elastoviscoplastic model is proposed. This model contains then two viscoplastic flow mechanisms using the Perzyna's viscoplasticity theory, Perzyna (1966), Owen (1982). Applications of such a model are particularly aimed at subsidence analysis of petroleum reservoirs and study of stability of underground constructions at great depth. Theoretical predictions of creep behaviour of rock are presented and compared with experimental results. The calibration of the model will be presented in the full length version.

## FORMULATION OF THE MODEL
The concept of multimechanisms in plasticity is used, Cambou (1988), Shao (1990). The present viscoplastic model is due to the extension of the elastoplastic model proposed by Shao and Henry (1990) for the time independent behaviour of porous rock. The total strain rate is divided into three components : elastic one, collapse plastic one and deviatoric plastic one :

$$\dot{\varepsilon}_{ij} = \dot{\varepsilon}_{ij}^{\,e} + \dot{\varepsilon}_{ij}^{\,c} + \dot{\varepsilon}_{ij}^{\,d}$$

The elastic strain is supposed time independent in the present study and determined by Hooke's law. According to the Perzyna's viscoplasticity theory, the onset of viscoplastic behaviour is governed by a scalar yield condition of the form :

$$F_c = f_c(\sigma_{ij}, \varepsilon_{ij}^{c}) - y_c(\varepsilon_{ij}^{c}) = 0$$

$$F_d = f_d(\sigma_{ij}, \varepsilon_{ij}^{d}) - y_d(\varepsilon_{ij}^{d}) = 0$$

It is assumed that viscoplastic flow occurs for values of $f_c > y_c$ and $f_d > y_d$ only. The collapse and deviatoric viscoplastic strain rates can be then written :

$$\dot{\varepsilon}_{ij}^{\,c} = \gamma_c < \phi_c(F_c) > \frac{\partial Q_c}{\partial \sigma_{ij}}$$

$$\dot{\varepsilon}_{ij}^{\,d} = \gamma_d < \phi_d(F_d) > \frac{\partial Q_c}{\partial \sigma_{ij}}$$

Where $\gamma_c$ and $\gamma_d$ are two fluidity parameters controlling the plastic flow rate. $Q_c$ and $Q_d$ can be considered as viscoplastic potentials. The notation $< >$ implies

$$< \phi(x) > = \phi(x) \quad \text{for} \quad x > 0$$

$$< \phi(x) > = 0 \quad \text{for} \quad x \leq 0$$

For the collapse flow mechanisms, we have used :

$$f_c = I_1$$

$$y_c = y_c^{o} + a \; Pa \; \xi_c^{\,n} \; e^{c.\xi_c}$$

$$Q_c = f_c$$

$$\phi_c = \left(\frac{f_c}{y_c} - 1\right)^{\alpha_c}$$

where $I_1$ is the first invariant of stress tensor, and a, n, c, $y_c{}^{o}$ and $\alpha_c$ are material parameters. $P_a$ is the atmospheric pressure. For the deviatoric flow mechanisms, the following relations have been taken :

$$f_d = \left(\frac{I_1^3}{I_3} - 27\right)\left(\frac{I_1}{P_a}\right)^m$$

$$y_d = \frac{y_d^r \, \xi_d^{\,u}}{(1+\xi_d)^u}$$

$$Q_d = I_1{}^3 - 27 I_3$$

$$\phi_d = a_2\left(\frac{f_d}{y_d}\right)^{b_2}\left(\frac{f_d}{y_d} - 1\right)^{\alpha_d}$$

where m, $y^d{}_r$, u, $a_2$, $b_2$ and $\alpha_d$ are material parameters. $I_3$ is the third invariant of stress tensor. Experimental results have shown that the fluidity of rock depends on hydrostatic stress. In this study, we have taken $\gamma_c = \gamma_d = \gamma$ and $\gamma$ is assumed to be dependent on $I_1$.

$$\gamma = \beta \left( \frac{I_1}{3\, P_a} \right)^t$$

On the other hand, parameter u depends generally on initial stress state of problem. In the above relations, $\xi_c$ and $\xi_d$ represent equivalent plastic strain defined by :

$$\xi_c = \int_0^T \sqrt{\dot{\varepsilon}_{ij}^{\,c}\ \dot{\varepsilon}_{ij}^{\,c}}\ dt$$

$$\xi_d = \int_0^T \sqrt{\dot{\varepsilon}_{ij}^{\,d}\ \dot{\varepsilon}_{ij}^{\,d}}\ dt$$

In order to take into account the tensile strength which can be sustained by rock, a transformation of stress coordinates will be performed according to Desai (1987) :

$$\sigma_{ij}^* = \sigma_{ij} + R\, P_a\, \delta_{ij}$$

where R is tensile strength parameter, and $\delta_{ij}$ Kronecker delta. After this transformation, $I_1$ and $I_3$ will be evaluated with $\sigma_{ij}^*$.

The constitutive model contains 15 parameters which can be completely estimated from available hydrostatic and triaxial creep tests. The associated calibration process of the model will be presented in the full length version.

## VERIFICATION OF THE MODEL

The first verifications of the model are now presented. It consists of simulation of creep tests. In figures 1 and 2 are presented comparisons between model predictions and experimental results for three hydrostatic creep tests under different stresses. It can be seen that the model describes correctly the creep behaviour of the rock. Simulations of two triaxial creep tests at different stress deviators are presented in figures 3 to 6 . We can also notice a good agreement between numerical predictions and the experiment but there are some differences for the axial strain.

## CONCLUSION

A viscoplastic model with two flow mechanisms for porous rock is proposed according to experimental investigations. Only the first verifications of the model, simulation of creep tests are presented. Further verifications of the model wille be presented in the full length version.

## REFERENCES


Cambou B., Jafari K. (1988) - "Modèle de comportement des sols non cohérents" - *Revue Française de Géotechnique*, n° 44, p. 43-55

Cristescu N. (1987) - "Elastic/viscoplastic constitutive equations for rock" - *Int. J. Rock Mech. Min. Sci & Geomech. Abstr.*, Vol. 24, n° 5, 271-282

Desai C.S., SALAMI M.R. (1987) - "A constitutive model and associated testing for soft rock" - *Int. J. Rock Mech. Min. Sci. & Geomech. Abstr.*, 24, n° 5, 297-307

Gioda G. (1981) - "A finite element solution of non-linear creep problems in rocks" - *Int. J. Rock Mech. Min. Sci. & Geomech. Abstr.*, Vol. 18, 35-46

Owen D.R.J., Hinton E. (1980) - "Finite element in plasticity - Theory and Practice" *Pineridge Press*, Swansea, U.K.

0wen D.R.J. (1982) - "Viscoplastic analysis of solids - Stability considerations" - *Recent advances in non linear computational mechanics*, Hinton et al (eds), 225-254

Perzyna P. (1966) - "Fundamental problems in viscoplasticity, *Recent advances in Applied Mechanics*, Academic Press, Vol. 9, 243-377

Shao J.F., Henry J.P. (1990) - "Validation of an elastoplastic model for chalk" - *Computers and geotechnics 9*, pp. 257-272

Toshihisa A., Fusao 0., Keisuke K. (1987) - "Viscoplastic constitutive model of soft rocks with strain softening" - *Constitutive laws for Engineering materials - Theory and applications*, Desai et al (eds), 659 - 665

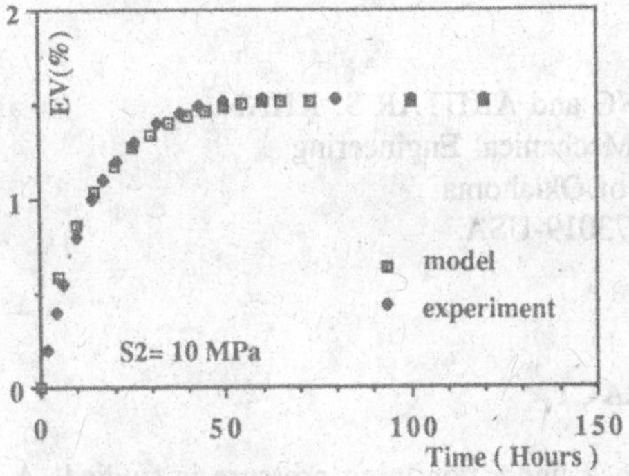


FIG1: isotropic creep test (S3=10MPa)

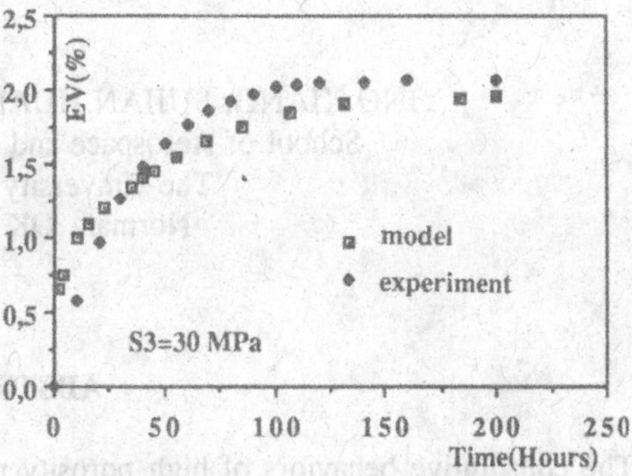


Fig 2: isotropic creep test (S3=30MPa)

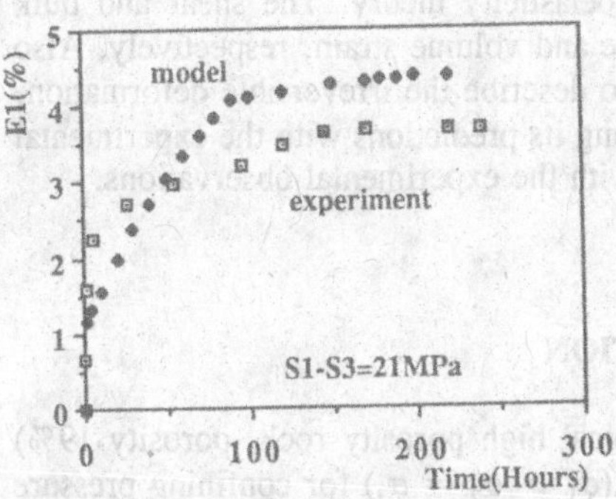


Fig3: Triaxial creep test (S3=20MPa)

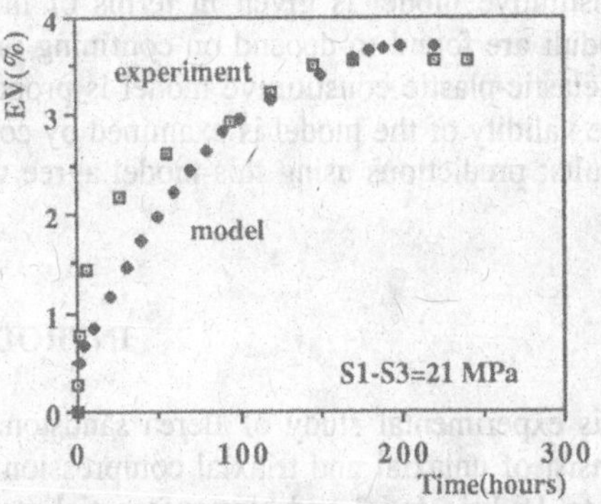


Fig4: triaxial creep test(S3=20MPa)

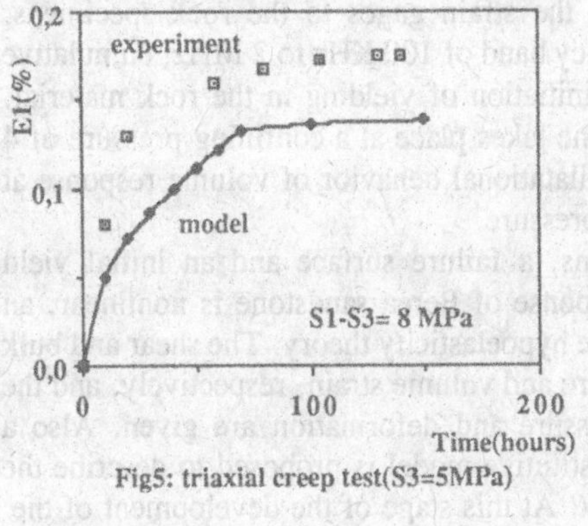


Fig5: triaxial creep test(S3=5MPa)

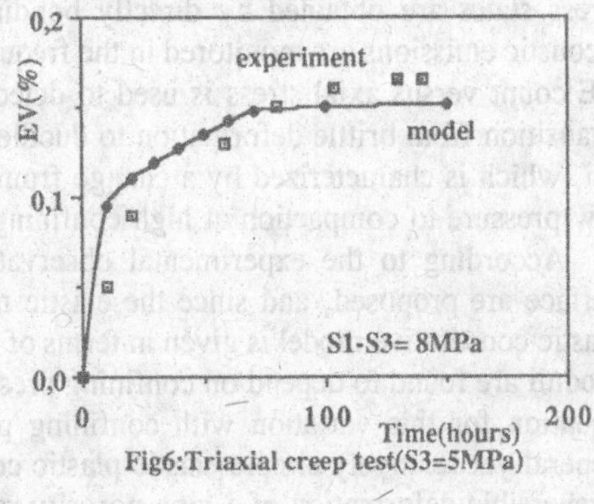


Fig6:Triaxial creep test(S3=5MPa)

# ELASTIC-PLASTIC BEHAVIOR OF HIGH POROSITY ROCK

YING XIANG, SUJIAN HUANG and AKHTAR S. KHAN
School of Aerospace and Mechanical Engineering
The University of Oklahoma
Norman, OK 73019-USA

## ABSTRACT

The constitutive behaviors of high porosity rock under confining pressure is studied. A failure surface and an initial yield surface are proposed based on the results, and an elastic constitutive model is given in terms of the hypoelasticity theory. The shear and bulk moduli are found to depend on confining pressure and volume strain, respectively. Also an elastic-plastic constitutive model is proposed to describe the irreversible deformation. The validity of the model is examined by comparing its predictions with the experimental results; predictions using this model agree well with the experimental observations.

## INTRODUCTION

This experimental study of Berea sandstone (typical high porosity rock, porosity 19%) consist of uniaxial and triaxial compression tests ($\sigma_1 > \sigma_2 = \sigma_3$) for confining pressure up to 15 ksi. Axial and circumferential strains versus axial stress as well as the failure stress states are obtained by directly bonding the strain gages to the rock specimens. Acoustic emissions are monitored in the frequency band of 100 KHz to 2 MHz; cumulative AE count versus axial stress is used to detect initiation of yielding in the rock material. Transition from brittle deformation to ductile one takes place at a confining pressure of 4 ksi, which is characterized by a change from dilatational behavior of volume response at low pressure to compaction at high confining pressure.

According to the experimental observations, a failure surface and an initial yield surface are proposed, and since the elastic response of Berea sandstone is nonlinear, an elastic constitutive model is given in terms of the hypoelasticity theory. The shear and bulk moduli are found to depend on confining pressure and volume strain, respectively, and the equation for this variation with confining pressure and deformation are given. Also a general yet relatively simple elastic-plastic constitutive model is proposed to describe the irreversible deformation of a high porosity rock. At this stage of the development of the model, an associated flow rule is used. The validity of the model is examined by comparing its predictions with the experimental results for different triaxial compression tests for Berea sandstone; predictions using this model agree well with the experimental observations.

## RESULTS AND DISCUSSION

The reader is referred to Xiang and Khan [3] for the experimental setup and detailed discussion of the experimental procedures and the results. Based on the experimental observations, a preliminary analysis of the test data has been carried out to determine: (i) Failure surface; (ii) Initial yield surface; (iii) Constitutive equation for elastic deformation; (iv) An elastic-plastic constitutive model.

**(i) Failure surface**

The failure surface proposed in this study is

$$(2J_2)^{1/2} = a_0 + a_1 I_1 + a_2 I_1^2 \tag{1}$$

here $I_1 = \sigma_{ii}$, $J_2 = (1/2)s_{ij}s_{ij}$. The material constants $a_0$, $a_1$, and $a_2$ are determined by the least square technique using the experimental data, and are given below:

$$a_o = 0.86662\ ksi, \quad a_1 = 0.622507, \quad a_2 = -0.0032261\ ksi \tag{2}$$

Fig.1 shows the comparison between the curve given by the proposed failure function and the experimental data.

**(ii) Initial Yield Surface**

The initial yield surface is determined by considering the deformation behaviors of distortion and volume change, and is given below

$$(2J_2)^{1/2} = b_0 + b_1 I_1 + b_2 I_1^2 \tag{3}$$

where $b_0$, $b_1$ and $b_2$ are material constants:

$$b_0 = 0.430258\ ksi, \quad b_1 = 0.384076, \quad b_2 = -0.0018614\ ksi^{-1} \tag{4}$$

Also the initial yield points are determined on the basis of the deviation form linearity of the cumulative AE counts versus stress curv. Fig.2 shows the determination of the yield point from AE signal in case of confining pressure of 8 ksi. Fig.1 gives the comparison between the curve given by the proposed initial yield function and the experimental data, as well as the result obtained from the AE signals.

**(iii) ELASTIC CONSTITUTIVE MODEL**

The experimental observation shows that bulk modulus K is dependent only on the volume strain $\epsilon_v$, but shear modulus G is dependent on the confining pressure P. The elastic behavior of Berea sandstone can be given by

$$s_{ij} = 2\left[G_s - (G_s - G_0)\, e^{-\alpha\left(\frac{p}{p_0}\right)}\right] e_{ij} \tag{5}$$

$$\sigma_m = K_s e_v - \left(\frac{1}{r}\right)\left[(K_s - K_0)\left(1 - e^{-r e_v}\right)\right] \tag{6}$$

here the quantities $s_{ij}$ and $e_{ij}$ are stress and strain deviators, respectively, and $G_0$, $G_s$, $\alpha$, as well as $k_0$, $k_s$, and r are material constants determined from the experimental data :

$$G_0 = 468\ ksi, \quad G_s = 1720.97\ ksi, \quad \alpha = -0.00041 \tag{7}$$

$$K_0 = 167.1\ ksi, \quad K_s = 2099.00\ ksi, \quad r = 221.87 \tag{8}$$

## (iv) AN ELASTIC-PLASTIC MODEL FOR HIGH POROSITY ROCK

The proposed yield function in the study is

$$Y_f = \sqrt{2J_2} - \alpha\left(b_0 + b_1 I_1 + b_2 I_1^2\right), \qquad 1 \le \alpha \le 1.8 \tag{9}$$

and an appropriate hardening parameter may be defined as

$$\alpha = \alpha_{max} - (\alpha_{max} - 1)\, e^{-F\xi} \tag{10}$$

here $\alpha_{max}$ and F are material constants and obtained from the experimental data as 1.8 and 14, respectively. An associated flow rule is used. The form of the incremental stress-strain relation can be obtained as

$$d\sigma_{ij} = C^{ep}_{ijkl}\, de_{kl} \tag{11}$$

in which $C^{ep}_{ijkl}$ = constitutive tensor composed of elastic and plastic parts and is expressed in terms of the elastic and plastic parameters. The experimentally observed stress-strain responses were back-predicted for the tests with confining pressures of 3 ksi and 10 ksi in Fig.3 and Fig.4, respectively; predictions using this model agree well with the test data. These results are described in detail in two forthcoming papers [1, 2].

## CONCLUSIONS

(1) The brittle-ductile transition in Berea sandstone occurred at C.P. of 4 ksi.
(2) AE activity is a good method for detecting the onset of inelastic deformation.
(3) Failure function is a parabolic curve in the $(2J_2)^{1/2}$ -$I_1$ space.
(4) An elastic constitutive model was proposed for the nonlinear elastic response.
(5) An elastic-plastic model for high porosity rock was developed.
(6) The model was used to simulate the behavior of Berea sandstone. Good agreement was shown between experimental stress-strain curves and the predictions using the model.

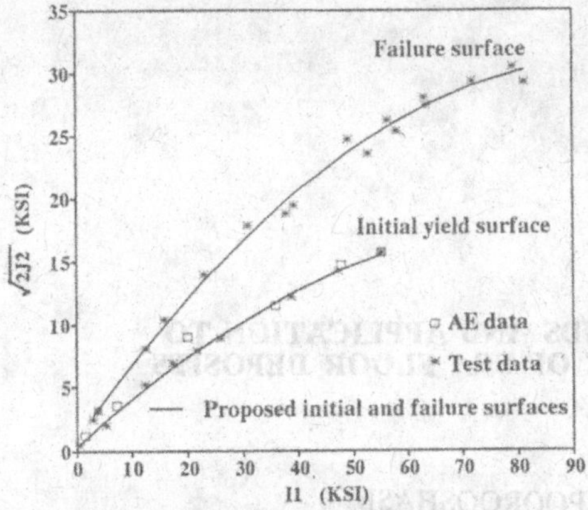


Fig.1 The initial and failure surfaces.

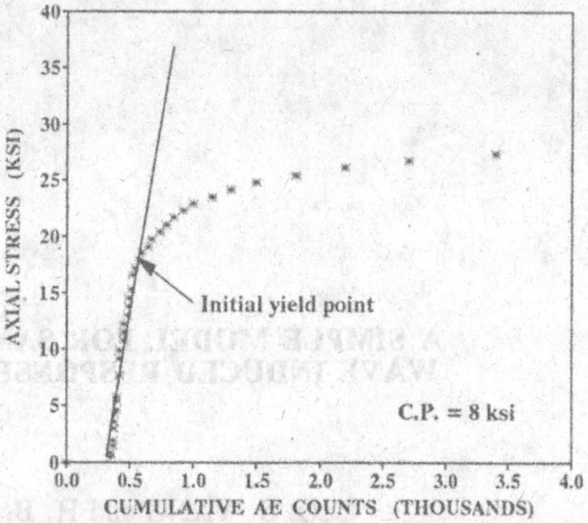


Fig.2 Enlarged cumulative AE counts versus axial stress for the determination of the initial yield point.

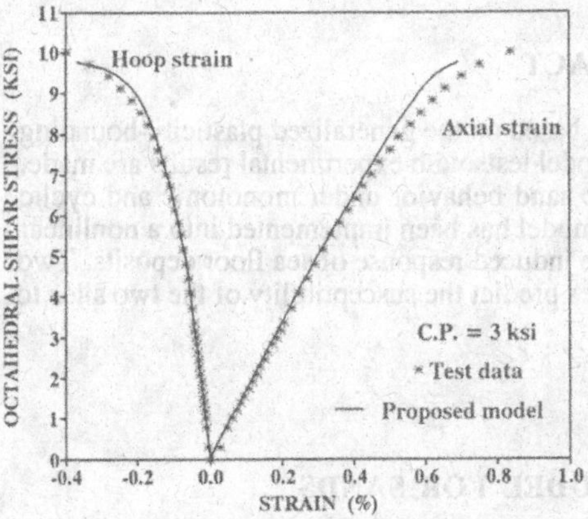


Fig.3 The comparison between the predicted and observed stress-strain curves (C.P. = 3 ksi).

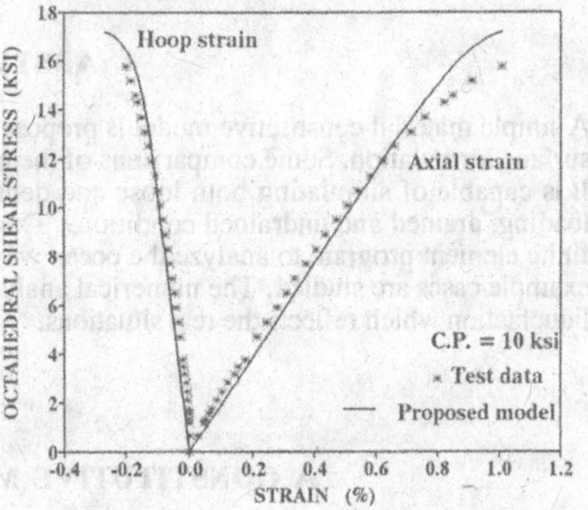


Fig.4 The comparison between the predicted and observed stress-strain curves (C.P. = 10 ksi).

## REFERENCES


1. Khan, A.S., Xiang, Y. and Huang, S., Behavior of Berea sandstone under confining pressure, Part I: yield and failure Surfaces, and nonlinear elastic response. Accepted for publication in International Journal of Plasticity, 1991.

2. Xiang, Y., Khan, A.S. and Huang, S., Behavior of Berea sandstone under confining pressure, Part II: elastic-plastic response. Accepted for publication in International Journal of Plasticity, 1991.

3. Xiang, Y. and Khan, A.S., Experimental study of yield and failure behaviors of Berea sandstone under confining pressure. Submitted to Experimental Mechanics, 1991.

# A SIMPLE MODEL FOR SANDS AND APPLICATION TO WAVE INDUCED RESPONSE OF SEA FLOOR DEPOSITS

Q. S. YANG and H. B. POOROOSHASB
**Civil Engineering Department, Concordia University, 1455 De Maisonneuve,W., Montreal, Canada H3G 1M8**

## ABSTRACT

A simple material constitutive model is proposed based on the generalized plasticity-bounding surface formulation. Some comparisons of the model tests with experimental results are made. It is capable of simulating both loose and dense sand behavior under monotonic and cyclic loading, drained and undrained conditions. The model has been implemented into a nonlinear finite element program to analyze the ocean wave induced response of sea floor deposits. Two example cases are studied. The numerical analyses predict the susceptibility of the two sites to liquefaction which reflects the real situations.

## A CONSTITUTIVE MODEL FOR SANDS

The classical concept of a yield surface implies a purely elastic stress range contrary to the reality for many soils. Some very important aspects of soil behavior, mainly relating to the cyclic response cannot be adequately described. This shows the necessity to develop constitutive laws within a more fundamental framework. Among the new concepts is that of the bounding surface originally introduced by Dafalias and Popov for metal plasticity[1]. The salient features of a bounding surface formulation are that plastic deformation may occur for stress states within the surface and the possibility of having a very flexible variation of the plastic modulus according to the generalized plasticity [2].

An unit vector **n** at stress space which determines both loading and unloading direction is required in generalized plasticity. If $\mathbf{n}^T d\sigma' > 0$, loading and $\mathbf{n}^T d\sigma' < 0$, unloading. The incremental stress-strain relations could be written in compact matrix notation in which the subscripts l and u refer to loading and unloading respectively,

$$d\sigma' = \mathbf{D}_{l/u}\, d\varepsilon\,, \qquad \mathbf{D}_{l/u} = \mathbf{D}^e - \frac{\mathbf{D}^e \mathbf{n}_{g\,l/u}\, \mathbf{n}^T \mathbf{D}^e}{H_{l/u} + \mathbf{n}^T \mathbf{D}^e \mathbf{n}_{g\,l/u}}$$

The material behaviour can be fully described if we can prescribe $\mathbf{D}^e$, **n**, $\mathbf{n}_{gl/u}$ and $H_{l/u}$ for all states of the material.

In bounding surface formulation, the bounding surface is functioning as the classical yield surface for virgin loading. However, if the actual stress point is within the bounding surface, **n**, $\mathbf{n}_{gl/u}$ and $H_{l/u}$ are defined by a suitable interpolation rule from the values of its 'image' on the bounding surface.

The proposed model is based on the generalized plasticity incorporating the bounding surface formulation with a non-associative flow rule [3]. It is also a simplified version of the two-surface model [4]. For comparison, these two models in the triaxial configuration are shown in Figure 1. The model in the general effective stress space has been implemented into a computer program MDTEST. Eight material parameters in total, $K_z$, $\kappa$, $\nu$, $\phi_f$, $\phi_{cr}$, A, B, $\gamma_0$ are used. A number of numerical model tests has been done to simulate the loose and dense sand behavior under monotonic and cyclic, drained and undrained conditions [3]. Only a few numerical tests are presented here to show the performance of the model.

Figure 2(a) is the results of cyclic drained test [5]. The plastic volumetric strain develops throughout the test, causing the soil to contract. A remarkable fact is that volumetric plastic strain developed during unloading is always of a contractive nature. The model predicted behavior as shown in Figure 2(b) reproduces the cyclic load-densification phenomena.

Figure 3 shows the influence of relative density on undrained behavior. The numerical simulations agree well with the triaxial compression tests performed by Castro [6]. The strength of loose sand after reaching a peak, drops continuously to almost zero and complete liquefaction is achieved. For dense sand, negative pore pressure occurs while mean effective stress increases after passing the phase transformation line.

Figure 4 shows the results of a strain controlled undrained test [7] in which a constant strain amplitude was applied. Progressive reduction in stress amplitude accompanies the generation of pore pressure. After a number of cycles the sample liquifies.

The main features of the model are its relative simplicity because of a vanishing yield surface while the reflecting plastic potential is retained implicitly and its capability of simulating several fundamental aspects of sand behavior which have already been recognized through laboratory investigations.

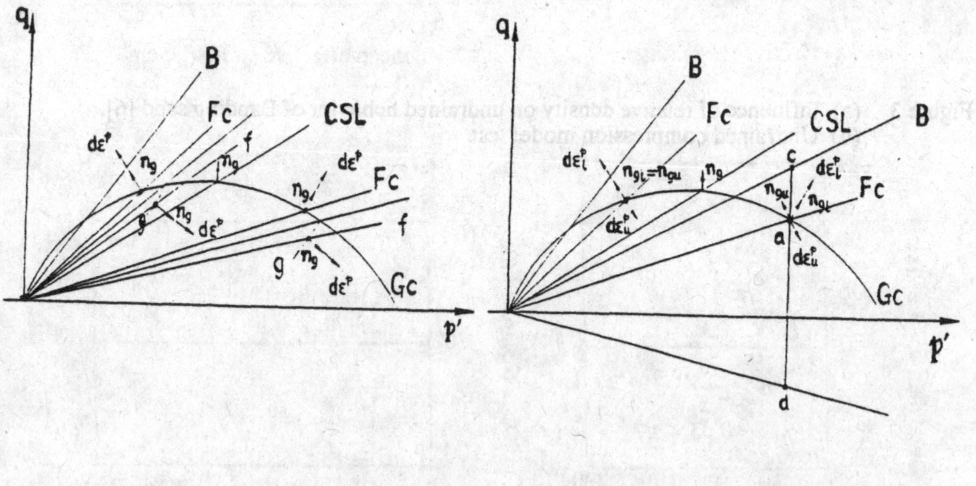


Figure 1 (a) Two-surface model with reflecting plastic potential [4].
(b) Simple model for sands in the triaxial plane [3].

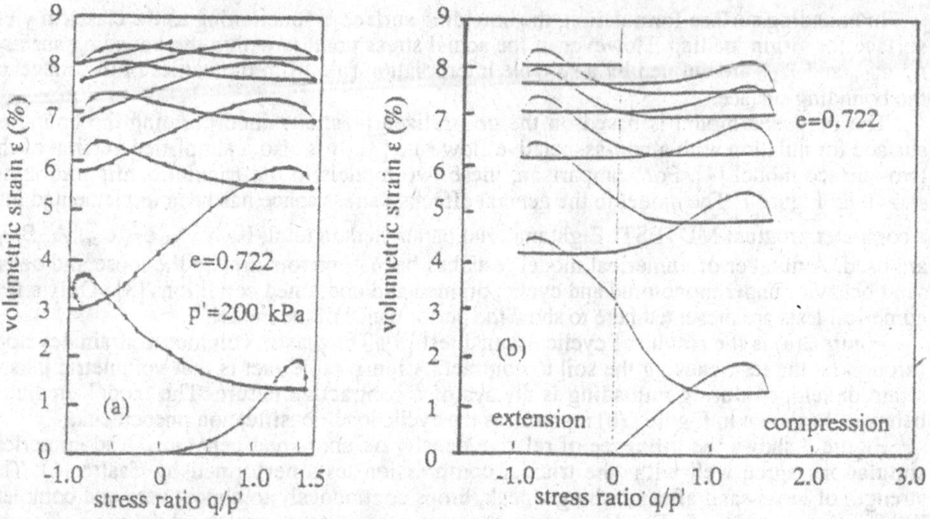


Figure 2 (a) Drained test on Fuji river sand [5].
(b) Cyclic drained triaxial model test ($\sigma_r$ = 200 kPa).

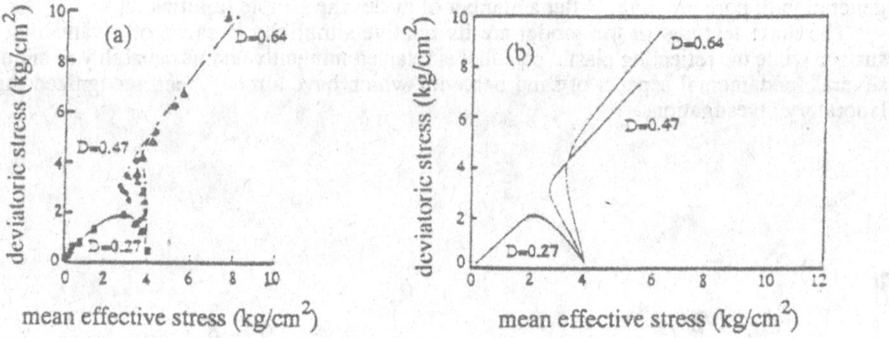


Figure 3 (a) Influence of relative density on undrained behavior of Banding sand [6].
(b) Undrained compression model test.

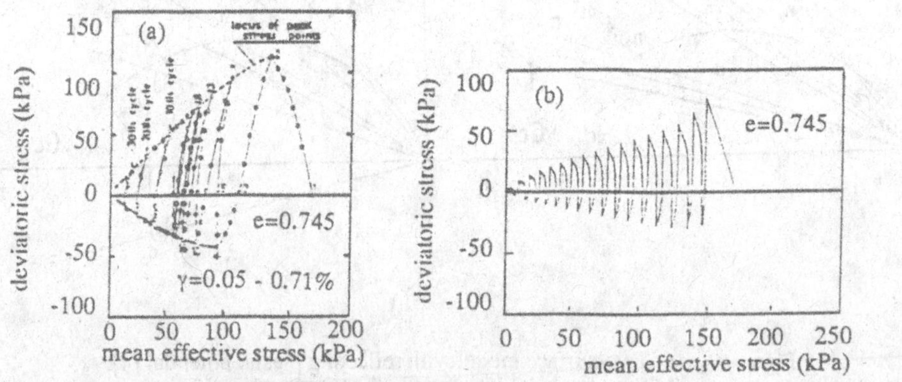


Figure 4 (a) Strain controlled undrained test results [7].
(b) Strain controlled cyclic undrained model test.

## RESPONSE OF SEAFLOOR DEPOSITS TO TRAVELLING WAVES

The sea floor deposit is assumed to be a saturated porous medium. The coupled governing equations are derived from the theory of mixture. The proposed constitutive model for sands has been implemented into a computer program PCPTRAV for study the travelling wave induced response. An artificial boundary is suggested to deal with the unbounded domain problem in the two phase coupled nonlinear finite element analysis. The seafloor is considered as being subjected to an ocean wave induced plane pressure wave travelling across the surface of the underlying deposits.

Two example cases are studied. The first example examines the seabed response to the wave train at a pipeline site in Lake Ontario where the deposit was measured to be loose and medium dense [8]. The results show decreasing of the mean and vertical effective stresses along with the wave propogation while the pore water pressure is increasing. The horizontal displacement increases noticeably along the direction of wave travel when liquefaction. This indicates a loss of lateral resistance of the soil system. The numerical analyses predict the evidence of liquefaction for both loose sand and medium dense sand in the circumstances of Lake Ontario [3],where liquefaction did occur. A single 3.05 m (10 ft) diameter steel pipeline failed several times, the buried pipeline floated to the surface, apparently because of liquefaction [8].

The second example investigates the seafloor behavior under a storm wave at the Ekofisk tank site in North Sea where the deposit was reported to be dense [9]. The analyses predict that the stress path is almost overlapping after 35 wave periods, indicating a situation generally referred to as cyclic mobility. There is no evidence of liquefaction in North Sea dense sand environment from numerical results [3], where the Ekofisk oil storage tank has successfully withstood 100-year design storm waves without experiencing wave-induced liquefaction [9]. The numerical analyses predict the susceptibility to liquefaction which reflects the realities.

## REFERENCES


1. Dafalias, Y. F. and Popov, E. P., A model of nonlinearly hardening materials for complex loading, Acta Mechanica, 1975, Vol. 21, pp. 173-192.

2. Zienkiewicz, O. C. and Mroz, Z., Generalized plasticity formulation and applications to geomechanics, in Mechanics of Engineering Materials, Editor, C. S. Desai and R. Gallagher, Wiley, Ch. 33., 1984, pp. 655-679.

3. Yang, Q. S., Wave Induced response of Sea Floor Deposites: A Simple Model for Sands and Nonlinear Analysis by FEM, Ph. D. thesis, Concordia University, Montreal, 1990.

4. Poorooshasb, H. B. and Pietruszczak, S., On yielding flow of sand; a generalized two-surface model, Computers and Geotechnics, 1985, Vol. 1, pp. 33-58.

5. Tatsuoka, F. and Ishihara, K., Drained deformation of sand under cyclic stress reversing direction, Soils and Foundations, 1974, Vol. 14, No. 3, pp. 51-65.

6. Castro, G., Liquefaction of Sand, Ph.D. thesis, Havard University, Cambridge, 1969.

7. Ishihara.,K., Tatsuoka, F. and Yasuda, S., Undrained deformation and liquefaction of sand under cyclic stresses, Soils and Foundations, 1975,Vol. 15, No. 1, pp. 29-44.

8. Christian, J. T., Taylor, P. K., Yen, J. K. C. and David, R. E., Large Diameter Underwater Pipeline for Nuclear Power Plant Designed Against Soil Liquefaction, Proceedings, Sixth Annual Offshore Technology Conference, Houston, Taxas, Paper No. OTC 2094, May, 1974, pp. 597-606.

9. Lee, K. L., and Focht, J. A., Liquefaction potential of Ekofisk Tank in North Sea, Journal of the Geotechnical Engineering Division, ASCE, 1975, Vol. 100, No. GT1, pp. 1-18.

# FINITE PLASTICITY

# PLASTIC DUAL POTENTIAL FOR METALS AND APPLICATION TO MINIMUM PLASTIC WORK PATH CALCULATIONS

F. BARLAT, K. CHUNG AND O. RICHMOND
Alcoa Laboratories, Alcoa Center, PA 15069, USA.

## ABSTRACT

In this work, an equation that defines the dual potential for plastically deforming metals is proposed. This potential can be identified with the work effective strain rate. The particular case of an isotropic FCC polycrystal is examined but extension to the case of anisotropic materials is outlined. Associated with a work-hardening curve, this equation completely describes the plastic behavior of metals. This definition is extremely useful for the calculation of minimum plastic work path, as illustrated by pure shear and simple shear cases.

## PLASTIC DUAL POTENTIAL

Calculations of minimum plastic work paths provide a useful basis for both design and analysis of metal forming processes. These calculations can be largely simplified when an explicit expression of the effective strain rate is known. This expression can be obtained from constitutive equations. Polycrystal models can be used to describe the plastic behavior of metals, but they do not provide a definition of the effective strain rate. Recently, Barlat et al. [1] proposed phenomenological yield functions that give an analytical description of the yield surface of textured polycrystals. However, these functions are expressed in stress space and their mathematical forms are not simple enough to give an explicit expression of the effective strain rate. Ziegler [2] and Hill [3] have shown that, based on the work-equivalent strain differential, a meaningful plastic dual potential can be associated to any convex yield function. Fortunier [4] has introduced the "flow polyhedron" as the dual potential of an FCC single crystal. Arminjon and Bacroix [5] have given an analytical description of the dual potential of BCC polycrystals with quadratic functions.

In this work, an analytical expression of the plastic dual potential for an FCC polycrystal is proposed. This potential $\psi$ can be identified with the work effective plastic strain rate $\dot{\bar{\varepsilon}}$:

$$\Psi = \left\{ \left(\frac{M}{3}\right)^{\mu} \frac{1}{1+2^{1-\mu}} \left( |2\dot{\varepsilon}_1 - \dot{\varepsilon}_2 - \dot{\varepsilon}_3|^{\mu} + |2\dot{\varepsilon}_2 - \dot{\varepsilon}_3 - \dot{\varepsilon}_1|^{\mu} + |2\dot{\varepsilon}_3 - \dot{\varepsilon}_1 - \dot{\varepsilon}_2|^{\mu} \right) \right\}^{1/\mu} = \dot{\bar{\varepsilon}} \quad (1)$$

where $\dot{\varepsilon}_i$ are principal strain rates, $\mu$ is a material coefficient and $M$ is a proportionality constant. The principal deviatoric stresses are generated by the gradient of the potential $\psi$. Using plastic incompressibility, the stresses can be expressed as follows:

$$S_i = \frac{\partial \Psi}{\partial \dot{\varepsilon}_i} = \frac{M^{\mu}}{3\left(1+2^{1-\mu}\right)}\left[2sign(\dot{\varepsilon}_i)|\dot{\varepsilon}_i|^{\mu-1} - sign(\dot{\varepsilon}_j)|\dot{\varepsilon}_j|^{\mu-1} - sign(\dot{\varepsilon}_k)|\dot{\varepsilon}_k|^{\mu-1}\right]\psi^{\frac{1-\mu}{\mu}} \quad (2)$$

Figure 1 shows that, when $\mu$=4/3, this phenomenological formulation is in very good agreement with the Taylor [6] / Bishop and Hill constitutive model [7] for polycrystals. In this figure, $M$ was chosen as the Taylor factor for an isotropic FCC polycrystal subjected to uniaxial tension ($M$=3.06) and the dual potential was represented for $\psi$=1.

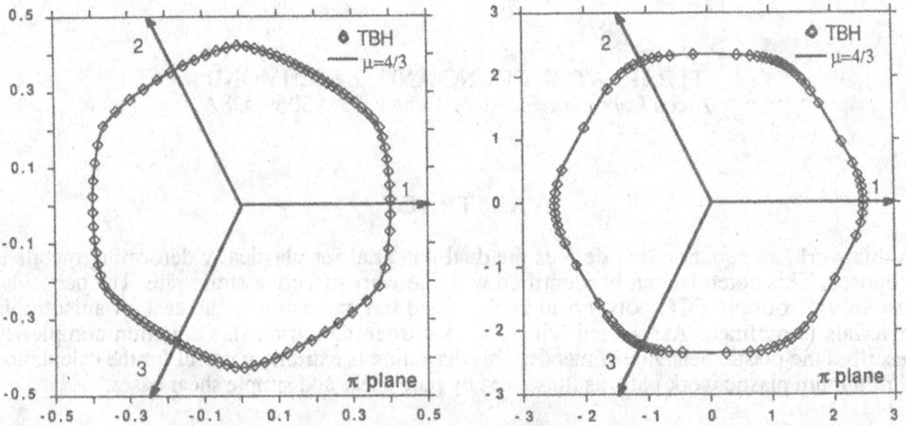


FIGURE 1. Plastic dual potential (left) and associated yield surface (right) calculated with the Taylor / Bishop and Hill polycrystal model and with Equations 1 and 2 for an isotropic FCC polycrystal.

Generalization of the phenomenological plastic dual potential for anisotropic materials exhibiting orthotropic symmetry can be obtained by considering the following matrix:

$$\mathbf{L} = \begin{bmatrix} \frac{c_3(\dot{\varepsilon}_{11}-\dot{\varepsilon}_{22})-c_2(\dot{\varepsilon}_{33}-\dot{\varepsilon}_{11})}{3} & c_6\dot{\varepsilon}_{12} & c_5\dot{\varepsilon}_{31} \\ c_6\dot{\varepsilon}_{12} & \frac{c_1(\dot{\varepsilon}_{22}-\dot{\varepsilon}_{33})-c_3(\dot{\varepsilon}_{11}-\dot{\varepsilon}_{22})}{3} & c_4\dot{\varepsilon}_{23} \\ c_5\dot{\varepsilon}_{31} & c_4\dot{\varepsilon}_{23} & \frac{c_2(\dot{\varepsilon}_{33}-\dot{\varepsilon}_{11})-c_1(\dot{\varepsilon}_{22}-\dot{\varepsilon}_{33})}{3} \end{bmatrix} \quad (3)$$

The six material coefficients $c_i$ are expressed in a given reference frame attached to the material. When the coefficients are equal to unity, the matrix reduces to the strain rate tensor. The eigenvalues of the matrix can be calculated and substituted in Eq. 1 and 2. When the coefficients are not simultaneously equal to one, they characterize the anisotropy of the material. Finding the eigenvalues of this matrix **L** and substituting in the phenomenological yield function leads to an expression of the plastic dual potential for anisotropic materials.

# MINIMUM PLASTIC WORK AND ITS APPLICATION

A deformation path to achieve a desired homogeneous deformation with a minimum amount of plastic work has been studied by Nadai [8] and Hill [9]. When this path corresponds to a path of minimum effective strain, as broadly assumed, the minimum work path is achieved when the principal material lines are fixed and, at the same time, the ratios of true strain rates along those principal material lines are constant. Later, Chung and Richmond [10] showed that, for such an optimum deformation path, the minimum effective strain can be easily calculated from the definition of the effective strain rate by replacing strain rate components with true strain components. Therefore, the minimum effective strain for the proposed dual potential in Eq. 1 becomes:

$$\bar{\varepsilon}_{min} = \left\{ \left(\frac{M}{3}\right)^{\mu} \frac{1}{1+2^{1-\mu}} \left( |2\varepsilon_1 - \varepsilon_2 - \varepsilon_3|^{\mu} + |2\varepsilon_2 - \varepsilon_3 - \varepsilon_1|^{\mu} + |2\varepsilon_3 - \varepsilon_1 - \varepsilon_2|^{\mu} \right) \right\}^{1/\mu} \quad (4)$$

where $\varepsilon_i$ are principal true strains. For the anisotropic materials proposed along with Eq. 3, $\varepsilon_i$ are the principal values of **L** whose strain rate components are replaced with true strain components. This simple relation between the minimum effective strain and effective strain rate is possible for general anisotropic materials that harden isotropically.

In Fig. 2, for demonstration purposes, minimum effective strains are calculated for the proposed isotropic materials that undergo simple shear deformation. The parameter x is the travel distance of a point originally located at (1,0). Then, the minimum effective strain, $\varepsilon_{min}$, which is the effective strain of pure shear in this case, and the effective strain for simple shear deformation, $\varepsilon_{sim}$, become:

$$\bar{\varepsilon}_{min} = \kappa \cdot \ln\left(\frac{x+\sqrt{x^2+4}}{2}\right), \quad \bar{\varepsilon}_{sim} = \kappa \cdot \frac{x}{2} \quad \text{where} \quad \kappa = M.\left(0.5+2^{-\mu}\right)^{-\frac{1}{\mu}}, \quad (5)$$

respectively. These effective strains, normalized by $\kappa$, are plotted in Fig. 2. The proportionality parameter $\kappa$ is dependent on the material coefficient $\mu$, and this relation is also represented in Fig. 2. The result is valid for the proposed isotropic materials which include Von Mises $(\kappa = 2/\sqrt{3}\ ,\ \mu = 2\ ,\ M = 1)$ as well as Tresca materials $(\kappa = 1\ ,\ \mu = 1\ ,\ M = 1)$.

The minimum plastic work provides a useful basis for design as well as for analysis of metal forming processes, as summarized by Chung et al. [11]. For design purposes, the *ideal forming* theory has been developed by Chung and Richmond [12] based on the earlier work by Richmond [13]. In this design theory, optimum strain distributions and ideal forming processes are calculated assuming that materials deform in minimum work paths. This application, especially rigid-plastic formulation, involves the calculation of the plastic work gradient; i.e.,

$$\frac{\partial w}{\partial \mathbf{u}} = \bar{\sigma} \frac{\partial \bar{\varepsilon}}{\partial \mathbf{u}} \quad (6)$$

where w is the plastic work per unit volume when materials deform in minimum work paths; $\sigma$ and **u** are the effective stress and displacement, respectively. For analysis purposes, a whole process is discretized into a finite number of small forming increments for updated Lagrangian formulations, and then the minimum work path is applied for each increment. Eq. 6 is applied incrementally in such applications. Consequently, the proposed anisotropic effective strain rate and its minimum effective strain in analytical forms provide a convenient mathematical basis for minimum work based formulations used in the design and analysis of anisotropic materials.

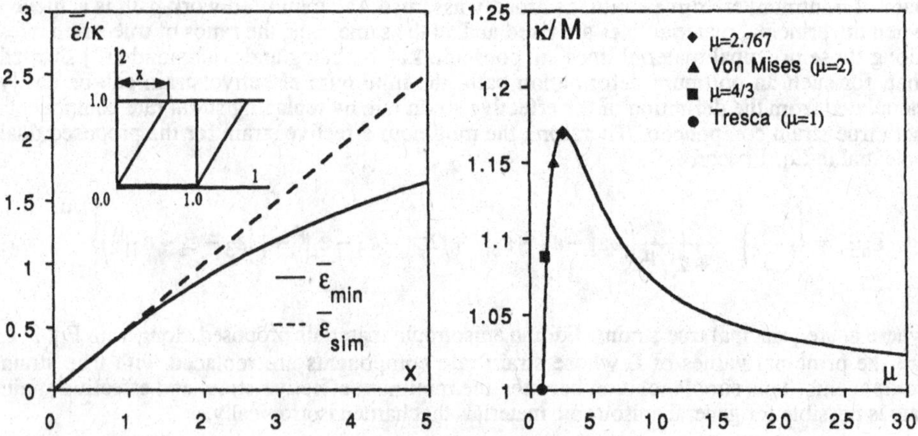


FIGURE 2. Comparison of effective strains between pure shear and simple shear (left) and dependence of the proportionality parameter κ on the material coefficient μ (right).

## REFERENCES


1. Barlat, F., Lege, D.J. and Brem, J.C. accepted for publication in Int. J. Plasticity, 1990.
2. Ziegler, H., An Introduction to Thermomechanics, North-Holland Publishing Company, Amsterdam, The Netherlands, 1977.
3. Hill, R., J. Mechanics and Physics of Solids, 1987, **35**, 22-33.
4. Fortunier, R., J. Mechanics and Physics of Solids, 1989, **37**, 779-790.
5. Arminjon, M. and Bacroix, B., accepted for publication in Acta Mechanica, 1990.
6. Taylor, G.I., J. Inst. Metals, 1938, **62**, 307-324.
7. Bishop, J.W.F. and Hill, R., Phil. Mag., 1951, **42**, 1298-1307.
8. Nadai, A., Theory of Flow and Fracture of Solids, McGraw-Hill, New York, 1963.
9. Hill, R., J. Mechanics and Physics of Solids, 1986, **6**, 1-8.
10. Chung, K. and Richmond, O., Unpublished work, 1990.
11. Chung, K., Richmond, O., Germain, Y. and Wagoner, R.H., NUMIFORM'89, Thompson, E.G. et al. eds, Belkema, Rotterdam, The Netherlands, 1989, 129-134.
12. Chung, K. and Richmond, O., Proc. Korea Sym. on Sci. & Tech., 1990, 2061-2065.
13. Richmond, O., Mechanics of Solid State, University of Toronto Press, 1963, 154-167.

# AN EXPERIMENTAL STUDY OF INTERNAL CONSTRAINTS BOUNDING ISOTROPIC AND ORTHOTROPIC DOMAINS FOR LARGE FINITE DEFORMATION IN CUBIC SINGLE CRYSTALS

JAMES F. BELL
The Johns Hopkins University
Baltimore, Maryland, USA

## ABSTRACT

Based upon a knowledge of all nine x-ray determined direction cosines between the crystal axes of cubic single crystals and the axial frame of rectangular specimens, a new kinematical analysis is described. The complex slip structure provides a general continuum theory and a specified bound separating domains of isotropic and orthotropic large finite deformation.

## INTRODUCTION

During a lecture before the Royal Society of London on February 23, 1923, G. I. Taylor described laboratory measurements that introduced a new field of study in the mechanics of solids. He measured all dimensions and angles between lateral and vertical sides and axes of a square cross-sectioned cubic single crystal at eight successive stages during elongation to fracture at an axial strain of 78.5%. From these data he identified the presence of an unstretched plane at large plastic deformation. On such a plane, he assumed that deformation occurs as simple shear resolvable in single slip in the direction of a specified face diagonal of the cube. Taylor and Elam found that such a resolved shear response function was reproducibly parabolic with **identical** numerical coefficients for tension or compression loading, for all initial crystallographic angles of the axis Z.

Thirty years later, when such tests were repeated upon zone-refined high purity cubic single crystals, it became evident that the stage III region of finite plastic deformation in which Taylor had assumed single slip was, in fact, a complex multiple slip domain.

Some years after this development, in order to address the implied dichotomy of indisputable global order found when calculations are made in single slip in a known complex slip domain, Nelson Hsu [4] extended Taylor and Elam's measurements to include the crystallographic location of the X and Y axes perpendicular to the sides of the rectangular specimen. His main findings were that in non-proportional loading, the initial complex of slips dominates even after a radical change in the direction of loading path, and that, in general, single slip does **not** characterize the response. In the present study, Hsu's x-ray data are re-phrased to provide the nine direction cosines relating the X, Y, Z specimen frame and the x, y, z crystal cubic frame. I have found that Taylor's unstretched plane provides a common, stable point of reference relating measured plastic deformation in the specimen frame to that in the cubic crystal axes frame. Projecting displacements in the X, Y, Z frame to the x, y, z frame, we obtain a unified **continuum** description of plastic deformation for cubic single crystals.

In what follows, $\eta$ is the angle between the specimen axis and the 101 direction for primary slip; $\phi$ is the angle between the specimen axis and a perpendicular to the unstretched plane. [The subscript o refers to initial values.] $\tau$ is Taylor's resolved shear stress in the 101 primary slip direction and $\gamma$ is Taylor's resolved shear strain. $E_Z$ is the nominal, axial tensile strain; $\sigma_Z$ is a corresponding engineering stress. For single slip,

$$\tau = \sigma_Z Cos\phi_o Cos\eta \tag{1}$$

$$\gamma = (Cos\eta / Cos\phi) - (Cos\eta_o / Cos\phi_o). \tag{2}$$

where $1 + E_Z = Sin\eta_o / Sin\eta = Cos\phi_o / Cos\phi.$ (3)

Unlike Taylor in 1923 [1] or Bell and Green in 1967 [2] who made x-ray measurements of $\eta$ and $\phi$ while the deformation is in progress, in the overwhelming majority of studies $\eta$ and $\phi$ for eqn (2) are obtained solely from knowledge of $E_Z$ and the initial angles $\eta_o$ and $\phi_o$ in eqn (3). Figure 1 shows Taylor's shear strain $\gamma$ when calculated using only initial values in eqn (3) (crosses). In this test #1027, circles are for a calculation where $\eta$ and $\phi$ are determined by x-ray diffraction Laue patterns as $E_Z$ increases. More important than the agreement between the two types of calculations, first reported by Bell and Green in 1967 [2], is the present discovery that the determination of Taylor's resolved shear strain $\gamma$ in cubic single crystals does **not require a knowledge of any crystallographic angles !** That is:

$$\gamma = 2\, E_Z. \tag{4}$$

That this is a general result is exhibited by the 82 tests of Fig. 2. The initial location of the Z axis covers the primary triangle. Each circle in Fig. 2(a) and 2(b) represents the final strain of an individual test in Hsu's study. In Fig. 2(a) are 23 proportional loading tests (axial tension). In Fig. 2(b) are 42 non-proportional tests (lateral compression followed by axial tension). The solid lines correspond to eqn (4). Figure 2(c) has a 99.99% purity test (#823) [9] with x-ray measured $\eta$ and $\phi$ to compare with the 99.6% purity test of Fig. 1. Figure 2(d) shows 9 tension tests of Bell and Green [2]. Figure 2(e) compares 8 tests of Bell and 4 of Taylor. Tests in both (d) and (e) proceed to a sufficiently large strain to reveal that eqn (4) is bounded at $\gamma_K = 71\%$. K divides domains of isotropic and orthotropic plasticity.

## KINEMATICS FOR THE ISOTROPIC DOMAIN

The angle between the 101 primary diagonal of the cube and the 100 or z direction is 45° while the angle between this z axis and the perpendicular to Taylor's unstretched plane is 54.7°. Since the crystal structure is preserved, the projection of Taylor's resolved shear stress and strain to the z or 100 crystal axis is Cos 54.7° × Cos 45° = $1/\sqrt{6}$. Two previously unexplained laboratory generalizations described in a monograph [3] and in numerous papers since the 1960's, become pertinent at this point:

$$\tau = (2/3)^r \beta_s(\gamma)^{1/2} \text{ and } T/\tau = \sqrt{6} = \gamma/\Gamma \text{ where } T = \sqrt{trace\, S^2} \text{ and } \Gamma = \sqrt{trace\, E^2}, \quad (5)$$

with the total stress **S** the deviatoric component of a stress $\boldsymbol{\sigma}$ referred to increments of material points, and **E** = **V** - **I** the strain component of the stretch **V** on the left side of the polar decomposition theorem. (See Bell [6] and particularly Bell [7] for details.)

Thus for the projection from the X, Y, Z frame to the x, y, z crystal frame,

$$\Gamma = \sqrt{2/3}\, E_Z \quad \text{and} \quad T = \sqrt{6}\, \sigma_Z \operatorname{Cos} \phi_o \operatorname{Cos} \eta\, . \quad (6)$$

From the nine direction cosines relating the X, Y, Z specimen frame with the x, y, z crystal frame, with z the edge of the cube making the smallest angle with the specimen axis Z, we determine the projected direction on the X-Y specimen plane of the 011 diagonal on the 001-010 crystal plane orthogonal to z [lines ending in arrows]. The direction of the maximum displacement, $\sqrt{(\Delta X)^2 + (\Delta Y)^2}$, occurs at $\psi = \tan^{-1} \Delta Y/\Delta X$ [lines ending in crosses]. For the proportional loading tests of Fig. 3(a), $\psi = \tan^{-1} \Delta Y/\Delta X$ coincides with the projected 011 diagonal. We find that the long known rotation of the specimen axis toward the 101 primary direction is accompanied by a counterclockwise rotation of the specimen plane. An example is shown in Fig. 3(c) that I have extracted from Taylor's "original test." In that same test Taylor recorded $\lambda$, the change of angle of the sides in the X-Y plane for each strain $E_Z$, from which, in 1991, I provide the data, Fig. 3(d). The rotation of the specimen plane is related to the axial strain by $2 \tan(\pi/2 - \lambda) = E_Z$.

Since direction cosines also provide projections of the 001 and 010 crystal axes on the X-Y plane [dashed lines, Fig. 3(a)], we may determine the corresponding strains from the expression $\Delta Z = \sqrt{2[(\Delta X)^2 + (\Delta Y)^2]}$ obtained from the averages of displacements. These measurements of 65 tests provide an internal constraint for isotropic deformation in cubic single crystals, the same as that measured for finite strain in polycrystals:

$$\text{trace } \mathbf{V} = 3 \quad \text{or} \quad \text{trace } \mathbf{E} = 0. \quad (7)$$

The coincidence of diagonals requires that the deformation in the crystal frame must be two-dimensional. More specifically, the measured internal constraint trace **V** = 3 requires that the response be **two-dimensional tension** in the 100-011 plane of the crystal cube. For the strain $E_1$ in the 100 or z direction we have $E_3 = -E_1$ in the 011 diagonal direction where $E_2 = 0$ in the direction perpendicular to the 100-011 plane. It follows that the deviatoric stress components for $\sigma_1$, $\sigma_2 = \sigma_1/2$, and $\sigma_3 = 0$ are $S_1 = \sigma_1/2$, $S_3 = -\sigma_1/2$ and $S_2 = 0$. For details see the parallel study of two-dimensional compression for a version of Bridgman's experiment of 1946 introduced by Bell and Florenz in 1969, and to measure all strains, further modified by Bell in 1982 [6]. For two-dimensional tension we have

$$dE_1 = 2S_1 dT/(2/3)^r \beta_o^2; \quad dE_2 = 0; \quad \text{and} \quad dE_3 = -2S_1 dT/(2/3)^r \beta_o^2 \tag{8}$$

where $\beta = (2/3)^{r/2}\beta_o$ are known material constants in a quantum structure for which the delineation of the source is part of the present study. (See BELL [3].) Tests [straight lines ending in circles in Fig. 3(b)] in which specimens were first compressed 3% to 5% in either the X or Y direction and then pulled in tension to large strain, have maximum displacements along an X or Y axis that indicate a change in the complex slip structure. Hsu grew his crystals in gangs, hence proportional and non-proportional tests can be compared for identical initial orientations. Leaving details to a forum where space is not dominant, we note here that integral changes in the mode index r are associated with this relocation of the maximum displacement.

## KINEMATICS FOR THE ORTHOTROPIC DOMAIN

Referring to the region $\gamma \geq 71\%$ ($E_{ZK} = 36\%$) in Fig. 2, eqn (4) is replaced by

$\gamma = \sqrt{2}\, E_Z + 0.208$ from which eqns (6) become (9)

$$\Gamma = \gamma/\sqrt{6} = [\, E_Z/\sqrt{3}] + 0.085\ ; \quad T = \sqrt{6}\, \sigma_Z \operatorname{Cos} \phi_o \operatorname{Cos} \eta\ . \tag{10}$$

Since $E_2 = 0$, the internal constraint of eqn (7) for the isotropic domain is, in incremental form, $dE_3 = -\,dE_1$. Analysis of the orthotropic domain for single crystals in Fig. 2(d) and 2(e), for $\gamma \geq 71\%$, provides the corresponding internal constraint $dE_3 = -\,(2/3)^2 dE_1$. Integrating, we have

$$E_3 = -\,(2/3)^{\Delta r}\, E_1 - 0.114 \quad \text{where } \Delta r = 2 \tag{11}$$

for substitution in the expression for generalized strain $\Gamma = \sqrt{E_1^2 + E_3^2}$.

For the two-dimensional compression in polycrystals studied earlier, in the orthotropic domain the two non-zero strain components also have mode indices that differ by $\Delta r = 2$. Thus the statements of eqn (8) for the isotropic domain become eqn (12) for the orthotropic domain:

$$dE_1 = 2\, S_1 dT/(2/3)^r \beta_o^2; \quad dE_2 = 0; \quad \text{and} \quad dE_3 = -2S_1 dT/(2/3)^{(r-2)} \beta_o^2. \tag{12}$$

From the theory of constraints, in the isotropic domain the volume gradually decreases to 0.958 at $\gamma_K = 71\%$. In the orthotropic domain the volume gradually recovers to a maximum of 0.996 at $E_1 = 50\%$ before again decreasing. I return now to Taylor and Elam's four tests on aluminum in the 1920's, the first of hundreds of such tests that include both isotropic and orthotropic domains. In Fig. 4, $\sigma_1^2$ vs $E_1$ and $\sigma_1^2$ vs $E_3$ plots are compared with eqns (8) and (12) (straight lines). To emphasize the generality of these results for cubic single crystals, Fig. 5 shows $E_i$ vs $E_j$ results for two-dimensional tension (circles) compared with results for a purely kinematical analysis of earlier measurements in two-dimensional compression. In Fig. 5 are data of Bell [6] (crosses) and of Khan and Wang [8] (triangles). In repeating my experiments of 1969, 1982, and 1988, Khan and Wang's strains are higher than mine or Bridgman's.

## REFERENCES


1. Taylor, G. T. & Elam, C. F., The Distortion of an Aluminium Crystal During a Tensile test, Taylor's Papers, 1923, **I**, 5, (1958) (Cambridge U.)
2. Bell, J. F. and Green, R. E., An Experimental Study of the Double Slip Deformation Hypothesis for Face-Centred Cubic Single Crystals, Phil. Mag., 1967, **15**, 469-476.
3. Bell, J. F., The Physics of Large Deformation of Crystalline Solids, Springer Tracts in Natural Philosophy, **14**, (1968) Springer.
4. Hsu, N. N-H., Experimental Studies of Latent Work Hardening of Aluminum Single Crystals, Ph.D Dissertation, 1969, (Johns Hopkins U., Baltimore, Md., USA).
5. Bell, J. F., The Experimental Foundations of Solid Mechanics, Handbuch der Physik, 1973, **VIa/1**, pp. 1-811, Springer. Reprinted (1984) Mechanics of Solids, **I.** Also Russian Translation (1984), NAUKA.
6. Bell, J. F., Plane Stress, Plane Strain, and Pure Shear at Large Finite Strain, Internat'l J Plasticity, 1988, **4,** No. 2, 127-148.
7. Bell, J. F., Material Objectivity in an Experimentally Based Incremental Theory of Large Finite Plastic Strain, Internat'l J Plasticity, 1990, **6,** No. 3, 293-314.
8. Khan, A. S. and Wang, X., An Experimental Study of Large Finite Plastic Deformation in Annealed 1100 Aluminum During Proportional and Nonproportional Biaxial Compression, Internat'l J Plasticity, 1990, **6,** No. 4, 485-504.
9. Bell, J. F., A Generalized Large Deformation Behaviour for Face-Centred Cubic Solids - High Purity Copper, Phil. Mag. 1964, **10**, 107-126.

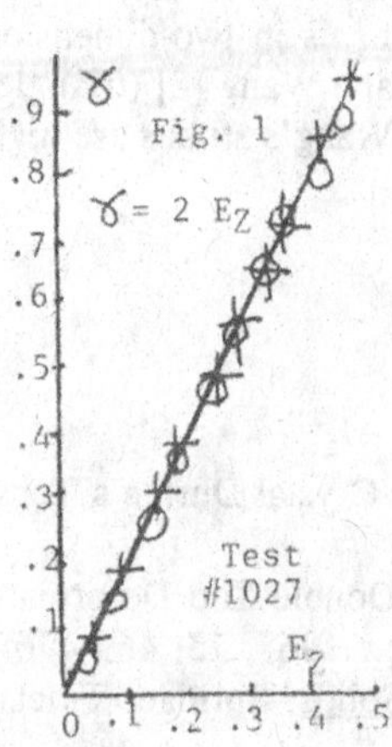
Fig. 1
γ = 2 E_Z
Test
#1027
E_Z
0 .1 .2 .3 .4 .5

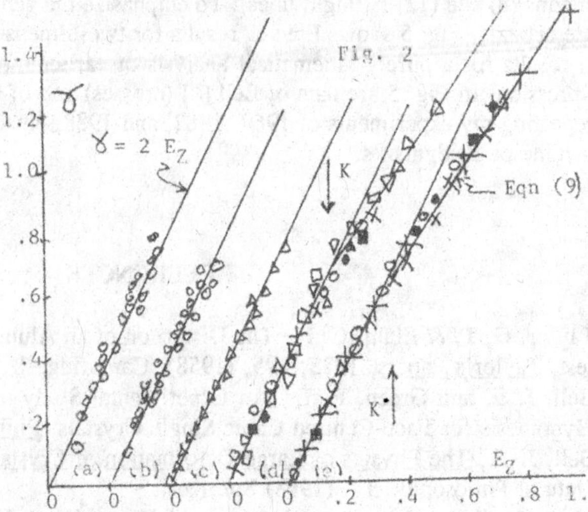
Fig. 2
γ
γ = 2 E_Z
K
Eqn (9)
K
(a) (b) (c) (d) (e)
E_Z

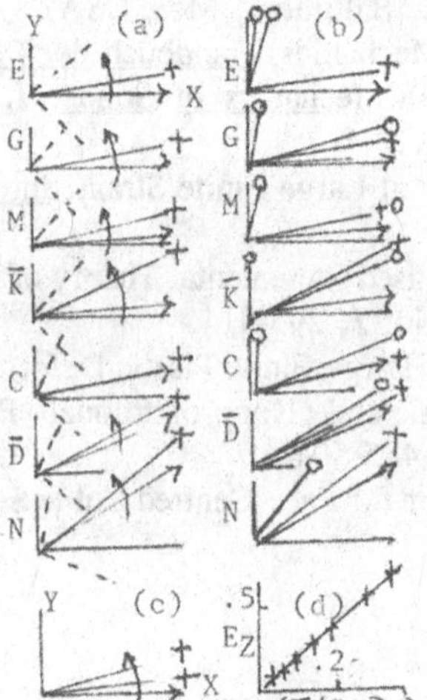
(a)
(b)
(c)
(d)
E_Z
tan(π/2-α)

Fig. 3

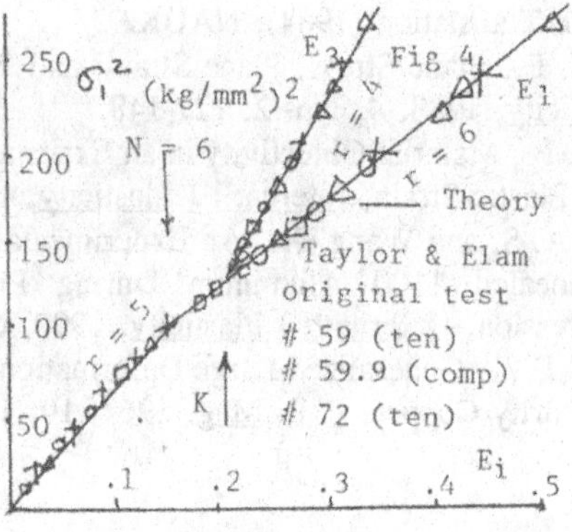
Fig 4
σ_1^2 (kg/mm^2)^2
N = 6
Theory
Taylor & Elam
original test
# 59 (ten)
# 59.9 (comp)
# 72 (ten)
K
E_i

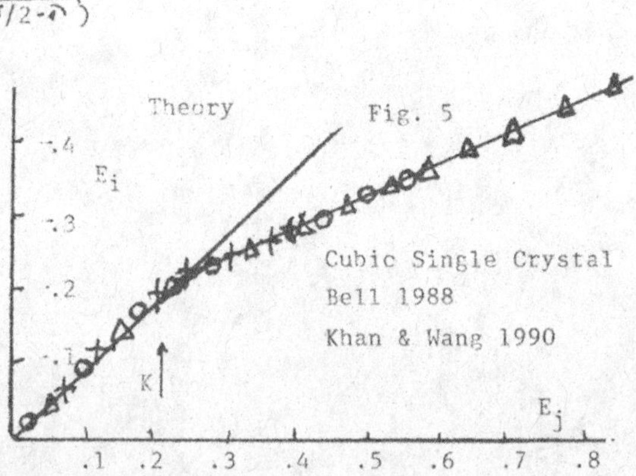
Theory
Fig. 5
E_i
Cubic Single Crystal
Bell 1988
Khan & Wang 1990
K
E_j

# RECENT DEVELOPMENTS IN FINITE RIGID PLASTICITY

J. CASEY
Department of Mechanical Engineering
University of California at Berkeley
Berkeley, CA 94720 , USA

## ABSTRACT

In the context of a strain-space formulation of finite plasticity, a constitutive theory is presented for rate-independent rigid-plastic materials. Both Eulerian and Lagrangian descriptions of the theory are discussed, and the nature of strain-hardening behavior is analyzed. Arbitrary objective rates are admitted. For special constitutive equations, a number of classical results, including the Hencky theorems, hold even at finite deformations.

## INTRODUCTION

An extensive classical literature on rigid plasticity exists, but it is only in recent years that finite kinematics, invariance requirements, and general constitutive relations have been incorporated into the subject. The present development grew out of the constitutive theory proposed in 1965 by Green and Naghdi [1] for finitely deforming elastic-plastic materials. The stress-space formulation utilized in [1] was replaced in [2] by a more satisfactory formulation relative to strain space. The latter leads in a natural way to the characterization of strain-hardening behavior [3]. Casey [4] developed a theory of finitely deforming rigid-plastic materials, by taking the rigid-plastic limits of elastic-plastic constitutive equations, and by then using these limits to motivate a set of *a priori* constitute relations for the rigid-plastic case. A Lagrangian strain-space formulation was employed in [4]. The equivalent Eulerian formulation was presented in [5], where the question of the choice of objective rates for stress and back-stress was treated in detail. Casey and Naghdi [5] showed that the theory can be formulated in such a way that it is form-invariant under arbitrary transformations of objective rates (see also [6], where a general rate-type constitutive equation is discussed). Casey and Chan [7] have considered special rigid-plastic materials in order to study strain-hardening behavior in more concrete terms, and have also given new proofs of the Hencky theorems.

## BASIC EQUATIONS

Let **X** and **x** denote the reference and current position vectors, respectively, of a particle in a three-dimensional deformable continuum, and represent the motion of the body by the smooth mapping $\mathbf{x} = \chi(\mathbf{X},t)$, where t is the current time. Define the fields

$$\mathbf{v} = \frac{\partial \chi}{\partial t} = \dot{\mathbf{x}}\,,\ \mathbf{F} = \frac{\partial \chi}{\partial \mathbf{X}}\,,\ J = \det\mathbf{F}\ (>0),$$
$$\mathbf{E} = \frac{1}{2}(\mathbf{F}^T\mathbf{F} - \mathbf{I})\,,\ \mathbf{L} = \frac{\partial \mathbf{v}}{\partial \mathbf{x}}\,,\ \mathbf{D} = \frac{1}{2}(\mathbf{L} + \mathbf{L}^T)\,, \qquad (1)$$

where **I** is the identity tensor, and observe that

$$\dot{\mathbf{E}} = \mathbf{F}^T\mathbf{D}\mathbf{F} \ . \tag{2}$$

Let **T** and **S**, respectively, denote the Cauchy and symmetric Piola-Kirchhoff stress tensors, and note that they are related through the Piola transformation

$$\mathbf{T} = \pi\{\mathbf{S}\} = \frac{1}{J}\,\mathbf{F}\mathbf{S}\mathbf{F}^T \ . \tag{3}$$

Under superposed rigid motions (s.r.b.m.) of the continuum, $\mathbf{F},\mathbf{E},\dot{\mathbf{E}},\mathbf{D},\mathbf{S},\dot{\mathbf{S}}$, and **T** are transformed objectively, i.e.

$$\begin{gathered}\mathbf{F}^+ = \mathbf{Q}\mathbf{F}\,,\ \mathbf{E}^+ = \mathbf{E},\ \dot{\mathbf{E}}^+ = \dot{\mathbf{E}},\ \mathbf{D}^+ = \mathbf{Q}\mathbf{D}\mathbf{Q}^T,\\ \mathbf{S}^+ = \mathbf{S},\ \dot{\mathbf{S}}^+ = \dot{\mathbf{S}},\ \mathbf{T}^+ = \mathbf{Q}\mathbf{T}\mathbf{Q}^T,\end{gathered} \tag{4}$$

where **Q** is a time-dependent proper orthogonal tensor representing the rotation in the superposed motion. The material time derivative of **T** is not objective, but a particular objective rate of **T** can be found by applying a Piola transformation to the material derivative of **S**:

$$\overset{t}{\mathbf{T}} = \pi\{\dot{\mathbf{S}}\} = \dot{\mathbf{T}} - \mathbf{L}\mathbf{T} - \mathbf{T}\mathbf{L}^T + \mathbf{T}\mathrm{tr}\mathbf{D},\ \overset{t}{\mathbf{T}}{}^+ = \mathbf{Q}\overset{t}{\mathbf{T}}\mathbf{Q}^T. \tag{5}$$

$\overset{t}{\mathbf{T}}$ is called the Truesdell rate of **T** . An arbitrary objective rate can be generated from $\overset{t}{\mathbf{T}}$ by an equation of the form [5]

$$\overset{a}{\mathbf{T}} = \overset{t}{\mathbf{T}} + B^{a,t}(\mathbf{T})[\mathbf{D}], \tag{6}$$

where $B^{a,t}$ is a fourth-order tensor which depends on **F**, **T** (and possibly other variables in the case of plasticity) and which acts linearly on **D** .

## CONSTITUTIVE THEORY : LAGRANGIAN FORM

For finitely deforming rigid-plastic materials, the plastic strain tensor is coincident with the Lagrangian strain tensor **E**, defined in $(1)_4$. As additional variables, we admit a scalar work-hardening parameter $\kappa$ and a back-stress tensor $\alpha_R$ belonging to the space of symmetric Piola-Kirchhoff stress tensors. We assume the existence of a yield function $f(\mathbf{S},\mathbf{E},\alpha_R,\kappa)$, such that for fixed values of $(\mathbf{E},\alpha_R,\kappa)$, the equation

$$f(\mathbf{S},\mathbf{E},\alpha_R,\kappa) = 0 \tag{7}$$

describes an orientable yield surface that separates stress space into two disjoint parts (the yield surface may be closed, but it is not necessarily so). Points in the region $f < 0$ are called *preplastic*, while points on the yield surface are called *plastic*. Loading criteria are defined by the strain-space conditions

$$\text{(a) } \dot{\mathbf{E}} = \mathbf{0} : \text{non–loading ; (b) } \dot{\mathbf{E}} \neq \mathbf{0} : \text{loading} . \tag{8}$$

It is assumed that during non-loading $\dot{\kappa} = 0$ and $\dot{\alpha}_R = \mathbf{0}$, whereas during loading

$$f = 0,\ \dot{\mathbf{E}} = \gamma\rho,\ \dot{\alpha}_R = H_R[\dot{\mathbf{E}}] = \gamma\beta_R,\ \dot{\kappa} = \gamma\lambda\ , \tag{9}$$

where

$$\gamma = \|\dot{\mathbf{E}}\| = (\mathrm{tr}\dot{\mathbf{E}}^2)^{1/2},\ \beta_R = H_R[\rho], \tag{10}$$

and the constitutive functions $\rho, \lambda, H^R$ depend on the same variables as f. The tensor $\rho$ has unit magnitude.

As a consequence of the consistency condition of plasticity, $\dot{f} = 0$ during loading, and hence

$$\hat{f} = \gamma\Gamma \tag{11}$$

where

$$\hat{f} = \frac{\partial f}{\partial \mathbf{S}} \cdot \dot{\mathbf{S}},\ -\Gamma = \rho \cdot \frac{\partial f}{\partial \mathbf{E}} + \beta \cdot \frac{\partial f}{\partial \alpha_R} + \lambda \frac{\partial f}{\partial \kappa}\ . \tag{12}$$

The quantity $\hat{f}$ is a measure of the rate at which the yield surface is locally expanding or contracting during loading. The rigid-plastic material is said to be hardening, softening, or perfectly plastic according as

$$\Gamma > 0,\ \Gamma < 0,\ \Gamma = 0, \tag{13}$$

respectively. During loading in a region of hardening or softening behavior, the flow and hardening rules in (9) can be expressed in the stress-space forms

$$\dot{\mathbf{E}} = \frac{\hat{f}}{\Gamma}\rho,\ \dot{\alpha}_R = \frac{\hat{f}}{\Gamma}\beta_R,\ \dot{\kappa} = \frac{\hat{f}}{\Gamma}\lambda\ . \tag{14}$$

## CONSTITUTIVE THEORY : EULERIAN FORM

The Eulerian description of the constitutive theory can be constructed from the foregoing Lagrangian description by the following procedure.

We utilize the Piola transformation $\pi$ to obtain an Eulerian representation $\alpha$ for the back-stress. Thus,

$$\alpha = \pi\{\alpha_R\},\ \alpha^+ = \mathbf{Q}\alpha\mathbf{Q}^T\ . \tag{15}$$

We define a yield function $f_*$ through the identities

$$f(\mathbf{S},\mathbf{E},\alpha_R,\kappa) = f(J\mathbf{F}^{-1}\mathbf{T}\mathbf{F}^{-T}, \frac{1}{2}(\mathbf{F}^T\mathbf{F} - \mathbf{I}), J\mathbf{F}^{-1}\alpha\mathbf{F}^{-T}, \kappa)$$

$$= f_*(\mathbf{T},\mathbf{F},\alpha,\kappa) . \tag{16}$$

As for the loading criteria, it is clear from (8 a,b), $(1)_3$ and (2) that

$$\text{(a)}\ \mathbf{D} = \mathbf{0} \text{ for non–loading ; (b) } \mathbf{D} \neq \mathbf{0} \text{ for loading .} \tag{17}$$

Let

$$\gamma_* = ||\mathbf{D}|| . \tag{18}$$

The flow rule $(9)_2$ can then be written in the form

$$\mathbf{D} = \gamma_*\xi, \tag{19}$$

where

$$\xi = \mathbf{F}^{-T}\rho\mathbf{F}^{-1}/||\mathbf{F}^{-T}\rho\mathbf{F}^{-1}|| \tag{20}$$

is a tensor of unit magnitude and depends on the same variables as $f_*$ . Clearly,

$$\frac{\gamma_*}{\gamma} = ||\mathbf{F}^{-T}\rho\mathbf{F}^{-1}|| > 0 . \tag{21}$$

The hardening rule $(9)_4$ can be written as

$$\dot{\kappa} = \gamma_*(\frac{\gamma\lambda}{\gamma_*}) = \gamma_*\lambda_* . \tag{22}$$

It is easy to show that (11) and $(12)_1$ can be expressed in the Eulerian forms

$$\hat{f} = \gamma_*\Gamma_*, \quad \hat{f} = \frac{\partial f_*}{\partial \mathbf{T}} \cdot \overset{t}{\mathbf{T}}. \tag{23}$$

The characterization of strain-hardening behavior in terms of $\Gamma_*$ is equivalent to the characterization (13).

To obtain the Eulerian form of the evolution equation for back-stress, we first apply the Piola transformation to both sides of $(9)_3$ to obtain

$$\overset{t}{\alpha} = H[\mathbf{D}] = \pi\{H_R[\dot{\mathbf{E}}]\} . \tag{24}$$

It then follows that

$$\overset{t}{\alpha} = \gamma_* \beta \,, \tag{25}$$

with

$$\beta = H[\xi] \,. \tag{26}$$

Upon transformation to an arbitrary objective rate a, $(24)_1$, (25), and (26) are form-invariant:

$$\overset{a}{\alpha} = \overline{H}[\mathbf{D}] = \gamma_* \overline{\beta},\ \overline{\beta} = \overline{H}[\xi] \,. \tag{27}$$

The functions $\overline{H}$ and $\overline{\beta}$ depend on $\mathbf{T},\mathbf{F},\alpha,\kappa$ and also on the choice of objective rate.

A characterization of strain-hardening behavior can be given in terms of an arbitrary objective rate. Thus, if we let

$$\hat{f}^a = \frac{\partial f_*}{\partial \mathbf{T}} \cdot \overset{a}{\mathbf{T}} \,, \tag{28}$$

it can be deduced that

$$\frac{\hat{f}^a}{\gamma_*} = \Gamma^a \,, \tag{29}$$

where $\Gamma^a$ is rate-independent. We can define a-*hardening*, a-*softening*, and a-*perfectly plastic behavior* according as

$$\Gamma^a > 0 \,, \Gamma^a < 0 \,, \Gamma^a = 0 \,, \tag{30}$$

respectively. The previous characterizations in terms of $\Gamma$ and $\Gamma_*$ are relative to the choices of the material derivative, and its Eulerian counterpart, the Truesdell rate, as objective rates.

## ST. VENANT – LÉVY – MISES MATERIALS

An important special case of the foregoing constitutive equations is the classical St. Venant - Lévy - Mises theory, which may be obtained as follows. Let $\mathbf{T}$ be decomposed into spherical and deviatoric parts:

$$\mathbf{T} = -p\mathbf{I} + \tau \,,\ p = -\frac{1}{3}\mathrm{tr}\mathbf{T} \,. \tag{31}$$

Specify the response functions $\lambda_*$ and $\beta$ to be zero, and suppose that the initial value of $\kappa$ is $K^2>0$, and that the initial value of $\alpha$ is zero. Also, suppose the yield function $f_*$ is given by

$$f_* = \frac{1}{2}\tau \cdot \tau - K^2 \,, \tag{32}$$

and let $\xi$ satisfy the normality relation

$$\xi = \frac{\partial f_*}{\partial \mathbf{T}} / \| \frac{\partial f_*}{\partial \mathbf{T}} \| = \tau / \| \tau \| \ . \tag{33}$$

Then, the flow rule (19) becomes

$$\mathbf{D} = \gamma_* \tau / \sqrt{2}\, K \ . \tag{34}$$

The flow is isochoric. Note that deformation is not restricted to be infinitesimal, and that the strain measure is the finite strain tensor **E**.

For the above class of materials, the function $\Gamma_*$ reduces to

$$\Gamma_* = -2\mathrm{tr}(\tau^3 - p\tau^2)/\| \tau \| \ . \tag{35}$$

The state of hardening depends only on the stress tensor. The manner in which the function $\Gamma_*$ varies in stress space determines the strain-hardening topography of the material.

For plane deformations of the foregoing materials, new proofs of a number of classical results are contained in [7].

## REFERENCES


1. Green, A.E and Naghdi, P.M., A general theory of an elastic-plastic continuum. Arch. Rational Anal., 1965, **18**, 251-281.

2. Naghdi, P.M. and Trapp, J.A., The significance of formulating plasticity theory with reference to loading surfaces in strain space. Int. J. Engng. Sci. , 1975, **13**, 785-797.

3. Casey, J. and Naghdi, P.M., On the characterization of strain-hardening in plasticity. J. Appl. Mech. , 1981, **45**, 285-286.

4. Casey, J., On finitely deforming rigid-plastic materials. Int. J. Plasticity , 1986, **2**, 247-277.

5. Casey, J. and Naghdi, P.M., On the relationship between the Eulerian and Lagrangian descriptions of finite rigid plasticity. Arch. Rational Mech. Anal. , 1988, **102**, 351-375.

6. Casey, J. and Naghdi, P.M., Eulerian versus Lagrangian descriptions of rate-type constitutive theories. In Constitutive Laws for Engineering Materials , eds. C.S. Desai, E. Krempl, G. Frantziskonis, and H. Saadatmanesh, ASME Press, NewYork, 1991, pp. 15-20.

7. Casey, J. and Chan, Y., Special results in finite rigid plasticity. Submitted for publication.

# ON THE APPLICATIOM OF THE INTEGRAL CONSTITUTIVE EQUATION FOR THE DESCRIPTION OF ANISOTROPIC HARDENING IN PLASTICITY

KANGHUA CHEN and ZHEN-BANG KUANG

Xi'an Jiaotong University
Xi'an, Shaanxi Province, 710049, P.R.China

## ABSTRACT

A new kind of endochronic with generalized metric tensor in an integral constitutive equation is proposed in this paper. This constitutive equation may be adapted to describe the anisotropic hardening in plasticity, i.e. it may be uesd to describe the translation, uniform expansion or contraction, distortion of the subsequent yield surface. The validity of the present theory is confirmed by good agreement with some previous experimental observations.

## INTEGRAL CONSTITUTIVE EQUATION OF AN ELASTIC–PLASTIC MEDIUM

The constitutive equation of a simple material at constant temperature and small deformation may be written as follows:[1][2]

$$\sigma = \Phi(\mathbf{e}(t-\tau);\mathbf{e}(t)) \tag{1}$$

where $\sigma$ is Cauchy stress tensor, **e** is Euler strain tensor, $\tau$ is time and t is the current time, $-\infty < \tau < t$. For rate–independent materials we can use an arc length s in the plastic strain space to take the place of time $\tau$.[2][3] Thus equation(1) may be reduced to:

$$\sigma = \Phi(\mathbf{e}(S-s);\mathbf{e}(S)) \tag{2}$$

$$ds = (d\mathbf{e}^p:\mathbf{P}:d\mathbf{e}^p)^{\frac{1}{2}} \tag{3}$$

where $s = s(\tau)$ called endochronic and $S = s(t)$. $\mathbf{P} = \mathbf{P}(\sigma, s)$ is the generalized metric tensor, which is dependent on $\sigma$ in the plastic strain space. It is emphasized that though **P** may be written as $\mathbf{P} = \mathbf{P}(s)$ due to that $\sigma$ is the funtional of **e**, but we let **P** be the evident funtion of $\sigma$ due to convenience to investigate the subsequent yield function.

Under some appropriate assumptions the continuous functional $\Phi$, and thus $\sigma$, can be approximated by a sum of mutiple integrals of the form[3][4]

$$\sigma = \int_0^S \mathbf{H}^{(1)}(S-s):d\mathbf{e}^p(s) + \int_0^S\int_0^S d\mathbf{e}^p(s_1):\mathbf{H}^{(2)}(S-s_1, S-s_2):d\mathbf{e}^p(s_2) \tag{4}$$

where $\mathbf{H}^{(1)}$ *and* $\mathbf{H}^{(2)}$ are fouth and sixth order tensors respectively. If the material is initially isotropic, then the appropriate symmetry group is the orthogonal group; if the material is plastic incompressible and the deformation is small, then $\mathrm{tr}(\mathbf{e}^p) = 0$. In this case the equation(3) can be reduced to:

$$\sigma' = \int_0^S f_1(S-s)d\mathbf{e}^p(s) + \int_0^S\int_0^S f_2(S-s_1, S-s_2)d\mathbf{e}^p(s_1).d\mathbf{e}^p(s_2) \tag{5}$$

where $\sigma' = \sigma - tr\sigma/3$.

Following reference [4] we let:

$$\left.\begin{aligned} f_1(s) &= h_0\delta(s) + h_1(s) \\ f_2(s) &= \beta_1\delta(s_1,s_2) + \beta_2\delta(s_1) + \beta_3\delta(s_2) + \beta_4(s) \end{aligned}\right\} \tag{6}$$

where $\delta(s_1,s_2)$is a Kronecker $\delta$ funtion with two variables, $\delta(s)$ is an usual Kronecker $\delta$, $h_0$, $h_1$, $\beta_1$, $\beta_2$, $\beta_3$and $\beta_4$are regular funtions. To investigate the subsequent yeild funtion we let:

$$ds^2 = de^p : de^p / K^2(s,\sigma) \tag{7}$$

where $K^2(s,\sigma)$ represents the anisotropic hardening funtion. Substituting equations (6) and (7) into (5) we get:

$$\beta_1 \frac{de^p}{ds} \cdot \frac{de^p}{ds} + \frac{de^p}{ds} \cdot (h_0\mathbf{I} + \mathbf{R}_2) = \sigma' - \mathbf{R} \tag{8}$$

$$\left.\begin{aligned} \mathbf{R} &= \int_0^S h_1(S-s)de^p(s) + \int_0^S\int_0^S \beta_4(S-s_1,S-s_2)de^p(s_1)\cdot de^p(s_2) \\ \mathbf{R}_2 &= \int_0^S [\beta_2(S-s) + \beta_3(S-s)]de^p(s) \end{aligned}\right\} \tag{9}$$

Equation (8) is a more general and complicated non–linear elastic–plastic incremental constitutive equation. If we let $\beta_1 = 0$ in equation (8) then we can get:

$$F = \frac{de^p}{ds} : \frac{de^p}{ds} - K^2(s,\sigma') = \frac{\sigma' - \mathbf{R}}{h_0\mathbf{I} + \mathbf{R}_2} : \frac{\sigma' - \mathbf{R}}{h_0\mathbf{I} + \mathbf{R}_2} - K^2(s,\sigma') = 0 \tag{10}$$

The above equation represents an yield funtion, which may be applied to anisotropic hardening. In the most of cases we can neglect the second term, i.e. the term of double integral, in equation (4), and then (10) becomes:

$$F = \frac{1}{h_0^2}(\sigma' - \mathbf{R}):(\sigma' - \mathbf{R}) - K^2(s,\sigma') = 0 \tag{11}$$

where $\mathbf{R}$ is given in equation $(9)_1$with $\beta_4 = 0$. In the following discussions we only use (11).

From the experimental results [5–7] it is found that along a definite distorted direction. the subsequent yield surface is separated in two parts: on the forward part the surface has high curvature and on the rear part the surface becomes flate (Fig.1).

Let the distorted direction be $\mathbf{v}$:

$$\mathbf{v} = \frac{\mathbf{V}}{|\mathbf{V}|} \quad or \quad v_{ij} = \frac{V_{ij}}{(V_{kl}V_{kl})^{\frac{1}{2}}} \tag{12}$$

How to select $\mathbf{v}$ is an important problem and is not fully solved. Usually define a distorted parameter $\xi$:

$$\xi = \frac{\sigma' - \mathbf{R}}{h_0} : \frac{\mathbf{V}}{|\mathbf{V}|} = \frac{(\sigma'_{ij} - R_{ij})V_{ij}}{h_0(V_{kl}V_{kl})^{\frac{1}{2}}} \tag{13}$$

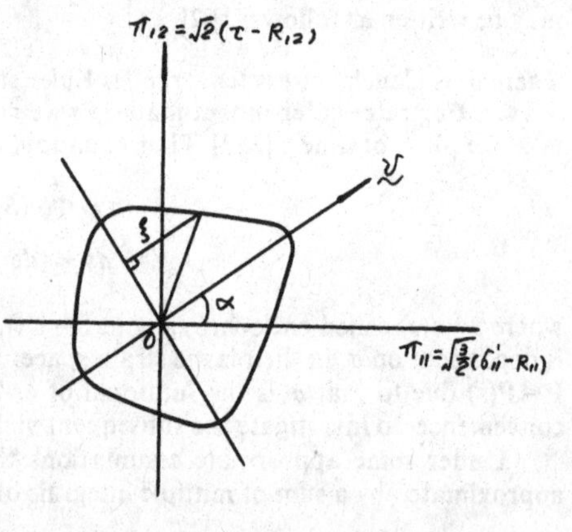


Fig.1 Distortion of yield surface

Let

$$K^2(s,\sigma') = B_0(s) - \sum_{n=1}^{N} B_n(s)\xi^{p_n} \tag{14}$$

$$B_0(0) = 1, \qquad \sum_{n=1}^{N} B_n(0) = 0 \tag{15}$$

So that from equations (11) and (14) we know that $h_0$ is the radius of the initial yield surface. $B_0(s)$, $B_n(s)$ and $p_n$ will be determined by experimental data. Substituting equation (14) into (11) we get:

$$F = \frac{1}{h_0^2}(\sigma' - \mathbf{R}):(\sigma' - \mathbf{R}) - B_0(s) + \sum_{n=1}^{N} B_n(s)\xi^{p_n} = 0 \quad (16)$$

The increment of endochronic is determined by equations (7) and (14).

## AN EXAMPLE

Phillips and Tang [5] had done experimental studies of yield surfaces of commercially pure aluminum at several temperatures. In their experiments the thin–walled tubes were loaded in combined tension and torsion. Their results showed that the subsequent yield surface due to prestressing is due to the superposition of a rigid body motion and a deformation in the direction of prestressing. The width of the yield curve in the direction of prestressing will decrease, but the width of yield curve perpendicular to the drection of prestressing will be constant. This result showed that $B_0(s)$ in equation (14) was equal to 1.

For the tension–torsion problem we have:

$$\left.\begin{aligned} &\sigma'_{11} = \frac{2}{3}\sigma, \quad \sigma'_{12} = \sigma_{12} = \tau, \quad \sigma'_{22} = \sigma'_{33} = -\frac{1}{3}\sigma, \quad \sigma'_{13} = \sigma'_{23} = 0 \\ &de^p_{11} = de^p, \quad de^p_{12} = d\gamma^p, \quad de^p_{22} = de^p_{33} = -\frac{1}{2}de^p, \quad de^p_{13} = de^p_{23} = 0 \\ &R_{11} \neq 0, \quad R_{12} \neq 0, \quad R_{22} = R_{33} = -\frac{1}{2}R_{11}, \quad R_{13} = R_{23} = 0 \end{aligned}\right\} \quad (17)$$

Let:

$$\pi_{11} = \sqrt{\frac{3}{2}}(\sigma'_{11} - R_{11}) = \sqrt{\frac{2}{3}}\sigma - \sqrt{\frac{3}{2}}R_{11}, \quad \pi_{12} = \sqrt{2}(\sigma_{12} - R_{12}) = \sqrt{2}(\tau - R_{12}) \quad (18)$$

where $\sigma$ is the tension stress and $\tau$ is shear stress.

In the following we will only discuss the case of proportional loading. In this case we let $V_{ij} = \pi_{ij}$. Assuming $\alpha$ is the angle between v and $\pi_{11}$. then we have:

$$\xi = \frac{1}{h_0}(\pi_{11} cos\alpha + \pi_{12} sin\alpha) \quad (19)$$

$$ds = \sqrt{\frac{3}{2}}[(de^p_{11})^2 + (de^p_{12})^2]^{\frac{1}{2}} / [B_0(s) - \sum_{n=1}^{N} B_n(s)\xi^{p_n}] \quad (20)$$

and

$$\left.\begin{aligned} R_{11} &= \int_0^S h_1(S-s)\frac{de^p_{11}}{ds}ds = cos\alpha\int_0^S h_1(S-s)[B_0(s) - \sum_{n=1}^{N} B_n(s)\xi^{p_n}]ds \\ R_{12} &= \int_0^S h_1(S-s)\frac{de^p_{12}}{ds}ds = sin\alpha\int_0^S h_1(S-s)[B_0(s) - \sum_{n=1}^{N} B_n(s)\xi^{p_n}]ds \end{aligned}\right\} \quad (21)$$

In deriving (21) the fact that the direction of $de^p$ coincides with $\pi$ was used. The yield funtion becomes:

$$F = \pi^2_{11} + \pi^2_{12} - h^2_0 B_0(s) + h^2_0 \sum_{n=1}^{N} B_n(s)\xi^{p_n} = 0 \quad (22)$$

For simplicity, we only discuss the pure tension problem. In this case we have:

$$\alpha = 0, \quad R_{12} = 0, \quad \xi = \pi_{11} / h_0$$

and

$$F = \pi^2_{11} + \pi^2_{12} - h^2_0 + B_2(s)\pi^2_{11} + \frac{1}{h_0}B_3(s)\pi^3_{11} \quad (23)$$

In equation (23) the result $B_0(s) = 1$ was used. The parameters $B_2$ and $B_3$ are determined by the experimental data obtained by [5] and the convexity of the yield surface.

Let $B_i(s)$ and $h_1$ in equation (9) take the following forms:

$$B_i(s) = b_i(1 - e^{-c_i s}) \qquad (i = 2,3); \qquad h_1 = f_1 g_1 e^{-g_1 s} \tag{24}$$

then we get:

$$b_2 = 9.9, \quad b_3 = -13.9 \quad c_2 = 613.2, \quad c_3 = 405.4; \quad f_1 = 12.2, \quad g_1 = 2060.0 \tag{25}$$

Finally we get:

$$F = \pi_{11}^2 + \pi_{12}^2 - 4 + 9.9(1 - e^{-613.2S})\pi_{11}^2 - 7.0(1 - e^{-405.4S})\pi_{11}^3 \tag{26}$$

$$dS = \sqrt{\frac{3}{2}}[(de_{11}^p)^2 + (de_{12}^p)^2]^{\frac{1}{2}} / 0.5(1 + e^{-1271.5S}) \tag{27}$$

Fig.2 gives the comparision of the subsequent yield surfaces between the calculated results from equation (26) and the experimental results obtained by [5] at temperature 70 F. It can be seen that these two results are coinsided each other. Using the same method we may research other experiments discussed in [5–7].

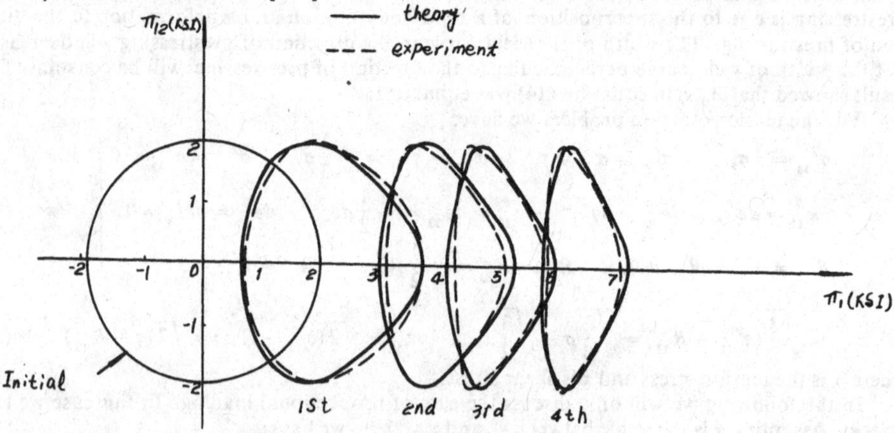


Fig.2 Comparision with experiment

## CONCLUSIONS

1. A new kind of endochronic which may be related to stress tensor $\sigma$ is introduced in this paper. A rough study showes that this kind of endochronic may be used to describe the anisotropic hardening for the subsequent yield function.
2. The integral constitutive equation with a new endochronic introduced in this paper may be applied to the case of combined loading. Obviously the further study is needed.

## REFERENCES

[1] Gurtin, M.E., An Introduction to Continuum Mechanics, Academic Press, 1981

[2] Kuang, Zhen–Bang, The Fundations of Non–Linear Continuum Mechanics, Xi'an Jiaotong University Publishing House, 1989 (in Chinese)

[3] Pipkin, A.C. and Rivlin, R.S., Mechanics of Rate–Independent Materials, in " Proceeding of the International Symposium on Laser–Physics and Applications " Bern(1964), Edited by Meyer, K.P., Brandli, H.P. and Dundliker, R.

[4] Kuang, Zhen–Bang, Acta Mechanica Solids Sinica, 3, 245–262 (1990)

[5] Phillips, A. and Tang, J.L., Int. J. Solids Struct. 8, 463–474 (1972)

[6] Helling, D.E., Miller, A.K. and Stout, M.G., J. Eng. Mat. Tech., 108, 313–320, (1986)

[7] Dvorak, G.J. etc., J. Mech. Phys. Solids, 36, 655–687 (1988)

# A Kinematical Justification of Some Constitutive Assumptions in Plasticity

Gianpietro Del Piero
Istituto di Meccanica, Università di Udine
Viale Ungheria 43, 33100 Udine, Italy

## Abstract

**The plastic deformation, usually considered as an internal variable, is given here a purely kinematical status by introducing a notion of deformation more general than the one used in Continuum Mechanics. With this approach, the additive decomposition of the total deformation into an elastic and a plastic part appears as the most natural one. It also comes out that the plastic deformation is isochoric if and only if it is reversible.**

## Simple Deformations and Structured Deformations

**In this communication I describe some new ideas on plastic deformation, originated from a study by D.R.Owen and myself, whose initial aim was the construction of a rational kinematics for fractured continua. With this aim in mind, we introduced a class of deformations, called simple deformations, in which the displacement vector is allowed to be discontinuous across some singular surfaces, interior to the body and not given a priori. When considering sequences of simple deformations, we realized that sometimes they converged to limit elements which did not belong to the class of simple deformations. Of basic importance are the two following examples.**

**1. Consider the unit cube. Cut it into horizontal slices of height 1/h and define the simple deformation $f_h$ as the deformation which leaves the first slice fixed and assigns to the $k^{th}$ slice a rigid translation relative to the $(k-1)^{th}$ slice, of amount 1/h and in a fixed direction a in the horizontal plane. When $h\rightarrow\infty$, the sequence $\{f_h\}$ converges to a pure shear in the plane determined by a and the vertical. On the other hand, since each slice does experience a rigid**

translation, at the interior points of each slice the deformation gradient $Df_h$ is equal to the identity.

2. Take again the unit cube, and subdivide it into cubic elements of side 1/h, with centers at the points with coordinates

$$(1) \qquad \left(\frac{2p-1}{2h}, \frac{2q-1}{2h}, \frac{2r-1}{2h}\right), \qquad p,q,r = 1,2,...h.$$

Let $\lambda,\mu,\nu$ be reals greater than one, and let $f_h$ be the piecewise rigid translation which sends each cubic element into the cube centered at the point

$$(2) \qquad \left(\lambda\frac{2p-1}{2h}, \mu\frac{2q-1}{2h}, \nu\frac{2r-1}{2h}\right).$$

For $h\rightarrow\infty$, the $f_h$ converge to the $C^1$ function which maps the point $(x_1,x_2,x_3)$ into the point $(\lambda x_1,\mu x_2,\nu x_3)$. But once more, just as in the preceding example, for each value of h the gradient $Df_h$ is equal to the identity I.

In both examples, a sequence of discontinuous deformations $f_h$ converges to a smooth deformation g, while the sequence of the derivatives converges to a limit G different from Dg. The $f_h$ in the first example describe slips along horizontal planes, of amount decreasing with h but with a number of slip planes increasing with h, in such a way that the total amount of slip is constant. It is natural to think to the limit element of the sequence as representing an infinite number of microscopic slips, whose macroscopic effect is a unit shear g. Furthermore, the fact that $G=I$ tells us that the material consituting the body does not undergo any deformation. Thus, we conclude that the inequality $Dg\neq G$ reveals the presence of microscopic slips, and that Dg and G measure the total (macroscopic) deformation and the deformation of the body as a continuum, respectively.

In the second example the $f_h$ describe a homogeneous spread of matter with creation of voids. The limit element in the sequence describes a continuum in which matter and voids are mixed in a way undetectable by macroscopic observation. This time, the inequality $Dg\neq G$ reveals the presence of voids and, as in the preceding example, the equality $G=I$ tells us that the matter is not deformed. In this example, the determinant of Dg is the macroscopic volume, det(G) is the volume actually occupied by the continuum, and $1 - (\det(G)/\det(Dg))$ is the void ratio.

The first example is appropriate to describe the plastic shear of a metal, and the second one to describe the dilatancy of a soil. Clearly, a complete description of these phenomena requires the formulation of constitutive hypotheses on the behavior of the stress across the discontinuity interfaces, i.e., across the planes of slip or separation. For one-dimensional examples of constitutive assumptions of this type see [1].

We call <u>structured deformations</u> the pairs (g,G) obtained as limits of sequences of simple deformations. As it is clear from the preceding examples, structured deformations carry more information than the classical deformations of Continuum Mechanics. It is worth noting that they succeed in describing in a purely kinematical way the "inelastic parts of the deformation" which are usually described by a set of "internal variables" and form the object of ad hoc constitutive assumptions. In what follows, we consider only limit functions g which are smooth, excluding thereby from our analysis the case of bodies undergoing macroscopic fractures. We also assume that G is continuous throughout the body.

## MAIN RESULTS

Let a be a unit vector and let $\mathcal{L}$ be a line segment interior to the body and parallel to a. For a given simpe deformation $f_h$, denote by $Z_h$ the set of the discontinuity points of $f_h$ in $\mathcal{L}$ and by $[f_h(z)]$ the discontinuity of $f_h|_{\mathcal{L}}$ at $z \in Z_h$. It is possible to prove that, if a sequence $\{f_h\}$ converges to a structured deformation (g,G), then

$$\lim_{n\to\infty} \sum_{z\in Z_h} [f_h(z)] = \int_{\mathcal{L}} (Dg(x)-G(x))a\, dx \quad (3)$$

This formula allows us to characterize the vector (Dg(x)-G(x))a as the <u>linear density of microfractures</u> at x in the direction a. By this way, the identity

$$Dg - I = (G - I) + (Dg - G) \quad (4)$$

determines the decomposition of the macroscopic displacement gradient Dg-I into the sum of a displacement gradient due to microfracture and one due to deformation without fracture.

Let now $\mathcal{S}$ be a planar surface in the body, with boundary $\mathcal{C}$. The integration of (4) along $\mathcal{C}$ and the use of Stokes' theorem yield [2]

$$(\operatorname{curl} G(x))^T n = \lim_{\operatorname{area}\mathcal{S}\to 0} \left(\frac{1}{\operatorname{area}\mathcal{S}} \int_{\mathcal{C}} (Dg(\xi)-G(\xi))t(\xi)\, d\xi\right). \quad (5)$$

This shows that the transpose of the curl of G measures the <u>area density</u> of the deformation due to microfracture. This result is a counterpart of a known result in the theory of continuous distributions of dislocations.

The formula (4) suggests the additive decomposition of the displacement gradient as the "natural" way for decomposing a finite deformation into an elastic and a plastic part. Of course, an equivalent multiplicative decomposition can be defined, but, in this context, a plastic deformation coming from a multiplicative decomposition looks more like a derived rather than like a primitive variable.

Another property of structured deformation relevant for Plasticity is the following. Let (g,G) be a structured deformation, and let it be the limit element of a sequence $\{f_h\}$ of simple deformations which are injective, in such a way that no interpenetration of matter is allowed at any stage of the limit process. Under these conditions, it is possible to prove that [3]

(6) $$\det(Dg) \geq \det(G).$$

If a structured deformation is reversible, then the same inequality applies between the inverses of Dg and G, and therefore (6) must hold as an equality. In other words, reversible plastic deformations are isochoric. This implies that the plastic deformation described in our Example 1, due to slips which are reversible by their own nature, has to be isochoric. This property is usually stated as a constitutive assumption, motivated by experiments; here, it is a direct consequence of non-interpenetration and reversibility. On the contrary, a plastic deformation occurring by dilatancy can be considered as irreversible, in that the incorporation of voids determines a change in the internal structure of the continuum, which cannot be eliminated by purely kinematical transformations.

## REFERENCES


1. Del Piero, G., and R.Sampaio, A Unified Treatment of Damage and Plasticity Based on a New Definition of Microfracture. In Proc. 4th Meeting Unilateral Problems in Structural Analysis, Capri 1989. Publ. Birkhäuser Verlag, 1991.

2. Del Piero, G., and D.R.Owen, New Concepts in the Mechanics of Fractured Continua, Proc. 10th National Congress AIMETA, Pisa 1990.

3. Del Piero, G., and D.R.Owen, in preparation.

# ON KINEMATIC HARDENING RULES

R. N. DUBEY, R. SAUVE and S. BEDI
University of Waterloo, Waterloo, Canada.

## ABSTRACT

One dimensional evolution rule is used as basis for generalization to three-dimensional finite deformation applications. It is shown that the strain-rate is not a flux and hence it should not be employed in the flow rule in Kinematic hardening theory.

## INTRODUCTION

For a specimen under uniaxial loading, the current stress may be expressed in terms of the yield stress, $\sigma_Y$, and a parameter $\alpha$:

$$\sigma = \sigma_Y + \alpha \tag{1}$$

where $\sigma_Y$ and $\alpha$ depend on plastic deformation. The yield stress $\sigma_Y$ in (1) is associated with strain-hardening or strain-softening, whereas $\alpha$ is a measure of kinematic hardening. In absence of kinematic hardening, yield stress in tension and compression is symmetric about a state of zero stress. Kinematic hardening shifts the point of symmetry to $\alpha$. Hence, it is also known as the shift or back stress.

The strain used as a measure of plastic deformation in unixial loading is

$$e = \ln(\lambda) \tag{2}$$

where $\lambda$ is the ratio of the final length of a line to its initial length.

The material derivative of (1) yields

$$\dot{\sigma} = \dot{\sigma}_Y + \dot{\alpha} \tag{3}$$

in which

$$\dot{\sigma}_Y = \frac{\partial \sigma_Y}{\partial e}\dot{e} > 0 \tag{4}$$

is the loading criterian, and

$$\dot{\alpha} = \frac{\partial \alpha}{\partial e}\dot{e} \tag{5}$$

is an example of evolution rule for back stress. The aim of the present work is to generalize (5) for three-dimensional finite deformation. For this purpose, use the following generalization of uniaxial stress,

$$\sigma_M = \sqrt{\frac{3}{2} s_{ij} s_{ij}} \tag{6}$$

in which $s_{ij}$ may be interpreted as deviatoric stress components of $\sigma_Y$. Hencky generalized the uniaxial strain in the form

$$e_H = \sqrt{\frac{2}{3} e_{ij} e_{ij}} \tag{7}$$

used in the deformation theory of plasticity. In this equation, $e_{ij}$ may be treated as components of $e_H$. Its material derivative yields

$$\dot{e}_H = \frac{2}{3} \frac{e_{ij} \dot{e}_{ij}}{e_H} \tag{8}$$

In the incremental theory of plasticity, the generalized strain is obtained from

$$de_M = \sqrt{\frac{2}{3} de_{ij} de_{ij}} \tag{9}$$

Note that $de_M$ may not be a differential. Nevertheless, the generalized strain is obtained from (9) by integration. A common practice is to replace the increments $de_{ij}$ by strain-rate $D_{ij}$, also known as the rate of deformation tensor. This procedure implicitly assumes $\dot{e}_{ij} = D_{ij}$. This assumption however, is not strictly correct because the strain-rate is not a flux, and its integration can not in general be identified with strain. For a verification of this statement, consider the simple shear.

## STRAIN-RATE AND RATE OF STRAIN

Choose a fixed reference axes, $x_i$, and suppose the deformation gradient of a body in simple shear is such that $F_{11} = F_{22} = 1$, $F_{12} = \gamma$, $F_{21} = 0$. The corresponding principal stretches and directions are obtained from

$$\lambda_1 = (1/\lambda_2) = \frac{\gamma + \sqrt{\gamma^2 + 4}}{2} \tag{10}$$

and

$$tan 2\beta = 2/\gamma \tag{11}$$

respectively. In the case of the logarithmic measure,

$$e_1 = -e_2 = \ln(\lambda_1) \tag{12}$$

may be treated as the principal components of Hencky's generalized strain,

$$e_H = \frac{2}{\sqrt{3}} e_1 \tag{13}$$

which is obtained from (7). Its $x_i$-components are

$$E_{11} = -E_{22} = \frac{\gamma}{\gamma^2 + 4} e_1 \tag{14}$$

$$E_{12} == \frac{2}{\gamma^2 + 4} e_1$$

It is easily shown that the material derivatives of normal components of strain in (14), $\dot{E}_{11}$ and $\dot{E}_{22}$, are not zero.

Use the chain-rule of differentiation to express the rate of deformation gradient in terms of the strain-rate and spin, $W_{ij}$:

$$\dot{F}_{ij} = (D_{ik} + W_{ij}) F_{kj} \tag{15}$$

The above equation used along with $\dot{F}_{11} = \dot{F}_{22} = \dot{F}_{21} = 0$, $\dot{F}_{12} = \dot{\gamma}$ results in

$$D_{11} = D_{22} = 0,\ D_{12} = \dot{\gamma}/2 \tag{16}$$

A comparison of (16) with the rate of (14) reveals that $\dot{E}_{ij} \neq D_{ij}$ and hence $E_{ij} \neq \int D_{ij} dt$. Thus, the equations

$$\dot{E}_{ij} = D_{ij},\ E_{ij} = \int D_{ij} dt \tag{17}$$

are not strictly correct. For small strains however, one can use

$$\dot{E}_{ij} \simeq D_{ij},\ E_{ij} \simeq \int D_{ij} dt \tag{18}$$

as an approximation for practical applications.

Substitute (16) in the rate-form of (9), which is $\dot{e}_M = \sqrt{\frac{2}{3} D_{ij} D_{ij}}$, to obtain

$$\dot{e}_M = \dot{\gamma}/\sqrt{3},\ e_M = \gamma/\sqrt{3} \tag{19}$$

It is not clear how to interpret $e_M$ obtained from (19) especially in view of (18). It would appear that the use of $e_M$ as generalized strain is not justified for three-dimensional finite deformation except perhaps for special loading histories.

## GENERALIZATION

It is proposed to use the generalized stress $\sigma_Y$ with components $s_{ij}$ and Hencky's generalized strain $e_H$ with components $e_{ij}$ . The rate of strain is obtained from (8). Choose an objective coaxial-rate, denoted by $(\ )^o$, such that

$$\begin{aligned} \dot{\sigma}_M &= \frac{3}{2} s_{ij} \dot{s}_{ij} = \frac{3}{2} s_{ij} s^o_{ij} \\ \dot{e}_H &= \frac{2}{3} e_{ij} \dot{e}_{ij} = \frac{2}{3} e_{ij} e^o_{ij} \end{aligned} \tag{20}$$

For small strains, the evolution rule (5) can be generalized to

$$\dot{\alpha}_{ij} = \frac{\partial \alpha}{\partial e} \dot{e}_{ij} \tag{21}$$

The evolution rules proposed by Melan, Prager, Kadashevitch and Novozhilov, Ziegler, Arutunayn and Vakulenko, Eisenberg and Phillips, Backhaus and Mroz, Shrivastava and Dubey (see [1] for more details) can be expressed in the form (21). It describes linear hardening if the coefficient of $\dot{e}_{ij}$ is constant. A non-linear kinematic hardening in the form

$$\dot{\alpha}_{ij} = c\, \dot{e}_{ij} - d\, \dot{e}_M\, e_{ij} \tag{22}$$

was proposed by Mroz et al in [1]. The aim of this work is to generalize (22) for application to finite deformation.

## GENERALIZED EVOLUTION RULE

Consider a generalization of the shift stress and assume that it is related to its components according to

$$\alpha = \sqrt{\frac{2}{3}\alpha_{ij}\,\alpha_{ij}} \tag{23}$$

The rate of the above equation can be expressed in the form

$$\dot{\alpha} = \frac{2}{3\,\alpha}\alpha_{ij}\dot{\alpha}_{ij} = \frac{2}{3\,\alpha}\alpha_{ij}\alpha^{o}_{ij} \tag{24}$$

An evolution rule in terms of $\alpha^{o}_{ij}$ is needed to find the back stress $\alpha$. Note that any one of the objective J-, Z- or E-rates can be used for $(\,)^{o}$.

The evolution rule (22) of Mroz et al is generalized for finite deformation to

$$\alpha^{o}_{ij} = c\,e^{o}_{ij} - d\,e_{ij}\dot{e} \tag{25}$$

The generalization yield stress

$$\sigma_Y = \sqrt{\frac{3}{2}(s_{ij} - \alpha_{ij})(s_{ij} - \alpha_{ij})} \tag{26}$$

was proposed by Mroz et al [1].

Assume that the yield stress is a function of plastic strain. Then, the consistency condition can be written in the form

$$(s_{ij} - \alpha_{ij})s^{o}_{ij} - (s_{ij} - \alpha_{ij})\alpha^{o}_{ij} = \frac{2}{3}\sigma_Y\sigma'_Y\dot{e} \tag{27}$$

In view of the difficulty in interpreting $e_M$, Hencky's generalized strain will be used for $e$ in (25,27). Since $D_{ij}$ is not a flux, an objective rate of strain will be used in the flow rule, which is written in the form

$$e^{o}_{ij} = \frac{3}{2}(s_{ij} - \alpha_{ij})\,\frac{(s_{kl} - \alpha_{kl})s^{o}_{kl}}{\sigma_Y^2 h} \tag{28}$$

where $h$ is a hardening parameter. Substitute (25, 28) in (27) to obtain $h$ in terms of $c$ and $d$ and the loading history:

$$h = c + \frac{(s_{ij} - \alpha_{ij})e_{ij}}{\sigma_Y\,e}\left(\frac{2}{3}\sigma'_Y - d\frac{(s_{ij} - \alpha_{ij})e_{ij}}{\sigma_Y}\right) \tag{29}$$

# References

[1] Mroz, Z., Shrivastava, H.P. and Dubey, R.N., "A Non-Linear Hardening Model and Its Application to Cyclic Loading", Acta Mechanica, Vol.25, pp 51-61, 1976.

# EQUIVALENT SYSTEMS FOR INELASTIC ANALYSIS OF MEMBERS SUBJECTED TO LARGE DEFORMATIONS

DEMETER G. FERTIS AND CHIN T. LEE
The university of Akron, Akron, Ohio 44325 U.S.A.

## ABSTRACT

The research here deals with the analysis of prismatic and nonprismatic flexible bars where their material is permitted to be stressed well beyond its elastic limit and all the way to failure. This condition of loading causes the modulus of elasticity of the material to vary along the length of the bar. Therefore, the analysis of flexible bars must take into consideration the above variation in the modulus of elasticity when large deflections and rotations are calculated. The analysis and mathematical formulation of this problem is based on the method of the equivalent systems which was developed by the first author of this research. The large deflections and rotations of the bar may be obtained for both elastic and inelastic ranges, all the way to failure, thus permitting to observe progressive deterioration of the bar's ability to resist load, stress, and deformation.

## INTRODUCTION

In the most general case, the nonlinear differential equation of the elastic line of a member with variable moment of inertia $I_x$, and variable modulus of elasticity $E_x$, that is subjected to large deflections and rotations, may be written as

$$\frac{y''}{[1+(y')^2]^{3/2}} = -\frac{M_x}{E_x I_x} \tag{1}$$

where $M_x$ is the bending moment at cross sections along the length of the member.

Fertis and Pallaki [1], and Fertis and Afonta [2,3], utilized Eq. (1) in order to obtain equivalent pseudolinear and equivalent nonlinear systems of constant stiffness $E_1 I_1$, in order to simplify the solution of such problems. Their solution, however, assumed that the modulus of elasticity $E$ remains elastic, and that the elastic limit of the material is not exceeded. If the rotations $y'$ in Eq. (1) are small, then this equation may be written as

$$y'' = -\frac{M_x}{E_x I_x} \tag{2}$$

This is a linear equation, but if the material of the member is permitted to be stressed beyond its elastic limit, a condition is created that results in a variable modulus of elasticity $E_x$ along the length of the member.

If both $E_x$ and $I_x$ are permitted to vary along the length of the member, the variable stiffness $E_x I_x$ may be expressed as

$$E_x I_x = E_1 I_1 \, g(x) f(x) \tag{3}$$

where $g(x)$ represents the variation of $E_x$ with respect to a reference value $E_1$, and $f(x)$ represents the variation of $I_x$ with respect to a reference value $I_1$. If $E$ and $I$ are both constant, then $f(x)=g(x)=1.00$.

For small deformation inelastic analysis, the computation of the function $g(x)$ in Eq. (3) may be carried out by using the methodologies of Timoshenko [4], Fertis and Keene [5], and Fertis and Lee [6]. These methodologies are based on the computation of a reduced modulus $E_r$ and the utilization of equivalent systems of constant stiffness $E_1 I_1$.

The research in this paper deals with the inelastic behavior of flexible prismatic or nonprismatic bars, where the material is permitted to be stressed well beyond its elastic limit, practically all the way to failure. Therefore, the large deflections and rotations of such flexible members will incorporate the effect of the variation in the modulus elasticity along their length. Various cases of loading and boundary conditions are investigated and the results are compared with the ones obtained by using the 4th Order Runge-Kutta method.

## INELASTIC ANALYSIS OF FLEXIBLE BARS

The nonlinear differential equation of a flexible bar in bending, Eq. (1), may be written as

$$y'' = -\frac{M_x Z_e}{E_x I_x} \tag{4}$$

where

$$Z_e = [1 + (y')^2]^{3/2} \tag{5}$$

By utilizing Eqs. (3) and (4) we write

$$y'' = -\frac{1}{E_1 I_1} \cdot \frac{M_x Z_e}{f(x) g(x)} \tag{6}$$

It is assumed here that both $E_x$ and $I_x$ are variable.

It may be proven here, see Fertis and Pallaki [1], and Fertis and Afonta [3], that the original bar of variable $E_x$ and $I_x$ may be replaced by an equivalent pseudolinear system of constant stiffness $E_1 I_1$ that has the same deflection curve, length, and boundary conditions as the original nonlinear variable stiffness bar. The moment diagram $M_e'$ of the pseudolinear equivalent system may be obtained from the equation

$$M_e' = \frac{M_x Z_e}{f(x)g(x)} \tag{7}$$

The equivalent shear force $V_e'$ and loading $W_e'$ of the pseudolinear system may be obtained from Eq. (7) by differentiation.

By substituting Eq. (7) into Eq. (6), we obtain

$$y'' = -\frac{M_e'}{E_1 I_1} \tag{8}$$

which is the differential equation of the pseudolinear system of constant stiffness $E_1I_1$. If it is possible to determine $M_e'$, then pseudolinear analysis may be used to solve Eq. (8). Integration of Eq. (8) yields

$$E_1 I_1 y' = \int M_e' dx + C_1 \tag{9}$$

where $C_1$ is the constant of integration. The inelastic analysis of flexible bars is illustrated here by the following example.

Consider a variable stiffness cantilever bar loaded as shown in Fig. (1). Its large deflection configuration is shown in the same figure. The moment $M_x$ at any $0 \leq x \leq L_o$ from the fixed end, is given by the expression

$$M_x = \frac{wLL_o}{2} - wLx + \frac{wx_o x}{2} \tag{10}$$

where $x_o$ is taken along the deformed length of the member as shown in Fig. (1).

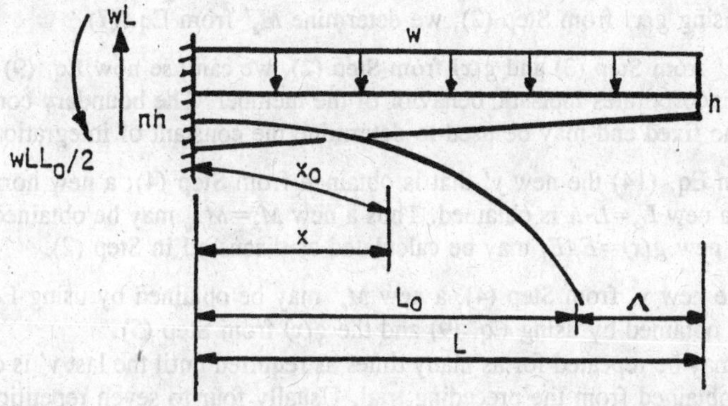


Figure 1. Variable Stiffness Cantilever Bar Loaded with Distributed Load *w*.

In Fig. (1) we note that for every length $x_o$ there corresponds a length $x$, where $0 \leq x \leq L_o$ with $L_o = L - \Lambda$. Thus we may write

$$x_o = x + \Lambda(x) \tag{11}$$

where $\Lambda(x)$ represents the horizontal displacement of $x_o$. We also know that

$$x_o = \int_o^x [1 + (y')^2]^{1/2} dx \tag{12}$$

Therefore Eq. (10) may be written as

$$M_x = \frac{wL}{2}(L-\Lambda) - wLx + \frac{wx}{2}[x + \Lambda(x)] \tag{13}$$

where $\Lambda$ is the horizontal displacement of the free end of the beam and $\Lambda(x)$ is the horizontal displacement at any $0 \leq \times \leq L_o$.

The inelastic analysis here may be carried out by a trial-and-error procedure and utilization of equivalent pseudolinear systems represented by Eq. (8). This procedure may be initiated as follows:

Step 1. A first approximation of $g(x)=E_r/E_1$, where $E_r$ is the reduced modulus, may be obtained by assuming $g(x)=1$ in Eq. (7), and proceed with the integrations in Eq. (9) to determine $y'$. The value of $\Lambda$ that is required in this integration may be determined from the equation

$$L = \int_o^{L-\Lambda} [1 + (y')^2]^{1/2} dx \tag{14}$$

by using a trial-and-error procedure. See Fertis and Afonta [3]. With known $\Lambda$, the values of $M_x$ at any $0 \leq \times \leq L_o$ can be determined from Eq. (13).

Step 2. With known $M_x$, we can start the procedure to determine the reduced modulus $E_r$, and consequently $g(x)=E_r/E_1$ as shown by Fertis, Taneja, and Lee [7].

Step 3. By using the values of $y'$ from Step (1) we can determine $Z_e$ from Eq. (5). Thus, with known $Z_e$ and using $g(x)$ from Step (2), we determine $M_e'$ from Eq. (7).

Step 4. By using $M_e'$ from Step (3) and $g(x)$ from Step (2), we can use now Eq. (9) to obtain a new $y'$ which incorporates inelastic behavior of the member. The boundary condition of zero rotation at the fixed end may be used to determine the constant of integration $C_1$.

Step 5. By using in Eq. (14) the new $y'$ that is obtained from Step (4), a new horizontal displacement $\Lambda$ and a new $L_o=L-\Lambda$ is obtained. Thus a new $M_x=M_{req}$ may be obtained from Eq. (13), and also a new $g(x)=E_r/E_1$ may be calculated as discussed in Step (2).

Step 6. By using the new $y'$ from Step (4), a new $M_e'$ may be obtained by using Eq. (7), and thus a new $y'$ is obtained by using Eq. (9) and the $g(x)$ from Step (5).

The procedure may be repeated for as many times as required until the last $y'$ is closely identical to the one obtained from the preceding trial. Usually four to seven repetitions are sufficient. The whole procedure may be easily computerized for convenience, and Simpson's rule may be used to carry out the required integrations. The utilization of pseudolinear analysis, that is $M_e'$, facilitates a great deal the solution and convergence of the above trial-and-error procedure, and consequently the solution of such types of problems.

Now that the correct values of $y'$ are known, the large deflections may be obtained by either (a) integrating Eq. (9) once and determining the constant of integration by satisfying the condition of zero deflection at the fixed end, or (b) by using the moment-area method, since $M_e'$ is known. See Fertis and Pallaki [1] and Fertis and Afonta [2,3]. Note that the rotation $\theta=\tan^{-1}(y')$.

Let it be assumed in Fig. (1) that $L=1000$ in. (25.4 m), n=2, h=3 in. (0.0762 m), and width b=3 in. (0.0762 m). The inelastic analysis for various values of the uniformly distributed load $w$ was carried out as discussed above, and by subdividing $L_o$ in each trial into forty segments. Table (1) shows the final values of $g(x)$, $f(x)$, $\Delta$, $y'$, $Z_e$, $M_{req}=M_x$, and $M_e'=Z_eM_x/f(x)g(x)$, for a load $w=3.5$ lb/in. (613.19 N/m). With known $M_e'$, deflections and rotations at any $0 \leq x \leq L_o$ can be determined as discussed in the preceding sections. In Table (2), the values of $\delta$, $\Lambda$, and $\theta$ at the free end of the bar, for various values of the distributed load $w$, are shown. The results obtained by using the 4th Order Runge-Kutta method are also shown in the same table for comparison purposes. These results are in close agreement for practical purposes. The material of the above member is Monel, and $E_r$ and $g(x)$ are determined by using a 3-line approximation of its stress-strain curve. This yields very accurate results for all practical purposes. The solution of other types of variable stiffness flexible members may be obtained in a similar manner.

## CONCLUSIONS

The research here proved that inelastic analysis of flexible bars with uniform or variable moment of inertia $I_x$ and variable modulus $E_x$ along the length of the bar, may be carried out by using the concept of pseudolinear equivalent systems and a reduced modulus $E_r$. The variation of $E_x$ along the length of the bar is caused by the applied load when it is large enough to stress the material of the bar beyond its elastic limit. The member becomes increasingly inelastic with increasing load, until its ultimate capacity to resist load is reached. This progressive deterioration of the flexible member's ability to resist load and deformation, may be used in practice to establish specific performance criteria in a given situation. The methodology proposed in this research may be used for any stress-strain curve variation of the material, since its utilization in the analysis involves the approximation of its shape with straight-line segments. The number of such segments that are usually required for reasonable accuracy is small; usually three to four segments are sufficient.

## REFERENCES


1. Fertis, D.G. and Pallaki, S. "Pseudolinear and Equivalent Systems for Large Deflections of Members." J. Engrg. Mech., (1989) ASCE, Vol. 115, No. 11.

2. Fertis, D.G. and Afonta, A.O. "Large Deflection of Determinate and Indeterminate Bars of Variable Stiffness." J. Engrg. Mech., ASCE, (1990), Vol. 116, No. 7.

3. Fertis, D.G., and Afonta, A.O. "Equivalent Systems for Large Deformation of Beams of Any Stiffness Variation". European Journal of Mechanics, A/Solids, 1991, in press.

4. Timoshenko, S. Strength of Materials, Part II, Advanced Theory and Problems, 3rd Ed., Robert E. Krieger Publishing Co., Hunting, NY. (1976)

5. Fertis, D.G., and Keene, M.E. "Elastic and Inelastic Analysis of Nonprismatic Members." J. Structural Engrg, ASCE, (1990) 16(2).

6. Fertis, D.G., and Lee, C.T. "Inelastic Analysis of Prismatic and Nonprismatic Members with Axial Restraints." International Journal of Mechanics of Structures and Machines, 1991, in press.

7. Fertis, D.G., Taneja, R., and Lee, C.T. "Equivalent Systems for Inelastic Analysis of Nonprismatic Members." Journal of Computers and Structures, Vol. 38, No. 1, pp. 31-39, 1991.

TABLE 1
Final Values for Strain Δ, $g(x)$, $y'$, $Z_e$, $M_{req}$ and $M_e'$ (1 in. = 0.0254 m, 1 kip-in. = 113.0 Nm)

| x (in.) (1) | f(x) (2) | Strain Δ (3) | $g(x)$ (4) | $y'$ Radians (5) | $Z_e$ (6) | $M_{req}=M_x$ (kip-in) (7) | $M_e'=Z_eM_x/f(x)g(x)$ (kip-in) (8) |
|---|---|---|---|---|---|---|---|
| 0. | 8.0 | $3.0964\times10^{-2}$ | 0.2281 | 0. | 1.0000 | 1398.40 | 766.35 |
| 40. | 7.53 | $1.8664\times10^{-2}$ | 0.3593 | $1.6745\times10^{-1}$ | 1.0427 | 1275.10 | 491.54 |
| 79.9 | 7.07 | $1.1414\times10^{-2}$ | 0.5559 | $2.7567\times10^{-1}$ | 1.1174 | 1156.40 | 329.01 |
| 119.9 | 6.61 | $8.1636\times10^{-3}$ | 0.7317 | $3.5351\times10^{-1}$ | 1.1955 | 1042.40 | 257.38 |
| 159.8 | 6.15 | $6.5636\times10^{-3}$ | 0.8537 | $4.1824\times10^{-1}$ | 1.2766 | 933.11 | 226.15 |
| 199.8 | 5.74 | $5.6136\times10^{-3}$ | 0.9327 | $4.7683\times10^{-1}$ | 1.3635 | 828.87 | 211.63 |
| 239.7 | 5.32 | $4.9136\times10^{-3}$ | 0.9823 | $5.5970\times10^{-1}$ | 1.4585 | 729.80 | 203.22 |
| 279.7 | 4.91 | $4.4636\times10^{-3}$ | 0.9993 | $5.8699\times10^{-1}$ | 1.5643 | 636.05 | 203.34 |
| 319.6 | 4.52 | $4.0506\times10^{-3}$ | 1.0000 | $6.4179\times10^{-1}$ | 1.6835 | 547.77 | 204.22 |
| 359.6 | 4.13 | $3.6488\times10^{-3}$ | 1.0000 | $6.9681\times10^{-1}$ | 1.8169 | 465.14 | 204.49 |
| 399.5 | 3.76 | $3.2428\times10^{-3}$ | 1.0000 | $7.5163\times10^{-1}$ | 1.9641 | 388.36 | 202.70 |
| 439.5 | 3.41 | $2.8340\times10^{-3}$ | 1.0000 | $8.0559\times10^{-1}$ | 2.1237 | 317.66 | 197.98 |
| 479.4 | 3.07 | $2.4243\times10^{-3}$ | 1.0000 | $8.5776\times10^{-1}$ | 2.2924 | 253.29 | 189.35 |
| 519.4 | 2.74 | $2.0169\times10^{-3}$ | 1.0000 | $9.0697\times10^{-1}$ | 2.4649 | 195.55 | 175.83 |
| 559.4 | 2.43 | $1.6162\times10^{-3}$ | 1.0000 | $9.5179\times10^{-1}$ | 2.6337 | 144.73 | 156.66 |
| 599.3 | 2.14 | $1.2291\times10^{-3}$ | 1.0000 | $9.9065\times10^{-1}$ | 2.7895 | 101.13 | 131.63 |
| 639.3 | 1.87 | $8.6494\times10^{-4}$ | 1.0000 | 1.0221 | 2.9218 | 65.05 | 101.49 |
| 679.2 | 1.62 | $5.3734\times10^{-4}$ | 1.0000 | 1.0449 | 3.0218 | 36.73 | 68.40 |
| 719.2 | 1.39 | $2.6508\times10^{-4}$ | 1.0000 | 1.0589 | 3.0846 | 16.37 | 36.23 |
| 759.1 | 1.19 | $7.4089\times10^{-5}$ | 1.0000 | 1.0650 | 3.1124 | 4.11 | 10.78 |
| 799.1 | 1.0 | 0. | 1.0000 | 1.0661 | 3.1175 | 0. | 0. |

TABLE 2
Values of End Displacements δ, Λ, and θ, with Increasing Distributed Load *W*. (1 in. = 0.0254 m, 1 lb/in. = 175.2 N)

| Load *W* (lb/in.) (1) | δ (in.) Eq. Sys. (2) | δ (in.) 4th Order Runge-Kutta (3) | Diff. % (4) | Λ (in.) Eq. Sys. (5) | Λ (in.) 4th Order Runge-Kutta (6) | Diff. % (7) | θ (Deg.) Eq. Sys. (8) | θ (Deg.) 4th Order Runge-Kutta (9) | Diff. % (10) |
|---|---|---|---|---|---|---|---|---|---|
| 1.0 | 150.08 | 150.98 | 0.60 | 12.40 | 12.55 | 1.21 | 12.90 | 13.15 | 1.94 |
| 1.5 | 219.14 | 219.57 | 0.20 | 27.70 | 27.95 | 0.90 | 18.93 | 19.09 | 0.85 |
| 2.0 | 281.85 | 282.98 | 0.40 | 48.40 | 48.87 | 0.97 | 24.53 | 24.84 | 1.26 |
| 2.5 | 349.89 | 351.96 | 0.59 | 75.90 | 76.73 | 1.07 | 30.39 | 31.17 | 2.57 |
| 3.0 | 450.59 | 452.61 | 0.49 | 126.20 | 128.58 | 1.89 | 38.18 | 39.54 | 3.56 |
| 3.5 | 565.39 | 569.40 | 0.71 | 200.90 | 206.92 | 0.51 | 46.83 | 48.65 | 3.89 |
| 4.0 | 652.16 | 646.31 | -0.90 | 273.60 | 272.23 | -0.50 | 53.75 | 53.88 | 0.24 |
| 4.5 | 713.35 | 706.31 | -0.99 | 336.50 | 333.18 | -0.99 | 59.03 | 57.91 | -1.89 |
| 5.0 | 757.40 | 747.97 | -1.25 | 390.00 | 385.13 | -1.25 | 63.13 | 60.95 | -3.45 |
| 5.5 | 791.49 | 780.21 | -1.43 | 435.90 | 444.09 | -1.88 | 66.44 | 64.07 | -3.57 |
| 6.0 | 817.50 | 793.87 | -2.89 | 475.50 | 469.56 | -1.25 | 69.13 | 66.44 | -3.89 |
| 6.5 | 838.10 | 811.91 | -3.12 | 510.10 | 495.36 | -2.89 | 71.38 | 68.51 | -4.02 |
| 7.0 | 854.72 | 830.02 | -2.90 | 540.40 | 524.24 | -2.99 | 73.27 | 70.24 | -4.14 |
| 7.5 | 864.98 | 840.85 | -2.79 | 567.31 | 549.61 | -3.12 | 76.71 | 72.96 | -4.89 |
| 8.0 | 874.83 | 848.59 | -2.99 | 591.31 | 578.76 | -2.12 | 78.90 | 74.96 | -4.99 |
| 8.5 | 885.61 | 857.30 | -3.20 | 613.21 | 594.89 | -2.99 | 80.21 | 75.99 | -5.26 |

# G. I. TAYLOR REVISITED: THE CONE OF UNEXTENDED DIRECTIONS IN DOUBLE SLIP

K. S. HAVNER
Department of Civil Engineering
North Carolina State University
Raleigh, NC 27695-7908, U.S.A.

## ABSTRACT

Recently developed general equations for the cone of unextended directions in arbitrary double slip are applied to the interpretation of finite distortion measurements from the classic series of papers (1923-27) by G. I. Taylor on cubic metal crystals. Measurements of comparable completeness are not readily found elsewhere in the literature. From the equations and the experimental data (including lattice positions from X-ray diffraction analysis), the respective amounts of finite double slip are calculated for several cases from comparisons of the theoretical and experimentally determined cones in stereographic projection. Misconceptions in the literature about the deformations of Taylor's crystals after the loading axis reached a symmetry line are specifically addressed.

## BACKGROUND

In [1] G. I. Taylor introduced his method of the cone of unextended material lines for the original determination (with C. F. Elam) of the experimental laws of finite deformation of f.c.c. metals, based upon X-ray diffraction analysis and complete distortion measurements at each of several stages throughout the tensile test of an aluminum crystal. At every stage the 'unstretched cone' was determined from these measurements in both the current and reference configurations, as illustrated for an extension of 30% in figures 7 and 8 of [1]. At each strain level through 40% the cone was found to be the degenerate form of two planes. Taylor and Elam readily identified the unique unstretched material plane and fixed material direction of simple shear that were common to all levels of distortion. Finally, by X-ray diffraction analysis of the changing lattice orientation, they were able to rigorously establish that the finite single slip took place on the (111) plane and in the [110] direction of greatest initial resolved shear stress.

Taylor's method for establishing the basic mechanics of finite distortion of crystals was further applied in [2]-[6], including extension to b.c.c. crystals in [3] and to f.c.c. crystals in compression in [4]-[6]. The early stages of finite deformation, sometimes to nominal strains as great as 60% (dependent upon initial crystal orientation), were consistently found to correspond to single slip and the degenerate case of two planes of unextended directions at each level of strain (only one of which, the actual slip plane, was fixed in the material). At larger strains, however, true cones of momentarily unstretched lines were found (see figs. 13 and 14 of [1] for example) that were judged in general to correspond to double slip, including the original slip plane and direction as one of the contributing slip systems (and commonly the dominant one).

In [5] Taylor developed equations for the unstretched cone in equal double slip on equally stressed slip-systems in axial compression and made comparisons with experimentally determined cones from experiments on aluminum crystals. In none of his papers, however, was the general and more common case of unequal double slip on unequally stressed systems rigorously analyzed for either tension or compression. [In [2] estimates were made of the relative amounts of arbitrary double slip in aluminum crystals in tension based upon the far simpler analysis of the cone of zero *rates* of extension.]

## SELECTED NEW RESULTS

General equations for the cone of unextended directions in finite double slip (in both tension and compression) have recently been developed [7], based upon earlier analyses of the kinematics of arbitrary (proportional) double slip in [8]. Here we apply these equations to the interpretation of several of the finite deformation experiments of Taylor and colleagues ([1]-[6]) for which complete distortion measurements are available. Such measurements are not readily found elsewhere in the literature.

In Fig. 1 is shown a comparison between experimentally and theoretically determined unstretched cones in the reference state for an aluminum crystal in an extension from 74.5% to 93.6% strain (nominal measure). At the beginning of this deformation interval the tensile axis has Euler angles $\theta_z = 33.75^\circ$ from [001] and $\phi_z = 138.3^\circ$ from [100], which position is equivalent to that of specimen 59 in [2] at the same stage. Thus, the axis already is rotated 3.3° beyond the symmetry line [001]-$[\bar{1}11]$ of equal resolved shear stresses between systems $(111)[\bar{1}01]$ and $(\bar{1}\bar{1}1)[011]$, henceforth called systems 1 and 2. System 1 was the primary system in single slip during the earlier stages of the deformation [2].

The experimental unstretched cone is determined from deformation measurements in table 1 of [2]. [It should be noted that the determinant of the deformation gradient calculated using these measurements gives a volume increase of 1.27% during the interval of 11% additional relative extension. As this increase seems rather large, there may be small errors in some of Taylor and Elam's reported values, making the equation of the unstretched cone [7] and its projection somewhat in error as well.] The theoretical cone is determined by iterative solution of the nonlinear equations (2.9) (with (2.33)) and (2.28) of [7] (corresponding to the initial axis position and the relative stretch $\lambda_z = 1.936/1.745$) to obtain the ratio $\alpha = \gamma_2/\gamma_1$ (and corresponding $\gamma$'s) that gives the best fit to the experimental locus. As indicated in the figure, this is $\alpha = 0.29$. (Taylor and Elam [2] gave a quite good estimate of one-third based upon an equation for the cone of zero stretch-rates in double slip.) Taking

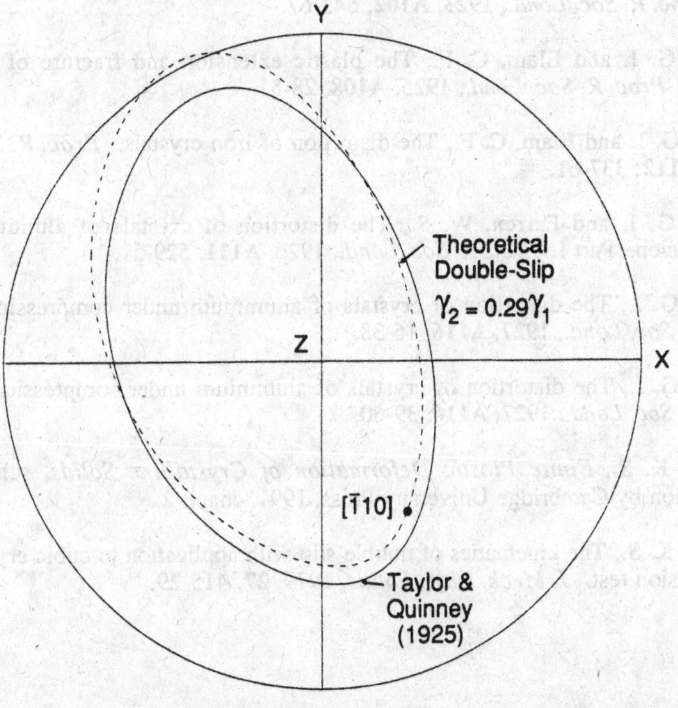


FIGURE 1. Tensile axis stereographic projections, in an aluminum crystal extended 74.5%, of the position of theoretical and experimental unstretched cones for extension from 74.5 to 93.6%.

into account the point already noted about the volume change, the experimental and theoretical loci of unextended directions may be judged to be in satisfactory agreement, differing primarily by a small relative displacement in the direction of their long axes. ($[\bar{1}10]$ is the intersection of the two slip planes.)

It is evident from this comparison that slip in the original system of Taylor and Elam's specimen 59 dominated during this interval of additional deformation. The tensile axis, which already had overshot the symmetry line before attaining a 74.5% extension, continued to rotate further into the spherical triangle of highest resolved shear stress on the secondary system. This behavior is representative of other specimens in [2] and was the first unambiguous evidence of 'overshooting' in f.c.c. crystals. Only rarely has this been recognized in the literature.

## REFERENCES


1. Taylor, G. I. and Elam, C. F., The distortion of an aluminium crystal during a tensile test. *Proc. R. Soc. Lond.*, 1923, A**102**, 643-67.

2. Taylor, G. I. and Elam, C. F., The plastic extension and fracture of aluminium crystals. *Proc. R. Soc. Lond.*, 1925, A**108**, 28-51.

3. Taylor, G. I. and Elam, C. F., The distortion of iron crystals. *Proc. R. Soc. Lond.*, 1926, A**112**, 337-61.

4. Taylor, G. I. and Farren, W. S., The distortion of crystals of aluminium under compression. Part I. *Proc. R. Soc. Lond.*, 1926, A**111**, 529-51.

5. Taylor, G. I., The distortion of crystals of aluminium under compression. Part II. *Proc. R. Soc. Lond.*, 1927, A**116**, 16-38.

6. Taylor, G. I., The distortion of crystals of aluminium under compression. Part III. *Proc. R. Soc. Lond.*, 1927, A**116**, 39-60.

7. Havner, K. S., *Finite Plastic Deformation of Crystalline Solids*, scheduled for publication by Cambridge University Press, 1991, chapt. 2.

8. Havner, K. S., The kinematics of double slip with application to cubic crystals in the compression test. *J. Mech. Phys. Solids*, 1979, **27**, 415-29.

# LARGE STRAIN EFFECTS DURING FREE-END TORSION OF COPPER BARS

JOHN J. JONAS* and LASZLO S. TOTH**
*Metallurgical Engineering, McGill University
3450 University Str., Montreal, H3A 2A7 CANADA
**Institute for General Physics, Eötvös University
1445 Budapest, P.O.Box 323, HUNGARY

## ABSTRACT

Experimental results are presented for the large strain (up to $\gamma$=12) free-end torsion of copper bars carried out between room temperature and 300°C. Crystal plasticity model simulations combined with microstructural investigations show that dynamic recrystallization is the mechanism responsible for shortening of the samples.

## INTRODUCTION

Copper bars generally lengthen when twisted in free-end torsion [1]. Shortening only takes place at high strains ($\gamma$>5) and when the temperature is raised above 200°C [2,3]. Copper wires containing <111> fibre textures also display shortening behavior, even at room temperature [4]. Such length changes are directly caused by the textures that are present, i.e. by the plastic anisotropy that develops and evolves during deformation [4,5]. In this paper, experimental as well as theoretical results are reported for the large strain (up to $\gamma$=12) behavior of copper bars. Evidence is presented for the occurrence of dynamic recrystallization (DRX), which is responsible for the shortening observed at large strains ($\gamma$>5) and elevated temperatures (>200°C).

## METHOD AND RESULTS

Specimens were machined from OFHC copper (99.95%) bars, which had been vacuum heat treated for a 1/2 hour at 550°C prior to testing. The tests were performed at room T, 125, 200 and 300°C in air. The torque, angular displacement and length change were recorded continuously during testing.

The experimental length change-shear strain curves display initial lengthening at all testing temperatures, but at 200 and 300°C, shortening is observed above shear strains of 10 and 5, respectively. The length change curve obtained at room temperature is identical to the one reported by Swift [1] but is extended up to $\gamma$=12.

The shear strain - shear stress curves exhibit stage III work hardening, followed by stage IV. At all temperatures, a maximum stress is attained which is followed by softening. The

TABLE 1
The ideal orientations of shear textures.
(hkl): plane of shear, [uvw]: shear direction.

| | A | $\bar{A}$ | B | $\bar{B}$ | C | $A_1^*$ | $A_2^*$ |
|---|---|---|---|---|---|---|---|
| (hkl) | $11\bar{1}$ | $\bar{1}\bar{1}1$ | $11\bar{2}$ | $\bar{1}\bar{1}2$ | 100 | $11\bar{1}$ | $1\bar{1}1$ |
| [uvw] | $1\bar{1}0$ | $\bar{1}10$ | $1\bar{1}0$ | $\bar{1}10$ | $0\bar{1}1$ | $2\bar{1}1$ | $\bar{2}\bar{1}1$ |

texture was examined by measuring pole figures and calculating orientation distribution functions (ODFs). The ideal components are listed in Table 1.

## DISCUSSION

According to the results of the ODF investigation, C is the strongest component below 200°C. This ideal orientation also exhibits an intensity maximum at medium strains. The behaviors of the C, $A_1$* and $A_2$* components are similar to those observed in the fixed-end case (see [6]). The variations in the strengths of these components are in accordance with crystal plasticity simulations [7], i.e. the predicted texture transition $A_2$* -> C -> $A_1$* is indeed observed. The relative strengths of the A/$\bar{A}$ and B/$\bar{B}$ components compared to the C increase as the testing temperature is increased. Above 125°C, the A/$\bar{A}$ component is the strongest. Its strengthening takes place mainly at the expense of the C component.

Texture development simulations were carried out for the free-end case using the rate sensitive polycrystal model [7]. The experimental length change curves were used to prescribe the deformation path of the polycrystalline aggregate. The predicted room temperature textures are in satisfactory agreement with the experimental results. However, at 300°C, the C component is still predicted to be the most intense, which is not in agreement with the experimental observations. The texture changes at elevated temperatures, therefore, cannot be accounted for solely by means of crystallographic slip. The occurrence of another mechanism, in this case dynamic recrystallization, is necessary to explain the changes in the texture intensities.

The preferred orientations observed at room T and 125°C are typical 'high stored energy' textures. This is because of the predominance of the C component, which displays the highest Taylor factor (√3) among all the ideal orientations [7]. This means that grains near the C position are expected to recrystallize dynamically earlier than any other ideal orientation. According to the oriented growth theory, those nuclei are favored in fcc materials which are rotated by 40° or 23° around a {111} pole of the parent grain [8]. In the present case, the parent grains are assumed to be in the C positions. The new grains are also subjected to torsional deformation. Using the rate sensitive deformation code [5], it was found that nuclei which are defined by the 23° rotation rotate to positions near the A/$\bar{A}$ and B/$\bar{B}$ components. These new orientations contribute appreciable components of continuous shortening during plastic deformation.

The above model for DRX in heavily twisted copper successfully explains the texture changes observed in the pole figures and in the ODF's. It is also consistent with the TEM observations, in which the orientations of the new grains are located near the A/$\bar{A}$ and B/$\bar{B}$ positions. It is even conceivable that DRX can take place at room temperature. According to Kvam [9], DRX is facilitated by an extreme supersaturation of vacancies at low homologous temperatures. Experimental observations of vacancy concentrations during the torsional deformation of Cu and Al [10] strongly support the validity of such a model. We can therefore conclude that the shortening phenomenon observed at elevated temperatures is associated with the occurrence of dynamic recrystallization.

## ACKNOWLEDGEMENTS

JJJ and LST are grateful to the Canadian Steel Industry Research Association for the partial support of this work. The help received from Dr. D. Daniel (McGill University, ODF analysis) and Mr. A.Ö. Kovács (Eötvös University, TEM) is also gratefully acknowledged.

## REFERENCES

1. Swift, H.W., Engineering, 1947, 163, 253-257.

2. Morozumi, F., Nippon Kokan Techn., 1965, Rep. no. 4, 67.

3. Hardwick, D. and Tegart, W.J.,McG, Mém. scient. Revue Métall., 1961, 58, 869-880.

4. Tóth, L.S., Szászvári, P., Kovács, I., and Jonas, J.J., Materials Science and Technology, 1991, in press.

5. Tóth, L.S, Jonas, J.J., Gilormini, P. and Bacroix, B., Int. J. Plasticity, 1990, 6, 83-108.

6. Montheillet, F., Cohen, M., and Jonas, J.J., Acta Metall., 1984, 32, 2077-2089.

7. Neale, K.W., Tóth, L.S., Jonas, J.J., Int. J. Plasticity, 1990, 6, 45-61.

8. Aust, K.T., and Chalmers, B., Metallurgical Transactions, 1970, 1, 1095-1104.

9. Kvam, E.P., Scripta Metall., 1989, 23, 1341-1346.

10. Haessner, F., and Schmidt, J., Scripta Metall., 1988, 22, 1917-1922.

# FINITE INELASTIC DEFORMATION ANALYSIS WITH UNIFIED CONSTITUTIVE MODELS

H.-P. HACKENBERG and F.G. KOLLMANN
Technische Hochschule Darmstadt
Fachgebiet Maschinenelemente und Maschinenakustik

## ABSTRACT

In this paper, finite inelastic deformation analysis is considered with the use of unified constitutive models. An attempt is made to obtain a constistent frame which allows the use of different unified constitutive models and which can also be used to extend models formulated for small elastic strains to finite deformations. The description is based on a manifold theoretical approach as given by MARSDEN AND HUGHES [4]. The implementation in a finite element code is done and numerical examples are shown.

## INTRODUCTION

Unified contitutive models have been proposed for small deformations by different authors (i.e. BODNER AND PARTOM [2]). Often, the equations of such models can be shown to have a common structure and they differ only in the choice of the flow, hardening and recovery functions. Other models have been proposed for or extended to finite deformations (BAMMAN AND AIFANTIS [1], BROWN, KIM AND ANAND [3]). However, for finite deformations the underlying structure of the equations generally differ, since there is still disagreement about the basic concepts for the phenomenological description of finite inelastic deformations. Basic issues are the multiplicative decomposition of the deformation gradient into an elastic and an inelastic part and the uniqueness of the associated intermediate configuration.

In this paper, the following concept for the description of large inelastic processes is employed. A basic assumption is the multiplicative decomposition of the deformation gradient into an elastic and an inelastic part. This allows to formulate deformation tensors, strain tensors and deformation rate tensors on the different configurations involved. Separate constitutive equations for the elastic and the inelastic part of the deformation are first given on the intermediate configuration, where elastic and inelastic parts of the strain and deformation rate tensors are completely decoupled. Through 'push-forward' with the elastic part of the deformation gradient, the constitutive equations can then be obtained on the actual configuration. The separation of the equations into volumetric and deviatoric parts is important in order to achieve a numerically efficient scheme. Approximations based on the assumption that elastic deformations are small are introduced. Special care has to be taken in order to correctly account for the plastic incompressibility constraint.

## KINEMATICS

The multiplicative decomposition of the deformation gradient $\mathcal{F}$ into an elastic and an inelastic part

$$\mathcal{F} = \hat{\mathcal{F}}_e \mathcal{F}_p \tag{1}$$

leads to definition of the deformation tensor $\tilde{\mathbf{b}}^\sharp$

$$\tilde{\mathbf{b}}^\sharp := \phi_*^e(\hat{\mathbf{G}}^\sharp) = \hat{\mathcal{F}}_e \hat{\mathbf{G}}^\sharp \hat{\mathcal{F}}_e^T \tag{2}$$

which is often called 'elastic' Finger tensor, since it is obtained through push-forward of the contravariant metric tensor $\hat{\mathbf{G}}^\sharp$ of the intermediate configuration with the elastic part of the deformation gradient. A definition of push-forward operators $\phi_*(.)$ and pull-back operators $\phi^*(.)$ can be found in MARSDEN AND HUGHES [4]. In correspondance with (2) the Finger tensor with respect to the total deformation can be defined as

$$\mathbf{b}^\sharp := \phi_*(\mathbf{G}^\sharp) = \mathcal{F}\,\mathbf{G}^\sharp\,\mathcal{F}^T \tag{3}$$

where $\mathbf{G}^\sharp$ is the contravariant metric tensor of the reference configuration. On the actual configuration the following decomposition of the Almansi strain tensor $\mathbf{e}^\flat$ into an elastic part $\mathbf{e}_e^\flat$ and an inelastic part $\mathbf{e}_p^\flat$ can be given

$$\mathbf{e}^\flat := \frac{1}{2}(\mathbf{g} - (\mathbf{b}^\sharp)^{-1}) = \mathbf{e}_e^\flat + \mathbf{e}_p^\flat \tag{4}$$

$$\mathbf{e}_e^\flat := \frac{1}{2}(\mathbf{g} - (\tilde{\mathbf{b}}^\sharp)^{-1}) \,. \tag{5}$$

The deformation rate tensor $\mathbf{d}^\flat$ is defined as

$$\mathbf{d}^\flat := \frac{1}{2}(\mathbf{g}\,\mathbf{l} + \mathbf{l}^T\mathbf{g}) = \mathcal{L}_v(\mathbf{e}^\flat) \tag{6}$$

where $\mathbf{l}$ is the velocity gradient, $\mathbf{g}$ the metric tensor on the actual configuration and $\mathcal{L}_v(.)$ denotes the Lie derivative [4]. From (4) and (6) one possible decomposition of the deformation rate tensor into elastic and inelastic parts can be obtained

$$\mathbf{d}^\flat = \mathbf{d}_e^\flat + \mathbf{d}_p^\flat \tag{7}$$

$$\mathbf{d}_e^\flat := \mathcal{L}_v(\mathbf{e}_e^\flat) \tag{8}$$

$$\mathbf{d}_p^\flat := \mathcal{L}_v(\mathbf{e}_p^\flat) \,. \tag{9}$$

It can be shown that the following relations hold

$$\mathbf{d}_p^\flat = \phi_*^e(\hat{\mathbf{D}}_p^\flat) = \phi_*^e({}^s(\hat{\mathbf{G}}\;\dot{\mathcal{F}}_p\;\mathcal{F}_p^{-1})) \tag{10}$$

$$\mathcal{L}_v(\tilde{\mathbf{b}}^\sharp) = -\,2\,\tilde{\mathbf{b}}^\sharp\,\mathbf{d}_p^\flat\,\tilde{\mathbf{b}}^\sharp \tag{11}$$

of which we will make use in the following.

## CONSTITUTIVE EQUATIONS

For the elastic part of the deformation, hyperelasticity is assumed. In particular, the following elastic constitutive model is used

$$\boldsymbol{\tau}^{\sharp} = K \ln J \, \mathbf{g} + \mu J^{-\frac{2}{3}} \operatorname{dev} \tilde{\mathbf{b}}^{\sharp} \tag{12}$$

in which the volumetric and deviatoric responses are decoupled [5]. Here $\boldsymbol{\tau}^{\sharp}$ is the Kirchhoff stress tensor, $J = \det \mathcal{F}$ and $K$ and $\mu$ are the bulk and shear modulus, respectively.

The inelastic constitutive equations are first formulated on the intermediate configuration. There the deformation rate tensor $\hat{\mathbf{D}}_p^{\flat}$ (for definition see (10)) has to be specified, which under the assumption of inelastic incompressibility is a deviatoric tensor. Therefore, the flow rule is assumed to depend on the deviatoric stress tensor on the intermediate configuration and some scalar and tensorial state variables. Making use of equations (10) und (11) allows formulating a flow rule on the actual configuration in terms of $\mathcal{L}_{\boldsymbol{v}}(\tilde{\mathbf{b}}^{\sharp})$. The condition of inelastic incompressibility gives

$$\mathcal{L}_{\boldsymbol{v}}(\tilde{\mathbf{b}}^{\sharp}) : (\tilde{\mathbf{b}}^{\sharp})^{-1} = 0 \,, \tag{13}$$

which shows that $\mathcal{L}_{\boldsymbol{v}}(\tilde{\mathbf{b}}^{\sharp})$ is generally not a deviatoric tensor. Next, some approximations are introduced, which are based on the notion of small elastic strains, which is a reasonable assumption for metals if the deformation rate is not too high. Further, the flow rule can be split into a volumetric and a deviatoric part

$$\operatorname{dev}(\mathcal{L}_{\boldsymbol{v}}(\tilde{\mathbf{b}}^{\sharp})) = -2\, \dot{\varphi}\,(I\!I, s, \theta)\, \mathbf{n}^{\sharp} \,, \tag{14}$$

$$\operatorname{tr}(\mathcal{L}_{\boldsymbol{v}}(\tilde{\mathbf{b}}^{\sharp})) = -2\, \dot{\varphi}\,(I\!I, s, \theta)\, \tilde{\mathbf{b}}^{\sharp} : \mathbf{n}^{\flat} \,. \tag{15}$$

Here, the flow direction is given through

$$\mathbf{n}^{\sharp} = \frac{3}{2} \frac{\operatorname{dev} \boldsymbol{\tau}^{\sharp}}{I\!I} \tag{16}$$

and $I\!I$ is the second invariant of the stress deviator

$$I\!I = \sqrt{\frac{3}{2} (\operatorname{dev} \boldsymbol{\tau}^{\sharp}) : (\mathbf{g} \operatorname{dev} \boldsymbol{\tau}^{\sharp} \mathbf{g})} \,. \tag{17}$$

For simplicity, it is assumed that the inelastic constitutive model considered here has only one scalar internal variable $s$, for which an evolution equation exists

$$\dot{s} = h(I\!I, s, \theta)\ \dot{\varphi}\,(I\!I, s, \theta) - r(I\!I, s, \theta) \,. \tag{18}$$

The flow function $\dot{\varphi}$, the hardening and recovery functions $h$ and $r$, respectively, depend on the second invariant $I\!I$, the internal variable $s$ and the temperature $\theta$. Additional scalar or tensorial internal variables are possible.

Equations (14), (15) and (18) can be used as a frame for the formulation of inelastic constitutive models at finite deformations. Existing constitutive models for large deformations like Anand's model [3] can be modified to fit into this frame. Also, these equations can be used to extend constitutive models formulated for small inelastic strains to finite deformations.

## NUMERICAL IMPLEMENTATION

The inelastic constitutive equations have to be integrated numerically. Since the differential equations are numerically stiff, implicit integration algorithms have to be used for stability requirements. Here, a backward implicit method with automatic time step control is used. Since hyperelasticity is considered (12), the resulting numerical scheme is incrementally objective [6]. A hybrid, mixed finite element model has been considered for the implementation of the inelastic model [6]. The numerical intergration scheme has to be consistently linearized to obtain quadratic convergence of the incremental procedure. **Figure 1** shows the result of a 60% upset forging of a cylindrical billet.

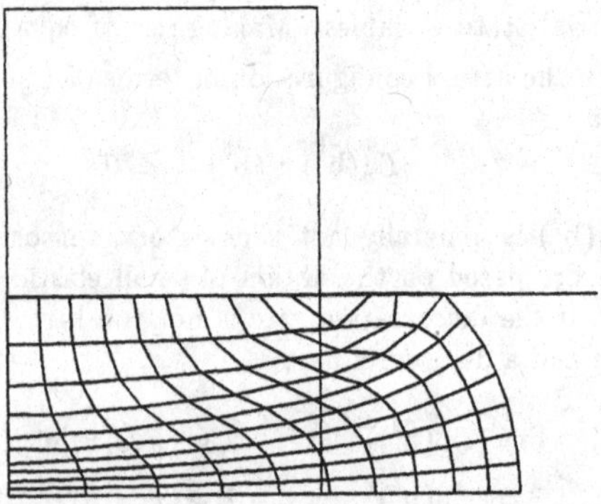

**Figure 1:** Axisymmetric upset forging example (60% height reduction)

## REFERENCES


[1] Bammann, D. J.; Aifantis, E. C.: A model for finite-deformation plasticity. *Acta Mechanica* **69**, 97–117 (1987).

[2] Bodner, S. R.; Partom, Y.: Constitutive Equations for elastic-viscoplastic strain-hardening materials. *Trans. ASME, J. Appl. Mech.* **42**, 385–389 (1975).

[3] Brown, S. B.; Kim, K. H.; Anand, L.: An internal variable constitutive model for hot working of metals. *Int. J. Plasticity* **5**, 95–130 (1989).

[4] Marsden, E.; Hughes, T. J. R.: *Mathematical foundations of elasticity.* Englewood Cliffs: Prentice Hall 1983.

[5] Simo, J. C.: A framework for finite strain elastoplasticity based on maximum plastic dissipation and the multiplicative decomposition: Part I. Continuum formulation. *Comp. Meths. Appl. Mech. Engrg.* **66**, 199–219 (1988).

[6] Simo, J. C.; Taylor, R. L.; Pister, K. S.: Variational and projection methods for the volume constraint in finite deformation elastoplasticity. *Comp. Meth. Appl. Mech. Engrg.* **51**, 177–208 (1985).

# THE ENDOCHRONIC PLASTIC BEHAVIOR OF BRASS UNDER TENSION-TORSION STRAIN PATH WITH CIRCULAR CORNERS

C.F.LEE
Dept. of Engineering Science, National Cheng-Kung University
Tainan, Taiwan, R.O.C.

## ABSTRACT

**In this paper, a method, based on simple loading data, is proposed to determine all associated material functions in the Endochronic plasticity without yield surface. Within data provided by simple shear stress-strain curve, the Kernel function $\rho(z)$ and material hardening function $f(\varsigma)$ of Brass are then determined and can be expressed:**

$$\rho(z) = \frac{2e^{-4.5z}}{z^{0.86}}$$

$$f(\varsigma) = 0.98 + 0.02e^{27.6\varsigma} \quad ; \quad \varsigma < 5.8 \times 10^{-2}$$

**Using the above functions, the incremental form of isotropic Endochronic plasticity is used to predict the plastic behavior of Brass during complex tension-torsional strain paths. The computational results agree very well with the experimental data. This show that the incremental form has excellent predictive capabilities. The theory can naturally predict that the (plastic) strain history has a gradually losing influence on the present material stress responses. The incremental form can predict exactly the stress relaxation during different tension-torsional strain paths and also predict the delay phenomenon of stress relaxation which is more severe as the radius of curvature of the strain trajectory increased. Consequently, the method proposed in the paper is practical and accurate.**

## INTRODUCTION

At 1978 Valanis [1] proposed an evolved form of the Endochronic theory of plasticity in which the plastic strain increment is used as a measure of the intrinsic time. In the theory, Valanis opens two avenues to investigate the plastic behavior of materials. The first avenue is to preserve the resulting yield surface and, hence, the material Kernel function can have a finite value at its origin. The second avenue is to shrink the size of resulting yield surface into zero and, hence, the material Kernel function possess a weak singularity at its origin. Along this line, methods of determining the Kernel function and computational methods associated with the theory become interest subjects of research which will be discussed in this paper.

For discussions, constitutive equations of the Endochronic Plasticity, under small strain, isotropic and plastically incompressible conditions, can be written as

$$\underset{\approx}{S} = 2\int_0^z \rho(z - z')\frac{\partial \underset{\sim}{e}^p}{\partial z'}dz' \tag{1}$$

$$\underset{\sim}{e}^p \equiv \underset{\sim}{e} - \underset{\sim}{S}/2\mu \tag{2}$$

$$dz = d\varsigma/f(\varsigma) \quad ; \quad d\varsigma \equiv \|d\underset{\sim}{e}^p\| \tag{3,4}$$

here the material function $f(\varsigma)$ represents hardening, softening an stable behavior. The material Kernel function $\rho(z)$ takes the following form

$$\rho(z) = \frac{\rho_o}{z^\alpha}G(z) \quad ; \quad G(0) = 1, \qquad 0 < \alpha < 1 \tag{5}$$

Determination of the material Kernel function $\rho(z)$

A systematic method of determining $\rho(z)$ from steady hysteresis loops of cyclic loading had been proposed by the author at 1987. The resulting $\rho(z)$ incorporated with the integration scheme of Gaussian Quadrature in eq.(1) can describe cyclic behavior of OFHC and 316 SS quite well [2]. However, most of experimental data available in public are under simple tension or shear. In such a case, the author proposes the following procedures: (1) As indicated everywhere (e.g. [1,2]), when the increment of $z$ after loading, unloading or reloading are small the resulting equation of eq.(1) predict a power form of stress-plastic strain relation whose exponent is equal to $1-\alpha$. Thus $\alpha$ can be determined. (2) From stress-(plastic) strain curve of simple loading, an approximate form of $\rho(z) = \sum_{i=1}^n A_i e^{-B_i z}$ can be used in eq.(1) to determine $A_i$ and $B_i$, when $z$ is away far enough from its origin. (3) Refer to eq.(5) a semi-log. plote of $z^\alpha \sum_{i=1}^n A_i e^{-B_i z}$ v.s. $z$ can be used to determine the form of $G(z)$. (4) Apply the resulting $\rho(z)$ into eq.(1) and then adjust $\rho_o$ to match with the data of simple loading. This procedure fixes the value of $\rho_o$.

An incremental form of the Endochronic Plasticity

In 1989 the author [3] proposed an incremental form of eq.(1) which is similar to the method proposed by Valanis and Fan. In the case, $\rho(z)$ is approximated by a finite sum of exponential decay functions and $\rho(o)$ is taken to be a very large number. After applying the Leibnitz's rule, eq.(1) can be resolved into an incremental form. Through several practice, the author finds that the computational precision and convergence depend strongly on the size of increment of $z$ and $\rho(o)$. To overcome this difficulty, the author proposes the following incremental form. From eq.(1) at $z = z_m$, one can write

$$\underset{\sim}{S}(z_m) = \underset{\sim}{S}(z_{m-1}, z_m) + \Delta\underset{\sim}{S}(z_{m-1}, z_m) \tag{6}$$

here

$$\Delta\underset{\sim}{S}(z_{m-1}, z_m) \equiv 2\int_{z_{m-1}}^{z_m} \rho(z_m - z')\frac{\partial \underset{\sim}{e}^p}{\partial z'}dz' \tag{7}$$

$$\underset{\sim}{S}(z_{m-1}, z_m) \equiv 2\int_0^{z_{m-1}} \rho(z_m - z')\frac{\partial \underset{\sim}{e}^p}{\partial z'}dz' \tag{8}$$

Let $\Delta z_m \equiv z_m - z_{m-1}$, which is small to allow one to make an approximation in eq.(7), i.e.,

$$\Delta\underset{\sim}{S}_m = 2\frac{\Delta \underset{\sim}{e}^p_m}{\Delta z_m}M(\Delta z_m) \tag{9}$$

where, by $y' \equiv z_m - z'$,

$$M(\Delta z_m) \equiv \int_0^{\Delta z_m} \rho(y')dy' \tag{10}$$

In eq.(5), if $z^*$ is large enough away from the singularity $z = 0$, then one can approximate $\rho(z)$, starting from $z^*$, by

$$\rho(z) \cong \sum_{r=1}^{n} C_r e^{-\alpha_r z} \quad , \quad z \geq z^* \tag{11}$$

In eq.(8), if $\Delta z_m \geq z^*$, then eq.(11) can be casted into eq.(8) to yield an excursive formula, i.e.,

$$\underset{\sim}{S}(z_{m-1}, z_m) = 2\sum_{r=1}^{n} \underset{\sim}{h_r}(z_{m-1})e^{-\alpha_r \Delta z_m} \tag{12}$$

here

$$\underset{\sim}{h_r}(z_{m-1}) = \underset{\sim}{h_r}(z_{m-2})e^{-\alpha_r \Delta z_{m-1}} + \frac{c_r}{\alpha_r}\frac{\Delta \underset{\sim}{e}^p_{m-1}}{\Delta z_{m-1}}(1 - e^{-\alpha_r \Delta z_{m-1}}) \tag{13}$$

and $\underset{\sim}{h_r}(0) = 0$. Denote the stress increment under the increment of strain $\Delta \underset{\sim}{e}_m$ as $\Delta \underset{\sim}{S}_m \equiv \underset{\sim}{S}(z_m) - \underset{\sim}{S}(z_{m-1})$. Hence an incremental form can be written as follow:

$$\Delta \underset{\sim}{S}_m = \frac{2\mu}{1+D^{-1}}\Delta \underset{\sim}{e}^p_m + \frac{2}{1+D}\sum_{r=1}^{n} \underset{\sim}{h_r}(z_{m-1})e^{-\alpha_r \Delta z_m} - \frac{1}{1+D}\underset{\sim}{S}(z_{m-1}) \tag{14}$$

here $D \equiv M(\Delta z_m)/\mu \Delta z_m$.

Eq. (14) does not contain the effects of singularity due to $\rho(z)$ and hence deserves a further research on its computational precision and convergence.

<u>Endochronic Plastic Behavior of Brass under Tension-Torsion Strain Path with Circular Corners.</u>

In this paper, the author uses Ohashi's experiments on 64 Brass under complex tension-torsion strain space with and without circular corners [4,5]. Following the procedures proposed in the determination of $\rho(z)$, the experimental data of simple shear is used to arrive

$$\rho(z) = 2e^{-4.5z}/z^{0.86} \tag{15}$$

and

$$f(\varsigma) = 0.98 + 0.02e^{27.6\varsigma} \quad ; \quad \varsigma < 5.8 \times 10^{-2} \tag{16}$$

In the application of eq.(14), $\rho(z)$ is approximated by 3 terms, i.e.,

$$\rho(z) \cong 21000e^{-12600z} + 1170e^{-535z} + 121e^{-37z} \quad ; \quad z^* = 5.0 \times 10^{-4} \tag{17}$$

The computational precision is quite well by using the convergence criterion on $\Delta z$ up to the 4 digits of significant number. The iterative numbers is about 15 in each increment of strain path.

The computational results by the incremental form of eq.(14) agree very well with the experimental data of strain paths with circular corners, e.g. Figs.1 & 2. These show that the incremental form can naturally predict that the (plastic) strain history has a gradually losing influence on the present material stress responses. The theory can also predict exactly the stress relaxation during different tension-torsion strain paths in which the delay phenomenon of stress relaxation is more severe as the radius of curvature of the strain trajectory increased. In conclusion, the incremental form and its associated computational method proposed in this

**of stress relaxation is more severe as the radius of curvature of the strain trajectory increased. In conclusion, the incremental form and its associated computational method proposed in this paper are practical and accurate, hence it deserves further research on its FEM applications.**

## REFERENCES


1. Valanis, K. C., Fundamental Consequence of a New Intrinsic Time Measure: Plasticity as a Limit of the Endochronic Theory, Arch. Mech., 1980, **32**, 171-191.
2. LEE, C. F., A Systematic Method of Determining Material Functions in the Endochronic Plasticity, J. Chinese Society of Mech. Eng., 1987, **8**, 419-430.
3. LEE, C. F., Endochronic Plastic Behavior of Perforated Thin Plate, Symposium of an Advances in Plasticity, ed. A. S. Khan and M. Tokuda, Pergamon Press, Oxford, 1989, pp.163-166.
4. Ohashi, Y., Effect of Curvatures of Strain Trajectory on the Plastic Behavior of Brass, J. Mech. Phys. Solids, 1981, **29**, 59-66.
5. Ohashi, Y., Effects of Complicated Deformation History on Inelastic Behavior of Metal, Memories of the Faculty of Engineering, Nagoya University, Japan, Vol.34, No.1, May 1982.


## FIGURES

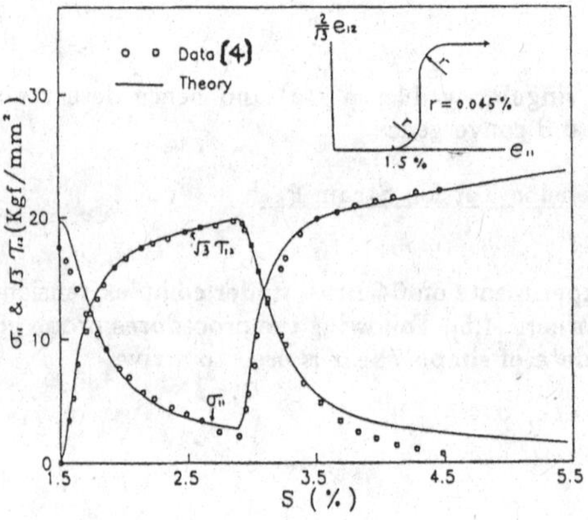


**Fig.1** Here $S = \frac{1}{2}\|\underset{\sim}{e}\|$ ; $r = 0.045\%$

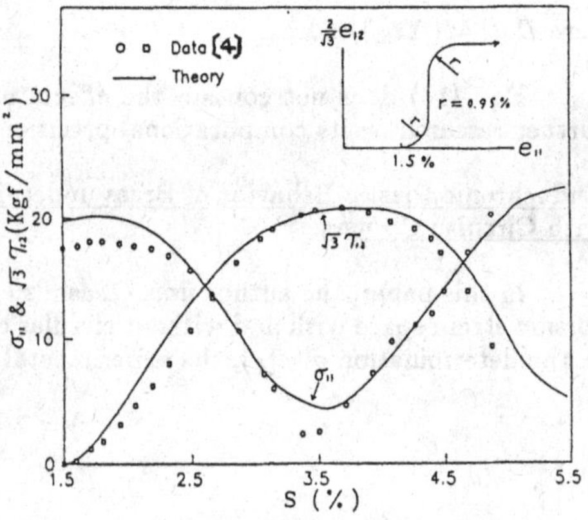


**Fig.2** Here $S = \frac{1}{2}\|\underset{\sim}{e}\|$ ; $r = 0.95\%$

# ON THE CHARACTERIZATION OF PLASTIC LOADING IN THE THEORY OF MATERIALS WITH ELASTIC RANGE

M. LUCCHESI* and P. PODIO-GUIDUGLI**

* Dipartimento di Scienze e Storia dell'Architettura, Università di Chieti
Viale Pindaro, 42 - 65100 Pescara, Italy

** Dipartimento di Ingegneria Civile Edile, Università di Roma 2,
Via E. Carnevale - 00173 Roma, Italy

## ABSTRACT

We show how a notion of plasticity index, *i.e.* an indicator of plastic loading situations, arises within our theory of materials with elastic range [1, 2]. This notion allows us to lay down some criteria for the choice of an evolution law for the back stress in the theory of v. Mises materials with hardening.

## BACKGROUND NOTIONS

The theory of materials with elastic range deals with elastic-plastic solids whose mechanical response to finite deformations is described by a frame-indifferent and isotropic constitutive functional delivering the Kirchhoff stress.

Deformation processes are called *histories*. Formally, a history is a continuous and piecewise continuously differentiable mapping F from [0,1] into the space $\mathrm{Lin}^+$ of second-order tensors with positive determinant; the value $F(\tau)$ of F at "time" $\tau$ is interpreted as the deformation gradient from a fixed reference configuration, at a fixed material point. For F a history and $\tau \in [0,1]$, the *$\tau$-section* of F is the history $F_\tau$ such that $F_\tau(\tau') = F(\tau\tau')$ for all $\tau' \in [0,1]$; a history G is said to be a *$\tau$-continuation* of F if there exists a $\tau$-section $G_\tau$ of G such that $G_\tau = F$.

Let $\mathfrak{D}$ be the set of all histories whose initial value F(0) belongs to Rot, the subspace of $\mathrm{Lin}^+$ consisting of all rotations, and let Sym be the space of all simmetric tensors. We denote by $\tilde{K} : \mathfrak{D} \rightarrow \mathrm{Sym}$ the *constitutive functional*, whose value $\tilde{K}(F)$ gives the Kirchhoff stress at the end of history F (the Kirchhoff stress is Cauchy stress times the determinant of the deformation gradient). For each $F \in \mathcal{D}$ and $\tau \in [0,1]$ fixed, we write $K(\tau)$ for $\tilde{K}(F_\tau)$ and interpret $K(\tau)$ as the Kirchhoff stress at time $\tau$ during history F.

For all $F \in \mathcal{D}$, let $\mathcal{E}(F)$ be the *elastic range* corresponding to F, namely, a closed, connected set $\mathcal{E}(F)$ of $Lin^+$, whose boundary is attainable from interior points only, and whose points are interpreted as the gradients of all deformations from the reference configuration to configurations which are elastically accessible from the current configuration [1]. An *elastic continuation* of F *up to* A is a $\tau$-continuation G of F such that $G(\tau') \in \mathcal{E}(F)$ for all $\tau' \in [\tau,1]$ and $G(1) = A$. We stipulate the constitutive hypothesis that each $F \in \mathcal{D}$ can be uniquely decomposed as $F(\tau) = V(\tau)P(\tau)$, with $V(\tau) \in Sym^+$, the space of all positive-definite elements of Sym, and with P an *unloaded history, i.e.*,a history such that $P(\tau) \in \mathcal{E}(F_\tau)$, $\det P(\tau) = 1$ and $\tilde{K}(G) = 0$ for each elastic continuation G of F up to $P(\tau)$. Moreover, we write

$$P(\tau) = R(\tau)S(\tau)\ , \ R(\tau) \in Rot\ , \ S(\tau) \in Sym^+\ , \tag{1}$$

for the right polar decomposition of P. Denoting time differentiation by a superposed dot, the symmetric and skew-symmetric parts of $\dot{F}F^{-1}$ are the *stretching* D and the *spin* W associated with history F; moreover, the symmetric part of $\dot{P}P^{-1}$ is the *plastic* stretching $D^P$, and $Z := \dot{R}R^T$.

We say that a history $F \in \mathcal{D}$ *satisfies the alternative* if, whenever $\tau \in ]0,1[$ and $F(\tau) \in \partial\mathcal{E}(F_\tau)$ then there exists $\varepsilon > 0$ such that

either

$$F(\bar{\tau}) \in \mathcal{E}(F_\tau) \text{ for each } \bar{\tau} \in [\tau, \tau+\varepsilon[\ , \tag{A$_1$}$$

or

$$F(\bar{\tau}) \notin \mathcal{E}(F_\tau) \text{ and } D^P(\bar{\tau}) \neq 0 \text{ for each } \bar{\tau} \in ]\tau, \tau+\varepsilon[\ . \tag{A$_2$}$$

The situations that may occur under $(A)_1$ are usually referred as *elastic unloading* and *neutral loading*. The situation under $(A)_2$ is known as *plastic loading*; during plastic loading we have $F(\bar{\tau}) \in \partial\mathcal{E}(F_{\bar{\tau}})$ for each $\bar{\tau} \in [\tau, \tau+\varepsilon[$ . We shall denote by $\mathcal{D}_a$ the collection of all histories in $\mathcal{D}$ that satisfy the alternative.

We assume that, for each $F \in \mathcal{D}$, the stress at time $\tau$ is completely determined by $V(\tau)$ through one and the same *structural mapping* $K^*$, a frame-indifferent, isotropic and symmetric-valued mapping defined over a neighborhood $\mathcal{N}$ of the identity tensor I in $Lin^+$, such that the restriction of $K^*$ to $\mathcal{N} \cap Sym^+$ is one-to-one and

$$K(\tau) = K^*(V(\tau))\ . \tag{2}$$

It then turns out that

$$\dot{K} + KW - WK = DK^*(V)[DV - VD^P]\ , \quad K(0) = 0\ , \tag{3}$$

with $DK^*$ the derivative of $K^*$ [2]. For $F \in \mathcal{D}$, $\mathcal{K}(F)$ is the *stress range* and $\mathcal{Y}(F) := \partial\mathcal{K}(F)$ the *yield surface* corresponding to F; we know from [2] that $F(\tau) \in \partial\mathcal{E}(F_\tau) \Leftrightarrow K(\tau) \in \mathcal{Y}(F_\tau)$. We write $K_0(\tau)$ for the traceless part of $K(\tau)$, and

$\mathrm{Sym}_0$ for the collection of all traceless symmetric tensors. We assume that (i) for all $F \in \mathfrak{D}$ there is a convex subset $\mathcal{K}_0(F)$ of $\mathrm{Sym}_0$ such that the stress range is the cylinder $\mathcal{K}(F) = \{\alpha I + \mathcal{K}_0(F),\ \alpha \in \mathbb{R}\}$ ; (ii) an associated flow rule holds, stating that $D^p(\tau)$, when different from zero, belongs to the normal cone of $\partial\mathcal{K}_0(F)$ at $K_0(\tau)$. We call the set $\mathcal{Y}_0(F) := \partial\mathcal{K}_0(F)$ the *yield locus*; a point $K_0 \in \mathcal{Y}_0(F)$ is *regular* if the outward unit normal $M(K_0)$ at $K_0$ is well defined; $\tilde{\mathcal{Y}}_0(F)$ denotes the set of all the regular points of $\mathcal{Y}_0(F)$.

## THE PLASTICITY INDEX

Let now $F \in \mathfrak{D}_a$ be fixed. For all $\tau \in\, ]0,1[$ and for all $\overline{\tau} \in [\tau,1[$ we write

$$F(\overline{\tau}) = \tilde{V}(\tau,\overline{\tau})\tilde{R}(\tau,\overline{\tau})S(\tau) \quad \text{and} \quad \tilde{Z}(\tau,\overline{\tau}) = \left(\frac{\partial}{\partial\overline{\tau}}\tilde{R}^T(\tau,\overline{\tau})\right)\tilde{R}^T(\tau,\overline{\tau}), \tag{4}$$

where $\tilde{R}(\tau,\overline{\tau})$ and $\tilde{V}(\tau,\overline{\tau})$ are the polar factors of $F(\overline{\tau})S(\tau)^{-1}$. We observe that $\tilde{R}(\tau,\tau) = R(\tau)$ and $\tilde{V}(\tau,\tau) = V(\tau)$; if $F(\tau) \in \partial\mathcal{E}(F_\tau)$ and situation $(A)_1$ of the alternative occurs, then $\tilde{Z}(\tau,\tau) = Z(\tau)$, but otherwise $\tilde{Z}(\tau,\tau) \neq Z(\tau)$; for simplicity, we write $\tilde{Z}(\tau)$ for $\tilde{Z}(\tau,\tau)$. For $F(\tau) \in \partial\mathcal{E}(F_\tau)$, let $K_0(\tau) \in \tilde{\mathcal{Y}}_0(F_\tau)$ and let $M(\tau)$ be the outward unit normal at $K_0(\tau)$; we call

$$\begin{aligned}\gamma(\tau) := M(\tau)\cdot\{&DK^*(V(\tau))[D(\tau)V(\tau)] + \\ &(W(\tau) - \tilde{Z}(\tau))K_0(\tau) - K_0(\tau)(W(\tau) - \tilde{Z}(\tau))\}\end{aligned} \tag{5}$$

the *plasticity index* of history $F$ at time $\tau$. This name is giustified by the following proposition [3].

**Proposition 1** *For each* $F \in \mathfrak{D}_a$ *and* $\tau \in\, ]0,1[$ *such that* $F(\tau) \in \partial\mathcal{E}(F_\tau)$ *and* $K_0(\tau) \in \tilde{\mathcal{Y}}_0(F_\tau)$,

$$(A)_1 \Rightarrow \gamma(\tau) \le 0\,,\quad \gamma(\tau) < 0 \Rightarrow (A)_1\,;\ (A)_2 \Rightarrow \gamma(\tau) \ge 0\,,\quad \gamma(\tau) > 0 \Rightarrow (A)_2\,. \tag{6}$$

A material with elastic range is said to be a v. Mises material if the structural mapping satisfies the Baker-Ericksen inequality and, for each history $F \in \mathfrak{D}$ and $\tau \in [0,1]$, the yield locus $\mathcal{Y}_0(F_\tau)$ is a sphere, whose center $C(\tau) \in \mathrm{Sym}_0$ is called the *back stress*: $\mathcal{Y}_0(F_\tau) = \{K_0 \in \mathrm{Sym}_0 \mid \|K_0 - C(\tau)\| = \rho(\tau)$.

**Proposition 2** *For each* $F \in \mathfrak{D}_a$ *and* $\tau \in\, ]0,1[$ *such that* $F(\tau) \in \partial\mathcal{E}(F_\tau)$ *and* $(A)_1$ *prevails,*

$$\dot{C}(\overline{\tau}) + C(\overline{\tau})Z(\overline{\tau}) - Z(\overline{\tau})C(\overline{\tau}) = 0\,. \tag{7}$$

Let

$$\tau \mapsto \zeta(\tau) := \int_0^\tau \|D^P(\tau')\|d\tau' \tag{8}$$

be the *Odqvist function*. In light of (8), the form of the associated flow rule is

$$D^P(\tau) = \dot{\zeta}(\tau)M(\tau)\,. \tag{9}$$

We accept the following hardening rules: (i) there is a material mapping $\Gamma$, such that, for $\dot{\zeta}(\tau) \neq 0$,

$$\dot{C}(\overline{\tau}) + C(\overline{\tau})\tilde{Z}(\overline{\tau}) - \tilde{Z}(\overline{\tau})C(\overline{\tau}) = \dot{\zeta}(\tau)\Gamma(\zeta(\tau), C(\tau), M(\tau)) \; ; \qquad (10)$$

(ii) there is a nondecreasing material function $\hat{\rho}$ such that $\rho(\tau) = \hat{\rho}(\zeta(\tau))$, for all $\tau \in [0,1]$. Let $\hat{\rho}' := d\hat{\rho}/d\zeta$; for each $F \in \mathfrak{D}_a$ and $\tau \in ]0,1[$ such that $F(\tau) \in \partial\mathcal{E}(F_\tau)$, we put

$$\xi(\tau) := \hat{\rho}'(\tau) + M(\tau)\cdot\{\Gamma(\zeta(\tau), C(\tau), M(\tau)) + DK^*(V(\tau))[V(\tau)M(\tau)]\} \; . \qquad (11)$$

**Proposition 3** *For each* $F \in \mathfrak{D}_a$ *and* $\tau \in ]0,1[$ *such that* $F(\tau) \in \partial\mathcal{E}(F_\tau)$,

$$(A)_1 \Rightarrow \dot{\zeta}(\tau) = 0 \quad and \quad (A)_2 \Rightarrow \dot{\zeta}(\tau) = \gamma(\tau)/\xi(\tau) \; . \qquad (12)$$

*Moreover,*

$$\gamma(\tau) \leq 0 \Rightarrow \dot{\zeta}(\tau) = 0 \quad and \quad \gamma(\tau) > 0 \Rightarrow \dot{\zeta}(\tau) = \gamma(\tau)/\xi(\tau) \; . \qquad (13)$$

With (13) and (9) we can, in principle, compute the plastic stretching during any history $F \in \mathfrak{D}_a$ for any v. Mises material with hardening.

Observe that, by (2) and the injectivity of $K^*$ on $\mathfrak{N} \cap Sym^+$, the mapping $\tau \mapsto V(\tau) = (K^*)^{-1}(K(\tau))$ is well defined. Consequently, (3) may be viewed as a first-order differential equation, parametrized by $D^p$, for the Kirchhoff stress mapping K: $\dot{K}(\tau) = \mathfrak{K}(K; D^p)$. Thus, if one could find reasonable hypotheses under which the nonlinear system (3), (5), (7), (9), (10), (11) and (13) be solvable for K at each given history $F \in \mathfrak{D}_a$, the constitutive behaviour of a v. Mises material with hardening would be completely accounted for. We are now in a position to justify the occurrence of $\tilde{Z}$ in formula (10). We have remarked that (i) $\tilde{Z}$ coincides with Z during any elastic continuation of a history (compare (10) and (7)). We offer here two other reasons for our selection: (ii) for each history one can compute $\tilde{Z}$ indipendently of the current plastic flow $\dot{\zeta}$; (iii) whenever $\gamma(\tau) > 0$, the plastic flow $\dot{\zeta}$, as computed on the basis of $(13)_2$ and the evolution equation (10) for C, is indeed positive-valued, as required by (8). We believe that whatever spin-like construct one would be willing to use in place of $\tilde{Z}$ in (10) must have properties (i) - (iii).

## REFERENCES


1. Lucchesi, M. and Podio-Guidugli, P., Materials with elastic range: a theory with a view toward applications. Part I. Arch. Rational Mech. Anal., 1988, 102, 23-43.
2. Lucchesi, M. and Podio-Guidugli, P., Materials with elastic range: a theory with a view toward applications. Part II. Arch. Rational Mech. Anal., 1990, 110, 9-42.
3. Lucchesi, M., Owen, D.R. and Podio-Guidugli, P., Materials with elastic range: a theory with a view toward applications. Part III. Approximate constitutive relations. To appear in Arch. Rational Mech. Anal., 1991.

# ROTATION PROBLEMS IN SIMPLE SHEAR DEFORMATION

KUNIO MIYAUCHI
Material Fabrication Laboratory
The Institute of Physical and Chemical Research
2-1 Hirosawa, Wako-Shi, Saitama-Ken, Japan 351-01

## ABSTRACT

According to Cauchy's theorem of moment equilibrium for all the arbitrary elements of deforming solid materials under loading, a balanced pair of shear stresses orthogonally crossing each other must always be present and is called pure shear stress. This assumption results in a very strange way of explanation that simple shear strain consists of pure shear strain and accompanying rotation because simple shear strain must be identical with pure shear strain. In a series of investigations into the dilemmas of the traditional assumption of shear stress and strain for equilibrium, this paper tries to make clear what is wrong in the treatment of the rotation of a deforming element. The limitation of present tensor analysis is also discussed.

## INTRODUCTION

Human beings have so far been very successful in the development of science and technology in almost all the fields. As for the mechanical physics of solids, there has been made a remarkable progress in the dynamics and mechanics of solids including rigid body mechanics. When we got into the problems of elastic and plastic deformation, it must have been done to very carefully examine what we can employ among the established laws of Newtonian mechanics. As the laws of force and moment equilibrium were undoubtedly valid in rigid body mechanics, A.-L. Cauchy simply applied this idea of equilibrium to the theorem of stress state in a deforming body in 1823 (1). Since then, his theorem has been playing a very important role in the simplification of complex formulas of deformation to lead us to a satisfactory solution in numerical analysis. The present FEM analysis based on his theorems of equilibrium is one of the marked technological achievements in elastic and plastic deformation.

As far as his theorem of moment equilibrium, however, is accepted, shear stress can be of pure shear only. We must here face the problem of two different shear strains, pure shear and simple shear, which must be generated by the single stress state, pure shear stress. It is widely known that simple shear strain is regarded as equal to pure shear strain accompanied by appropriate rotation. The author has already pointed out the dilemma of his theorem of moment equilibrium from the geometrical, phenomenological and mechanical points of view (2) - (6), while he has met with no reasonable objection or counter explanation against his doubts either.

This paper tries to re-examine and more clearly understand the character of rotation which is employed to make simple shear identical to pure shear.

## CHARACTER OF ROTATION IN PLASTIC DEFORMATION

In the past public discussions on simple shear deformation, they say that the rotation of an element which is given pure shear strain can make itself simple shear strain, while no additional deformation takes place, because it is like the rotation of a rigid body. Before we consider the present topic, it may be worthwhile reminding us of some fundamentals of rigid body mechanics.

The motion of a rigid body may be divided into three types; straight displacement, curved displacement like the orbital motion of the moon, and rotation like the earth's on its axis. They may appear at the constant or varying velocity. A free rigid body can stand still, or can continue a straight motion or rotation at a constant velocity, unless any other force exerts on the body. There must be present a certain external force whenever the rotation starts, changes its speed or stops. Even a free rigid body can move under such limited conditions as described above.

We know that the rotation of any deforming element takes place only when the exerting external forces change. Otherwise, the element keeps its standing position even under stationary loading, in the absence of creep. This is a proof that neither elastic nor plastic behavior of any deforming element is possible independently of the surrounding conditions. If someone proposes an imaginary rotation to superimpose pure shear strain on simple shear strain, it will be more meaningless or more unreal (4). As it is obvious that the deforming element is completely different from the free rigid body with respect to the surrounding conditions, the rotation of the deforming element cannot simply be adopted to prove the identity of pure shear and simple shear.

## POSSIBLE SIMPLE SHEAR DEFORMATION

There is another excuse for Cauchy's theorem that, although pure shear stress is the sole existence, simple shear strain can directly be caused by the mechanical anisotropy of deforming materials, as is well known for single crystals in the mode of symmetrical stress like the uniaxial tensile test. Simple shear strain is here exactly different from pure shear strain phenomenologically and geometrically.

In general, polycrystalline metals have some anisotropic nature to cause asymmetric deformation until the texture reaches to its stable orientation under the given stress mode. No rotation of the texture but symmetric deformation takes place in the state of stable orientation. We have had no evidence that the element subjected to pure shear strain rotates to produce simple shear strain in the ordinary mode of deformation. The author has shown experimentally that simple shear strain can directly be obtained in any directions in polycrystalline metal sheets by his planar simple shear test (2)(3). He has also proved that no rotation of a deforming element can freely occur even in the local absence of moment equilibrium caused by simple shear stress as shown in Fig. 1 (5).

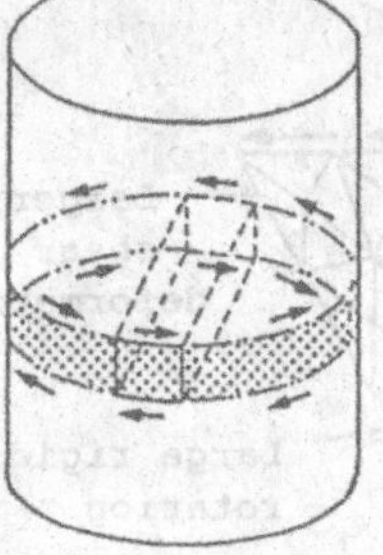
Total Moment equilibrium is
satisfied in any imaginary round
plate as well as in an imagined
inner bar.

Figure 1. An example ofthe local loss of moment equilibrium in the state of simple shear stress which satisfies the whole moment equilibrtium.

The combination of pure shear strain in different directions can make simple shear strain without any approximation as shown in Fig. 2. This evidence also proves the contradiction of the present theory of shear deformation.

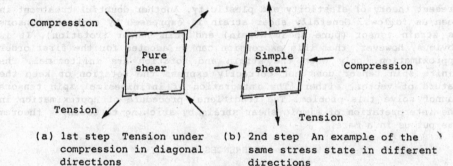

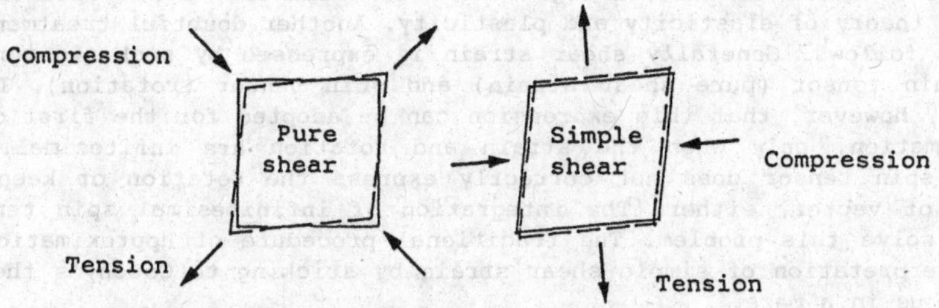
Compression
Pure
shear
Tension
Simple
shear
Compression
Tension

(a) 1st step Tension under compression in diagonal directions (b) 2nd step An example of the same stress state in different directions

Figure 2. An example of deforming process which produces simple shear strain without rigid body rotation.

## ROTATION OF DEFORMING ELEMENTS IN SIMPLE SHEAR

The torsion of a round bar or tube is one of the typical examples of simple shear deformation. W. A. Backofen dealt with the rotation of textures in torsion which was little seen with the inceasing strain from a plastic shear strain of 3.95 up to 5.25 and did not take place at all in reversed plastic torsion from 5.25 back to zero (7). There have been made some efforts to give a reasonable explanation of the relation between pure shear and simple shear with respect to the rotation of deforming elements and/or principal stress axis (8). No one seems to have been successful on this matter, yet.

As the outer geometry of a round bar does not change at all during torsion, it is natural to assume that the stress mode does not change though the stress level varies with the given strain. As long as no change of the stress mode is present, it is unreasonable to assume any rotation of principal stress axis (4)(6).

According the FEM analysis, we can see geometrical mesh changes in a specimen in simple shear and two different mesh rotations ; one is like rigid body rotation for the coordinate system corresponding to the external forces and another derived from the decomposition of simple shear strain as shown in Fig. 3.

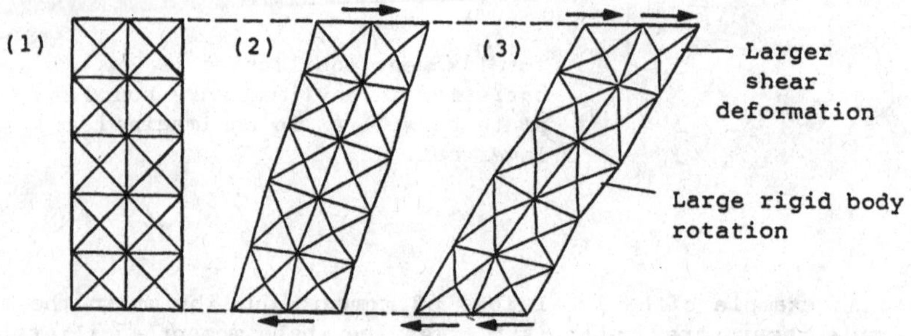


Fig. 3. Complex behavior of deforming elements in the macroscopic mode of simple shear under plane strain condition.

## LIMITATION OF STRAIN AND SPIN TENSORS

The rotation of a deforming element is carelessly dealt with in the present theory of elasticity and plasticity. Another doubtful treatment is seen as follows. Generally shear strain is expressed by such two tensors as strain tensor (pure shear strain) and spin tensor (rotation). It is obvious, however, that this expression can be adopted for the first order approximation, only when the strain and rotation are infitesimal. The finite spin tensor does not correctly express the rotation or keep the nature of vector, either. The integration of infinitesimal spin tensors cannot solve this problem. The traditional procedure of approximation in the interpretation of simple shear strain by sticking to Cauchy's theorem has put us in a maze.

## REFERENCES


1. Cauchy, A.-L., Recherches sur l'Equilibre et le Mouvement Interieur des Corps Solides ou l'Elastiques ou Non Elastiques. Bull. Soc. Philomath. Paris, 1823, 9-13 = Oeuvres (2) **2**, 300-4.
2. Miyauchi, K., A Proposal of a Planar Simple Shear Test in Sheet Metals. Sci. Papers Inst. Phys. Chem. Res., 1984, **78**, 27-40.
3. Miyauchi, K., Deforming Behavior of Sheet Metals in Planar Simple Shear Deformation. ibid., 1987, **81**, 27-38.
4. Miyauchi, K., On Simple Shear Deformation., ibid., pp. 57-67.
5. Miyauchi, K., Mechanical Response in Elastic and Plastic Deformation III (Rotation in Shear Deformation). Proc. 40th Japanese Joint Conf. Tech. Plasticity, Niihama, 1989, pp. 441-4.
6. Miyauchi, K., On the Equilibrium Equations in the Constitutive Laws, ed. F. Jianhong and S. Murakami, International Academic Publishers (Pergamon Press), Beijing, 1989, pp. 832-6
7. Backofen, W. A., The Torsion Texture of Copper. J. Metals, 1950, **188**, 1454-9.
8. Drucker, D. C., Appropriate Simple Idealizations for Finite Plasticity. PLASTICITY TODAY: Modelling, Methods and Applications, ed. A. Sawczuk and G. Bianchi, Elsevier Applied Science Publishers, London, 1983, pp. 47-59.

# MICROSCOPIC ASPECTS OF FINITELY DEFORMING INELASTIC MATERIALS

P. M. NAGHDI
Department of Mechanical Engineering
University of California at Berkeley
Berkeley, CA 94720, USA

## ABSTRACT

First a rapid review is provided of some important issues in finite plasticity and of a justifiable basis for adopting the strain-space formulation. This is followed by a brief account of the structure of the constitutive equations in both Lagrangian and Eulerian formulations of the theory, and of some results pertaining to constitutive restrictions derived from a physically plausible work assumption. The remainder of the lecture is devoted to microstructural aspects of finite plasticity. The approach taken is to introduce one or more additional vector-valued kinematical fields--called directors--intended to capture such physical properties as microcrack growth in brittle materials, and slip and related phenomena in ductile materials. These kinematical fields must necessarily be accompanied by appropriate balance laws, as well as constitutive equations. The character of the resulting dynamical field equations will, in general, differ from one class of microstructural phenomena to another.

## INTRODUCTION

Although the discussion that follows will focus mainly on the formulation of purely mechanical theories of inelastic behavior which incorporate microstructural effects, the first part of this lecture (as a point of reference) highlights the main features and advantages of the strain-space formulation of the existing continuum theory of elastic-plastic materials. An exposition of this development can be found in [1, Secs. 4-5], where a list of the original references on the subject can also be found. As remarked in [1, Sec. 8], once a fairly general structure of a continuum theory possessing sufficient flexibility is known, this can serve as a building block for the construction of a more general theory which could include microstructural effects. In other words, the point of view taken here is that (macroscopic) continuum theories should have a fairly broad base and a general outlook that allow for the inclusion of additional microscopic ingredients. The additional ingredients can be incorporated into the continuum theory in a variety of ways but an approach which appears natural to this writer is the use of director fields, together with associated balance laws of the director theory, as a basis for generating physically well-motivated additional kinematical (and kinetical) macroscopic variables which represent the main microscopic feature(s).

In order to indicate more specifically the nature of the additional ingredients mentioned in the preceding paragraph, in the next section we present an extended summary of the main features of a (macroscopic) dynamical theory of brittle materials in [2] that incorporates the growth of microcracks and their influence upon the material response, which is assumed to be perfectly elastic in the absence of cracks. The theory constructed in [2] represents an idealized characterization of inelastic behavior in the presence of crack growth which accounts for energy dissipation without explicit use of macroscopic plasticity and related effects, and the class of materials addressed by the theory are geological materials such as rocks and concrete, among others.

Due to space limitation, attention here is confined only to a theoretical development for microcrack growth in brittle materials, but the basic discussion here can be used with profit to motivate the construction of other macroscopic theories (for example, as in crystal plasticity) which must include microstructural effects.

## AN EXTENDED SUMMARY

We begin with a discussion of some background information on crack growth at the microscopic level and then motivate the choice of new independent variables, in addition to those of ordinary classical continuum mechanics. For this purpose, we need to consider two distinct scales of physical modeling, namely the *macroscopic* and *microscopic* scales. The former represents the scale of bulk or "smeared" response of the medium and the latter refers to the scale on which microcracks are observed.

By way of additional background, we note that the term "microcracking" refers to a particular type of inelastic behavior commonly found in brittle materials: Such materials contain a distribution of small cracks or voids which can be seen only at a microscopic level. When a part of the material (on the microscopic level) is subjected to surface forces, "damage" commences to accumulate at a macroscopic level and takes the form of the growth of the microcrack ranging from the cracking of a few isolated grains to the complete comminution of regions of the material. In the context of the development in [2], a crack on the microscale is defined as a void in a material which is much narrower in one direction, say in the direction of a unit vector $\mathbf{n}^*$, than in directions perpendicular to[1] $\mathbf{n}^*$. The area of the projection of the volume of a crack on the surface which is normal to $\mathbf{n}^*$ is taken as representing the predominant feature of the crack and will be simply referred to as the "crack area." Changes in crack area can then occur in two possible ways, i.e.,

[1] Throughout the paper, symbols with added asterisks such as $\mathbf{n}^*$ are used to designate microscopic variables.

(i) a crack may be stretched as the (microscopic) bounding surface which bounds the crack is stretched such that this surface is always material--we refer to such motion as *crack deformation;* and

(1)

(ii) a crack may grow in excess of that described in (i). We refer to such motions as *crack growth.* Crack growth is the consequence of the effects of actual fracture processes in which material bonds are destroyed.

It is assumed that in the absence of microcracks, the macroscopic and microscopic theories become coincident. Moreover, the microscopic effects arising from the microcrack growth are represented here (on the macroscopic scale) by two additional independent variables, together with additional conservation laws and constitutive equations. A macroscopic theory of this kind, in the absence of microcracks, must necessarily include classical nonlinear elasticity. As will be seen presently, one of the additional variables is a vector-valued variable $\bar{d}$ [see Eq. (3) below] whose magnitude is a measure of the area of the cracks in the microscopic description and whose direction represents the direction of the orientation of the cracks at the microscopic level. The other is a scalar-valued variable representing the crack density defined as the limiting value of the number of cracks per unit volume in a reference configuration at the microscopic level.

In the rest of this section, we motivate the development of the macroscopic theory based on the usual formulation of the theory of directed media. Thus, we recall that in the usual macroscopic description of materials, a body $\mathcal{B}$ bounded by a closed surface $\partial\mathcal{B}$ is regarded to consist of material points (or particles) X. Further, in the context of directed media, we suppose that each material point X is endowed with an additional kinematical vector field--called a *director* --whose magnitude (or length) remains unaltered under superposed rigid body motions. We define two sufficiently smooth vector functions $\chi$ and $\mathcal{D}$ which, respectively, assign the place **x** and director **d** to each material point in the current configuration $\kappa$ of $\mathcal{B}$ at time t, i.e.,

$$\mathbf{x} = \chi(\mathbf{X}, t) \quad , \quad \mathbf{d} = \mathcal{D}(\mathbf{D}, t) \ , \tag{2}$$

where **X** and **D** are the reference values of the material point and the director in the reference configuration $\kappa_o$ of $\mathcal{B}$. Since D depends on X, the right-hand side of $(2)_2$ can be expressed as a different function of **X** and t. The pair of functions $\{\chi, \mathcal{D}\}$ will be referred to as a *process.* The function $\chi$ is called a motion and an explicit physically relevant identification of $\mathcal{D}$ will be made presently (after Eq. (3) below).

In order to make suitable identification between the variables (2) in the macroscopic description and the corresponding variables on the microscopic scale, we recall from [2] that a microscopic material point in an arbitrary microscopic volume $\mathcal{S}^*$ of $\mathcal{B}$ is designated by X* and is identified by position vectors $\mathbf{x}^*$ and $\mathbf{X}^*$ respectively in the current and reference configurations $\boldsymbol{\kappa}^*$ and $\boldsymbol{\kappa}_o^*$ occupied by $\mathcal{S}^*$. The places $\mathbf{x}$ and $\mathbf{X}$ in $(1)_1$ are identified with the location of the center of mass of $\mathcal{S}^*$ in $\boldsymbol{\kappa}^*$ and $\boldsymbol{\kappa}_o^*$, respectively. [For additional discussion concerning the relationship between description of motion on the macroscopic and microscopic scales, see Sec. 2 and Appendix A of [2].] Due to the possible irreversible growth of cracks, an arbitrary process is not necessarily reversible on the microscopic level of scaling. This is because the microscopic bounding surfaces (including the boundary surfaces of the microcracks) of the body are altered by the fracture processes and hence the adjoining microscopic particles do not return to their original state upon reversal of the deformation. In this connection, we introduce the notion of a reversible process which in the present context can be defined on the microscopic scale as follows: Given that the body is assumed to be elastic in the absence of the cracks, a *reversible process* of the body is any process in which every microscopic particle contained in the body can be returned to its reference state simply by reversing the motion $\boldsymbol{\chi}$ of the body. It then follows that during such reversible processes, the cracks in the body can undergo only crack deformation as defined in (i) of (1) above. Thus, translated into macroscopic terms, a reversible process is any process for which the reversal of $\boldsymbol{\chi}$ in $(2)_1$ implies the reversal of $\mathcal{D}$ in $(2)_2$ to its reference value $\mathbf{D}$. For an arbitrary process, we can identify the irreversible part of the process by reversing $\boldsymbol{\chi}$ in $(2)_1$ and allowing $\mathbf{d}$ to take its value $\bar{\mathbf{d}}$ representing the irreversible crack growth (see (ii) of (1)). For convenience we denote the resulting configuration by $\bar{\boldsymbol{\kappa}}$, i.e., one in which $\mathbf{x} = \mathbf{X}$ but $\bar{\mathbf{d}}$ is not necessarily equal to $\mathbf{D}$. It should be noted here that the concepts of reversible and irreversible processes, as defined here, are applicable not only to the body as a whole but also to any subset of the body.

The deformation gradient $\mathbf{F}$ and its determinant J are defined by $\mathbf{F} = \frac{\partial \boldsymbol{\chi}}{\partial \mathbf{X}}$, $J = \det \mathbf{F}$ and the Lagrangian strain tensor is given by $\mathbf{E} = \frac{1}{2}(\mathbf{F}^T\mathbf{F} - \mathbf{I})$, where $\mathbf{F}^T$ is the transpose of $\mathbf{F}$ and $\mathbf{I}$ the identity tensor. We assume that $(2)_1$, but not $(2)_2$, is invertible for a fixed value of t so that the Jacobian of the transformation associated with $(2)_1$ does not vanish; and, for definiteness, we require $J > 0$. Also, the ordinary particle velocity $\mathbf{v}$ and director velocity $\mathbf{w}$ are defined by $\mathbf{v} = \dot{\mathbf{x}}$, $\mathbf{w} = \dot{\mathbf{d}}$, where a superposed dot denotes material time differentiation with respect to t holding $\mathbf{X}$ fixed.

We introduce now two additional kinematical variables. One of these is defined as

$$\bar{\mathbf{d}} = \bar{\mathcal{D}}(\mathbf{D}, t) = \frac{1}{J}\mathbf{F}^T\mathbf{d} = \frac{1}{J}\mathbf{F}^T\mathcal{D}(\mathbf{D}, t) \; , \tag{3}$$

and the other is the crack density

$$n = n(\mathbf{X}, t) \tag{4}$$

per unit volume in the reference configuration $\boldsymbol{\kappa}_o$. The definition (3), which is motivated on the basis of microscopic considerations, relates $\bar{\mathbf{d}}$ to the director $\mathbf{d}$ (and hence the function $\bar{\mathcal{D}}$

to $\mathcal{D}$ in $(2)_2$) and represents the composition of the inverse of the adjugate of the deformation gradient **F** and the director **d**. It is easily seen that the values of both d and $\bar{d}$ coincide with D in the reference configuration $\kappa_0$ where F = I. An explicit interpretation of $\bar{d}$ in (3) and hence also of **d** on the basis of microscopic considerations is given in [2]. However, we note that the basic kinematical variables here are (E, $\bar{d}$, n).

The appropriate balance laws here, aside from being in Lagrangian form, consist of those ordinarily adoped in classical continuum mechanics and two others, which are associated with the kinematical variables n and $\bar{d}$. These balance laws may be stated in words as:

rate of change of crack number = rate of production of new cracks (5)

and

rate of change of momentum associated with crack growth = all forces arising from (and maintaining) crack growth, (6)

where (6) represents a momentum-like balance law for the velocity $\bar{w} = \dot{\bar{d}}$.

There is considerable support in the literature on microcrack growth in brittle materials (see, for example, the review article by Kranz [3], especially his Figs. 2 and 4 and Table 1) for stipulating that

crack growth tends to occur along a surface which is normal to the particular principal direction of the strain E in a macroscopic theory for which the associated eigenvalue is maximum. (7)

Let $a^3$ refer to the principal direction of E which possess a maximum eigenvalue $\beta^{(3)}$. Recalling that $\bar{w}$ may be identified with the rate of crack growth and in line with the stipulation (7), we require that $\bar{w}$ be constrained to be parallel to $a^3$ by the constraint $\bar{w} = \alpha a^3$ , where $\alpha = \alpha(\mathbf{X}, t)$. Using this constraint condition, a class of constitutive equations is developed in [2] for microcrack growth which accounts for energy dissipation associated with the rate of increase of kinetic energy due to fracturing processes in a nonmaterial region arising from formation of new cracks and the growth of the preexisting ones. This is followed by further discussions of constitutive equations, including a development of the conditions for initiation of microcracking, together with illustrative examples for uniform extensive and compressive straining of a microcracking material [2, Secs. 5-6].

## REFERENCES

1. Naghdi, P.M., A critical review of the state of finite plasticity. J. Appl. Math. Phys. (ZAMP), 1990, **41**, 315-394.

2. Marshall, J.S., Naghdi, P.M. and Srinivasa, A.R., A macroscopic theory of microcrack growth in brittle materials. Phil. Trans. R. Soc. Lond., 1991, Ser. A (in press).

3. Kranz, R.L., Microcracks in rocks: a review. Tectonophysics, 1983, **100**, 449-480.

# A SURVEY OF SOME MATHEMATICAL DESCRIPTIONS OF DEFORMATIONS AND STATES FOR ELASTIC–PLASTIC MATERIALS

DAVID R. OWEN
Department of Mathematics
Carnegie Mellon University
Pittsburgh, PA 15213, USA

## INTRODUCTION

In this paper I wish to give a brief account of some recent mathematical studies of constitutive equations in plasticity, restricting my attention to constitutive equations for isothermal response and to studies that rest on the concept of a simple material, either in the original form [1] or in a newer form [2], both due to Noll. Essential ingredients in all of these approaches are a concept of deformation and a concept of state or, alternatively, of history of deformation.

## STUDIES THAT EMPLOY HISTORIES OF DEFORMATION

The idea that the present value of stress should depend upon the past history of deformation occurs in the earliest studies of viscoelastic materials, but it was incorporated explicitly for elastic–plastic materials only in the mid 1960's in Pipkin & Rivlin's study of rate–independent materials [3]. Specifically, Pipkin and Rivlin studied materials whose present stress response is a rate–independent functional of the past history of deformation gradient. They called such a material elastic–plastic if to each history of deformation gradient there is an elastic range, i.e., a set of deformation gradients such that the stress response functional is path–independent on continuations of the given history that remain in that set. Pipkin and Rivlin showed that the work done by the stresses on each history also possesses this property of local path–independence, and they introduced a derivative of the work functional that measures the hysteresis associated with deformations that change the elastic range. Their analysis shows that non–negativity of this derivative implies that the work done in small deformation cycles is non–negative and, for the case of infinitesimal deformations, that the yield surface in stress space is convex and the increments of plastic deformation are normal to the yield surface. Pipkin and Rivlin also took advantage of the fact that notions of material symmetry and frame–indifference have direct and simple mathematical forms in the context of materials with memory, and they determined restrictions that follow from rate–independence, together with material symmetry and frame–indifference.

Subsequent steps in the development of the ideas of Pipkin and Rivlin appeared in articles by Owen [4] and Šilhavý [5], in which the concept of elastic range was further refined, the concept of plastic or permanent deformation was placed on a precise footing, and transformation laws for both under changes in frame and changes in reference configuration were studied. It was shown also that this point of view provides a rational basis for two important ingredients in classical theories of plasticity: flow rules, i.e., constitutive equations governing the velocity–strain associated with plastic deformations, and decompositions of the total velocity–strain into elastic and plastic parts. Although Pipkin and Rivlin introduced explicitly a yield functional, no such functional was needed in the developments in the articles [4], [5]. The absence of a yield functional served there to clarify the logical structure of the framework under study but it was never advocated that plasticity could or should be developed entirely without yield functions or functionals.

An important step in the further development of plasticity based on stress functionals was a closing of the substantial gap between the theory presented in the articles [3]–[5] and the standard engineering theories of plasticity. The major part of this step was accomplished in two articles [6] by Lucchesi & Podio–Guidugli, where the introduction of a yield surface and a dissipation axiom, laid down only after careful refinement of the earlier theory and whose consequences were studied in full mathematical rigor, provided yield conditions, flow rules, and formulas for stress–rate that are essential ingredients in engineering theories. Thus, the work of Lucchesi and Podio–Guidugli provided a background theory for engineering theories of plasticity, and what then remained was to show explicitly how one can recover not only the standard constitutive equations of engineering plasticity, but also constitutive equations intermediate in some sense to the background theory [6] and to the standard engineering theories. This step was accomplished by Lucchesi, Owen & Podio–Guidugli in the recently completed article [7] in which they developed a systematic approximation scheme based on four nested classes of histories of deformation. Passage from each class to the next smaller is made by placing further restrictions on the elastic shear, the elastic shearing, the elastic stretching, the total stretching, or the total spin. For each class, approximate formulas for the stress–rate, yield condition, and rate of plastic deformation are derived, and the important classical equations of Prandtl–Reuss and of infinitesimal plasticity are recovered through approximation as one passes to the smallest two classes of histories. These developments as well as the implications of this methodology in non–isothermal contexts, currently under study by Lucchesi & Šilhavý, are the subject of a separate session of this Symposium and will not be elaborated upon here.

## STUDIES THAT EMPLOY EXPLICITLY THE CONCEPT OF STATE

In 1972, Noll introduced his "New Theory of Simple Materials" [2] in which he identified the concept of "state" as a natural replacement for the notion of "history of deformation gradient" in his first theory [1]. One of Noll's reasons at the time for changing his point of view was the absence of a fully developed treatment of plasticity within his first theory. The subsequent progress of the first theory in describing plasticity [5]–[7] notwithstanding, the classical theories of plasticity can be described directly without the use of histories of deformation, and this fact alone suggests that approaches without the use of histories should be pursued. Noll's new theory has lead to new treatments of plasticity, including some unpublished work of Noll, as well as articles by Del Piero [8], and Šilhavý & Kratochvíl [9]

Noll's new theory describes a material element in terms of the deformation processes that the element can undergo and the corresponding changes in state that can accompany such processes. The states of a material element need not have any special structure, but the deformations, or (intrinsic) configurations, encountered in each process are inner products on the tangent space of the body manifold at the point under consideration. (Although Noll's description is intrinsic, i.e., does not employ any placements of the body in a given Euclidean space, it is useful to think of configurations as values of the left Cauchy–Green strain tensor $F^TF$ computed in terms of a given reference placement.) The constitutive equations of the material element are obtained through specification of a state transformation function, whose values represent the final state obtained when the element in a given initial state undergoes a given process, and through specification of both a stress function, whose value at a state is the (intrinsic) stress for the element in that state, and a configuration function, whose value at a state is the configuration of the element in that state. Thus, specification of a material element in a given state determines, at least, the stress and configuration of the element; in general, the state carries additional information that is not assigned an explicit mathematical structure, a feature that renders Noll's theory quite general. Among the various features of Noll's theory, two particular ones stand out: the notion of material symmetry and the notion of relaxed state. Noll observes that each invertible linear mapping A of the tangent space of a material element induces natural transformations on the stresses, the configurations, and, hence, on the processes of the element. If the mapping A also determines a one–to–one correspondence $\iota_A$ from the state space of the element onto itself such that corresponding processes produce corresponding changes in state, and corresponding states produce corresponding stresses and configurations, then the linear mapping A is called a material automorphism of the element. Noll shows that the permutation of states $\iota_A$ is uniquely determined by the material automorphism A, and the resulting collection of material automorphisms forms a group, the symmetry group of the material element. Given a state of the material element, a subgroup of the symmetry group is obtained by collecting together all the material automorphisms A such that the permutation $\iota_A$ of states leaves the given state fixed, and the material isomorphisms with this property are called material autoomorphisms for the given state. Each material autoomorphism for a given state determines a change in reference placement that cannot be detected in processes of the element that start at that state, and, thus, the symmetry group for a given state corresponds to the notion of symmetry group introduced by Noll in his first theory of simple materials. A second important concept in Noll's second theory is the notion of relaxed state. Given any state, Noll assumes that one can freeze indefinitely the configuration at the value corresponding to that of the given state and recover in the limit of large times a limiting state called the relaxed state determined by the given state. Noll identifies elastic materials as those for which the states can be put in one–to–one correspondence with the set of configurations of the element, and he calls a material element semi–elastic if the relaxed states can be put in one–to–one correspondence with the set of configurations. Thus, for a semi–elastic material element, the process of freezing the element at a particular configuration starting at a given state results in a relaxed state that is independent of the given state. In this sense, a semi–elastic material has a fading memory and, therefore, cannot describe the permanent memory of an elastic–plastic material.

With this background from Noll's new theory of simple materials at hand, I can give some indication of recent progress in describing elastic–plastic materials within the framework of Noll's theory. Del Piero [8] described a class of material elements each of whose states is a pair consisting of configuration and stress. Elastic ranges are introduced as pieces of a given partition of the state space on which the stress is determined by the deformation. Transitions between elastic ranges are accomplished by first moving to the boundary of a given elastic range, and hence to the boundary of the state space, moving along the boundary and subsequently entering the desired elastic range. In a natural way, Del Piero obtains an evolution equation for the stress in the form of a differential equation in which the stress–rate is subject to a switching rule that is invoked when the current state reaches the boundary of the state space and would otherwise move out of the state space. He presents a qualitative study of the initial–value problem for this evolution equation. The result is a material element in which every state is a relaxed state, and the structure of the evolution law of the elements is compatible with many classical models of plasticity, including those with kinematical hardening.

The two–part study of plasticity by Šilhavý & Kratochvíl [9] is a comprehensive treatment of various ingredients essential to descriptions of elastic–plastic materials. Noteworthy in this work is the use of the notion of material automorphism to define what is meant by the plastic deformation of a material element in a given state. Moreover, Kratochvìl and Šilhavý introduce a notion of elastic range that requires that material response be locally path–independent only in the limit of rapid deformation processes. Their systematic and novel use of material automorphisms permits the natural identification of material symmetries associated with the elastic response of the element for processes that remain in elastic ranges. Thus, in a complete and natural way, they describe material elements whose states can be reached through elastic deformations followed by plastic deformations, and, if appropriate, through additional structural changes associated with work– or strain–hardening. Kratochvìl and Šilhavý discuss the evolution of plastic deformations and structural parameters, and obtain laws that retain the main features of flow rules in classical theories of plasticity.

## REFERENCES


1. Noll, W., A mathematical theory of the mechanical behavior of continuous media. *Arch. Rational Mech. Anal.*, 1958, 2, 197–226.
2. Noll, W., A new mathematical theory of simple materials. *Arch. Rational Mech. Anal.*, 1972, 48, 1–50.
3. Pipkin, A. C., and Rivlin, R. S., Mechanics of rate–independent materials. *ZAMP*, 1965, 16, 313–326.
4. Owen, D. R., A mechanical theory of materials with elastic range. *Arch. Rational Mech. Anal.*, 1968, 37, 85–110.
5. Šilhavý, M., On transformation laws for plastic deformations of materials with elastic range. *Arch. Rational Mech. Anal.* 1977, 63, 169–182.
6. Lucchesi, M., and Podio–Guidugli, P., Materials with elastic range: A theory with a view toward applications. Part I. *Arch. Rational Mech. Anal.*, 1988, 102, 23–43. Part II. *Arch. Rational Mech. Anal.*, 1990, 110, 9–42.
7. Lucchesi, M., Owen, D. R., and Podio–Guidugli, P., Materials with elastic range: A theory with a view toward applications. Part III. *Arch. Rational Mech. Anal.*, (to appear).
8. Del Piero, G., On the elastic–plastic material element. *Arch. Rational Mech. Anal.*, 1975, 59, 111–129.
9. Šilhavý, M., and Kratochvíl, J., A theory of inelastic behavior of materials. Part I. *Arch. Rational Mech. Anal.*, 1977, 65, 97–129; Part II. *Arch. Rational Mech. Anal.*, 1977, 65, 131–152.

# TENSILE DEFORMATION OF TI-6AL-4V ALLOY SHEET UNDER TRANSFORMATIONAL SUPERPLASTICITY CONDITION

KINJI SATO TOMIO KAJIKAWA KAZUHIDE UGAJIN
Science University of Tokyo Faculty of Science and Technology
Department of Mechanical Engineering
Noda-shi, CHIBA-KEN 278 JAPAN

## INTRODUCTION

Ti-6Al-4V alloy has excellent characteristics , but it has the problem of being difficult to work at room temperature. Superplastic flow is produced under low stress, does not decreace dislocation density, and does not induce elastic residual strain. Therefore it is expected that super-plasticity may be used to solve this problem . This report is studies tensile deformation of sheet under transformational superplasticity.

## EXPERIMENTAL PROCEDURE

Tensile deformation experiments were performed using Ti-6Al-4V alloy sheet under thermal cycling and loading. The experimental loading conditions were static, constant stress and constant strain rate. Strain begins to increase at the 800°C start of the $\alpha \rightarrow \beta$ phase trans-formation , and the strain increment becomes maximum near the $\beta$ phasetransformation point at 980°C. In this time when thermal amplitude and cycleare varied, their influences are considered.

## EXPERIMENTAL RESULTS AND DISCUSSION

### 1 TENSILE DEFORMATION UNDER STATIC LOADING CONDITION

### INFLUENCE OF INITIAL STRESS ON STRAIN

The relation between nominal strain and loading time under conditions of initial stress and thermal cycles is shown in Fig.1 . The figure shows that the values and increment of strain are very different for varying initial stress. Strain differs by a factor of two between 12MPa and 13MPa and the latter ruptures . Other cycles display similar phenomina . Although the initial stress is different, rupture is produced by a difference of

1MPa. When initial stress increases, the deformation rate increase and in early steps rupture occurs and values of strain decrease.

## INFLUENCE OF THERMAL CYCLES ON STRAIN

The period of thermal cycles is an important factor , as shown in the relation between last nominal strain and initial stress given in Fig.2 . Strain usually increase as cycles becomes longer and the converse is aiso true. This effect is shown above initial stress of 12MPa, which is taken to be large strain.

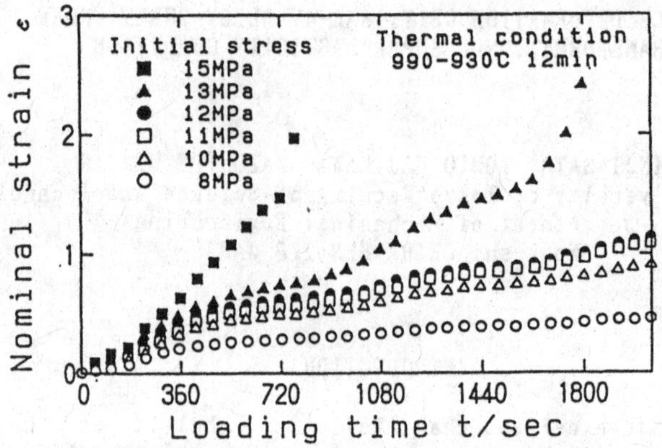


Figure 1. Relation between nominal strain ε and loading time t

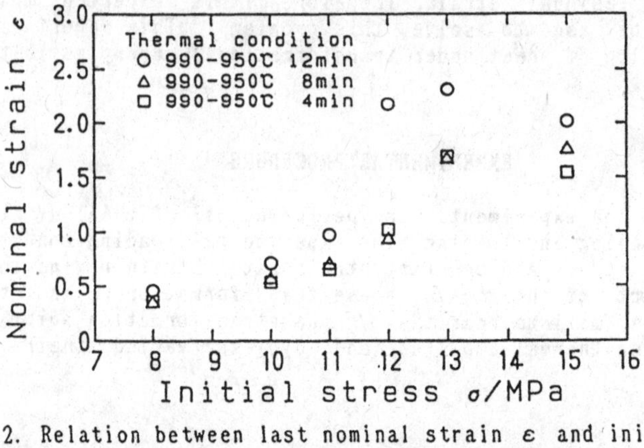


Figure 2. Relation between last nominal strain ε and initial stress σ

## 2 TENSILE DEFORMATION UNDER CONSTANT STRESS CONDITION

In conditions of constant tress the influence of amplitude change of thermal cycles on strain is considered . When thermal cycles of the same rate of heating and cooling are given under a constant stress of 18MPa, the relation between nominal strain and loading time is as shown in Fig.3. Unlike the static loading condition, the amount of strain increases as the period of thermal cycles decreases. A difference of approximately 20% between 8min and 4min is observed . It is considered that , because load decreases in response to strain, strain rate should be large initially to obtain larger strains.

When thermal cycles of different rates of heating and cooling are given under constant stress of 18MPa, the relation between nominal strain and loading time becomes that shown in Fig.4. Strain becomes large at heating and cooling time of 4min-2min. Similarly, strain decreases as the heating-cooling cycles change from 6min-4min, 4min-6min, 6min-2min.

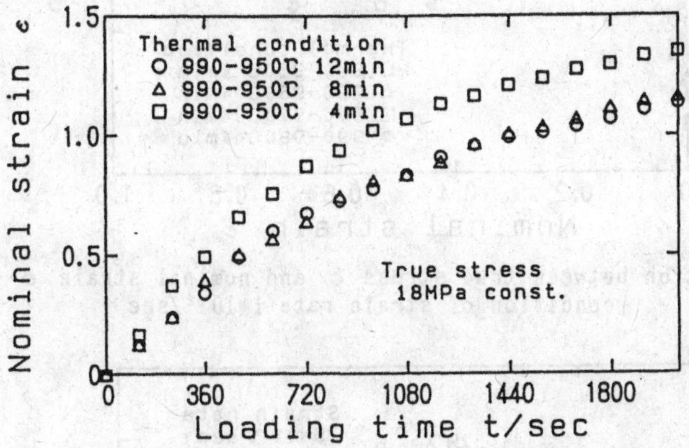


Figure 3. Relation between nominal strain $\varepsilon$ and loading time t under condition of constant stress 18MPa

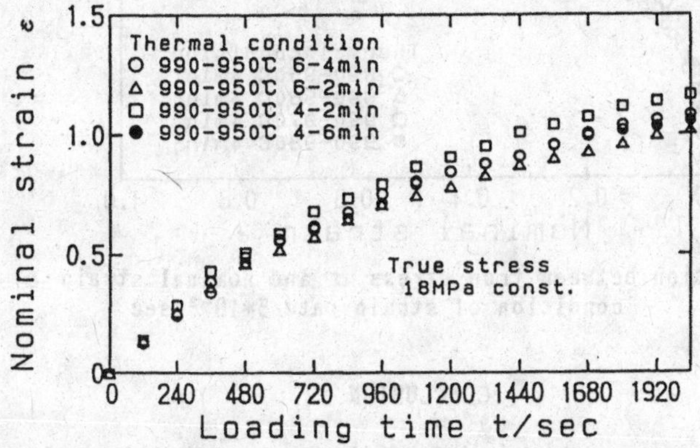


Figure 4. Relation between nominal strain $\varepsilon$ and loading time t under condition of constant stress 18MPa

## 3 CONSTANT STRAIN RATE CONDITION

Relations between true stress and nominal strain are shown in Fig.5 and 6 for a specified thermal cycle and strain rates of $1*10^{-2}sec^{-1}$, $5*10^{-3}sec^{-1}$, respectively. From the results it is noted that total elongation isaffected by the amount of strain at the ultimate tensile strength. Thatpoint occurs under large strain for conditions of wide amplitude andtensile strength shown in Fig.5 and 6 show reverse tendencies. It isconsidered that the point changes depending on whether the phase trans-formation is $\alpha \rightarrow \beta$ or $\beta \rightarrow \alpha$.

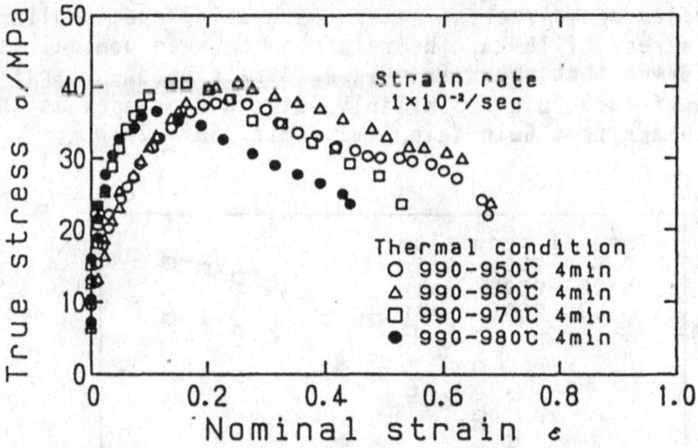


Figure 5. Relation between true stress $\sigma$ and nominal strain $\varepsilon$ under condition of strain rate $1*10^{-2}$/sec

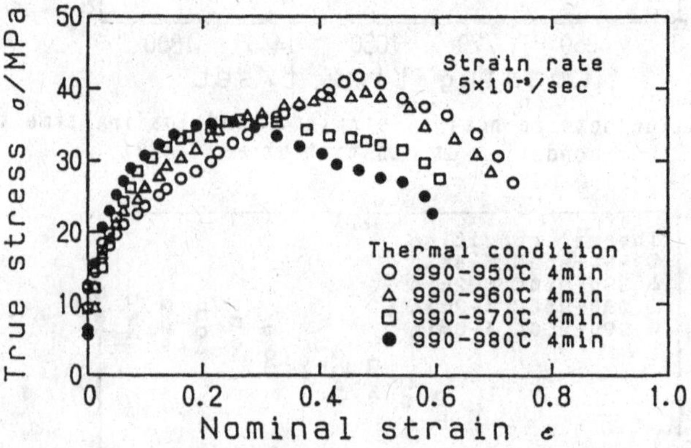


Figure 6. Relation between true stress $\sigma$ and nominal strain $\varepsilon$ under condition of strain rate $5*10^{-3}$/sec

## CONCLUSION

The following are related to deformation under transformational superplasticity.

(1) Under static load condition, in a thermal cycle , larger strain was obtained in heating process than in cooling process . The magnitude of strain was dependent on the initial stress. The maximum strain value was obtained for the initial stress of 13MPa.

(2) Under constant stress condition , the magnitude of strain was directly proportional to the magnitude of strain rate. The maximum strain value was obtained for a constant stress of 18-20MPa.

(3) For the case of constant strain rate, tensile strength was recorded in the case of largest thermal cycle amplitude over shortest period, total elongation in this case was largest as compared to other case.

# ON THE LOADING CONDITIONS AND THE DECOMPOSITION OF DEFORMATION

CHARALAMPOS TSAKMAKIS
Technische Hochschule Darmstadt, Institut für Mechanik
Hochschulstrasse 1, 6100 Darmstadt, Germany

## ABSTRACT

In this paper various definitions of loading criteria for elastic-plastic materials are investigated. A loading function is introduced, which allows equivalent stress space and strain space formulations of all loading conditions.

## INTRODUCTION

In an earlier paper [1], which presents an overview of basic concepts concerning the numerical implementation of rate-independent plasticity theories, HUGHES introduces the socalled "non-classical" loading and unloading conditions in a stress space formulation. On the other hand NAGHDI et al. (see e.g. [2,3]) have demonstrated the "advantages" defining loading conditions in a strain space formulation. In the present paper these various definitions are analyzed and compared with the classical ones.

In the case of small deformations the classical loading criteria are given by

$$\Lambda \gtreqless 0 \quad \text{with the loading function} \quad \Lambda = \frac{\partial \varphi}{\partial \mathbf{T}} \cdot \dot{\mathbf{T}} \quad .$$

Here $\varphi=\varphi(\mathbf{T},\mathbf{q})$ is the yield function, $\dot{\mathbf{T}}$ is the material time derivative of the stress tensor and $\mathbf{q}$ represents a set of internal variables, which are assumed to be constant during elastic processes.

## THE RIGID PLASTIC SOLID WITH ISOTROPIC HARDENING

We start to investigate finite deformations of rigid plastic solids and assume plastic incompressibility, i.e. $\det \mathbf{F} = 1$, where $\mathbf{F}$ is the deformation gradient. Linear isotropic hardening is introduced in the v.Mises yield function:

$$f(\mathbf{S},k) = \frac{1}{2}\,\mathbf{S}' \cdot \mathbf{S}' - \frac{1}{3}\,k^2 \,, \qquad k = k_0 + k_1\, s$$

( $\dot{s} = \sqrt{\frac{2}{3}\,\mathbf{D}\cdot\mathbf{D}}$ , $k_0, k_1$ : material constants , $\mathbf{S}'$: deviator of the weighted Cauchy stress tensor $\mathbf{S}$ ). Because of $\mathbf{S} = \mathbf{F}\tilde{\mathbf{T}}\mathbf{F}^T$, $\mathbf{C} = \mathbf{F}^T\mathbf{F}$ and $\mathbf{E} = (\mathbf{C}-\mathbf{1})/2$, ($\mathbf{1}$ : unit tensor of second order), we have

$$f(\mathbf{S},k) = \frac{1}{2}\left(\mathbf{C}\tilde{\mathbf{T}}\mathbf{C}\cdot\tilde{\mathbf{T}} - \frac{1}{3}(\tilde{\mathbf{T}}\cdot\mathbf{C})^2\right) - \frac{1}{3}\,k^2 \overset{\text{def}}{=} \tilde{f}(\tilde{\mathbf{T}},\mathbf{C},k) \;.$$

The associated flow rule is given by

$$\mathbf{D} = \lambda \frac{\partial f}{\partial \mathbf{S}} \quad \text{or, equivalently} \quad \dot{\mathbf{E}} = \lambda \frac{\partial \tilde{f}}{\partial \tilde{\mathbf{T}}} \quad \text{with} \quad \frac{\partial \tilde{f}}{\partial \tilde{\mathbf{T}}} = \mathbf{F}^T \frac{\partial f}{\partial \mathbf{S}}\,\mathbf{F} \;.$$

$\mathbf{D}$ is the symmetric part of the velocity gradient $\mathbf{L}$. The $\lambda$-factor is determined from the consistency condition. For this model a usual generalization of the classical loading function is

$$\Lambda = \frac{\partial f}{\partial \tilde{\mathbf{T}}} \cdot \dot{\tilde{\mathbf{T}}} = \frac{\partial f}{\partial \mathbf{S}} \cdot \overset{\nabla}{\mathbf{S}} \,, \qquad \left( \overset{\nabla}{\mathbf{S}} = \dot{\mathbf{S}} - \mathbf{L}\mathbf{S} - \mathbf{S}\mathbf{L}^T \right) \;. \tag{1}$$

In the special case of uniaxial loading the matrices of $\mathbf{F}$ and $\mathbf{S}$ have diagonal form: $F_{11} = l/l_0 = \exp(s)$, $F_{22} = F_{33} = \exp(-s/2)$; $S_{11} = \sigma$, $S_{22} = S_{33} = 0$. Thus we obtain for this example $\Lambda = 2\sigma\dot{s}(k_1 - 2k_0 - 2k_1 s)/3$, and we observe that for $s \geq (k_1 - 2k_0)/2k_1$ the loading condition is not fullfilled, i.e. $\Lambda \leq 0$.

In a more general calculation NAGHDI et al. (see e.g. [2,3]) have shown, that for an elastic-plastic solid with arbitrary hardening properties loading criteria defined with eq. (1), fail to hold in general. Alternatively they proposed a corresponding definition of a loading function in strain space formulation as primary. However, the rigid plastic solid has no elastic range and no strain space formulation is possible. In the following, a modification of the classical loading conditions is presented by which this difficulty can be avoided.

## A MODIFIED LOADING FUNCTION

We consider once more the classical yield function $\varphi = \varphi(\mathbf{T},\mathbf{q})$ and define a loading function as follows:

$$\Lambda = \left[\frac{\partial \varphi}{\partial \mathbf{T}} \cdot \dot{\mathbf{T}}\right]_{\mathbf{q} = \text{const.}} \qquad (2)$$

Because of the structure of the rate-independent plasticity theory $\mathbf{q}$ = const. implies that the plastic strain is constant and $\dot{\mathbf{T}}$ in the loading function $\Lambda$ is recognized to be the trial rate of the stress in [1].

By means of an appropriate elasticity law we are able to introduce a yield function g in the strain space formulation i.e.,

$$\varphi(\mathbf{T},\mathbf{q}) = \varphi(\mathbf{T}(\mathbf{E} - \mathbf{E}_P),\mathbf{q}) \overset{\text{def}}{=} g(\mathbf{E},\ldots) \quad .$$

Now, it is not difficult to see that a loading function in strain space formulation, i.e.,

$$\Lambda = \frac{\partial g}{\partial \mathbf{E}} \cdot \dot{\mathbf{E}} ,$$

where all internal variables in $\Lambda$ are held constant, is equivalent to the loading function in the stress space formulation defined by eq. (2). Thus, we see obvious relations between the definitions of HUGHES [1] and NAGHDI et al. [2,3] including the definition of this paper.

For the uniaxial loading of the rigid plastic solid considered above the modified loading condition (see eq. (2)) yields

$$\Lambda = \frac{\partial \tilde{f}}{\partial \tilde{\mathbf{T}}} \cdot \dot{\tilde{\mathbf{T}}} \Big|_{\mathbf{E} = \mathbf{E}_P = \text{const.}} = \frac{\partial f}{\partial \mathbf{S}} \cdot \overset{\nabla}{\mathbf{S}} \Big|_{\mathbf{E} = \mathbf{E}_P = \text{const.}} = \frac{2}{3} \sigma \dot{\sigma} \quad 0 \; .$$

Therefore the loading condition is now satisfied, in contrast to the previous situation.

## THE DECOMPOSITION OF THE DEFORMATION AND RELATED LOADING CONDITIONS

The modification of the classical loading function, defined in this paper, can be incorporated in a more general concept to represent elastic-plastic material properties. The analytical and geometrical motivation for this concept is based on stress and strain measures, which are dual to each other in the sense of the stress power and the complementary stress power [4].

In particular there are two different families of strain tensors with corresponding dual stresses. Within each particular family the multiplicative decomposition of the deformation gradient into elastic and plastic parts implies additive decompositions of the corresponding strain and strain rate tensors. On the basis of the associated dual stress rates the modified loading criteria apply to both families of dual stress and strain measures, in such a way that full "invariance requirements" (see e.g. [5]) are satisfied.

## REFERENCES


1. Hughes, T.J.R., Numerical implementation of constitutive models: Rate-independent deviatoric plasticity. Workshop on Theoretical Foundations for Large Scale Computations of Nonlinear Material Behavior, Northwestern University, Evanston, IL (1983), pp. 29-63.

2. Naghdi, P.M. and Trapp, J.A., The significance of formulating plasticity theory with reference to loading surfaces in strain space. Int.J. Engng Sci., 1975, 13, pp. 785-797.

3. Naghdi, P.M. and Casey, J., On the characterization of strain-hardening in plasticity. J. App. Mech., 1981, 48, pp. 285-295.

4. Haupt, P. and Tsakmakis, Ch., On the application of dual variables in continuum mechanics. Continuum Mech. Thermodyn., 1989, 1, pp. 165-196.

5. Casey, J. and Naghdi, P.M., A remark on the use of the decomposition $\mathbf{F} = \mathbf{F}_e \mathbf{F}_p$ in plasticity. ASME J. Appl. Mech., 1980, 47, pp. 672-675.

# ANALYSIS OF DAMAGE AND PLASTICITY FOR LARGE DEFORMATION WITH APPLICATION IN SIMPLE SHEAR

GEORGE Z. VOYIADJIS* and PETER I. KATTAN**
*Professor and **Research Associate
Department of Civil Engineering
Louisiana State University
Baton Rouge, LA 70803, USA

## ABSTRACT

A coupled theory of continuum damage mechanics and finite strain plasticity is formulated in the Eulerian reference system. A linear transformation is shown to exist between the effective deviatoric Cauchy stress tensor and the total Cauchy stress tensor. In addition, an effective elasto-plastic stiffness tensor is derived that includes the effects of damage. The problem of finite simple shear is investigated. It is noticed that the resulting differential equations are solved numerically using a Runge-Kutta-Verner fifth order and sixth order method. The results for the stress, backstress and damage variables are compared with a previous damage theory by the authors, as well as with an undamaged plasticity model.

## CONSTITUTIVE MODEL

The basic principles of continuum damage mechanics appeared for the first time in 1958 when Kachanov [1] introduced the concept of effective stress. In the simplest case of uniaxial tension, the effective stress is defined to be the stress in a fictitious deformed but undamaged sample that is mechanically equivalent (in terms of the total force acting on it) to the actual deformed damaged sample. In the general case of three-dimensional deformation and damage, the nature and definition of the effective stress become more complex. In the formulation that follows, the Eulerian reference system is used. A linear transformation between the Cauchy stress tensor σ and the effective Cauchy stress tensor σ is introduced as follows [2,3]:

$$\sigma_{ij} = M_{ijkl}\, \sigma_{kl} \tag{1}$$

where $M_{ijkl}$ are the components of the fourth-rank linear operator called the damage effect tensor that depends on the damage variable $\phi$. Rewriting the above equation for the deviatoric stress $\tau$, one obtains

$$\tau_{ij} = N_{ijkl}\, \tau_{kl} \tag{2}$$

where $N_{ijkl}$ are the components of a fourth-rank tensor given by

$$N_{ijkl} = M_{ijkl} - \frac{1}{3} M_{rrkl}\, \delta_{ij} \tag{3}$$

Equation (2) represents a linear transformation between the effective deviatoric Cauchy stress tensor $\tau$ and the Cauchy stress tensor $\sigma$. However, in this case the linear operator N is not simply the damage effect tensor M, but a linear function of M as seen by equation (3).

In the following derivation, it is assumed that the elastic strains are small compared to the plastic strains. In addition, it is assumed that an elastic strain energy function exists such that a linear relation can be used between the Cauchy stress tensor $\sigma$ and the engineering elastic strain tensor $\varepsilon'$. Using these assumptions, one obtains the desired linear relationship between the elastic strain tensor $\varepsilon'$ and its effective counterpart $\bar{\varepsilon}'$

$$\bar{\varepsilon}'_{kl} = M^{-T}_{klmn}\, \varepsilon'_{mn} \tag{4}$$

The constitutive model to be developed here is based on a von Mises type yield function $f(\tau,\alpha,\kappa,\phi)$ in the undamaged configuration that involves both isotropic and kinematic hardening through the evolution of the plastic work $\kappa$ and the backstress tensor $\alpha$, respectively (see reference [4] for more details). The corresponding yield function $f(\tau,\alpha,\kappa,0)$ in this configuration is given by

$$f = \frac{3}{2} (\bar{\tau}_{kl} - \bar{\alpha}_{kl})(\bar{\tau}_{kl} - \bar{\alpha}_{kl}) - \bar{\sigma}^2_o - c\bar{\kappa} = 0 \tag{5}$$

where $\sigma_o$ and c are material parameters denoting the uniaxial yield strength and isotropic hardening, respectively. The plastic flow in the undamaged configuration is described by the associated flow rule. The following nonlinear transformation equation for the plastic part of the spatial strain rate is then obtained:

$$\bar{d}''_{ij} = X_{ijkl}\, d''_{kl} + Z_{ij} \tag{6}$$

where the tensors X and Z are given by

$$X_{ijkl} = \frac{a_2}{a_1} M^{-1}_{ijkl} \tag{7}$$

$$Z_{ij} = 3\, \frac{a_3}{a_1} M^{-1}_{ijkl}\, N_{klrs}\, N_{mnrs}\, (\sigma_{mn} - \beta_{mn}) \tag{8}$$

One also obtains the following transformation equation for $d'$:

$$\bar{d}'_{ij} = \mathring{M}^{-T}_{ijmn}\, \varepsilon'_{mn} + M^{-T}_{ijkl}\, d'_{kl} \tag{9}$$

Finally, after several algebraic manipulations, one obtains the desired inelastic constitutive relation

$$\mathring{\sigma}_{ij} = \bar{D}_{ijkl}\, d_{kl} + \bar{G}_{ij} \tag{10}$$

where the effective elasto-plastic stiffness tensor D and the additional tensor G are given by

$$\bar{D}_{ijkl} = O^{-1}_{pqij}\, D_{pqmn}\, X_{mnkl} \tag{11}$$

$$\bar{G}_{ij} = O^{-1}_{pqij}\, D_{pqmn}\, Z_{mn} \tag{12}$$

and the fourth-rank tensor O is given by a lengthy expression (see reference [5] for details).

The effective elasto-plastic stiffness tensor is equation (11) is the stiffness tensor including the effects of damage and plastic deformation. It is derived in the actual current configuration of the deformed and damaged body. Equation (10) can now be used in finite element analyses.

## FINITE SIMPLE SHEAR

The problem of finite simple shear in the $x_1$-direction with finite plastic deformation and damage is solved using the proposed model. Substituting the relevant expressed for $d_{ij}$ of this problem into the constitutive equation (10) and simplifying the resulting equations, one obtains:

$$\mathring{\sigma}_{ij} = k\, \bar{D}_{ij12} + \bar{B}_{ij} \tag{13}$$

where k is a constant representing shearing strain.

In order to solve the resulting system of differential equation, the IMSL subroutine DIVPRK (a version of the IMSL routine IVPRK that uses double precision) is used. The method of solution used in this subroutine is a Runge-Kutta-Verner fifth order and sixth order method. For the numerical calculations, a cylindrical bar made of aluminum alloy 2024-T4 is used. The material properties for this problem are E = 10,600 ksi, $\nu = 0.25$ and $\sigma_o = 56$ ksi. The isotropic and kinematic hardening parameters c and b are, respectively, 138 ksi and 73 ksi. The shear strain $\gamma$ is incremented by controlling the time t with k = 1. The numerical tolerance used in the differential equation solver is taken to be 0.01. The damage parameters used are $\mu = 0.5$ and $\beta = 1 \times 10^{12}$.

The distributions of the backstress $\alpha_{11}$ is shown vs. $\gamma$ in Figure 1 for $\omega = 0.1$. It is clear that all the curves produced are monotonically increasing. However, it is noticed that the curves produced by the previous theory do not show the same degree of smoothness of the other curves. This can be mainly attributed to the unrealistically simplifying assumptions used in the previous damage model.

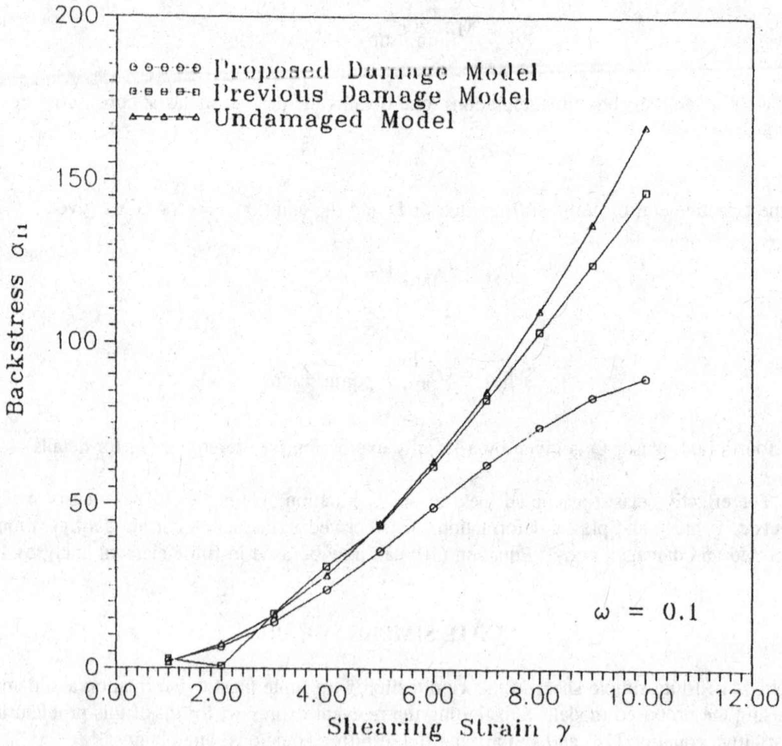


Figure 1. Variation of backstress $a_{11}$ vs. shearing strain $\gamma$ for $\omega = 0.1$.

## REFERENCES


1. Kachanov, L.M., On the creep fracture time. Izv Adad. Nauk USSR Otd. Tekh., 1958, **8**, 26-31.

2. Sidoroff, F., Description of anisotropic damage application to elasticity. in IUTAM Colloquium on Physical Nonlinearities in Structural Analysis, Springer-Verlag, Berlin, 1981, pp. 237-244.

3. Murakami, S., Notion of continuum damage mechanics and its application to anisotropic creep damage theory. J. Engineering Materials and Technology, 1983, **105**, 99-105.

4. Voyiadjis, G.Z. and Kattan, P.I., Eulerian constitute model for finite strain plasticity with anisotropic hardening. Mechanics of Materials Journal, 1989, **7**, 4, 279-293.

5. Voyiadjis, G.Z. and Kattan, P.I., A coupled theory of damage mechanics and finite strain elasto-plasticity. Part II: Damage and Finite Strain Plasticity. International Journal of Engineering Science, 1990, **28**, 6, 505-524.

# ON SUBSEQUENT YIELD SURFACES AFTER FINITE SHEAR PRE-STRAINING

XINWEI WANG, AKHTAR S. KHAN, AND HUIGENG YAN
The University of Oklahoma
Norman, Ok 73019, USA

## ABSTRACT

The subsequent yield surface for an annealed OFHC copper after finite shear pre-loading is experimentally investigated. It was found that when large offset strains or back extrapolation yield stresses were used, the corresponding subsequent yield surfaces were partially or completely outside the von Mises loading surface. On the other hand, the subsequent yield surface determined by the back extrapolation method was close to the Tresca loading surface.

## INTRODUCTION

Researches on metal plasticity have been focused on the yield surface and hardening law for many years [1]. Difficulties in defining the yield point for materials without sharp yield point are still one of the major concerns, since the behaviors of the yield surface, such as expansion, translation, distortion, cross-effect, and Bauschinger effect, strongly depend on the definition used in obtaining the yield surface.

In literature, three definitions were generally adopted to define a yield surface in multi-axial stress space. One is so-called the departure from the linearity. The definition is obvious to use but an exact value is very difficult to obtain experimentally. Thus, an alternative definition but essentially the same one is the very small offset strain definition (say, $\epsilon_{off} < 20\,\mu\epsilon$). The second definition is the so-called offset definition. A von Mises equivalent offset strain is generally adopted. However, for an anisotropic material, this equivalent strain is not a good measure. Thus, a third definition, called the backward extrapolation method, is recommended. The method requires that the deformation be large enough in order to extrapolate in the backward direction and intersect with the elastic line. In this paper, the offset and the back extrapolation methods are used. Eight tubular specimens were used to define the subsequent yield surface in $\sigma$-$\sqrt{3}\tau$ space or $\sigma$-$2\tau$ space. One specimen was used on one loading path, so that large incursions were made into the strain hardening range to make possible the use of large offset strain and back extrapolation methods.

## EXPERIMENTAL DETAILS

The material used in this study was OFHC copper. Thin-walled tubular specimens, approximately 1.2 in. (30.48 mm) long in test section, with an inside diameter of 0.375 in. (9.53 mm) and an outside diameter of 0.450 in. (11.43 mm), were used for all the experiments reported here. All specimens were annealed in an oxygen free atmosphere at 1100°F (593.3°C) for an hour and then furnace cooled to room temperature before performing a test.

The testing machine was a dead weight type with a capability of combined tension-torsion loadings. Details of the machine were described elsewhere [2]. A few modifications were made over the past in the experimental measurements, for example, data acquisition was completed by using a micro-computer.

The loading paths in the stress space ($\sigma$-$\sqrt{3}\tau$ space) for all tests are shown in Fig. 1a, where $\alpha$ is the angle between the $\sqrt{3}\tau$ axis and the second loading path. During the first stage, i.e., the torsional pre-loading and partial unloading stage, the angles of twist were measured by using an Angular Displacement Transducer (ADT). While in the second loading stage, the shear strains and the axial strains were measured by bonded strain gages (KFD-1-C1-11) to obtain higher sensitivity.

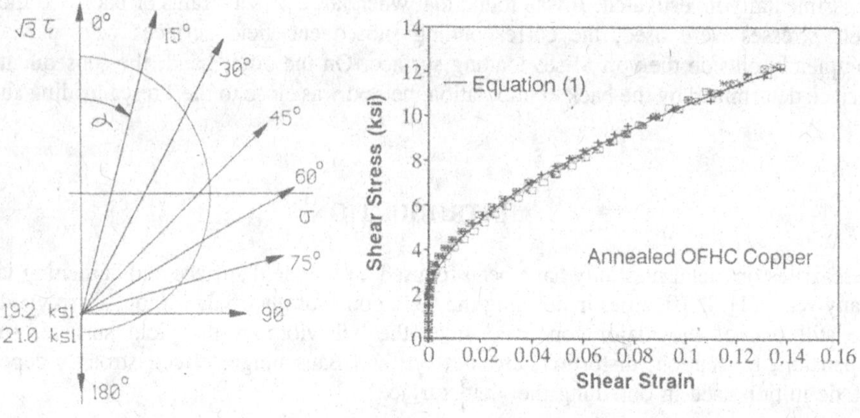


(a) (b)

Figure 1 (a) Various Subsequent Loading Paths in the $\sigma$-$\sqrt{3}\tau$ Space; (b) Shear Stress - Shear Strain Diagrams in the Preloading Stage

## EXPERIMENTAL RESULTS AND DISCUSSIONS

Figure 1b shows the shear stress Hxshear strain relationship for the pre-loading stage at an approximate loading rate of 17 psi per second (117 kpa per second). Although small specimen to specimen variations would be expected, all shear stress-shear strain curves, denoted by various symbols, are very close to each other. Thus it can be stated that every specimen used in the present investigation was fairly close to each other from material point of view. From Fig. 1b, it can be seen that the shear yield stress is approximately 1.75 ksi (12.07 kpa). The solid line in Fig. 1b represents Eq. 1, described in detail elsewhere [3],

$$\tau = 28.8\, e^{\left(\frac{0.02\gamma}{(\gamma+0.002)^{1.6}}\right)} \gamma^{0.5} \quad (ksi) \tag{1}$$

where $\gamma$ is the engineering shear strain and $\tau$ is the shear stress in unit of ksi.

The subsequent yield surface and its variation with the offset strains in the $\sigma$-$\sqrt{3}\tau$ and $\sigma$-$2\tau$ spaces are shown in Figs. 2 and 3, respectively. Symbols represent the experimentally determined data and lines are smoothed curves, each corresponding to a different level of plastic strain. Dotted lines denote either the von Mises or the Tresca loading surfaces. The offset von Mises and Tresca equivalent plastic strains, $\epsilon_{voff}$ and $\epsilon_{Toff}$, are computed by the following equation,

$$\epsilon_{voff} = \sqrt{\epsilon_p^2 + \frac{\gamma_p^2}{3}}\,, \qquad \epsilon_{Toff} = \sqrt{\epsilon_p^2 + \frac{4\gamma_p^2}{9}}\,, \tag{2}$$

where $\epsilon_p$ and $\gamma_p$ are the plastic axial and shear strains, respectively. These strains are based on the partially unloaded configuration.

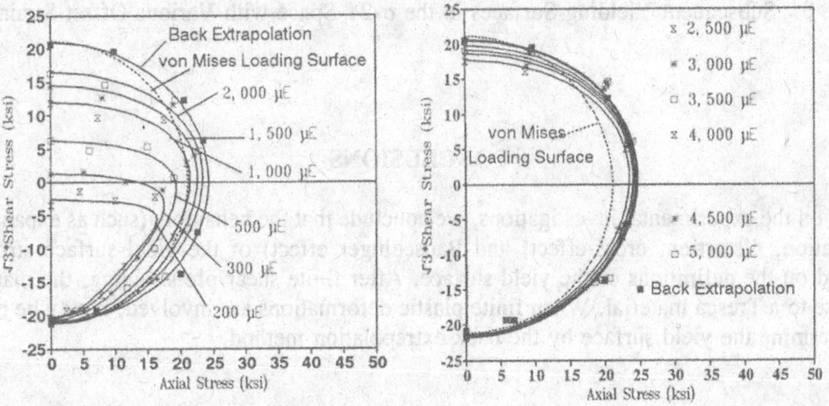


Figure 2. Subsequent Yielding Surfaces in the $\sigma$-$\sqrt{3}\tau$ Space with Various Offset Strains

As can be seen that the subsequent yield surfaces are translated in the preloading direction, expanded, and distorted with a rounded nose in the preloading direction. These results agree with the results in the existing literature. However, it is found that when large offset strains were used, the corresponding surfaces were no longer completely within the von Mises loading surface.

For comparisons, the yield surface determined by the backward extrapolation method is also included in Figs. 2 and 3, represented by the filled squares. It is interesting to note that the subsequent yield surface determined by the backward extrapolation method is almost completely outside the von Mises loading surface and the round nose in the preloading direction is no longer present. In this case, little Bauschinger effect and strong cross effect are observed. On the other hand, the subsequent yield surface determined by the backward extrapolation method is approximately the Tresca loading surface. In other words, the OFHC copper is a Tresca material after large torsional loading.

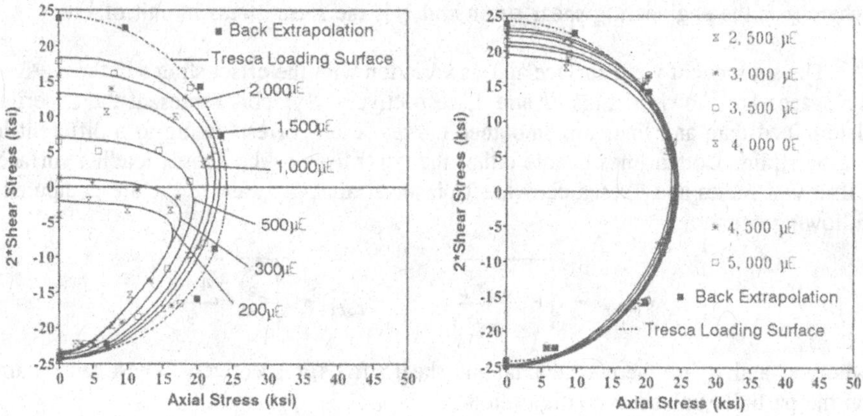


Figure 3. Subsequent Yielding Surfaces in the $\sigma$-$2\tau$ Space with Various Offset Strains

## CONCLUSIONS

Based on the experimental investigations, we conclude that the behaviors (such as expansion, translation, distortion, cross-effect, and Bauschinger effect) of the yield surface strongly depend on the definitions of the yield surface. After finite shear pre-straining, the material is close to a Tresca material. When finite plastic deformations are involved, it may be better to determine the yield surface by the back extrapolation method.

## REFERENCES

1. Ikegami, K., An Historical Perspective of the Experimental Study of Subsequent Yield Surfaces for Metals - Parts 1 & 2, J. Soc. Mat. Sci., 1975, **24**, 491-505 & 709-719.

2. Khan, A.S. and Parikh, Y., Large Deformation in Polycrystalline Copper Under Combined Tension-Torsion, Loading, Unloading and Reloading or Reverse Loading: A Study Of Two Incremental Theories Of Plasticity," Int. J. Plasticity, 1986, **2**, 379.

3. Wang, X. and Khan A.S., Hardening Functions For Finite Plastic Strains, to be submitted to J. Engng. Mater. Tech., 1991.

# ON OSCILLATORY SHEAR STRESS IN SIMPLE SHEAR

WEI WU and DIMITRIOS KOLYMBAS
Institute of Soil Mechanics and Rock Mechanics, Karlsruhe University, Germany

## ABSTRACT

Why the shear stress may oscillate for large simple shear deformation is the objective of the present paper. It is shown that the shear stress oscillates when the difference between the two normal stress components varies its sign periodically. If this sign variation is supressed in a constitutive equation, oscillatory shear stress can be avoided. Contrary to the results obtained by Dienes it is shown that the polar stress rate proposed by Green and McInnis fails to account for the rotation at large shear deformation properly.

## INTRODUCTION

In studying hypoelastic constitutive equations in simple shear deformation, Truesdell [1] found that the shear stress can be oscillatory and called this behaviour *instable.* Truesdell's paper failed to raise the attention until a retreatment of this problem followed by Dienes [2]. Dienes showed that the oscillatory shear stress can be avoided using the polar stress rate proposed by Green and McInnis [3]. Dienes paper gave rise to vivid discussions concerning various stress rates in large simple shear deformation. A general review of the literature can be found in [4]. In most of the papers the resulting differential equations are solved either analytically or numerically. Despite the many publications the mechanism of the oscillatory shear stress remains unclear and the question why the shear stress oscillates and how it can be avoided remains unanswered. In the present paper, we attempt to give a clear answer to these questions. The polar stress rate is also briefly discussed. Unless stated otherwise the notations are of Truesdell and Noll [1].

## ROTATIONAL PART OF JAUMANN STRESS RATE

Among various objective stress rate tensors, the stress rate tensor proposed by Jaumann, called Jaumann stress rate,

$$\overset{\circ}{\mathbf{T}} = \dot{\mathbf{T}} - \mathbf{WT} + \mathbf{TW} \tag{1}$$

is widely used in formulating incremental constitutive equations. It can be seen from Eq. (1) that the Cauchy stress rate $\dot{\mathbf{T}}$ consists of two parts, namely the constitutive part $\overset{\circ}{\mathbf{T}}$ and the rotational part $\mathbf{WT} - \mathbf{TW}$. Apparently, the constitutive part does not induce oscillatory response. We can leave the constitutive part aside and consider only the rotational part. For a simple shear deformation, $x_1 = X_1 + X_2 f$, $x_2 = X_2$, $x_3 = X_3$, it can be shown that the spin

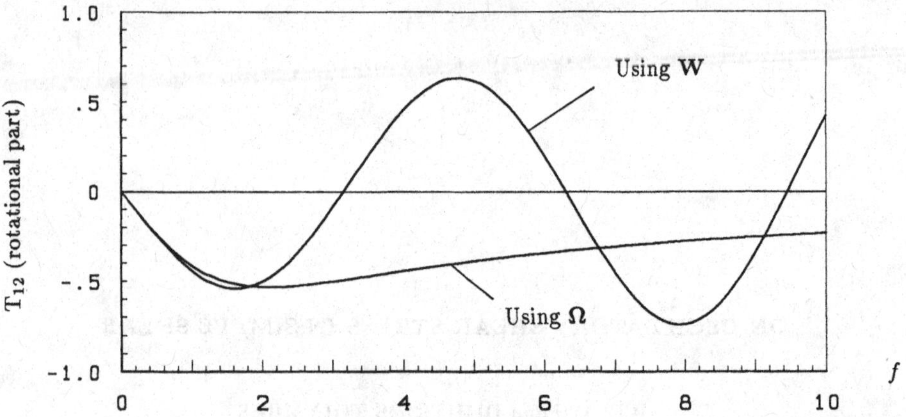


Figure 1: Shear response of the rotational part

tensor **W** can be written out in the following matrix form

$$\mathbf{W} = \frac{1}{2}\begin{pmatrix} 0 & \dot{f} & 0 \\ -\dot{f} & 0 & 0 \\ 0 & 0 & 0 \end{pmatrix} \tag{2}$$

It is then straightforward to get the rotational part in the following matrix form

$$\mathbf{TW} - \mathbf{WT} = \begin{pmatrix} -2\mathrm{T}_{12}\dot{f} & (\mathrm{T}_{11} - \mathrm{T}_{22})\dot{f} & 0 \\ (\mathrm{T}_{11} - \mathrm{T}_{22})\dot{f} & 2\mathrm{T}_{12}\dot{f} & 0 \\ 0 & 0 & 0 \end{pmatrix} \tag{3}$$

In deriving (3) the stresses with the index 3 are taken to be zero. If the shear stress oscillates, the rate of shear stress $\dot{\mathrm{T}}_{12}$ must vary its sign during the shear deformation, e.g. $\dot{\mathrm{T}}_{12} > 0$ for $t_i < t < t_{i+1}$; $\dot{\mathrm{T}}_{12} < 0$ for $t_{i+1} < t < t_{i+2}$ etc. From Eq. (3) it can be seen that this is only possible if in the same time intervals we have $\mathrm{T}_{11} > \mathrm{T}_{22}$ and $\mathrm{T}_{11} < \mathrm{T}_{22}$ and so on, since $\dot{f}$ is assumed to be always positive. The above reasoning can be explained with the following example in a more comprehensive manner. Suppose a simple shear is calculated with the following initial condition

$$\mathbf{T} = \begin{pmatrix} \mathrm{T}_{11} & 0 & 0 \\ 0 & \mathrm{T}_{22} & 0 \\ 0 & 0 & 0 \end{pmatrix} \tag{4}$$

with $\mathrm{T}_{11} > \mathrm{T}_{22}$. According to Eq. (3) we have at the initial stage of the calculation, $0 < t < t_1$, $\dot{\mathrm{T}}_{11} < 0$, $\dot{\mathrm{T}}_{22} < 0$ and $\dot{\mathrm{T}}_{12} > 0$. With continuing deformation $\mathrm{T}_{22}$ increases and $\mathrm{T}_{11}$ decreases. At a certain time instant $t = t_1$, we have $\mathrm{T}_{11} = \mathrm{T}_{22}$. Afterwards, for $t > t_1$, we have $\mathrm{T}_{11} < \mathrm{T}_{22}$ and according to Eq. (3) $\dot{\mathrm{T}}_{12} < 0$. This process will be repeated during further shear deformation and results in an oscillatory shear response for large shear deformation. Figure 1 shows the numerical result of Eq. (3).

## HYPOELASTIC CONSTITUTIVE EQUATIONS

From last section we know why oscillatory shear stress in large shear deformation may occur. It can be seen from Eq. (1) that the total response is obtained by adding the constitutive part to the rotational part. Whether oscillation of the shear stress occurs in the total response depends not only on the rotational part but also on the constitutive part. It was shown by Dienes that oscillatory shear stress is obtained with the following simple hypoelastic constitutive equation

$$\overset{\circ}{\mathbf{T}} = h\mathbf{D} \tag{5}$$

where $h$ is a scalar constant. It can be easily shown by writing out Eq. (5) in matrix form that the sign variation of $T_{11} - T_{22}$ is not influenced. As an example, let us consider the following constitutive law, which consists of one term from the representation theorem of a bilinear tensorial function of two symmetric second rank tensors,

$$\overset{\circ}{\mathbf{T}} = \mathbf{TD} + \mathbf{DT} \tag{6}$$

The matrix form of Eq. (6) can be shown to be

$$\mathbf{TD} + \mathbf{DT} = \begin{pmatrix} 2T_{12}\dot{f} & (T_{11} + T_{22})\dot{f} & 0 \\ (T_{11} + T_{22})\dot{f} & 2T_{12}\dot{f} & 0 \\ 0 & 0 & 0 \end{pmatrix} \tag{7}$$

The total response can be obtained by adding Eq. (7) into (1).

$$\dot{\mathbf{T}} = \mathbf{TD} + \mathbf{DT} + \mathbf{WT} - \mathbf{TW} = \begin{pmatrix} 4T_{12}\dot{f} & 2T_{22}\dot{f} & 0 \\ 2T_{22}\dot{f} & 0 & 0 \\ 0 & 0 & 0 \end{pmatrix} \tag{8}$$

It can be seen from Eq. (8) that the sign variation of $\dot{T}_{12}$ is supressed and no oscillatory shear stress occurs. It can be concluded that if we have a constitutive equation which contains (6), then no oscillatory shear stress can be obtained. This is the case for the hypoplastic constitutive equation in [5].

## POLAR STRESS RATE

In studying generalized hypoelastic constitutive equations a stress rate tensor different from Jaumann's was proposed by Green and McInnis [2]. The only difference from Jaumann stress rate tensor lies in that the skew–symmetric tensor, $\mathbf{\Omega}$, defined by

$$\mathbf{\Omega} = \dot{\mathbf{R}}\mathbf{R}^T \tag{9}$$

is used in the rotational part in Eq. (1) instead of the spin tensor **W**. The rotation tensor **R** in (9) is orthogonal and follows from polar decomposition of the deformation gradient. After Dienes a monotonic shear stress is obtained for Eq. (5) using $\mathbf{\Omega}$. This was at first rather puzzling for us, since the reason for the oscillatory response lies in the skew–symmetric structure of the spin tensor **W**. The tensor $\mathbf{\Omega}$ can be easily proved to be skew–symmetric. As a consequence, we expect also an oscillatory shear stress for $\mathbf{\Omega}$. That no oscillation of the shear stress acctually occurs becomes clear after $\mathbf{\Omega}$ is written out in the following matrix form.

$$\mathbf{\Omega} = \frac{2}{4 + f^2} \begin{pmatrix} 0 & \dot{f} & 0 \\ -\dot{f} & 0 & 0 \\ 0 & 0 & 0 \end{pmatrix} \tag{10}$$

Comparing Eq. (2) with (10) reveals that while the spin tensor **W** remains constant the skew–symmetric tensor $\mathbf{\Omega}$ becomes smaller for large shear deformation, e.g. $\mathbf{\Omega} \approx 0.01\mathbf{W}$ for $f = 10$. This tendency can be also observed in Figure 1. Theoretically, $\mathbf{\Omega}$ vanishes for infinitely large shear deformation, so that we have $\overset{\circ}{\mathbf{T}} \approx \dot{\mathbf{T}}$. Owing to the fact that oscillation of shear stress is avoided by using $\mathbf{\Omega}$, many invetigators advocate the use of the polar stress rate. According to our analyses, however, the polar stress rate should be discarded, since the rotation is not accounted for in $\mathbf{\Omega}$ for large shear deformation. In this case, it is no wonder that no oscillatory shear stress is obtained. It is worthwhile to point out that **W** depends only on the instantaneous configuration whereas $\Omega$ depends on both the instantaneous and the reference configuration.

## EXPERIMENTAL EVIDENCE

It is a widespread opinion that the oscillatory shear response is unplausible. Many attempts have been made to avoid the oscillatory shear response. In fact, the plausibility of such a response should be judged through experiments. Nevertheless, large homogeneous simple shear deformation can be hardly realized in the laboratory, since the specimen is prone to instabilities, e.g. in form of localized deformation. In [6] some experiments conducted on glas beads with a rotational shear apparatus were reported. In these experiments the shear moment oscillates after it arrives at a maximal value. Despite the inhomogeneity of deformation the experiments provide an insight into the behaviour at large shear deformation.

## CONCLUSIONS

- The mechanism of oscillatory shear stress in large simple shear is due to the sign variation of $T_{11} - T_{22}$. If this variation is supressed, then oscillation of shear stress can be avoided.
- The polar stress rate fails to predict the rotation at large shear deformation. It is therefore clear in this case why oscillation of the shear stress ceases to occur.
- Oscillation of the shear stress is not necessarily unplausible, because it can be observed in laboratory experiments.

## REFERENCES

[1] Truesdell, C. and W. Noll (1965), *The Nonlinear Field Theories of Mechanics*, in: S. Flügge, ed., *Encyclopedia of Physics* III/1, Springer, Berlin.

[2] Dienes, J. K. (1979), On the analysis of rotation and stress rate in deformed bodies, *Acta Mechanica*, **32**, 217–232

[3] Green, A. E. and B. C. McInnis (1967), Generalized hypo–elasticity, *Proc. Roy. Soc. Edinburgh*, **A 67**, Part III, 220–230

[4] Atluri S. N. (1984), On constitutive relations at finite strain, *Computer Methods in Applied Mechanics and Engineering*, **43**, 137–171

[5] Wu, W. and D. Kolymbas (1990), Numerical testing of the stability criterion for hypoplastic constitutive equations, *Mechanics of Materials*, **9**, 245–253

[6] Gebhard, H. (1982), Scherversuche an leicht verdichteten Schüttgütern unter besonderer Berücksichtigung des Verformungsverhaltens, *Fortschritt Berichte VDI*, Reihe 3, Nr. 68

# DUAL NON-EUCLIDEAN NORMS AND NORMALITY THEOREM IN PLASTICITY

WEI H. YANG
*The University of Michigan*
*Ann Arbor, Michigan 48109, USA*

## ABSTRACT

Functional analysis and the mathematical theory of plasticity both enjoyed flourishing but independent development in the 1950s. Yet little cross fertilization of ideas between mathematicians and engineers in those respective fields took place until almost three decades later. Recent advances in computational approach to optimization have provided a catalyst to combine certain ideas of functional analysis and plasticity, resulting in some new formulations and methods of solutions for more general and complex problems. In this paper, we present the duality of norms in non-Euclidean spaces in connection with a generalized Hölder inequality. These mathematical concepts offer a new explanation to the normality relation in plasticity deduced originally from a physical principle. Applications of these ideas to minimax theorems in plasticity and subsequent optimization problems are discussed.

## INTRODUCTION

In the theory of plasticity, the normality relation between the strain rate matrix and the gradient of the convex yield function of the stress matrix was derived from a physical argument namely non-negative dissipation of the plastic deformation. This constitutive model may seem to have closed the chapter on theoretical research concerning flow rules. Even before the discovery of this normality relation, Prandtl and Reuss in 1930 conjectured from their great insight a flow rule which not only foretold a part of the modern theory but was also verified experimentally. Now a mathematical argument applied to the inner product of the stress and strain rate matrices inspires a Hölder type inequality which independently infers normality. A general theorem is established for this inequality in terms of paired non-Euclidean vector and matrix norms. This generalized Hörder inequality provides the fundamental duality relation in plasticity. This is a significant addition to the constitutive model of plasticity. All known minimax theorems (duality) in plasticity are the consequences of the said inequality which can now apply to new materials and produce new solutions. Convexity of a yield function, normality of a flow rule and now duality of stress and strain rate matrices complete a mathematical model of plasticity.

The familiar Cauchy-Schwarz inequality in the Euclidean space $R^n$ can be stated for the inner product of vectors,

$$|\mathbf{x}^t\mathbf{y}| \le \|\mathbf{x}\|_2 \|\mathbf{y}\|_2, \tag{1}$$

where $\mathbf{x}$, $\mathbf{y}$ are vectors in $R^n$ and $\|\cdot\|_2$ denotes the Euclidean norm. The equality holds if

$$\mathbf{y} = \alpha\mathbf{x}, \qquad \alpha \in R \tag{2}$$

or the two vectors are co-linear. Inequality (1) is often used in the bounding process of mathematical analysis. A "sharp" bound that includes the equality case is vitally important in the field of functional analysis [1] and its applications. Obviously, a function that is bounded above has a finite supremum. The maximum of a function is contained in the range of a sharp upper bound function. Therefore a search for the least upper bound (supremum) will recover the maximum of the original function. This indirect method of finding the maximum of a function will fail if its upper bound function is not sharp. The sharpness condition (2) for the Cauchy-Schwarz inequality does not hold in a general real vector space $R^n$ when non-Euclidean norm measures are used.

A sharpness condition for a pair of dual vectors defined in non-Euclidean spaces does exist and arises from certain specific pairing other than the co-linearity of the two vectors. The pairing depends on the norm measures defined. This fact is well known in a sub-class of non-Euclidean spaces called $l_p$-spaces in which the family of Minkowski norms [2] for a vector $\mathbf{x} \in R^n$ is defined by

$$\|\mathbf{x}\|_p = \left(\sum_{i=1}^{n} |x_i|^p\right)^{\frac{1}{p}}, \qquad 1 \le p \le \infty \tag{3}$$

which, called a $p$-norm, includes the Euclidean norm ($p = 2$) as a special case. A pair of norms, $\|\cdot\|_p$ and $\|\cdot\|_q$, in this family are said to have a dual relation if $\frac{1}{p} + \frac{1}{q} = 1$, $1 \le p, q \le \infty$.

The Hölder inequality holds for any pair of such dual norms that

$$|\mathbf{x}^t\mathbf{y}| \le \|\mathbf{x}\|_p \|\mathbf{y}\|_q \tag{4}$$

for all vectors $\mathbf{x}$, $\mathbf{y} \in \mathbf{R}^n$. For the equality to hold, $\mathbf{x}$ and $\mathbf{y}$ must satisfy the relation [3] such that

$$\mathbf{y} \propto \nabla\|\mathbf{x}\|_p \quad \text{or} \quad \mathbf{x} \propto \nabla\|\mathbf{y}\|_q \tag{5}$$

where $\nabla$ is the gradient operator. When operated on a $C^1$ function of $n$ variavles, it produces a gradient vector in $R^n$. A gradient vector is normal to the level sets (contours) of the function. Hence, the sharpness condition (5) is also called the normality relation. To avoid the issue of differentiability of the $p$-norm functions, we exclude from our discussion the limiting cases, $p, q = 1, \infty$, even they are equally valid in the context of normality. The Cauchy-Schwarz inequality (1) is a special cases of (4) and its sharpness condition (2) is a special case of (5).

The normality relation between dual vectors is not restricted to the family of Minkowski norms. General non-Euclidean norms have a dual structure for pairing vectors that participate in an inequality similar to that of Hölder.

## GENERALIZED HÖLDER INEQUALITY

All norms share a common property as convex, non-negative and homogeneous functions of degree one. We shall consider a pair of norms, $\|\cdot\|_{(p)}$ and $\|\cdot\|_{(d)}$, which are said to have a dual relation implied by their subscripts, $(p)$ for primal and $(d)$ for dual. The parentheses on the subscripts prevent confusion with the Minkowski norms. One may arbitrarily define a primal norm as long as it satisfies the conditions: $(i)$ $\|\mathbf{x}\| > 0 \;\forall\, \mathbf{x} \neq 0$, $(i')$ $\|\mathbf{x}\| = 0$ if and only if $\mathbf{x} = 0$, $(ii)$ $\|\alpha\mathbf{x}\| = |\alpha|\|\mathbf{x}\|$, $(iii)$ $\|\mathbf{x}+\mathbf{y}\| \leq \|\mathbf{x}\| + \|\mathbf{y}\|$ for all $\alpha \in R$ and $\mathbf{x}, \mathbf{y} \in R^n$. The part $(i')$ is omitted for the definition of a seminorm.

Usually, the definition of a primal norm (or seminorm) arises naturally from a specific application. A dual norm must be matched to the primal norm so that the generalized Hölder inequality may be stated as a theorem.

**Theorem:** *For any two vectors $\mathbf{x}, \mathbf{y} \in R^n$ where $\|\mathbf{x}\|_{(p)}$ is given as a properly defined primal norm (or seminorm), there exists a dual norm (or seminorm) $\|\mathbf{y}\|_{(d)}$ such that the inequality*

$$|\mathbf{x}^t\mathbf{y}| \leq \|\mathbf{x}\|_{(p)}\|\mathbf{y}\|_{(d)} \tag{6}$$

*holds in general and the case of equality is attained when*

$$\mathbf{y} = \|\mathbf{y}\|_{(d)} \nabla \|\mathbf{x}\|_{(p)}. \tag{7}$$

*is chosen.*

The theorem, of course, covers the original Hölder inequality (4) as a special case. A proof of the theorem can be found in [3]. This theorem is fundamental to variational formulations of plasticity. A method of constructing dual norms from a given primal norm is also discussed in the reference quoted above including non-smooth norms.

## APPLICATION TO PLASTICITY

The concept of a norm is not restricted to vectors. There are matrix norms [4], function norms and operator norms [5]. We shall give an example of matrix norms for which the theorem above applies. In plasticity, the natural primal variable is the stress denoted by $\sigma \in R^{3\times 3}$, a real symmetric $3 \times 3$ matrix representing the state of force per unit area at a point in the material. Since there exists no real material that is infinitely strong, the strength of a material can be modeled by an inequality, $\|\sigma\| \leq \sigma_0$, bounding the matrix $\sigma$ where $\sigma_0$ is a constant, the strength of a material. The specific form of the norm is called the yield function derived from experimental data. The von Mises yield function [6] is among the best known. To shorten the discussion but still keep all the essentials relevant to the intended discussion in this paper, the von Mises yield function is presented in a subspace where the stress is a $2 \times 2$ symmetric matrix

$$\sigma = \begin{bmatrix} \sigma_{xx} & \sigma_{xy} \\ \sigma_{xy} & \sigma_{yy} \end{bmatrix} \tag{8}$$

which represents a state of plane stress in a sheet of material located in the $(x, y)$-plane of a Cartesian coordinate. The von Mises yield criterion,

$$\|\sigma\|_{\mathsf{v}} = \sqrt{\sigma_1^2 - \sigma_1\sigma_2 + \sigma_2^2} \leq \sigma_0 \tag{9}$$

is written here in terms of the eigenvalues $\sigma_1$ and $\sigma_2$ of the plane stress matrix in (8) where $\|\cdot\|_{\mathsf{v}}$ denotes the von Mises norm. It is a non-Euclidean norm outside the class of Minkowski norms. One can not find this norm in a mathematics book since it is derived from a physical origin. This definition is generalized, for broader applications, by introducing a parameter $\beta$ into (9). We define the $\beta$-norm yield criterion [7] as

$$\|\sigma\|_{(\beta)} = \sqrt{\sigma_1^2 - \beta\sigma_1\sigma_2 + \sigma_2^2} \leq \sigma_0 \quad -2 < \beta < 2 \tag{10}$$

where the range of $\beta$ between $-2$ and 2 is chosen so that the definition (10) satisfies the conditions of a norm. The $\beta$-norm function may also be expressed in terms of the components of $\sigma$.

The $\beta$-norm so defined represents a parametric family which may fit a wider range of plastic behavior of materials. It even includes the matrix norm ($\beta = 0$) known as the Frobenius norm. For other values of $\beta$ in the range, the norm is non-Euclidean. The case $\beta = 1$ is the von Mises norm. All members in the $\beta$-family are valid norms.

The dual variable in plasticity is the strain rate $\epsilon$, also a $2 \times 2$ symmetric matrix in the context of plane stress. The inner product $\sigma : \epsilon$ appears frequently in the science of mechanics. From a physical principle of non-negative dissipation, engineers had reached a conclusion that the plastic strain rate matrix is a constant multiple of the matrix $\nabla\|\sigma\|$ (a gradient vector arranged in the form of a matrix). This is known as the "normality theorem" in plasticity [8]. This conjecture and the theorem presented in this paper reach the same conclusion.

Using the generalized Hölder inequality (6), we can establish

$$|\sigma : \epsilon| \leq \|\sigma\|_{(\beta)} \|\epsilon\|_{(-\beta)} \tag{11}$$

for the $\beta$-norm family where

$$\|\epsilon\|_{(-\beta)} = \frac{2}{\sqrt{4 - \beta^2}} \sqrt{\epsilon_1^2 + \beta\epsilon_1\epsilon_2 + \epsilon_2^2} \tag{12}$$

where $\epsilon_1$ and $\epsilon_2$ are the eigenvalues of the strain rate matrix $\epsilon \in R^{2\times 2}$. The inequality (11) for the $\beta$-family is a sub-class of the generalized Hölder inequality. For other primal norms applied to plasticity. (eg. Tresca, Johenson, Rankine, Coulomb etc. [9]), special inequalities and their dual norms can be constructed. Since these inequalities are sharp, maximizing the left side and minimizing the right side lead to the same value. This is related to the classical upper and lower bound theorems in plasticity [6], but the minimax approach [10] is more mature in theory and more powerful in practice.

## A REMARK

One of the intellectual rewards from research is to discover the relation of seemingly different topics and to reach conclusions that broaden view points and offer new possibilities. Although the inequality theorems of non-Euclidean norms in mathematics and the normality theorem in plasticity are classical and well-known in their respective fields, their relation had not been recognized, at least not to its full extend. When the elegant mathematical and physical reasonings behind these theorems are brought side by side, a deeper insight can be added to each.

## REFERENCES


1. Goffman, C. and Pedrick, G., First Course in Functional analysis, Prentice-Hall 1965.
2. Birkhoff, G. and MacLane, S., A Survey of Modern Algebra, MacMillian 1953.
3. Yang, W. H., "On generalized Hölder inequality", Nonlinear Analysis, **16**, (5), p489-498, 1991.
4. Householder, A. S., On Norms of Vectors and Matrices, ORNL Report No. 1759, 1954.
5. Rall, L. B., Computational Solutions of Nonlinear Operator Equations, John-Wiley p13-14, 1969.
6. Hill, R., The Mathematical Theory of Plasticity, Oxford at Clarendon Press, 1950.
7. Yang, W. H., "A duality theorem for plastic plates", Acta Mechanica, **69**, p177-193, 1987.
8. Drucker, D. C., "A more fundamental approach to plastic stress strain relations", Proceedings, 1st U. S. Congr. Appl. Mech. p487-491, 1952.
9. Yang, W. H., "A useful theorem for constructing convex yield functions", J. Appl. Mech. **47**, p301-303, 1980.
10. Matthies, H., Strang, G. and Christiansen, E., "The saddle point of a differential program", Energy Methods in Finite Element Analysis (eds. Glowinski etc.) Wiley, 1979.

# DYNAMIC PLASTICITY AND VISCO-PLASTICITY

# EVALUATION OF CORRECT FLOW EQUATIONS FOR INTERNAL VARIABLE CONSTITUTIVE MODELS

STUART BROWN, VIVEK DAVE, and PRATYUSH KUMAR
Department of Materials Science and Engineering
Massachusetts Institute of Technology
77 Massachusetts Avenue, Room 8-106
Cambridge, Massachusetts 02139

## ABSTRACT

Although many constitutive models for rate-dependent deformation have been proposed, there is little fundamental justification for the particular functional forms used to characterize viscoplastic flow. Power law forms, although convenient analytically, do not necessarily represent physical processes, particularly given internal variable constitutive models that decouple evaluation of microstructure from constant structure flow behavior. Here, we present material data for two FCC systems, very high purity lead and 1145 aluminum, using constant structure experiments to probe directly the appropriate form of the flow equation at constant microstructures. The data represents conditions of moderate strain rates ($10^{-3}$ to $10^{0}$ sec$^{-1}$) and high to very high homologous temperatures ($0.5T_m < T < 0.99T_m$).

## INTRODUCTION

The unambiguous characterization of flow behavior is a particular requirement of internal variable constitutive models that simultaneously track the evolution of structure with rate-dependent flow. Although these internal variable models are employed extensively for predicting the evolution of deformation and structure, considerable debate remains about the proper material functions that comprise these models. Internal variable models of the form

$$\frac{d\epsilon_{ij}^{vp}}{dt} = \hat{f}_{ij}(\sigma_{kl}, T, s_1, \ldots, s_m) \qquad (1)$$

$$\frac{ds_n}{dt} = \hat{g}_n(\sigma_{kl}, T, s_1, \ldots, s_m), \quad 1 \leq n \leq m \qquad (2)$$

require very specific experiments to characterize unambiguously the material functions (or equations) $\hat{f}_{ij}(\ldots)$ and $\hat{g}_n(\ldots)$.

The following data describe constant structure experiments performed at moderate homologous temperatures on a commercial purity 1145 aluminum alloy and at high homologous temperatures on high purity lead. We then describe the data analysis performed with the experimental results to evaluate constant structure rate dependence.

## EXPERIMENTAL PROCEDURE

The constant structure experiments were performed with an Instron servohydraulic test machine. Strain feedback was provided for the aluminum experiments using a Capacitec high temperature extensometer designed so that the frequency response of the extensometer exceeded 250 Hertz. Data collection rates were normally 2500 samples per second to avoid aliasing difficulties.

A thermocouple was attached to each specimen to assure accurate temperature control. Temperature uniformity was held within 2 degrees Celsius. All tests were performed at constant true strain rates, the control signal to the servoelectric provided by a Keithley digital function generator.

## DATA ANALYSIS

Stress/strain data was obtained from the lead constant-strain-rate tests by first subtracting the effect of machine compliance and then digitally smoothing the data. Strain data for the aluminum was obtained directly using an extensometer. A local regression smoothing technique was used to reduce noise in the data for both the lead and the aluminum.

The constant structure rate dependence was determined under the assumption that first order constitutive behavior could be characterized by equations (1) and (2). The form of this evolution equation is commonly used in internal variable formulations [7,4, 2,5,6], frequently for analytic and computational convenience. The different structural processes can operate independently, however, and there is insufficient experimental data to settle unambiguously on another functional form.

Previous investigations indicate that static recovery rates at higher strain rates are a much lower that dynamic recovery rates, and can usually be neglected[1]. Similarly, the period of the oscillations of the stress/strain data at constant true strain rates indicate that the rate of change in structure due to "dynamic" recrystallization is significantly less than the hardening rate. The result is that over short transients at high strain rates, the primary mechanisms for structural evolution are dynamic recovery and hardening. Since both mechanisms require dislocation motion, we assume they are proportional to the imposed strain rate. The stress responses to different changes in strain rate therefore should be compared at an identical plastic strain offset from the point of the strain rate change. This is illustrated schematically in Figure 1.

Figure 2 presents the variation of the constant structure **power law** exponent determined from the strain rate change data as a function of plastic strain offset. Note that the power law rate dependence for the lead decreases from a value of approximately 25 to the value of 6 over a relatively small strain offset range. Similarly, the 1145 aluminum demonstrates a much higher rate dependence than is normally attributed to metals at these homologous temperatures (greater than 0.6 $T_m$).

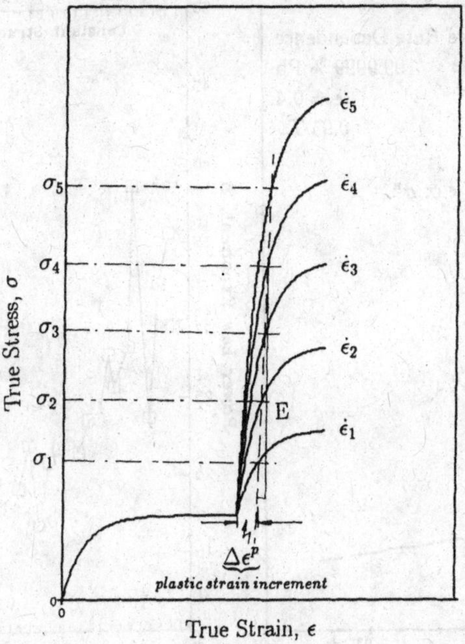


Figure 1: Schematic of plastic strain offset criterion used to select stress corresponding to strain rate.

## CONCLUSIONS

Internal variable constitutive models require greater attention to the appropriate flow equation that represents rate dependence at a given structure. Given the larger power law exponent, certainly a power law dependence with an exponent of 4 is incorrect.


## ACKNOWLEDGEMENT

Support for this work was provided by the U.S. National Science Foundation (Grant No. DMR-8806901).


## REFERENCES


[1] S B Brown, K H Kim, and L Anand. An internal variable constitutive model for the hot working of metals. *International Journal of Plasticity*, 5(2):95–130, 1989.

[2] K S Chan, U S Lindholm, S R Bodner, and K P Walker. A survey of unified constitutive theories. In *Nonlinear Constitutive Relations for High Temperature Application - 1984*, pages 1–23, NASA, 1985.

[3] U F Kocks, A S Argon, and M F Ashby. *Progress in Materials Science: Thermodynamics and Kinetics of Slip*. Pergamon, Oxford, New York, 1975.

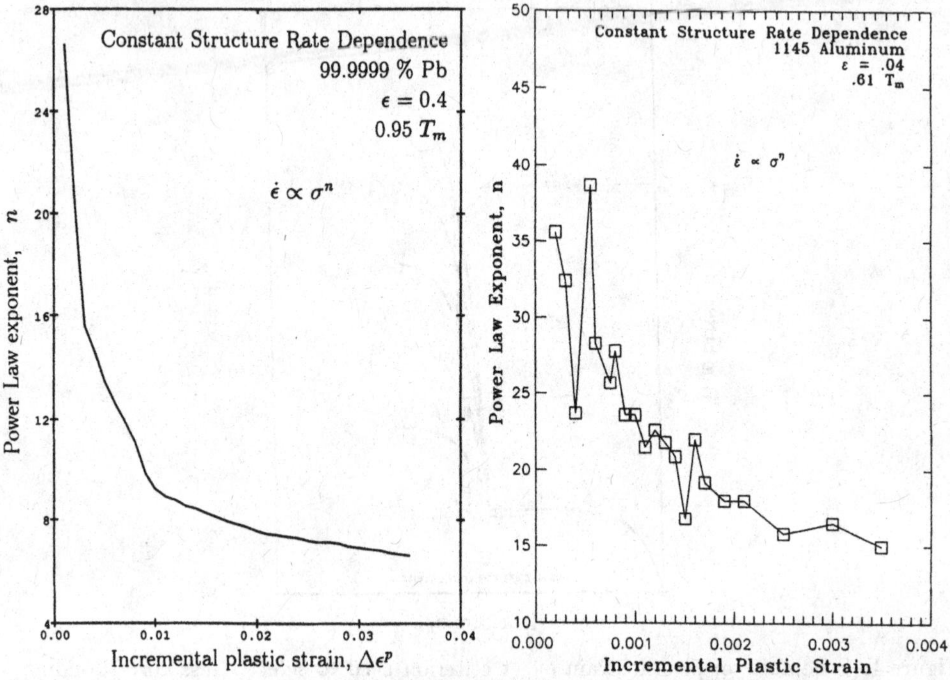


Figure 2: Constant structure rate dependence of 1145 aluminum and high purity lead as a function of plastic strain offset assuming power law behavior.

[4] E Krempl, J J McMahon, and D Yao. Viscoplasticity based on overstress with a differential growth law for the equilibrium stress. In *2nd Symposium on Nonlinear Constitutive Relations for High Temperature Applications, NASA Conference Publication 2369*, pages 25–50, 1985.

[5] W D Nix, J C Gibeling, and D A Hughes. Time-dependent deformation of metals. *Metallurgical Transactions*, 16A:2215–2226, December 1985.

[6] F B Prinz and A S Argon. The evolution of plastic resistance in large strain plastic flow of single phase subgrain forming metals. *Acta Metallurgica*, 32-7:1021–1028, 1984.

[7] D C Stouffer, V G Ramaswamy, J H Laflen, R H Van Stone, and William R. A constitutive model for the inelastic multiaxial response of rene 80 at 871C and 982C. *Journal of Engineering Materials and Technology*, 112:241–246, April 1990.

# SOME REMARKS ON THE CONCEPT OF VISCOPLASTIC MODELS OF THE OVERSTRESS-TYPE

O.T. BRUHNS
Institute of Mechanics I, Ruhr-University Bochum,
Universitätsstraße 150, D-4630 Bochum, FRG

## ABSTRACT

A concept is discussed which enables to develope rate-dependent models on the base of approved elastic-plastic constitutive equations with the goal to separate rate-dependent from rate-independent functions. This allows both the application of an appropriate strategy to evaluate the material functions and the attainment of an exact limiting value in the case of quasi-static processes. Ordinary viscoplastic models of overstress-type show purely elastic response in this limit whereas elastic-plastic behaviour should be predicted.

## INTRODUCTION

In many problems of dynamic loading of structures the analysis has to take into consideration that especially in the range of inelastic deformations different regimes can be observed with different prevailing process velocities.

A typical problem of this kind is the impact of a projectile against a steel target. The target plate fails by "plugging" due to shear band formation and by "spalling" due to cracks which are nucleated by tensile waves reflected on the rear side of the plate. Within this zone of failure the whole range of strain rates is observed. Close to a shear band and close to the tip of a crack e.g. the deformations are related with high strain rates up to $10^5$ sec$^{-1}$. In some distance, however, the srain rate may decrease to the values of quasi-static deformations. Since modern structures are made from alloys that exhibit a strong rate-dependency especially at elevated temperatures, the constitutive relations which have been developed to describe the behaviour of the alloys must be able to reproduce these facts. Processes related with high velocities therefore must be described by a rate -dependent viscoplasticity whereas the same constitutive relations for vanishing velocities must turn over to a rate-independent plasticity.

Prior to establishing constitutive relations that can fulfill these conditions, a proper physical interpretation of the phenomena observed during inelastic deformations is needed. This means that the different mechanisms that can explain the dissipation implied by internal changes of the material are to be analysed. It is known from appropriate microscopic investigations [1] that in a temperature, strain -rate spectrum in general different regions can be observed reflecting different mechanisms of inelastic deformations. These are:

i) athermal mechanisms characterized by a yield stress relatively insensitive to the strain rate,
ii) thermally activated dislocation motion characterized by a more markedly temperature and rate sensitivity of the yield stress and
iii) damping of dislocation motion at very high strain rates related with an extreme strain rate sensitivity of the yield stress.

Deformations in region (i) at low strain-rates are normally described by classical rate-independent theories of plasticity subject to appropriate yield and loading conditions. Deformations in (ii) and (iii), on the other hand, at moderate and high strain rates can be modelled by rate-dependent relations of the overstress -type [2-4]. Our intention is now to describe the behaviour of a material in the entire spectrum of strain rates and temperatures. We therefore combine both formulations with the demand that in the limit of vanishing velocities the classical rate-independent relation will be gained. We note further that several attempts of Perzyna [2, 4] to show that this limit also can be deduced from rate-dependent viscoplasticity seem to be questionable.

## FUNDAMENTAL RELATIONS

Let us consider a representative volume element $\Delta V$ as a polycrystal. The motion of this element may be defined by averaging values over $\Delta V$, e.g. the strain rate as average effect of the different mechanisms that contribute to the motion. A continuum theory thus is characterized by the additive decomposition of strain rate tensor $\mathbf{D}$ according to

$$\mathbf{D} = \mathbf{D}_e + \mathbf{D}_i = \mathbf{D}_e + \mathbf{D}_a + \mathbf{D}_t \tag{1}$$

where $\mathbf{D}_e$ and $\mathbf{D}_i$ are elastic and inelastic parts, respectively, and $\mathbf{D}_i$ contains contributions $\mathbf{D}_a$ of athermal dislocation motion and $\mathbf{D}_t$ of thermally activated dislocation motion. Only for simplicity here the contribution of the damping processes is neglected. Additional variables are introduced in order to describe the microstructural changes and to formulate necessary yield and loading conditions. The identification of these variables is based on an approximate homogenization method which has been explained in [5]. Let

$$\mathcal{Q} = \{\mathbf{X}_a, \mathbf{X}_t, \varkappa\} \tag{2}$$

be the set of internal variables and further u be a process variable, where $\mathbf{X}_a$ and $\mathbf{X}_t$ are different kinematic hardening tensors related to the athermal and thermally activated dislocation motion, respectively. $\varkappa$ is an isotropic hardening parameter related to the total dislocation density N, and $u = N_{mob}/N$ is the quotient of mobile and total dislocation densities.

For the description of nonisothermal processes the specific free enthalpy g is introduced as a function

$$g = g(\sigma, \Theta; \mathcal{Q}) , \tag{3}$$

of the external variables (true) stress $\sigma$ and temperature $\Theta$, and the internal variables while the process variable does not alter the energy of the solid.

Under the aforementioned assumptions the constitutive relations comprehend

i) the state function for the specific free enthalpy, from which the reversible strains and the specific entropy are deduced,

ii) evolution laws for the internal variables $\boldsymbol{\mathcal{Q}}$ and the process variable u, and

iii) flow rules for the different parts $\mathbf{D}_a$, $\mathbf{D}_t$, $\mathbf{D}_d$ of $\mathbf{D}_i$ and the inelastic spin $\mathbf{W}_i$.

According to the different mathematical nature of time-dependent and time-independent relations, the latter are subject to yield and loading conditions while the former are only subject to "yield conditions".

The decomposition (1) is motivated by the introduction of a so-called intermediate configuration [6]. The evolution laws and flow rules are first formulated for variables with respect to this intermediate configuration and thereafter transformed to the actual configuration. This conversion includes multiplications with the reversible part $\mathbf{F}_e$ of the deformation gradient and is simplified in the present case since in metals only small reversible changes of the shape occur.

Thus the reversible part of the strain rate tensor becomes

$$\mathbf{D}_e = \frac{1}{2G}\left\{ \overset{\circ}{\sigma} - \frac{\nu}{1+\nu}\,\mathrm{tr}(\overset{\circ}{\sigma})\,\mathbf{1} \right\} + \alpha_t \dot{\Theta}\,\mathbf{1} \tag{4}$$

for an instantaneously isotropic material. Herein G (shear modulus), $\nu$ (Poisson's ratio) and $\alpha_t$ (thermal expansion coefficient) are the material parameters of the elastic material. Furthermore the symbol $\overset{\circ}{(\cdot)}$, e.g.

$$\overset{\circ}{\sigma} := \dot{\sigma} - (\mathbf{W}-\mathbf{W}_i)\,\sigma + \sigma\,(\mathbf{W}-\mathbf{W}_i) \tag{5}$$

expresses a special objective time derivative (Jaumann-type rate) introduced in the intermediate configuration [7, 8]. Further the dislocation induced inelastic strain rates become

$$\mathbf{D}_a = \left[\gamma_a(u,\ldots)\right]_a \langle LC_a \rangle\, \mathbf{N}_a \quad \text{if} \tag{6$_1$}$$

$$\left.\begin{aligned} F_a &= (\sigma' - \mathbf{X}_a)\cdot(\sigma' - \mathbf{X}_a) - g_a(\varkappa, \Theta, u) \geq 0 \ \text{and} \\ LC_a &= 2(\sigma' - \mathbf{X}_a)\cdot \overset{\circ}{\sigma} - \frac{\partial g_a}{\partial \Theta}\,\dot{\Theta} > 0 \end{aligned}\right\} \tag{6$_2$}$$

$$\left.\begin{aligned} \mathbf{D}_t &= \gamma_t(u,\ldots)\left[\Phi_t\right]_t \mathbf{N}_t \quad \text{if} \\ F_t &= (\sigma' - \mathbf{X}_t)\cdot(\sigma' - \mathbf{X}_t) - g_t(\varkappa, \Theta, u) \geq 0 \end{aligned}\right\} \tag{7}$$

for a v. Mises material, where

$$\mathbf{N}_t = \frac{\sigma' - \mathbf{X}_a}{\|\sigma' - \mathbf{X}_a\|} \qquad \mathbf{N}_a = \frac{\sigma' - \mathbf{X}_t}{\|\sigma' - \mathbf{X}_t\|} \tag{8}$$

are the normals to the surfaces $F_t$ = const. and $F_a$ = const., respectively. The square brackets here and in what follows mean

$$[x]_y = \begin{cases} x & \text{if } F_y \geq 0, \\ 0 & \text{otherwise} \end{cases} \tag{9}$$

and $\langle \cdot \rangle$ denotes a Maccauly bracket with respect to the loading condition $LC_a > 0$. The $F_y \geq 0$ are different yield conditions, and the prime $(\cdot)'$ denotes the deviatoric part of a tensor. The material functions $\gamma_y$, $\Phi_y$ and $g_y$ have been introduced as in usual theories of plasticity and viscoplasticity, respectively, except an additional dependency of some of these functions on the process variable u. Moreover different functions have been introduced for the description of the different mechanisms that contribute to the motion. Furthermore the process variable u as function of an appropriate process velocity controls the time-dependence and -independence, respectively, of the relations. I.e. e.g. function $\gamma_t(u,...)$ in (7) vanishes for a vanishing process velocity (quasi-static processes).

The evolution of the hardening of the material is here described in a somewhat simplified form; this means, e.g. that only for simplicity any recovery terms in the hardening laws and moreover some coupling effects of minor interest have been neglected

$$\overset{\circ}{\mathbf{X}}_a = c_a(\varkappa)\,\mathbf{D}_a\,, \quad \overset{\circ}{\mathbf{X}}_t = c_t(\varkappa)\,\mathbf{D}_t\,, \quad \dot{\varkappa} = k(\varkappa)\,\boldsymbol{\sigma}\cdot\mathbf{D}_i \tag{10}$$

where $c_a(\cdot)$, $c_t(\cdot)$ and $k(\cdot)$ are functions of the average dislocation density.

The evolution of the process variable u follows the second order differential equation

$$\ddot{u} + c\dot{u} + \frac{c^2}{4}(u - u_{eq}) = 0\,; \qquad \text{with } c = c_0(\alpha + v_p) \tag{11}$$

where $v_p := \max\{\|\mathbf{D}\| + \chi\dot{\Theta}\}$ is a sufficient measure to describe the process velocity and $u_{eq}(\cdot)$ is the equilibrium value of u which for different values of $v_p$ can be determined from experiments [9].

This concept is compared with an alternative procedure where additionally the stresses have been decomposed into their athermal and thermally activated parts

$$\boldsymbol{\sigma} = \boldsymbol{\sigma}_a + \boldsymbol{\sigma}_t \tag{12}$$

and where instead of u the static stresses $\boldsymbol{\sigma}_a$ have been introduced as process variables. We finally note that according to (12) compared with the static stresses $\boldsymbol{\sigma}_t$ can be interpreted as (dynamic) overstresses.

## REFEENCES


1. Rosenfield, A.R. and G.T. Hahn, *Trans. Amer. Soc. Metals*, 1966, **59**, 962-980
2. Perzyna, P., *Advances in Applied Mechanics Vol. 11*, Academic Press, 1971, 313-354
3. Perzyna, P., *Mechanical Properties at High Rates of Strain*, The Institute of Physics, 1974, 138-153
4. Perzyna, P., *Arch. Mechanics*, 1980, **32**, 403-420
5. Bruhns, O.T. and H. Diehl, *Arch. Mechanics*, 1989, **41**, 427-460
6. Lehmann, T., *The Constitutive Law in Thermoplasticity , CISM Courses and Lectures No. 281*, ed. T. Lehmann, Springer-Verlag, 1984, 379-463
7. Dafalias, Y.N., *Mech. Materials*, 1984, **3**, 223-233.
8. Dafalias, Y.N., *J. Appl. Mech.*, 1985, **52**, 865-871.
9. Shioiri, J. and K. Sakino, *Loading and Dynamic Behaviour of Materials*, eds. C.Y. Chiem, H.D. Kunze and L.W. Meyer, DGM, 1988, 793-800

# A UNIFIED CONSTITUTIVE MODEL WITH DISLOCATION DENSITIES AS INTERNAL VARIABLES

Y. ESTRIN and H. MECKING
Technical University Hamburg–Harburg,
P.O.Box 901052, 2100 Hamburg 90, F.R.G.

## ABSTRACT

A microstructurally founded unified elastic–viscoplastic constitutive model is exposed. Developed some years ago, the model has been extended to account for a number of features of mechanical response of metallic materials. These refer to multiaxial loading, cyclic loading, and dynamic strain aging. The model is presented both in its most extended form and in an arbridged version.

## INTRODUCTION

Several years ago, we proposed a unified elastic–viscoplastic constitutive model without a yield criterion [1] which was based on microstructural considerations. In contrast to many constitutive models, that one was very simple in its form and contained an unusually small number of adjustable parameters. The relation of these parameters to the underlying physical processes could easily be established. However, with only one internal variable included, the original version of the model was suited primarily to monotonic uniaxial loading. Additional features had to be introduced to make the model competitive in a broader range of applications. New developments of the model, such as generalization to the case of multiaxial loading [2], adaptation to cyclic loading conditions [3], and including dynamic strain aging effects [4,3] will be summarized below.

The basics of the model are as follows:

(i) The tensor of total strain rate $\dot{\epsilon}_{ij}$ is represented as the sum of the elastic ($\dot{\epsilon}_{ij}{}^{e}$) and plastic ($\dot{\epsilon}_{ij}{}^{p}$) parts.
(ii) The elastic part is given by the time derivative of the Hooke's law.
(iii) The plastic part is expressed, according to the Prandl–Reuss flow rule, through the deviatoric stress tensor $\sigma = (3J)^{1/2}$ and the corresponding equivalent plastic strain rate $\dot{\epsilon}^{p} = (2/\sqrt{3})(D_2^p)^{1/2}$. Here $J_2$ and $D_2^p$ denote the second invariants of the respective tensors.
(iv) A power–law relation between $\dot{\epsilon}^{p}$ and $\sigma$ is adopted [5,1]. In the most extended version of the model, this relation contains three microstructure–related

internal variables: the densities of mobile (X) and forest (Y) dislocations and the effective solute concentration $C_s$ on mobile dislocations temporarily arrested at localized obstacles. All three variables are scalars and enter the model in nondimensional form. $C_s$ is normalized by saturation solute concentration on dislocations, while the dislocation densities have been normalized in such a way that the initial values of X and Y are equal to unity.
(v) Evolutionary equations for the internal variables X, Y, and $C_s$ are based on considering dislocation storage and annihilation processes, as well as dynamic strain aging kinetics.
(vi) A special recipe for integrating the evolution equations for X and Y takes into account partial remobilization of forest dislocations upon stress reversal.

## CONSTITUTIVE MODEL: UNABRIDGED FORM

The complete set of equations comprising the constitutive model reads as follows

$$\dot{\epsilon}_{ij} = \dot{\epsilon}^e_{ij} + \dot{\epsilon}^p_{ij} \tag{1}$$

$$\dot{\epsilon}^e_{ij} = [(1+\nu)/E]\,[\delta_{ik}\delta_{jl} - (\nu/(1+\nu))\delta_{ij}\delta_{kl}]\dot{\sigma}_{kl} \tag{2}$$

$$\dot{\epsilon}^p_{ij} = (3/2)(\dot{\epsilon}^p/\sigma)s_{ij} \tag{3}$$

$$\dot{\epsilon}^p = \xi\,(\sigma/\sigma_0)^m\, X\,(Y + \beta C_s)^{-m/2} \tag{4}$$

$$dX/d\epsilon^p = [- C - C_1Y^{1/2} - C_3X + C_4\, Y/X] \cdot q \tag{5}$$

$$dY/d\epsilon^p = [C + C_1Y^{1/2} - C_2Y + C_3X] \tag{6}$$

$$C_2 = C_{20}\,(\dot{\epsilon}_p/\dot{\epsilon}_0)^{-1/n} \tag{7}$$

$$C_s = 1 - \exp\,[- C_n\,(t_a/t_d)^{2/3}] \tag{8}$$

$$dt_a/d\epsilon^p = 1/\dot{\epsilon}^p - t_a/(\psi X Y^{-1/2}) \tag{9}$$

Here $d\epsilon^p = [(2/3)\, d\epsilon^p_{ij}\, d\epsilon^p_{ij}]^{1/2}$ is the equivalent plastic strain–increment. E and $\nu$ denote Young's modulus and Poisson's ratio, respectively; $\delta_{ij}$ is the Kronecker delta. The unabridged model contains 12 *temperature independent* parameters:

$$C,\ C_1,\ C_{20},\ C_3,\ C_4,\ \xi,\ \dot{\epsilon}_0,\ \sigma_0,\ \beta,\ \psi,\ q,\ \text{and}\ C_n$$

and 3 *temperature dependent* parameters:

$$m,\ n,\ \text{and}\ t_d\,.$$

The exponents m and n are inversely proportional to the absolute temperature, while the temperature dependence of the characteristic solute diffusion time $t_d$ is given by a Boltzmann–type expression. The above division into temperature dependent and temperature independent parameters is applicable for the low temperature range (up to about 2/3 of the melting temperature). For high temperatures, n should be considered constant, while $\dot{\epsilon}_0$ assumes a Boltzmann–type temperature dependence. The physical significance of the parameters is explained in previous publications, cf. [1–4]. In the present brief summary, the meaning of only most important of them is explicitly mentioned. The range of variation of all parameters can be evaluated from physical considerations.

In integrating the evolution equations for X, Y and $C_s$ (or $t_a$), the initial value of $t_a$ has to be specified in addition to those of X and Y (which are equal to unity). If quasi–steady state deformation conditions may be assumed for the prehistory of the material, then $dt_a/d\epsilon^p = 0$, and the initial value of $t_a$ is given by $\psi/\dot{\epsilon}^p$. Here $\dot{\epsilon}^p$ is the plastic strain rate at the end of pre–straining. If static aging took place before straining, $t_a$ should be identified with the static aging time.

A special feature is introduced for cyclic loading. The initial values of X and Y for the (i+1)–th half–cycle are given, respectively, by $X_i^E + \varphi q(\Delta Y)_i$ and $Y_i^E - \varphi(\Delta Y)_i$, where $(\Delta Y)_i$ is the increment of Y in the i–th half–cycle. A half–cycle refers to deformation between two stress reversals. The parameter q equals the ratio of the forest and the mobile dislocation density, while the quantity $\varphi$ denotes the fraction of the last increment in Y that is remobilized upon stress reversal. Thus $0 \leq \varphi \leq 1$. The inclusion of this remobilization effect introduces a possibility of accounting for an asymmetry with respect to the direction of stress (the Bauschinger effect).

The parameter C is structure sensitive. It is inversely proportional to the grain size or – in the case of particle–strengthened materials – to the particle spacing. If the structure is changing in the process of straining, the corresponding variation of C may be included [3].

## CONSTITUTIVE MODEL: ABRIDGED FORM

For materials not exhibiting dynamic strain aging, the constitutive model is simplified by setting $\beta$ equal to zero and dropping eqs. (8) and (9). The number of adjustable parameters is then reduced by five. In the case of monotonic loading a further simplification is possible. Equation (5) is dropped, the coefficient $C_3$ in eq. (6) is set equal to zero, and the product $\xi X$ is replaced by a constant $\hat{\epsilon}$ which, for practical reasons, may be identified with $\dot{\epsilon}_0$. The set of constitutive equations is then comprised of eqs. (1) through (3), together with the equations

$$\dot{\epsilon}^p = \dot{\epsilon}_0 \, (\sigma/\sigma_0)^m \, Y^{-m/2} \tag{4'}$$

and

$$dY/d\epsilon^p = C + C_1 Y^{1/2} - C_2 Y \tag{6'}$$

where $C_2$ is given by eq. (7).

This is a very "economical" version of the constitutive model proposed, the total number of adjustable parameters being as low as seven. All of them can be relatively easily identified using conventional uniaxial tests [1,3].

Examples of validating the model by successfully applying it (both in the full and in the abridged form) to describe the mechanical response of various materials [3] will be discussed in the talk.

## REFERENCES

1. Estrin, Y., and Mecking, H., A Unified Phenomenological Description of Work Hardening and Creep Based on One–Parameter Models, Acta Metall., 1984, 32, 57–70.

2. Estrin, Y., and Mecking, H., An Extension of the Bodner–Partom Model of Plastic Deformation, Intl. J. of Plasticity, 1986, 2, 73–85.

3. Estrin, Y., A Versatile Unified Constitutive Model Based on Dislocation Density Evolution, Proc. of the Conference on High–Temperature Constitutive Modeling: Theory and Applications, ASME, Atlanta, Dec. 1991, A. Freed and K. Walker, Eds. (to be published).

4. McCormick, P.G., and Estrin, Y., Proc. of the TMS Annual Meeting, New Orleans, 1991 (in press).

5. Kocks, U.F., Laws for Work–Hardening and Low–Temperature Creep, ASME J. Engg. Mater. Techn., 1976, 98, 76–85.

# PREDICTION AND VERIFICATION OF MATERIAL BEHAVIOR USING THE CHABOCHE VISCOPLASTIC CONSTITUTIVE THEORY

D. L. JONES* and A. M. EL-ASSAL**

*Department of Civil, Mechanical and Environmental Engineering
The George Washington University, Washington, DC 20052

** Department of Mechanical Engineering Technology
Banha Institute of Technology, Banha, Egypt

## ABSTRACT

The Chaboche viscoplasticity theory has been widely used to predict various types of time-dependent material behavior for many engineering materials. The authors have developed a procedure for determining the Chaboche material parameters from four tensile tests at different strain rates. Tensile tests have been performed on Type 316 stainless steel to obtain the material parameters, and then predictions were made for many types of inelastic material response. This paper presents predictions of the creep and stress relaxation behavior of Type 316 stainless steel, which were then compared with subsequent experimental results.

## INTRODUCTION

The Chaboche viscoplastic constitutive model [1] is a yield-based model that has been developed and enhanced considerably since it was introduced in 1977 [2-4]. It is a unified model in that it is capable of modelling many types of inelastic material response, including fatigue cycling with hold times, dual-rate tensile testing, creep, and stress relaxation. The authors have previously shown [5] that the material constants can be evaluated using data from four tensile tests at different strain rates. This paper presents predictions of both creep and stress relaxation using the material parameters obtained from tensile test data. The predictions are compared with creep and stress relaxation tests also performed by the authors.

## THE CHABOCHE VISCOPLASTIC MODEL

The uniaxial form of the Chaboche model can be described as follows. It is based on the linear decomposition of the strain rate tensor into elastic (recoverable) and inelastic (nonrecoverable) components, namely

$$\dot{\epsilon} = \dot{\epsilon}' + \dot{\epsilon}'' , \qquad (1)$$

where the superimposed dot represents time differentiation. The elastic portion of the strain rate is related to the stress rate through Hooke's law. The inelastic strain rate is formulated in terms of a yield function, f, which is based on the von Mises yield criterion, as

$$\dot{\epsilon}'' = 0 \qquad \text{for } f \le 0, \tag{2}$$

$$\dot{\epsilon}'' = (f/k)^n \qquad \text{for } f > 0. \tag{3}$$

where n and k are material constants and the inelastic strain rate tensor is assumed to be directed normal to the yield surface. The yield function is given by

$$f = |\sigma - Y| - R \quad , \tag{4}$$

where Y is kinematic hardening variable, R is the isotropic hardening variable, and $\sigma$ represents the applied stress. The evolution of the kinematic and isotropic hardening variables are given by

$$\dot{Y} = c\ (a\dot{\epsilon}'' - Y|\dot{\epsilon}''|) - \gamma(|Y|)^m\ \mathrm{sgn}(Y), \tag{5}$$

$$\dot{R} = b\ (q - R)\ \dot{p} \quad , \tag{6}$$

where the quantities c, a, $\gamma$, m, b, and q are material constants, and sgn(Y) is the sign of the kinematic hardening variable. The accumulated inelastic strain, p, is given by

$$p = \int_0^t |\dot{\epsilon}''|dt \quad . \tag{7}$$

The model parameters have been evaluated by the authors for Type 316 stainless steel under room temperature conditions [5], where it was assumed that the thermal recovery term, $\gamma(|Y|)^m$ sgn(Y) in Eq. (5), is insignificant. Thus, $\gamma$ was set to zero, with the remaining parameters having the following values:

| | |
|---|---|
| E (modulus of elasticity) | = 200 GPa |
| n (inelastic strain rate exponent) | = 4.55 |
| k (overstress parameter) | = 85.2 MPa $\sec^{-1/n}$ |
| b (isotropic hardening parameter) | = 21.3 |
| q (saturation value of isotropic hardening | = 436 MPa |
| c (kinematic hardening exponent) | = 843 |
| a (saturation value of kinematic hardening) | = 111 MPa |

## RESULTS AND DISCUSSION

All tests were performed using a commercial servo-hydraulic testing system that was controlled by a personal computer through an IEEE 488 interface. The tensile tests were performed under strain control so that the selected strain rates could be maintained during the elastic and inelastic portions of the test. For the creep tests, a dual-ramp stress controlled signal was generated by the computer and used to control the test system. The second stress-rate was set to zero, thus establishing the creep boundary condition. Different initial stress rates and hold stresses were used for the creep tests. The stress relaxation tests were also performed by the use of dual ramp functions, with strain as the controlled parameter. Different ramp rates and hold strains were also used in these tests.

Output data were collected in analog format using an x-y recorder and in digital format with the test system computer. Standard tensile test specimens of commercial quality Type 316 stainless steel were purchased from Laboratory Devices Company. The specimens were made according to ASTM E 8 and the testing followed this standard as closely as possible

The equations of the Chaboche model were incorporated into a suitable program that utilized a fourth-order Runge-Kutta numerical integration procedure to solve for the desired values as a function of time.

The predicted creep response for Type 316 stainless steel for a hold stress of 448 MPa is presented in Fig. 1, along with experimental results for the same load rate and hold stress values. Both results exhibit primary creep and secondary creep with steady-state creep rates close to zero. The creep rates agreed rather well, but the magnitude of the experimental creep strain was nearly four times as great as the predicted value. This result suggests the need for some refinement or optimization of the material parameters. However, an optimization was not performed because one of the objectives of this study was to determine how well tensile test data alone could be used to predict creep response.

The stress relaxation response for the same material was predicted for a hold strain of 0.015 and is shown in Fig. 2, along with experimental results for the same conditions. The agreement between the predicted and experimental results is much better for the stress relaxation behavior than for the creep response. The final predicted stress differed about eight percent from the experimental results. The general shapes of the curves were similar, as was the total amount of stress relaxation.

## CONCLUSIONS

The Chaboche viscoplastic constitutive model has been used to predict creep and stress relaxation behavior for Type 316 stainless steel for comparison with experimental results. Agreement between the predicted and experimental curves was better for the stress relaxation than for the creep response. The stress relaxation comparisons may have been better because both the stress relaxation and the tensile tests used to determine the material parameters were performed in strain control. A number of additional predictions and comparisons will be included in the presentation and the full paper.

## REFERENCES

1. Chaboche, J.L., Viscoplastic constitutive equations for the description of cyclic and anisotropic behavior of metals, Bulletin De L'Academie Polanaise des Sciences, 1977, **XXV**.

2. Eftis, J., Abdel-Kader, M.S., and Jones D.L., Comparisons between the modified Chaboche and Bodner-Partom viscoplastic constitutive theories at high temperature, Int. J. Plasticity, 5, 1989, pp. 1-27.

3. Eftis, J., and Jones, D.L., Life prediction for a structural material under cyclic loads with hold times using a viscoplastic constitutive model, Final Sci. Rpt, AFOSR Grant 83-0066, Washington, DC, Dec., 1984.

4. Abdel-Kader, M.S., Eftis, J., and Jones, D.L., An extension of the

Chaboche theory to account for rate-dependent initial yield, SECTAM XIII Proceedings, Univ. of S. Carolina, 1986, pp. 263-269.

5. El-Assal, A.M. and Jones D.L., Evaluation of the Chaboche viscoplastic constitutive theory from a series of monotonic tensile tests for strain-rate sensitive materials, Mech. Eng. & Tech. MET-4, Cairo, 1989.

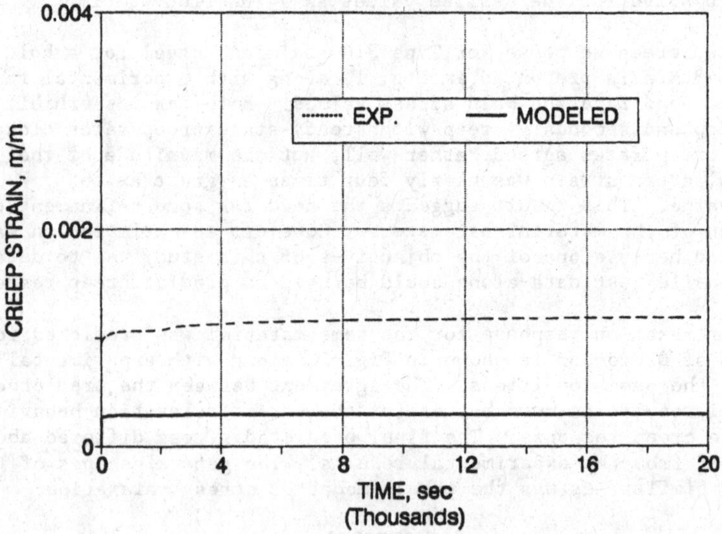


Fig. 1. Creep of Type 316 stainless steel at a hold stress of 448 MPa.

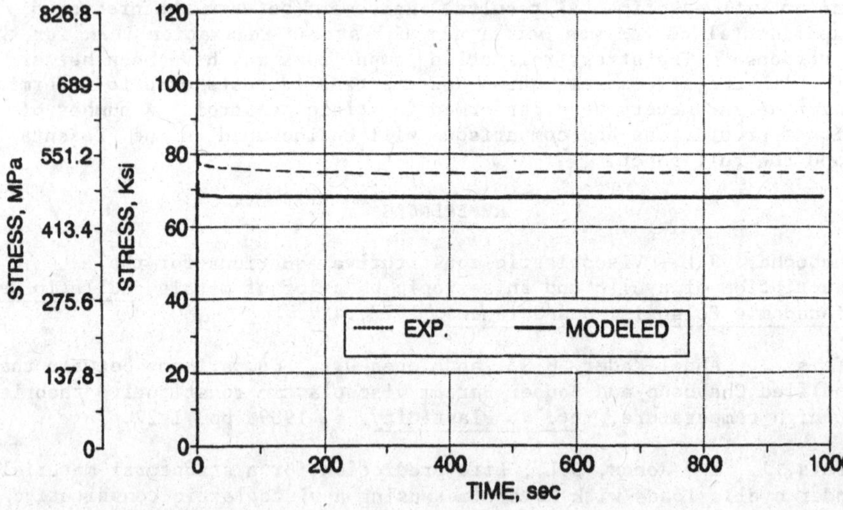


Fig. 2. Stress relaxation of Type 316 stainless steel at a hold strain of 0.015.

# MODELING OF THE DYNAMIC DEFORMATION AND FRACTURE OF THE TENSILE BAR

J.A. NEMES* and J. EFTIS†

*Mechanics of Materials Branch, Materials Science and Technology Division,
Naval Research Laboratory, Washington, D.C., 20375, U.S.A.

†Department of Civil, Mechanical, and Environmental Engineering,
George Washington University, Washington, D.C., 20052, U.S.A.

## ABSTRACT

A recently developed viscoplastic-damage constitutive theory is used to describe the deformation behavior and ductile fracture of a cylindrical bar subject to dynamic tensile loading that produces strain rates of the order $10^2$ - $10^4$. In this simulation the localization of inelastic deformation is allowed to occur naturally, rather than to have it induced at a particular location along the gage length of the bar. It appears that location of localization is significantly influenced by the dynamic nature of the loading.

## INTRODUCTION

The role of inertia and material rate dependence on the dynamic tension test has been of interest since the work of Malvern [1] on plastic wave propagation in a long bar subjected to impact. The propagation of waves along the length of the bar establishes a heterogeneous deformation field from the onset of deformation, which affects the subsequent localization. As in the quasi-static tensile test, localization in ductile metals results in the formation of a neck in the bar with accompanying high triaxial tensile stresses leading to void nucleation, growth, and eventually coalescence to final fracture.

Regazzoni *et al.*, [2] used a one dimensional finite-difference approach with an elasto-viscoplastic constitutive model to establish the heterogeneous deformation fields for different strain-hardening and strain-rate behavior. Two-dimensional numerical calculations were performed by Tvergaard and Needleman [3] to analyze the deformation and fracture of round tensile bars at quasi-static rates considering a small initial geometric inhomogeneity to induce localization. Recently Tvergaard and Needleman [4] studied ductile rupture in notched bars under impulsive tensile loading. In this paper high rate deformation and fracture of smooth tensile bars of high purity copper is considered, where no a priori determination of localization position is made. Instead, localization is permitted to occur naturally as a result of the heterogeneous deformation fields induced by the dynamic nature of the deformation and the actual geometry of the bar.

## VISCOPLASTIC-DAMAGE CONSTITUTIVE MODEL

The rate of deformation is decomposed into elastic, $\underset{\sim}{D}^e$, and viscoplastic, $\underset{\sim}{D}^p$, components. The elastic strains are assumed to be small. However because of the presence of voids, the elastic properties of the material are taken to be degraded as a function of the void volume fraction, $\xi$, according to the model proposed by MacKenzie [5]. The rate of elastic deformation for the voided solid is given in terms of an objective time rate of the Cauchy stress tensor $\underset{\sim}{T}$ and the rate of plastic deformation is taken to have the form proposed by Perzyna [6], whereby

$$\underset{\sim}{D}^{e} = \frac{1}{2\bar{\mu}}\left[\overset{\nabla}{\underset{\sim}{T}} - \frac{\bar{\nu}}{1+\bar{\nu}}\left(\mathrm{tr}\,\overset{\nabla}{\underset{\sim}{T}}\right)\underset{\sim}{1}\right] \quad , \quad \underset{\sim}{D}^{p} = \frac{\gamma_0}{\left(\frac{I_2}{I_2^s} - 1\right)^m}\left[\frac{J_2' + n\xi J_1^2}{\left[q + (\kappa_0 - q)\,e^{-\beta\epsilon^p}\right]^2\left[1 - n_1\xi^{\frac{1}{2}}\right]^2} - 1\right]^{m_1}\frac{1}{\kappa_0}\left(2n\xi J_1\,\underset{\sim}{1} + \underset{\sim}{T}'\right) \qquad , (1)$$

where $\gamma_0$ is a material viscosity constant, $\kappa_0$ is a parameter related to the quasi-static yield stress, q, $\beta$, and $n_1$ are material hardening and softening parameters ,with m and $m_1$ as additional material parameters. $I_2$ is the norm of the square root of the second invariant of $\underset{\sim}{D}'$, $I_2 = \left\|\left(II_{\underset{\sim}{D}'}\right)^{\frac{1}{2}}\right\|$ , and $I_2^s$ is the value of $I_2$ at the quasi-static rate of deformation. Because of the compressibility of the voids, the plastic deformation of the voided solid will also be compressible, consequently the yield function is taken to have both a deviatoric part and a dilatational part. Therefore both $J_1$, the first invariant of the stress tensor, and $J_2'$, the second invariant of the stress deviator tensor appear in (1) with n serving as a material weighting parameter.

Increase in the void volume fraction in polycrystalline metals is attributed to void nucleation and void growth. Under conditions of high mean stress, such as occurring from impact, a threshold stress, thermally-activated type mechanism is appropriate for describing void nucleation [7]. During tensile bar deformation, however, mean stresses that develop in the neck region are usually insufficient to overcome the void nucleation threshold stress. Thus the void nucleation occurs from mechanisms that are strain driven, as discussed by Curran *et al.* [8], who proposed a general void nucleation model. Here, however, a simplified approach is used initially in which only the void growth from an initial void volume fraction $\xi_0$ is considered. The rate of void growth for the porous solid can be expressed by the set of relations [6,9]

$$\dot{\xi} = \frac{1}{\eta}\,g(\xi)\,F(\xi,\xi_0)\,(\sigma - \sigma_G) \qquad , \sigma > \sigma_G \,, \qquad (2)$$

where

$$F(\xi,\xi_0) = \frac{\sqrt{3}}{2}\,\xi\left(\frac{1-\xi}{1-\xi_0}\right)^{\frac{2}{3}}\left[\xi - \left(\frac{\xi}{\xi_0}\right)^{\frac{2}{3}}\right]^{-1} \quad , \quad \sigma_G = \frac{1}{\sqrt{3}}(1-\xi)\ln\left(\frac{1}{\xi}\right)\left[2q + (\kappa_0 - q)\,F_1(\xi,\xi_0)\right] \,,$$

$$F_1(\xi,\xi_0) = \exp\left[\frac{2}{3}\beta\,\frac{(\xi_0 - \xi)}{\xi(1-\xi_0)}\left(\frac{1-\xi}{1-\xi_0}\right)^{-\frac{2}{3}}\left(\frac{\xi_0}{\xi}\right)^{-\frac{2}{3}}\right] + \exp\left[\frac{2}{3}\beta\,\frac{(\xi_0 - \xi)}{(1-\xi_0)}\left(\frac{1-\xi}{1-\xi_0}\right)^{-\frac{2}{3}}\right] \,. \qquad (3)$$

In these expressions $\sigma = (1/3)J_1$ is the mean stress, $\sigma_G$ is the void growth threshold mean stress, $\eta$ is a material parameter, and $g(\xi)$ is a material function to account for void interaction (coalescence). A local criteria for ductile fracture is defined by attainment of a critical value $\xi = \xi_F$ for the void volume fraction. Computationally, elements are deleted when the void volume reaches a value slightly less than $\xi_F$.

Determination of the viscoplastic material parameters requires a knowledge of the rate-dependent stress-strain response of the material extending to large strains, where influence of specimen geometry is minimized. Extensive experiments on copper under uniaxial compression have been performed by Follansbee and Kocks [10], from which the viscoplastic material parameters can be obtained. Good agreement between the computed and experimental response over the range of rates is obtained. The material parameters used for the homogeneous response and subsequent simulation of the dynamic tensile test are given in the table below.

## SIMULATION OF THE TENSILE TEST

The geometry of the tensile bar considered in the simulations along with the finite element discretization is shown in Fig. 1. Due to symmetry along the length only one half of the bar is modelled. The finite element mesh shown contains approximately 1000 elements. More refined meshes containing 1600 and 4000 elements were also used in the study. Specified velocities were applied to the upper nodal points indicated in the figure using a

TABLE I Material Parameters

| | | | | | |
|---|---|---|---|---|---|
| $\mu$ | = | $4.84 \times 10^4$ MPa | $\gamma_0$ | = | $2.18 \times 10^5\ s^{-1}$ |
| K | = | $14.00 \times 10^4$ MPa | $I_2^s$ | = | $1.21 \times 10^{-4}\ s^{-1}$ |
| q | = | 202 MPa | $\kappa_0$ | = | 23.1 MPa |
| $\beta$ | = | 5.34 | n | = | 0.25 |
| $\rho_R$ | = | 8.93 gm/cm$^3$ | m | = | 0.167 |
| $\xi_0$ | = | $3 \times 10^{-4}$ | $\xi_F$ | = | 0.32 |
| $\eta$ | = | 120 Poise | $n_1$ | = | 1.77 |
| g | = | $e^{\alpha\xi}$ | $\alpha$ | = | 20 |
| $m_1$ | = | 2 | | | |

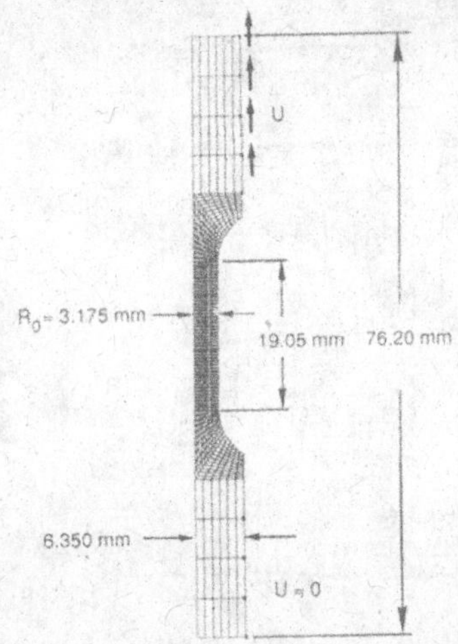


Fig. 1 Geometry and finite element discretization of the tensile bar.

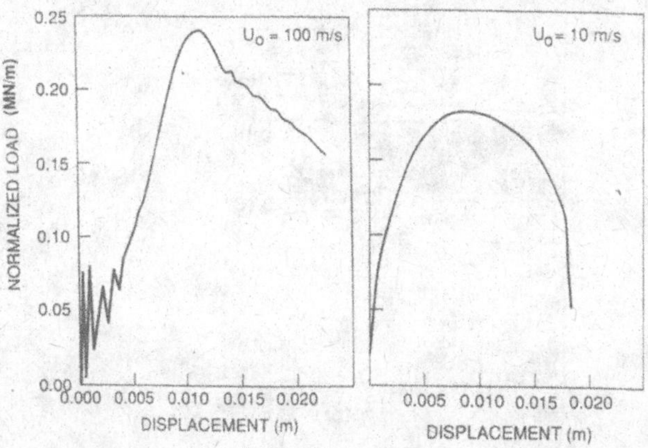


Fig. 2 Computed load-displacement histories at two loading rates.

ramp history, where the applied velocity $U= U_0t/t_0$ for $t< t_0$ and $U = U_0$ for $t\geq t_0$. The lower nodal points indicated in the figure have displacements set to zero. Simulations were conducted using a value of 50 μs for $t_0$ and values of U equal to 10, 25, 50, and 100 m/s, which result in nominal strain rates in the gage section of 500, 1250, 2500, and 5000 $s^{-1}$, respectively.

The computations showed that the position of localization along the length of the gage section is strongly affected by loading rate, thus indicating the effect of the heterogeneous deformation field established from the dynamic loading. At a nominal rate of 5000 $s^{-1}$ localization occurred 7.04 mm above the specimen centerline compared to 5.38 mm for 2500 $s^{-1}$. At 1250 $s^{-1}$ two necks formed symmetrically 3.31 mm from the centerline. Only at the lowest rate considered, 500 $s^{-1}$, did localization occur at the centerline of the gage section, indicating, perhaps, negligible effects of inertia. This is also apparent from the computed load-displacement curves at the highest and lowest rates considered, shown in Fig. 2. In Fig. 2a dynamic effects are clearly visible from the oscillations in the loading history while at the lower rate such dynamic behavior is absent. The elongation at instability is determined by considering the change of length in the gage section at the point of maximum load. Thus $e = \Delta l/l$ at maximum load is computed and shown in Fig. 3, where it compares well to data from Regazzoni and Montheillet [11].

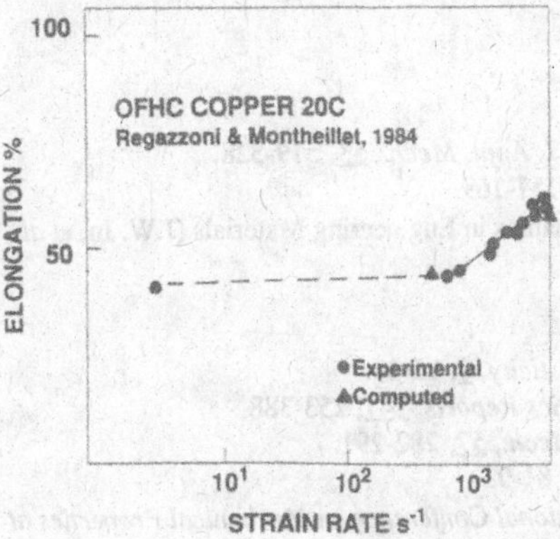


Fig. 3 Elongation at tensile instability.

Subsequent to localization induced by the dynamic loading the triaxial tensile stress that develops in the necked region is sufficient to overcome the void growth threshold stress defined by (3), thus initiating the void growth process. The void growth begins along the centerline of the specimen and than propagates outward. The final fracture condition, showing deleted elements and void volume contours for the 5000 $s^{-1}$ strain rate is shown in Fig. 4. Figure 5 shows contours of the extremely large values of equivalent plastic strain. From the computation the area at fracture is approximately 21% for all the rates considered compared to a value of 10% observed experimentally. The elongation at fracture is computed and shown compared to the data of Regazzoni and Montheillet [11] in Fig. 6. Although there is not strong quantitative agreement, the trend of increasing elongation with increasing rate is seen in the computations as well as the

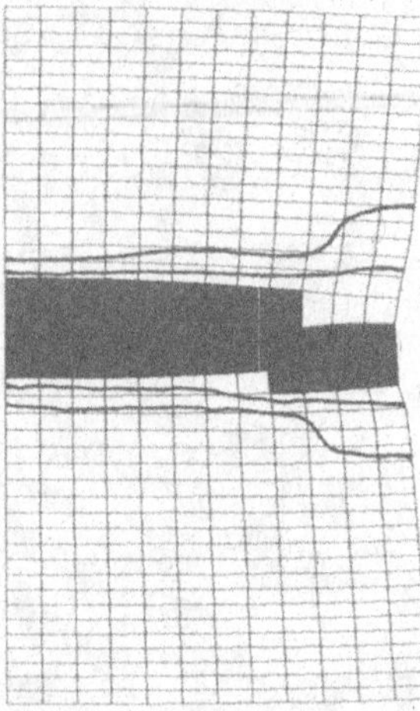


Fig. 4 Deleted elements and void volume fraction contours at fracture.

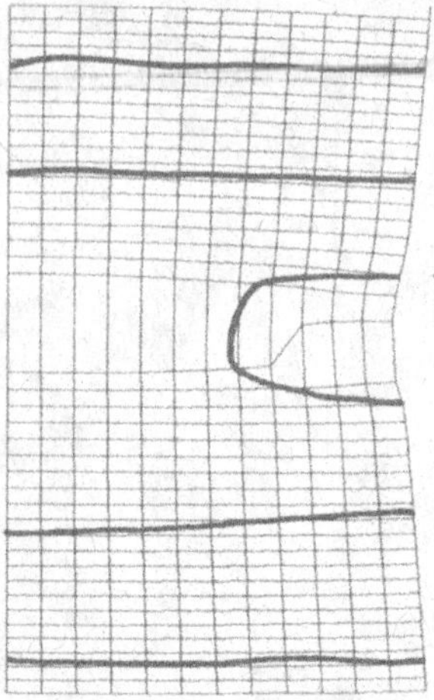


Fig. 5 Contours of equivalent plastic strain at fracture.

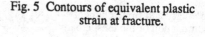

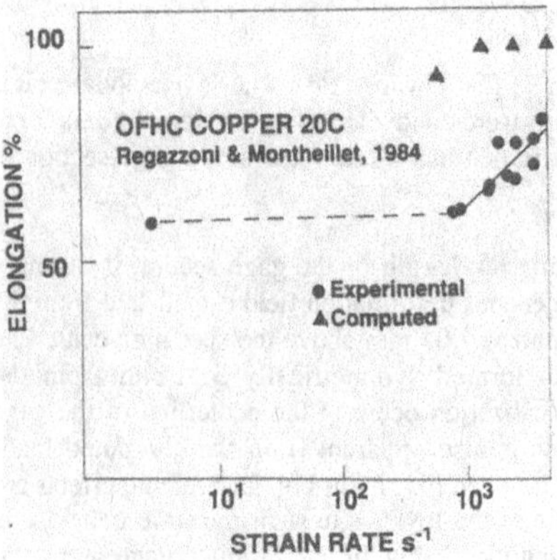


Fig. 6 Elongation at fracture.

data. It is expected that inclusion of a strain driven void nucleation term would reduce the elongation at fracture, which would be more consistent with the experimental data.

## REFERENCES


1. Malvern, L.E. (1951), *J. Appl. Mech.*, 18, 203-208.
2. Regazzoni, G., J.N. Johnson and P.S. Follansbee (1986). *J. Appl. Mech.*, 53, 519-528.
3. Tvergaard, V. and A. Needleman (1984). *Acta metall.*, 32, 157-169.
4. Tvergaard, V. and A. Needleman (1990) In: Damage Mechanics in Engineering Materials (J.W. Ju, *et al.*, Eds.), pp. 117-128, ASME, New York.
5. MacKenzie, J.H. (1950). *Proc. Phys. Soc.*, 63, 2-11.
6. Perzyna, P. (1986). *Int. J. Solids Struct.*, 22, 797-818.
7. Eftis, J., J.A. Nemes and P.W. Randles (1991). *Int. J. Plasticity*, 7, 15-39.
8. Curran, D.R., L. Seamen,and D.A. Shockey (1987). *Physics Reports*, 147, 253-388.
9. Nemes, J.A., J. Eftis and P.W. Randles (1990). *J. Appl. Mech.*, 57, 282-291.
10. Follansbee, P.S. and U.F. Kocks (1988), *Acta metall.*, 36, 81-93.
11. Regazzoni, G. and F. Montheillet (1984). In: *Third International Conference on Mechanical Properties at High Rates of Strain*, (J. Harding, Ed.), pp. 63-70, Institute of Physics, New York.

# COMPARATIVE VISCOPLASTIC FE-CALCULATIONS OF A NOTCHED SPECIMEN UNDER CYCLIC LOADINGS

J. OLSCHEWSKI, R. SIEVERT, A. BERTRAM
Bundesanstalt für Materialforschung und -prüfung (BAM), Berlin, FRG

## ABSTRACT

In this paper, results from comparative FE-calculations using the unified models of Chaboche and Bodner-Partom are presented. The material constants in the constitutive models are obtained from uniaxial data of the Ni-base superalloy IN 738 LC at 850 °C. The finite element model of a circumferentially notched specimen is employed to determine the predictive capabilities of both models under inhomogeneous stress-strain states. The calculated stress- and strain distributions in the net section of the specimen show no significant differences in the high hardening regime.

## INTRODUCTION

The design and analysis of structural components to operate at elevated temperatures, such as gas turbine engines, require an accurate prediction of the nonlinear stress-strain response of high-temperature materials subjected to cyclic loading conditions. For that purpose, two different unified constitutive models based on the proposals of Chaboche [1] and Bodner-Partom [2] have been selected to predict the material behaviour of the particle hardened Ni-base alloy IN 738 LC. The calibration of these models is based solely on uniaxial hot tension, LCF and creep tests at the service temperature of 850 °C. For the validation of the constitutive models under consideration an experimental program with a broad variety of complex loading processes has been carried out. In particular, the ability of the selected constitutive models to describe the material response under multiaxial loading conditions have been demonstrated in [3] by comparing the calculated results with those obtained in experiments with thin- walled tubular specimens subjected to proportional and non-proportional LCF- loadings.

The main objective of this paper is to demonstrate the predictive capabilities of unified models based on different constitutive assumptions in the case of nonhomogeneous states of stresses and strains as can be found in the net section of an circumferential notched specimen under various displacement controlled cyclic loadings.

## CONSTITUTIVE MODELS

The complete set of constitutive equations which has been used in this study is given in Table 1 below. The material constants presented in Table 2 have been calculated from the experimental data by using an optimization procedure based on the Levenberg-Marquardt algorithm. Details concerning the identification process have been omitted here due to the limited space.

TABLE 1
Summary of the Chaboche and Bodner-Partom model

| Chaboche model | Bodner-Partom model |
|---|---|

A1: strain decomposition

$$E = E_e + E_i$$

A2: Hooke' s law

$$S = \overset{<4>}{C}_e \,..\, E_e$$

A3: flow rule

$$\dot{E}_i = \sqrt{3/2} \left\langle \frac{J_2(S'-X) - R}{K} \right\rangle^n \frac{S'-X}{\|S'-X\|} \qquad \dot{E}_i = \sqrt{2}\, D_o \exp\left[ -\frac{1}{2} \left( \frac{Z}{J_2(S')} \right)^{2n} \right] \frac{S'}{\|S'\|}$$

$$J_2(-) = \sqrt{3/2}\,\|(-)\|\,, \quad \dot{p} = \sqrt{2/3}\,\|\dot{E}_i\|$$

A4: hardening rules

$$Z = Z^I + Z^D$$

- isotropic hardening

$$\dot{R} = (Q_o - R)\,\dot{p}\,,\ R(p=0) = k \qquad \dot{Z}^I = m_1 (K_1 - Z^I)\, \dot{W}_i - A \left( \frac{Z^I - K_2}{K_1} \right)^r$$

$$Z^I(t=0) = K_o$$

- kinematical/directional hardening

$$\dot{X} = c \left( a \frac{S'-X}{J_2} - \Phi(p) X \right) \dot{p} - d \left( \frac{J_2(X)}{a} \right)^r \frac{X}{J_2(X)} \qquad \dot{B} = m_2 \left( K_3 \frac{S}{\|S\|} - B \right) \dot{W}_i - A \left( \frac{\|B\|}{K_1} \right)^r \frac{B}{\|B\|}$$

$$\Phi(p) = \Phi_\infty - (\Phi_\infty - 1)\, e^{-\omega p} \qquad Z^D = B \,..\, \frac{S}{\|S\|}$$

TABLE 2
Material constants for IN 738 LC at 850 °C

$E$ = 149 650 MPa, $\nu_e$ = 0.33

| | | | |
|---|---|---|---|
| $K$ | = 1150 | $D_o$ | = $2.45\ 10^6\ s^{-1}$ |
| $n$ | = 7.7 | $n$ | = 0.289 |
| $k$ | = 153 MPa | $K_o$ | = $4.18\ 10^5$ MPa |
| $Q_o$ | = 0.0 | $K_1$ | = $3.76\ 10^5$ MPa |
| $a$ | = 311 MPa | $K_2$ | = $3.07\ 10^5$ MPa |
| $b$ | = 317 MPa | $K_3$ | = $1.54\ 10^5$ MPa |
| $c$ | = 201 | $m_1$ | = 0.581 |
| $\Phi_\infty$ | = 1.1 | $m_2$ | = 0.344 |
| $\omega$ | = 0.04 | $A$ | = $4.59\ 10^3$ MPa/s |
| $d$ | = $2.27\ 10^{-2}$ MPa/s | $r$ | = 5.4 |
| $r$ | = 4.8 | | |

## FE-CALCULATIONS

Both sets of unified constitutive equations have been incorporated into the ADINA finite element computer code [4]. As an example for a structural component, a cylindrical circumferentially notched bar with a stress concentration factor of $K_t = 2.5$ has been used. The finite element mesh consists of 70 axisymmetric isoparametric 8-node-elements which give rise to 462 degrees of freedom. The notched bar has been cyclicly loaded under displacement controlled conditions. Figures 1 to 4 show some results concerning the material response in the vicinity of the notch root. In Figure 1 the notch opening stress distribution ahead of the notch root is plotted for the first and fifth cycle. It can be seen that the material exhibits no significant hardening effect and that the stress maximum calculated with Bodner-Partom's model is slightly higher than that based on Chaboche's model. The radial notch root displacement is compared in

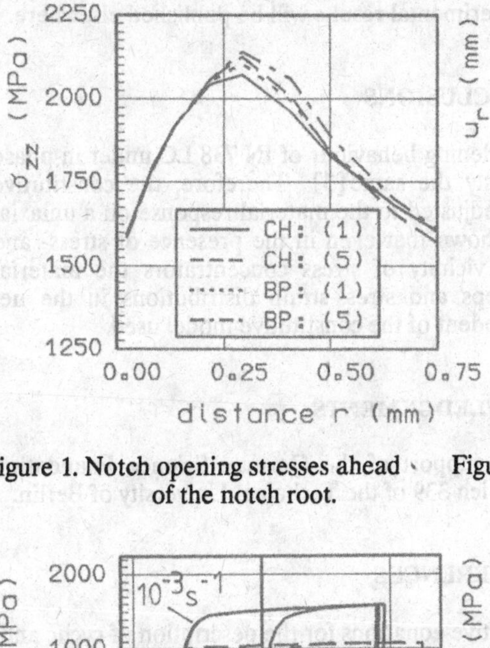


Figure 1. Notch opening stresses ahead of the notch root.

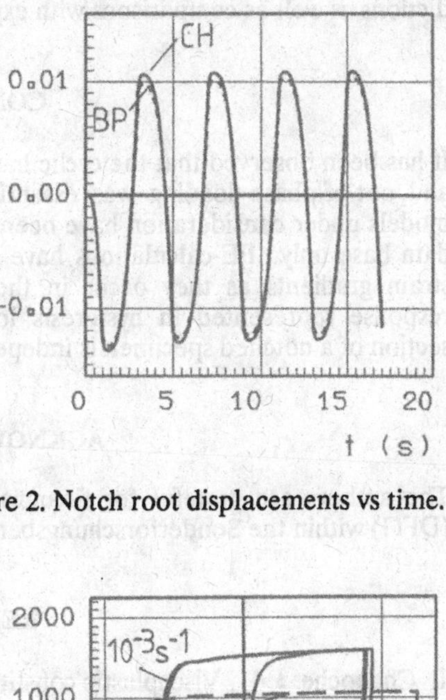


Figure 2. Notch root displacements vs time.

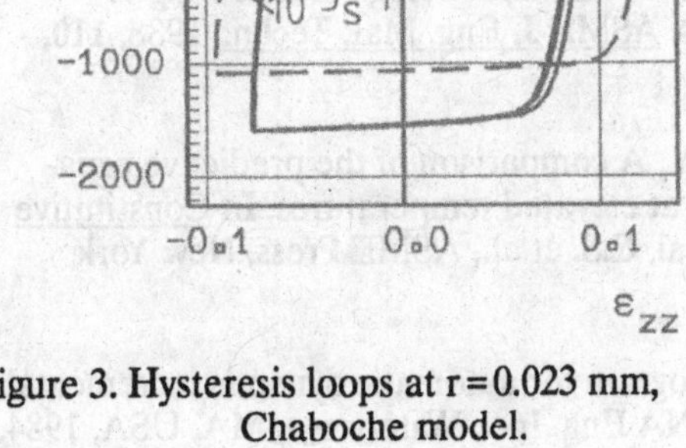


Figure 3. Hysteresis loops at r = 0.023 mm, Chaboche model.

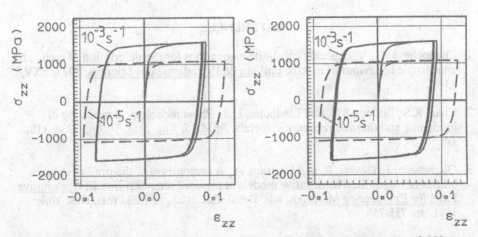

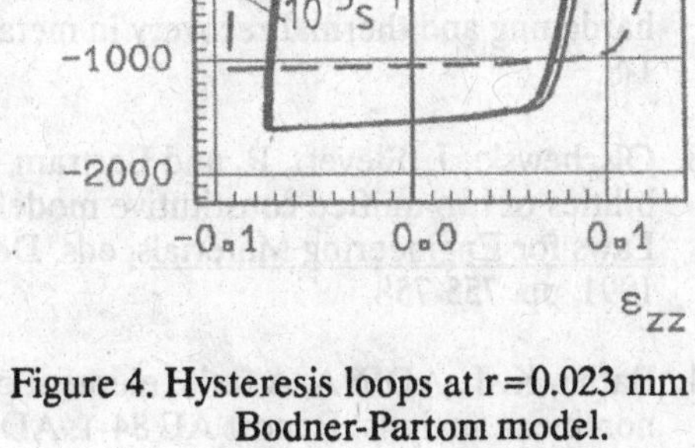


Figure 4. Hysteresis loops at r = 0.023 mm Bodner-Partom model.

Figure 2. Again, there is no transient hardening effect at all and the differences calculated by both models are negligible. Nevertheless, this displacement reflecting the hardening state of the material at the notch root can be used to validate constitutive models. The material response close to the notch root (r=0.023 mm) represented by notch opening stress-strain hysteresis loops is shown in Figures 3 and 4. The plotted hysteresis loops are calculated with respect to two loading rates which correspond to far field strain rates of about $10^{-3}\,s^{-1}$ and $10^{-5}\,s^{-1}$. It can be observed that in the high hardening regime both models predict the same material response including strain rate effects. But, in the low hardening regime caused by a lower load amplitude the calculated notch root hysteresis loops differ from each other. Here, the width of the hysteresis loops based on the Bodner-Partom model is larger than those calculated with the Chaboche model.

More detailled results showing the influence of the load amplitude on the model predictions as well as comparisons with experimental results will be published elsewhere.

## CONCLUSIONS

It has been observed that the cyclic hardening behaviour of IN 738 LC under in-phase and out-of-phase loading was essentially the same [3]. Therefore, the constitutive models under consideration have been adjusted to the material response on a uniaxial data base only. FE-calculations have shown that even in the presence of stress- and strain gradients as they occur in the vicinity of stress concentrators the material response represented in hysteresis loops and stress-strain distributions in the net section of a notched specimen is independent of the constitutive model used.

## ACKNOWLEDGEMENTS

The authors are grateful for financial support of the German Science Foundation (DFG) within the Sonderforschungsbereich 339 of the Technical University of Berlin.

## REFERENCES

1. Chaboche, J.-L., Viscoplastic constitutive equations for the description of cyclic and anisotropic behaviour of metals. Bulletin de L'Academie des Sciences, 1977, **XXV**, 33-42.

2. Chan, K.S., Bodner, S.R. and Lindholm, U.S., Phenomenological modeling of hardening and thermal recovery in metals. ASME J. Eng. Mat. Techn., 1988, **110**, 1-8.

3. Olschewski, J., Sievert, R. and Bertram, A., A comparison of the predictive capabilities of two unified constitutive models at elevated temperatures. In Constitutive Laws for Engineering Materials, eds. Desai, C.S. et al., ASME Press, New York, 1991, pp. 755-758.

4. Bathe, K.-J., ADINA, A finite element program for automatic dynamic incremental nonlinear analysis. Report AE 84-1, ADINA Eng. Inc., Watertown, MA, USA, 1984.

# DYNAMIC DUCTILE FAILURE UNDER MULTI-AXIAL LOADING

A.M. RAJENDRAN, S.J. BLESS, AND D.J. GROVE
UNIVERSITY OF DAYTON RESEARCH INSTITUTE
DAYTON, OHIO 45469-0120

## ABSTRACT

Dynamic ductile failure under a three dimensional strain state was modeled using a recently developed spall model. We considered a plate impact configuration in which a flyer plate impacts the base of a solid right circular cone. At high velocity impact, a complex spall pattern was generated in the cone. The pattern consisted of radial cracks at the base and tunnel (longitudinal) cracks inside. The failure model simulated spall pattern matched reasonably well with the experimental results.

## INTRODUCTION

Recently Rajendran, Dietenberger, and Grove [1] developed a three dimensional spall model (RDG) and successfully predicted spall in a thick plate target due to a planar impact by a thin plate. In this geometry, spall is induced under a one dimensional strain state. Model constants for several metals were determined and reported in reference 2. The main objective of this paper is to describe spall under a three dimensional strain state using the RDG model constants determined from the a conventional plate impact experiment. For this purpose, we considered a right angle cone whose base was impacted by a thin circular disk. The EPIC-2 finite element code was used in the numerical simulation. To simulate the spall failure, specialized subroutines describing the RDG model have been added to this code.

## THE RDG MODEL

The RDG model is a state variable and continuum mechanics based failure model. The material is assumed to be intact until the generation of voids. The Bodner-Partom viscoplastic model [3] describes the intact material behavior. Void nucleation is modeled as a Gaussian distribution about a mean threshold stress. The voids make the material porous and plastically compressible. Therefore, a pressure dependent yield function is used to describe the plastic flow in the porous material. Complete failure is assumed when the void volume fraction reaches the theoretical value of one. Details of this model can be found in references 1 and 4.

## EXPERIMENTS

From the quasi-static and SHB stress-strain curves at different strain rates, the Bodner-Partom constants were determined. The RDG model constants were calibrated by matching the computed stress history with the test data through trial and error simulation of the plate impact test using the one dimensional strain option in the EPIC-2 code. The final computed stress history is compared with the stress gauge data in Figure 1.

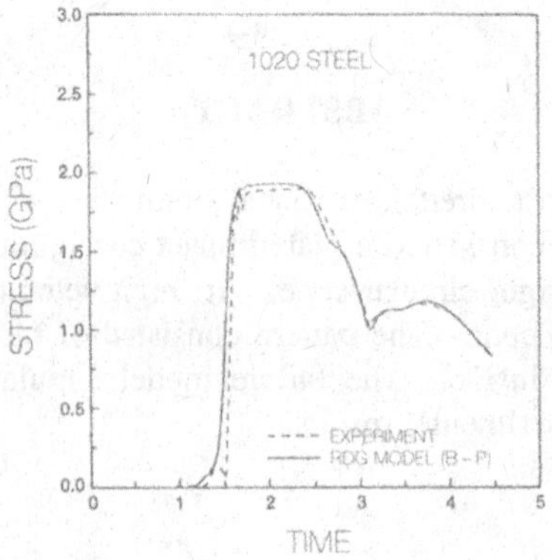


Figure 1. Comparison of test and simulated stress histories.

To further validate the ability of the RDG model in modeling spall, we considered an experimental configuration which involved the impact of a 2 mm thick circular disk of 38 mm diameter on the base of a right circular cone. The cone height was 25 mm. Both the flyer and target were 1020 steel. The experimentally observed spall patterns inside the cone target are shown in Figure 2.

It can be seen that four separate fracture systems developed in the target. A one dimensional spall-type fracture developed parallel to the cone surfaces. This fracture was apparently the first one to develop and it is first manifest at about midheight. The second fracture system consisted of distributed cavities near the axis. The third system below midheight toward the base is roughly a cylindrical failure region about 12 mm in diameter symmetric to the cone axis. The fourth system consisted of radial cracks as shown in Figure 3. Between the radial cracks there were voids. This failure is believed to have occurred last, as a result of surface motion induced by wave reflection from the flanks of the cone.

Figure 2. Spall patterns in the cone.

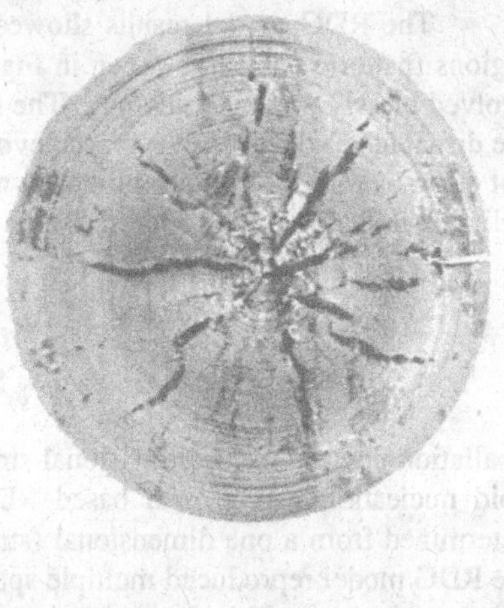

Figure 3. Radial cracks in the cone.

## RESULTS

We used the EPIC-2 finite element code to simulate the cone impact experiment. In the simulation, voids nucleated and grew almost to completion during the first seven microseconds. Results, in terms of 1) time histories of the pressure and damage for an element inside a spall zone and 2) the void volume fraction contours, are shown in Figures 4 and 5 respectively. At a location inside the first fracture region, the pressure is initially compressive and becomes tensile at t= 3 microseconds due to release wave interactions. High tensile loading (negative pressure) generates voids in the initially intact material and rapid void growth occurs leading to spallation as can be seen in Figure 4.

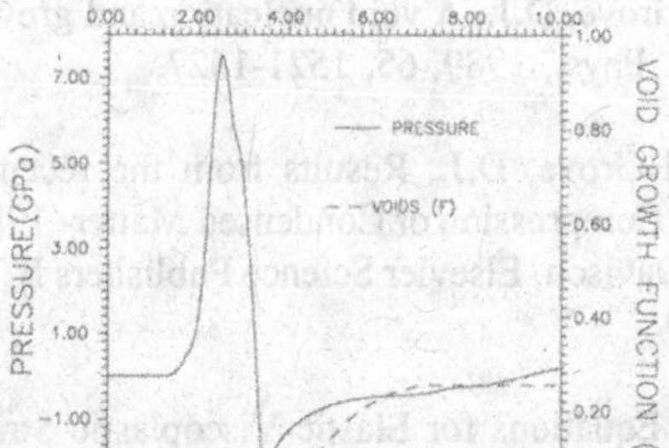


Figure 4. Pressure and damage time history.

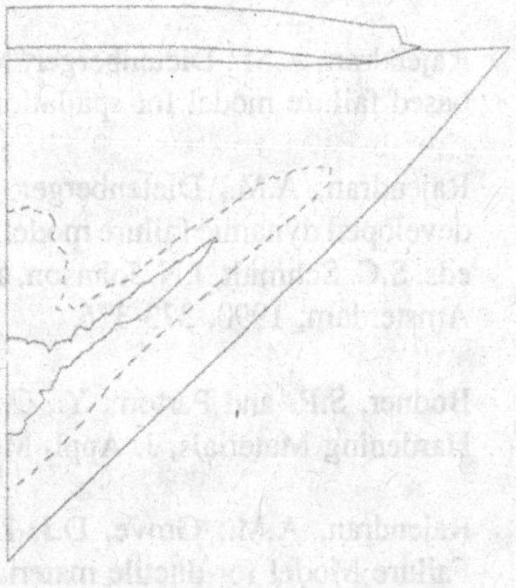

Figure 5. Void contours. Dotted line: - 5%, solid line: 15%.

The RDG model results showed evolution of the experimentally observed fracture regions (patterns) as can be seen in Figure 5. The spall region parallel to the cone surface evolved clearly in the simulation. The model also reproduced the second region containing the distributed voids. However, the evolution and location of the cylindrical spall zone was not clear. From the contour plot shown in Figure 5, we can observe the not-so-well-formed cylindrical spall zones around the axis of symmetry.

## SUMMARY

Spallation under a three dimensional strain state was modeled using the recently developed, void nucleation and growth based RDG failure model [1]. The model constants were determined from a one dimensional (strain) plane plate impact test. Using those constants, the RDG model reproduced multiple spall regions in a cone target under a three dimensional stress-strain state. This demonstration greatly increases our confidence in the generality of this model, even when calibrated from one dimensional experiments.

## ACKNOWLEDGMENTS

This work was supported by Air Force contract number F33615-86-C-5064 and monitored by Dr. Theodore Nicholas of the Wright Laboratories at the Wright Patterson Air Force Base, Ohio. William H. Cook, Eglin Air Force Base, FL partially supported this work. In addition, this work was supported in part by a grant from the Ohio Supercomputer Center, Columbus, Ohio.

## REFERENCES

1. Rajendran, A.M., Dietenberger, M.A. and Grove, D.J., A void nucleation,and growth based failure model for spallation. J. Appl. Phys., 1989, 65, 1521-1527.

2. Rajendran, A.M., Dietenberger, M.A. and Grove, D.J., Results from the recently developed dynamic failure model. In Shock Compression of Condensed Matter- 1989, eds. S.C. Schmidt, J.N. Johnson, and L.W. Davison, Elsevier Science Publishers B.V., Amsterdam, 1990, 373-376.

3. Bodner, S.R. and Partom, Y., Constitutive Equations for Elastic-Viscoplastic Strain Hardening Materials, J. Appl. Mech., 1975, 42, 385-389.

4. Rajendran, A.M., Grove, D.J, Dietenberger, M.A., and Cook, W.H., A dynamic Failure Model for ductile materials, Technical Report, UDR-TR-90-109, University of Dayton, Dayton, OH 45469, August 1990.

# DAMAGE AND VISCOPLASTICITY EFFECTS IN THE INDENTATION OF A POLYCRYSTALLINE SOLID

A.P.S.SELVADURAI and M.C.AU
Department of Civil Engineering
Carleton University
Ottawa, Ontario, Canada, K1S 5B6

## ABSTRACT

This paper examines a problem related to the dynamic indentation of layer by a rigid punch. The layer consists of a polycrystalline material, such as ice, which exhibits elasticity with damage, delayed elasticity, non-linear creep and viscoplasticity. The finite element technique is used to evaluate the force exerted on the punch due to steady indentation.

## INTRODUCTION

The subject of continuum damage mechanics has been applied quite extensively to the study of the mechanics of progressive failure and degradation in a variety of artificial and natural materials. Accounts of developments in this area are given by Kachanov [1], Krajcinovic [2], and Lemaitre and Chaboche [3]. In such modelling, damage is interpreted as a reduction in the stiffness of the material due to the generation of micro-cracks and other voids. In generalized treatments, it is assumed that continuum damage processes can be described by tensorial measures and such damage can occur with elastic, plastic and viscoplastic behaviour of materials. This paper examines the application of damage mechanics to the study of the mechanics of a polycrystalline material such as ice. In general, the constitutive response of a temperature and rate sensitive material such as ice can be characterized by elastic, delayed elastic, non-linear creep and visco-plastic phenomena. Damage processes can occur to varying degrees in each constitutive response depending upon the rate of loading, the microstructural features of the ice and stress state. In this study we examine the influence of brittle elastic damage on the time-dependent creep and viscoplastic processes encountered during the steady indentation of a layer of polycrystalline ice by a rigid punch. A finite element technique is used to evaluate the dynamic interaction between the rigid punch and the layer of finite thickness which is underlain by a rigid base.

## CONSTITUTIVE MODELLING

The constitutive behaviour of polycrystalline material such as ice has to take into consideration a variety of time and rate dependent phenomena. For example at low rates of strain a material such as ice can experience continuous creep whereas at high rates of strain ice can experience brittle fracture [4,5]. The constitutive response of a polycrystalline material such as ice can only be formulated by recourse to experimental data which investigate the influence of loading rates. Recently Karr and Choi [6] have postulated a three-dimensional constitutive model which takes into consideration, damage processes in the elastic, delayed elastic and non-linear creep phenomena. The latter responses are also assumed to be indirectly influenced by the measure of elastic damage. Of particular interest in this model is the provision for intergranular and intragranular damage evolution. In this study, the basic model of Karr and Choi [6] is extended to include effects of viscoplasticity. We assume that the incremental strain rate $d\dot{\epsilon}_{ij}$ is composed of the following:

$$d\dot{\epsilon}_{ij} = d\dot{\epsilon}_{ij}^{(el)} + d\dot{\epsilon}_{ij}^{(de)} + d\dot{\epsilon}_{ij}^{(v)} + d\dot{\epsilon}_{ij}^{(vp)} \tag{1}$$

The incremental elastic constitutive relationships governing $d\dot{\sigma}_{ij}$ and $d\dot{\epsilon}_{ij}^{(e)}$ in the damaged material can be written as:

$$d\dot{\sigma}_{ij} = \overline{K}_{ijk\ell} d\dot{\epsilon}_{k\ell}^{(el)} \tag{2}$$

where

$$\overline{K}_{ijk\ell} = K_{ijk\ell} + C_1 \left(\delta_{ij} D_{k\ell} + \delta_{k\ell} D_{ij}\right) + C_2 \left(\delta_{ik} D_{j\ell} + \delta_{jk} D_{i\ell}\right) \tag{3}$$

In (2) $C_{ijk\ell}$ is the elasticity tensor of the undamaged material, $D_{ij}$ is the symmetric damage tensor and $C_1$ and $C_2$ are constants. A single system of parallel microcracks is defined on a plane given by the unit normals $n_i$. A measure of the effective area is $\omega_i = \omega n_i$. If there are $m$ systems of parallel micro-cracks:

$$D_{ij} = \int_{\theta_1}^{\theta_2} \int_{\varphi_1}^{\varphi_2} \omega n_i n_j \sin\phi \, d\theta \, d\varphi \tag{4}$$

with the orientation of the micro-crack system is given within the range $\theta_1 \leq \theta \leq \theta_2$ and $\varphi_1 \leq \varphi \leq \varphi_2$. The effect of damage on $d\dot{\epsilon}_{ij}^{(de)}$, $d\dot{\epsilon}_{ij}^{(v)}$ and $d\dot{\epsilon}_{ij}^{(vp)}$ occurs through an effective stress $\overline{\sigma}_{ij}$ given by:

$$\overline{\sigma}_{ij} = R_{ijk\ell} \sigma_{k\ell} \tag{5}$$

where $R_{ijk\ell} = R_{ijk\ell}\left(K_{ijk\ell}, \overline{K}_{ijk\ell}\right)$. With the aid of (5) we have

$$d\dot{\epsilon}_{ij}^{(de)} = A\left[\frac{3}{2} K \, d\,\overline{\sigma}_{ij}^{D} - d\epsilon_{ij}^{(de)}\right] \tag{6}$$

$$d\dot{\epsilon}_{ij}^{(v)} = \frac{3}{2}\, K_n \left[\frac{3}{2}\, \overline{\sigma}_{\ell k}^{D}\, \overline{\sigma}_{\ell k}^{D}\right]^{\left(\frac{n-1}{2}\right)} d\dot{\overline{\sigma}}_{ij}^{D} \tag{7}$$

$$d\dot{\epsilon}_{ij}^{(vp)} = \gamma < \Phi(F) > \frac{\partial F}{\partial \overline{\sigma}_{ij}} \tag{8}$$

where $\overline{\sigma}_{ij}^{D}$ is the effective stress deviator tensor, $A$, $K$, $K_n$ and $n$ are material constants; $\gamma$ is the fluidity parameter, $\Phi(F)$ is a monotonically increasing function for $F > 0$, $F$ is the yield function; $F = F\left(\overline{\sigma}_{ij},\ \epsilon_{ij}^{(vp)}\right) - F_0 = 0$ and $F_0$ is the uniaxial yield stress. The convention $< >$ implies that $< \Phi(F) >= \Phi(F)$ if $F > 0$; $< \Phi(F) >= 0$ if $F \leq 0$. Examples of the viscoplastic yield function are given by Perzyna [7]. The description of the evolution of damage is achieved by specifying the dependence of $\omega$ on $\epsilon_{ij}$.

For the intergranular type the damage evolution law suggested by Karr and Choi [6] takes the form:

$$\frac{d\omega^{(1)}}{d\epsilon_{nn}^{(de)}} = \beta\left[\alpha_1 - \omega^{(1)}\right] \tag{9}$$

where $\omega^{(1)} = 0$ for $d\epsilon_{nn}^{(de)} \leq d\epsilon_{cr}^{(de)}$; $\alpha_1 = \exp\left(\gamma_1 \sigma_{nn}\right)$ for $\sigma_{nn} \geq 0$; $\alpha_1 = 1$ for $\sigma_{nn} < 0$; similarly

$$\beta_1 = \begin{cases} \zeta_1 \log_{10}|\dot{\epsilon}_{nn}^{(de)}| + \eta_1\ ; & \text{for } |\dot{\epsilon}_{nn}^{(de)}| > 10^{-5} \\ -5\zeta_1 + \eta_1 & \text{for } |\dot{\epsilon}_{nn}^{(de)}| < 10^{-5} \end{cases} \tag{10}$$

For intra-granular cracks:

$$\frac{d\omega^{(2)}}{d\epsilon_{nn}} = \beta_2\left[\alpha_2 - \omega^{(2)}\right] \tag{11}$$

and similar constraints apply to $\alpha_2$ and $\beta_2$.

## THE INDENTATION PROBLEM

The constitutive model presented in the previous section has been incorporated in a finite element formulation of the problem. The finite element technique is used to examine the influence of material damage and viscoplasticity on the steady indentation of a polycrystalline ice layer, underlain by a rigid smooth boundary, by a rigid punch. The contact between the punch and the ice layer is assumed to be smooth and the contact between the rigid boundary and the layer is considered to be bonded. The plane strain problem is examined by discretizing the domain with quadrilateral elements. Appropriate mesh discretization is achieved in the vicinity of the indenting punch to accommodate the anticipated high stress gradients. Standard procedures of time integration are adopted in the dynamic analysis of indentation. The Figure 1 shows some typical results derived from the finite element study for the time history of the force induced on the indentor due to constant rate of penetration of the indentor. The results illustrate the distinction in the dynamic response derived by incorporating viscoplastic effects in the damage process.

## REFERENCES


[1] Kachanov, L.M., *Introduction to Continuum Damage Mechanics*, Martinus Nijhoff Publ., The Netherlands (1986).

[2] Krajcinovic, D., Continuum damage mechanics revisited; Basic concepts and definitions, *J. Appl. Mech.*, 52 (1985), pp.829-834.

[3] Lemaitre, J. and Chaboche, J.L., Mechanics of Solid Materials, Cambridge University Press, Cambridge, 1990.

[4] Jones, S.J., The confined compressive strength of polycrystalline ice, *J. Glaciol.*, 28 (1982), pp.171-176.

[5] Gold, L.W., Failure processes in columnar grained ice, National Research Council of Canada Tech Paper 369 (1972).

[6] Karr, D.G. and Choi, K., A three-dimensional constitutive damage model for polycrystalline ice, *Mech. Materials*, 8 (1989), pp.55-66.

[7] Perzyna, P., Fundamental problems in viscoplasticity, Advances in Applied Mech., 9 (1966), pp.244-368.


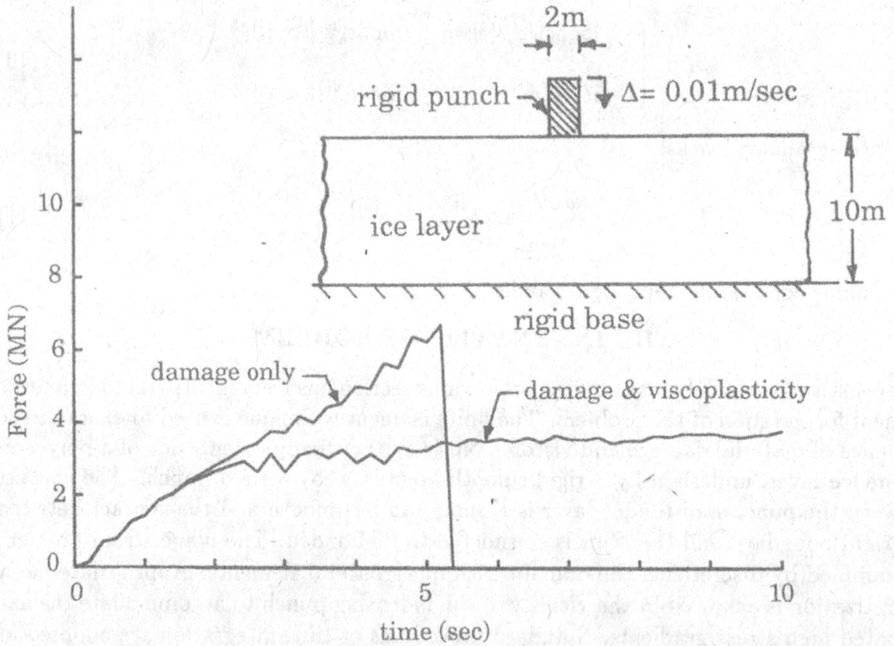


Figure 1. Dynamic indentation of ice layer. ($\dot{\Delta}$ = penetration rate of indentor, $F$ = induced contact force on indentor).

# THEORY AND FINITE ELEMENT COMPUTATION OF FINITE ELASTO–VISCO–PLASTIC STRAINS

ERWIN STEIN & CHRISTIAN MIEHE
Institut für Baumechanik und Numerische Mechanik,
University of Hanover, Germany

## INTRODUCTION

This paper presents a constitutive model of finite strain elasto–visco–plasticity suitable for metals which is highly effective for numerical implementation. Noval aspects of the proposed formulation are (i) the derivation of *power–type* associative viscoplastic evolution equations based on a penalty augmentation of the viscoplastic dissipation and (ii) a particular *logarithmic* elastic model suitable for metal elasticity. The combination of (i) and (ii) allows a direct implementation of integration algorithms used in linear viscoplasticity.

## A CONSTITUTIVE MODEL OF VISCOPLASTICITY

**Kinematical Description.** Let $\mathcal{B} \subset \mathbb{R}^3$ be the reference configuration of the body and $\varphi(\boldsymbol{X},t) : \mathcal{B} \times \mathcal{I} \rightarrow \mathbb{R}^3$ the deformation map at time $t \in \mathcal{I} \subset \mathbb{R}_+$. Within a neighborhood $\mathcal{N}(\boldsymbol{X})$ of every $\boldsymbol{X} \in \mathcal{B}$ we consider the standard multiplicative decomposition $\boldsymbol{F} = \boldsymbol{F}^e \boldsymbol{F}^{vp}$ of the tangent map $\boldsymbol{F} := T_X \varphi_t(\boldsymbol{X})$ in an elastic part $\boldsymbol{F}^e$ and a viscoplastic part $\boldsymbol{F}^{vp}$. $\boldsymbol{F}^{vp-1}$ defines *locally* a macro–stress–free *plastic intermediate configuration* and is regarded as *given* by a constitutive assumption.

Within a spatial setting of finite strain viscoplasticity it will suffice, see SIMO & MIEHE [1990], to consider as strain measure the elastic Finger tensor

$$\boldsymbol{b}^e := \boldsymbol{F}^e \boldsymbol{F}^{e\,T} \tag{1}$$

defined on the current configuration $\varphi_t(\mathcal{B})$. Time differentiation of (1) gives

$$\dot{\boldsymbol{b}}^e = \boldsymbol{l}\boldsymbol{b}^e + \boldsymbol{b}^e \boldsymbol{l}^T + \mathcal{L}_v \boldsymbol{b}^e \tag{2}$$

where $\boldsymbol{l} := \dot{\boldsymbol{F}}\boldsymbol{F}^{-1}$ is the spatial velocity gradient and $\mathcal{L}_v \boldsymbol{b}^e := \boldsymbol{F}\dot{\boldsymbol{C}}^{vp-1}\boldsymbol{F}^T$ the Lie derivative or Oldroyd rate of $\boldsymbol{b}^e$. $\boldsymbol{C}^{vp-1} := [\boldsymbol{F}^{vp\,T}\boldsymbol{F}^{vp}]^{-1}$ is the inverse plastic right Cauchy–Green tensor defined on the reference configuration $\mathcal{B}$.

**Free Energy, Stresses and Viscoplastic Dissipation.** Let $\psi$ denote the change in free energy within the isothermal viscoplastic process from the reference configuration to the current configuration. $\psi$ is assumed to be given by the constitutive expression

$$\psi = \hat{\psi}(\boldsymbol{b}^e, \alpha) \tag{3}$$

in terms of a *free energy function* $\hat{\psi}(\boldsymbol{b}^e, \alpha)$: $\mathbf{R}^6 \times \mathbf{R}_+ \to \mathbf{R}_+$ satisfying $\hat{\psi}(\mathbf{1}, 0) = 0$. Here $\alpha \in \mathbf{R}_+$ is a strain–like internal variable which describes the stored free energy due to micro–stress fields caused by dislocations and point deffects. We remark that the proposed spatial formulation (3) is automatically restricted to *isotropic response* due to the objectivity demand $\hat{\psi}(\boldsymbol{b}^e, \cdot) = \hat{\psi}(\boldsymbol{Q}\boldsymbol{b}^e\boldsymbol{Q}^T, \cdot)\ \forall\ \boldsymbol{Q} \in SO(3)$ where $SO(3)$ is the special orthogonal group.

The Clausius Planck inequality reads for the isothermal case

$$\mathcal{D}^{vp} := \boldsymbol{\tau} \cdot \boldsymbol{d} - \dot{\psi} \geq 0 \tag{4}$$

where $\boldsymbol{\tau}$ denotes the Kirchhoff stress tensor and $\boldsymbol{d} := \mathrm{sym}(\boldsymbol{l})$ the rate of deformation tensor. Time differentiation of (3) gives

$$\dot{\psi} = \left[\partial_{\boldsymbol{b}^e}\hat{\psi}\, \boldsymbol{b}^e\right] \cdot \left[2\boldsymbol{l} + (\boldsymbol{\pounds}_{\boldsymbol{v}}\boldsymbol{b}^e)\boldsymbol{b}^{e-1}\right] + \partial_\alpha\hat{\psi} \cdot \dot{\alpha}\ . \tag{5}$$

Observe that $\partial_{\boldsymbol{b}^e}\hat{\psi}$ *commutes* with $\boldsymbol{b}^e$ as a result of the isotropy assumption. Thus one obtains by insertion of (5) in (4)

$$\mathcal{D}^{vp} = \left[\,\boldsymbol{\tau} - 2\partial_{\boldsymbol{b}^e}\hat{\psi}\,\boldsymbol{b}^e\,\right] \cdot \boldsymbol{d} + 2\partial_{\boldsymbol{b}^e}\hat{\psi}\,\boldsymbol{b}^e \cdot \left[\,-\tfrac{1}{2}(\boldsymbol{\pounds}_{\boldsymbol{v}}\boldsymbol{b}^e)\boldsymbol{b}^{e-1}\,\right] - \partial_\alpha\hat{\psi} \cdot \dot{\alpha} \geq 0\,. \tag{6}$$

The standard argument of Rational Thermodynamics leads to the constitutive equations

$$\boldsymbol{\tau} = 2\,\partial_{\boldsymbol{b}^e}\hat{\psi}(\boldsymbol{b}^e, \alpha)\,\boldsymbol{b}^e \qquad \text{and} \qquad \beta := \partial_\alpha\hat{\psi}(\boldsymbol{b}^e, \alpha) \tag{7}$$

for the Kirchhoff stresses $\boldsymbol{\tau}$ and the stress–like internal variable $\beta$ conjugate to $\alpha$ (*definition*). Insertion of (7) in (6) gives the expression

$$\mathcal{D}^{vp} = \boldsymbol{\tau} \cdot \left[\,-\tfrac{1}{2}(\boldsymbol{\pounds}_{\boldsymbol{v}}\boldsymbol{b}^e)\boldsymbol{b}^{e-1}\,\right] + \beta \cdot \left[\,-\dot{\alpha}\,\right] \geq 0 \tag{8}$$

for the viscoplastic dissipation which is identical with the formulation propsed by SIMO & MIEHE [1990].

**Associative Viscoplastic Evolution Equations.** Associated viscoplastic evolution equations result from the minimization of a penalty functional based on an augmentation of the viscoplastic dissipation, see e.g. SIMO & HUGHES [1989] and MÜLLER–HOPPE & STEIN [1990]. Following this strategy we write formally the viscoplastic dissipation (8) as an inner product of *thermodynamical forces* $\boldsymbol{f}$ and *thermodynamical fluxes* $\boldsymbol{e}$

$$\mathcal{D}^{vp} = \hat{\mathcal{D}}^{vp}(\boldsymbol{f}, \boldsymbol{e}) = \boldsymbol{f} \cdot \boldsymbol{e} \geq 0 \tag{9}$$

where $\hat{\mathcal{D}}^{vp}(\boldsymbol{f}, \boldsymbol{e})$: $\mathbf{R}^{6+1} \times \mathbf{R}^{6+1} \to \mathbf{R}_+$ is the viscoplastic dissipation function based on the definitions $\boldsymbol{f} := \{\boldsymbol{\tau}, \beta\}$ and $\boldsymbol{e} := \{-\tfrac{1}{2}(\boldsymbol{\pounds}_{\boldsymbol{v}}\boldsymbol{b}^e)\boldsymbol{b}^{e-1}, -\dot{\alpha}\}$. As a main characteristic of viscoplasticity we introduce an *elastic domain* $\mathbf{E}_f$ in the space of the thermodynamical forces

$$\mathbf{E}_f := \{ \, \boldsymbol{f} \in \mathbf{R}^{6+1} \mid \hat{\phi}(\boldsymbol{f}) \le 0 \, \} \tag{10}$$

where $\hat{\phi}(\boldsymbol{f})$: $\mathbf{R}^{6+1} \to \mathbf{R}$ is a convex *flow criterion function.* Consider furthermore the penalty functional $\hat{\mathcal{P}}^{vp}(\boldsymbol{f}, \boldsymbol{e})$: $\mathbf{R}^{6+1} \times \mathbf{R}^{6+1} \to \mathbf{R}$ defined as

$$\hat{\mathcal{P}}^{vp}(\boldsymbol{f},\boldsymbol{e}) \;:=\; -\hat{\mathcal{D}}^{vp}(\boldsymbol{f},\boldsymbol{e}) \;+\; c\,\hat{\gamma}[\,\hat{\phi}^{+}(\boldsymbol{f})\,] \;. \tag{11}$$

Here $c \in \mathbf{R}_{+\backslash 0}$ is the penalty parameter, $\hat{\gamma}(\phi^{+})$: $\mathbf{R}_{+} \to \mathbf{R}_{+}$ a monotonically increasing $C^1$ *penalty function* satisfying $\hat{\gamma}(0) = 0$ and $\hat{\phi}^{+}(\boldsymbol{f})$: $\mathbf{R}^{6+1} \to \mathbf{R}_{+}$ defined by

$$\hat{\phi}^{+}(\boldsymbol{f}) \;:=\; \begin{cases} 0 & \text{if } \hat{\phi}(\boldsymbol{f}) \le 0, \\ \hat{\phi}(\boldsymbol{f}) & \text{if } \hat{\phi}(\boldsymbol{f}) > 0 \end{cases} \tag{12}$$

the *viscoplastic loading function.* We now consider the minimization problem

$$\overset{*}{\boldsymbol{f}} = \operatorname{Arg}\left\{ \underset{\boldsymbol{f} \in \mathbf{R}^{6+1}}{\operatorname{Min}} \hat{\mathcal{P}}^{vp}(\boldsymbol{f}, \overset{*}{\boldsymbol{e}}) \right\}. \tag{13}$$

where $\overset{*}{\boldsymbol{f}}$ is the minimizer of $\hat{\mathcal{P}}^{vp}$ at *fixed* $\boldsymbol{e} = \overset{*}{\boldsymbol{e}}$. The inversion of equation (13) determines the actual thermodynamical fluxes $\overset{*}{\boldsymbol{e}}$ in terms of the actual thermodynamical forces $\overset{*}{\boldsymbol{f}}$

$$\overset{*}{\boldsymbol{e}} = \lambda\, \partial_{\boldsymbol{f}} \hat{\phi}(\overset{*}{\boldsymbol{f}}) \qquad \text{with} \qquad \lambda := c\,\frac{d}{d\phi^{+}}[\,\hat{\gamma}(\,\hat{\phi}^{+}(\overset{*}{\boldsymbol{f}}))\,] \ge 0 \tag{14}$$

which results in the *associative viscoplastic evolution equations*

$$\left.\begin{aligned} -\tfrac{1}{2}\,(\boldsymbol{\mathcal{L}_v b^e})\,\boldsymbol{b}^{e-1} &= \lambda\, \partial_{\boldsymbol{\tau}} \hat{\phi}(\boldsymbol{\tau},\beta) \\ -\dot{\alpha} &= \lambda\, \partial_{\beta} \hat{\phi}(\boldsymbol{\tau},\beta) \end{aligned}\right\} \text{with} \quad \lambda := c\,\frac{d}{d\phi^{+}}[\,\hat{\gamma}(\,\hat{\phi}^{+}(\boldsymbol{\tau},\beta))\,] \ge 0 \tag{15}$$

where the parameter $c \in \mathbf{R}_{+\backslash 0}$ plays the role of an additional material parameter.

**Model Problem: $J_2$–Theory of Viscoplasticity.** Observe that in the associative case the constitutive viscoplastic equations are completely determined by *three fundamental constitutive functions*: the *free energy function* $\hat{\psi}$, see (3), the *flow criterion function* $\hat{\phi}$ in (10) and the *penalty function* $\hat{\gamma}$ in (11). As a model problem we propose the following explicit formulation of these functions which are suitable for metal–viscoplasticity:

$$\left.\begin{aligned} \hat{\psi}(\boldsymbol{b}^e,\alpha) &= \{\, \tfrac{1}{2}\kappa[\ln J^e]^2 \;+\; \tfrac{1}{4}\mu\, \operatorname{tr}[\ln \bar{\boldsymbol{b}}^e]^2 \,\} \;+\; \{\, \tfrac{1}{2} h\, \alpha^2 \,\} \\ \hat{\phi}(\boldsymbol{\tau},\beta) &= \|\, \operatorname{dev} \boldsymbol{\tau} \,\| - \sqrt{\tfrac{2}{3}}\,[\beta + y_0] \\ \hat{\gamma}(\phi^{+}) &= \frac{1}{\epsilon+1}[\,\phi^{+}\,]^{\epsilon+1} \end{aligned}\right\}. \tag{16}$$

Here the free energy function contains a decoupled macroscopic and microscopic elastic part, $\hat{\psi} = \hat{\psi}^e_{macro}(\boldsymbol{b}^e) + \hat{\psi}^e_{micro}(\alpha)$, where the latter one is associated with linear strain hardening characterized by a hardening modulus $h \in \mathbf{R}_{+}$. The macroscopic part contains a decoupled volumetric and isochoric contribution, $\hat{\psi}^e_{macro}(\boldsymbol{b}^e) =$

$\hat{\psi}^{e,vol}_{macro}(J^e) + \hat{\psi}^{e,iso}_{macro}(\bar{b}^e)$ formulated in terms of the logarithms of $J^e := \sqrt{\det b^e}$ and $\bar{b}^e := J^{e-2/3} b^e \in SL(3)$, where $SL(3)$ is the special linear group with unit determinant. $\kappa \in \mathbf{R}_{+\backslash 0}$ denotes the macroscopic bulk modulus and $\mu \in \mathbf{R}_{+\backslash 0}$ the macroscopic shear modulus. The flow criterion function $\hat{\phi}$ is the standard $J_2$ v. Mises function formulated in terms of the Kirchhoff stresses where $y_0 \in \mathbf{R}_+$ is the initial flow stress. The penalty function $\hat{\gamma}$ with exponent $\epsilon \in \mathbf{R}_{+\backslash 0}$ characterizes a power law typically used in metal–viscoplasticity, see e.g. MÜLLER–HOPPE & STEIN [1990].

The exploitation of (7) results in the constitutive equations for the Kirchhoff stresses $\boldsymbol{\tau}$ and the stress–like internal variable $\beta$

$$\left.\begin{aligned} \boldsymbol{\tau} &= \kappa[\ln J^e]\mathbf{1} + \mu[\ln \bar{\boldsymbol{b}}^e] \\ \beta &= h\,\alpha \end{aligned}\right\} . \tag{17}$$

where we have used the relationship $\operatorname{tr}[\ln \bar{\boldsymbol{b}}^e] = \ln[\det \bar{\boldsymbol{b}}^e] = 0$. The constitutive model is completed by the associated viscoplastic evolution equations. Exploitation of (15) yields

$$\left.\begin{aligned} -\tfrac{1}{2}(\mathcal{L}_v \boldsymbol{b}^e)\boldsymbol{b}^{e-1} &= \lambda \boldsymbol{n} \\ \dot{\alpha} &= \lambda\sqrt{\tfrac{2}{3}} \end{aligned}\right\} \quad \text{with} \quad \left\{\begin{aligned} &\boldsymbol{n} := \operatorname{dev}\boldsymbol{\tau}\,/\,\|\operatorname{dev}\boldsymbol{\tau}\| \\ &\lambda := c\,[\phi^+]^\epsilon \geq 0\ . \end{aligned}\right. \tag{18}$$

The flow rule $(18)_1$ is *exact volume perserving*, see SIMO & MIEHE [1990]. Observe that the set (16)–(18) of constitutive equations describes for $y_0 = h = 0$ a finite strain version of the classical Norton exponential law of creep for the isochoric part of the deformation. Setting in addition $\epsilon = 1$ the isochoric part of the deformation behaves like a linear Newton fluid where $\eta := c^{-1}$ is the linear viscosity.

## SUMMARY

We proposed a particular constitutive model of isotropic associated finite strain viscoplasticity which contains a *power–type* viscoplastic flow rule formulated in terms of the Lie derivative of $\boldsymbol{b}^e$, see (18), and an elastic model in terms of the *logarithm* of the elastic Finger tensor, $\ln \boldsymbol{b}^e$, see $(16)_1$. This model allows a direct implementation of integration algorithms used in linear viscoplasticity. We discuss the algorithmic setting and finite element implementation in the extended version of this paper.

## REFERENCES

Müller–Hoppe, N. ; Stein, E. [1990], "On Finite Elastoviscoplastic Strains in Spatial Description", Formulation, Numerical Analysis and Implementation", to appear in *Journal of Mechanics and Physics of Solids.*

Simo, J.C.; Hughes, T.J.R. [1989], "Elastoplasticity and Viscoplasticity — Computational Aspects", to appear in Springer Verlag Series in Applied Mathematics.

Simo, J.C.; Miehe, C. [1990], "Associated Coupled Thermoplasticity at Finite Strains: Formulation, Numerical Analysis and Implementation", to appear in *Computer Methods in Applied Mechanics and Engineering.*

# GENERATED FORCE AND DEFORMATION AT A CONICAL IMPACT-END OF ALUMINUM BAR

SHINJI TANIMURA
Department of Mechanical Engineering,
University of Osaka Prefecture
Mozu-Umemachi, Sakai, Osaka 591, Japan

## ABSTRACT

A brief survey of recent studies on local plastic deformation during collision of two bodies and on generation of impact force at the contact part is made. Presented are recently obtained experimental results, when aluminum bars with a circular truncated cone end, collide axially, perpendicularly, against a plane wall at a comparatively low speed. It is confirmed that the proposed theoretical relations can be used to predict the impact force generated at the contact part where pulse shape sensitively varies with the shape of the contact part of a colliding body, as well as with the size of the body and the collision velocity.

## INTRODUCTION

To design a machine or a structure which prevents fracture or large deformation under impact loading caused by a collision, it is fundamentally important to predict the distribution of the generated impact force at the region of contact.

In many cases of collisions, for example, between machine parts, or machine to machine or to structure, as well as collisions of flying object against a structure or machine, the speed of the collisions is lower than a few 10 m/s. Plastic deformation usually occurs at the contact part, its behaviour sensitively affects the generation of the impact forces. To measure such impact forces directly and accurately, a new method was devised by Tanimura [1, 2]. Using this method, generation of the impact force and local plastic deformation at the contact part were investigated using aluminum bars of various sizes whose conical ends were made to collide against the sensing plane wall [3]. Theoretical relations were then formulated by which the impact force generated at a contact part can be calculated [4, 5].

In this paper, a brief survey of recent studies on the local plastic deformation at the contact part of such collisions and the generation of impact forces is made. New results are presented, which have been obtained by carrying out experiments in which aluminum bars with circular truncated cone ends were made to collide axially, perpendicularly, against the devised sensing plane wall. It was confirmed that the derived theoretical relations can be used to predict fairly well the impact force generated at the contact part whose pulse shape varies sensitively with the shape of the contact part of a colliding body, the size of the body and with the collision velocity.

## THEORETICAL RELATIONS

### (a) Relation Taking no Account of the Stress Wave in a Colliding Body

Experiments in which aluminum and pure iron bars whose ends were of varying shapes of truncated cones were collided axially, perpendicularly, against a plane wall established the following phenomena [3, 4, 5];

(1) The contact part during collision keeps the form of circular truncated cone as shown by the shadowed portion in FIGURE 1.

(2) The conical angle of the forming circular cone, $\theta'$, depends only on the conical angle of the truncated circular cone before the collision, $\theta$, so that the relation $\theta' = \theta'(\theta)$ is held which is not dependent on the length of the colliding bar, diameter of the impact end of the cone, $d_1$, and collision velocity $v_0$, excepting the initial time of contact of the collision.

(3) During the collision, the mean stress $\sigma_2$ on the boundary face of diameter $d_2'$ shown in FIGURE 1 does not depend on L, $\theta$, $d_1$ and $v_0$, it behaves like a material constant of the colliding body.

Following the experimental evidence as mentioned above (2),

$$d_1'/2 = xk - hp \;, \quad d_2'/2 = xk \tag{1}$$

hold, where $k=\tan(\theta/2)$, $p=\tan(\theta'/2)$, and symbols $x$, $d_1'$,......, are as shown in FIGURE 1. When we put $\gamma(x) \equiv h/x$, and $\gamma'$ denotes the $\gamma$ when $x_0=0$, the $\gamma'$ is expressed by

$$\gamma' = \frac{k}{p}\{1 - (1 - \frac{p}{k})^{1/3}\} \; . \tag{2}$$

Because of $\theta' = \theta'(\theta)$ as shown above (2), we can see that the value of $\gamma'$ depends only on the value of $\theta$.

The impact force generated at the contact part, F(t), can be expressed by

$$F(t) = \pi(d_1'/2)^2\sigma_1 \; , \tag{3}$$

where $\sigma_1$ is the mean nominal stress on the contact end face of the diameter $d_1'$. Assuming $d_1'^2\sigma_1 = d_2'^2\sigma_2$, F(t) is given by

$$F(t) = \pi^2k^2\sigma_2x^2 \; . \tag{4}$$

The x-t relation has been derived by Tanimura et al. [4] as follows;

$$t = \int_{d_1/2(k-\gamma'p)}^{x} \frac{dx}{\sqrt{(a-bx^{-3})^{-2}[v_0^2(a-bx_0^{-3})^2 + 2K\{\frac{a}{3}(x_0^3 - x^3) - b\cdot\log\frac{x_0}{x}\}]}} \;, \tag{5}$$

where $v_0$ is the collision velocity (initial velocity of the colliding body), $a \equiv 1-\gamma'$, $b \equiv (2/3+4p/9k+5p^2/9k^2)x_0^3$, $K \equiv \pi k^2\sigma_2/M$, and M is the mass of the colliding body. Using relations (4) and (5), we can easily predict the impact force F(t) generated at the contact part with the help of a simple desk-top computor, when the size of the colliding body (M), the material parameter ($\sigma_2$) and the shape of the impact end $(\theta, x_0)$ are given.

**(b) Relation Taking Account of Stress Waves in a Colliding Body**

When a comparatively long bar collides axially, perpendicularly, against a plane wall the generated stress wave at the contact part is propagated into the bar, gets reflected at the free end of the bar, and is again reflected at the impact end. These reflections are repeated back and forth. The generated impact force $F_n(t)$, which was affected by the unloading wave reflected the nth time at the free end of the colliding bar, has been derived by Tanimura et al. [5] as follows;

$$\frac{dF_n(t)}{dt} = \{\frac{2v_0}{3A}F_n(t) - \frac{2c_0}{3AS_0E}F_n^2(t) - \frac{4c_0}{3AS_0E}\sum_{l=0}^{n-1} F_n(t)F_l(t - \frac{2(n-1)/L}{c_0})\}/F_n^{1/2}(t), \tag{6}$$

$$F_n(\frac{2nL}{c_0}) = F_{n-1}(\frac{2nL}{c_0}),$$

where t is the elapsed time from the initiation of the collision, $S_0$ is the cross section of the colliding bar, E is the Young's modulus, $c_0$ is the elastic wave velocity of the bar, and A is a coefficient whose value is determined by the shape of the contact part (k and p), the mean stress $\sigma_1$ on the contact face and the dynamic yield stress $\sigma_y$ of the colliding bar.

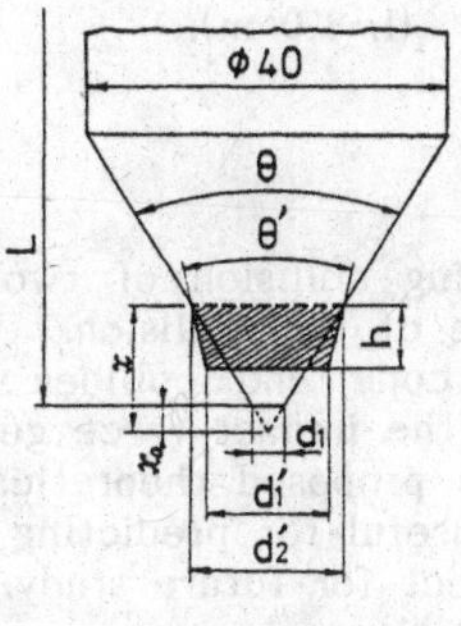


FIGURE 1. Shape of impact end.

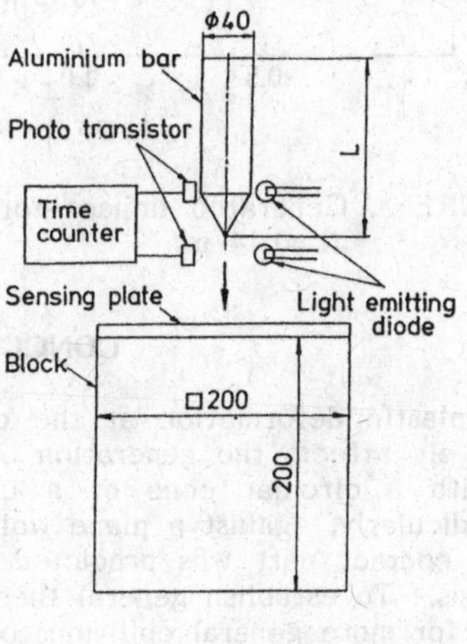


FIGURE 2. Schematic diagram of collision testing apparatus.

## EXPERIMENTAL RESULTS AND DISCUSSIONS

An aluminum bar whose end was a circular cone or truncated cone (frustum) which was collided axially, perpendicularly, against the impact surface of a sensing plate, devised by Tanimura [3, 4, 5], is shown in FIGURE 2. The sensing plate was set in close contact with a steel block. On the reverse side of the impact surface of the sensing plate, pits were bored in which small strain gages were cemented.

A typical record of the generated impact force when the aluminum bar of 40 mm diameter and 0.14 m length was collided at a velocity of 10.15 m/s against the impact surface is shown by the dotted curve in FIGURE 3. The solid curve in the figure is the calculated result from Eqs. (4) and (5) where $\sigma_2$=100 MPa was chosen as a material parameter.

Another typical record of the impact force when an aluminum bar of 1.0 m length collided with the impact face after a drop from 0.2 m height is shown by the dotted curve in FIGURE 4. The solid curve is the calculated result from Eq. (6).

The other calculated results of the impact forces when the impact conditions were varied, have also agreed fairly well to the experimental records as shown in FIGURES 3 and 4.

It is, therefore, confirmed that the theoretical relations of Eqs. (4) and (5) can be used to predict the generated impact force for the collision of a comparatively short bar, and that Eq. (6) can be used well for the collision of a comparatively long bar.

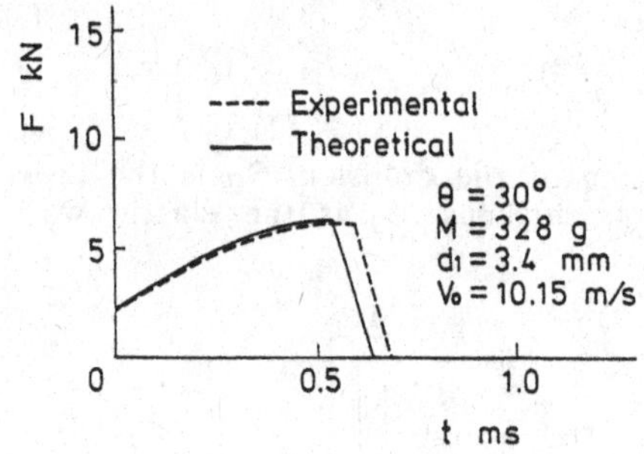


FIGURE 3. Generated impact force (L=0.14 m).

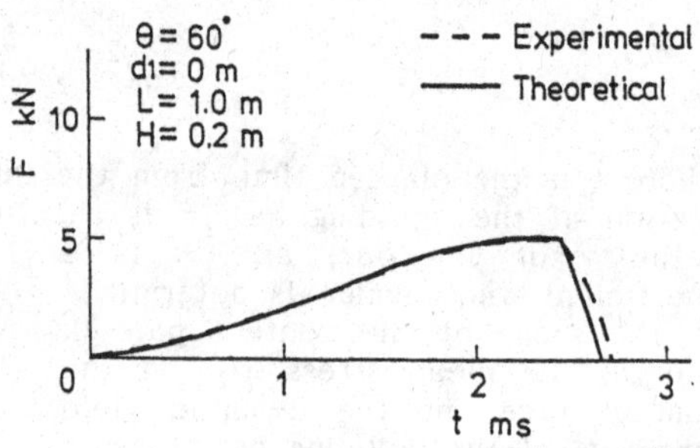


FIGURE 4. Generated impact force (L=1.0 m).

## CONCLUDING REMARKS

Local plastic deformation at the contact region during collision of two bodies sensitively affects the generation of the impact force of the collision. When a bar with a circular cone or a circular truncated cone end collides axially, perpendicularly, against a plane wall at a low speed the impact force generated at the contact part was predicted fairly well by the proposed theoretical relationships. To establish general theoretical relations, useful for predicting impact forces for more general collisions of bodies, is a subject for future study.

## REFERENCES

1. Tanimura, S., A new method for measuring impulsive force at contact parts. Exp. Mech., 1984, **24**, 271-276.
2. Daimaruya, M., Naitoh, M., Kobayashi, H. and Tanimura, S., A sensing-plate method for measuring force and duration of impact in elastic-plastic impact of bodies. Exp. Mech., 1989, **29**, 268-273.
3. Tanimura, S. and Aiba, M., On the deformation and impulsive force at a protrudent impact end in collision of an aluminium bar against a wall. Trans. JSME (in Japanese), 1984, **50**, 2009-2016.
4. Tanimura, S., Chuman, Y. and Akaishi, Y., Generated force and deformation at a conical end of aluminium bar due to impact. Trans. JSME (in Japanese), 1986, **52**, 1142-1147.
5. Tanimura, S. Sasaki, Y., Kaizu, K., Higashi, K. and Isuzugawa, K., Generated force and deformation at a conical end of aluminum bar due to impact (Theory). Trans. JSME (in Japanese), 1990, **56**, 2255-2261.

# A CONSTITUTIVE EQUATION DESCRIBING STRAIN HARDENING, STRAIN RATE SENSITIVITY, TEMPERATURE DEPENDENCE AND STRAIN RATE HISTORY EFFECT

SHINJI TANIMURA* and KOICHI ISHIKAWA**
* Department of Mechanical Engineering,
University of Osaka Prefecture
Mozu-Umemachi, Sakai, Osaka 591, Japan
** Osaka Municipal Technical Research Institute
Morinomiya, Joto-ku, Osaka 536, Japan

## ABSTRACT

A constitutive equation with which strain hardening, strain rate sensitivity, temperature dependence and strain rate history effect can be described covering a wide range of strain rates is proposed. To describe the strain hardening and the strain rate history effect, a new idea which takes into account the plastic work in a manner of fading memory is introduced. It is illustrated that material behaviors observed not only in constant strain rate tests but also in jump tests can be represented fairly well by the proposed equation.

## INTRODUCTION

It is well known that the flow stress depends not only on the current strain, strain rate and temperature but also on the history of strain rate and temperature (for example, [1]). Quite a number of studies have been done on the constitutive equation describing strain rate history effect (for example, [2, 3]).

One of the authors of this study had proposed a constitutive equation for elastic/viscoplastic body covering a wide range of strain rates [4, 5]. In this paper, the constitutive equation is extended to describe the strain hardening and the strain rate history effect by introducing the plastic work into "back stress" and by taking into account the plastic work in a manner of fading memory. The obtained constitutive equation is first compared with the monotonic stress-strain curves for 1100-0 aluminum at constant strain rates. The response in the strain rate jump tests is investigated by changing the parameters included in the constitutive equation.

Although the extended constitutive equation includes a few more parameters, these parameters can be evaluated comparatively easily from experimental results by using a desk-top computor.

## ANALYTICAL FORMULATION

### Fundamental Constitutive Equation

In simple torsion of a thin walled tube, the constitutive equation proposed by Tanimura [4] is reduced to Eqn. (1).

$$\dot{\gamma} = \frac{\dot{\tau}}{2G} + \eta^*(\tilde{e}^P)\exp\{\frac{-D(\tilde{e}^P, T)}{\tau - \tau^*(\tilde{e}^P, T)}\} , \quad (1)$$

where $\tilde{e}^P = \int_0^t \frac{2}{\sqrt{3}}|\dot{\gamma}^P|dt' \equiv \frac{2}{\sqrt{3}}\tilde{\gamma}^P$, $D(\tilde{e}^P, T) = D_0(T) + K(T)(\tilde{e}^P)^n$,

$\gamma$ (tensor definition) is shear strain, $\tau$ is shear stress, G is the shear modulus, $\eta^*(\tilde{e}^P)$ is a pre-exponential factor, $\tau^*(\tilde{e}^P,T)$ is the back stress due to the work hardening, $\tilde{e}^P$ is the integrated inelastic strain, T is absolute temperature, $D(\tilde{e}^P,T)$ is a parameter which exhibits the strain rate sensibility, t' is the time variable and n denotes the index of work hardening. For aluminum, $K(T) \propto T$ or $K(T) \propto T^{1/2}$ is approximately held. It may be approximately assumed that at constant temperature, $D_0(T)=0$ and $n=1/2$ for aluminum.

When we deal with the behavior at a constant temperature, $D(\tilde{e}^P,T)=K_0 \times(\tilde{e}^P)^{1/2}$, where $K_0$ is a constant.

The constitutive equation proposed by Kocks et al. [6] is rewritten as follows;

$$\dot{\gamma}^P = \eta^*\exp[-\frac{\Delta F}{kT}\{1-(\frac{\tau_t}{\hat{\tau}})^p\}^q] , \quad (2)$$

where $\eta^* = \eta_0^*(\frac{\tau_t}{\mu})^2$, $\tau_t = \tau - \tau^*$, $0 \le p \le 1$, $1 \le q \le 2$,

$\dot{\gamma}^P$ is inelastic shear strain rate, $\tau_t$ is the effective stress, $\eta_0^*$ is a frequency, $\mu$ is the shear modulus, $\Delta F$ is the Helmholtz total free energy, k is Bolzmann's constant, T is absolute temperature, and $\hat{\tau}$ is the flow stress at absolute zero.

### Expansion of the Fundamental Constitutive Equation

To describe strain hardening and the strain rate history effect, we assume that the strain hardening and the strain rate history effect are mainly caused by $\tau^*$. Thus $\tau^*$ is a function of the plastic work, which is taken account of in a manner of fading memory. As a possible form, $\tau^*$ can be expressed as follows;

$$\tau^*(\int dW_p') = \tau_0^* + H\int_0^{\tilde{\gamma}^P} [1-h+h\cdot\exp\{-(\tilde{\gamma}^P-\tilde{\gamma}^{P\prime})/C\}]2\tau' d\tilde{\gamma}^{P\prime} , \quad (3)$$

where $dW_p'=2\tau' d\tilde{\gamma}^{P\prime}$, $\tau'$ is the stress variable, $\tilde{\gamma}^{P\prime}$ is the strain variable, $\tilde{\gamma}^P$ is the current strain, $\tau_0^*$ is the value of $\tau^*$ at the initial state, $\tilde{\gamma}^P=0$ (which ordinarily corresponds to t=0), H is a parameter with which the plastic work is transformed into the increment of $\tau^*$, and h and C are parameters which are related to the fading memory.

Assuming Mises yield condition, an expanded form in the uni-axial stress state of Eqn. (1) becomes as follows;

$$\sigma(\tilde{\varepsilon}^p)=\sqrt{3}\tau_0^* + \sqrt{3}H\int_0^{\tilde{\varepsilon}^p} [1-h+h\cdot\exp\{-\sqrt{3}(\tilde{\varepsilon}^p-\tilde{\varepsilon}^{p\prime})/C\}]\sigma' d\tilde{\varepsilon}^{p\prime} + \frac{\sqrt{3}K_0(\tilde{\varepsilon}^p)^{1/2}}{\ln\{\eta^*/(\sqrt{3}\dot{\varepsilon}^p/2)\}}, \quad (4)$$

where $\sigma$ is the current stress, $\tilde{\varepsilon}^p$ is the current strain, $\dot{\varepsilon}^p$ is the strain rate, $\sigma'$ is the stress variable and $\tilde{\varepsilon}^{p\prime}$ is the strain variable.

Eqn. (2) can be expanded in the same manner as Eqn. (1).

## CALCULATED RESULTS AND DISCUSSION

Here, we discuss the calculated results from Eqn. (4).

### Constant Strain Rate Test

The stress-strain curves which were re-expressed on the basis of the experimental results obtained for 1100-0 aluminum by Lindholm [7] are shown in broken curves in FIGURE 1. These are the curves at the strain rates of $10^{-3}$, $10^0$ and $10^3\ s^{-1}$. The parameters in Eqn. (4) were chosen by fitting calculated stress-strain curves to the experimental ones. $\tau_0^*=11.5$ MPa and $\eta^*=10^8\ s^{-1}$ were assumed, and the other parameters were as follows; H=106, h=0.9968, C=0.006 and $K_0$=345 MPa. The solid curves in FIGURE 1 represent the curves calculated for the set of these values of the parameters. It can be seen that the calculated curves coincide fairly well with experimental ones.

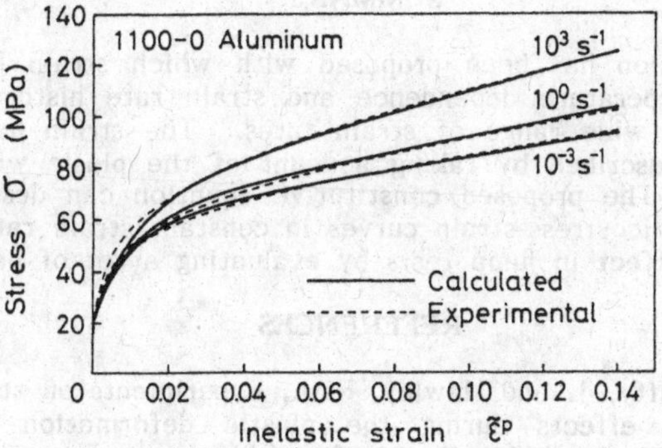


FIGURE 1. Experimental and calculated stress-strain curves for 1100-0 aluminum at constant strain rates.

### Strain Rate Jump Test

The calculated curves obtained from Eqn. (4) when the strain rate jumps rapidly from $10^{-3}$ to $10^3\ s^{-1}$ at the strain of 0.02, 0.04 or 0.08 are shown in FIGURE 2. These curves are calculated by using the same values of parameters which are given above. The dotted curves are the ones at constant strain rates of $10^{-3}\ s^{-1}$ and $10^3\ s^{-1}$. It is observed that the jumped curves gradually approach the monotonic stress-strain curves at the constant strain rates. The chained curve in FIGURE 2 illustrates the change of "back stress", $\sqrt{3}\,\tau^*$, before and after jumping of the strain rate at the strain of 0.08. It is observed that the change of $\sqrt{3}\tau^*$ is accompanied by that of the strain rate. In the case of the decreasing strain rate jump, the same behavior as the reported experimental results can be simulated by using the same set of the parameters. The material behavior under the some types of strain rate jump tests can also be shown with the other sets of parameters in FIGURE 3. As the value of the parameter, C, becomes smaller, the response of the stress-strain curve becomes more rapid.

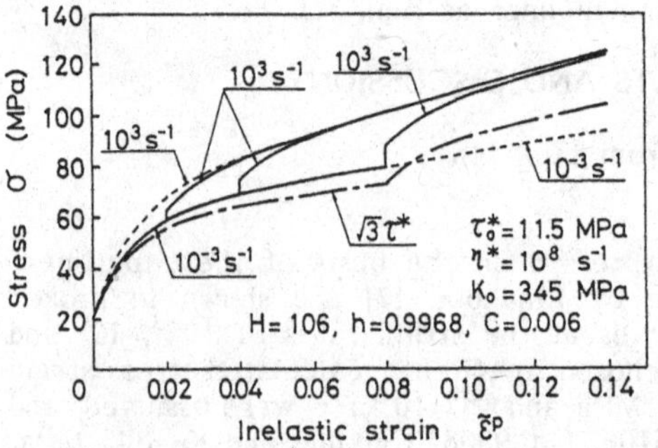


FIGURE 2. Calculated stress-strain curves in strain rate change test.

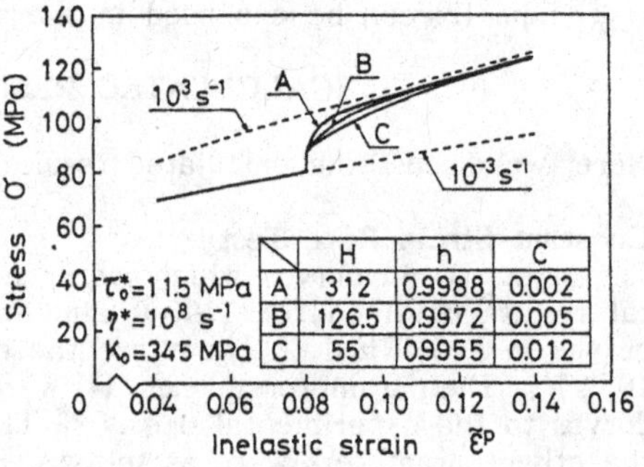


FIGURE 3. Response under some types of strain rate change test.

## SUMMARY

A constitutive equation has been proposed with which strain hardening, strain rate sensitivity, temperature dependence and strain rate history effect can be described covering a wide range of strain rates. The strain hardening and the history effect are described by taking account of the plasic work in a manner of fading memory. The proposed constitutive equation can describe fairly well not only the monotonic stress-strain curves in constant strain rate tests but also strain rate history effect in jump tests by evaluating a set of parameters.

## REFERENCES

1. Senseny, P.E., Duffy, J. and Hawley, R.H., Experiments on strain rate history and temperature effects during the plastic deformation of close-packed metals. J. Appl. Mech., 1978, **45**, 60-66.
2. Bodner, S.R. and Merzer, A., Viscoplastic constitutive equation for copper with strain rate history and temperature effects. J. Engng. Mater. Tech., 1978, **100**, 388-394.
3. Klepaczko, J. and Duffy J., Strain rate history effects in bcc metals. Tech. Rep. Dev. Eng. Brown Univ. CME 79-23742/2 MRLE-128, 1981.
4. Tanimura, S., A practical constitutive equation covering a wide range of strain rates. Int. J. Engng. Sci., 1978, **17**, 997-1004.
5. Tanimura, S. and Igaki, H., A constitutive equation for elastic/viscoplastic media covering a wide range of strain rates. J. Soc. Mater. Sci., Japan, 1980, **29**, 137-142 (in Japanese).
6. Kocks, U.F., Argon, A.S. and Ashby, M.F., Thermodynamics and Kinetics of Slip, Pergamon Press, New York, 1975.
7. Lindholm, U.S., Some experiments in dynamic plasticity under combined stress. In Mechanical Behavior of Materials under Dynamic Loads, ed. U.S. Lindholm, Springer-Verlag, Berlin, 1968, pp. 77-95.

# SOME BASIC PROPERTIES OF CONSTITUTIVE MODELS IN PLASTICITY AND VISCOPLASTICITY

CHARALAMPOS TSAKMAKIS and PETER HAUPT
University of Kassel, Institute of Mechanics
Mönchebergstr. 7
3500 Kassel, Germany

## ABSTRACT

A particular nonradial stress-controlled homogeneous process is analyzed, for which experimental data are known from the literature. First, we apply a rate-independent plasticity theory, which is based on the concept of a yield surface. Then, we investigate the case that the diameter of the yield surface is zero, which corresponds to a vanishing elastic region. Finally, we calculate the given process on the basis of a constitutive model of rate-dependent viscoplasticity. The numerical results suggest, that in view of a representation of nonradial behavior there exist some qualitative differences between the predictions, which are based on constitutive models of plasticity and viscoplasticity.

## CONSTITUTIVE MODELS OF RATE INDEPENDENT PLASTICITY

### Rate-independent plasticity

The following analysis is restricted to small deformations. Second order tensors are denoted by boldface letters. The inner product of two tensors **A** and **B** is $\mathbf{A}\cdot\mathbf{B} = \mathrm{tr}(\mathbf{A}\mathbf{B}^T)$, where $\mathbf{B}^T$ is the transpose of **B**. $\mathbf{A}'$ denotes the deviator of **A**, **1** the unit tensor and $(\ )^{\cdot}$ the material time derivative of a field ( ).

We assume the additive decomposition of the linearized total strain tensor **E** into elastic and plastic parts, a linear and isotropic elasticity relation, a v. Mises yield function and (for the case of loading in the plastic range) the associated flow rule:

$$\mathbf{E} = \mathbf{E}_e + \mathbf{E}_p \tag{1}$$

$$\mathbf{T} = 2\mu\left[\mathbf{E}_e + \frac{\nu}{1-2\nu}\,\mathrm{tr}(\mathbf{E}_e)\mathbf{1}\right] \tag{2}$$

$\mu$ and $\nu$ are the shear modulus and Poisson's ratio, respectively.

$$f(\mathbf{T},\mathbf{X}) = \tfrac{1}{2}(\mathbf{T}-\mathbf{X})' \cdot (\mathbf{T}-\mathbf{X})' - \tfrac{1}{3}k^2 \tag{3}$$

$$\dot{\mathbf{E}}_p = \lambda(\mathbf{T}-\mathbf{X})' \tag{4}$$

$\mathbf{X}$ is the center of the yield surface and $k$ the yield stress. The factor $\lambda$ is determined from the consistency condition $\dot{f} = 0$.

**Kinematic hardening**

The following hardening model has been proposed by Chaboche and Rousselier [1]:

$$\dot{\mathbf{X}} = \dot{\overline{\mathbf{X}}} + c\,\dot{\mathbf{E}}_p - b\,\dot{s}(\mathbf{X} - \overline{\mathbf{X}}), \qquad \dot{\overline{\mathbf{X}}} = \overline{c}\,\dot{\mathbf{E}}_p - \overline{b}\,\dot{s}\,\overline{\mathbf{X}} \tag{5}$$

The variable $s$ is the accumulated plastic strain, defined by $\dot{s}(t) = \sqrt{\tfrac{2}{3}\dot{\mathbf{E}}_p \cdot \dot{\mathbf{E}}_p}$ .
The special case $\overline{c} = \overline{b} = 0$ corresponds to the Armstrong/Frederick model [2].

**Vanishing elastic range: k = 0**

The case of a vanishing elastic region corresponds to a vanishing yield stress, i. e., $k = 0$. This implies $\dot{f} \equiv 0$ or $f \equiv 0$ . Thus, we always have plastic loading; a case of neutral loading does not occur. For a vanishing elastic region the stress deviator $\mathbf{T}'$ equals the kinematic hardening variable $\mathbf{X}$, and no flow rule is necessary: In fact, we infer from the hardening model (5), that the plastic strain rate is immediately related to the stress, $\mathbf{X} = \mathbf{T}'$, which itself is related to the elastic strain through the deviatoric part of eq. (2). The decomposition (1) of the strain into elastic and plastic parts completes this special set of constitutive equations: For a given strain process $\mathbf{E}(t)$ equations (1), (2) and (5) are sufficient to determine the functions $\mathbf{T}(t)$, $\mathbf{E}_e(t)$ and $\mathbf{E}_p(t)$. However, in this case we have a system of implicit differential equations, which cannot be solved explicitely with respect to the time derivatives. This has to be observed in view of the numerical integration.

## CONSTITUTIVE MODELS OF RATE DEPENDENT VISCOPLASTICITY

A representative constitutive model of viscoplasticity is defined as follows: In the decomposition (1) we replace the plastic strain $\mathbf{E}_p$ by the inelastic strain $\mathbf{E}_i$ and assume instead of eq. (4) for the inelastic strain rate the flow rule

$$\dot{\mathbf{E}}_i = \Phi(\langle F \rangle)\,\frac{\partial f}{\partial \mathbf{T}},$$

with $\quad F = \dfrac{f-k}{r} \quad$ and $\quad f = \sqrt{\tfrac{3}{2}(\mathbf{T}-\mathbf{X})' \cdot (\mathbf{T}-\mathbf{X})'}$ .

$k$ and $r$ are material parameters and $\Phi$ is a material function with $\Phi(0) = 0$. The symbol $\langle\ \rangle$ means that $\langle F \rangle = F$ or $\langle F \rangle = 0$ if $F$ takes positive or negative values, as it is possible in contrast to rate-independent plasticity. More specifically, we assume $\Phi(\alpha) = \alpha^m$ with a real valued parameter $m$, such that we have

$$\dot{\mathbf{E}}_i = \left\langle \frac{f-k}{r} \right\rangle^m \frac{3}{2f}(\mathbf{T}-\mathbf{X})' \ . \tag{6}$$

The constitutive equations are now eqs. (1), (2), (6) and the hardening model (5).

## STRESS CONTROLLED PLANE LOADING PROCESSES

Figure 1 shows a nonradial loading path (**O-A-B-C-O**) and a stress strain curve from experiments on AISI 304, which have been taken from the literature [3]. Figures 2 and 3 illustrate the response of different constitutive models of rate-independent plasticity. We observe, that at the point **A** a discontinuous change of the stress increment occurs (there is a very short way through the elastic range), which gives rise to a jump of the slope of the stress strain curve (see Figures 2 and 3). This particular shape of the curve is unrealistic in comparison with the experimental curve and it is a characteristic property of the rate-independent constitutive theory; in fact, it occurs even if the radius of the yield surface is set to zero (Figure 3). The curvature of the observed stress strain curve near the point **A** must be understood to be a consequence of the rate-dependence of the real material. Indeed, Figure 4 shows, that the response of the viscoplastic model yields a stress strain curve with a more realistic shape in the neighborhood of the point **A**. As a concluding conjecture we presume, that in certain non-radial loading situations only a constitutive theory of rate-dependent material behavior can imply a good qualitative agreement between theory and experiment.

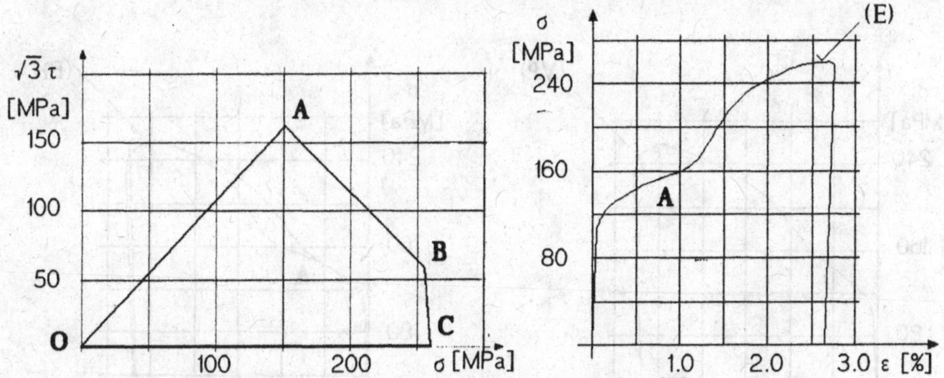


Figure 1. Nonradial loading path and experimental results (E)

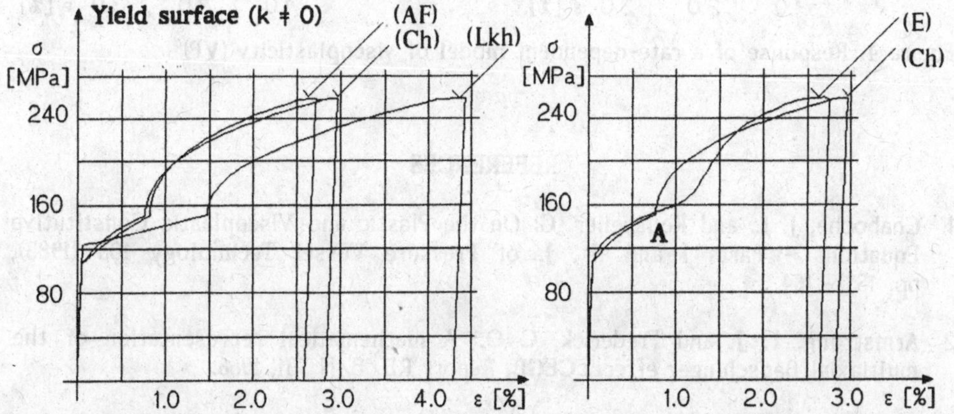


(AF) = Armstrong-Frederick, (Ch) = Chaboche, (Lkh) = Linear kinematic hardening, (E) = Experiment

Figure 2. Response of different models of rate-independent plasticity (k ≠ 0)

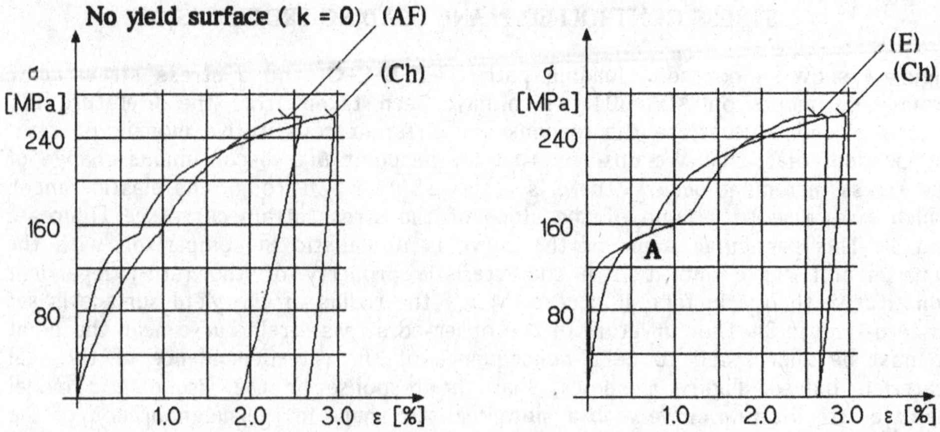


(AF) = Armstrong-Frederick , (Ch) = Chaboche, (E) = Experiment

Figure 3. Response of different models of rate-independent plasticity (k = 0)

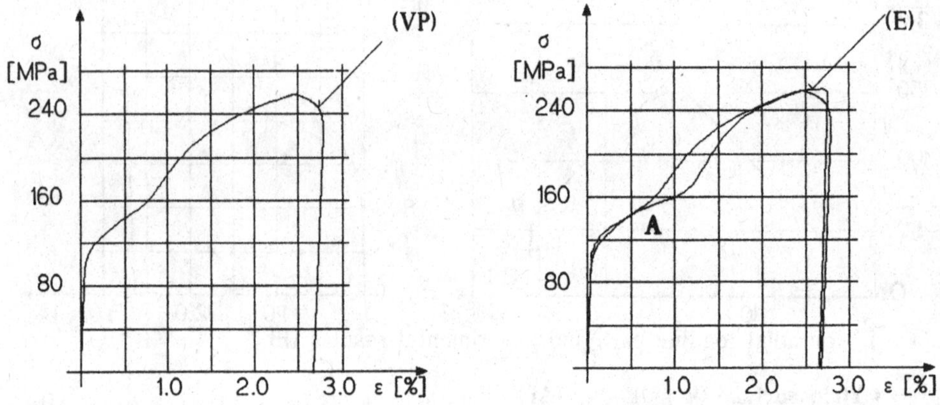


Figure 4. Response of a rate-dependent model of viscoplasticity (VP)

## REFERENCES

1. Chaboche, J. L. and Rousselier, G: On the Plastic and Viscoplastic Constitutive Equations - Parts I and II , J. of Pressure Vessel Technology 105 (1983), pp. 153 - 164

2. Armstrong, P. J. and Frederick, C. O.: A mathematical representation of the multiaxial Bauschinger effect. CEGB, Report RD/B/N 731, 1966.

3. Bruhns, O.T. and Müller, R: Some Remarks on the Application of a Two-Surface Model in Plasticity. Acta Mechanica 53 (1984), pp. 81 - 100.

# YIELDING OF MILD STEEL AFTER HYDROSTATIC PRESSURIZATION

F. YOSHIDA, M. ITOH and M. OHMORI
Department of Mechanical Engineering,
Hiroshima University
Higashi-Hiroshima 724, Hiroshima, Japan

## ABSTRACT

A sharp yield point of mild steel is absent or substantially reduced in specimens subjected to high hydrostatic pressure prior to tension. The effect of high hydrostatic pressurization on the yielding behavior in the subsequent uniaxial tension is investigated by means of FEM simulation. A unit-cell model of an elastic-viscoplastic matrix with an elastic inclusion is used for the analysis. This analysis shows that the yield stress reduces markedly, only when the matrix has a sharp yield point and abrupt yield-drop characteristics, and there exists an elastic inhomogeneity between the matrix and the inclusion.

## INTRODUCTION

Mild steel normally shows a sharp yield point in tension at atmospheric pressure. This yield is absent or substantially reduced in specimens subjected to high hydrostatic pressure prior to tension[1]. Bullen et al.[1] explained that the disappearance of the yield point is due to free dislocation created during pressurization around elastic inhomogeneities, such as particles of second phase or voids.

The aim of this paper is to show how the plastic deformation is created around the inhomogeneities of the specimen during the pressurization, and to discuss the effect of the plastic deformation zone on the disappearance of the yield point in the subsequent uniaxial tension by means of FEM simulation. A unit-cell model of an elastic-viscoplastic matrix with an elastic inclusion is used for the analysis. For the matrix of the unit-cell, Hahn's constitutive equation[2] which describes the sharp yield drop based on dislocation multiplication and velocity characteristics is employed. Three different matrices, with or without a sharp yield drop, having different strain hardening, are examined. The effects of volume fraction and rigidity of the inclusion on the yielding are also studied.

## ANALYSIS

Mild steel is modeled as a cubic unit-cell of elastic-viscoplastic matrix with a cylindrical elastic inclusion, as illustrated in Fig. 1. The stress/strain response of the matrix is described by Hahn's constitutive equation[2] as

$$\dot{\bar{\varepsilon}}^p=0.5bf[\rho_o+C(\bar{\varepsilon}^p)^a](2\tau_o)^{-n}(\bar{\sigma}-q\bar{\varepsilon}^p)^n.$$

Here, $\bar{\sigma}$, $\bar{\varepsilon}^p$ and $\dot{\bar{\varepsilon}}^p$ are the effective stress and effective plastic strain and its rate; b, the Burgers vector; $\rho_o$, an average density of free dislocation in a virgin material; f, a fixed fraction of the dislocation density; $\tau_o$, the resolved shear stress corresponding to unit velocity; q, a strain hardening coefficient [$\Delta\bar{\sigma}=q\bar{\varepsilon}^p$]; and n, an exponent of dislocation velocity [velocity=$(\tau/\tau_o)^n$]. This constitutive equation can describe a sharp yield point and the subsequently abrupt yield drop based on dislocation multiplication and velocity characteristics.

Three different matrices, of which uniaxial stress-strain curves are illustrated in Fig.2., were used for the calculation. Matrices I and II have sharp yield points and the yield drops, but matrix III does not, and the difference in strain hardening after the yield drops are seen between I and II. FEM simulation was performed for the hydrostatic pressurization ($P_x=P_y=P_z$) of the unit-cell and the subsequent uniaxial tension under the generalized plane strain condition.

## RESULTS AND DISCUSSIONS

Figure 3(a) shows macroscopic stress-strain curves of the unit-cell with matrix I in uniaxial tension at atmospheric pressure after some degrees of hydrostatic pressurization. The yield point of the unit-cell is subsatantially reduced with increasing pressure because of the formation of the yielding zone around the inclusion under high hydrostatic pressure, as illustrated in Fig. 4. Emphasis is that if a matrix has no yield-drop characteristics such as matrix III, the pressurization will only cause the strain hardening of the matrix, and it will lead to an increase of flow stress of the unit-cell in the subsequent tension, as illustrated in Fig.3(b).

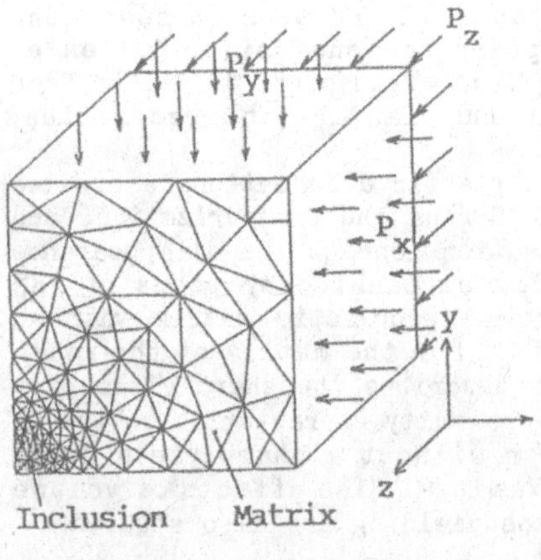


Fig. 1 Unit-cell and FEM mesh.

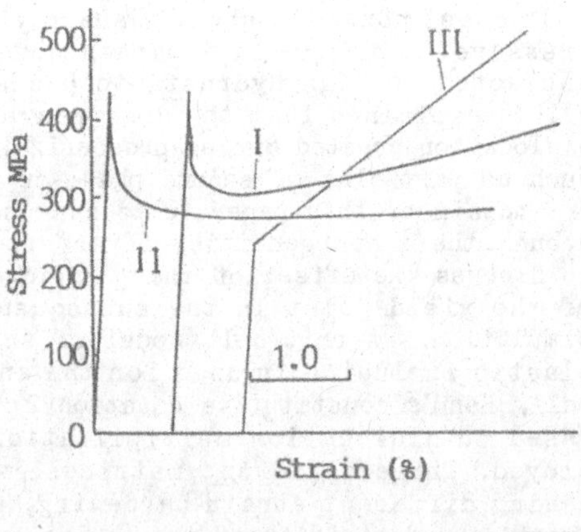


Fig.2 Stress-strain curves of the matrices.

I; $\rho_o=10^{-3}$ mm$^{-2}$, q=3.7*10$^3$ MPa

II; $\rho_o=10^{-3}$ mm$^{-2}$, q=9.8*10$^2$ MPa

III; $\rho_o=10^3$ mm$^{-2}$, q=9.8*10$^3$ MPa

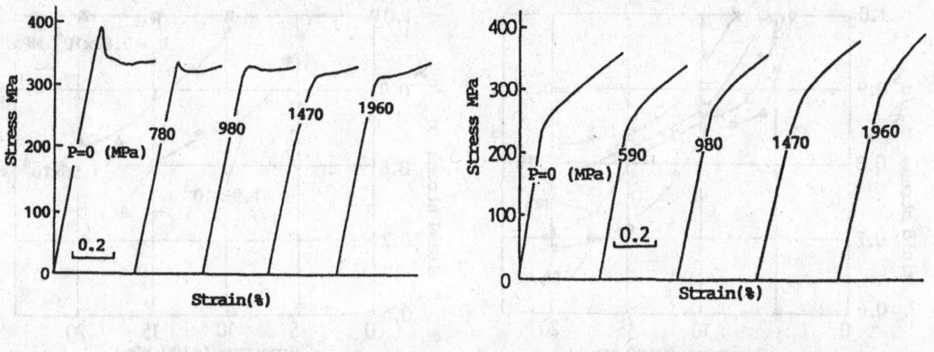


(a) Matrix I ($V_f$=1.0%) (b) Matrix III ($V_f$=3.0%)

Fig. 3 Stress-strain curves of the unit-cells in uniaxial tension after hydrostatic pressurization.

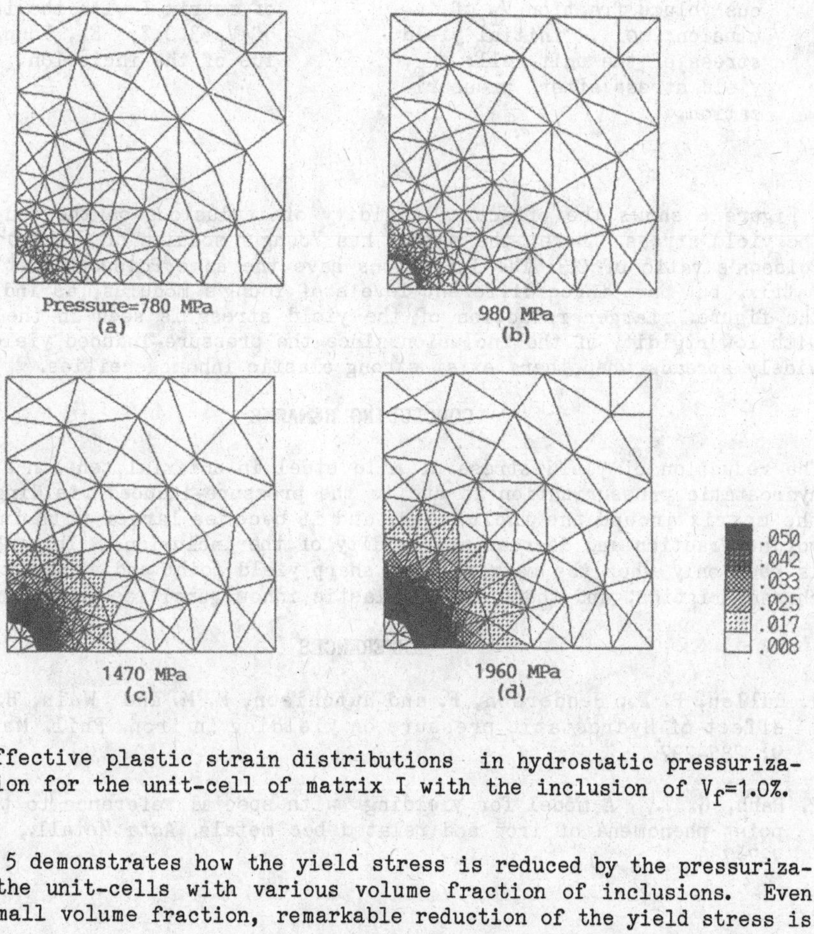


Fig. 4 Effective plastic strain distributions in hydrostatic pressurization for the unit-cell of matrix I with the inclusion of $V_f$=1.0%.

Figure 5 demonstrates how the yield stress is reduced by the pressurization for the unit-cells with various volume fraction of inclusions. Even for the small volume fraction, remarkable reduction of the yield stress is observed.

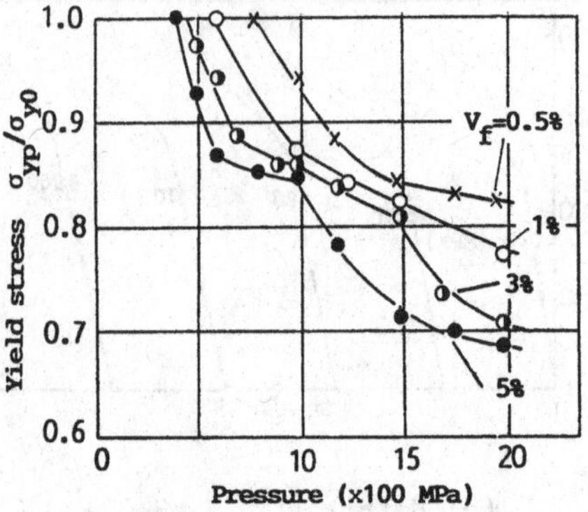


Fig. 5 Reduction of yield stress by hydrostatic pressurization for the unit-cells with various volume fraction $V_f$ of inclusion; $\sigma_{yo}$, initial yield stress of the unit-cell; $\sigma_{yp}$, yield stress after pressurization.

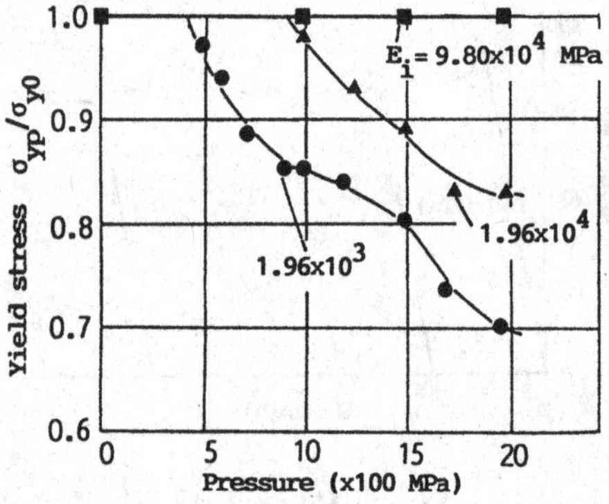


Fig. 6 Effect of the rigidity of inclusion on the reduction of yield stress for the unit-cells of matrix I with the inclusions of $V_f$=3.0 %; $E_i$, Young's modulus of the inclusion.

Figure 6 shows the effect of rigidity of inclusions on the reduction of the yield stress. Here, the matrix has Young's modulus of $2.06*10^5$ MPa and Poisson's ratio of 0.3. The inclusions have the same Poisson's ratio as the matrix, but have three different levels of Young's modulus, as indicated in the figure. Larger reduction of the yield stress is seen in the material with low rigidity of the inclusion since the pressure-induced yielding zone widely spreads when there exist strong elastic inhomogeneities.

## CONCLUDING REMARKS

The reduction of yield stress of mild steel in uniaxial tension after high hydrostatic pressurization is due to the pressure-induced yielding zone in the matrix around the inclusions, and it becomes larger with increasing volume fraction and decreasing rigidity of the inclusions. This phenomenon is seen only when the matrix has a sharp yield point and abrupt yield-drop characteristics, and there exist elastic inhomogeneities in the material.

## REFERENCES

1. Bullen, P. F., Henderson, F. and Hutchison, M. M. and Wain, H. L., The effect of Hydrostatic pressure on yielding in iron. Phil. Mag., 1964, **9**, 285-297

2. Hahn, G. T., A model for yielding with special reference to the yield-point phenomena of iron and related bcc metals. Acta Metall., 1962, **10**, 727-738

# CYCLIC PLASTICITY

# LIMIT RESPONSE IN CYCLIC PLASTICITY

SAMIR AKEL*, GAZ DE FRANCE* and QUOC SON NGUYEN**

Direction des Etudes et Techniques Nouvelles
361, Av. du Pdt Wilson BP 33 - 93211 La Plaine St Denis Cedex, FRANCE

ECOLE POLYTECHNIQUE - Laboratoire de Mécanique des Solides
91128 Palaiseau Cedex, FRANCE

## ABSTRACT

The objective of this communication is to present a numerical investigation on the computation of the cyclic response of elastic-plastic structures under cyclic loading. First, we investigate a cyclic approach method based upon a direct research of cyclic solutions by iterative procedure ensuring the periodic response at the local scale and at every step of iteration. This method may be a priori interesting since one deals only with cyclic responses of stress and strain.

In the case of repetitive rolling loads, the iterative procedure leading directly to the limit response can be extremely accelerated when noting the fact that this limit response is not only cyclic but also stationnary. Illustartion were made on axisymmetric structure.

## CONSTITUTIVE EQUATIONS

Elastic-plastic constitutive law can be expressed in the following incremental form:

$$\mathring{\sigma} = E.(\mathring{\varepsilon} - \mathring{\varepsilon}^p) \qquad \text{where} \quad (\mathring{\ }) = d/dt$$

$$\mathring{\varepsilon}^p = \lambda.\partial f/\partial\sigma \qquad \lambda \geq 0,\ \lambda.f = 0$$

and $f(\sigma, \varepsilon^p, \ldots) \leq 0$ the plastic criterion.

## ASYMPTOTIC ANALYSIS

In many practical applications, it may happen that both the loading and the initial state are not known with sufficient precision. The knowledge of the whole evolution in that case has only limited interest. If the asymptotic behavior of the load is given, it may be interesting to know the asymptotic behavior of the solution, whatever be the initial conditions of the structure. Shakedown theorems give a satisfactory answer to this question and represent the basis of the analysis of the cyclic bahavior of stuctures. Three principal features may occur: elastic shakedown, plastic shakedown and ratcheting. In all the three cases, the knowledge of the limit response in stress and strain is usefull to predict the reliability of the structure with respect to the phenomenon of strain ratcheting and to the oligocyclic fatigue or damage analysis.

## CYCLIC APPROACH

Since cyclic plasticity is our main interest, here after, we shall only deal with time periodic parameters ($\sigma$, $\varepsilon$, $\varepsilon^P$). Let F(t) be a cyclic loading path of period T. Consider a temporal discretization of [0,T] to $0 < t_1 <\ldots< t_n = T$ and $Q_i$ the external corresponding load at time $t_i$. The cyclic approach can be resumed as following:

i) **Equilibrium equations (global step)** : at each increment $t_i$, one must satisfy the global equilibrium of the structure by solving n linear systems:

$$K.U_i = Q_i + Q_i^P \quad \text{for} \quad i = 1,\ldots,n$$

where K is the stiffness, $U_i$ the displacement field and $Q_i^P$ the equivalent nodal loading du to initial strain. Note that the resolution is done in terms of total and not incremental variables as in the classical step by step plastic resolution. This provides n couples ($\sigma_i^e$, $\varepsilon_i$) statically and kinematically admissible. Moreover, at this level, one notices that a complete cycle of loading was accomplished, despite the fact that the constitutive law was not taken into account yet.

ii) **Constitutive equations (local step)** : for the above n couples, we search for n triplets ($\sigma_i$, $\varepsilon_i$, $\varepsilon_i^P$) related to ($\sigma_i^e$, $\varepsilon_i$) by normality rule, and where :

$$\sigma_{i+1} - \sigma_i = E.(\varepsilon_{i+1} - \varepsilon_i - \varepsilon^P_{i+1} + \varepsilon^P_i)$$

The n resulting triplets are plastically, kinematically admissible and periodic.

To reach the limit response, an iterative procedure combining i and ii is made. This method was illustrated by a plate with a crack in tension. Figure 1 shows the limit response at the crack tip obtained by the classical step by step method and by the direct cyclic approach.

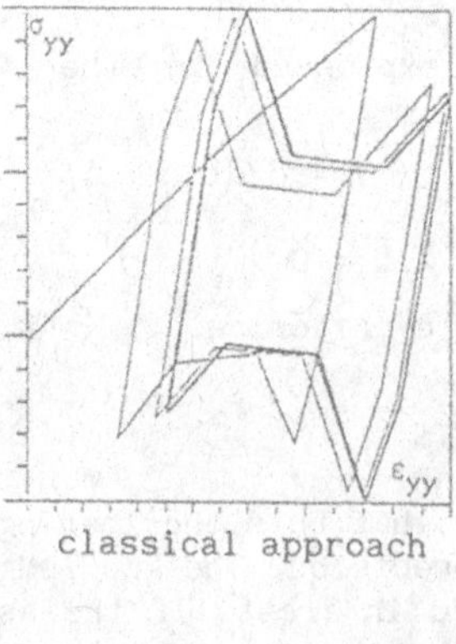

classical approach

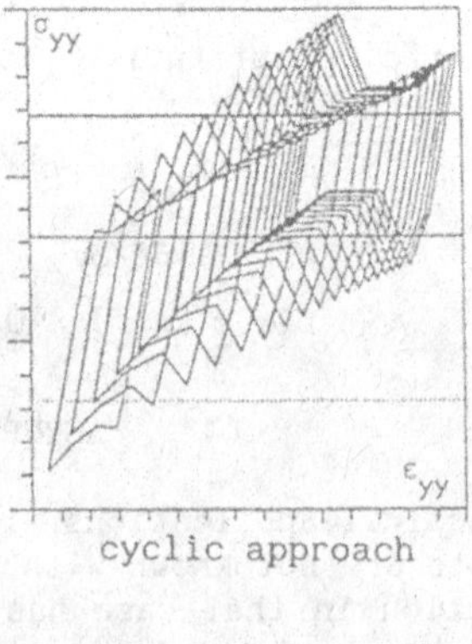

cyclic approach

figure 1

## STATIONNARY CASE

For certain cyclic loadings, the limit response may show, beside its periodic aspect, a stationnary one; repetitive rolling loads constitute a typical example : loaded moving wheel, "dudgeonning"... Transient regime presenting no interest, we shall only be concerned by computing directly the limit response in permanent regime.
Since incremental plasticity is written in terms of time derivative, the permanent regime assumption in small strains induces the following changes in the flow rule:

$$\dot{\varepsilon}^p = - V.\text{grad}\ \varepsilon^p = \lambda.\partial f/\partial\sigma = \frac{1}{H}.<-\partial f/\partial\sigma.E.(V.\text{grad}\ \varepsilon)>.\partial f/\partial\sigma$$

where H describes hardening, < > denotes the positive part, and V the time-independant loading velocity. *Du to small strain assumption, stream lines are parallel to* V. For axisymmetrical structures under rolling load in the circumferential direction, the above equation becomes :

$$\varepsilon^p_{,\theta} = -\frac{1}{H}.<-\partial f/\partial\sigma.E.\varepsilon_{,\theta}>.\partial f/\partial\sigma \qquad (1)$$

Whence, along a stream line, we get:

$$\varepsilon^p(\theta) = \varepsilon^p(0) - \int_0^\theta \frac{1}{H}.<-\partial f/\partial\sigma.E.\varepsilon_{,\phi}>.\partial f/\partial\sigma.d\phi$$

so that $\varepsilon^p$ can be computed at any point of the structure if $\varepsilon$ and $\varepsilon^p(0)$ for each stream line are known.

## STATIONNARY APPROACH

Spatial discretization of equation (1) leads to :

$$\varepsilon^p_{i+1} = \varepsilon^p_i - \frac{1}{H}.<-\partial f/\partial\sigma.E.(\varepsilon_{i+1} - \varepsilon_i)>.\partial f/\partial\sigma$$

where i and i+1 are two adjacent points on the same stream line. This prescribes a particular mesh in a finite element analysis in order to respect the stream lines. $\varepsilon^p(0)$ being an unknown, an iterative cyclic procedure can be performed by using :

$$\varepsilon^p_k(0) = \varepsilon^p_{k-1}(2.\pi) \qquad \text{for iteration " k "}$$

du to shakedown theorems ($\varepsilon^p_0(0) = 0$ for example). This iterative procedure can be accelerated when combined to the classical plastic one as in figure 2; the same decomposition into global and local steps being maintained.

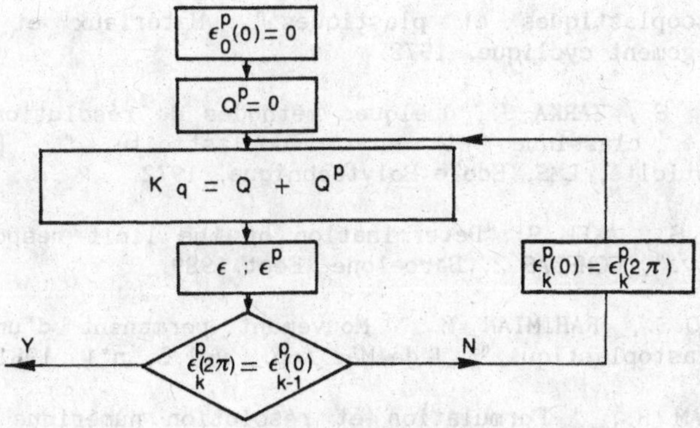

figure 2

The example of a thick infinite cylinder under internal rolling load was examined by this approach. The stationnary cyclic limit response is given in figures 3 and 4 in terms of equivalent plastic strain and displacement field.

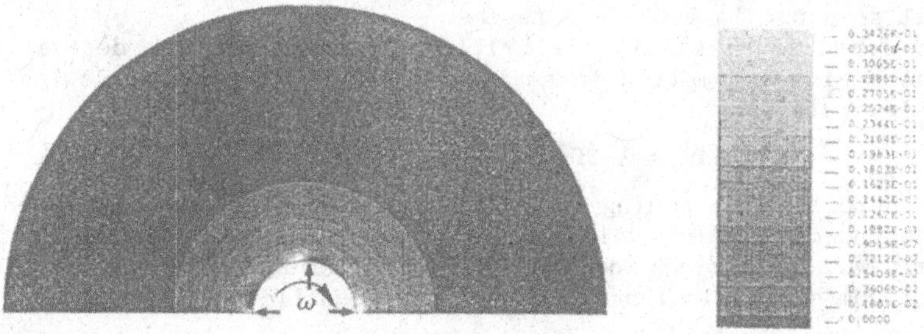

figure 3 (iso - $\varepsilon^{p}_{eq.}$ )

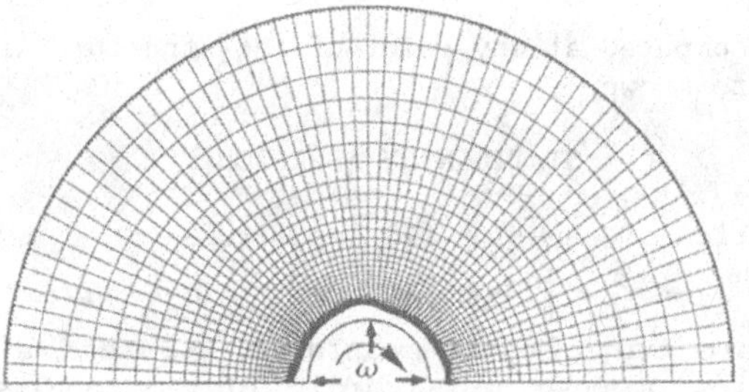

figure 4 (displacements)

**REFERENCES**

[1] BOISSE P.,"Nouvel algorithme à grand incrément de temps pour le calcul des structures plastiques.", Thèse Paris VI, 1987.

[2] HALPHEN B., "Accomodation et adaptation des structures élastoviscoplastiques et plastiques.", Matériaux et Structures sous chargement cyclique, 1978.

[3] NGUYEN Q. S., ZARKA J.,"Quelques méthodes de résolution numérique en plasticité classique et en viscoplasticité ", Plasticité et Viscoplasticité, LMS, Ecole Polytechnique, 1972.

[4] NGUYEN Q.S., AKEL S.,"Determination of the limit response in cyclic plasticity.", COMPLAS 2, Barcelone, Sept.1989.

[5] NGUYEN Q.S., RAHIMIAN M.," Mouvement permanent d'une fissure en milieu élastoplastique.", J.de Méc.App., vol.5, n°1, 1981.

[6] MAITOURNAM H., " Formulation et résolution numérique des problèmes thermoviscoplastiques en régime permanent.", Thèse ENPC 1987.

# A THREE DIMENSIONAL CYCLIC CONSTITUTIVE LAW FOR METALS WITH A SEMI-DISCRETE MEMORY VARIABLE

S. ANDRIEUX, S. TAHERI
Département Mécanique et Modèles Numériques
Etudes et Recherches ,Electricité de France
1, Avenue du Général de Gaulle, 92141 Clamart Cedex, FRANCE

**ABSTRACT**

The extension to 3D situations of an uniaxial cyclic constituive law with a discrete memory variable is presented. Attention is focused on the precise definition of the extended 3D corresponding variables. The discrete nature of the memory leads to discontinuity problems for some loading paths, a modification is then proposed which uses a differential evolution law. For large enough uniaxial cycles, the uniaxial law is nevertheless recovered. Lastly, an incremental form of the implicit evolution problem is given.

## THE UNIAXIAL LAW

The study of cyclic elastoplastic constitutive laws is, at the moment, focused on non proportional loadings, but for uniaxial loadings some problems remain, as for example the ability for a law to describe simultanously ratcheting in non symmetrical load-controlled test, elastic and plastic shakedown in symmetrical and non symmetrical ones. This is the reason why it has been proposed in [1] a law which, in addition to previous phenomena, describes the cyclic hardening in a pushpull test, the cyclic softening after overloading and also the dependence of cyclic strain stress curve on the history of loading. These are the usual properties of a 316 stainless steel at room temperature.

This law uses, besides the usual plasticity variables, an internal discrete memory variable: namely the plastic strain at the last unloading, denoted by $\varepsilon^p{}_n$, and the maximal past stress $\sigma_p$. All these macroscopic variables are deduced from a microscopic analysis, which will be now quickly summed up [2].

When the cross slipping of dislocations is possible for a metal (easy cross slip for pure Al and Cu, difficult cross slip for a 316 stainless steel) the microscopic structure is characterised at low cyclic amplitude by permanent slip bounds and at higher amplitude by cell structure, whose mean size decreases with an increasing amplitude of loading. But when the amplitude of loading decreases again, the cell structure is stable (at room temperature). The cell structure seems also to be detected for monotonic loading. As before the cell size decreases when the maximal stress or strain increases. <u>We suppose that the mean cell size is determined by the maximal stress supported by the material in its history</u>. The asymptotic form of the curves showing mean cell size in fonction of the amplitude loading suggests furthermore to make the hypothesis that there exists a minimal cell size which depends only on the material and not on the loading. During cycling, dislocations pile up on the obstacles (walls), stabilization is obtained when the numbers of dislocations created and anihilated are equal. At stabilized state the plastic deformation is created by active dislocations which sweep away the cell volume and then are anihilated by dislocations of opposite sign. Once a stabilized state is obtained if the maximal stress or the amplitude of loading increases, smaller cells will be obtained, meaning that new obstacles are created on which dislocations have to be piled up again to obtain a new stabilized state. Usually more than one cycle is needed to get a new stabilised state, as macroscopicaly showed by a pushpull test. Different experiments show that at room temperature ratcheting is obtained at a nearly fixed maximal stress, independently of the amplitude of loading. After some cycles, there is a constant gap between each traction or compression curve. Increasing of a small quantity the maximal stress leads to a plateau: the metal becomes practically perfectly plastic (for a 316). This suggests to use a cyclic ultimate stress parameter $S_r$ in relation with minimal cell size.

We define now the macroscopic variables in relation with the previous microscopic analysis:

- $\varepsilon^p$ usual plastic part of the deformation tensor, related to the gliding of dislocations ,
- $\lambda$ cumulated plastic strain, related to the density of dislocations,
- $\sigma_p$ maximal past absolute value of stress supported by the material in his history, related to the actual mean size of cells, this variable is used partly as $S_r-\sigma_p$, the ratcheting stress $S_r$ being a material parameter,
- $\varepsilon^p_n$ plastic deformation at the last unloading point, here the significant variable is the difference $\varepsilon^p-\varepsilon^p_n$, on stabilised cycles, it measures the plastic deformation amplitude. Note the discrete nature of this variable.

The uniaxial constitutive law is now simply described by an elastoplastic model which yield function combines isotropic and kinematic hardening:

$$F\ (\sigma,\ \varepsilon^p,\ \lambda,\ \sigma_p,\ \varepsilon^p_n) = \left|\ \sigma_D - X\ (\varepsilon^p,\ \lambda,\ \sigma_p)\ \right| - R\ (\lambda,\ \sigma_p,\ \left|\varepsilon^p_n - \varepsilon^p\right|\ )$$

The evolution equations for internal variables $\sigma_p$ and $\varepsilon^p_n$ follow directly from their definition, whereas the usual normality and consistency relations are used for the remaining variables $\varepsilon^p$ and $\lambda$. The next figures illustrate two features of this law obtained for simple choice of the yield function F [1].

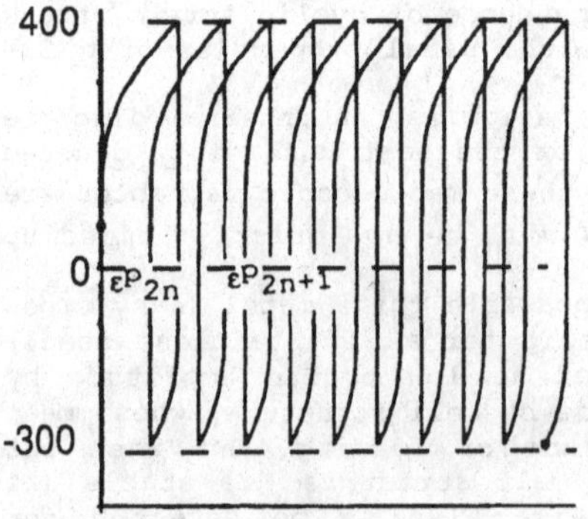

Fig 1. Rachetting

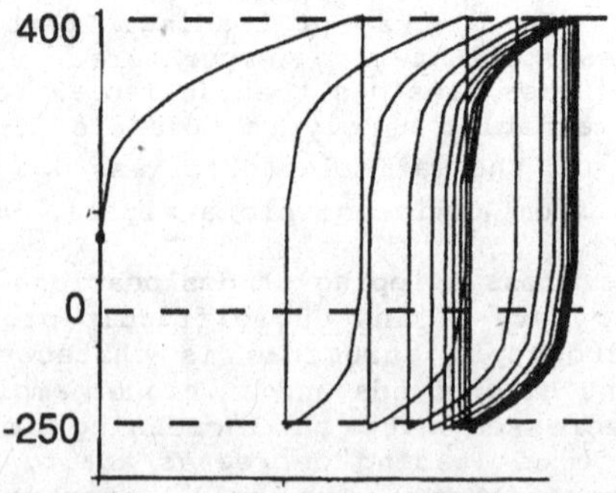


Fig 2. Plastic Shakedown in non symmetrical loading

## THREE DIMENSIONAL LOADINGS, SEMI-DISCRETE MEMORY AND MAXIMAL PAST STRESS

The extension to three dimensionnal situations of the previous iniaxial law encounters two main difficulties. First, choices have to be made on the variables themselves: Is the small number of internal variables sufficient for 3D situations? To this first question, although some answers can be gained by extra microscopic analysis, we choose here, for the sake of simplicity (as frequently done) the deviatoric part of the tensors and to keep the uniaxial law's general form.

The second class of difficulties arise from extension of the definitions and evolution equations of the memory variables $\sigma_p$ and $\varepsilon^p{}_n$. As a matter of fact, the uniaxial loadings histories are very poor (there is no tangent loadings, cycling is only defined by two extreme values etc), furthermore in applications incremental time integration is rarely performed because advantage is taken of algebraic relations between extremal states. Intrinsic definitions and more precise evolutions laws are then needed in the 3D case.

**Definition and Evolution of $\sigma_p$**

This variable is defined as the maximal past deviatoric norm of the stress experienced by the material (the norm is denoted by $|\sigma_D|$). With initial value $\sigma_p{}^0$ of $\sigma_p$, the precise definition is then:

$$\sigma_p(t) = \max_{u \in [0,t]} \left(\sigma_p^0, |\sigma_D(u)|\right)$$

leading to the evolution equation for $\sigma_p$ (H is the Heavyside function):

$$\dot{\sigma}_p = H\left(|\sigma_D| - \sigma_p\right) \frac{\sigma_D \cdot \dot{\sigma}}{|\sigma_D|}$$

**Definition and Evolution Law of $\varepsilon^p{}_n$**

Two questions arise from the definition of the evolution law of $\varepsilon^p{}_n$. One is common with cyclic memory variables: do the stress-strain curve in a partial elastic unloading and reloading, fit the initial monotonic curve? In other words, can the description of the material behavior admit some undershooting of the monotonic stress-strain curve after an elastic unloading? The answer to this question is not straightforward and can depend on the material and perhaps on the kind of stress, so that we do not address to this problem here. The second question, more important from a physical point of view, is the requirement of continuity of the stress-strain curve with respect to very small unloadings. With full discrete memory, this requirement is generally not fullfiled: any unloading, even as small as possible, leads to an (discontinuous) evolution of the memory variable which induces in turn a discontinuity on the value of the yield function F. This last discontinuity can finally cause the violation of the yield condition $F \leq 0$. For 3D loading pathes, this problem is of primary importance because "micro-unloadings" can result from changes of direction of the loading path in the stress space.

To overcome this last difficulty, we modify the discrete evolution law for $\varepsilon^p{}_n$ to a semi discrete one (the word semi-discrete is used because of the saturation of the memory ensuing from the definition of the evolution). Starting from the discrete model rewrited here as:

$$\Delta\varepsilon^p{}_n = \varepsilon_n^{p+} - \varepsilon_n^{p-} = \varepsilon^p - \varepsilon_n^{p-}, \quad \text{if } F = 0, \ \frac{\partial F}{\partial \sigma}\dot{\sigma} \leq 0$$

we introduce a scalar differential evolution equation together with a consistency condition ensuring the fullfilment of the yield condition:

$$\left\{\begin{array}{ll} \dot{\varepsilon}^p{}_n = \alpha\left(\varepsilon^p - \varepsilon_n^p\right) & \text{if } F = 0, \ \frac{\partial F}{\partial \sigma}\dot{\sigma} \leq 0 \\ \alpha \geq 0, \quad \alpha F = 0 & F \leq 0 \end{array}\right.$$

It can be seen that (with appropriate "generalized hardening conditions" on the yield function F):

i) the yield condition ($F \leq 0$) is never violated,

ii) the continuity with respect to the chronology parameter is restored

iii) the memory shows a "saturation effect": when, during the unloading, the value of $\varepsilon^p{}_n$ reaches $\varepsilon^p$, then $\varepsilon^p{}_n$ stays at this value and the unloading becomes purely elastic (no internal variable evolution)

iv) for uniaxial cycling loadings, the discrete memory is recovered between two successive unloadings (provided the cycle is large enough).

## THE IMPLICIT INCREMENTAL PROBLEM

The evolution equations of the remaining variables $\lambda$ and $\varepsilon^p$ are deduced from the usual normality and consitency relations. It is easily seen that when $|\sigma_D|$ does not reach $\sigma_p$ the law reduces (in loading evolutions) to the standard platicity laws. Non standard flow rules are obtained when $\sigma_p$ varies during loading.(but normality for $\dot{\varepsilon}^p$ still remains)
To write down extensively the proposed constitutive law, we present hereafter it implicit incremental form which will be used in computations. We denote by e the "state" ($\sigma$, $\varepsilon^p$, $\lambda$, $\sigma_p$, $\varepsilon^p{}_n$), by A the elasticity tensor, and by $<x>^+$ the positive part of x.

$$\begin{cases} \Delta\sigma + \Delta\lambda \ A \dfrac{\partial F}{\partial \sigma}(e+\Delta e) = \Delta\varepsilon \\ \Delta\varepsilon^p - \Delta\lambda \ \dfrac{\partial F}{\partial \sigma}(e+\Delta e) = 0 \\ \Delta\varepsilon^p_n = \left[1 - \left\langle 1 - \Delta\alpha \right\rangle^+\right] \ (\varepsilon^p - \varepsilon^p_n) \\ \Delta\sigma_p = \left\langle R\ (e+\Delta e) + \left| X\ (e+\Delta e) \right| - \sigma_p \right\rangle^+ \end{cases}$$

with consistency relations

$$F(e + \Delta e)\ \Delta\lambda = 0, \qquad F(e + \Delta e)\ \Delta\alpha = 0, \qquad F \leq 0$$

$$\Delta\lambda \geq 0 \ , \ \Delta\alpha \geq 0 \ , \ \Delta\lambda\ \Delta\alpha = 0$$

Implicit Incremental Formulation for the 3D Constitutive Law

The coherence of these equations and the conditions on F (generalized hardening conditions) ensuring existence and unicity of the response ($\Delta\sigma$, $\Delta\varepsilon^p$, $\Delta\lambda$, $\Delta\sigma_p$, $\Delta\varepsilon^p{}_n$) for a given increment $\Delta\varepsilon$, have been established and will be reported elsewhere.

## CONCLUSION

The mean features of the proposed 3D cyclic law are the following:

i) it exhibits two surfaces: the usual yield surface together with a ratcheting surface $\sigma_p = S_r$,

ii) it presents, because of the semi-discrete nature of the memory variable, a decomposition of the unloading in two phases. First, unloading begins with elastic response but continuous actualisation of the memory variable $\varepsilon^p{}_n$, once the memory variable $\varepsilon^p{}_n$ is saturated by the value of the plastic deformation at beginning of unloading, the unloading is purely elastic.

Let us quote lastly that the 3D law keeps the properties listed at the beginning of the paper for the uniaxial one.

## REFERENCES

[1] Taheri,S.,(1989). Une Loi de Comportement Uniaxial Cyclique Avec Variable à Memoire Discrète. 9° Congres Francais De Mécanique Metz France.

[2] Taheri,S.,(1990). Une Loi de Comportement Uniaxial En Elastoplasticité Pour le Chargement Cyclique, Note Interne E.D.F/D.E.R/HI-71/6812.

# RATE-DEPENDENT PLASTIC DEFORMATION-EXPERIMENTS AND CONSTITUTIVE MODELLING

F.ELLYIN*, Z.XIA and K.SASAKI**

* Department of Mechanical Engineering, University of Alberta, Canada T6G 2G8
** Department of Mechanical Engineering II, Hokkaido University, Sapporo 060, Japan

## INTRODUCTION

Experiments show that some metals exhibit rate-dependent behaviour in room temperature. The rate-sensitivity generally increases with increasing temperature. Tests on the rate-dependent behaviour have been performed mostly on specimens subject to uniaxial or combined axial-torsional loading. Even for these simple loading forms, there has been very few published data regarding the effect of strain history on a material's hardening response. Recently a series of uniaxial and biaxial tension-compression cyclic tests was performed in our laboratory. The tests included proportional and nonproportional loading paths with different strain-rate histories. The results of these tests will be presented in the next section.

A rate-dependent constitutive model recently proposed by Ellyin and Xia (1991) will be outlined in the third section. The predictions of the model are compared with the experimental result, and the need for further investigation are discussed in this paper.

## EXPERIMENTAL PROCEDURE AND RESULTS

Thin-walled circular cylindrical specimens of stainless steel type 304, were subjected to uniaxial and biaxial cyclic loading under strain-controlled condition. The strain-path for the biaxial loading is shown in Fig. 1. Each test comprised of three levels of strain-rates applied in a predetermined order. In each step 50 cycles were applied before switching to another one in order to ensure stable material response at each strain-rate level.

For uniaxial tests the imposed strain amplitude was $\Delta\varepsilon/2 = \pm 0.3\%$ with strain rates ranging from $6 \times 10^{-3}$ to $6 \times 10^{-4}$ to $6 \times 10^{-5}$ $s^{-1}$. The strain amplitude for biaxial tests was $\Delta\varepsilon_a/2 = \Delta\varepsilon_t/2 = \pm 0.2\%$ with strain-rates varying from $4 \times 10^{-3}$, to $4 \times 10^{-4}$ to $4 \times 10^{-5} s^{-1}$. For the sake of convenience we will use the letters F to denote the highest rate, S the lowest, and M the intermediate strain rates.

A typical result indicating the influence of strain-rate history for the uniaxial loading case is shown in Fig. 2. It is seen that a transient hardening occurs during a few initial cycles. Thereafter, the response is stabilized, and an abrupt change of strain-rate may cause a change in the size of the stress-strain loop. However, this change in strain hardening is achieved instantaneously following the change in the strain-rate without a transient regime. For tests with F-M-S sequence, we observe a decrease in the stress amplitude with decreasing strain-rate. However, no appreciable change is observed with the S-M-F sequence. That is, for increasing sequence of strain-rates, the abrupt changes do not cause a corresponding change in the stress amplitude. Finally, for the F-S-F sequence, there is a decrease in the stress amplitude due to decrease in the strain-rate from F to S, and a slight increase in stress amplitude when returning to the initial fast strain-rate.

Figure 3 shows the biaxial stress-strain response for the first 10 cycles at a fast strain-rate of $4 \times 10^{-3}$ $s^{-1}$. It is noted that the transient response is again limited to a few initial cycles. A rather interesting phenomenon is the difference in response between axial and circumferential directions (cf. Figs. 3a and 3b). Referring to Fig. 1, it is seen that at first the material is loaded to point A and subsequently the loading path follows a square diamond shape ABCDA. At point A, the material has already experienced different plastic deformation in two directions, and this initial anisotropy influences the subsequent response.

In the case of non-proportional biaxial tests, the same trend as the uniaxial tests (Fig. 2) is observed with respect to strain-history. That is, once a material is stabilized under a lower strain-rate, the stress-strain response is not appreciably affected by a jump to a faster strain-rate. A possible explanation is that at a given stress level, the plastic strain is greater at a lower strain-rate compared to that at a higher strain-rate. Thus, the dislocation cells formed at a lower strain-rate are more stable and subsequent change to a higher strain-rate would not affect the stabilized cell structure and consequently the stress-strain response.

## CONSTITUTIVE MODEL AND PREDICTED RESULTS

Recently, Ellyin [1], Ellyin and Xia [2-6] have presented a hierarchy of constitutive models each appropriate to a class of applications. Here we will outline a rate-dependent elastoplastic constitutive law. The model is based on three types of hypersurfaces, viz., yield, stress memory, and strain memory surfaces. The yield surface specifies the locus of elastic regime and is expressed by:

$$\phi_y = f_y(\underset{\sim}{\sigma} - \underset{\sim}{\alpha}) - q^2 = 0 \tag{1}$$

where $\underset{\sim}{\alpha}$ and q specify the centre and radius of the current yield surface. The size of the yield surface is rate- and history-dependent, i.e.

$$q = q(\dot{\varepsilon}_{eq}, \ell_p) \tag{2}$$

where $\dot{\varepsilon}_{eq} = [2(\dot{\underset{\sim}{e}} \cdot \dot{\underset{\sim}{e}})/3]^{1/2}$, and $\dot{\underset{\sim}{e}} = \dot{\underset{\sim}{\varepsilon}} - \underset{\sim}{1}\, tr(\dot{\underset{\sim}{\varepsilon}})/3$, $\ell_p = \int [2(d\underset{\sim}{\varepsilon}^p \cdot d\underset{\sim}{\varepsilon}^p)/3]^{1/2}$ and a superscript dot denotes the time rate $\partial/\partial t$.

The stress and strain memory surface are related to the maximum equivalent stress and strain level experienced by the material during its previous history,

$$\phi_m^\sigma = f_m^\sigma(\underset{\sim}{\sigma}) - R^2(\sigma_{eq,\max}) = 0 \tag{3}$$

$$\phi_m^\varepsilon = f_m^\varepsilon(\underset{\sim}{\varepsilon}) - C^2(\varepsilon_{eq,\max}) = 0. \tag{4}$$

The evolution of the yield surface is specified with respect to the plastic loading types, viz. monotonic loading (ML) where the stress memory surface expands with the movement of the yield surface and the two surfaces remain tangent at the loading point. The second type of loading is termed reloading (RL) for which the current loading point is inside the stress memory surface.

The stress-strain relation is given by:

$$\dot{\underset{\sim}{\varepsilon}} = \dot{\underset{\sim}{\varepsilon}}^e + \dot{\underset{\sim}{\varepsilon}}^p \tag{5}$$

where $\dot{\underset{\sim}{\varepsilon}}^e$ is specified by the generalized Hooke's law in its rate form and

$$\dot{\underset{\sim}{\varepsilon}}^p = g\left(\frac{\partial f_y}{\partial \underset{\sim}{\sigma}}\right)\left(\frac{\partial f_y}{\partial \underset{\sim}{\sigma}} \cdot \dot{\underset{\sim}{\sigma}}\right) \tag{6}$$

where g is a hardening modulus related to the instantaneous tangent modulus, $E_t = E_t(\sigma_{eq}, \ell_p, \dot{\varepsilon}_{eq})$. Calculation of the tangent modulus is also dependent on the loading types (ML or RL).

The transient hardening is reflected in a change in the elastic regime and in the tangent modulus. The description of transient hardening and details regarding determination of material constants or function relations can be found in [5].

Figure 4 shows the predicted results for the biaxial loading. The predicted transient response for the first four cycles is slightly different from the experimental observation shown in Fig. 3. However, the stable response is very close to the experimental data. The model also predicts the effect of strain-rate history for the sequence F-M-S in agreement with the experimental data. However, it predicts almost the same stable hysteresis loops for sequences S-M-F and F-S-F, which are not in conformity with Fig. 2 observations. Efforts are under way to remedy this situation.

REFERENCES


1. Ellyin, F., "An Anisotropic Hardening Rule for Elastoplastic Solids Based on Experimental Observations", J. Applied Mechanics, Vol. 56, 1989, pp. 499-507.
2. Ellyin, F. and Xia, Z., "Elastoplastic Stress-Strain Relation Based on a New Anisotropic Hardening Model", in: Yield, Damage and Failure of Anisotropic Solids, EGF5, J.P. Boehler (ed.), Mech. Engng. Publ. London, 1990, pp. 155-169.
3. Ellyin, F. and Xia, Z., "A Rate-Independent Constitutive Model for Transient Non-Proportional Loading", J. Mechanics and Physics of Solids, Vol. 37, 1989, pp. 71-91.
4. Xia, Z. and Ellyin, F., "Nonproportional Multiaxial Cyclic Loading - Experiments and Constitutive Modelling", J. Applied Mechanics, 1991 (in press).
5. Ellyin, F. and Xia, Z., "A Rate-Dependent Inelastic Constitutive Model, Part I: Elastic Plastic Flow", ASME J. Engng. Mater. Technol. (to appear).
6. Xia, Z. and Ellyin, F., "A Rate-Dependent Inelastic Constitutive Model, Part II: Creep Deformation Including Prior Plastic Strain Effects", ASME J. Engng. Mater. Technol. (to appear).

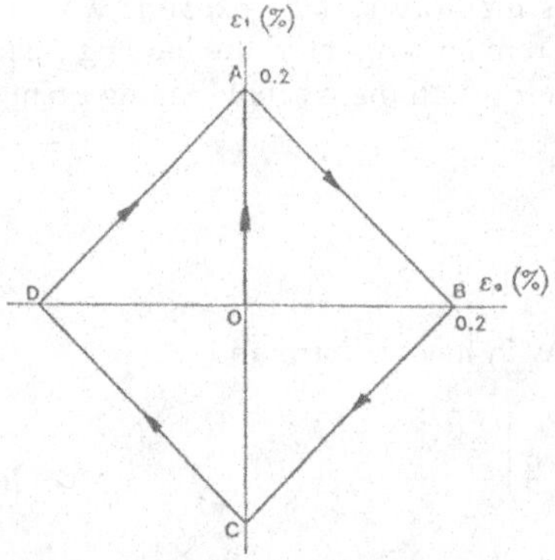


Fig. 1 The strain-path for the biaxial cyclic loading

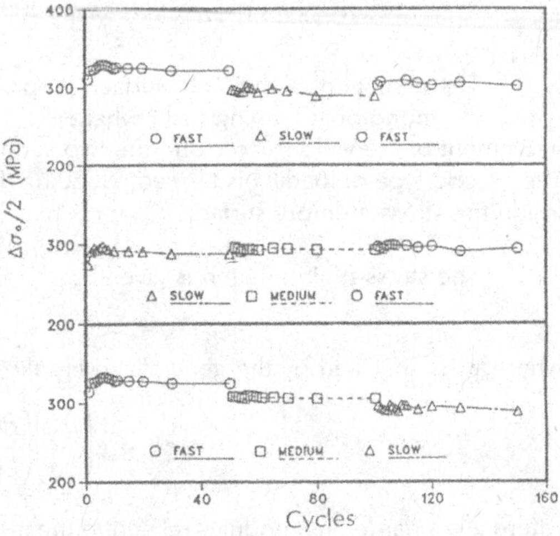


Fig. 2 Stress amplitude versus number of cycles for the uniaxial cyclic tests

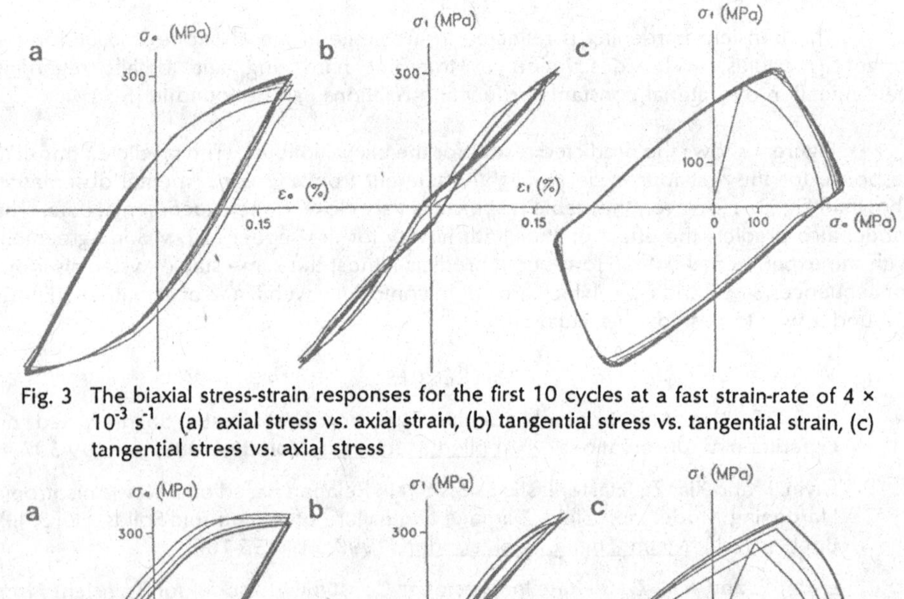


Fig. 3 The biaxial stress-strain responses for the first 10 cycles at a fast strain-rate of $4 \times 10^{-3}\ s^{-1}$. (a) axial stress vs. axial strain, (b) tangential stress vs. tangential strain, (c) tangential stress vs. axial stress

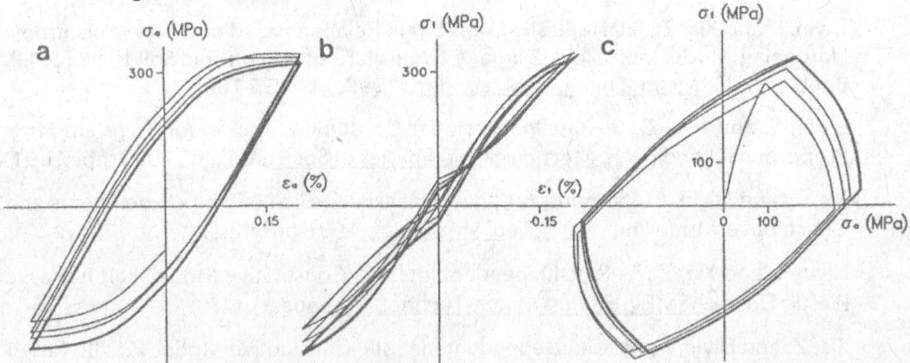


Fig. 4 The predicted results for the biaxial cyclic loading.
(a) axial stress vs. axial strain, (b) tangential stress vs. tangential strain, (c) tangential stress vs. axial stress

# RATCHETTING STRAINS IN CYCLIC PLASTIC BENDING

**S.J. HARVEY, D. ELCOCK, A.P. TOOR**
**COVENTRY POLYTECHNIC, DEPARTMENT OF COMBINED ENGINEERING,**
**PRIORY STREET, COVENTRY, CV1 5FB.**

## ABSTRACT

The incremental downward collapse of cantilevers subjected to cyclic horizontal deflections and a steady sustained vertical load is examined. The material behaviour of elemental fibres in the beams, undergoing push-pull cycles, is modelled using an appropriate power law. An interative elemental fibre analysis allows material properties to be updated with each increment of strain and the position of the neutral surface to be located. The method allows the ratchetting deflections to be accurately predicted when steady state cyclic conditions are achieved.

## INTRODUCTION

When engineering components are subjected to cyclic plastic strains, the presence of a superimposed or follow up load results in incremental or ratchetting strains with each cycle. The effect is cummulative and failure of the component can be caused by gross-deformation before fatigue failure can occur. The ratchetting process has been observed and reported by many researchers. Coffin [1] in his work on fatigue of uniaxial specimens observed incremental growth under combined cyclic and steady loads. Several problems of incremental collapse have also been examined by Edmunds and Beer [2] and Bree [3]. Interest in cyclic plastic bending was promoted by two major developments. Firstly, the development of plastic methods of structural design, with limit analysis and minimum weight design, resulted in the acceptance of plastic strains within some structural elements. Since the loads on some structures would also be cyclic, this could result in incremental collapse or shakedown, depending on the magnitude of the applied loads. Secondly, the need to obtain material design data for components subjected to low cycle or high strain fatigue resulted in plastic cyclic bending tests being used for high strain ranges, because of the instability problems associated with push-pull tests. Since a beam in bending can be considered to be made up of elemental fibres undergoing push-pull cycles it is possible to generate quite large cyclic plastic strains. If fatigue failure is defined as crack initiation at the outer fibre of the beam it should correspond to the fatigue life of a component in push-pull at the same strain level.

Conversely it can be shown that the cyclic moment curvature relationship for the beam can be predicted from cyclic push-pull data [4]. The problem of biaxial plastic bending is more complex and has not been extensively examined. The movement of the neutral surface can lead to elements of the beam unloading with plastic strain reversal. The problem is simplified if the direction of the applied moment is assumed to be fixed [5]. In this investigation a cantilever beam is considered to be made up of elemental fibres undergoing push-pull cycles when subjected to cyclic horizontal deflections with a sustained steady vertical load. This loading system results in ratchetting downward deflections with each cycle of horizontal deformation.

## MATERIALS PROPERTIES AND SPECIMENS

In order to predict the ratchetting deflections it will be necessary to model the cyclic material properties of each elemental fibre with each cycle, and throughout each cycle, and also account for any cyclic hardening or softening.

The model used is the analysis as shown in Fig. 1. The cross-section of the beam is divided into a number of elemental fibres, then as moments Mx and My are applied, the stress and deformation of each elemental fibre can be identified and its position on the monotonic or cyclic stress strain curve established. One of the simplest relationships to use in modelling the plastic stress-strain curve is the power law.

$$\sigma = \sigma_0 \epsilon^n \qquad (1)$$

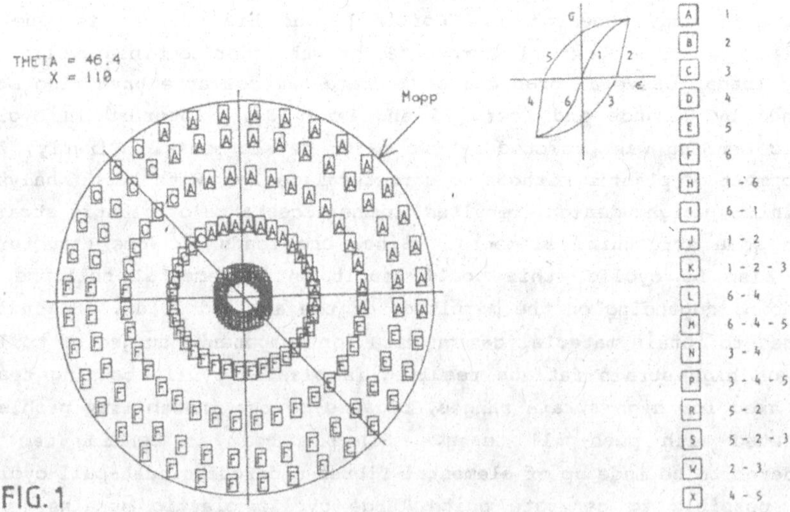


FIG. 1

This relationship gives a good fit for the cyclic stress-strain curve after the first quarter cycle, because it models a bauschinger effect quite well. However, its use for the first quarter cycles leads to significant errors. Two test materials were used in this investigation. A low carbon steel (EN32B) and an 18/8 stainless steel (T321). Two types of cantilever specimens were used, namely a) square beams 12 mm x 12 mm and b) circular beams, 11.5 mm diameter. The test procedures are fully described elsewhere [5]. Both materials cycled to a steady state condition after a few cycles in push-pull tests. These tests showed the dependence of $\sigma_0$ and $n$ on the cyclic strain range and the constants were determined for the steady state cyclic condition. The constants for both materals rapidly approach constant values for cyclic strains greater than 1.2% [6].

## THEORETICAL APPROACH

With reference to Fig. 1. the basic approach was to (a) Apply vertical constant load (b) Increment horizontal load from zero to maximum value (c) Set strain value at onset of plastic deformation (d) Calculate co-ordinates of element with respect to centroid of beam (e) Estimate initial strain distribution (f) Evaluate stress using equation (1) (g) Calculate internal bending moment and compare with applied bending moment (h) Adjust estimated value of maximum strain, and repeat until internal moment equals applied moment. This procedure is repeated for other sections of the beam. Usually three to five iterations were required, depending on the initial estimation, to establish the strains. When the horizontal force is increased, the material data is updated, using equation (1), and based on the previous elemental load history. Using the evaluated strains, the curvature is determined and the deflections are then calculated.

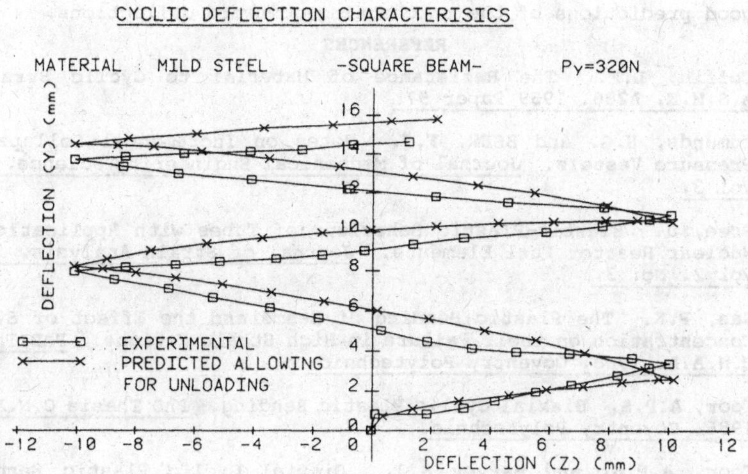


FIG.2

## RESULTS

The validity of the theoretical approach was tested by calculating the horizontal cyclic load-deflection characteristics, without a vertical load. The predictions compared very well with experimental results for square and round sections for both materials [6]. Any errors were consistent with our inability to model the material behaviour more accurately with a simple power law. When a sustained vertical load is applied at the end of the cantilever the neutral surface rotates with increasing horizontal loading. This causes no problems in the elastic region, but once plastic strains are induced some. elemental fibres unloaded and are reloaded in the opposite direction, with a subsequent bauschinger effect. Particular elemental fibres can be identified together with their position within the stress-strain hysteresis curve, Fig. 1. The prediction of horizontal cyclic load-deflection characteristics were again very good [6]. However, a more sensitive measure of the approach adopted is to compare actual and theoretical vertical deflections. A typical result is shown in Fig. 2. where it can be seen that the method gives good predictions both within a cycle and with each subsequent cycle.

## CONCLUSIONS

The theoretical approach adopted appears to work quite well. It is very dependent on the ability to model the cylic material behaviour accurately. The power law used is not suitable for the first quarter cycle but gives good results for subsequent cycles. Generally the power law is more appropriate for the larger plastic strains and more complex laws would be necessary if there is an extensive elastic region existing in the cyclic stress-strain curve. The non-radial biaxial bending problem creates difficulties because of the unloading and re-loading of individual elements during a bending cycle. However the method adopted gives good predictions of the ratchetting downward deflections.

## REFERENCES


1. Coffin, L.F. The Resistance of Material to Cyclic Strain. A.S.M.E. A286. 1959 Paper 57.

2. Edmunds, H.G. and BEER, F.J. Notes on Incremental Collapse of Pressure Vessels. Journal of Mechanical Engineering Science. 1961 Vol 3.

3. Bree, J. Elastic-Plastic Behaviour of Tubes with Application to Nuclear Reactor Fuel Elements. Journal of Strain Analysis. 1967. Vol 2. No. 3.

4. Das, P.K. The Plastic Bending of Beams and the Effect of Strain Concentration on their Failure in High Strain Fatigue. PhD Thesis C.N.A.A. 1968. Coventry Polytechnic.

5. Toor, A.P.S. Biaxial Cyclic Plastic Bending. PhD Thesis C.N.A.A.. 1986. Coventry Polytechnic.

6. Toor, A.P.S. and Harvey S.J. Biaxial Cyclic Plastic Bending. Third International Conference on Biaxial/Multiaxial Fatigue. Stuttgart. 1989 Vol 1. Paper 17.

# HIGH TEMPERATURE PLASTIC BEHAVIOR OF IN738LC UNDER LCF LOADING

F. JIAO*, W. CHEN**, J. ZHU* and R. P. WAHI*
* Hahn-Meitner-Institut Berlin GmbH, Glienicker Str. 100, 1000 Berlin 39, FRG: ** Institut für Metallforschung, Technische Universität Berlin, Hardenberg Str. 36, 1000 Berlin 12, FRG

## ABSTRACT

Fully reversed (R=-1), total axial strain controlled LCF tests were performed on button head type specimens at 850°C. Tests were conducted at a constant strain rate of $10^{-3}$ $s^{-1}$ using different total strain amplitudes from 0.4% to 1.0%. The Transmission Electron Microscopic (TEM) observations of the deformed specimens show that the γ' precipitates are sheared by dislocations producing superlattice stacking faults in the precipitates at the lower strain amplitudes and at the higher strain amplitudes they are overcome by Orowan process. The results of the cyclic plasticity are discussed in the light of microscopic mechanisms.

## INTRODUCTION

The nickel base alloy IN738LC is a commonly used material for stationary gas turbine blades. Low cycle fatigue and subsequent cyclic crack propagation at elevated temperatures (700 - 900°C) are important damage mechanisms for turbine blades and disks. In the design of such components engineers are very much interested in the labortory test results on crack initiation and propagation obtained under creep and LCF loading, respectively. The crack initiation unter LCF loading might occur due to the accumulation of deformation debris on surface, grain boundaries and carbides. Therefore, it is important to study the high temperature plastic behavior of nickel base superalloys under LCF loading

The fatigue properties of IN738LC have been studied in detail [1]. However, very few systematic microstructural investigations have been undertaken to understand the mechanical behavior. In this study, the deformation microstructure of IN738LC after LCF testing at 850°C was studied in detail. The behavior under monotonic loading is reported elsewhere [2].

## MATERIALS AND EXPERIMENTAL METHODS

Specimens of IN738LC supplied by Thyssen Guß AG, Germany were machined from investment castings after the thechnical heat treatment. This treatment produced a bimodal γ' distribution having average size of 400 and 80 nm and a volume fraction of about 20% each. The procedure of the heat treatment and the chemical composition are given in [3]. All the LCF tests were carried out under fully reversed total strain controll, at a constant strain rate of $10^{-3}$ $s^{-1}$, and total strain amplitudes from 0.4% to 1.0% at 850°C.

Discs for TEM-foil preparation were sectioned perpendicular to the direction of applied stress. Details of specimen preparation are reported elsewhere [2]. The foils were examined in a Philips EM400 microscope operated at 120KV.

## RESULTS

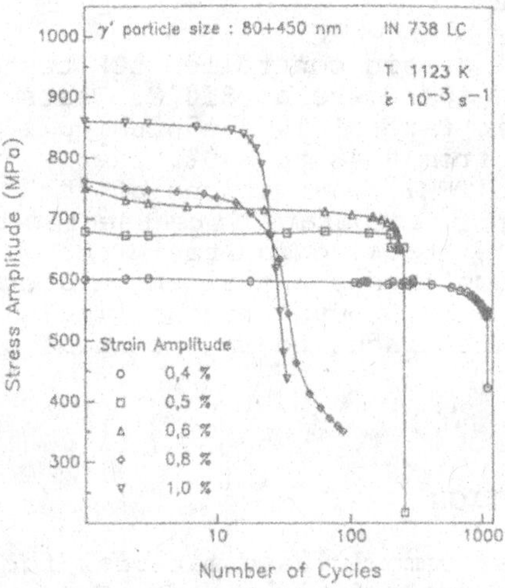


Fig. 1. Cyclic stress response curves

Cyclic stress response curves are presented in Fig. 1. The material shows a slight softening at a very slow rate under all testing conditions. The reasons for the observed softening behavior will be discussed later in the light of microstructural observations. Finally, near the end of the test the stress drops off very rapidly due to crack propagation.

The microstructural changes as a function of strain amplitudes are given in Fig. 2. The analysis of the microstructure at the smaller strain amplitudes ($<$0.6%) shows that γ' precipitates are sheared by {111}<112> slip [4]. This causes the formation of super intrinsic/extrinsic stacking fault (SISF/SESF) in γ' (seeFig. 2a). All Burgers vectors in the dislocation configuration shown in Fig. 2b were analysed under two beam condition. The results of the analysis show that the fault lies on $(1\bar{1}\bar{1})$ and the Burgers vectors of dislocation 1, 2, 3, 4 and 5 are $a/6[\bar{1}1\bar{2}]$ (Aβ), a/6[211] (Dβ), $a/6[12\bar{1}]$ (βC), $a/2[\bar{1}0\bar{1}]$ (AD) and a/2[110] (DC), respectively (Fig. 2b). As reviewed in ref. [4], different models have been proposed for SISF/SESF formation. The present configuration can be explained by the Condat's model [5]. In this model the fault formation is caused by 1/2<110> dislocations which enter γ' precipitates and produce APB in γ'. Then a shockley partial loop is nucleated in the APB zone, grows and transforms APB to SISF/SESF. According to this model the analysed configuration can be formed by groups of AD and DC moving together. A schematic representation is shown in Fig. 3. At the larger strain amplitudes ($\geqslant$0.8%), dislocation rings were observed only around the small γ' particles. No dislocation within the precipitates were seen (see Fig. 2c). These observations indicate that the deformation occurs only in matrix, and the small particle can be overcome by dislocation following Orowan mechanism.

No indication of a change in γ' morpholoy was found in this study, perhaps due to relatively short testing time. However, a significant coarsening of large γ' particles and dissolution of small γ' particles in the same material were observed in creep tests [6]

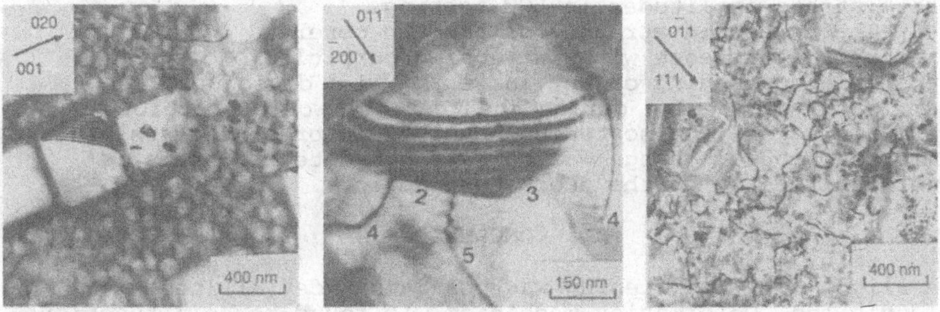


(a) (b) (c)

Fig. 2. Deformation microstructures at strain amplitude of (a) and (b) 0.4%, and (c) 0.8%.

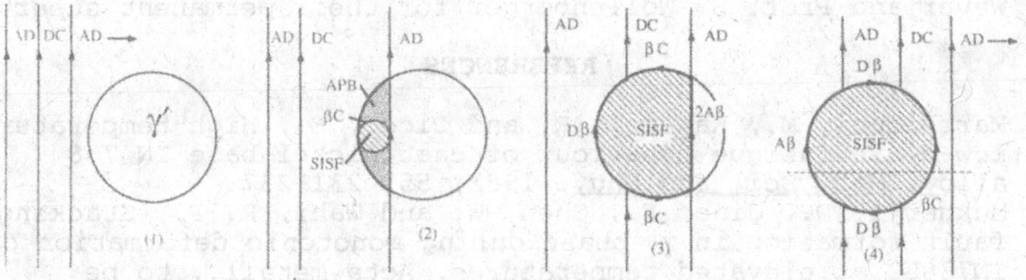


Fig. 3. Schematic reprensentation of dislocation reaction leading to the configuration shown in Fig. 2b.

## DISCUSSION

Two deformation mechanisms viz. shearing by formation of stacking fault in γ' phase and dislocation bowing, in different regimes of strain amplitudes respectively, were identified by TEM observation. In this material containing a monomodal γ' distribution, dislocation bowing was not observed under the same testing conditions [7]. This difference may be due to the fact that the local volume fraction in the material with bimodal γ' distribution is about 25% as compared to about 40% in the material with monomodal distribution. The interparticle spacing in the former material is correspondingly larger (smaller Orowan stress) than in the latter material. A theoretical calculation of threshold stress for shearing process involving stacking fault formation in γ' phase according to Condat's model has been performed in [4]. The result shows the threshold stress decreases with increasing particle spacing. An estimate of the threshold stresses for both the observed mechanisms [4] shows that for the present γ' morphology the Orowan stress is higher than the threshold stress for the shear mechanism. This suggests that the Orowan mechanism is possible at high strain amplitudes which lead to high stress amplitudes in LCF testing.

The threshold stresses of the two observed deformation mechanisms decrease with increasing particle spacing. A possible coarsening of the large γ' and a dissolution of smaller γ' particles during the test [6] would therefore explain the slight softening observed in present tests. Such a microstructural change was however not detected. This is perhaps related to the short test periods.

## CONCLUSIONS

A slight cyclic softening was observed under all test conditions. Deformation mechanism was found to be a function of strain amplitude. At low strain amplitudes, the γ' precipitates were sheared by {111}<112> slip. At high strain amplitudes, the Orowan process is operative.

***Acknowledgements***-The work was supported by Deutsche Forschungsgemeinschaft (SFB 339). The authors are grateful to Prof. H. Wever and Prof. H. Wollenberger for their permanent supert.

## REFERENCES

1. Marchionni, M., Ranucci, R. and Picco, E., High temperature low-cycle fatigue behaviour of cast nickel-base IN 738 alloy, Mat. Sci. and Eng., 1982, 55, 231-237.
2. Mukherji, D., Jiao, F., Chen, W. and Wahi, R. P., Stacking fault formation in γ' phase during monotonic deformation of IN738LC at elevated temperatures, Acta metall., to be published, 1991.
3. Chen, W., Frohberg, G. and Wever, H., Gefügeeinfluß auf das machanische Verhalten von IN738LC bei einachsiger Wechsel- und Kriechbeanspruchung in Hinblick auf Plastizität und Schädigung. Forschungsbericht: TP B3, SFB 339, (Feller, H. G., Ed.), Technische Universität Berlin, 1990.
4. Jiao, F., Elektronenmikroskopische Untersuchung des Hochtemperatur-Verformungsmechanismus an der Nickelbasis-Superlegierung IN738LC, thesis, TU Berlin, 1991.
5. Condat, M. and Decamps, B., Shearing of γ' precipitates by single a/2<110> matrix dislocations in a γ/γ' Ni-base superalloy, Sripta Met., 1987, 21, 607-612.
6. Henderson, P. J. and McLean M., Microstructural changes during the high temperature creep af a directionally solidified nickel-base superalloy, Scripta Met., 1985, **19**, 99-104.
7. Jiao, F., Chen, W., Mukherji, D., Zhu, J. and Wahi, R. P., Deformation behavior and microstructural evolution in IN738LC under LCF loading, ICM-6, kyoto, 1991.

# A NEW MODEL FOR A REPRESENTATION OF HARDENING PROPERTIES IN CYCLIC PLASTICITY

MARC KAMLAH, CHARALAMPOS TSAKMAKIS, PETER HAUPT
Institut für Mechanik, Universität-Gesamthochschule Kassel, 3500 Kassel, BRD

## ABSTRACT

Proposals are made, how cyclic plasticity can be described by using evolution laws of an Armstrong-Frederick-type regarding to a suitable transformed arc length.

## INTRODUCTION

Kinematic hardening rules are an important part of the formulation of time-independent plasticity. A very simple assumption is the linear kinematic rule of Prager. However, the deformation history usually has to be modelled by using nonlinear relations, such as multi-surface models, like the Dafalias-Popov model, the Armstrong-Frederick model or the

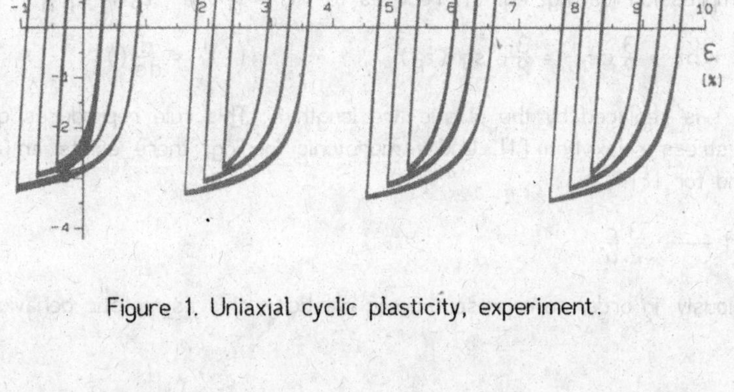

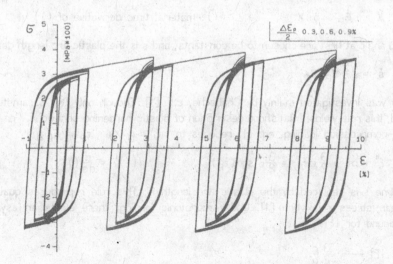

Figure 1. Uniaxial cyclic plasticity, experiment.

Chaboche model, c.f. [1]. This simple models yield satisfying results.

The phenomena of cyclic plasticity have been investigated by many researchers. Typical cyclic tension-compression experiments with steel are reported in [1] and [2] (c.f. Fig. 1), in which the mean strain and the strain amplitude are controlled. After i) increasing respectively ii) decreasing the strain amplitude at fixed mean strain, cyclic hardening respectively softening is observed. The stabilized stress amplitude doesn't depend on the mean strain. After iii) changing the mean strain, a cyclic relaxation of the mean stress takes place. Finally, a strain history effect can be observed, which has been interpreted as a maximum plastic strain memory [1]. In this paper, only the first three processes are considered.

Many methods have been developed to describe cyclic plasticity, which use e.g. additional loading conditions, discrete parameters or updating procedures. In order to represent uniaxial cyclic plasticity, the Armstrong-Frederick kinematic hardening rule is formulated in the present paper with regard to a transformed arc length. The corresponding transformation of the plastic arc length is influenced by an additional internal variable, which undergoes a continuous evolution.

## THE ELASTIC-PLASTIC CONSTITUTIVE MODEL

The **linearized GREEN's strain tensor $\mathbf{E}$** is usually decomposed into elastic and plastic parts, i.e. $\mathbf{E} = \mathbf{E}_e + \mathbf{E}_p$. The **CAUCHY stress tensor** is related to $\mathbf{E}_e$ by the generalized Hooke's law. The rate of $\mathbf{E}_p$ is given by the **associated flow rule (normality rule)** in combination with a **VON MISES** yield function. In this paper, kinematic hardening, but no isotropic hardening, is incorporated in the yield function.

### The kinematic hardening rule

A simple evolution equation for kinematic hardening, which initially was proposed by Armstrong and Frederick (c.f. [1]), is given by

$$\dot{\mathbf{X}} = c\dot{\mathbf{E}}_p - b\dot{s}\mathbf{X} \quad , \qquad \left(\dot{(\,)} : \text{material time derivative of } (\,)\right) \tag{1}$$

where b and c at first are chosen to be constants, and s is the **plastic arc length** defined by

$$\dot{s} := \sqrt{\frac{2}{3}\dot{\mathbf{E}}_p \cdot \dot{\mathbf{E}}_p} \quad . \tag{2}$$

Eq. (1) was investigated mainly by Chaboche, c.f. [1]. Though only two parameters are involved, this rule yields a satisfying description of plastic hardening behaviour. For uniaxial tension-compression loading, eq. (1) reduces to ( $\xi := \frac{2}{3} X_{11}$ , $\varepsilon_p := E_{p\,11}$ )

$$\xi' + b\xi = \frac{3}{2} c\,\varepsilon_p' = \frac{3}{2} c \,\operatorname{sgn}(\dot{\varepsilon}_p) \quad , \qquad \left((\,)' := \frac{d}{ds}(\,)\right) \tag{3}$$

when time t is replaced by the plastic arc length s. This rule reproduces qualitatively the mean stress relaxation [1]. Under monotonic loading, there exists an asymptotic upper bound for $|\xi|$:

$$|\xi| \longrightarrow \frac{3}{2}\frac{c}{b} \tag{4}$$

Obviously, in order to represent cyclic plasticity, this asymptotic behaviour can be influenced by assuming, that b is not a constant. Marquis (c.f. [1]) chose b to be a function of the plastic arc length, i.e. $b = \hat{b}(s) = b_1 + b_2 e^{-b_3 s}$, which describes a certian class of cyclic loading processes. For more general loading processes, it is proposed here, that b is a function of e.g. an additional internal variable p, i.e. $b = \hat{b}(p)$, of which the evolution equation has to be chosen skillfully. Every function $\hat{b}(p) > 0$ defines a transformation of the plastic arc length s to a new arc length z via

$$\dot{z} = \hat{b}(p)\dot{s} \quad , \tag{5}$$

with regard to which the evolution equation of $\xi$ is of the type (3) with $b \equiv 1$.

In the next section, a function $\hat{b}(p)$ and evolution equations for p will be proposed. The aim of this paper is to show that cyclic plasticity effects can be reproduced qualitatively by such methods, so that only the dependence of $\xi$ upon $\varepsilon_p$ (instead of $\sigma$ upon $\varepsilon$) is considered.

**Kinematic Hardening with a modified arc length**

Assuming $\hat{b}(p) = b/(1+\alpha p)$, where b and $\alpha$ are constants, the evolution equation for $\xi$ is given by

$$\xi'(s) = \frac{3}{2} c \,\mathrm{sgn}(\dot{\varepsilon}_p) - \frac{b}{1+\alpha p}\, \xi(s) \quad . \tag{6}$$

Generalizing eq. (4), the upper bound for $|\xi|$ can now be regarded to be a linear function of p. The definition of the evolution equation for p, when combined with eq. (6), should repesent the above mentioned three effects of cyclic plasticity. A series of possibilities will be discussed in the following.

First, consider

$$p = \frac{1}{s_0} \int_{s-s_0}^{s} |\varepsilon_p(\sigma)|\, d\sigma \quad , \quad s_0 \equiv \text{const.} \quad , \tag{7}$$

where s is the current value of the plastic arc length. The product $s_0 p$ is the area under $|\varepsilon_p|$ in $[s-s_0, s]$. If the loading process has zero plastic mean strain and constant plastic strain amplitude in $[s-s_0, s]$, p is nearly (depending on the magnitude of $s_0$) proportional to the plastic strain amplitude. If the amplitude changes in $[s-s_0, s]$ (at fixed zero mean strain), p will adjust itself continiously. Eqs. (6) and (7) exhibit cyclic hardening and softening, but the stress amplitude of the stabilized hysteresis loops depends on the plastic mean strain.

Because of this, in a further step, eq. (7) should be replaced by

$$p = \frac{1}{s_0} \int_{s-s_0}^{s} |\varepsilon_p(\sigma) - \varepsilon_p^M(\sigma)|\, d\sigma \quad , \tag{8}$$

where $\varepsilon_p^M$ represents the plastic mean strain, and only the oscillating part $\varepsilon_p - \varepsilon_p^M$ of $\varepsilon_p$ is considered. The plastic mean strain can be defined by

$$\varepsilon_p^M(s) = \frac{1}{s_1} \int_{s-s_1}^{s} \varepsilon_p(\sigma)\, d\sigma \quad , \quad s_1 \equiv \text{const.} \tag{9}$$

A more satisfying formulation of eqs. (8) and (9), using a continuously fading memory (along the plastic arc length), would be

$$p = \frac{1}{s_0} \int_0^s e^{-\frac{s-\sigma}{s_0}} |\varepsilon_p(\sigma) - \varepsilon_p^M(\sigma)| \, d\sigma \quad , \quad \varepsilon_p^M(s) = \frac{1}{s_1} \int_0^s e^{-\frac{s-\sigma}{s_1}} \varepsilon_p(\sigma) \, d\sigma \qquad (10)$$

( $s_0$, $s_1$ : constants ), which is equivalent to

$$p' + \frac{1}{s_0} p = \frac{1}{s_0} |\varepsilon_p(s) - \varepsilon_p^M(s)| \quad , \quad \varepsilon_p^{M\prime}(s) + \frac{1}{s_1} \varepsilon_p^M(s) = \frac{1}{s_1} \varepsilon_p(s) \qquad (11)$$

( $p(0) = 0$, $\varepsilon_p^M(0) = 0$ ). The three above cited effects i), ii) and iii) of cyclic plasticity are represented qualitatively by eqs. (6) and (11). On the other hand problems occur because of the need to introduce a somewhat arbitrary mean value of $\varepsilon_p$ into the functional p, since $\varepsilon_p^M$ doesn't adjust itself spontaneously to the currently prescribed mean strain.

As a consequence, p should be a functional of a quantity, which is a measure for the plastic strain amplitude, but on the other hand, doesn't require one to consider its mean value. Because of the mean stress relaxation property of $\xi$ governed by eq. (3), $|\xi|$ fulfills both of these conditions. Therefore, the evolution equation

$$p' + \frac{1}{s_0} p = \frac{1}{s_0} |\xi(s)| \quad , \quad p(0) = 0 \quad , \quad s_0 \equiv \text{const.} \qquad (12)$$

is proposed, which together with eq. (6) describes the three cyclic effects i), ii) and iii) without the above mentioned problems. This is shown in Fig. 2, where the result of a numerical integration of eqs. (6) and (12) is presented. These proposals will be a basis for further development.

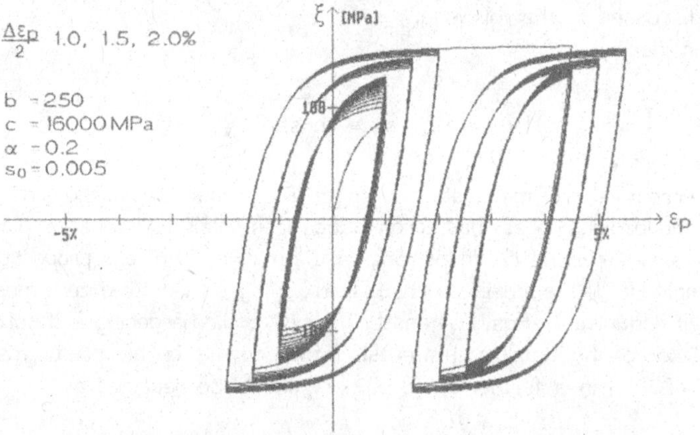


Figure 2. Numerical integration of eqs. (6) and (12)

## REFERENCES


1. CHABOCHE, J.-L., Time-independent Constitutive Theories for Cyclic Plasticity. Int. J. Plasticity, 1986, **2**, 149-188
2. PAPE, A., Zur Beschreibung des transienten und stationären Verfestigungsverhaltens von Stahl mit Hilfe eines nichtlinearen Grenzflächenmodells. Diss. Bochum 1988

# NONLINEAR KINEMATIC HARDENING RULE WITH CRITICAL STATE FOR ACTIVATION OF DYNAMIC RECOVERY

N. OHNO and J.-D. WANG
*Department of Mechanical Engineering, Nagoya University*
*Chikusa-ku, Nagoya, 464-01 Japan*

## ABSTRACT

*A kinematic hardening rule formulated in a hardening/dynamic recovery format is examined for simulation of ratchetting behavior. This rule, characterized by decomposition of the kinematic hardening variable into components, is based on the assumption that each component has a critical state for activation of its dynamic recovery. It is capable of predicting much less accumulation of ratchetting strain under uniaxial and multiaxial loadings than the Armstrong and Frederick rule; however, if ratchetting strain is negligible, they may give nearly the same predictions. It is shown that the present rule simulates well ratchetting experiments of Modified 9Cr-1Mo steel by Tanaka et al. and a nonproportional experiment of OFHC copper by Lamba and Sidebottom.*

## INTRODUCTION

Simulation of ratchetting behavior remains difficult, because it is concerned with secondary deformation accumulating cycle by cycle in the direction of nonzero mean stress. In fact, most of existing constitutive models have poor capability of predicting it under uniaxial and multiaxial loading conditions [e.g., 1, 2].

The nonlinear kinematic hardening rule of Armstrong and Frederick [3] is well known. However, it also gives poor predictions with respect to ratchetting behavior; it usually overpredicts ratchetting strain accumulation under both uniaxial and multiaxial loading conditions [e.g., 1, 2]. Nevertheless, the concept employed in this rule, i.e., hardening and dynamic recovery of kinematic hardening, is simple and physically sound. Modifications of the Armstrong and Frederick rule were thus discussed in several works; for example, Burlet and Cailletaud [4] made a modification effective in multiaxial ratchetting behavior, and Chaboche [5] introduced a threshold of dynamic recovery.

In the present paper, a kinematic hardening rule with a critical state for activation of dynamic recovery is examined in comparison with ratchetting experiments by Tanaka et al. [6] and a nonproportional experiment by Lamba and Sidebottom [7].

## KINEMATIC HARDENING RULES EXAMINED

Since metallic materials deformed inelastically may have various internal stresses acting on dislocations, we assume that the kinematic hardening variable $\alpha$ consists of components: $\alpha = \alpha_1 + \cdots + \alpha_N$.

Dislocation pile-up to obstacles can be a micromechanism for kinematic hardening, and piled-up dislocations can recover mobility due to cross slip. This dynamic recovery however may have a threshold resulting from the resistance to be overcome for cross slip. It is thus suggested that dynamic recovery of $\alpha_i$ is activated when the magnitude of $\alpha_i$, $\bar{\alpha}_i = (3/2)^{1/2}\|\alpha_i\|$, attains a critical value $r_i$. Let us represent the critical state of dynamic recovery by a surface

$$f_i = \bar{\alpha}_i^2 - r_i^2 = 0. \tag{1}$$

Then, an evolution equation of $\alpha_i$ is

$$\dot{\alpha}_i = \zeta_i[(2/3)r_i\dot{\varepsilon}^p - \mathrm{H}(f_i)\langle n : k_i\rangle \alpha_i \dot{p}], \tag{2}$$

where $\zeta_i$ is a constant, $\dot{\varepsilon}^p$ denotes inelastic (or plastic) strain rate, H stands for the Heaviside step function, $\langle\ \rangle$ indicates the Macauley bracket, $n = \dot{\varepsilon}^p / \|\dot{\varepsilon}^p\|$, $k_i = \alpha_i / \|\alpha_i\|$, and $\dot{p} = (2/3)^{1/2}\|\dot{\varepsilon}^p\|$. In eq. (2), $\langle n : k_i\rangle$ is necessary in the dynamic recovery term, because $\alpha_i$ remains located on the surface $f_i = 0$ when $\bar{\alpha}_i = r_i$ and $\dot{\varepsilon}^p : \alpha_i \geq 0$.

We may allow nonlinear activation of dynamic recovery inside the surface $f_i = 0$; i.e., dynamic recovery of $\alpha_i$ becomes significant nonlinearly, as $\bar{\alpha}_i$ approaches $r_i$. Equation (2) then may take a form

$$\dot{\alpha}_i = \zeta_i[(2/3)r_i\dot{\varepsilon}^p - (\bar{\alpha}_i / r_i)^{m_i}\langle n : k_i\rangle \alpha_i \dot{p}], \tag{3}$$

where $m_i$ (»1) is a material constant. This equation is reduced to eq. (2), when $m_i = \infty$.

According to the Armstrong and Frederick rule [3], on the other hand, dynamic recovery of $\alpha_i$ is proportional to $\alpha_i$:

$$\dot{\alpha}_i = \zeta_i[(2/3)r_i\dot{\varepsilon}^p - \alpha_i \dot{p}]. \tag{4}$$

To simulate multiaxial ratchetting behavior, Burlet and Cailletaud [4] modified eq. (4) as

$$\dot{\alpha}_i = \zeta_i[(2/3)r_i\dot{\varepsilon}^p - \mu_i \alpha_i \dot{p} - (2/3)^{1/2}(1-\mu_i)(n : \alpha_i)\dot{\varepsilon}^p], \tag{5}$$

where $\mu_i$ is a constant. Incidentally, eq. (4) with nonlinearity of $\alpha_i$ in the dynamic recovery term was examined for modeling of ratchetting behavior by Chaboche [5].

If we do not consider isotropic hardening for simplicity, $\dot{\varepsilon}^p$ can be expressed in terms of effective stress $s_{eff}$ $(= s - \alpha)$:

$$\dot{\varepsilon}^p = (3/2)A\bar{s}_{eff}^{\,n-1} s_{eff}, \tag{6}$$

where $\bar{s}_{eff} = (3/2)^{1/2}\|s_{eff}\|$, $A$ and $n$ are material constants, and **s** denotes the stress deviator.

## COMPARISON WITH EXPERIMENTS

First we are concerned with multiaxial ratchetting experiments of Modified 9Cr–1Mo steel at 550°C by Tanaka et al. [6] (Fig. 1(a)). Although this material exhibits strain softening, it was not taken into account in the simulations shown in Figs. 1(b) to 1(d). The present rule simulates well the experiments, while the Armstrong and Frederick rule highly overpredicts the experiments. The Burlet and Cailletaud rule gives better simulation than the Armstrong and Frederick rule; however, it is not so good as the present rule.

Lamba and Sidebottom [7] performed a nonproportional experiment of OFHC copper, resulting in the stress response shown in Fig. 2(a). The present rule and the Armstrong and Frederick rule provided nearly the same simulations (Figs. 2(b) and 2(c)), which are in good agreement with the experiment; in other words, if effects of ratchetting behavior are not marked, these two rules may have little difference in simulation. The Burlet and Cailletaud rule, on the other hand, was found to be unsuccessful (Fig. 2(d)).

*Acknowledgment* — The authors are grateful to Dr. E. Tanaka and Professor S. Murakami of Nagoya University as well as to Dr. K. Iwata of PNC for their kindness in contributing some of their experimental data.

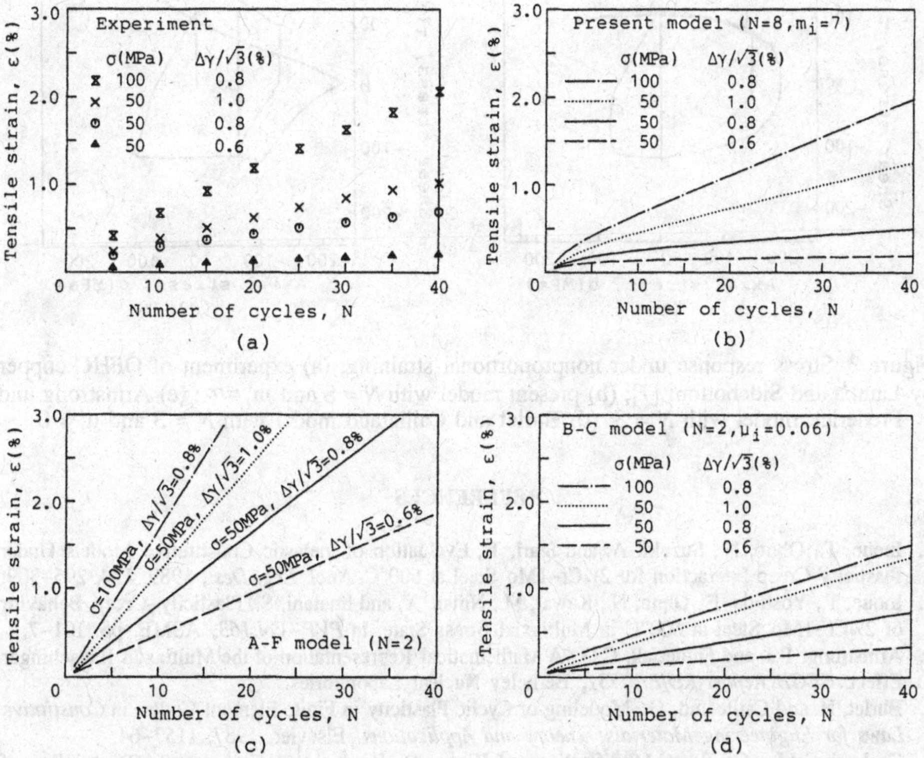


Figure 1. Accumulation of tensile strain under constant tensile stress combined with cyclic torsional straining; (a) experiment of Modified 9Cr–1Mo steel at 550°C by Tanaka et al. [6]; (b) present model with $N$ = 8 and $m_i$ = 7; (c) Armstrong and Frederick model with $N$ = 2; (d) Burlet and Cailletaud model with $N$ = 2 and $\mu_i$ = 0.06.

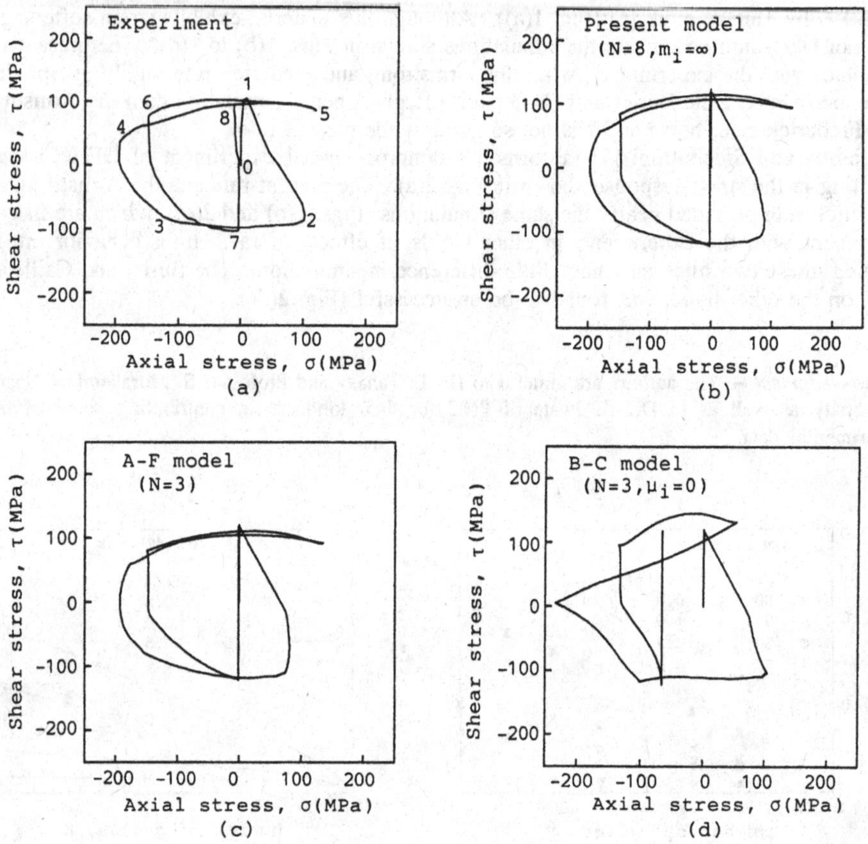


Figure 2. Stress response under nonproportional straining; (a) experiment of OFHC copper by Lamba and Sidebottom [7]; (b) present model with $N = 8$ and $m_i = \infty$; (c) Armstrong and Frederick model with $N = 3$; (d) Burlet and Cailletaud model with $N = 3$ and $\mu_i = 0$.

## REFERENCES


1. Inoue, T., Ohno, N., Suzuki, A. and Igari, T., Evaluation of Inelastic Constitutive Models Under Plasticity Creep Interaction for 2¼Cr–1Mo Steel at 600°C. *Nucl. Eng. Des.*, 1989, **114**, 295–309.
2. Inoue, T., Yoshida, F., Ohno, N., Kawai, M., Niitsu, Y. and Imatani, S., Plasticity–Creep Behavior of 2¼Cr–1Mo Steel at 600°C in Multiaxial Stress State. in *PVP–Vol.163*, ASME, pp. 101–7.
3. Armstrong, P.J. and Frederick, C.O., A Mathematical Representation of the Multiaxial Bauschinger Effect. *CEGB Report RD/B/N/731*, Berkeley Nuclear Laboratories.
4. Burlet, H. and Cailletaud, G., Modeling of Cyclic Plasticity in Finite Element Codes, in *Constituive Laws for Engineering Materials: Theory and Applications*. Elsevier, 1987, 1157–64.
5. Chaboche, J.L., On Some Modifications of Kinematic Hardening to Improve the Description of Ratchetting Effects. to appear in *Int. J. Plasticity*.
6. Tanaka, E., Murakami, S., Mizuno, M., Yamada, H. and Iwata, K., Inelastic Behavior of Modified 9Cr–1Mo Steel and Its Unified Constitutive Model. to be presented at ICM–6, 1991, Kyoto, Japan.
7. Lamba, H.S. and Sidebottom, O.M., Cyclic Plasticity for Nonproportional Paths, *ASME J. Eng. Mat. Tech.*, **100**, 96–111.

# A DAMAGE CUMULATION LAW UNDER NON PROPORTIONAL CYCLIC LOADING WITH OVERLOADS FOR THE PREDICTION OF CRACK INITIATION

S. TAHERI
Département de Mécanique et Modèles Numériques
Etudes et Recherches ,Electricité de France
1, Avenue du Général de Gaulle, 92141 Clamart Cedex, FRANCE

**ABSTRACT**

For a sequence of cyclic loadings of constant amplitude and containing overloads, we propose a method for damage cumulation in non proportional loading. This method uses as data cyclic stabilized states at non proportional loading and initiation or fatigue curve in uniaxial case. For that, we take into account the mean cell size and we define a stabilized uniaxial state cyclicaly equivalent to a non proportional stabilized state through a family of cyclic strain stress curves (C-S-S-C). Although simple assumptions like linear damage function and linear cumulation are used, we obtain a sequence effect for difficult cross slip materials as 316 stainless steel, but the Miner rule for easy cross-slip materials. We show that for the second case, in a load controled test, the non proportional loading is less damaging than the uniaxial one for the same equivalent stress, while the result is opposite in a strain controlled test.

## 1 INTRODUCTION

The most important law used in damage cumulation is the Miner law. The cumulated damage is given by:$D=\Sigma_i(n_i/N_f^i)$. $N_f^i$ is the number of cycles to failure (or initiation) at amplitude $\Delta\varepsilon_i$ (resp. $\Delta\sigma_i$) and $n_i$ is the number of cycles at this amplitude. Failure (initiation) is obtained when D = 1. In the following, $N_f$ is the number of cycles to initiation, but for small amplitudes it may be replaced by the number of cycles to failure without an important error. To get a sequence effect (a high law cycling damages differently from a law-high cycling), non linear cumulation and non linear damage functions have been used. This seems complicated. We are going to show that with a C-S-S-C depending on load history a linear damage function and a linear damage cumulation will give us a cumulation damage method which takes into account the sequence effect.

## 2 A DAMAGE CUMULATION MODEL IN UNIAXIAL LOADING

For materials with easy cross-slip (Cu, mild steel) C-S-S-C is independent of the load history (for Cu when $\Delta\varepsilon>0.002$), while for difficult cross-slip materials it depends on monotonic or cyclic prehardening. Fig.1 shows the result of an experiment on a 316 stainless steel at 20 Celsius's degrees. The curve A shows a C-S-S-C obtained by an increasing incremental test on a single sample (pushpull test), B a decreasing pushpull test from point M, and C an increasing one from point N. C and B are practicaly superposed. This shows that the loading history memory of a small cyclic amplitude is cleared by a higher amplitude, and that the memory of the high amplitude is perfectly conserved for smaller ones.

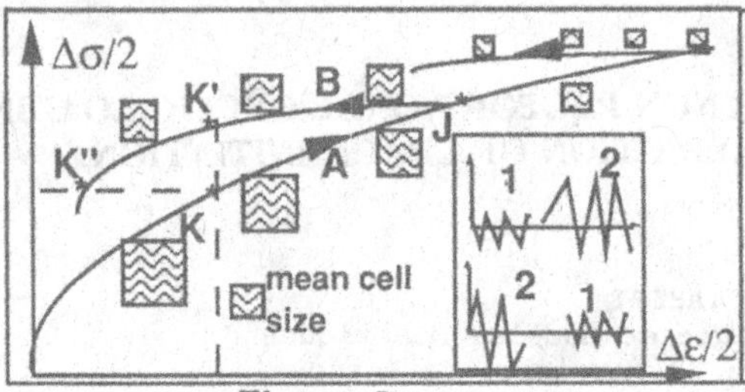


Figure 2.

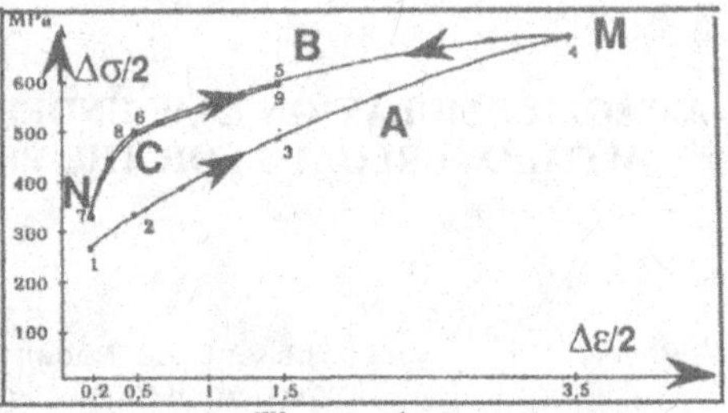


Figure 1.

We consider a sequence of constant amplitude loading, 1 and 2 (fig 2), where we suppose that the number of cycles $n_1$ (resp. $n_2$) is sufficient to get a stabilized state and this number is negligible with respect to the number of cycles to initiation.

For a strain-controlled-test, stabilization at load 1 corresponds to the point K. The damage, using a linear damage function is: $d_1=n_1/N_f(\Delta\varepsilon_1)$. Applying then the load 2, the stabilized point is designed by J. The damage is $d_2=n_2/N_f(\Delta\varepsilon_2)$ so the total damage with a linear cumulation is: $D^{\varepsilon}_{(1-2)} = n_1/N_f(\Delta\varepsilon_1)+n_2/N_f(\Delta\varepsilon_2)$. Now we apply first 2 and then 1. The points of stabilization are J and K'. The damage cumulation is: $D^{\varepsilon}_{(2-1)} = n_1/(\tilde{N}_f(\Delta\varepsilon_1) +n_2/N_f(\Delta\varepsilon_2)$. $\tilde{N}_f(\Delta\varepsilon_1)$ is the number of cycles to initiation for the amplitude $\Delta\varepsilon_1$ but with cyclic prehardening as far as stabilization at point J. We compare points K and K', strain amplitudes are identical while the stress amplitude is higher in K' than in K so obviously :$D^{\varepsilon}_{2-1} > D^{\varepsilon}_{1-2}$. This result is usually accepted in literature[1], a high-low cycling is more damaging than a low-high cycling, but for a load controlled test we obtain an opposite result. For a stress-controlled-test the sequence 1-2 gives the points K and J while the sequence 2-1 gives the points J and K". The damages are respectively:

$D^{\sigma}_{(1-2)} = n_1/N_f(\Delta\sigma_1)+n_2/N_f(\Delta\sigma_2)$ and $D^{\sigma}_{(2-1)} = n_1/(\tilde{N}_f(\Delta\sigma_1)+n_2/N_f(\Delta\sigma_2)$

The same analysis as before shows that :$D^{\sigma}_{(2-1)}<D^{\sigma}_{(1-2)}$. We may remark that $D^{\varepsilon}_{(1-2)}$ is overestimated by $D^{\varepsilon}_{(2-1)}$ and $D^{\sigma}_{(2-1)}$ is overestimated by $D^{\sigma}_{(1-2)}$ In the literature, it is not easy to find initiation or fatigue curves with cyclic prehardening. So we propose to use these overestimatings as a **simplified method**. So in a stress controlled test $\tilde{N}_f$ may be replaced by $N_f$, the number of cycles to initiation without any prehardening (Wöhler curve) In this case a **conservative response is obtained**. In a strain controlled test $\tilde{N}_f$ may be replaced by $N_f^{max}$ where $N_f^{max}$ is the Manson-Coffin curve obtained after a cyclic prehardening at a maximal strain amplitude.

## 3 MICROSTRUCTURAL ANALYSIS.

In Fig.1 the difference between curves A and B comes from the difference of the mean cell sizes due to the different amplitudes at which they have been created[1]. The superposition of B and C comes from the stability of the mean cell size obtained at a great amplitude for smaller one. Beside we suppose that a minimal mean cell size exists[1]. So we may suppose that for the amplitudes of loading greater than that for creating the minimal cell

size, the C-S-S-C is single. It is obvious that if the minimal cell size is reached at a very low amplitude, C-S-S-C is independent of prehardening for in service loadings. This may explain the uniquness of C-S-S-C for pure Cu or mild steel. This analysis **may be extended to the non proportional loading**. For non proportional loading, we suppose that[1] for the same equivalent stress or strain (in Mises sense) the cell size is smaller for a non proportional loading than for an uniaxial one. So we may suppose that the limit cell size is obtained for smaller equivalent amplitudes in non proportional loading. For easy cross slip material as pure copper[2] or mild steel[4], in uniaxial case and non proportional case the C-S-S-C and microscopic structures are similar. This may be explained by the fact that the minimal cell size is reached for the usual amplitude of loading ($\Delta\varepsilon^p$ >0.002 for pure Cu). So in the uniaxial and non proportional loading, we have the same cell size and so that the same C-S-S-C. Finally:
easy cross slip ⇔ C-S-S-C independent of prehardening ⇔cell size stabilized (at minimal size)⇔{non proportional C-S-S-C ⇔ uniaxial C-S-S-C}
For a 316 stainless steel, the limit cell size is not reached at usual amplitudes, the microstructure and C-S-S-C depend on prehardening. But as shown on fig. 2 (curve B), the cell size is stable on curve B for all amplitudes smaller than the amplitude of point J. So for a metal defined by curve B we are in the same situation as for pure copper. So we may suppose that for all states on the curve B, the uniaxial and the non proportional C-S-S-C are superposed. This brings us to the following definition. We define a non proportional stabilized state as cyclically equivalent to an uniaxial stabilized state when the mean size of cell structure are identical in both cases. Practically this means that for a 316 steel a non proportional stabilized state defined by ($\Delta\varepsilon^1{}_{eq}$, $\Delta\sigma^1{}_{eq}$), (point N fig. 3), is equivalent to a uniaxial stabilized state defined by $\Delta\varepsilon_{11} = \Delta\varepsilon^1{}_{eq}$ and $\Delta\sigma_{11} = \Delta\sigma^1{}_{eq}$ with a cyclic prehardening as far as stabilization at the point $m_1$. It is obvious that with such a definition for the case of a single C-S-S-C, a cyclically equivalent state has the the same meaning as a equivalent state defined by the Mises criterion.

## 4 A DAMAGE CUMULATION MODEL IN NON PROPORTIONAL LOADING

When C-S-S-C is single, uniaxial and non proportional C-S-S-C are superposed. A stabilized limit state ($\Delta\varepsilon_{eq}$, $\Delta\sigma_{eq}$) corresponds to the uniaxial point $\Delta\varepsilon_{11} = \Delta\varepsilon_{eq}$, $\Delta\sigma_{11} = \Delta\sigma_{eq}$ on the uniaxial C-S-S-C. For these metals, fatigue curves are not very different for uniaxial and non proportional experiments[3]. For them, we suppose so that the damage created at each cycle at a stabilized state are identical in uniaxial case and in non proportional case. For a 316 stainless steel fatigue curves are very different in uniaxial and non proportional loadings. The analysis of §3 suggests us to suppose that the damage created in non proportional loading for each cycle at stabilized state ($\Delta\varepsilon_{eq}$, $\Delta\sigma_{eq}$) is identical to the one obtained at each cycle of a cyclicaly equivalent uniaxial stabilized state. A method of damage cumulation may so be proposed analogous to the uniaxial one through the following example. We take a sequence of cyclic loading at constant amplitude. The stabilized states are designed successively by ($\Delta\varepsilon^1{}_{eq}$, $\Delta\sigma^1{}_{eq}$), ($\Delta\varepsilon^2{}_{eq}$, $\Delta\sigma^2{}_{eq}$), ($\Delta\varepsilon^3{}_{eq}$, $\Delta\sigma^3{}_{eq}$). For each state the number of cycles are $n_1$, $n_2$, $n_3$. We bring the above points on an uniaxial C-S-S-C diagram (fig.3). These points are on the curves $B_1$, $B_2$, $B_3$ so the damage is:

$$D = n_1/N_{f_1}^{m_1} + n_2/N_{f_2}^{m_2} + n_3/N_{f_3}^{m_3}$$

where $N_{f_1}^{m_1}$ is the number of cycles to initiation in an uniaxial case for the amplitude $\Delta\varepsilon_{11} = \Delta\varepsilon_{eq}$ or $\Delta\sigma_{11} = \Delta\sigma_{eq}$, prehardened as far as stabilization at $m_1$. Through this analysis, as explained in §2, during strain controlled test a non proportional loading is more damaging than an equivalent uniaxial loading. This is verified for a 316[1]. But the result is opposite for a stress controlled test: a non proportional loading is less damaging than the uniaxial equivalent one. For the same reason as explained in §2. **a simplified method** analogous to that of §2 may be proposed[1].

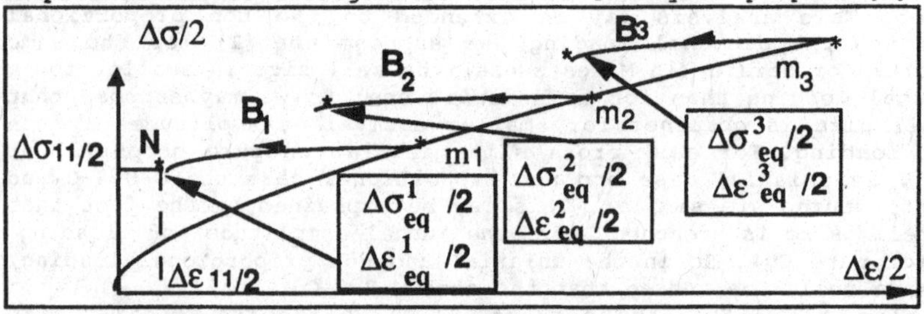


Figure 3.

## 5 DAMAGE CUMULATION MODEL TAKING INTO ACCOUNT THE OVERLOADS

If we suppose that the mean cell size is determined by the maximal stress supported in the load history, the C-S-S-C after an overload at $\sigma_{max}$ or $\varepsilon_{max}$ (fig 4.).will be placed (fig 5.) between the C-S-S-C without any prehardening and the C-S-S-C prehardened cyclicaly as far as stabilization for $\Delta\sigma=2\sigma_{max}$ or $\Delta\varepsilon=2\varepsilon_{max}$ [1]. For a single overload, when C-S-S-C is not single, in a load controlled test damaging at each cycle for a prehardened material (point P") may be overestimated by the damage for a virgin material (point K). In a strain controlled test we get a different result: the damage at P' may be overestimated by the damage at K'

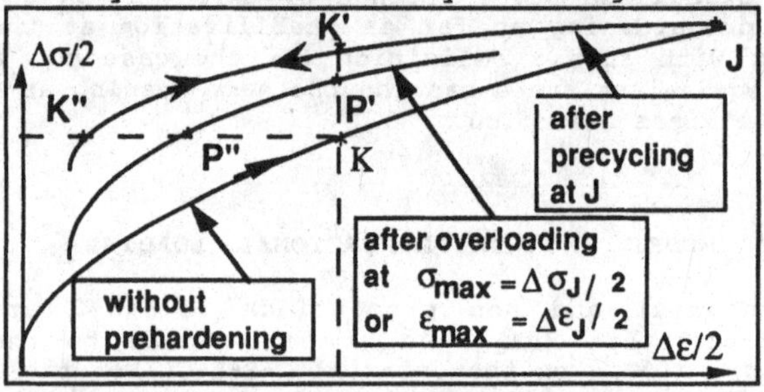


Figure 5.

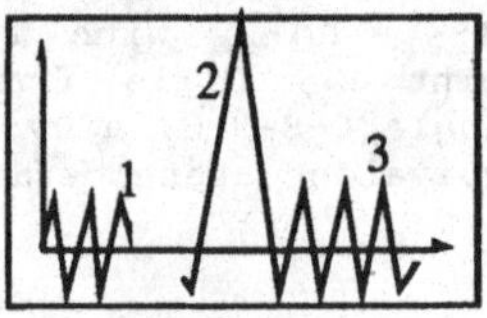


Figure 4.

That means: for a load controlled test, by the use of the initiation curve without prehardening, a **conservative** result is obtained if we neglect the overload, while for a strain controlled a **non conservative** answer is obtained. In the case of a load controlled test for a periodic overload [1], we show that, when the overload frequency is low, the life time increases, while for a high frequency it decreases.

## REFERENCES


1. TAHERI, S.,.Loi simplifiée de cumul de dommages sous chargement cyclique radial ou non radial avec surcharges. Note EDF HI-71/7169. 1990

2. DOONG,S.H., SOCIE, D.F., Deformation mecanisms of metals under complex non proportional cyclic loadings 3th Int. Conference on Biaxial and Multiaxial Fatigue, April 3-6 1989 Stuttgart.F.R.G

3. DOQUET,V., PINEAU,A. Multiaxial low cycle fatigue behaviour of a mild steel. 3th International Conference on Biaxial and Multiaxial Fatigue, April 3-6 1989 Stuttgart F.R.G..

# NUMERICAL PREDICTIONS OF BAUSCHINGER CURVES AND CYCLIC LOADING CURVES WITH THE LIN'S POLYCRYSTAL MODEL

H. TAKAHASHI[1] and N. ONO[2]

[1] Dept. of Mechanical Systems Engng.,Yamagata Univ.,Yonezawa JAPAN.

[2] Dept. of Precision Engng.,Tohoku Univ.,Sendai JAPAN.

## ABSTRACT

Bauschinger curves and cyclic loading curves of polycrystal aluminium are predicted numerically with the Lin's polycrystal model. Bauschinger effect is taken into account in the constitutive law of single crystal by introducing the backlash model. This model is based on the author's experimental findings in torsion tests of polycrystal aluminium that stress reversal yields transient region of non-workhardening. The calculated results show the good agreements with the experiments.

## INTRODUCTION

Some attempts have been made to predict Bauschinger curves of polycrystal metals. Hutchinson (1) and Ono et al.(2) calculated the Bauschinger curves with the KBW model assuming the Taylor's isotropic law for single crystal workhardening. However, the Taylor's law does not account for Bauschinger effect of single crystal. Budiansky & Wu(3) assumed the Ishlinsky-Prager's kinematic hardening law for yielding function of single crystal. Weng(4) also proposed the combined hardening law of isotropic and kinematic. Kratochvil & Tokuda(5) proposed the stress memory hardening law. All of these anisotropic hardening models are intended to account for stress decreasing due to stress reversal.

Hasegawa et al.(6) showed that cell structures developed during prestraining are partially dissolved by stress reversal and new cell structures consisting of opposite sign dislocations are formed. During resolution and reformation of cell structures, the flow stress is kept constant. The plateau at the Bauschinger curve has been observed by Takahashi et al.(7) for aluminium at large strain, by Miyauchi(8) for steel and by Christoudoul et al.(9) for copper.

The kinematic hardening model can not explain the stagnation of workhardening in Bauschinger curves. Hess & Sleeswyk(10) proposed the polarity model where piled dislocations may be annihilated by moving dislocations of opposite polarity. This model can explain well the stagnation. However, it can not take account for reversible process in reloading or cyclic loading which are found by the author(11). The present paper proposes the backlash model for single crystal workhardning and calculates cyclic loading curves of aluminium polycrystal with the Lin's polycrystal model.

## BACKLASH MODEL FOR SINGLE CRYSTAL WORKHARDENING

The author(7) found that stress reversal yields a transient region of nonworkhardening. After stagnating, workhardening restarts as if the material forgets the stress reversal. Since this feature resembles backlash in gear contact, the author(11) has proposed the backlash model to predict cyclic loading behavior in large strain. Now, we introduces the concept of backlash into the constitutive law of single crystal.

Suppose two kinds of strains for each slip system, reversible strain $\gamma_{rev}^{(r)}$ and effective strain for workhardening $\gamma_{eff}^{(r)}$ where the upper suffix (r) denotes slip system. The yielding shear stress of crystal is assumed as

$$k = F \cdot [\sum_{r} \gamma_{eff}^{(r)}]^{n} \quad ( F=32\ \mathrm{Mpa},\ \ n=0.24 ). \tag{1}$$

Here we assume the condition for $\gamma_{eff}^{(r)}$ to increase by the following.

$$\zeta^{(r)} \gamma_{rev}^{(r)} = \gamma_{b}^{(r)}, \qquad \zeta^{(r)} = \mathrm{sign}(\tau^{(r)} \text{or } d\gamma^{(r)}) \tag{2}$$

where $\gamma_{b}^{(r)}$ is the backlash strain and given by the following.

$$\gamma_{b}^{(r)} = \begin{cases} \gamma_{eff}^{(r)} & \text{for } \gamma_{eff}^{(r)} < \gamma_0 \\ \beta( \gamma_{eff}^{(r)} - \gamma_0 ) + \gamma_0 & \text{for } \gamma_{eff}^{(r)} > \gamma_0 \end{cases} . \tag{3}$$

with $\gamma_0$ =0.01 and $\beta$=0.4. Under the workhardening condition of eqn (2) the both of $\gamma_{eff}^{(r)}$ and $\gamma_{rev}^{(r)}$ increase with the following equations.

$$d\gamma_{eff}^{(r)} = |\ d\gamma^{(r)}|, \qquad d\gamma_{rev}^{(r)} = \zeta^{(r)} d\gamma_{b}^{(r)} . \tag{4}$$

Once the loading direction is reversed, $\zeta^{(r)}$ is changed and we find

$$\zeta^{(r)} \gamma_{rev}^{(r)} < \gamma_{b}^{(r)} \tag{5}$$

During this transient region the effective strain $\gamma_{eff}^{(r)}$ could decrease and the reversible strain $\gamma_{rev}^{(r)}$ of previous sign is annihilated. The evolution law is assumed as follow.

$$\begin{cases} d\gamma_{eff}^{(r)} = |\ d\gamma^{(r)}|\ (\zeta^{(r)}\gamma_{rev}^{(r)} / \gamma_{b}^{(r)} + \lambda)/2 \\ d\gamma_{rev}^{(r)} = d\gamma^{(r)} \end{cases} \tag{6}$$

where $\lambda$ is an irreversible term valid for small strain and assumed as

$$\lambda = \begin{cases} 1 & \text{for } \gamma^{(r)}_{\mathrm{eff}} < \gamma_0 \\ \exp(-\eta(\gamma^{(r)}_{\mathrm{eff}} - \gamma_0)) & \text{for } \gamma^{(r)}_{\mathrm{eff}} > \gamma_0 \end{cases} \quad (\eta = 100). \tag{7}$$

When $\gamma^{(r)}_{\mathrm{eff}}$ increases infinitely, $\lambda$ vanishes and then the process of eqn (6) becomes purely reversible because $\zeta^{(r)}|\, d\gamma^{(r)}| = d\gamma^{(r)}_{\mathrm{rev}}$. It should be noted that stress decreasing just after stress reversal comes from decreasing of $\gamma^{(r)}_{\mathrm{eff}}$ due to $\zeta^{(r)}\gamma^{(r)}_{\mathrm{rev}} < 0$. As the deformation proceeds in the reverse direction, $|\gamma^{(r)}_{\mathrm{rev}}|$ decreases at first and then incereases again up to $\gamma^{(r)}_b$ where workhardening starts again.

## NUMERICAL PREDICTIONS WITH LIN'S POLYCRYSTAL MODEL

Each crystal has cubic shape and numerous blocks consisting of 27 crystals having different orientaions are piled up as illustrated in Fig.1 . The shear strain increment $d\gamma^{(r)}$ in each crystal are determined by the simplex method based on the Lin's polycrystal model, see Takahashi(12).

While the experimental curves are obtained by torsional test, that is simple shear, the calculation are carried out for pure shear. The macroscopic shear stress and plastic shear strain are denoted by $\mathrm{T}$ and $\Gamma$, respectively. Figure 2 shows the Bauschinger curves of the experiments and the calculations where

$$\overline{\mathrm{T}} = |\mathrm{T}|, \quad \overline{\Gamma} = \int |d\Gamma|. \tag{8}$$

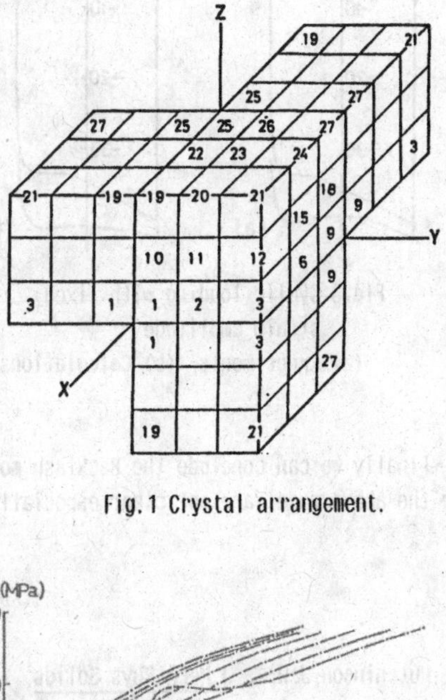


Fig.1 Crystal arrangement.

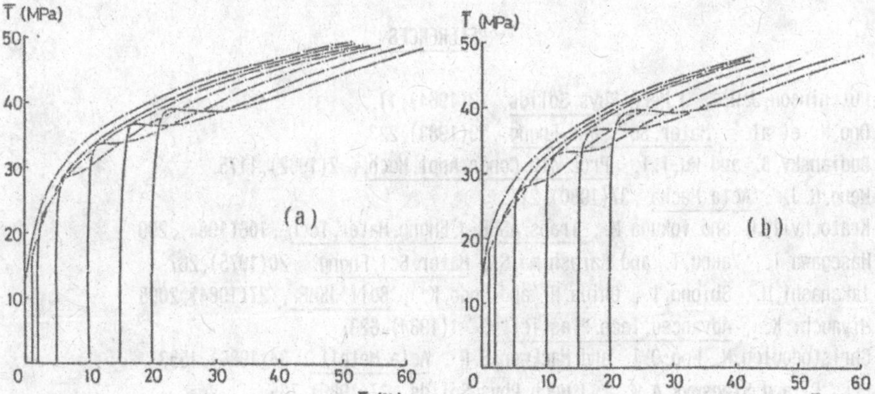


Fig.2 Bauschinger curves (a) Experiments, (b) Calculations.

Figure 3 shows the cyclic loading curve with fixed strain amplitude. The saturation of hysterysis loop can be realized almost after two cycles. Figure 4 shows the cyclic loading curve with increasing strain amplitude.

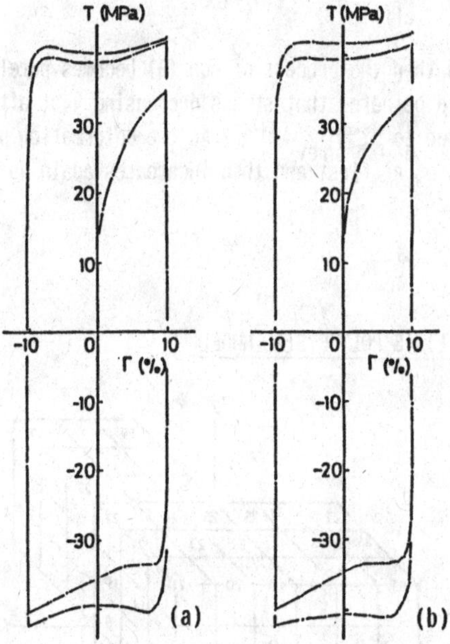


Fig.3 Cyclic loading with fixed strain amplitude
(a) Experiments, (b) Calculations.

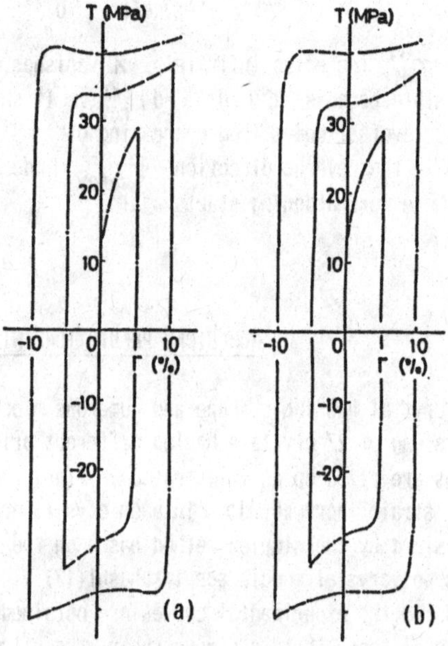


Fig.4 Cyclic loading with increasing strain amplitude
(a) Experiments, (b) Calculations.

Finally we can conclude the Backlash model works well. However, the physical explanations for the above model are not clear especially for the term $\lambda$ in eqn (6).

## REFERENCES


1. Hutchinson,J.W. , J.Mech.Phys.Solids, 12(1964),11.
2. Ono,N. et al. , Mater.Sci. and Engng.,59(1983),223.
3. Budiansky,B. and Wu,T.T., Proc.4th Congr.Appl.Mech., 2(1962),1175.
4. Weng,G.J. , Acta Mecha.,37(1980),217.
5. Kratochvil,J. and Tokuda,M., Trans.ASME.J.Engng.Mater.Tech.,106(1984),299.
6. Hasegawa,T., Yakou,T. and Karashima,S., Mater.Sci.Engng., 20(1975),267.
7. Takahashi,H., Shiono,I., Chida,N. and Endo,K. , Bull.JSME., 27(1984),2095 .
8. Miyauchi,K. , Advanced Tech.Plasticity, 1(1984),623.
9. Christodoulou,N, Woo,O.T. and MacEwen,S.R., Acta Metall., 34(1986),1553 .
10. Hess,F. and Sleeswyk,A.W. , J.Mech.Phys.Solids, 37(1989),735.
11. Takahashi,H. and Shiono,I. , Int.J.Plasticity, 7(1991),199.
12. Takahashi,H., Int.J.Plasticity, 4(1988),231.

# ANISOTROPIC HARDENING MODEL AND ITS APPLICATION TO CYCLIC LOADING

W. TRAMPCZYNSKI and Z. MROZ
Institute of Fundamental Technological Research
Warsaw, Poland

## ABSTRACT

The experimental study of evolution rules of two state variables: back stress $\alpha_{ij}$ and yield stress R is first discussed. It is shown that back stress evolution is affected by the maximal prestress and asymptotic value of yield stress depends on strain amplitude. The constitutive model is presented and its prediction for cyclic loading program is compared with experimental data.

## EXPERIMENTAL STUDY OF EVOLUTION RULES

Numerous models of plastic hardening use the concept of back stress tensor specifying the centre of the yield surface and its varying size is associated with the isotropic hardening. The evolution of back stress represents the varying Bauschinger anisotropy associated with elastic microstress evolution induced by inhomogeneous plastic deformation within the representative macro element. The isotropic hardening or softening is associated with the evolving dislocation structure during cyclic or variable loading involving loading, unloading and reloading phenomena. Neglecting complex effects associated with shape variation of the yield surface, the evolution rules for both kinematic and isotropic hardening can be tested experimentally by carrying out properly controlled cyclic loading tests. The results of such tests, cf.[1,2] obtained for two structural steels are briefly discussed here. It was assumed that the material obeys the Huber–Mises yield condition and the yield surface translates and expands or contracts. The points symmetrically situated with respect to the loading point on this surface were identified by performing unloading tests and using the specified offset value ($\varepsilon^p = 0.05\%$) as a definition of the yield point. The following properties of evolution were established.

i) The evolution of the back stress $\alpha_{ij}$ for cyclic loading with specified plastic strain amplitude attains the steady state practically after the first cycle and then it changes but

slightly to reach a stable loop. Shape of this loop is similar to the stress loop and no decay is observed, cf.Fig.3.
ii) The memory of maximal prestress is exhibited in the evolution of the back stress $\alpha_{ij}$, in both monotonic and cyclic loading tests. The prior deformation of larger value of stress or plastic strain affects the steady state reached subsequently for lower value of stress, cf.Fig.4.
iii) The isotropic hardening parameter $R$ representing the size of the yield surface does not exhibit the memory effect of prior prestress. Its steady state evolution depends only on the plastic strain amplitude (during the proportional straining), cf.Fig.4.
iv) The steady state evolution reached in one cyclic loading program is followed by a transient evolution before reaching the subsequent steady state corresponding to a new loading program.

Figures 2-5 present typical experimental results obtained for different loading histories of tubular specimens (of 21CrMoV57 steel) subjected to torsion tests at room temperature. The continuous lines correspond to the first cycle response and broken lines correspond to steady state response.

## CONSTITUTIVE MODEL

This model utilizes the concept of memory surface specified in the space of the back stress, cf.[3-6]. Consider the yield condition

$$S_{ij} - \alpha_{ij} = n_{ij}R(H) \tag{1}$$

where $S_{ij}$ is the stress deviator, $\alpha_{ij}$ is the back stress tensor representing kinematic hardening, $R$ is the yield stress representing isotropic hardening, $n_{ij}$ is the unit "vector" of directions in $S_{ij}$ space and $H$ is parameter describing history effect. Besides the yield surface, introduce the memory surface in the space of back stress

$$\varphi(\alpha_e,\ \alpha_m) = \alpha_e - \alpha_m \ \leq 0 \tag{2}$$

where

$$\alpha_e = (3/2\alpha_{ij}\alpha_{ij}) \tag{3}$$

and

$$\alpha_m = \sup \alpha_e(t-s) \quad 0 < s < t \tag{4}$$

During the initial loading process when $\alpha_{ij}$ grows from zero, the maximal value of $\alpha_e$ is equal to the actual value and then $\varphi(\alpha_e,\ \alpha_m) = 0$ (Fig.1a). On the other hand, when $\alpha_e$ decreases after reaching $\alpha_m$, the back stress state is represented by the interior of the memory surface and $\varphi(\alpha_e,\ \alpha_m) < 0$. When $\varphi(\alpha_e,\ \alpha_m) = 0$, $\dot{\alpha}_e > 0$ we have the $\alpha_e$ – loading process and when $\varphi(\alpha_e,\ \alpha_m) < 0$ or $\varphi = 0$, $\dot{\alpha}_e < 0$ we the have $\alpha_e$ – unloading process (reorientation process - Fig.1b).

Assume the following representation of the evolution rule for $\alpha_{ij}$

$$\dot{\alpha}_{ij} = \dot{\alpha}_{ij}^{(l)} + \dot{\alpha}_{ij}^{(r)} \tag{5}$$

Here $\dot{\alpha}_{ij}^{(l)}$ denotes the back stress rate associated with the loading process and $\dot{\alpha}_{ij}^{(r)}$ denotes the rate of $\alpha_{ij}$ due to reorientation process occurring for $\alpha_e < \alpha_m$.

$$\dot{\alpha}_{ij}^{(l)} = C^{(l)}\dot{\varepsilon}_{ij}^{p}, \quad C^{(l)} = C[(\alpha_f - \alpha_m)/R_o] \tag{6}$$

and $C$, $\alpha_f$, $R_o$, $m$ are the material constants
($\alpha_f$ denotes the asymptotic value of $\alpha_e$ reached in the loading process and $R_o$ denotes the initial yield limit - Fig.1a). Further we have:

$$\dot{\alpha}_{ij}^{(r)} = C^{(r)}\lambda_p\beta_{ij}^{r} \tag{7}$$

where:

$$\begin{aligned}
C^{(r)} &= D(\overset{*}{S}_d\,/R_o) \\
S_d &= [3/2(\overset{*}{\alpha}_{ij} - \alpha_{ij})(\overset{*}{\alpha}_{ij} - \alpha_{ij})]^{1/2} \\
\overset{*}{\alpha}_{ij} &= \alpha_m \dot{e}_{ij} \\
\dot{e}_{ij} &= \dot{\varepsilon}_{ij}^{p}/(2/3\dot{\varepsilon}_{kl}^{p}\dot{\varepsilon}_{kl}^{p})^{1/2} \\
\dot{\lambda}_p &= (2/3\dot{\varepsilon}_{ij}^{p}\dot{\varepsilon}_{ij}^{p})^{1/2} \\
\beta_{ij}^{r} &= (\overset{*}{\alpha}_{ij} - \alpha_{ij})/[3/2(\overset{*}{\alpha}_{kl} - \alpha_{kl})(\overset{*}{\alpha}_{kl} - \alpha_{kl})]^{1/2}
\end{aligned}$$

and $D$, $R_o$, $n$ are the material constants.
Thus the back stress reorientation occurs toward the image back stress $\overset{*}{\alpha}_{ij}$ lying on the memory surface $\varphi = 0$ and parallel to the instantaneous plastic strain rate (Fig.1b).

The isotropic hardening evolution is specified by

$$R = A[(R_a - R)/R_o]^k \dot{\lambda}_p \tag{8}$$

where $A$, $R_a$, $R_o$, $k$ are the material constants and $\dot{\lambda}_p$ is specified by eq.7.
The asymptotic value of $R_a$ is specified by

$$R_a = [R_o + [B(1 - 1/(1 + A_p))^l]](1 + F\omega) \tag{9}$$

where $B$, $l$, $F$ are the material constants.
($A_p$ – denotes the plastic strain cyclic amplitude and $\omega (0 \le \omega \le 1)$ is the parameter describing the cycle shape in the $\gamma_{xy}^{p}$ - $\varepsilon_{x}^{p}$ plane. Thus $A_p$ represents the amplitude dependent hardening evolution and $\omega$ represents hardening due to non proportional loading).

Next the model is applied to simulate experimental stress - strain curves and evolution curves. The identification of material parameters is obtained from loading-unloading tests. For 21CrMoV57 steel the list of parameters is as follows:
$C = 800$, $\alpha_f = 36$, $R_o = 20$, $m = 3$, $D = 3800$, $n = 2.5$, $A = 560$, $k = 1$, $B = 11$, $l = 0.6$, $F = 0.25$. Some of verificated results are shown in Figs. 2-5. It is seen that the present description provides a fair agreement with experimental data.

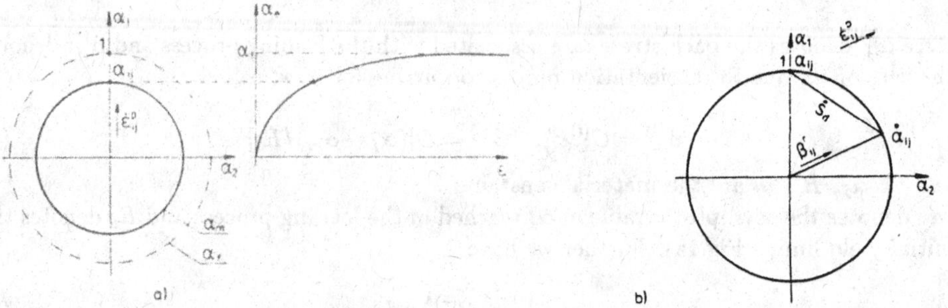

Figure 1. The evolution of back stress $\alpha_{ij}$ (a - loading process; b - reorientation process)

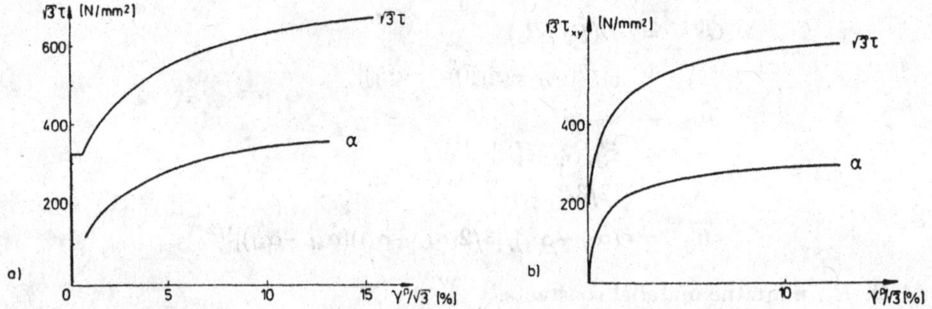


Figure 2. Experimental (a) and predicted (b) stress-strain and back stress-strain curves for monotonic torsion

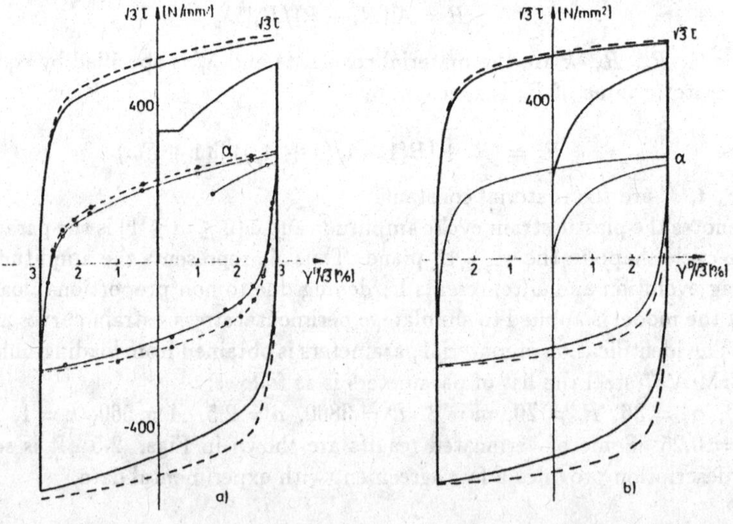


Figure 3. Experimental (a) and predicted (b) stress-strain and back stress-strain curves for cyclic torsion

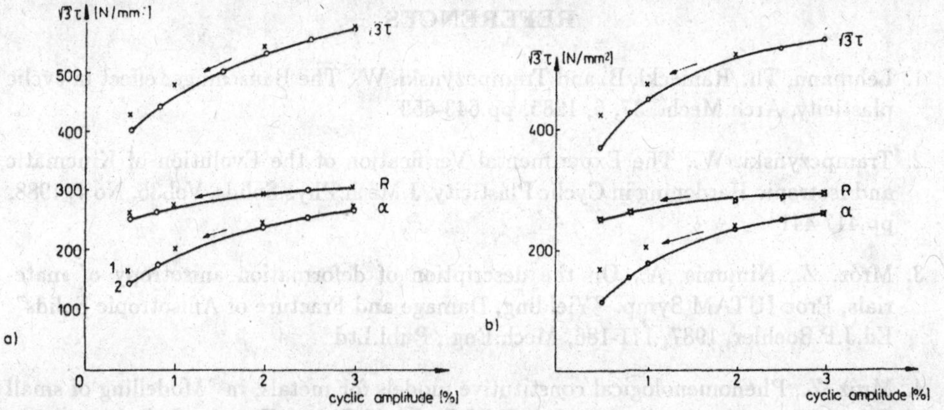


**Figure 4. Comparison between skeleton points for stress, $\alpha$ and $R$ values in the case of increasing (o) and decreasing (x) (after cyclic torsion $\varepsilon^p = + - 3\%$) plastic strain amplitudes; a - experiment, b - prediction.**

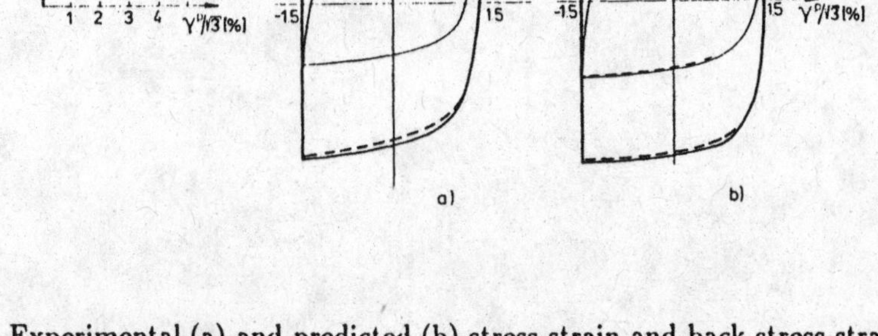

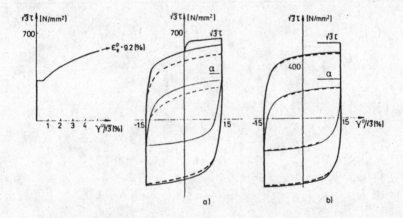

**Figure 5. Experimental (a) and predicted (b) stress-strain and back stress-strain curves for cyclic torsion after plastic prestrain $\varepsilon^p = 9.2\%$**

## REFERENCES

1. Lehmann, Th. Raniecki, B. and Trąmpczyński, W., The Bauschinger effect in cyclic plasticity, Arch.Mech., 37, 6, 1985, pp.643-659

2. Trąmpczyński, W., The Experimental Verification of the Evolution of Kinematic and Isotropic Hardening in Cyclic Plasticity, J.Mech.Phys.Solids, Vol.36, No 4, 1988, pp.417-441

3. Mróz, Z., Nimunis, A., On the description of deformation anisotropy of materials, Proc IUTAM Symp. "Yielding, Damage and Fracture of Anisotropic Solids", Ed.J.P.Boehler, 1987, 171-186, Mech.Eng., Publ.Ltd

4. Mróz, Z., Phenomenological constitutive models for metals, in "Modelling of small deformations of polycristals" 1986, Ed.J.Zarka, H.Gittus, Elsevier Sc.Publ, 293-344

5. Mróz, Z. and Trąmpczyński, W., On the Creep Hardening Rule for Metals with a Memory of Maximal Prestress, Int.J.Solids Struct., 20, 1984, pp.467

6. Mróz, Z., Trąmpczyński, W. and Hayhurst, D., Anisotropic creep hardening rule for metals and its application to cyclic loading, Int.Journ.of Plasticity, Vol.4, 1988, pp.279-299

# CONSTITUTIVE MODEL FOR CYCLIC PLASTICITY WITH RATCHETTING EFFECTS

GEORGE Z. VOYIADJIS and SRINIVASAN M. SIVAKUMAR

Department of Civil Engineering
Louisiana State University
Baton Rouge, Louisiana 70803, USA

## ABSTRACT

A robust kinematic hardening rule is proposed which appropriately blends the deviatoric stress rate rule and the Tseng-Lee rule in order to satisfy both the experimental observations made by Phillips et al. [1-3] and the nesting of the yield surface to the limit surface. A more general expression for the plastic modulus is proposed. In order to model a more sophisticated and extended memory of discrete events of unloading-reloading and to model ratchetting properly, an additional surface called the memory surface is introduced. An additional parameter $\zeta$ is introduced that reflects the dependence of the plastic modulus on the angle between the deviatoric stress rate tensor and the direction of the limit backstress relative to the yield backstress.

## INTRODUCTION

Phillips and co-workers [1-3] observed that the motion of the center of the yield surface in the deviatoric stress space is along the direction of the deviatoric stress rate. McDowell [4] showed that the deviatoric stress rate rule provides more accurate description of the motion of the yield surface compared to other proposed rules. But since the deviatoric stress rate rule does not ensure nesting of yield and limit surfaces in a multiple surface plasticity model, it is proposed here to appropriately blend the deviatoric stress rate and Tseng-Lee [5] rules to satisfy both nesting and movement of yield surface as observed in experiments.

Further, enhancement of the material model is made by decomposition of the generalized plastic modulus into two components. The additional component being attributed to attaining nearly complete closure of each subcycle obtained from partial reverse loading. A pair of memory surfaces is introduced that describe the maximum stress amplitude encountered in the recent past.

The motion of the limit surface is modified to allow ratchetting in the general framework of loading cases. The motion of the memory surface is coupled with the motion of the limit surface such that the center of memory is made to coincide with the center of the limit surface.

## PROPOSED KINEMATIC HARDENING RULE

The proposed rule makes use of the form proposed by Voyiadjis and Kattan [6] given by:

$$\overset{\circ}{\boldsymbol{\alpha}} = \|\overset{\circ}{\boldsymbol{\alpha}}\| \boldsymbol{\nu} \tag{1}$$

where $\boldsymbol{\nu}$ is expressed in terms of $\boldsymbol{l}$, the deviatoric stress rate tensor direction and $\boldsymbol{n}$, the direction normal to the yield surface tensor. Using the consistency condition, the magnitude of $\overset{\circ}{\boldsymbol{\alpha}}$ is obtained.

Phillips et. al [1-3] observed that yield surface moves in the direction of the deviatoric stress rate except when there is a requirement of tangency of yield surface to the limit surface (nesting of surfaces) to be satisfied. Moreover, McDowell [4] has found that the deviatoric stress rate direction of the yield surface provides superior correlation of the backstress rate direction of the yield surface compared to other rules. However, this rule does not ensure the nesting condition. Therefore, a kinematic hardening rule for von Mises yield and limit surfaces prescribed by a deviatoric stress rate which provides nesting of the two surfaces can describe the translation of the yield surface more accurately than by a Mroz type rule [4].

Considering the above observations, an appropriate blending of the deviatoric stress rate law and a kinematic hardening rule which satisfies the nesting of the two surfaces seems to be a more appropriate solution. Since Tseng-Lee rule provides superior correlation compared to Mroz type rule [4], it is proposed here to blend appropriately the deviatoric stress rate rule and Tseng-Lee rule in order to satisfy both nesting and movement of the yield surface in accordance to experimental observations. The blending is done such that as the yield surface approaches the limit surface, the domination of the Tseng-Lee rule increases over the deviatoric stress rate rule. At a point away from the limit surface, the deviatoric stress rate dominates. The process of blending is done two steps first using the parameter, $\Delta$ which denotes the nearness of the yield surface to the limit surface and then for the parameter, $\zeta$ which accounts for the direction of deviatoric stress rate. Both the parameters are normalised for a robust formulation.

After blending using both the parameters $\Delta$ and $\zeta$, the backstress rate direction is obtained as:

$$\boldsymbol{\nu} = \frac{\{(\Phi_o'\Phi_o + \Phi_o'\Phi_1 a)q + \Phi_1'\}\,\boldsymbol{l} + (\Phi_o'\Phi_1 bq)\boldsymbol{n}}{\|\{(\Phi_o'\Phi_o + \Phi_o'\Phi_1 a)q + \Phi_1'\}\,\boldsymbol{l} + (\Phi_o'\Phi_1 bq)\boldsymbol{n}\|} \quad or \quad \boldsymbol{\nu} = \frac{A\boldsymbol{l} + B\boldsymbol{n}}{\|A\boldsymbol{l} + B\boldsymbol{n}\|} \tag{2}$$

where

$$A = (\Phi_o'\Phi_o + \Phi_o'\Phi_1 a)q + \Phi_1' \quad and \quad B = \Phi_o'\Phi_1\, b\, q \tag{3}$$

$\Phi'_o$ and $\Phi'_1$ are blending functions related to the direction of limit backstress relative to the yield backstress and $\Phi_o$ and $\Phi_1$ are related to the nearness parameter. The functions $\Phi_o$ and $\Phi_1$ or $\Phi'_o$ and $\Phi'_1$ are related to each other as:

$$\Phi_o = 1 - \Phi_1 \quad and \quad \Phi'_o = 1 - \Phi'_1 \qquad (4)$$

q is the normalization factor used during the first step of blending. a and b are the coefficients of $l$ and $n$ respectively for the Tseng-Lee rule.

## MEMORY SURFACE

Experimental observations show that materials exhibit a more sophisticated and extended memory of discrete events of unloading-reloading compared to the behavior characterized by the model described above.

An attempt is made here to describe the material memory by means of a pair of additional surfaces called the memory surfaces. The two surfaces represent the material memory in two directions of the previously occured maximum stresses. The surfaces are updated as and when the maximum stress changes adhering to certain rules of updating and erasure. An influence factor is attached to each of the memory surfaces. The influence factor is the maximum (normalized to one) when the stressing direction coincides with the direction of the maximum stress affecting the memory surface. This factor dissipates as the direction tends to become perpendicular to the maximum stress direction.

A typical response with and without the memory surfaces is depicted in Figures 1(a) and 1(b). The near closure of the subcycles is due to the presence of the memory surface.

## PLASTIC MODULUS

The behavior of the material may be captured by assuming additive decomposition of the generalized plastic modulus H into two components $H_1$ and $H_2$ ,

$$H = H_1 + H_2 \qquad (5)$$

In this work, a more general expression for the plastic modulus, $H_1$ is proposed. The expression is given by,

$$H = H^* [1 + g(\delta_{in}, \zeta) * f(\delta, \delta_{in})] \qquad (6)$$

Note that here the proximity parameters $\delta_{in}$ and $\delta$ are defined as proposed by Tseng-Lee [5]. $H^*$ is the limiting plastic modulus. In order to predict the ratchetting behavior better, the motion of the limit surface is adjusted so that for a high accumulated strain, the motion goes on diminishing asymtotically to a fixity for a kinematic hardening. Figure 2 shows an example solved for a ratchetting with a non-zero mean stress.

The plastic modulus component, $H_2$ can be expressed as a summation of the components due to the two memory surfaces. $H_2$ can be expressed as:

$$H_2 = f(\bar{\delta_{in}}, \bar{R}) \frac{\bar{\delta}}{\bar{\delta_{in}} - \bar{\delta}} \qquad (7)$$

where $\bar{R}$ is the size of the memory surface and $\bar{\delta_{in}}$ and $\bar{\delta}$ are the proximity parameters for the memory surface.

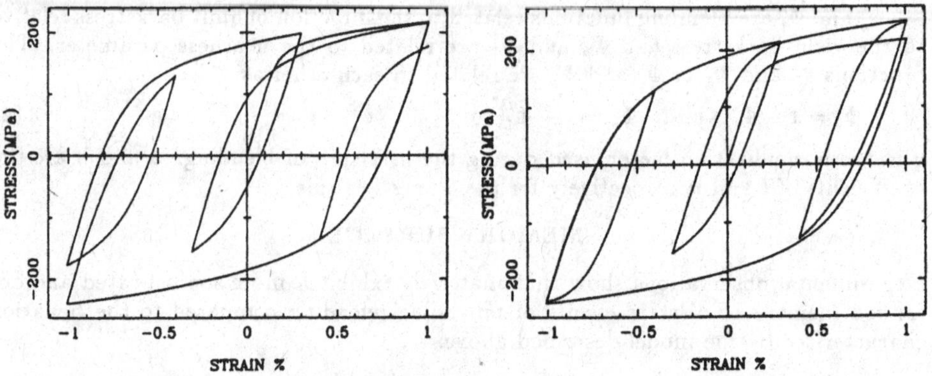


(a) (b)

**Figure 1. A Uniaxial Response (a) without (b) with memory surface.**

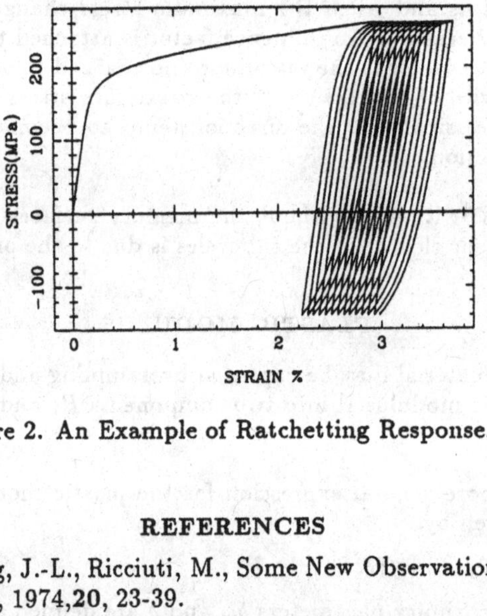


**Figure 2. An Example of Ratchetting Response.**

## REFERENCES


1. Phillips, A., Tang, J.-L., Ricciuti, M., Some New Observations on Yield Surfaces, Acta Mechanica, 1974,**20**, 23-39.

2. Phillips, A., Moon, H., An Experimental Investigation concerning Yield Surfaces and Loading Surfaces, Acta Mechanica, 1977, **27**, 91-102.

3. Phillips, A., Lee, C.-W., Yield Surfaces and Loading Surfaces. Experimental and Recommendations, Int. J. of Solids and Structures, 1979, **15**, 715-729.

4. McDowell, D.L., An Evaluation of Recent Developments in Hardening and Flow rules for Rate-Independent, Nonproportional Cyclic Plasticity, ASME Journal of Applied Mechanics, 1987, **54**, 323-334.

5. Tseng, N.T., and Lee, G.C., Simple Plasticity Model of the Two-surface type, ASCE J. Engg. Mech., 1983, **109**, 795-810.

6. Voyiadjis, G.Z. and Kattan, P.I., Eulerian Constitutive Model for Finite Deformation Plasticity with Anisotropic Hardening,Mech. Mater., 1989,**7**, 279-293.

# ANISOTROPIC HARDENING LAW OF PLASTICITY WITH CROSS AND ROTATIONAL EFFECTS

OSAMU WATANABE
Institute of Engineering Mechanics
University of Tsukuba
Tsukuba, Ibaraki 305, JAPAN

## ABSTRACTS

**Subsequent yield surfaces are known to be distorted along a loading direction, but for some materials, such as brass or stainless steel, a yield surface may be distorted along a direction perpendicular to the loading direction, which is called the cross hardening. Also, rotations of the yield surface have been experimentally investigated for a loading path curved with a corner. Based on the internal time concept, the present author[1] reported an anisotropic hardening law to describe accurately distortions of the yield surface with only two parameters of deformations. The present paper will describe a theoretical treatment for the cross and rotational hardenings for the subsequent yield surface in a linear vector space with its axes employing deviatoric stresses. The present theory for isotropic, kinematic, anisotropic and rotational hardenings wil be presented in a unified and systematic manner, and will be confirmed by a comparison of brass for biaxial stress field problem.**

## THEORY

The present consideration will be confined to small displacement and infinitesimal strain. If we employ three-dimensional Cartesian coordinate system with axis $x_i$ (i=1,2,3) and its base vector $e_i$, and adopt a dyadics notation for tensors of the second order, then the deviatoric stress s and plastic strain $e^p$ can be written by $s=s_{ij}e_ie_j$ and $e^p=e^p_{ij}e_ie_j$, respectively. In order to develop the anisotropic hardening model relating to geometrical representation, we will employ the following six dimensional linear vector space with the axes employing deviatoric stress components $s_{ij}$ as

$$s_1=s_{11},\quad s_2=s_{22},\quad s_3=s_{33},\quad s_4=\sqrt{2}s_{12}=\sqrt{2}s_{21},\quad s_5=\sqrt{2}s_{23}=\sqrt{2}s_{32},\quad s_6=\sqrt{2}s_{31}=\sqrt{2}s_{13}. \tag{1}$$

Similarly, we can define a linear vector space based on deviatoric plastic strains. If the coordinate axes for a deviatoric stress space coincides with those for a deviatoric plastic strain space, then stress increments can be related to plastic strain increments for an arbitrary loading histories. The coordinate axes in the linear vector space are denoted by 1, 2, ·····,6 axis with unit base vector taken as $b_i$ (i=1, 2, ·····,6).

### Description of yield surface

We introduce an intrinsic time z to describe plastic deformation history. The incremental time dz is assumed by using the internal time measure $d\zeta$ and its function $f(\zeta)$ as

$$dz=\frac{d\zeta}{f(\zeta)},\quad \text{with}\quad d\zeta=\|de^p\|=\sqrt{de^{p2}}, \tag{2}$$

where $f(\zeta)$ is non-negative and $f(0)=1$. Fig.1 shows the kinematics of the yield surface in the deviatoric stress space, where the point S on the initial yield surface moves to the point s on the subsequent yield surface through the plastic deformation from the internal time 0 to z expressed by

$$s=s(z,\frac{de^p}{dz}[0,z], S). \tag{3}$$

where $de^p/dz[0,z]$ denotes the history of plastic strains from internal time 0 to the present time z. We assume that the surface expands initially ($S^{(0)}\rightarrow S^{(1)}$), then moves ($S^{(1)}\rightarrow S^{(2)}$), deforms ($S^{(2)}\rightarrow S^{(3)}$), and finally rotates ($S^{(3)}\rightarrow S^{(4)}$) expressed by

$$s=Q\bullet(gS+r^*+q^*), \tag{4}$$

where g is a measure of change in size (isotropic hardening), $r^*$ is a location of the yield surface center (kinematic hardening), $q^*$ is a measure of change in shape (anisotropic hardening with cross effect), and Q is a measure of rotation (rotational hardening). Solving the initial stress S yields

$$S=\frac{Q^t}{g}\bullet(s-r-q), \quad \text{with } r = Q\bullet r^*, \quad q = Q\bullet q^*, \tag{5}$$

where the notation $(\ )^t$ means transpose. The vectors r and q are kinematic and anisotropic hardenings after rotational effects. The quantities g, $r^*$ and $q^*$ can be specified as,

$$g=g(z), \quad r^*=r^*(z,\frac{de^p}{dz}[0,z]), \quad q^*=q^*(z,\frac{de^p}{dz}[0,z], S)=q^*(z,\frac{de^p}{dz}[0,z], \hat{s}), \quad \text{with } \hat{s}=s-r. \tag{6}$$

The matrix Q is assumed to rotate the yield surface center r around the origin O assumed by

$$Q=Q(z,\frac{de^p}{dz}[0,z], r[0,z]). \tag{7}$$

Yield criteria of von Mises is adapted for the subsequent yield surface as

$$F=\frac{1}{S_y^0}\|S\|-1=\frac{1}{S_y^0 g}\|s-r-q\|-1=0, \tag{8}$$

where $S_y^0$ represents the size of the initial yield surface.

## Anisotropic hardening with cross effect

First of all, we shall show the anisotropic hardening before rotational effects, which is designated by $q^*$ in Eq.(6). We carry out transformation of the coordinate system by setting a new axis to the direction of the increment of plastic strain $de^p$. The axes for the new coordinate system are denoted by 1', 2', 3'$\cdots$, 6' axis, where the principal axis corresponding to the incremental plastic strain is denoted by 1' axis. If the unit base vectors in the new system are denoted by $b_i'$, then these vectors $b_i$ are related to $b_i$ in the former system by using the cosine of direction $A_{ij}$ as

$$b_i' = A_{ij}\, b_j, \quad A_{1j}= de_j^p / d\zeta, \tag{9}$$

where the coefficient $A_{1j}$ for the 1' axis can be determined as shown in Fig.2. The transformation matrix $A_{ij}$ is orthogonal. If we denote the quantities in the new coordinate system by the notation ( )', the transformation between the two frames are given by

$$\hat{s}'=A\bullet\hat{s}, \quad dr^{*\prime}=A\bullet dr^*, \quad dq^{*\prime}=A\bullet dq^*. \tag{10}$$

The deformation $dq_1^{*\prime}$ in the 1' axis is assumed to be proportional to the power law of the stress

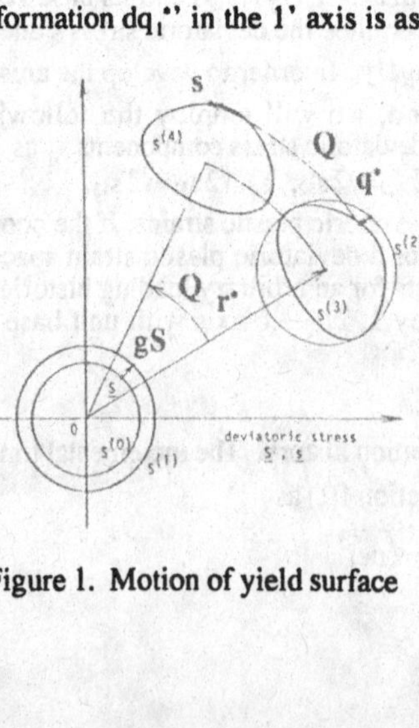


Figure 1. Motion of yield surface

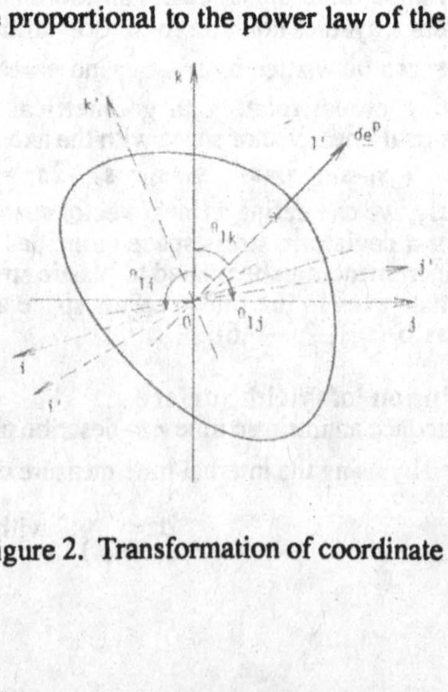


Figure 2. Transformation of coordinate system

$(\hat{s}'_1/S^o_y)$ in the loading direction. We consider cases for power number m=0, 1 and 2.

Case for power number m=0: The present case corresponds to the kinematic hardening $r^*$ expressed in the present 1', 2', 3', ⋯, 6' axes as

$$dr^{*\prime}_1 = 2\mu_o \rho^{(0)}(0) d\zeta; \quad dr^{*\prime}_j = 0, \quad (j=2,3,\cdots,6), \tag{11}$$

where $\mu_o$ is shear modulus, and $\rho^{(0)}(0)$ is a material constant for kinematic hardening. Thus, we have kinematic hardening in the former coordinate system by using Eqs.(10) and (11) as

$$dr^*_i = A_{ji}\, dr^{*\prime}_j = A_{1i}\, dr^{*\prime}_1 = 2\mu_o \rho^{(0)}(0) de^p_i. \tag{12}$$

Case for power number m=1: The cross hardening can be incorporated in this mode. The modes $dq_i^{(1)*\prime}$ are assumed for the linear hardening rule as

$$dq_1^{(1)*\prime} = 2\mu_o\, \rho^{(1,n)}(0) \left(\frac{\hat{s}'_1}{S^0_y}\right) d\zeta; \quad dq_j^{(1)*\prime} = 2\mu_o\, \rho^{(1,c)}(0) \left(\frac{\hat{s}'_j}{S^0_y}\right) d\zeta \quad (j=2, 3,\cdots,6), \tag{13}$$

where $\rho^{(1,n)}(0)$ and $\rho^{(1,c)}(0)$ are material constants. The expression $dq_i^{(1)*}$ in the former coordinate system can be obtained by substituting Eq.(13) into Eq.(10) as

$$dq_i^{(1)*} = A_{ji}\, dq_j^{(1)*\prime} = 2\mu_o\, \rho_c^{(1)}(0) \left(\frac{\hat{s}_i}{S^0_y}\right) d\zeta + 2\mu_o\{\rho_n^{(1)}(0) - \rho_c^{(1)}(0)\} \left(\frac{\hat{s}'_j}{S^0_y}\frac{de^p_j}{d\zeta}\right) de^p_i. \tag{14}$$

It is seen that the above equation excludes the undetermined coefficients $A_{ij}$ (i=2,3,⋯,6; j=1,2,⋯,6). The distortion $q_i^{(1)*}$ can be decomposed into two terms of the normal hardening $q_i^{(1,n)*}$ in the principal direction and the cross hardening $q_i^{(1,c)*}$ in the other axes as

$$q_i^{(1)*} = q_i^{(1,n)*} + q_i^{(1,c)*} \quad \text{where } q_i^{(1,n)*} = R_{ij}^{(1,n)} \frac{\hat{s}_j}{S^o_y f}, \quad q_i^{(1,c)*} = R^{(1,c)} \frac{\hat{s}_i}{S^o_y f} - R_{ij}^{(1,c)} \frac{\hat{s}_j}{S^o_y f}. \tag{15}$$

The evolutional equations for coefficients are as follows:

$$\dot{R}_{ij}^{(1,n)} = 2\mu_o\, \rho^{(1,n)}(0)\, \dot{e}^p_i \dot{e}^p_j, \quad \dot{R}^{(1,c)} = 2\mu_o\, \rho^{(1,c)}(0), \quad \dot{R}_{ij}^{(1,c)} = 2\mu_o\, \rho^{(1,c)}(0)\, \dot{e}^p_i \dot{e}^p_j. \tag{16}$$

Case for power number m=2: Since the distortion mode for m=2 is non-symmetric with respect to directions perpendicular to the loading direction, no cross hardening may not occur in this mode. The distortion $dq^{(2)*\prime}$ can be assumed as follows:

$$dq_1^{(2)*\prime} = 2\mu_o\, \rho^{(2)}(0) \left(\frac{\hat{s}'_1}{S^0_y}\right)^2 d\zeta; \quad dq_{ij}^{(2)*\prime} = 0 \quad (j=2,3,\cdots,6). \tag{17}$$

The evolutional equation for the coefficient is given by

$$q_i^{(2)*} = R_{ijk}^{(2)} \frac{\hat{s}_j}{S^o_y f} \frac{\hat{s}_k}{S^o_y f}, \quad \text{where } \dot{R}_{ijk}^{(2)} = 2\mu_o\, \rho^{(2)}(0)\, \dot{e}^p_i \dot{e}^p_j \dot{e}^p_k. \tag{18}$$

## Rotational hardening

The evolutional equation for rotation rate is assumed by

$$\dot{Q}_{kl} = \dot{\theta}_{kl} = \frac{\rho_\theta}{S^o_y} (r_k \dot{e}^p_l - r_l \dot{e}^p_k). \tag{19}$$

The right-hand side of Eq.(19) expresses the angular momentum around the origin O, if we regard the plastic strain increment ($de^p_k$, $de^p_l$) and the yield surface center ($r_k$, $r_l$) as a load increment and a lever, respectively, and $\rho_\theta$ is material constant for rotation. We employ a linear hardening rule for rotation based on the assumption that the rotation is independent of the direction. Differentiating Eq.(5) with respect to the internal time z yields

$$\dot{r} = \dot{Q}\cdot r^* + Q\cdot \dot{r}^*, \quad \dot{q} = \dot{Q}\cdot q^* + Q\cdot \dot{q}^*. \tag{20}$$

If we set the present configuration as a reference one, we can assume the followings:

$$r^* = r, \quad q^* = q, \quad Q = I. \tag{21}$$

The final constitutive equation for r and q considering nonlinear effects can be expressed by

$$r_i = \sum_k r_i^{(k)} \quad \text{where } \dot{r}_i^{(k)} = 2\mu_o\, \rho^{(0,k)}\, \dot{e}^p_i - \alpha^{(0,k)}\, r_i^{(k)} + \dot{Q}_{i\beta} r_\beta^{(k)}, \tag{22}$$

$$R_{ij}^{(1,n)} = \sum_k R_{ij}^{(1,n,k)} \quad \text{where } \dot{R}_{ij}^{(1,n,k)} = 2\mu_o\, \rho^{(1,n,k)}\, \dot{e}^p_i \dot{e}^p_j - \alpha^{(1,n,k)}\, R_{ij}^{(1,n,k)} + \dot{Q}_{i\beta} R_{\beta j}^{(1,n,k)}, \tag{23}$$

$$R_{ij}^{(1,c)} = \sum_k R_{ij}^{(1,c,k)} \quad \text{where } R_{ij}^{(1,c,k)} = 2\mu_o\, \rho^{(1,c,k)}\, \dot{e}^p_i \dot{e}^p_j - \alpha^{(1,c,k)}\, R_{ij}^{(1,c,k)} + \dot{Q}_{i\beta} R_{\beta j}^{(1,c,k)}, \tag{24}$$

$$\dot{R}^{(2)}_{ijk}=\sum_{m} \dot{R}^{(2,m)}_{ijk} \quad \text{where } \dot{R}^{(2,m)}_{ijk} = 2\mu_o\, \rho^{(2,m)}\, \dot{e}^p_i \dot{e}^p_j \dot{e}^p_k - \alpha^{(2,m)}\, R^{(2,m)}_{ijk} + \dot{Q}_{i\beta} R^{(2,m)}_{\beta jk}. \tag{25}$$

The equation for $R^{(1,c)}$ in Eq.(16) will not be affected by rotation, since $R^{(1,c)}$ is a scaler.

## TENSION-TORSION PROBLEM

We shall compare the experiment[2] of brass alloy, where a loading history consists of two straight paths with a corner as shown in Fig.3. The bent angle are selected as two kinds of $\phi = 90$ and 135 degrees. The final shape of the yield surface is observed to show congruence regardless of the bent angle after the second plastic strain greater than 2.5 percents. We approximate the final configuration by using the present theory as shown in Fig.4. Fig.5 shows the transient shape distortion in the deviatoric stress non-dimensionalized by the stress $\bar{s}^c_1$ just after the second path. Though we can see some discrepancies in the transient process, the overall agreements between the theoretical prediction and the experimental results are observed. Fig.5 also shows stress path, calculated from the associated flow rule of the plastic potential theory. The yield surface will be symmetric with respect to the loading direction $de^p$, and ceases to rotate at the final stage. The accumulated rotation angle $\theta_r$ will be depicted in the same figure.

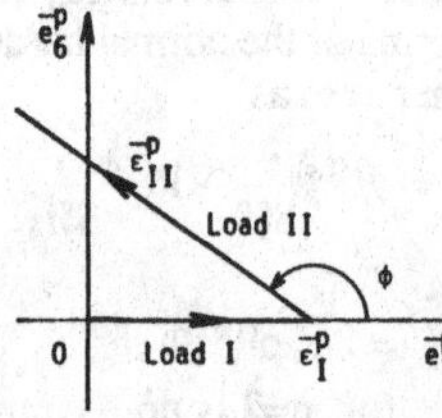


Figure 3. Strain path with a corner

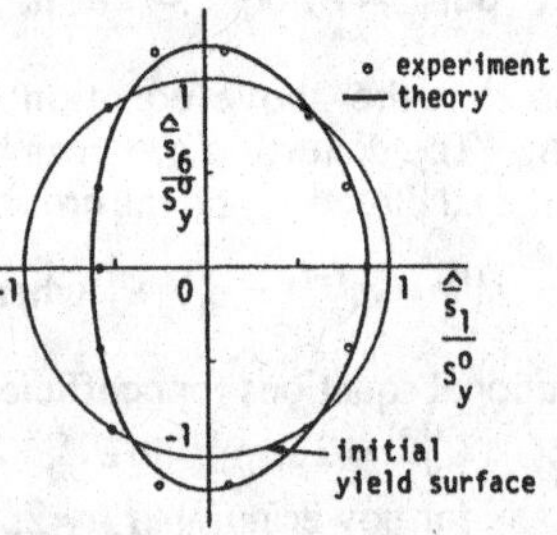


Figure 4. Final shape of yield surface

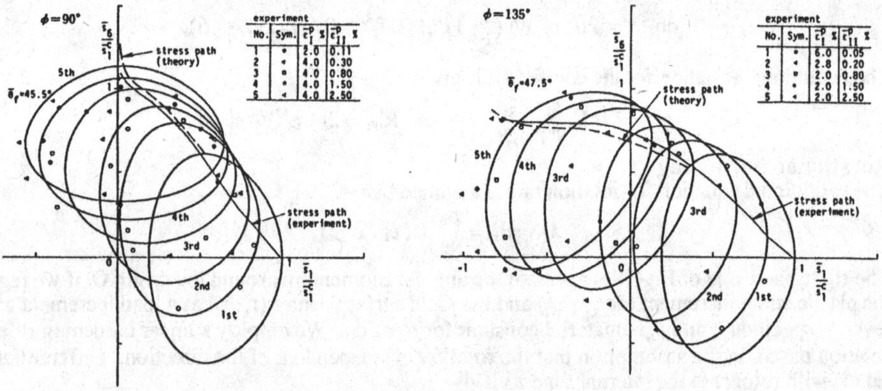


Figure 5. Comparison of experiment and theory

## REFERENCES

1. Watanabe, O., Anisotropic Hardening Law of Plasticity Using an Internal Time Concept (Deformation of Yield Surface). JSME International Journal, 1987, 30-264, pp.912-920.

2. Ikegami, K., Experimental Plasticity on the Anisotropy of Metals, Proc. Euromech Colloquim, ed. Boehler, J. P., 1979, 115, pp.201-242.

# POLYMERS, POLYMER COMPOSITES, PHASE TRANSITIONS AND SUPERALLOYS

# EVOLUTION OF PLASTIC ANISOTROPY IN AMORPHOUS POLYMERS DURING FINITE STRAINING

ELLEN M. ARRUDA AND MARY C. BOYCE
Mechanical Engineering Department
Massachusetts Institute of Technology
Cambridge, MA 02139, USA

## ABSTRACT

The large strain deformation response of amorphous polymers results primarily from orientation of the polymer chains during plastic straining. Through mechanical testing the nature of the evolution of chain orientation under different conditions of temperature and state of strain is developed. A rubber elasticity spring system which is capable of capturing the state of deformation dependence of hardening is used to develop a tensorial internal state variable model of the evolving anisotropic polymer response. This fully three-dimensional constitutive model is shown to be successfully predictive of our uniaxial compression, plane strain compression and uniaxial tension experiments on polycarbonate and polymethylmethacrylate over a range of strain rates and temperatures.

## INTRODUCTION

Amorphous polymers are capable of withstanding plastic straining on the order of 100%. During such deformations these materials become highly anisotropically strengthened. The developing anisotropy is due to orientation of the polymer molecular chains which depends on the state of deformation and temperature. This orientation-induced strain hardening is a well-known characteristic used in polymer processing to strengthen polymer components but it is not well characterized nor understood.

An existing framework for modelling large strain polymer deformation which accounts for the yield and post yield behavior with full three-dimensionality is adapted herein to predict the state of deformation dependent response of these polymers. The ability of several rubber elasticity spring systems to model the plastic strain anisotropy is considered.

## EXPERIMENTS

### Mechanical Testing

Constant strain rate uniaxial compression and plane strain compression tests were conducted on polycarbonate (PC) and polymethylmethacrylate (PMMA) over a range of strain rates and temperatures. Details of the testing procedure are given elsewhere [1]. Under isothermal, room temperature conditions the states of deformation response was found to be vastly different for plane strain versus uniaxial compression as seen in Figures 1 and 2 for PMMA and PC, respectively. The pressure dependent yield is followed by strain softening and then

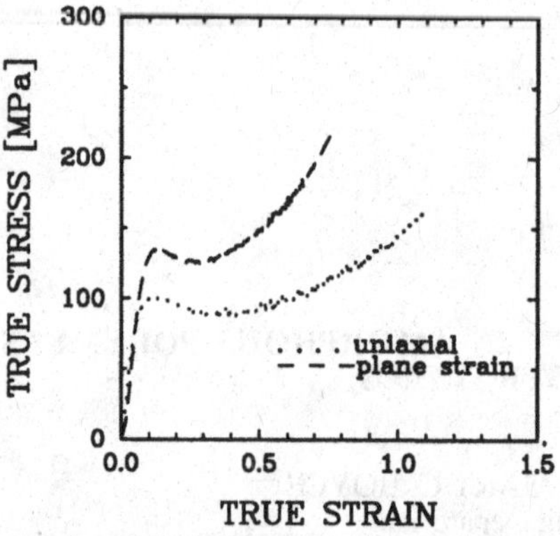


Figure 1. Uniaxial compression and plane strain compression response of PMMA at -.001/s and 23°C.

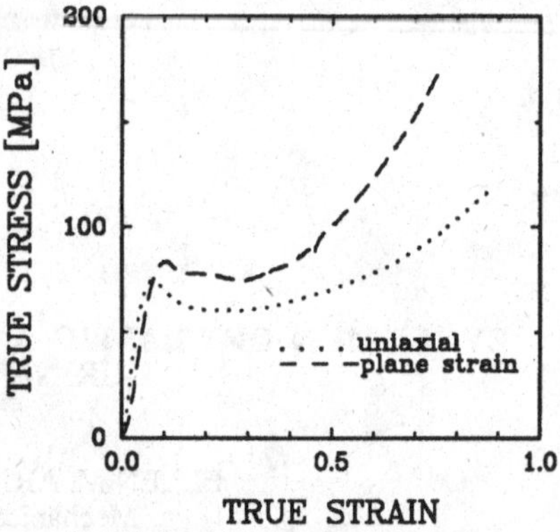


Figure 2. Uniaxial compression and plane strain compression of PC at -.01/s and 23°C.

strain hardening for both deformation states and in both materials. Failure occurs in these materials due to locking of the molecular chains with splitting occurring at roughly 45° to the principal orientation axis or plane. Manifested in the stress-strain response curves as asymptotically increasing stress with increased strain, locking is characteristic of both materials in either deformation scheme. The strain hardening evolution prior to locking accumulates more rapidly for plane strain versus uniaxial compression. The contrast is due to uniaxial alignment of the polymer chains in plane strain deformation and planar orientation of the chains in uniaxial compression.

The temperature dependent nature of the polymer response is shown in Figure 3 for PMMA. During isothermal testing at various temperatures the uniaxial compression results show temperature dependent yield and strain softening followed by temperature dependent strain hardening. The amount of hardening can be compared for 23°C and 75°C curves by noting that at a strain of 100% for instance, the higher temperature response showed hardening of 25 MPa as measured from the strain softened value of 35 MPa at ~40% strain to 60 MPa at 100% strain, while a similar measure of 55 MPa hardening is obtained at 23°C. PC shows similar temperature dependent behavior; temperature dependent yield is seen to occur in Figure 4 along with 80 MPa strain hardening at 23°C versus 40 MPa at 125°C for 100% strain.

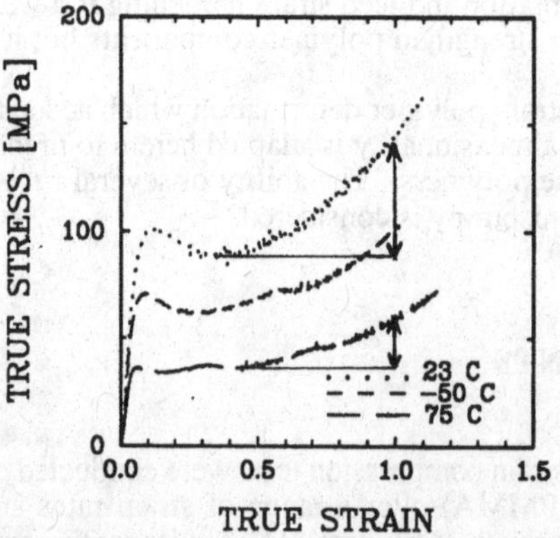


Figure 3. Effect of test temperature on the uniaxial compression response of PMMA at -.001/s.

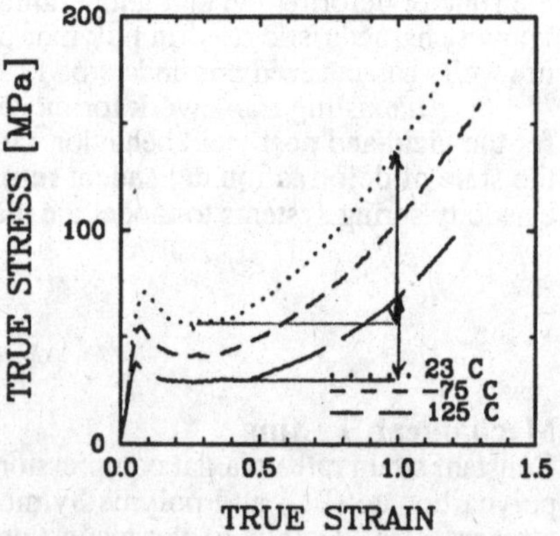


Figure 4. Effect of test temperature on the uniaxial compression response of PC at -.001/s.

# MODELLING

The model of Boyce, Parks and Argon has been shown to be predictive of the uniaxial response of polymers [1] over a range of temperatures and strain rates. The mechanics of large strain deformation have been worked out in that paper. In their model two resistances to plastic flow act in parallel to account for both isotropic and anisotropic resistances observed for these materials. Argon [2] developed an expression for the plastic shear strain rate which ensues once isotropic barriers to chain segment rotation have been overcome:

$$\dot{\gamma}^p = \dot{\gamma}_o \exp\left[\frac{-A\,s_o}{\Theta}\left(1-\left(\frac{\tau}{s_o}\right)^{5/6}\right)\right] \qquad (1)$$

where $\tau$ is the applied shear stress, $s_o$ is the athermal shear yield strength, $\Theta$ is the temperature, A is inversely proportional to Boltzmann's constant and $\dot{\gamma}_o$ is proportional to the attempt frequency.

The anisotropic strain hardening response to plastic deformation occurs as a result of molecular chain orientation and had been modelled using the Wang and Guth three-chain rubber elasticity spring [3]. Here, various schemes have been considered for the rubber elasticity element including the three-chain element of Wang and Guth [3], the four-chain model of Flory and Rehner [4] modified for a non-Gaussian network (Treloar [5]) and a new eight-chain element. Schematics of these rubber elasticity spring systems appear in Figure 5. In the mathematical model of each system the stress-stretch relation for each individual chain is given by:

$$\sigma_i = N\,k\Theta\,\mathcal{L}^{-1}\left(\frac{\lambda_i}{\sqrt{N}}\right)\lambda_i \qquad (2)$$

where $k$ is Boltzmann's constant, N is a statistical parameter related to the network locking stretch, $\lambda_i$ is the chain stretch and $\mathcal{L}^{-1}$ is the inverse Langevin function which asymptotically increases in value as its argument approaches 1.0. All three rubber elasticity systems considered here were taken to be incompressible. It will be shown that the existing three- and four-chain models investigated either fail to fully capture the network response of the polymer required to account for the state of deformation dependence on hardening or do not possess the symmetry required of the principal strain space. The proposed eight-chain system is found to provide the desired network response and cubic symmetry.

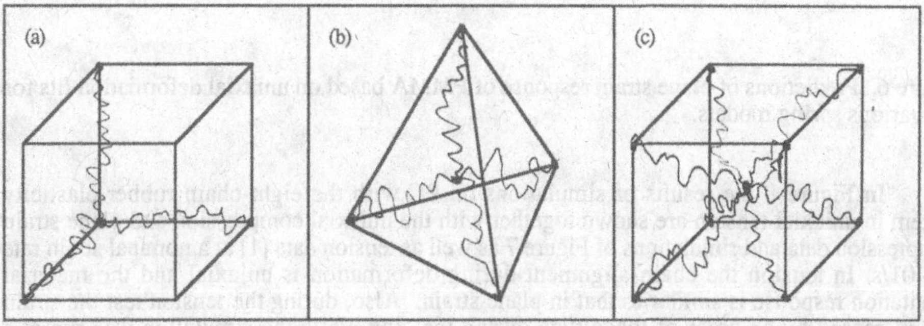


Figure 5. Rubber elasticity spring systems: (a) three chain, (b) four chain and (c) eight chain.

## RESULTS

The results of the uniaxial compression data were used to fit the model constants in order to predict the plane strain compression response of PC and PMMA. Four schemes were implemented for predicting the anisotropic network hardening response and all are compared for the isothermal, room temperature responses of these materials in Figures 6 and 7. The three-chain system is dependent upon a single chain stretch behavior with the behavior of the remaining chains arising solely through incompressibility constraint, i.e. a complete network-like behavior is not exhibited by this system and therefore the state of deformation dependence on hardening is not accurately captured. Two four-chain schemes were compared, these differed only in the orientation of the four chains with respect to the major principal strain of the deformation. Orientations along and perpendicular to the largest principal strain differed in their ability to capture the state of deformation response of the polymers; both methods showed some success, but the initial anisotropy of the four chain network was found to unjustly influence the ability of this model to predict this behavior. The eight-chain scheme was found to be far superior to these previous models. In Figure 6 the results on PMMA are shown to be extremely well predictive of the plane strain response. The results on PC in Figure 7 share the same amount of success as the PMMA results in modelling the state of deformation response of these materials.

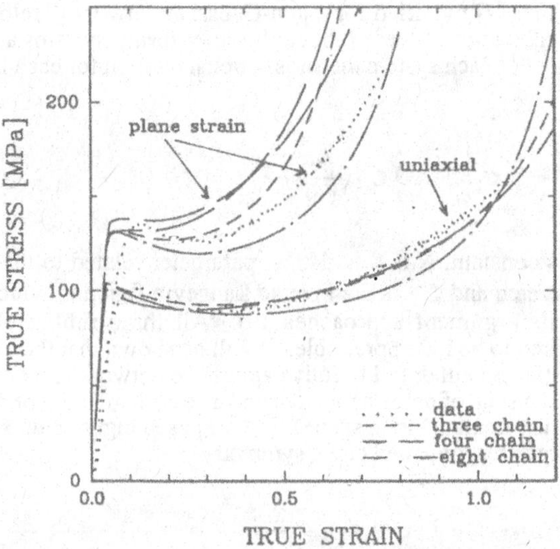


Figure 6. Predictions of plane strain response of PMMA based on uniaxial deformation fits for various spring models.

In Figure 8 the results of simulations on PC with the eight-chain rubber elasticity system in uniaxial tension are shown together with the uniaxial compression and plane strain compression data and simulations of Figure 7 as well as tension data [1] at a nominal strain rate of -.01/s. In tension the chain alignment during deformation is uniaxial and the material orientation response is similar to that in plane strain. Also, during the tension test the strain rate increased by an order of magnitude during yielding while the simulation was run at a constant true strain rate of -.01/s. Again the eight-chain model successfully predicts hardening.

The room temperature data were also used to predict the response of these materials at elevated temperatures. The results of the simulations on PMMA at 50°C and 75°C appear in Figure 9. The Argon model of the thermally activated isotropic resistance to plastic flow

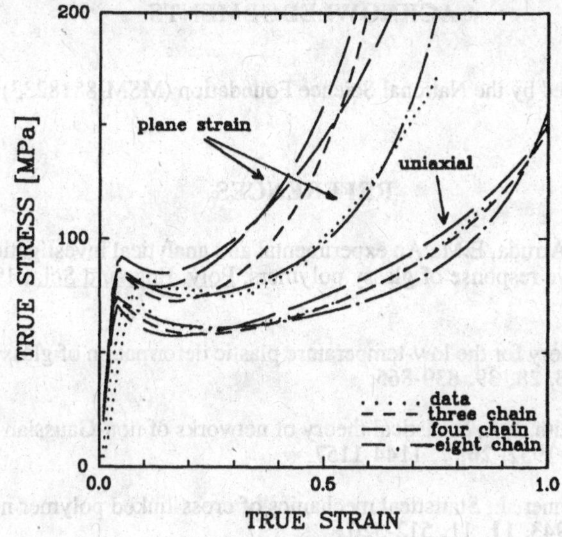


Figure 7. Predictions of plane strain response of PC based on uniaxial fits for various spring models.

predicts the temperature dependent yield behavior quite well. There is no account of the temperature dependence on strain hardening in the current model and therefore the model prediction begins to diverge from the experimental data at large strains for elevated temperature tests. The elevated temperature hardening curves show more gradual hardening than do the room temperature curves, this behavior was addressed in the discussion following Figures 3 and 4. Additional work to characterize the evolving plastic anisotropy as a function of temperature is in progress. The results of measurement of birefringence evolution with magnitude of strain, temperature and state of deformation will be used to augment the current tensorial internal variable model of the development of anisotropy with plastic deformation.

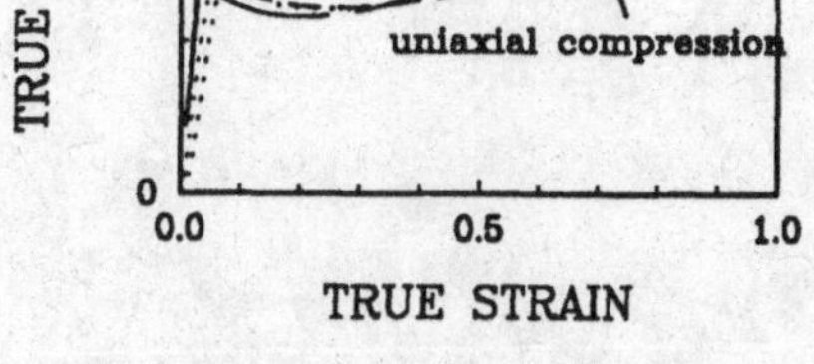

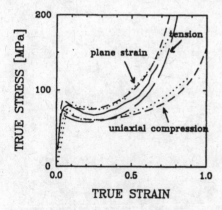

Figure 8. PC data and results of eight chain simulations in uniaxial compression, plane strain compression and tension at 23°C and -.001/s.

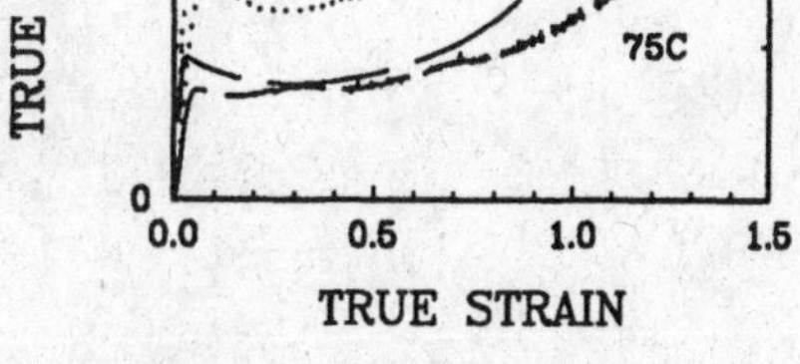

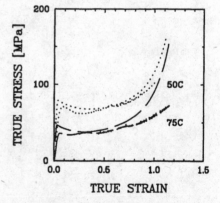

Figure 9. Data and results of simulations in uniaxial compression at different temperatures for PMMA at -.001/s.

## ACKNOWLEDGEMENTS

This work is sponsored by the National Science Foundation (MSM-8818233) and the M.I.T. Bradley Foundation.

## REFERENCES

1. Boyce, M.C. and Arruda, E.M., An experimental and analytical investigation of the large strain compressive response of glassy polymers. Poly. Eng. and Sci., 1990, **30**, 20, 1288-1298.

2. Argon, A.S., A theory for the low-temperature plastic deformation of glassy polymers. Phil. Mag., 1973, **28**, 39, 839-865.

3. Wang, Y.Y. and Guth, E.J., Statistical theory of networks of non-Gaussian flexible chains. J. Chem. Phys., 1952, **20**, 7, 1144-1157.

4. Flory, P.J. and Rehner, J., Statistical mechanics of cross-linked polymer networks. J. Chem. Phys., 1943, **11**, 11, 512-520.

5. Treloar, L.R.G., The Physics of Rubber Elasticity, Clarendon, London, 1985.

# EFFECTS OF INITIAL ANISOTROPY ON THE FINITE STRAIN DEFORMATION BEHAVIOR OF GLASSY POLYMERS

ELLEN M. ARRUDA, HARALD QUINTUS-BOSZ[†] and MARY C. BOYCE
Mechanical Engineering Department
Massachusetts Institute of Technology
Cambridge, MA 02139, USA
[†]Mechanical Engineering Department
Stanford University
Palo Alto, CA 94303, USA

## ABSTRACT

A polymeric product generally possesses a residual state of orientation incurred due to processing. The pre-oriented polymer exhibits a highly anisotropic mechanical behavior as characterised by a yield stress and post yield deformation which are both direction dependent. In this paper, polycarbonate (PC) has been subjected to various magnitudes of pre-orientation via uniaxial compression. The subsequent anisotropic yield and post yield behavior is then experimentally determined, also by compression. A constitutive model of glassy polymer deformation proposed by two of the authors is found to be successfully predictive of the prominent features of the observed experimental behavior.

## INTRODUCTION

The effects of state of strain on the yield and post-yield behavior of glassy polymers have previously been investigated through constant strain rate uniaxial and plane strain compression tests on polycarbonate (PC) and polymethylmethacrylate (PMMA) [1]. Strain hardening occurs in these polymers in a manner strongly dependent on the state of strain and is a direct result of the nature of developing molecular orientation with plastic straining. Upon removal of the loads polymeric materials relax somewhat and a portion of the plastic strain is recovered, but the remainder of the molecular orientation and the strengthening it produces is locked into the material. Polymer components in service often possess uncharacterised anisotropic strengths due to orientation during processing. In order to predict the response of such materials one must characterise the nature of the evolution of anisotropic orientation with state of deformation and temperature during straining and develop a constitutive model which accurately accounts for this evolution process. A current model has been shown elsewhere in these proceedings to be successfully predictive of the state of deformation dependence on the anisotropic hardening of PC and PMMA [1]. The present investigation is concerned with the ability of the proposed model to predict the subsequent anisotropic response of uniaxially oriented PC specimens.

## EXPERIMENTS

Uniaxial compression specimens were prepared by compression molding bars from PC pellets supplied by Mobay, subjecting the molded bars to an anneal above 150°C and machining cubes of length 12.7 mm on edge. These specimens were compressed at a constant strain rate of -.005/s to various magnitudes of strain and unloaded at the same strain rate. Pre-oriented cubic specimens were machined from the uniaxially compressed materials; the specimen sizes were limited by the final heights of the initially compressed specimens but were generally 6.5 mm to 7.5 mm in axial dimension. The pre-oriented specimens were tested in uniaxial compression both along and normal to the initial compression direction.

Three sets of pre-oriented conditions and subsequent deformations were chosen for consideration because they represented low, moderate and high levels of locked-in orientation prior to recompression. The measured principal stretch ratios of the three cases were $\lambda = .614$, $\lambda = .577$ and $\lambda = .508$ corresponding to plastic strains of $\varepsilon = -.488$, $\varepsilon = -.550$ and $\varepsilon = -.677$, respectively. Plastic incompressibility was assumed and used to determine the normal stretches for each case.

## MODELLING

Two resistances to plastic deformation in glassy polymers have previously been identified and incorporated by Boyce, Parks and Argon into their fully three-dimensional elastic-viscoplastic constitutive model of the response of glassy polymers under deformation [2]. In their model use is made of Argon's following expression for the shear stress $\tau$ required to overcome the molecular barriers to deformation at a given shear strain rate $\dot{\gamma}^p$ and temperature $\Theta$

$$\tau = s_o\left[1 - \frac{\Theta}{A s_o} ln\left(\frac{\dot{\gamma}_o}{\dot{\gamma}^p}\right)\right]^{6/5} \tag{1}$$

where $s_o$ is the athermal shear resistance and A is inversely proportional to Boltzmann's constant [3]. A second resistance, that due to chain orientation, has been modelled by Arruda and Boyce [1] as an eight chain rubber elasticity network system where each chain is modelled as a Langevin spring giving a chain stress-stretch relation of the form

$$\sigma_i = n \, k \Theta \, \mathcal{L}^{-1}\left(\frac{\lambda_i}{\sqrt{N}}\right) \lambda_i \tag{2}$$

where n is the number of statistical elements in a chain, $k$ is Boltzmann's constant, N is a statistical parameter related to the network locking stretch, $\lambda_i$ is the chain stretch and $\mathcal{L}^{-1}$ is the inverse Langevin function which asymptotically increases in value as its argument approaches 1.0. The success of this system in predicting the anisotropic strain hardening response of isotropic polymers within the Boyce, Parks and Argon framework is the subject of a companion paper in these proceedings [1]. The various material constants needed are fit to the model using uniaxial compression data as shown in Figure 1.

Simulation of the pre-oriented response requires an input to the model of the current plastic stretch tensor locked into the material due to prior deformation giving the polymer an internal backstress tensor which imparts an anisotropy to the plastic resistance. As a result of this initial plastic stretch tensor the subsequent anisotropic resistance to continued plastic deformation is affected. The initial plastic straining also acts to strain soften the material and therefore the isotropic resistance is lowered, this resistance is taken to be the strain softened value. The pre-oriented state is determined from the height after relaxation of the initially compressed specimens. Details of the framework for incorporating such an internal state are given in Boyce et. al. (1989) [4].

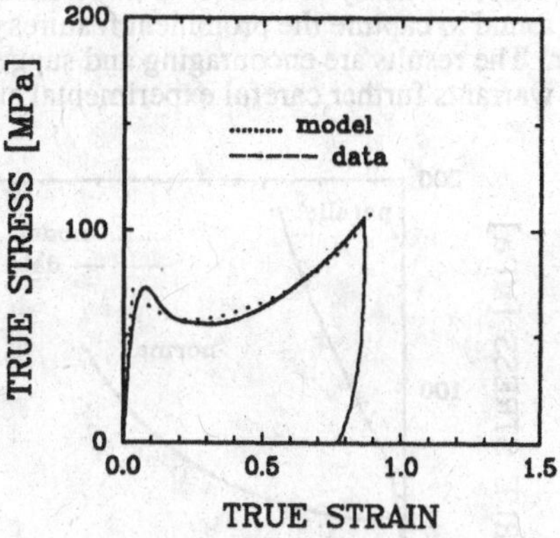


Figure 1. Uniaxial compression data and simulation to determine material constants for PC at 25°C and -.005/s.

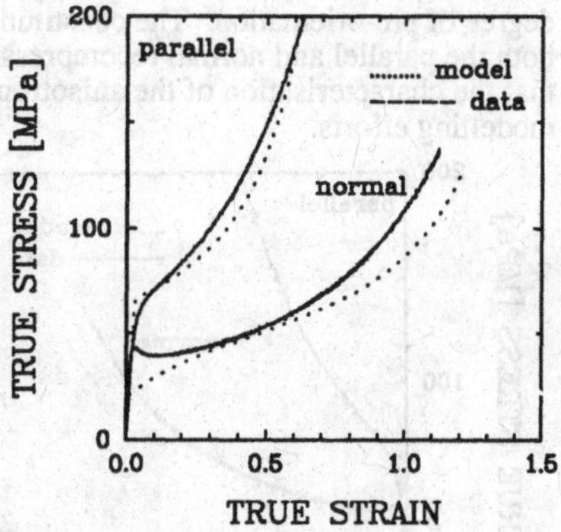


Figure 2. Effect of pre-orientation by $\lambda = .614$ on the response of PC at 25°C and -.005/s along the original compression direction and along a perpendicular direction.

## RESULTS

In Figure 1 the response of an isotropic PC specimen shows an isotropic resistance to plastic deformation of 73 MPa followed first by strain softening to an evolved isotropic resistance of 56 MPa and then anisotropic strain hardening as chains preferentially orient in a plane perpendicular to the loading direction. This specimen had been unloaded from a strain of -.85. Elastic and anelastic recovery occurred during unloading, in this case the specimen recovered to -.77 strain upon unloading and additional recovery to -.68 strain occurred prior to remachining of this specimen for pre-oriented studies.

Each of Figures 2 - 4 contains data and results of the simulations on pre-oriented specimens. In each case results of the simulations along the initial compression direction which is perpendicular to the chain orientation and results of compression parallel to the chain orientation direction are presented. In Figure 2 the response of a specimen containing an initial stretch of $\lambda = .614$ is plotted along with the results of the simulation. For recompression along the initial compression direction the material responds by first overcoming the isotropic barrier to plastic deformation at the strain softened value of 56 MPa as evidenced by a slowly evolving change in slope, followed by rapid strain hardening. For compression along a direction of previous chain orientation the response is markedly different from that obtained along the initial compression direction. The material responds to a reduced resistance to deformation as a result of prior chain alignment in this direction where the yield stress is initially 45 MPa and softens to 40 MPa. The model captures the reduced resistance effect but predicts 35 MPa for this event. The subsequent strain hardening follows that of the initially isotropic specimen but is offset from the isotropic response by a strain of -0.13. This offset strain is the strain required to undo the original state of orientation. The simulation is found to predict a similar response, however the offset strain is computed to be -0.22.

The trends in the experimental results and corresponding model predictions are similar for the higher degree of initial pre-orientation as in Figures 3 and 4. When recompressing in the same direction the material is observed to first acknowledge its passage through the isotropic resistance prior to rapidly reaching the strain hardened state. When tested normal to the orientation direction, a reduction in the yield stress is observed followed by delayed strain hardening with respect to the initially isotropic response; the delay increases with increasing degree of pre-orientation. The constitutive model is found to capture the prominent features of both the parallel and normal recompression behavior. The results are encouraging and suggest that the characterisation of the anisotropic behavior warrants further careful experimental and modelling efforts.

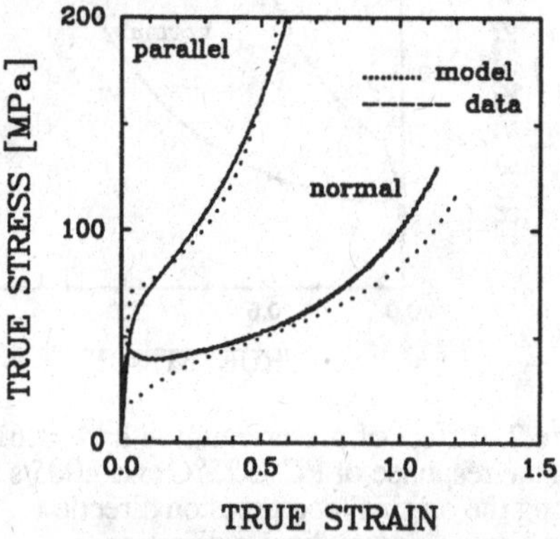


Figure 3. Effect of pre-orientation by $\lambda = .577$ on the response of PC at 25°C and -.005/s along the original compression direction and along a perpendicular direction.

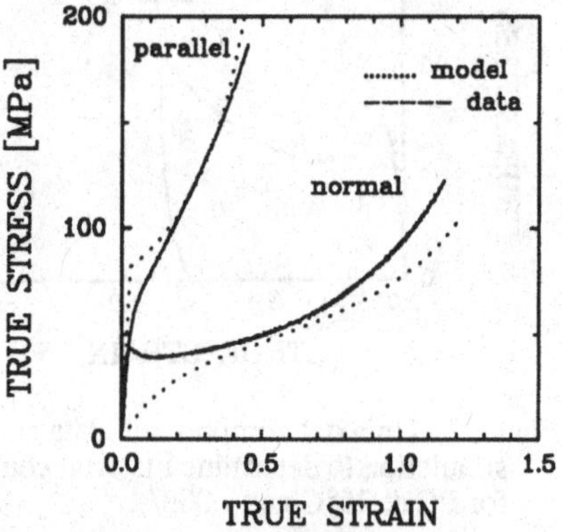


Figure 4. Effect of pre-orientation by $\lambda = .508$ on the response of PC at 25°C and -.005/s along the original compression direction and along a perpendicular direction.

## ACKNOWLEDGEMENTS

This work is sponsored by the National Science Foundation (MSM-8818233) and the M.I.T. Bradley Foundation.

## REFERENCES

1. Arruda, Ellen M. and Boyce, Mary C., Evolution of plastic anisotropy in amorphous polymers during finite straining. Proceedings of Plasticity '91, The Third International Symposium on Plasticity and its Current Applications, Grenoble, France, 1991

2. Boyce, M.C. and Arruda, E.M., An experimental and analytical investigation of the large strain compressive response of glassy polymers. Poly. Eng. and Sci., 1990, **30**, 20, 1288-1298.

3. Argon, A.S., A theory for the low-temperature plastic deformation of glassy polymers. Phil. Mag., 1973, **28**, 39, 839-865.

4. Boyce, M.C., Parks, D.M. and Argon, A.S., Plastic flow in oriented polymers. Int. J. of Plasticity, 1989, **5**, 593-615.

# INELASTIC ANISOTROPIC BEHAVIOUR OF LONG FIBRE REINFORCED POLYMER MATRIX COMPOSITES

A.H. CARDON, R. BROUWER, L. SCHILLEMANS, I. DAERDEN
Composite Systems and Adhesion Research Group of the Free University Brussels
COSARGUB
V.U.B. (TW-KB), Pleinlaan 2, B-1050 Brussels, Belgium

## ABSTRACT

The viscoelastic nature of the polymer matrix induces a viscoelastic behaviour of the polymer matrix composite. If this composite is unidirectional reinforced by long fibres, or by short fibres oriented in one direction, the mechanical behaviour is anisotropic, generally of the orthotropic or even transverse isotropic groups.
Time dependent behaviour will be changed with the environmental variations such as temperature changes or moisture uptake and diffusion.
The time dependent behaviour may be global : creep- and/or relaxation effects of a unidirectionally reinforced lamina ; or internal, stress transferts between viscoelastic and elastic components.Those stress transferts occur in a unidirectional reinforced lamina between matrix and fibres, but also between lamina in a laminate function of the combination in the different layers of viscoelastic and elastic dominated directions.
Those stress transferts at higher bondings must be completed by those resulting from matrix cracking, fiber debonding, fiber breaking and delaminations.
A complete limit analysis must consider viscoplastic behaviour and must be completed by a continuum damage analysis.
A thermodynamic based formulation such as proposed by R.A. Schapery, [1], can include those different aspects of stress redistributions in a composite laminate and give a good foundation for the analysis of the limitbehaviour of a composite structural component under general, mechanical and environmental, loading history.

## INTRODUCTION

The thermomechanical behaviour of a unidirectional reinforced composite lamina can be studied within the framework of the constitutive equations proposed by R.A. Schapery :

$$q_m = -\frac{\partial G_R}{\partial Q_m} + \frac{\partial \hat{Q}_n}{\partial Q_m} \int_0^t \Delta S_{np} (\psi - \psi') \frac{\partial [\hat{Q}_p / a_\sigma]}{\partial u} du$$

$m,n,p$ : 1,... 6
$q_m$ : components of the strain tensor
$\hat{Q}_n$ : generalised stresses
$Q_m$ : components of the stress tensor
$\Delta S_{np}$ : transient compliances

$$\psi = \int_0^t \frac{d\zeta}{a_\sigma} \; ; \; \psi' = \psi(t = u) \text{ reduced time scales}$$

$G_R$ : constant term in a series development of the Gibbs energy.

In the case of a plane stress state those equations were developed f.ex. by R. Brouwer, [2].

The final expressions involve 4 nonlinearizing functions. Those nonlinearizing functions may depend on the temperature, on some invariants of the stress state and on some expressions integrating the environmental effects.

Under conditions of perfect adhesion between fibres and matrix the viscoelastic nature of the matrix induces by relaxation the transfert of stresses to the fibres. This type of stress transfert occur also between lamina as function of the stacking sequence. Elements of such an analysis were given by A. Cardon, [3].

## INTERPHASE

In a composite system not only the bulk properties of the phases are important for the global thermomechanical behaviour but also the interaction between the phases as influenced by the manufacturing processes. The interactions between phases can be included by the properties of the interphase region. The properties of those interphase regions are not of easy access for direct experimental measurements, see f.ex.[4]. The basic component for the study of the behaviour of a laminate and/or a structural component is the unidirectional reinforced lamina. The thermomechanical properties of those basic elements under biaxial loading conditions of relative humidity combined with short term test results under different loading histories will give the starting elements of the prediction of the long term behaviour of laminates and structural components.

## LIMIT BEHAVIOUR

The prediction of the limit behaviour of a structural component under a general, mechanical and environmental, loading history is the central problem for safe structural applications.

Such a limit analysis can be achieved by the use of specific limit criteria or by the prediction of damage evolution. Global limit criteria are very useful for isotropic materials and even for anisotropic materials where the anisotropy is the macroscopic consequence of a physical behaviour on microscopic scale.

For composite materials, where the different phases can be identified on macroscopic level, even if their bulk properties are not easy to be measured, a limit criterium has to take into account some "structural" effects in the material behaviour, f.ex. compression. In those situations the prediction of the damage growth, after identification of the elementary damage modes, is probably a more convenient method for the prediction of the life time and the general durability analysis.

## CONCLUSIONS

The analysis of the limit behaviour of long fibre reinforced composite structural components must start from a thermodynamic based constitutive model, such as those proposed by R.A. Schapery and co-workers, where viscoelasticity and viscoplasticity can be included.

In those models damage can be included by hidden variables and the damage evolution can be computed.

The results of the predicted limit behaviour must be compared with real time data obtained on structural composite components under different loading histories and environmental variations, if we want to define safe design codes without overdimensioning of the components.

## REFERENCES


1. Schapery, R.A., On a Thermodynamic Constitutive Theory and its Application to Various Nonlinear Materials. In Thermoinelasticity, ed. B.A. Boley, Springer Verlag, Wien-New York, 1970, pp. 259-285.

2. Brouwer, R., Hiel, C., Cardon, A.H., Nonlinear Viscoelastic Behaviour of a Polymeric Matrix Composite in Biaxial Loading Conditions. In Rheology of Anisotropic Materials, ed. C. Huet, D. Bourgain, S. Richemond, Ed. CEPADVES, Toulouse, 1986, pp. 121-130.

3. Cardon, A.H., Global and Internal Time Dependent Behaviour of Polymer Matrix Composites. In Inelastic Deformation of Composite Materials, ed. G.J. Dvorak, Springer Verlag, 1991, pp. 489-499.

4. Cardon, A.H., De Wilde, W.P., Verheyden, M., Van Hemelrijck, D., Schillemans, L., Boulpaep, F., Interphase Region in Composite Systems : Analysis, Characterisation and Influence on Long Term Behaviour". In Interfaces in New Materials, ed. P. Grange, Elsevier Applied Science (to be published).

# A CONSTITUTIVE FRAMEWORK FOR LARGE VISCOELASTIC DEFORMATIONS OF ELASTOMERS

YANNIS F. DAFALIAS
Department of Civil Engineering, University of California, Davis, CA 95616, USA
and Division of Mechanics, National Technical University, Athens, 10682, HELLAS

## ABSTRACT

A specific large viscoelastic deformations constitutive framework for elastomers is presented in an Eulerian formulation. Within it, the predominant elastic part of the deformation can be described by a number of well known theories, while the viscous contribution comes for the evolution of an internal stress where the concepts of the effectively unstressed configuration and the internal spin are used. By comparison with experiments, the model is shown to capture salient features of the material response under cyclic large simple shear.

## INTRODUCTION

The mechanical response of polymeric materials at large deformations is bounded by the two extreme modes of behavior depending on the internal structure of these materials and loading conditions. On the one extreme, glassy polymers deform primarily in a plastic/viscoplastic mode with very small superposed elastic deformations, and on the other extreme rubber-like materials or elastomers deform elastically according to what is known as rubber elasticity. In between there is a plethora of responses where elastic, viscous and plastic elements contribute to the total large deformation of the material. In this work attention will be focused on a particular class of elastomeric response, where the primary mode of large deformation is elastic, accompanied by a small viscous contribution.

Large deformation viscoelasticity of polymers, in general, has been studied traditionally by the method of hereditary integrals extending the classical approach of linear viscoelasticity, properly modified to account for non-linear geometrical and material characteristics [1]. The fact that in the present case elasticity is the primary contributor to the large deformation, suggests an alternative approach where the vast knowledge of rubber-elasticity is supplemented by the introduction of internal variables whose evolution attributes the viscous aspect of the response. Along these lines the works in [2,3,4] in a Lagrangian setting can be mentioned. The present study presents an Eulerian formulation extending the original suggestion made by Dafalias [5] to include compressibility, and achieves a successful simulation of experimental data under cyclic loading in simple shear.

## THE CONSTITUTIVE FORMULATION

Rubber elasticity is based on the concept of an affine deformation of the molecular chain network at equilibrium. Yet, such affine deformation does not occur instantaneously upon application of a load or deformation, but during the transient period towards equilibrium the molecular chain network changes configuration and in the process intermolecular stresses which developed as resistance to applied strain relax, exhibiting a viscous-like response. It is proposed to introduce the concept of an evolving internal stress variable $\alpha$ to account for the above phenomenon, and write the constitutive law as

$$\sigma - \alpha = J^{-1} \sum_{i=1}^{3} \lambda_i \frac{\partial w}{\partial \lambda_i} (\mathbf{n}_i \otimes \mathbf{n}_i) \tag{1}$$

where $\sigma$ is the Cauchy stress, $\lambda_i$ are the principal stretches along the principal unit directions $\mathbf{n}_i$, w is the strain energy density and $J = \lambda_1\lambda_2\lambda_3$. Eq. (1) shows that zero elastic strains ($\lambda_i = 1$) occur not when $\sigma = \mathbf{0}$, but when $\sigma - \alpha = \mathbf{0}$ according to the concept of the effectively unstressed configuration [6]. It is now necessary to specify w and the law of evolution of $\alpha$.

While many ways exist to describe the elastic response, here the proposition by Ogden [7] is adopted, thereby

$$\lambda_i \frac{\partial w}{\partial \lambda_i} = \mu (\lambda_i^n - 1) - \frac{B}{\beta} (J^{-\beta} - 1) \tag{2}$$

(no sum over i) introducing the elastic constants $\mu$ (shear "modulus"), n, B and $\beta$, the latter two addressing the compressibility of the material.

The rate equation of evolution of $\alpha$ is proposed to be

$$\overset{\circ}{\alpha} = \dot{\alpha} - \omega\alpha + \alpha\omega = \frac{2}{3} h_\alpha \mathbf{D}' - c_r \alpha \tag{3a}$$

$$\omega = \Omega^E - \Omega^I \tag{3b}$$

$$\Omega^I = \eta_r (\alpha\mathbf{V} - \mathbf{V}\alpha) \tag{3c}$$

where $\mathbf{D}'$ is the deviatoric part of the rate of deformation, $\Omega^E$ is the spin of the Eulerian strain ellipsoid and $\Omega^I$ is an internal spin, depending on the degree of non-coaxiality between $\alpha$ and the left stretch tensor $\mathbf{V}$ ($\mathbf{F} = \mathbf{VR}$ according to the polar decomposition of the deformation gradient $\mathbf{F}$). Eqs. (3) introduce the "hardening" modulus $h_\alpha$ associated with the development of the internal stress, the recovery coefficient $c_r$ associated with a relaxation process, and the internal spin coefficient $\eta_r$. The definition of the constitutive spin $\omega$ according to Eq. (3b) reflects the tendency of the internal stress $\alpha$ to become coaxial with $\mathbf{V}$, since $\alpha$ is associated with the transient state of the molecular chain network which tends to deform affinely at equilibrium.

## APPLICATION AND COMPARISON WITH EXPERIMENTS

The model is used for the simulation of large simple shear $\gamma$ along $x_1$ on the $x_1$-$x_2$ plane. The simple shear kinematics in combination with the governing Eqs. (1)-(3) and a lengthy but straightforward algebra yield finally:

$$\sigma_{11} - \alpha_{11} = \mu \left( \frac{\lambda^{n+1} + \lambda^{-(n+1)}}{\lambda + \lambda^{-1}} - 1 \right) \quad ; \quad \sigma_{12} - \alpha_{12} = \mu \frac{\lambda^{n} - \lambda^{-n}}{\lambda + \lambda^{-1}} \tag{4}$$

$$\frac{d\alpha_{11}}{d\gamma} = \frac{2}{\gamma^2 + 4} \alpha_{12} - \frac{c_r}{\dot{\gamma}} \alpha_{11} - \frac{\eta_r}{\dot{\gamma}} \frac{2\gamma}{(\gamma^2 + 4)^{1/2}} (2\alpha_{11} - \gamma\alpha_{12})\alpha_{12} \tag{5a}$$

$$\frac{d\alpha_{12}}{d\gamma} = \frac{1}{3} h_\alpha - \frac{2}{\gamma^2 + 4} \alpha_{11} - \frac{c_r}{\dot{\gamma}} \alpha_{12} + \frac{\eta_r}{\dot{\gamma}} \frac{2\gamma}{(\gamma^2 + 4)^{1/2}} (2\alpha_{11} - \gamma\alpha_{12})\alpha_{11} \tag{5b}$$

with $\alpha_{22} = -\alpha_{11}$, a corresponding equation for $\sigma_{22} - \alpha_{22}$ not given for simplicity, and $\lambda = (1/2)\,[\gamma + (4 + \gamma^2)^{1/2}]$. Eqs. (5) can be solved by elementary numerical procedures, and the values of $\alpha_{11}$, $\alpha_{12}$ used in Eq. (4) to obtain $\sigma_{11}$ and $\sigma_{12}$ as functions of $\gamma$. With the values of the model constants specified as $\mu = 52$ psi, $n = 2.5$, $h_a = 280$ psi, $c_r = c_d\dot{\bar{\varepsilon}} + c_s = 10$ (with $\dot{\bar{\varepsilon}} = |\dot{\gamma}|/\sqrt{3} = 0.289\ \text{sec}^{-1}$) and $\eta_r = 0.1$, the experimental data obtained by Kelly and Celebi [8] as shown in Fig. 1(a) were simulated by the solution of Eqs. (4) and (5) as shown in Fig. 1(b) in terms of the stabilized stress-strain loops. It is seen that this preliminary simulation captures some of the salient feature of the material response quite well. Particularly interesting is the simulation of the initial reduction and subsequent slight increase of the effective shear modulus (average loop slope) with increasing strain amplitude, without the need to introduce any damage mechanism as in [4]. This results from the level of "saturation" of the internal stress $\alpha$ with amplitude, and the increased stiffening of the elastic response.

## ACKNOWLEDGEMENT

The support by the NSF small grant for exploratory research No. MSS-8918531 with program director Dr. O.W. Dillon is acknowledged. Mr. H.W. Cho provided the plots of Fig. 1.

## REFERENCES

1. Christensen, R.M., A nonlinear theory of viscoelasticity for application to elastomers. ASME, J. Appl. Mech., 1980, **47**, 762-768.

2. Green, M.S. and Tobolsky, A.V., A new approach to the theory of relaxing polymeric media. J. Phys. Chem., 1946, **14**, 80-92.

3. Lubliner, J., A model of rubber viscoelasticity, Mech. Res. Comm., 1985, **12**, 93-99.

4. Simo, J.C., On a fully three-dimensional finite-strain viscoelastic damage model: formulation and computational aspects. Comp. Meth. Appl. Mech. Eng., 1987, **60**, 153-173.

5. Dafalias, Y.F., Constitutive model for large viscoelastic deformations of elastomeric materials. Mech. Res. Comm., 1991, **18**, 61-66.

6. Dafalias, Y.F., Issues on the constitutive formulation at large elastoplastic deformations, part 2: kinetics. Acta Mech., 1988, **73**, 121-146.

7. Ogden, R.W., Large deformation isotropic elasticity: on the correlation of theory and experiment for compressible rubberlike solids. Proc. R. Soc. Lond., 1972, A. **328**, 567-583.

8. Kelly, J.M. and Celebi, M., Verification testing of prototype bearings for a base isolated building. Report UCB/SESM-84/01, Dept. Civil Eng., U.C. Berkeley, March 1984.


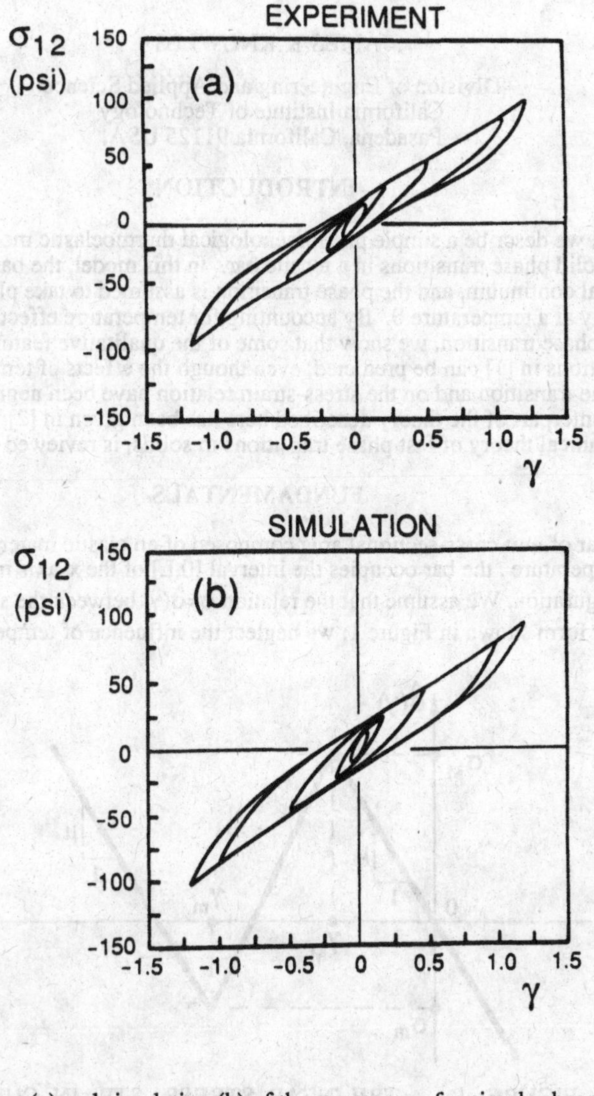


Fig. 1a,b. Experiment (a) and simulation (b) of the $\sigma_{12}$-$\gamma$ curve for simple shear cyclic loading.

# CONTINUUM MODELING OF PHASE TRANSITIONS IN SOLIDS

JAMES K.KNOWLES

Division of Engineering and Applied Science
California Institute of Technology
Pasadena, California 91125 USA.

## INTRODUCTION

In this paper, we describe a simple phenomenological thermoelastic model for stress-induced solid-solid phase transitions in a tensile bar. In this model, the bar is treated as a one-dimensional continuum, and the phase transition is assumed to take place quasi-statically and isothermally at a temperature $\theta$. By accounting for temperature effects solely in the kinetics of the phase transition, we show that some of the qualitative features of the experimental observations in [1] can be predicted, even though the effects of temperature on the nucleation of the transition and on the stress-strain relation have been neglected. A purely mechanical counterpart of the theory described here has been given in [2]; the latter work, as well as a mechanical theory of fast phase transitions in solids, is reviewed in [3].

## FUNDAMENTALS

Consider a bar of unit cross-sectional area composed of an elastic material; suppose that at a reference temperature , the bar occupies the interval [0,L] of the x-axis in a stress-free reference configuration. We assume that the relation $\sigma=\hat{\sigma}(\gamma)$ between the stress $\sigma$ and strain $\gamma$ has the trilinear form shown in Figure 1; we neglect the influence of temperature on $\hat{\sigma}(\gamma)$.

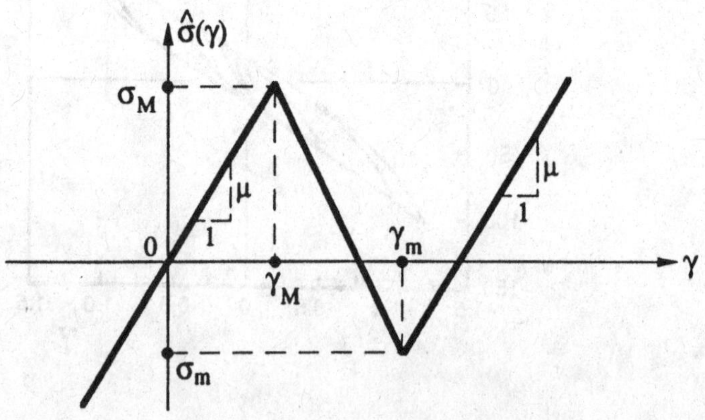


FIGURE 1. TRILINEAR STRESS - STRAIN CURVE .

Equilibrium in the absence of body force requires that the stress be constant along the bar. For the stress response function of Figure 1, the constancy of stress implies that $\gamma(x)$ is also constant if the applied stress $\sigma$ exceeds $\sigma_M$ or is less than $\sigma_m$. If $\sigma$ is *between* $\sigma_m$ and $\sigma_M$, however, the strain need be only *piecewise* constant. When $\sigma$ lies in the latter range, we suppose that, for x>s, the strain is given by $\gamma = \sigma/\mu$, corresponding to the first rising branch of the stress-strain curve (or "phase 1" of the material), while for x<s, $\gamma=\gamma_m + (\sigma - \sigma_m)/\mu$, a strain on the final rising branch ("phase 3"). One could view the declining branch of the curve in Figure 1 as an unstable phase, or "phase 2". Thus for $\sigma_m<\sigma<\sigma_M$, there is an equilibrium mixture of phases 1 and 3 in the bar; s/L is the volume fraction of phase 3.

The overall elongation $\Delta$ is the integral of $\gamma(x)$ from x=0 to x=L. Thus at an applied stress $\sigma$, the *macroscopic response* of the bar is given by

$$\Delta = \begin{cases} \sigma L/\mu\,, & \text{for } \sigma < \sigma_m\,, \\ \sigma L/\mu + (\gamma_m - \sigma_m/\mu)\, s\,, & \text{for } \sigma_m \le \sigma \le \sigma_M\,,\ 0 \le s \le L, \\ [\gamma_m + (\sigma - \sigma_m)/\mu]L\,, & \text{for } \sigma > \sigma_M\,. \end{cases} \tag{1}$$

This relation between stress $\sigma$ and elongation $\Delta$ involves the "internal variable" s that locates the interface between the the high- and low-strain phases.

In a *quasi-static motion* of the bar, s and $\sigma$ in (1) are replaced by functions of time t: s=s(t), $\sigma=\sigma(t)$. Then $\Delta=\Delta(t)$ as well. When $\sigma_m \le \sigma(t) \le \sigma_M$ in such a motion, it follows from results established in [2] that the *dissipation rate*, defined to be the difference between the rate of work done on the ends of the bar by the stress $\sigma$ and the rate of energy storage, is $D(t)=f(t)\dot{s}(t)$, where f(t), defined in [2], is called the *driving force* associated with the moving interface. If the process is assumed to be isothermal, the second law of thermodynamics requires that $D(t) \ge 0$. The explicit form of f can be obtained for the trilinear material using results from [2]; one finds that f(t)>0 for $\sigma(t) > \sigma_0$, f(t)<0 for $\sigma(t)<\sigma_0$ and f(t)=0 if $\sigma(t)=\sigma_0$, where the *Maxwell stress* $\sigma_0$ is such that the line $\sigma=\sigma_0$ cuts off triangles of equal area from the peak and the valley of the stress-strain curve in Figure 1. It then follows from the second law that s(t) cannot decrease (increase) if the applied stress $\sigma(t)$ is greater (less) than $\sigma_0$; if $\sigma=\sigma_0$, the phase boundary may move in either direction.

In a displacement-controlled process, the history of $\Delta$ is given, and the stress $\sigma(t)$ is to be determined. Because s(t) is unknown, (1) fails to provide a determinate macroscopic response in such a process. This indeterminacy is to be remedied as in [2] by supplying additional constitutive information in the form of a nucleation criterion for the initiation of the phase transformation and a kinetic relation that controls the rate at which the transformation proceeds. *Nucleation* was assumed to occur at a critical value of driving force f, while the *kinetic law* was assumed to have the form $\dot{s}=\Phi(f)$, where $\Phi$ is determined by the material.

In order to state the specific kinetic relation to be used here, we first introduce the Gibbs potential $G(\gamma, \sigma)$ defined by

$$G(\gamma, \sigma) = \int_0^{\gamma} \hat{\sigma}(\gamma')\, d\gamma' - \sigma\gamma\,, \tag{2}$$

For the trilinear material of Figure 1, the function $G(\cdot\,, \sigma)$ is convex with a single minimum for any fixed $\sigma$ that is either greater than $\sigma_M$ or less than $\sigma_m$. But if $\sigma_m<\sigma<\sigma_M$, $G(\cdot\,, \sigma)$ has two minima separated by a maximum, and thus represents a "two-well potential" that correponds to a two-phase material; the left (right) minimum occurs at a phase-1 (phase-3)strain.

The macroscopic rate at which the phase transition in the bar takes place is measured by $\dot{s}(t)$. As in classical thermal activation theory [4], we take this rate to be determined by the energy barriers $b_{13}(\sigma)$ and $b_{31}(\sigma)$ and the temperature $\theta$; $b_{13}(\sigma)$ is the the difference between the value of G at its local maximum and the value at the phase-1 minimum; $b_{31}(\sigma)$ is the corresponding barrier for the phase-3 minimum. One can find these barriers explicitly for the trilinear material. As the kinetic relation, we then take

$$\dot{s} = \nu(\theta)\left\{\exp[-\,kb_{13}(\sigma)/\theta] - \exp[-\,kb_{31}(\sigma)/\theta]\right\}, \tag{3}$$

where k is a constant, $\theta$ is the absolute temperature, and $\nu(\theta)>0$ is a materially-determined function of temperature. The stress $\sigma$ can be expressed in terms of the driving force f with the help of results in [2], so that the kinetic relation (3) can be written in the form $\dot{s} = \Phi(f)$ at each temperature, as assumed in [2,3]; moreover, one can show that $\Phi(f)f\geq 0$, so that the dissipation rate $D=f\dot{s}$ is automatically non-negative. Finally, we assume that nucleation of the phase 1→phase 3 transition occurs at the Maxwell stress $\sigma_0$, regardless of the temperature.

Consider now a quasi-static process in which the elongation $\Delta(t)$ of the bar is increased from zero at a given constant rate $\rho$. From (1), it follows that the bar remains in phase 1 with the stress given by $\sigma(t) = \mu\Delta(t)/L$ until $\sigma(t)=\sigma_0$, at which time a phase transition is nucleated and a phase boundary enters the bar at x=0. At this instant, the kinetic law (3) takes over and controls the evolution of the phase boundary x=s(t) and the resulting macroscopic response; indeed, eliminating s(t) between (1) and (3) yields a differential equation for the unknown stress $\sigma(t)$ as a function of time t after nucleation at the fixed temperature $\theta$.

## RESULTS

Let $\theta_0 = k\,\sigma_M\gamma_M/2$ and $T=\theta/\theta_0$ be a reference temperature and a dimensionless temperature, respectively, and set $\psi(T)=\nu(\theta)/\nu(\theta_0)$. Let $\delta = \Delta/(L\gamma_M)$ be the dimensionless elongation, $\tau=\sigma/\sigma_M$ the dimensionless stress and $\lambda=\rho/[\nu(\theta_0)\gamma_M]$ the dimensionless elongation rate. We also introduce dimensionless material constants $p=\gamma_m/\gamma_M$ and $q=-\sigma_m/\sigma_M$. With the help of these quantities, one obtains from (1), (3) a differential equation for the dimensionless stress $\tau$ as a function of dimensionless elongation $\delta$. Indeed, one finds

$$\frac{d\tau}{d\delta} = 1 - \frac{p+q}{\lambda}\,\psi(T)\left\{\exp\left[-\frac{p+q}{1+q}\frac{(\tau-1)^2}{T}\right] - \exp\left[-\frac{p+q}{1+q}\frac{(\tau+q)^2}{T}\right]\right\}, \tag{4}$$

together with the initial condition $\tau=(1/2)(1-q)$ when $\delta=(1/2)(1-q)$, corresponding to nucleation at the Maxwell stress. We consider this "loading" initial value problem on the interval $(1/2)(1-q) \leq \delta \leq \delta_1$. Upon achieving the elongation $\delta_1$, the bar is to be *unloaded* at a constant (dimensionless) elongation rate $\lambda$ until the stress vanishes. During this unloading stage, (4) holds with $\lambda$ replaced by $-\lambda$, with a new initial condition at $\delta=\delta_1$ that assures the continuity of $\tau(\delta)$ at $\delta=\delta_1$.

Some of the qualitative possibilities afforded by the model as regards macroscopic response in the above quasi-static, grip-controlled loading-unloading program have been illustrated by solving the initial value problem for (4) - and its counterpart for unloading - for various choices of $\psi$. Two representative cases are: (a) $\psi(T)=T^2$, (b) $\psi(T)=1$ . Case (b) corresponds to a temperature-independent "pre-exponential" factor $\nu$ in (3). In each case, the response has been calculated, for fixed values of $\lambda$ , p and q, at each of three different values of T. The results for cases (a) and (b) are shown in Figures 2 and 3, respectively. The response curves in Figure 2 are qualitatively similar in two respects to measured load-elongation curves for Ag-Cd single crystals reported in [1]: at the lowest temperature, there is permanent deformation, and at the highest temperature, there is virtually no hysteresis. In Case (a), the size of the hysteresis loop varies monotonically with temperature, whereas in Case (b), this is not the case. The experimental results in [1] do *not* appear to confirm that the hysteresis diminishes monotonically with increasing temperature. Finally, we note that the present model would *not* be adequate to describe experiments in which the *temperature* is cycled at fixed stress, as in experiments on indium-thallium single crystals carried out in [5].

The support of the U.S. Office of Naval Research is gratefully acknowledged.

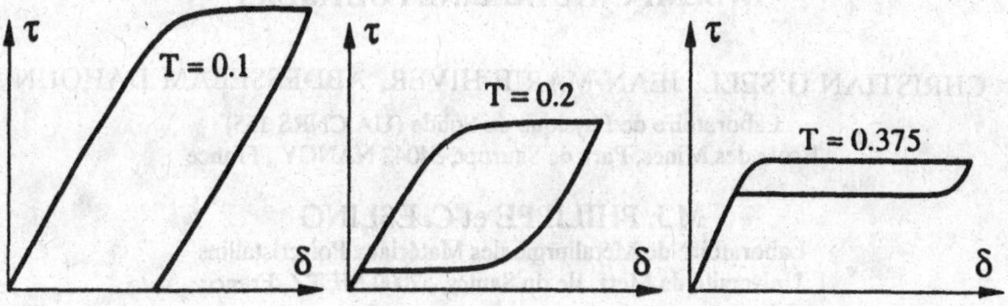


FIGURE 2. MACROSCOPIC RESPONSE, (a) $\Psi(T) = T^2$.

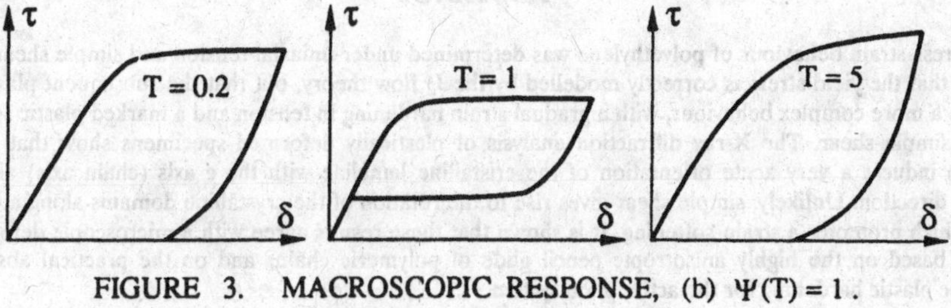


FIGURE 3. MACROSCOPIC RESPONSE, (b) $\Psi(T) = 1$.

## REFERENCES

1. Krishnan, R.V. and Brown, L.C., Pseudo-elasticity and the strain-memory effect in an Ag-45 at. pct. Cd alloy, Metallurgical Transactions, 1973, **4**, 423-429.

2. Abeyaratne, R. and Knowles, J.K., On the dissipative response due to discontinuous strains in bars of unstable elastic material, International Journal of Solids and Structures, 1988, **24**, 1021-1044.

3. Abeyaratne, R. and Knowles, J.K., Nucleation, kinetics and admissibility criteria for propagating phase boundaries, to appear in Proceedings of the Workshop on Shock Induced Transitions and Phase Structures in General Media, ed. R. Fosdick, Institute of Mathematics and its Applications, University of Minnesota.

4. Weiner, J. H., Thermal activation and tunneling phenomena in solids, Proceedings of the Sixth U.S. National Congress of Applied Mechanics, ed. G. Carrier, ASME Publishers, New York, 1970, pp. 62-77.

5. Burkart, M.W. and Read, T.A., Diffusionless phase change in the indium-thallium system, Journal of Metals, Transactions of the AIME, 1953, **197**, 1516-1524.

# ANISOTROPIC PLASTIC HARDENING IN SEMI-CRYSTALLINE POLYMERS

CHRISTIAN G'SELL, JEAN-MARIE HIVER, ABDESSELAM DAHOUN
Laboratoire de Physique du Solide (UA CNRS 155)
Ecole des Mines, Parc de Saurupt, 54042 NANCY , France

M.J. PHILIPPE et C. ESLING
Laboratoire de Métallurgie des Matériaux Polycristallins
Université de Metz, Ile du Saulcy, 57000 METZ, France

## ABSTRACT

**The stress-strain behaviour of polyethylene was determined under uniaxial tension and simple shear. It was shown that the yield stress is correctly modelled by the $J_2$ flow theory, but that the subsequent plastic flow follows a more complex behaviour, with a gradual strain hardening in tension and a marked plastic softening under simple shear. The X-ray diffraction analysis of plastically deformed specimens show that uniaxial tension induces a very acute orientation of the cristalline lamellae with the c axis (chain axis) along the tensile direction. Unlikely, simple shear gives rise to the rotation of the crystalline domains along a bimodal path which promotes a strain-softening. It is shown that these results agree with a microscopic deformation model based on the highly anisotropic pencil glide of polymeric chains and on the practical absence of intrinsic plastic hardening for the active glide system.**

## INTRODUCTION

It is now broadly estabished that the structure of most semi-crystalline polymers is constituted by lamellar single crystal arranged radially within spherulites, and interconnected by amorphous chains. When the material is subjected to plastic deformation, the macroscopic response is then primarily dependent on the microscopic mechanisms of the individual lamellae and to the distribution of their orientations. It has been acknowledged that these mechanisms control some industrial solid-state forming processes (stretching, thermoforming...). Although the case of polyethylene has received a particular attention in the past, most workers have focussed their attention on tension, compression [1] and rolling [2,3], the crystalline textures and the flow behaviour of the material was not precisely assessed for deformation pathes of lower symmetry. It is therefore the aim of this work to examine the effect of plastic deformation on the crystal orientation distribution in simple shear by comparison with uniaxial tension.

## MATERIAL AND METHODS

A commercial grade of high-density polyethylene of molecular weight $M_w = 172000$ g/mole was extruded in thick plates, and subsequently annealed under vacuum at 120 °C during 24 h in order to relax internal stresses. The degree of crystallinity is thus equal to about 71 %, with a spherulite size of a few microns in diameter. The crystalline structure is orthorhombic, with a=0.74 nm, b=0.494 nm and c=0.254 nm as lattice parameters.

The mechanical tests were performed at room temperature by using an original system based on the utilization of a digital video strain tranducer (4,5). In simple tension, the true stress and strain are measured in real time in the neck which forms at yield in the center of the hourglass-shaped specimen ; the system controls the servo-hydraulic machine in such a way that the true strain-rate is locally constant up to rupture. Similarly, with the simple shear system a single shear band is formed along a notch of a special specimen and the true stress and strain are locally controlled within the band at a constant shear rate.

As for the microstructural investigation, specimens deformed to well controlled tensile and shear strains were analysed by X-ray goniometry . In order to obtain complete pole figures of the crystal orientations, this analysis was performed in transmission on small cylinders carefully machined out of the calibrated part of the specimens. The incident beam was Cu-K$\alpha_1$ and the diffracting planes were (200), (110) and (011).

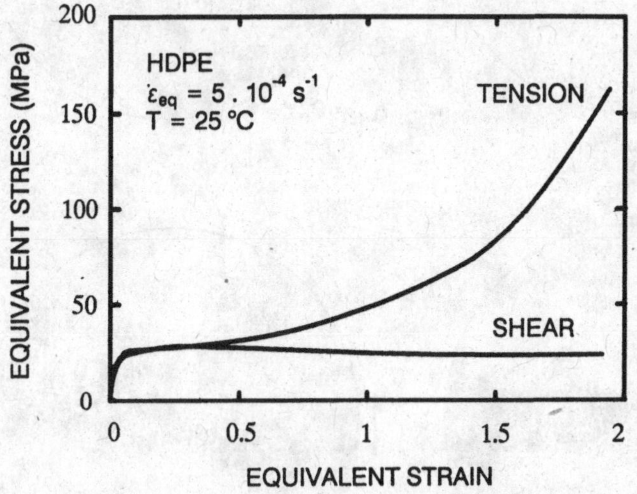


Figure 1. Equivalent stress/strain curves obtained from experimental tests in uniaxial tension and under simple shear.

## EXPERIMENTAL RESULTS

From the tensile and shear stress/strain curves obtained directly from the mechanical tests, the plastic behaviour of the polyethylene was displayed in Figure 1 in terms of Mises work conjugated quantities : i) in tension $\sigma = F_T \times P / A$ vs. $\varepsilon = \ln(A_o/A)$ where P is the load, A (resp. $A_o$) the current (resp. initial) cross-section and $F_T$ the Bridgman stress triaxiality factor taking into account the curvature of the neck, ii) in simple shear $\tau \times \sqrt{3}$ vs. $\gamma / \sqrt{3}$, where $\tau$ and $\gamma$ are the local shear stress and strain. In both cases, the equivalent strain rate was identical : $\mathring{\varepsilon}_{ij} = (2/3\ \mathring{\varepsilon}_{ij}\ \mathring{\varepsilon}_{ij}) = 5\ 10^{-4}\ s^{-1}$. It can be observed that for moderate strains (< 0.4) the equivalent stress/strain curves are almost superimposed, which indicates that, in this range of deformation, the material presents a Mises isotropic behaviour. However, for larger higher levels of plastic deformation, the two curves diverge, with a progressively increasing strain-hardening in uniaxial tension and a significant stress softening under simple shear (the stress drop continues until $\gamma$ is equal to about 3 and then the stress increases again).

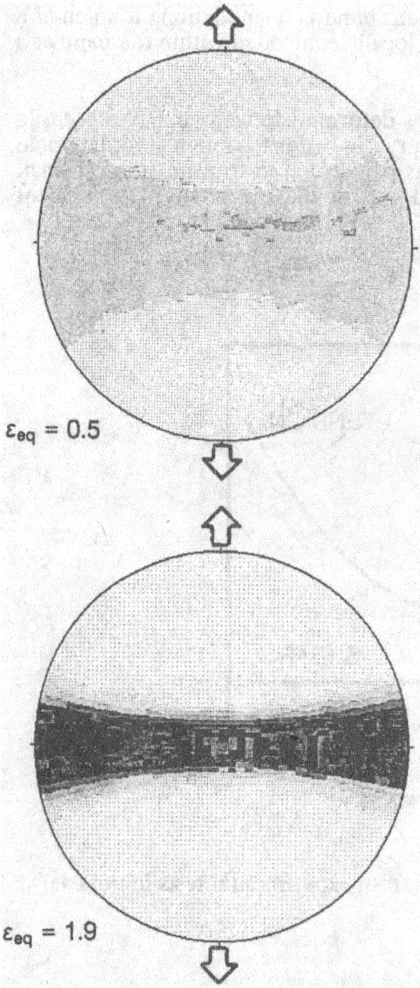


Figure 2. (200) pole figures obtained at increasing strains in tension

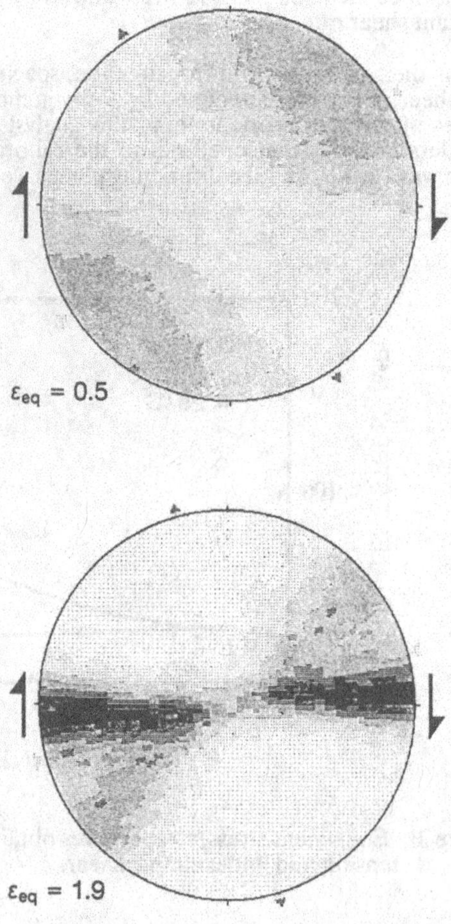


Figure 3. (200) pole figures obtained at increasing strains under shear

The pole figures displayed in Figure 2 confirm that the texture evolution or the crystallites are very different for the two loading paths followed in the mechanical tests. In this short paper, we have chosen to present the (200) pole figures only. For the case of uniaxial tension, the one notes that the (100) crystallographic plane normals rotate progressively towards the plane perpendicular to the tensile axis, indicating that the **c** = [001] vector (direction of the chains in the crystals) turns towars the tensile axis. This is the orientation distribution commonly referred to as the "fiber texture". Conversely, in shear, the (200) pole figures displayed in Figure 3 show the development of a transient bimodal texture : the majority of the [100] directions get oriented close to the plane perpendicular to the shear direction, that is with the **c** axis along the shear direction. The rest of the [100] axes stabilize for a while along an orientation intermediate between the latter and the shear axis, before rejoining the first set at very high shear strains..

## DISCUSSION

The basic ingredient of the model corresponds to the plastic mechanism of the crystalline lamellae. According to previous works [6-8], the polymer yields when dislocations begin to glide along the macromolecular chains densely packed within the crystallites. Because of the topological constraint imposed by the very high energy of the covalent bonds along the chains (relatively to the weak intermolecular Van-der-Waals bonds), the glide plane must be of the {hk0} type. The orthorhombic structure promotes the {100} planes as the most favorable ones for glide, although the {010} and {110} planes have a reasonable density as well. As for the glide direction, the <001> chain axis is much more probable than any transverse direction because of the low thickness of the lamellae (≈ 10 nm) with respect to their lateral size. It should be then conceived that plastic flow in the polyethylene results on this kind of pencil glide in the lamellae, and that the rubberlike amorphous layers act as a continuous soft medium which accomodates the rotations induced by the deformation.

Originally, the distribution of the crystal orientation is globally isotropic and the plastic deformation is initiated in those sectors of the spherulites where the Schmid factor is maximum. In a first approximation, the constitutive equation is of the $J_2$ Mises type. As the deformation proceeds, the lamellae which undergo plastic glide tend to rotate and this evolution differs according to the loading geometry and to their original position. In uniaxial tension the **c** chain axis turns towards the tensile axis, giving rise to the development of the fiber texture with lower and lower average Schmid factors and to a high strain hardening. Under simple shear, by contrast, the rotation of the lamellae differs according to their position in a given spherulite. A transient bi-modal orientation distribution occurs in which only a fraction of crystallites turns immediately towards the highly glissile orientation with the **c** axis along the macroscopic shear direction. The others will eventually rejoin this orientation, but with some delay. On the whole, however, the texture development in shear leads to a plastic softening. The ultimate hardening which is observed at very high strains results on the limited capability of the interlamellar tie chains to unfold under stress.

## REFERENCES


1. Krause, S.J. and Hosford, W.F., J. Polym. Sci., Polym. Phys. Ed., 1989, **27**, 1853-65
2. Krause, S.J. and Hosford, W.F., ibid, p 1867-86
3. Point, J.J., Homes, G.A., Gezovich, D. and A. Keller, J. Mater. Sci., 1969, **4**, 908-18
4. G'Sell, C., Rev. Phys. Appl., 1988, **23**, 1085-1101
5. G'Sell, C., Boni, S. and Shrivatava S., J. Mater. Sci., 1983, **18**, 903-18
6. Haudin J.M., in "Plastic deformation of amorphous and semi-crystalline materials", ed.B. Escaig & C. G'Sell, Les Editions de Physique, Les Ulis, France, 1982, 291-311
7. Young, R.J., Bowden, P.B., Ritchie, J.M. and Rider, J.G., J. Mater. Sci., 1973, **2**, 23
8. Frank, F.C., Keller, A. and O'Connor, A., Phil. Mag., 1958, **3**, 64-74

# MODELING OF THE PSEUDO-ELASTIC BEHAVIOR OF POLYCRYSTALLINE SHAPE MEMORY ALLOYS CU-ZN-AL

CHRISTIAN LEXCELLENT and PIERRE VACHER
Laboratoire de Mécanique Appliquée associé au C.N.R.S.
Route de Gray - La Bouloie - 25030 BESANCON CEDEX

## ABSTRACT

Pseudoelastic behaviour of Cu Zn Al shape memory alloys is studied by means of tensile and tensile compressive test. Electrical resistance measurements performed "in situ" allow the martensitic fraction present in the alloy to be known. So, the formulation of the pseudoelastic deformation rate is established.

## INTRODUCTION

The behavior in pure transformation plasticity of polycrystalline alloys Cu-Zn-Al is studied by means of uniaxial isothermal tests in loading/ unloading tensile tests (systematic discharge at 1 %, 2 %...) and in tension-compression.
Electrical resistance measurements performed in situ allow the evolution of the martensite fraction present in the alloy to be known [1]. Acoustic emission recordings, though they allow the stress transformation thresholds to be detected (austenite ⇔ martensite), yield only qualitative information the remainder of the time [2]. The complex problem of coupled transformation plasticity, confronted by Onodera and Tamura [3], Romero and Ahlers [4] for shape memory alloys, has been examined in another article [5]. Here, in reference to models existing in the literature, the pseudo-elastic behaviour of S.M.A. will be studied.

### Phenomenological approach

### Materials and techniques

The polycrystalline S.M.A. Cu-Zn-Al have been fabricated by the company "Trefimetaux".

| % weight | Cu | Zn | Al | $M_F$ | $M_S$ | $A_S$ | $A_F$ |
|---|---|---|---|---|---|---|---|
| R 205 | 67,93 | 28,07 | 4,0 | -110,5 | -98 | -94 | -91 |
| R 232* | 70,17 | 25,63 | 4,0 | 6 | 15 | 17 | 19,5 |
| R 244 | 69,64 | 26,21 | 4,09 | -10 | 1 | 5,5 | 11 |

* without additive constituents

The standard thermal treatment consists in obtaining a $\beta$ phase (10 mn at 850°C), by means of a water quench (at $T-A_S \approx 100$°C) followed be a return to ambient temperatures.

Experimental results

For the loading/unloading tests, the stress threshold of phase transformation is determined by the curve $\sigma_{11} \leftrightarrow \varepsilon_{11}$.

$$A \rightarrow M \quad \sigma_0 = b(T-M_s) \quad (1) \quad \text{with } b = 4 \text{ MPa. } ^\circ C^{-1} \quad \text{(Fig. 1)}$$

The electrical resistance measurements allow the evolution kinetics of the martensite volume fraction to be determined, in agreement with Koistinen and Marburger [6] and Tanaka [7].

$$A \rightarrow M \quad z = 1 - \exp(-a_c < \sigma_{11} - \sigma_0 >) \quad (2)$$

$$M \rightarrow A \quad z = z_0 \ (n \ \%) \exp(-a_d < \sigma'_0 - \sigma_{11} >) \quad (3)$$

with $\sigma'_0 = b(T-A_s)$.

The comparison between the measurements of $\varepsilon^{PE}$ (pseudo-elastic strain) and of z (figure n° 2) allow the following expression to be obtained :

$$\varepsilon^{PE} = \gamma(T)\, z \quad (4)$$

Since z evolves from 0 to 1, $\gamma(T)$ constitutes the limit values of $\varepsilon^{PE}$ at the temperature T. $\gamma$ is proportional to $(T-M_s)$ which agrees with Ono [8] (fig. 3).
The largest absolute value of the phase transformation threshold in compression or tension has already been predicted by Patel and Cohen, [9] and will be integrated in the model (fig. 4).

**Model of pseudoelastic behavior**

First of all, the isothermal loading/unloading tests will be identified. The validity of the maximum work principle extended to polycrystals [10] allows the associated transformation law to be obtained [11] :

$$d\underline{\varepsilon}^{PE} = a(T,z,\ldots)\ \frac{\partial F}{\partial \underline{\sigma}} \left( \frac{\partial F}{\partial \underline{\sigma}} : d\underline{\sigma} + \frac{\partial F}{\partial T} dT \right) \quad (5)$$

where F yield function of the polycrystal.

Two yield functions will be distinguished for the polycrystal : $F_1$ for $A \rightarrow M$ and $F_2$ for $M \rightarrow A$ (see Bertram [12].

$$A \rightarrow M \quad F_1 = < \overline{\sigma} - b(T-M_s) > \quad (6)$$

The choice of the function $a_1(T)$ sous la forme :

$$a_1(T) = a'_1(T)\,(1-z) \quad (7)$$

coupled with the kinetic of z ($A \rightarrow M$) (expression 2) allows the probabilistic model of Wang [13] to be applied thus yielding :

$$\boxed{d\underline{\varepsilon}^{PE} = \frac{3}{2}\, \frac{\text{dev}\, \underline{\sigma}}{\overline{\sigma}}\, \gamma(T)\, dz} \quad (8)$$

The same procedure is performed for M → A :

$$M \rightarrow A \quad F_2 = < b(T-A_s) - \bar{\sigma} > \qquad (9)$$

The choice of the function $a_2(T)$ in the form :

$$a_2(T) = a'_2(T)\, z \qquad (10)$$

coupled with the kinetic of z (M → A) (expression 3) leads to the same expression as (8) for $\underline{d\varepsilon^{PE}}$.
In loading/unloading, the experimental observations of the proportionality between $\varepsilon^{PE}$ and z are well accounted for in the model.
However, the fact that the exponential formulation for the kinetic of z fails to accurately describe the beginning of the reversion of the martensite platelets has an effect on the model (Figure n° 5).
By slightly modifying the charge function ($F_1$ and $F_2$) of the polycrystal, the tension-compression tests can be simulated (fig. n° 6) :

$$\bar{\sigma} \rightarrow A\,\bar{\sigma} + B\,\sigma_m \qquad (11)$$

with at the same temperature $\gamma_c(T) < \gamma_T(T)$.

## CONCLUSION

Given this experimental strategy, the modeling of the 1D loading/unloading and tension compression tests is acceptable. Meanwhile, a verification of equation (8) by biaxial and triaxial tests is necessary.
Finally, the time effects observed in creep and relaxation tests must be integrated.

## REFERENCES


1. Vacher, P. and Lexcellent, C., Rev. Phys. A 25, 1990, pp. 783-797
2. Lexcellent, C. and Vacher, P., EuroMech Coll. 263 : "The effect of phase transformation in solids, on constitutive lows", Vienna Technical University, 1990, July 2-4
3. Onodera, H. and Tamura, I., N.S.F. US/Japan Seminar, 1979, pp. 12-24
4. Romero, R. and Ahlers, M., Phil. Mag. A 59, 1989, pp. 1103-1112
5. Lexcellent, C. and Vacher, P., VII<sup>ème</sup> Symposium Franco Polonais : "Nouveaux aspects de la mécanique des matériaux élastoplastiques", Radziejowice Pologne 2-7 Juillet 1990
6. Koistinen, D.P. and Marburger, R.E., Acta Met. 7, 1959, pp. 59-70
7. Tanaka, K., Kobayashi, S. and Sato, Y., Int. J. Plasticity, 1986, 2, pp. 59-72
8. Ono, N., Mat. Trans. JIM 31, 1990, 5, pp. 381-385
9. Patel, J.R. and Cohen, M., Acta Met. 1, 1953, pp. 531-538
10. Mandel, J., Plasticité classique et viscoplasticité, Udine (1971)
11. Patoor, E., Eberhardt, A. and Berveiller, M., Acta Met. 35, 1987, 11, pp. 2779-2789
12. Bertram, A., Nuclear Eng. and Design, 1982, 74, pp. 173-182
13. Wang, Z.G., ICRS2, Nancy, 25-27 November 1988, pp. 529-538

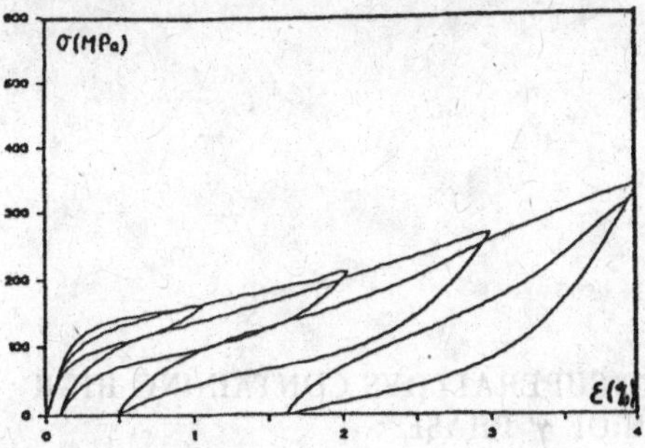


Fig.1 : Loading-unloading test

(R205, T=-70°c, $\dot{\sigma}$=0.55 MPa s$^{-1}$)

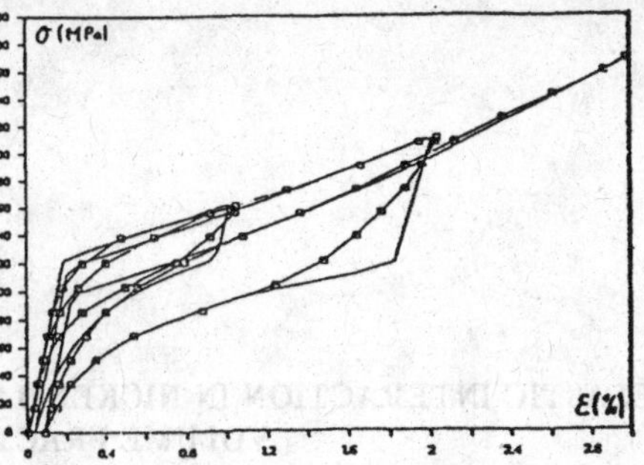


Fig.5 : Non linear modelization of

test fig.1

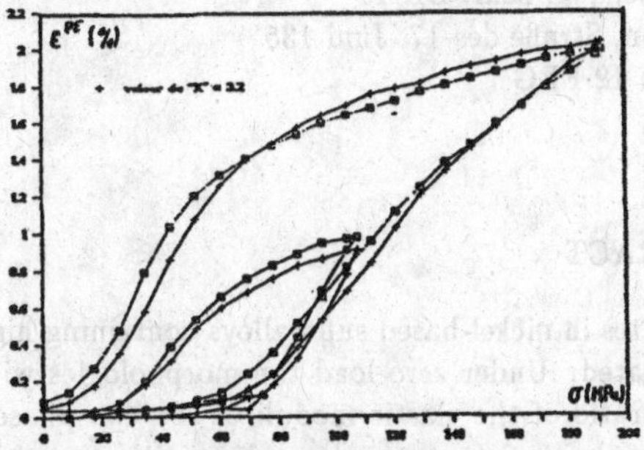


Fig.2 : Evolution of pseudoelastic deformation and volumic fraction of martensite z with applied stress (R205, T=-80°c, $\dot{\sigma}$=0.55 MPa s$^{-1}$)

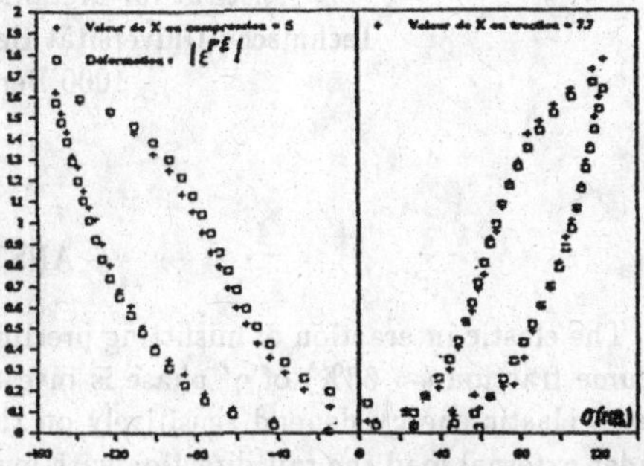


Fig.4 : Evolution of $\epsilon^{PE}=\gamma z$ with applied stress σ (tensil-compressive test) (R244, T=24°c, $\dot{\epsilon}$=10$^{-4}$s$^{-1}$, ±2%)

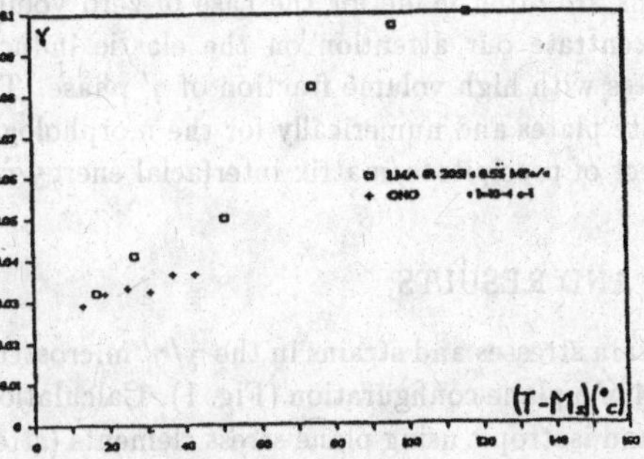


Fig.3 : Evolution of γ(T) with (T-Ms)

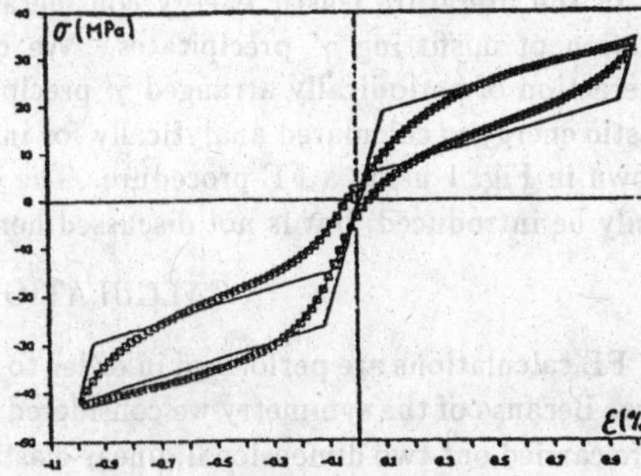


Fig.6 : Linear modelization of tensil-compressive test (R232, $\dot{\epsilon}$=0.01% s$^{-1}$, T=24°c, ±1%)

# ELASTIC INTERACTION IN NICKEL-BASED SUPERALLOYS CONTAINING HIGH VOLUME FRACTION OF $\gamma'$ PHASE

L. MULLER, U. GLATZEL and M. FELLER-KNIEPMEIER
Institut für Metallforschung, Sekr. BH 18
Technische Universität Berlin, Straße des 17. Juni 135
1000 Berlin 12 FRG

## ABSTRACT

The elastic interaction of misfitting precipitates in nickel-based superalloys containing high volume fraction ($\approx$ 63%) of $\gamma'$ phase is investigated. Under zero load the morphologies with lowest elastic energy depend sensitively on the ratio of the elastic moduli of the two phases. Under external load the raft direction with lowest elastic free energy depends sensitively on the misfit, the ratio of the elastic moduli and not on the volume fraction. The local distribution of the elastic energy varies significantly with morphology, but only slightly with the $E$-moduli ratio. The observed morphologies for SRR 99 can only be explained with $E_{\gamma'} > E\gamma$ at high temperatures.

## INTRODUCTION

In the literature elastic energy considerations are often made for the case of zero volume fraction of misfitting $\gamma'$ precipitates. We concentrate our attention on the elastic induced interaction of periodically arranged $\gamma'$ precipitates with high volume fraction of $\gamma'$ phase. The elastic energy is calculated analytically for infinite plates and numerically for the morphologies shown in Fig. 1 using a FE-procedure. The effect of precipitate/matrix interfacial energy can easily be introduced, but is not discussed here.

## CALCULATIONS AND RESULTS

FE calculations are performed in order to obtain stresses and strains in the $\gamma/\gamma'$ microstructure. Because of the symmetry we considered 1/4 of a plane configuration (Fig. 1). Calculations were carried out two dimensional, linear-elastic and isotropic using plane stress elements (stress in the third dimension is zero). Elastic constants and thermal expansion coefficients for the $\gamma$ and $\gamma'$ phase of SRR 99 were taken from table 1 in [1]. The ratio of the elastic moduli $E_{\gamma'}/E\gamma$ is $85MPa/92MPa = 0.924$ and $\nu \approx 0.403$. We also made calculations with the inverse E-moduli ratio 92MPa/85MPa because in general the ratio of the E-moduli between $\gamma$ and $\gamma'$ phase is experimentally not clear at high temperatures. The constrained misfit between $\gamma'$ and $\gamma$ phase, $\delta_c = -2.1 \cdot 10^{-3}$ (lattice parameter of $\gamma'$ phase smaller than that of the matrix) at 1253 K, was taken from measurements of mean dislocation distances in misfit networks by Link and Feller-Kniepmeier [2, 3]. We also assume that the morphological changes occur at constant volume fraction of the two phases at $T = 1253\,K$. The area fraction $A_f = A\gamma'/(A\gamma + A\gamma') = 324/441$ of the two-dimensional calculations represents an approximate volume fraction $V_f \approx A_f^{3/2} = 63\%$.

Under the condition that no external stress is acting on the system the total elastic energy $U$ and the components in the two phases $U^{\gamma}$ and $U^{\gamma'}$ have been calculated.

$$U = U^{\gamma} + U^{\gamma'} = \frac{1}{2} \cdot \sum_{\gamma,\gamma'} \int_V \sum_{i=1}^{3} \sum_{j=1}^{3} \sigma_{ij}\, \epsilon_{ij}\, dV \tag{1}$$

$\sigma_{ij}$ are the components of the stress tensor, $\epsilon_{ij}$ of the strain tensor, $V = l^3$ the volume with $l$ being the system length.

Fig. 2 displays the elastic energy and the distribution of the energy between the $\gamma$ and $\gamma'$ phase of the experimentally relevant morphologies sphere, butterfly distorted cube, infinite plate and cube under zero load. For the $E$-moduli ratio $E_{\gamma'}/E\gamma = 0.924$, spheres have got the lowest and cubes the highest value of $U$. The energy value of the infinite plates is also the self energy of the other morphologies. The interaction energy is the deviation from the energy value of the plates. For equal $E$-moduli and Poisson ratio the total elastic energy $U$ is independent of the morphology for the isotropic case.

The major part of the elastic energy is concentrated in the small matrix channels. Only 10% of the total elastic energy for cubic, 22% for butterfly, 30% for plate and sphere morphology is stored in the $\gamma'$ phase (Fig. 2b). The local elastic energy density fluctuates strongly in the matrix phase for $\gamma'$ spheres, intermediately for $\gamma'$ cubes and distorted cubes, whereas it does not fluctuate for infinite $\gamma'$ plates (Fig. 2c).

For the inverse $E$-moduli ratio $E_{\gamma'}/E\gamma = 1.082$ the sequence of the morphologies with minimum elastic energy is quite different(Fig. 3a). Now cubes are elastically more favoured. But the distribution of the energy between the two phases changes only change slighthly.

For [001] external load the elastic free energy $F_{el}$ of 1/4 of the configuration with edge length $l$ is calculated

$$F_{el} = U - P \tag{2}$$

P is the potential elastic energy of the system:

$$P = \int_{\partial V} \sigma_a\, \epsilon_a\, l\, dA = \sigma_a\, \epsilon_a\, V \tag{3}$$

$\sigma_a$ representing the external load in [001]-direction, $\sigma_a > 0$ tensile load and $\epsilon_a$ the strain in [001]-direction.

For infinite plates parallel and perpendicular to the stress axis, one can calculate analytically the difference $\Delta F_{el}$ of their elastic free energies $F_{el}^{||}$ and $F_{el}^{=}$ (in plane stress):

$$\Delta F_{el} = F_{el}^{||} - F_{el}^{=} = \frac{l_\gamma l_{\gamma'}}{l_{\gamma'} E_{\gamma'} + l_\gamma E_\gamma} \cdot \left[ \frac{1}{2} \sigma_a^2 \,(1 - \nu^2)\, \frac{(E_\gamma - E_{\gamma'})^2}{E_\gamma E_{\gamma'}} - \delta\, \sigma_a\, (1 + \nu)\, (E_{\gamma'} - E_\gamma) \right] \tag{4}$$

$\delta$ is the misfit, $\nu = \nu_\gamma = \nu_{\gamma'}$ the Poisson ratio and $l_i$ the width of phase i (i=$\gamma$,$\gamma'$, see Fig. 1).

The sign of $\Delta F_{el}$ in equation (4) allows to outline a map (Fig. 3b) indicating for various values of $E_{\gamma'}/E\gamma$ and $\sigma_a/(E_{\gamma'} \cdot \delta)$ which infinite plate configuration is the most stable. When a moderate stress is applied the configuration with the lowest $F_{el}$ depends on the second term of equation (4) and $\Delta F_{el} \sim \delta\sigma_a(E_{\gamma'} - E\gamma)$ depends sensitively on the sign of the misfit, the direction of the applied stress and the difference of the elastic moduli. Only for unphysically

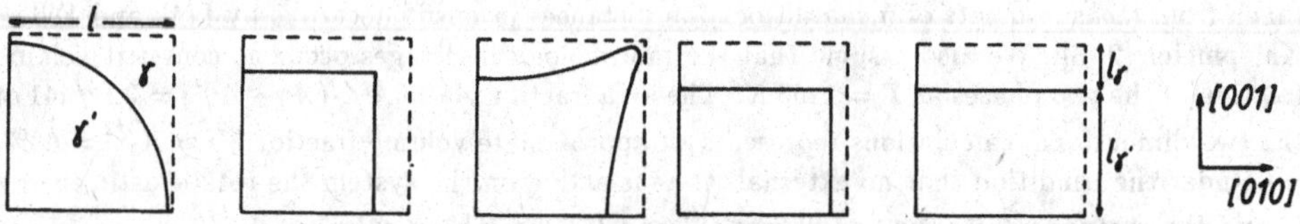


Fig. 1 One quater of the configuration of spheres, cubes, butterfly distorted cubes, finite rectangulars and infinite plates

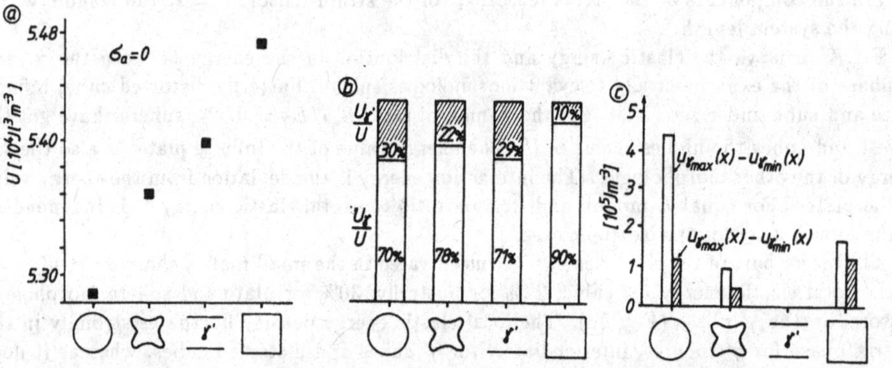


Fig. 2 a) Elastic energy U ($E_{\gamma'}/E_\gamma = 0.924$) b) Partition of U between $\gamma$ and $\gamma'$ phase c) Fluctuation of the local elastic energy density $u(x)$, $u^i_{max} - u^i_{min}$, for the phases i=$\gamma$, $\gamma'$.

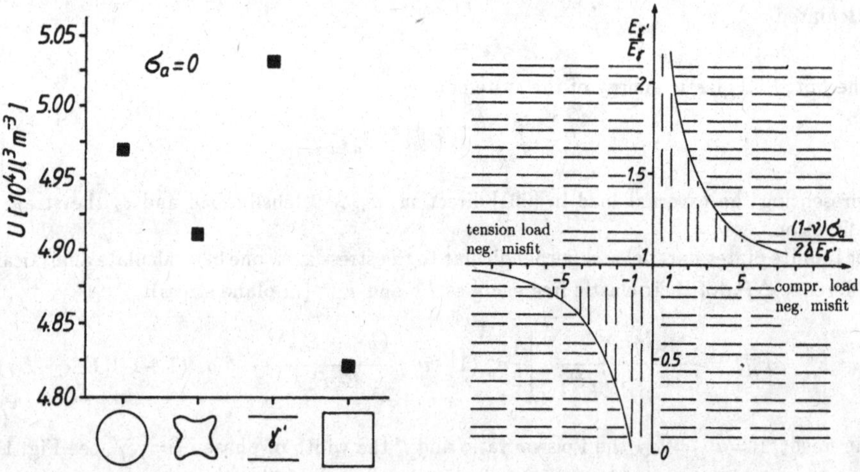


Fig.3 a) Elastic energy U ($E_{\gamma'}/E_\gamma = 1.082$) b) Stability diagram high external stresses (here $\sigma_a \sim 9000\,MPa$) the first term of equation (4) becomes important leading to plates perpendicular to the applied stress. The raft direction is independent of the volume fraction.

The infinite plate configurations take the extremal value of $F_{el}$ for our parameter set. In the isotropic case, for equal elastic constants, $F_{el}$ is independent of the particle shape.

## DISCUSSION

The morphology with the lowest $F_{el}$ depends sensitively on the sign of misfit, the direction of external load and the ratio of the elastic moduli. The morphologies predicted by minimizing the elastic energy under zero load agree with experimental observations for SRR 99 and other negative misfit alloys [4, 5, 2, 3] only with $E_{\gamma'} > E_\gamma$. Under this condition cubes are elastically more favoured than butterfly distorted cubes and spheres for $l \approx 0.5\mu m$. Under zero load and an $E$-moduli ratio of $E_{\gamma'}/E_\gamma = 0.924$ a periodic arrangement of $\gamma'$ spheres is predicted. Considering interfacial energy spheres are even more favoured for all sizes in contradiction to the experimental facts.

The $E$-moduli ratio of typical $\gamma$ and $\gamma'$ material is in general not clear at high temperatures. The $E$-modulus of pure $Ni_3Al$ is less than that of $\gamma$ materials for all temperatures [6]. Gayda and MacKay measured for a special $\gamma'$ material a higher $E$-modulus than for $\gamma$ material at high temperature [7]. There seems to be a need for a wider investigation of the material parameters of $\gamma$ and $\gamma'$ single crystals at high temperatures in order to verify the theoretical predictions.

For an external applied load we calculate that the $\gamma'$ morphologies with the lowest $F_{el}$ are rafts perpendicular to the applied tensile stress for the parameter set of SRR 99 ($E_{\gamma'} > E_\gamma$) in accordance with Gayda and MacKay's Monte Carlo simulation [7] (elastically anisotropic,2D) and Pineau's analytical calculations (isotropic, 3D) [8] for plate and needle configurations in an infinite matrix. Therefore the observed raft direction in SRR 99 can only be predicted with $E_{\gamma'} > E_\gamma$.

Glatzel and Feller-Kniepmeier [1] postulated that $\gamma'$ precipitates grow into the direction of minimum von Mises stresses. For an incompressible liquid (Poisson ratio $\nu = 0.5$) the local energy density is proportional to the square of the von Mises stresses. We show that the inversion of the elastic moduli influnence drastically the sequence of the morphologies with minimum elastic energy, but does not change significantly local energy densities. Therefore our elastic energy considerations do not support the above ad-hoc rule.

## CONCLUSION

Making two-dimensional elasic isotropic calculations we conclude for single crystal nickel-based superalloys with high volume fraction of $\gamma'$ phase:

The morphological changes in SRR 99 and other negative misfit alloys can only be explained with $E_{\gamma'} > E_\gamma$ at high temperature.

The distribution of the elastic energy in the microstructure varies significantly with the morphology, but not with the $E$-moduli ratio.

## REFERENCES


1. U. Glatzel and M. Feller-Kniepmeier, Scripta Met. 23, 1839 (1989)
2. M. Feller-Kniepmeier and T. Link, Z. Metallkunde 76, 283 (1985)
3. M. Feller-Kniepmeier and T. Link, Met. Trans. 20 A, 1233 (1989)
4. S. K. Tien and S. M. Copley, Met. Trans. 2, 543 (1971)
5. R. A. Mac Kay and L. S. Ebert, Scripta Met. 17, 1217 (1983)
6. H.A. Kuhn and H.G. Sockel, Phys. Stat. Sol. (a) 119, 93(1990)
7. J. Gayda and R. A. MacKay, Scripta Met. 23, 1835 (1989)
8. A. Pineau, Acta Met. 24, 559 (1976)

# STRAIN-INDUCED TRANSFORMATION PLASTICITY IN METASTABLE AUSTENITIC STEELS

D. M. PARKS and R. G. STRINGFELLOW
Massachusetts Institute of Technology, Cambridge, MA 02139, USA

## ABSTRACT

We have recently developed [1] a constitutive model of strain-induced transformation plasticity, the phenomenon by which metastable austenite transforms into martensite in association with on-going plastic slip in each phase. The kinetics of the transformation, and the effects of the transformation on mechanical behavior, are highly sensitive to temperature, extent of plastic strain, and the triaxiality of stress. Basic features of the model are reviewed. The model has been incorporated into the ABAQUS finite element program. We review applications of the model to two problems of elastic-plastic deformation: tensile necking and crack tip blunting. These solutions give insight into the remarkable levels of transformation toughening ($\sigma_y \geq 1300$ MPa; $K_{Jc} \geq 380$ MPa$\sqrt{\text{m}}$) which have been measured in these materials [2].

## MODEL

Within a certain range of temperatures, the transformation of austenite to martensite occurs subsequent to slip-based yielding in the parent austenite. This phenomenon has been termed *strain-induced transformation plasticity (SITP)*. Olson and Cohen [3] developed a model for strain-induced martensite evolution based on the assumption that nucleation of martensite occurs predominantly at shear-band intersections produced by slip in the austenite. Following [1], the rate of increase in the volume fraction of martensite, $\dot{f}$, is proportional to the rate of increase in the number of martensitic embryos per unit austenite volume, $\dot{N}_a$:

$$\dot{f} = (1 - f)\,\bar{v}_a \dot{N}_a\,, \tag{1}$$

where $\bar{v}_a$ is the average volume per martensitic unit. $N_a$ equals the number of shear-band intersections per unit volume, $N^I = \hat{N}^I(\Theta, \gamma_a)$, times the probability of transformation, $P = \hat{P}(\Theta, \Sigma)$, where $\Theta$ is a normalized temperature, $\gamma_a$ is the plastic shear strain in the austenite, and $\Sigma$ measures the triaxiality of the stress state ($\Sigma = -p/\sqrt{3}\bar{\tau}$, where $p$ is pressure and $\bar{\tau}$ is equivalent shear stress). The nucleation probability $P$ is cast in the form of a cumulative distribution function,

$$P(g) = \frac{1}{\sqrt{2\pi}} \int_{-\infty}^{g} exp\left[-\frac{1}{2}\left(\frac{g' - \bar{g}}{s_g}\right)^2\right] dg'\,, \tag{2}$$

where $\bar{g}$ and $s_g$ are the dimensionless mean and standard deviation of a probability distribution function, $g$, given by

$$g = g_0 - g_1\Theta + g_2\Sigma \,, \tag{3}$$

where $g_0$, $g_1$ and $g_2$ are dimensionless constants. Under isothermal conditions, $\dot{f}$ can be expressed as:

$$\dot{f} = A_f(P,\gamma_a)\dot{\gamma}_a + B_f(\gamma_a)\dot{\Sigma} \,. \tag{4}$$

The parameter $g$ (a normalized measure of the chemical driving force for transformation) determines the extent of transformation under different conditions: $g$ decreases with increasing $\Theta$ but increases with increasing $\Sigma$. Considering the high gradients of $\Sigma$ found near a crack tip, accounting for the $\Sigma$-dependence of the transformation process is essential for understanding (transformation) toughness enhancement.

We also proposed an isotropic model for inelastic deformation in SITP [1]. Its principal feature is a self-consistent decomposition of inelastic straining into the sum of terms due to martensite nucleation (shape strain) and due to slip in both phases. The model fits available tensile and compressive martensite evolution data and tensile stress-strain data [1, 3, 4] well.

## RESULTS

The model has been implemented as a material subroutine in the finite element code ABAQUS. As an illustration of the stress-state sensitivity of transformation, we consider homogeneous deformation under five states of stress, each characterized by a different level of $\Sigma$, Fig. 1. The evolution of $f$ with equivalent plastic strain, $\overline{\gamma}^p$, (bottom of Fig. 1) differs widely over the range of $\Sigma$ between plane-strain tension and compression. Curves at the top of Fig. 1 show how dramatically the formation of a harder martensite phase alters the equivalent stress/strain behavior. The large shape strain produced by the rapid transformation in plane strain tension causes the early strain-softening in that curve.

We applied the model to analyses of tensile necking and crack tip blunting. In both of these problems we compare behavior of a transforming (T) material with that of an identical, but nontransforming (NT), austenite. Material parameters are taken from [4].

The tensile necking problem is based on a round cross section of initial radius $R_0$ containing a small imperfection at one axial location. Fig. 2 shows the distribution of scalar fields $\overline{\gamma}^p$, $\Sigma$, and $f$ for T and NT materials, each loaded to a reduced section radius of $R/R_0 = 0.734$. Because of the increased $\Sigma$ at the neck center, $P$ is locally enhanced. The contours of $f$ are a direct result of the combined effect of the strain and $\Sigma$ fields. Fig. 3 shows the radial variation of $\overline{\gamma}^p$, $\Sigma$ and $f$ across the minimum section at the same neck strain of $R/R_0 = 0.734$. Because of the gradient of $\Sigma$, a corresponding gradient in transformation (and transformation hardening) raises $\overline{\tau}$ at the neck center, compared to the NT case, lowering the ratio $\Sigma$. The gradient hardening redistributes $\overline{\gamma}^p$ across the section more uniformly, lowering the curvature of the neck profile in accordance with experimental observations [4]. At lower temperatures, extreme transformation hardening leads to propagation of the neck down the length of the specimen (cold drawing) and significant increases in ductility. Fig. 4 shows the strain history of $f$ and $\overline{\tau}$ for a material point at the neck center. For reference, we have also plotted the corresponding curves for load histories of constant $\Sigma$ in uniaxial and plane strain tension. As $\Sigma_{\text{neck}}$ increases beyond $1/3$, $f_{\text{neck}}$ evolves more rapidly than in uniaxial tension, eventually crossing the reference plane strain history as $\Sigma$ exceeds $1/\sqrt{3}$. The increasing $\Sigma$ loading history shows a steadier hardening rate than either of the reference (constant $\Sigma$) curves.

We next consider the plane strain blunting in mode I small-scale yielding of an initially sharp crack. The analysis is patterned after McMeeking [5], with the asymptotic linear elastic $K_I$-field displacements applied as remote boundary conditions.

Fig. 5a shows the extent of transformation near the blunted crack. Because $\overline{\gamma}^p$ decays rapidly ahead of the crack tip, the extent of the transformed region is limited; the peak level of $f$ occurs nearer to the tip than the peak in $\Sigma$, Fig 5b. Fig 6 shows pairs of

$(\overline{\tau}, \overline{\gamma}^p)$ at ligament locations ahead of the crack; they trace the stress/strain history of a material point as $K_I$ increases. Locations of some of these points are given in terms of normalized distance ahead the crack tip, $(R/b)$. Transformation dramatically increases the hardening rate in the blunting zone directly ahead of the crack tip.

## DISCUSSION

Locally high strain *and* high triaxiality levels at the center of a tensile neck or ahead of a crack tip preferentially promote formation of local zones of enhanced martensite content, causing substantial local increases in (transformation) hardening. In the absence of transformation, high strength, low hardening, steels often fail by a process of shear instability. Enhanced transformation hardening can *delocalize* deformation, stabilizing flow and retarding the fialure process.

Increases in tensile ductility with transformation (*e.g.*, [4]) conform with this simple notion. More significantly, recent fracture toughness experiments [2] further support this concept. For NT materials, low toughness and classic "zig-zag" patterns of crack propagation associated with flow localization were observed. With crack tip transformation, however, the crack often branched, with substantial crack blunting (high toughness) prior to advance of one of the branches. The inhibition of shear localization and the deflection of the crack front away from the forward direction appear to be major factors contributing to the observed high toughnesses.

## REFERENCES


1. Stringfellow, R., Parks, D. and Olson, G., submitted to *Acta Metall.*, 1991.
2. Stavehaug, F., Ph.D. Thesis, MIT, Cambridge, MA, USA, 1990.
3. Olson, G. and Cohen, M., *Met. Trans.* **6A**, 791, 1975.
4. Young, C.–C., Ph.D. Thesis, MIT, Cambridge, MA, USA, 1988.
5. McMeeking, R., *J. Mech. Phys. Solids*, **25**, 357, 1977.


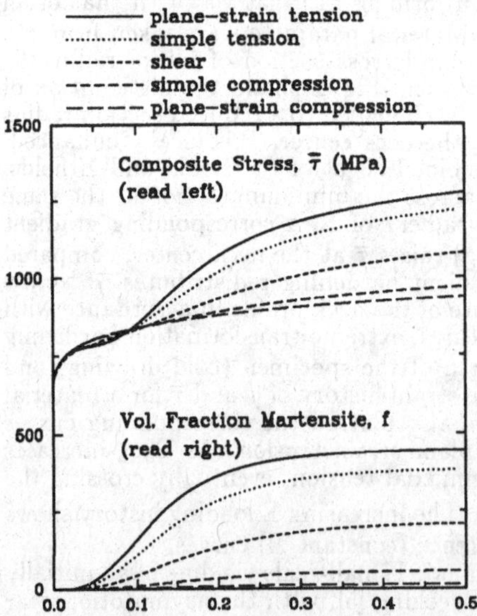


Figure 1 One element test results. Model data taken from Young [4].

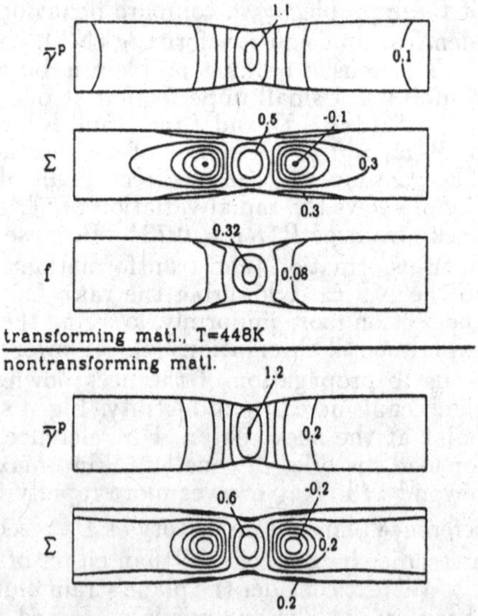


Figure 2 Comparison of scalar fields for $\overline{\gamma}^p$, $\Sigma$ and $f$ at equal minimum section reductions, $R/R_0 = 0.734$.

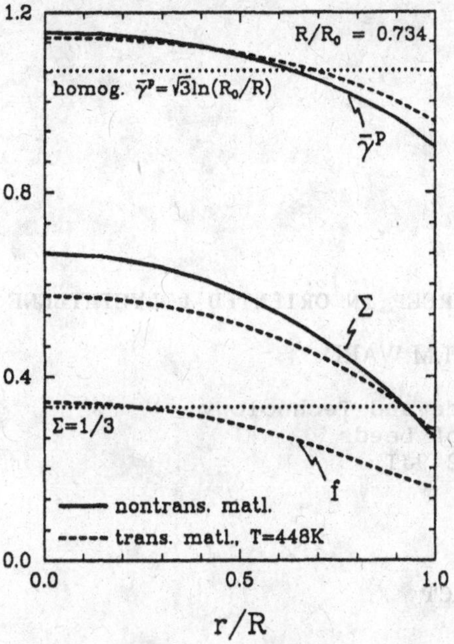


Figure 3 Variation of $\bar{\gamma}^p$, $\Sigma$ and $f$ across the min. section at reduction $R/R_0 = 0.734$.

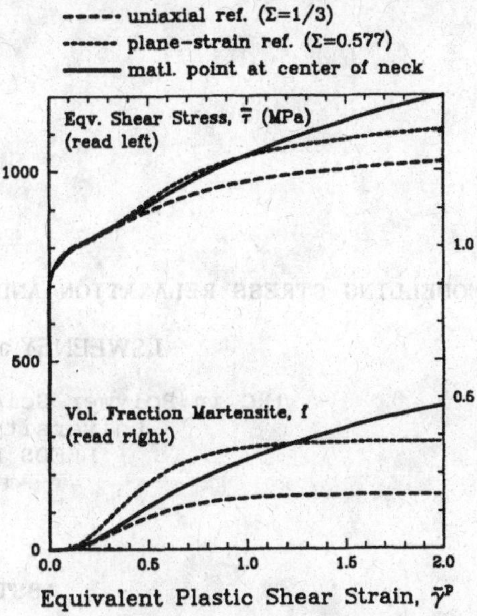


Figure 4 Comparison of the history of $f$ and $\bar{\tau}$ at the center of the neck along with reference uniaxial and plane strain curves.

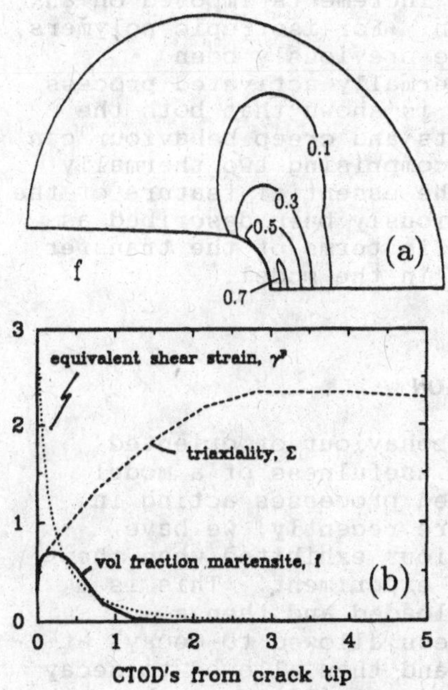


Figure 5 (a). Contours of f. (b). Variation of scalar fields $\bar{\gamma}^p$, $\Sigma$, and $f$ ahead of the crack tip.

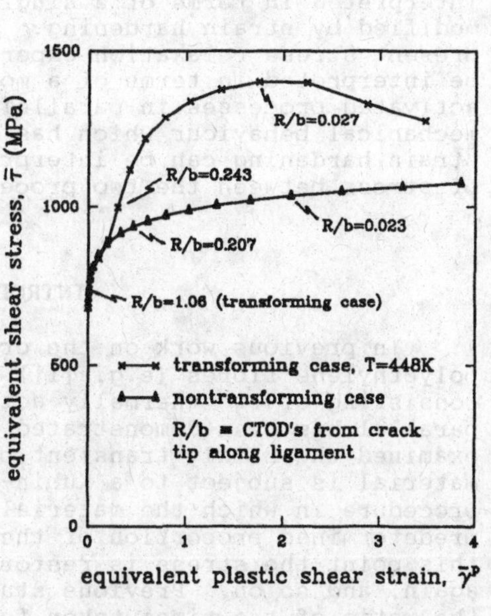


Figure 6 Pairs of $(\bar{\tau}, \bar{\gamma}^p)$ at nodal locations ahead of and approaching the crack.

# MODELLING STRESS RELAXATION AND CREEP IN ORIENTED POLYETHYLENE

J.SWEENEY and I.M.WARD

IRC in Polymer Science and Technology
University of Leeds
LEEDS LS2 9JT
UK

## ABSTRACT

The stress relaxation behaviour of high modulus oriented polyethylene fibre has been studied with regard to the response to successive small strain increments imposed on an initial relatively large deformation. For isotropic polymers, the results of such experiments have previously been interpreted in terms of a single thermally activated process modified by strain hardening. It is shown that both the present stress relaxation experiments and creep behaviour can be interpreted in terms of a model comprising two thermally activated processes in parallel. The essential feature of the mechanical behaviour which has previously been described as strain hardening can be interpreted in terms of the transfer of stress between the two processes in the model.

## INTRODUCTION

In previous work on the creep behaviour of oriented polyethylene fibres (e.g. [1]) the usefulness of a model consisting of two thermally activated processes acting in parallel has been demonstrated. More recently, we have examined the mainly transient behaviour exhibited when the material is subject to a Kubin-type experiment. This is a procedure in which the material is loaded and then a predetermined proportion of the stress allowed to decay; at this point the stress is restored, and then allowed to decay again, and so on. Previous studies (e.g. [2]) have shown that the ratio of the times taken for the stress to decay at successive steps is approximately constant. The two-process model and the step stress relaxation experiment are illustrated respectively in figures 1 and 2.

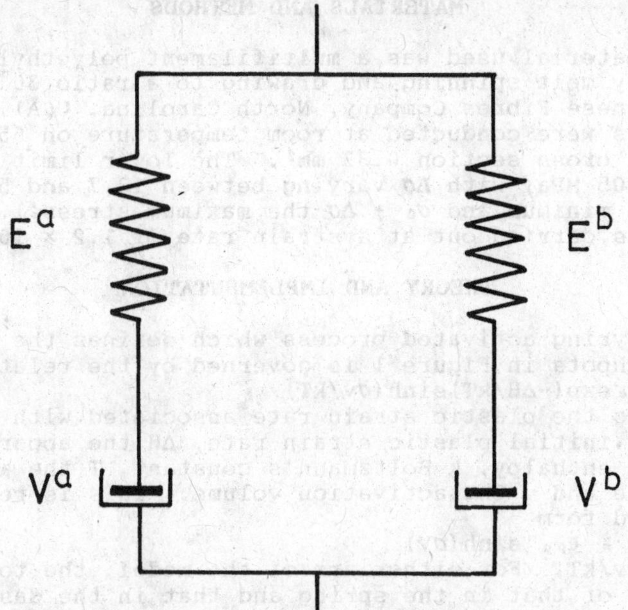


FIGURE 1. Schematic of two-process model.

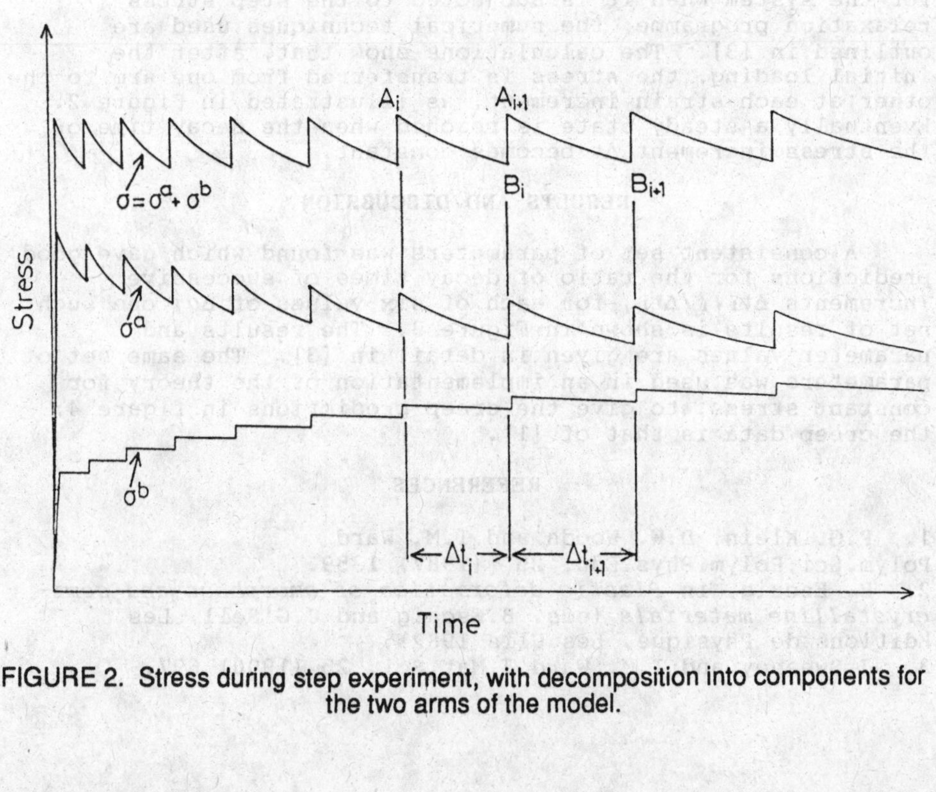


FIGURE 2. Stress during step experiment, with decomposition into components for the two arms of the model.

## MATERIALS AND METHODS

The material used was a multifilament polyethylene fibre, prepared by melt spinning and drawing to a ratio 30 (Alathon 7050, Celanese Fibres Company, North Carolina, USA). Experiments were conducted at room temperature on 65 mm gauge lengths of cross section 0.33 $mm^2$. The lower limit of stress was $\sigma_0$ = 205 MPa, with $\Delta\sigma$ varying between 13.7 and 58.4 MPa ($\sigma_0$ is the minimum and $\sigma_0 + \Delta\sigma$ the maximum stress.). The fast loading was carried out at a strain rate of $1.2 \times 10^{-3}$ $s^{-1}$.

## THEORY AND IMPLEMENTATION

The Eyring activated process which defines the behaviours of the dashpots in Figure 1 is governed by the relation

$$\dot{\epsilon}_p = \dot{\epsilon}_{po} \exp(-\Delta H/kT) \sinh(\sigma v/kT)$$

where $\dot{\epsilon}_p$ is the plastic strain rate associated with the stress $\sigma$, $\dot{\epsilon}_{po}$ the initial plastic strain rate, $\Delta H$ the apparent activation enthalpy, k Boltzmann's constant, T the absolute temperature and v the activation volume. This is rewritten in abbreviated form

$$\dot{\epsilon}_p = \dot{\epsilon}_{po}' \sinh(\sigma V) \qquad (1)$$

where V = v/kT. For either arm of the model, the total strain is the sum of that in the spring and that in the dashpot. Hence the total strain rate for arm a is given by

$$\dot{\epsilon}^a = \dot{\epsilon}_{po}^{a}{}' \sinh(\sigma^a V^a) + \dot{\sigma}^a/E^a \qquad (2)$$

and a similar expression applies for $\dot{\epsilon}^b$ and $\sigma^b$. The total stress $\sigma$ is given by $\sigma = \sigma^a + \sigma^b$. This quantity is calculated for the system when it is subjected to the step stress relaxation programme; the numerical techniques used are outlined in [3]. The calculations show that, after the initial loading, the stress is transferred from one arm to the other at each strain increment, as illustrated in Figure 2. Eventually a steady state is reached when the decay time of the stress increment $\Delta t$ becomes constant.

## RESULTS AND DISCUSSION

A consistent set of parameters was found which gave good predictions for the ratio of decay times of successive increments $\Delta t_{i+1}/\Delta t_i$, for each of six values of $\Delta\sigma$; one such set of results is shown in Figure 3. The results and parameter values are given in detail in [3]. The same set of parameters was used in an implementation of the theory for constant stress, to give the creep predictions in figure 4; the creep data is that of [1].

## REFERENCES


1. P.G. Klein, D.W. Woods and I.M. Ward Polym.Sci.Polym.Phys.Edn. **25** (1987) 1359.
2. B. Escaig, in *Plastic deformation of amorphous and semi-crystalline materials* (eds. B.Escaig and C.G'Sell, Les Editions de Physique, Les Ulis 1982).
3. J.Sweeney and I.M. Ward J.Mat.Sci. **25** (1990) 697.

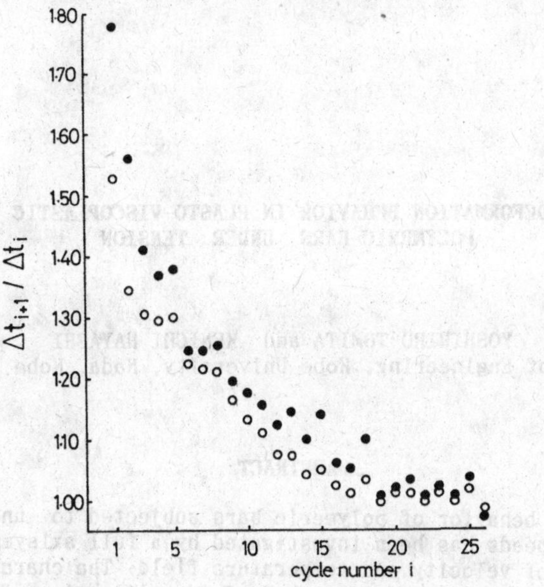


FIGURE 3. Decay time ratios when $\sigma_o$ = 205 MPa and $\Delta\sigma$ = 58 MPa; (●) observed (O) predicted.

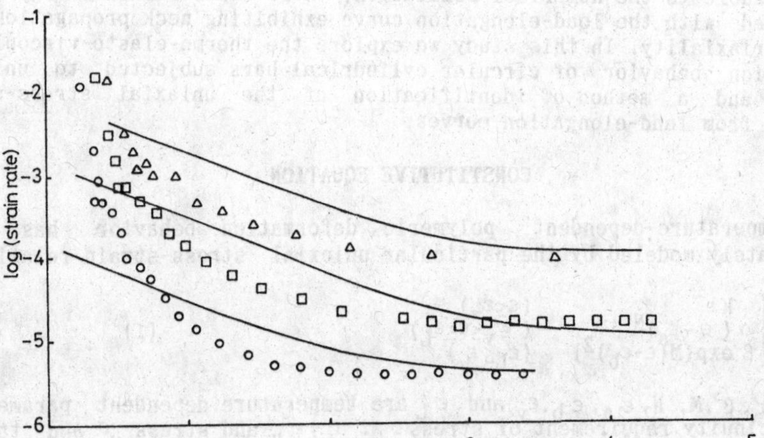


FIGURE 4. Predictions (lines) of creep behaviour at (O) 150, (□) 190 and (Δ) 280 MPa.

# DEFORMATION BEHAVIOR IN ELASTO-VISCOPLASTIC POLYMERIC BARS UNDER TENSION

YOSHIHIRO TOMITA and KENICHI HAYASHI
Faculty of Engineering, Kobe University, Nada, Kobe, Japan

## ABSTRACT

The deformation behavior of polymeric bars subjected to uniaxial tension with different speeds has been investigated by a full axisymmetric finite-element method of velocity and temperature field. The characteristics of the deformation behavior and a method of identification of the uniaxial stress-strain relation from load-elongation curves are explored.

## INTRODUCTION

Due to the strain rate and temperature sensitivity of polymeric materials, the instability propagation behaviors manifest different features associated with the boundary conditions applied[1,2]. Furthermore, although the determination of the uniaxial stress-strain relation[3] is indispensable to the numerical simulation, there still remain some problems associated with the load-elongation curve exhibiting neck propagation and stress triaxiality. In this study we explore the thermo-elasto-viscoplastic deformation behavior of circular cylindrical bars subjected to uniaxial tension and a method of identification of the uniaxial stress-strain relation from load-elongation curves.

## CONSTITUTIVE EQUATION

The temperature-dependent polymeric deformation behavior has been approximately modeled by the particular uniaxial stress-strain relation

$$\sigma_o = \begin{cases} K\varepsilon & (\varepsilon \leq \varepsilon_y) \\ \alpha(\varepsilon-\varepsilon_a)^N & (\varepsilon_y \leq \varepsilon \leq \varepsilon_L) \\ \beta \exp\{M(\varepsilon-\varepsilon_b)^2\} & (\varepsilon_L \leq \varepsilon), \end{cases} \tag{1}$$

where $K$, $\alpha$, $\beta$, $M$, $N$, $\varepsilon_a$, $\varepsilon_b$, $\varepsilon_y$, and $\varepsilon_L$ are temperature dependent parameters. The continuity requirement of stress $\sigma$ at $\varepsilon = \varepsilon_y$ and stress $\sigma$ and tangent modulus $d\sigma/d\varepsilon$ at $\varepsilon = \varepsilon_L$ determines $\alpha = K\varepsilon_y/(\varepsilon_y - \varepsilon_a)^N$, $M = N/\{2(\varepsilon_L - \varepsilon_b)(\varepsilon_L - \varepsilon_a)\}$, $\beta = \alpha(\varepsilon_L - \varepsilon_a)^N/\exp\{M(\varepsilon_L - \varepsilon_b)^2\}$. Consequently, the independent parameters reduce to six. The uniaxial stress-strain relations for different temperatures and the temperature dependency of the parameters are depicted in Fig.1 (a). The uniaxial stress-strain relation (1) is simply generalized to strain-rate-sensitive materials by

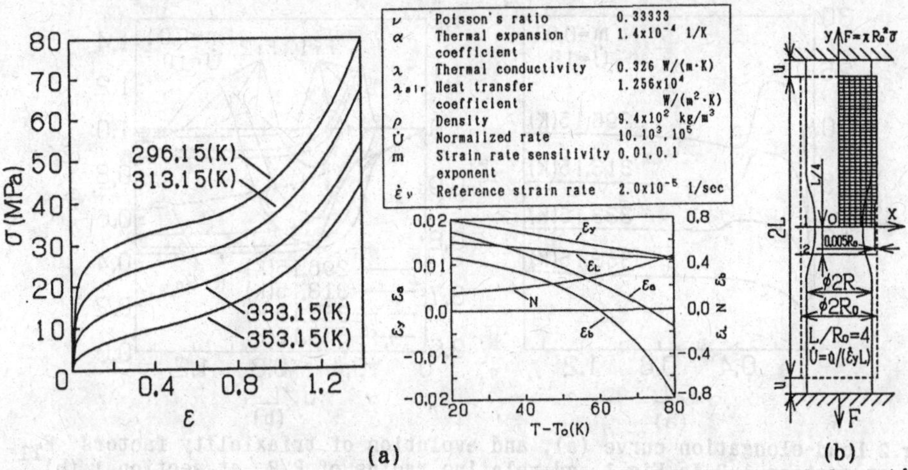


| | | |
|---|---|---|
| $\nu$ | Poisson's ratio | 0.33333 |
| $\alpha$ | Thermal expansion coefficient | $1.4\times10^{-4}$ 1/K |
| $\lambda$ | Thermal conductivity | 0.326 W/(m·K) |
| $\lambda_{air}$ | Heat transfer coefficient | $1.256\times10^{4}$ W/(m²·K) |
| $\rho$ | Density | $9.4\times10^{2}$ kg/m³ |
| $\dot{U}$ | Normalized rate | 10, $10^{3}$, $10^{5}$ |
| m | Strain rate sensitivity exponent | 0.01, 0.1 |
| $\dot{\varepsilon}_y$ | Reference strain rate | $2.0\times10^{-5}$ 1/sec |

Fig.1 Uniaxial true stress-natural strain curves for isothermal deformation (a) and analytical model and finite element discretization(b).

$$\sigma = \sigma_o(1+\dot{\varepsilon}/\dot{\varepsilon}_y)^m, \qquad (2)$$

where $\sigma_o$ is defined by Eq.(1) and m is the strain rate sensitivity exponent. The relation (2) is generalized to the three-dimensional constitutive equation relating stress rate ,strain rate, and temperature change [4].

## METHOD OF ANALYSIS AND COMPUTATIONAL MODEL

The numerical procedure employs the well-established finite element method[4] for thermo-elasto-viscoplastic deformation problem. Due to the symmetry of the deformation, one-half of the bar with finite-element discretization shown in Fig.1(b) is investigated. Each quadrilateral shown in the vertical section consists of four crossed triangular elements. Heat generated by viscoplastic work is assumed to be discharged through convection to air at the side surface. The remaining surfaces are assumed to be adiabatic boundaries. In all calculations reported here, the employed material and computational parameters are summarized in Fig.1(a) and the temperature dependency of the elasticity modulus and specific heat are expressed by $E=1045-10.63(T-T_o)$(MPa), and $c=1.55-25.1/(T-T_o-108)$(KJ/KgK), respectively. T is the absolute temperature and $T_o$=273.15K is the reference temperature.

## RESULTS AND DISCUSSION

Figure 2(a) shows the load versus elongation under isothermal deformation with $\dot{U}$=10 under four different temperatures. Except for the high-temperature case (353.15K), the deformations are essentially the same in that the force attains the maximum and then drops to the local minimum and with further straining, asymptotically approaches the steady-state value. Figure 2(b) shows the evolutions of the radius of the specimen $R/R_o$ and the triaxiality factors defined by $F_T=\tilde{\sigma}_{yy}/\tilde{\bar{\sigma}}$ at the cross sections 1 and 2 shown in Fig.1. $\tilde{\sigma}_{yy}$ and $\tilde{\bar{\sigma}}$ are the average axial stress and representative stress acting on the specific cross section, respectively. Since the deformation states are different, the two steady-state triaxiality factors have some

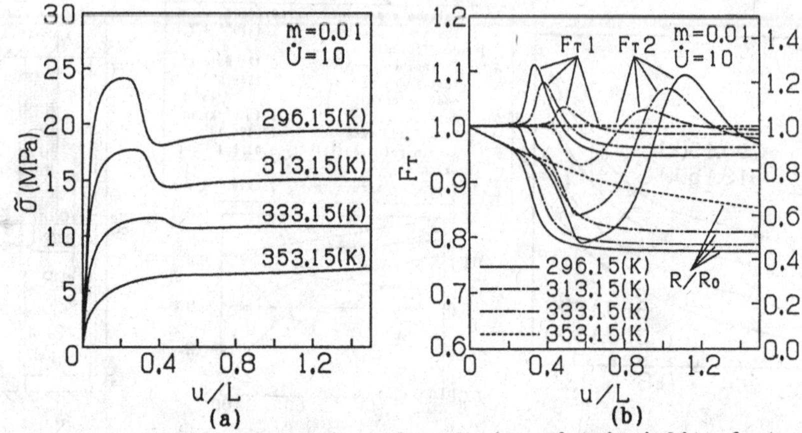


Fig.2 Load-elongation curve (a), and evolution of triaxiality factors $F_{T1}$, $F_{T2}$ at section 1,2 in Fig.1 and relative radius of $R/R_0$ at section 1 (b) for quasi-static isothermal deformation.

discrepancy. The triaxiality factor at the cross section far removed from section 1 behaves in much the same way as that of $F_{T2}$. Relative radius of the specimen $R/R_0$ at section 1 decreases uniformly as the deformation proceeds and then drops with the neck formation.

Next, the effect of the strain rate and temperature sensitivity has been investigated for the strain rate sensitivity parameter m=0.01, 0.05 and for the nominal deformation rate U=10, $10^3$, $10^5$(the corresponding strain rate is 0.0002, 0.02, 2/s) with material constants shown in Fig.1. Figure 3 shows the load versus elongation, the evolution of the triaxiality factor and the relative radius. An adiabatic response in which ninety-five percent of the viscoplastic work is converted to heat and causes a corresponding rise in temperature without heat transfer inside the specimen is also depicted in this figure as (AD). Results from the adiabatic case show a completely different feature than the conductive case (with heat transfer) which has a relatively low deformation rate. Corresponding deformation behaviors exhibit not any more steady-state. As discussed in ref.[1], increase in the strain rate sensitivity m and the deformation rate

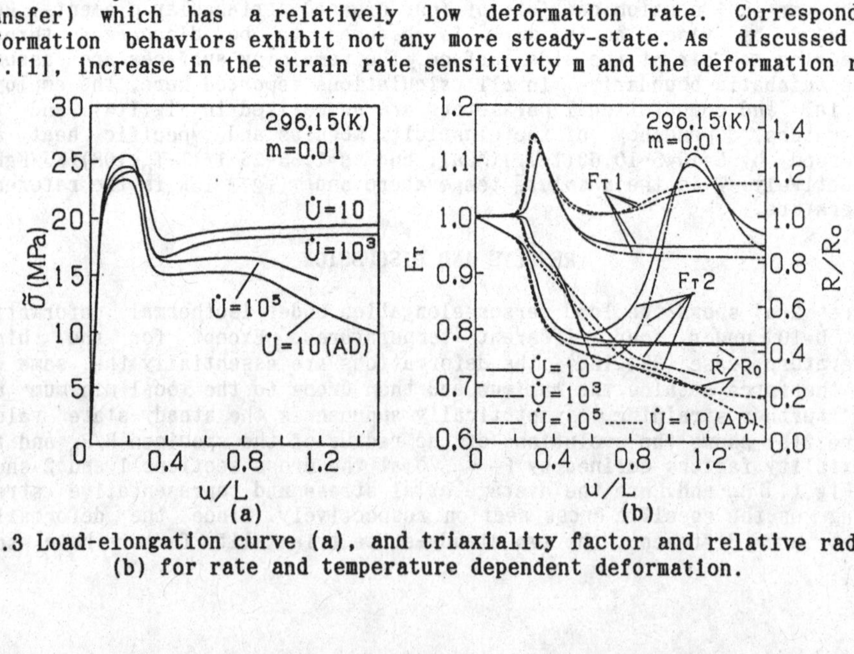


Fig.3 Load-elongation curve (a), and triaxiality factor and relative radius (b) for rate and temperature dependent deformation.

may not cause a substantial change in the elongation at which load maximum occurs. Furthermore, since the strain rate sensitivity stabilizes the deformation, after maximum load point, load-elongation curves exhibit gentle decreases accompanying low triaxiality factors for the material with higher m value. The influence of the rate of deformation on the triaxiality factors can be observed in their values at the later stages of deformation. According to the present results, the effect of heat induced by irreversible work is substantial. Due to the thermal softening, steady-state neck propagations may not be observed in high rate of deformation and in adiabatic deformation. The temperature distribution clearly captured the propagation of the heat source, which in turn obstructs the local heating and causes the gentle distribution of the heat to the unnecked parts for relatively slow deformation. A different picture has been seen in cases for a high rate of deformation and adiabatic deformation, where the high temperature area is fixed in a specific area.

## IDENTIFICATION OF UNIAXIAL STRESS-STRAIN RELATION

Consider the procedure of identification of the uniaxial stress-strain relation from the load-elongation curve shown in Fig.2(a). The uniaxial stress-strain curve is tentatively assumed to be expressed by Eq.(1). Then, there are six independent parameters: $K$, $\varepsilon_y$, $\varepsilon_L$, $\varepsilon_a$, $\varepsilon_b$, and $N$. The remaining parameters are determined by the continuity requirement. Furthermore, $K$ and $\varepsilon_y$ are the elasticity modulus and yield strain, respectively. They can be determined from the load elongation curve or directly by experiment for the elastic range.

Here, written symbolically, the load is the function of elongation $u$ and independent parameters $x_i$ which respectively stand for $x_1=\varepsilon_L$, $x_2=\varepsilon_a$, $x_3=\varepsilon_b$, and $x_4=N$ as $\tilde{\sigma}=\tilde{\sigma}(u,x_i)$. The independent parameters $x_i$ are determined such that the predicted load-elongation curve coincides with the given one. The real value of the parameters $x_i$ is expressed by the sum of the initial estimations of $x_{io}$ and perturbation $\Delta x_i$ as $x_i=x_{io}+\Delta x_i$. Substituting $x_i$ into $\tilde{\sigma}(u,x_i)$ and taking the Taylor expansion with respect to $\Delta x_i$ yields the linear expression. In determination of $\Delta x_i$, the least squares method is employed with the specific values of $\tilde{\sigma}$ and $\partial\tilde{\sigma}/\partial x_i$ at the elongation $u$ corresponding to the load maximum and minimum, and the $u/L=$ 0.6, 0.8, 1.2. The calculation is performed by introducing trial values $x_{io}$ and obtaining the value $\Delta x_i$. Due to the nonlinearity of the problem, some trial and error in determining the estimated values is required. While rather poorly estimated values are introduced, quite gentle and monotonical convergence is obtained. A detailed discussion will appear in the full-length version of this paper.

## REFERENCES

1. P.Tugcu and K.W.Neale, Analysis of neck propagation in polymeric fibers including the effect of viscoplasticity. Trans. ASME. J. Eng.Mat. Tech., 1988, **110**, 395-400.

2. P.Tugcu and K.W.Neale.,Cold drawing of polymers with rate and temperature dependent properties, Int. J. Mech. Sci., 1990, **32**, 405-416.

3. G.G'Sell and J.J.Jonas, Determination of the plastic behaviour of solid polymers at constant true strain rate, J. Mat. Sci., 1979, **14**, 583-591.

4. Y.Tomita, A.Shindo and T.Sasayama, Plane strain tension of thermo-elasto-viscoplastic blocks, Int. J. Mech. Sci., 1990, **32**, 613-622.

# THERMODYNAMIC CONSIDERATIONS AND THERMAL EFFECTS

# ON THE THERMODYNAMICS OF STRESS RATE IN THE EVOLUTION OF BACK STRESS IN VISCOPLASTICITY

ALAN D. FREED*, JEAN-LOUIS CHABOCHE** and KEVIN P. WALKER***

* National Aeronautics and Space Administration
Lewis Research Center
Cleveland, OH 44135, USA

** Office National d'Etudes et de Recherches Aérospatiales
92322 Châtillon, Cedex, France

*** Engineering Science Software, Inc.
Smithfield, RI 02917, USA

## ABSTRACT

A thermodynamic foundation using the concept of internal state variables is presented for the kinematic description of a viscoplastic material. Three different evolution equations for the back stress are considered. The first is that of classical, nonlinear, kinematic hardening. The other two include a contribution that is linear in stress rate. Choosing an appropriate change in variables can remove this stress rate dependence. As a result, one of these two models is shown to be equivalent to the classical, nonlinear, kinematic hardening model; while the other is a new model—one which seems to have favorable characteristics for representing ratchetting behavior. All three models are thermodynamically admissible.

## INTRODUCTION

Let us consider a complementary (Gibbs) free energy for the thermodynamic potential function, *i.e.*

$$\psi = \Psi[T, \sigma_{uv}, \beta_{uv}] \quad , \quad \beta_{ii} = 0\,, \tag{1}$$

where square brackets $[\cdot]$ are used to denote 'function of'. The temperature $T$ and stress $\sigma_{ij}$ are external variables whose variations can, in principle, be controlled by an observer. However, the evolution of the internal state variable $\beta_{ij}$—associated with kinematic hardening—cannot be controlled by an observer; it is a material response. This variable must therefore evolve as a function of state. Furthermore, if this kinematic variable is to remain invariant under transformation to another kinematic variable, then its evolution will, in general, also depend upon the rates of change in the external variables [1]. For simplicity, let us consider the following equation for the evolution of internal state, *i.e.*

$$\dot{\beta}_{ij} = \chi_{ij}[T, S_{uv}, \beta_{uv}] + (N/2H)\,\dot{S}_{ij}\,, \tag{2}$$

where $S_{ij} = \sigma_{ij} - 1/3\,\sigma_{kk}\delta_{ij}$ is the deviatoric stress with $\delta_{ij}$ representing the Kronecker delta, and where $N$ and $H$ are material constants. In general, a term that is linear in $\dot{T}$ may also be introduced into (2) [2], but it is ignored in this paper for the purpose of simplification. Notice that $\beta_{ij}$, and therefore $\dot{\chi}_{ij}$, are deviatoric by definition. The function $\dot{\chi}_{ij}$ describes the irreversible contribution to the evolution of $\beta_{ij}$.

Given (1) and (2), one can derive [1,2] relationships that govern the entropy,

$$S = \frac{-\partial\Psi}{\partial T}\,, \tag{3}$$

the strain,

$$\varepsilon_{ij} = \frac{-\partial\Psi}{\partial\sigma_{ij}} - \frac{N}{2H}\frac{\partial\Psi}{\partial\beta_{ij}} + \varepsilon^{p}_{ij} \quad , \quad \varepsilon^{p}_{ii} = 0\,, \tag{4}$$

and the intrinsic dissipation,

$$\sigma_{ij}\dot{\varepsilon}^{p}_{ij} - \frac{\partial\Psi}{\partial\beta_{ij}}\,\dot{\chi}_{ij} \geq 0\,, \tag{5}$$

where $\varepsilon^{p}_{ij}$ is the plastic strain, which is deviatoric. It evolves according to a separate evolution equation (13).

## VISCOPLASTIC THEORY

If we assume that the complementary free energy (1) is given by

$$\begin{aligned}\Psi \;=\; & \overbrace{\frac{-1}{4\mu}S_{ij}S_{ij} - \frac{1}{18\kappa}\sigma_{ii}\sigma_{jj} - \alpha\,\Delta T\,\sigma_{ii} - S_0\Delta T - C\left\{T - T_0\left(1 + \ln\left[\frac{T}{T_0}\right]\right)\right\}}^{\text{thermoelastic contribution}} + \\ & + \underbrace{H\beta_{ij}\beta_{ij} + N\left(\frac{N}{4H}S_{ij}S_{ij} - \beta_{ij}S_{ij} + \Lambda[S_{uv},\beta_{uv}]\right)}_{\text{viscoplastic contribution}}\,,\end{aligned} \tag{6}$$

then we obtain—from (3) and (4)—the constitutive equations for an isotropic Hookean material ; they are:

$$\Delta S = \alpha\,\sigma_{ii} + (C/T)\Delta T\,, \tag{7}$$

$$\sigma_{ii} = 3\kappa\,(\varepsilon_{ii} - \alpha\,\Delta T\,\delta_{ii})\,, \tag{8}$$

and

$$S_{ij} = 2\mu\,(E_{ij} - \varepsilon^{p}_{ij})\,, \tag{9}$$

provided that

$$\frac{\partial\Lambda}{\partial S_{ij}} = \frac{-N}{2H}\frac{\partial\Lambda}{\partial\beta_{ij}}\,. \tag{10}$$

Here $\mu$ and $\kappa$ are the elastic shear and bulk moduli, $\alpha$ is the coefficient of thermal expansion, $C$ is the specific heat, $S_0$ and $T_0$ are the initial values of entropy and temperature with $\Delta S = S - S_0$ and $\Delta T = T - T_0$, and $E_{ij} = \varepsilon_{ij} - 1/3\,\varepsilon_{kk}\delta_{ij}$ is the deviatoric strain. The first three terms in the viscoplastic contribution to $\Psi$ are introduced to remove unwanted cross products that would otherwise appear in (9) because of (4) [2]. One also obtains

$$\frac{\partial\Psi}{\partial\beta_{ij}} = 2H\left(\beta_{ij} - \frac{N}{2H}S_{ij} - \frac{\partial\Lambda}{\partial S_{ij}}\right) = 2H\left(\chi_{ij} - \frac{\partial\Lambda}{\partial S_{ij}}\right)\,, \tag{11}$$

which defines the thermodynamic force conjugate to $\beta_{ij}$. As a consequence, (5) becomes

$$\sigma_{ij}\dot{\varepsilon}^{p}_{ij} - 2H\left(\chi_{ij} - \frac{\partial\Lambda}{\partial S_{ij}}\right)\dot{\chi}_{ij} \geq 0\,, \tag{12}$$

which describes the intrinsic dissipation properties of our material. The function $\Lambda$, which is constrained by (10), is introduced into $\Psi$ to affect the intrinsic dissipation (12); its form is model dependent.

For the evolution of plastic strain, we shall consider that

$$\dot{\varepsilon}^p_{ij} = \tfrac{1}{2}\,\|\dot{\varepsilon}^p\|\,\frac{S_{ij} - B_{ij}}{\|S - B\|}\,, \tag{13}$$

where the back stress,

$$B_{ij} = 2H\beta_{ij} = 2H\chi_{ij} + NS_{ij}\,, \tag{14}$$

accounts for kinematic behavior, with $H > 0$ being its modulus and $0 \leq N < 1$. The norms used are defined by $\|I\| = \sqrt{\tfrac{1}{2} I_{ij} I_{ij}}$ where $I_{ij}$ is any deviatoric stress-like quantity, and by $\|J\| = \sqrt{2 J_{ij} J_{ij}}$ where $J_{ij}$ is any deviatoric strain-like quantity. These von Mises norms are scaled for shear.

## Model I

The classical, nonlinear, kinematic hardening model [3,4] has an evolution of internal state described by

$$\dot{\chi}_{ij} = \dot{\varepsilon}^p_{ij} - \frac{H\chi_{ij}}{L}\,\|\dot{\varepsilon}^p\|\,, \tag{15}$$

where $L$ is the limiting state for the back stress $B_{ij}$. There is no stress rate term in the evolution of $\beta_{ij}$ for this model, *i.e.* $N = 0$ in (2), and consequently

$$\dot{B}_{ij} = 2H\left(\dot{\varepsilon}^p_{ij} - \frac{B_{ij}}{2L}\,\|\dot{\varepsilon}^p\|\right) \tag{16}$$

describes its evolution for the back stress. Because $N = 0$, the last three terms in the complementary free energy (6) do not contribute to the dissipation in this model; its intrinsic dissipation (12) is therefore given by

$$\left(\|S - B\| + \frac{\|B\|^2}{L}\right)\|\dot{\varepsilon}^p\| \geq 0\,, \tag{17}$$

and it is always satisfied. Hence, Model I is thermodynamically admissible.

## Model II

This model uses the same description for the irreversible evolution of internal state that Model I uses, *viz.* (15). In addition, it assumes that $0 < N < 1$, which introduces a reversible attribute to the evolution of internal state. As a result,

$$\dot{B}_{ij} = 2H\left(\dot{\varepsilon}^p_{ij} - \frac{B_{ij} - NS_{ij}}{2L}\,\|\dot{\varepsilon}^p\|\right) + N\dot{S}_{ij} \tag{18}$$

describes this model's evolution for the back stress. Model II reduces to Model I when $N = 0$.

As Lubliner [1] discussed in his paper, one can always (in principle) transform from an internal state variable whose evolution contains terms that are linear in the external variable rates, to another internal state variable where there are no external variable rates present in the evolution equation. This is accomplished in our case by considering the linear transformation

$$X_{ij} = \frac{2H}{1-N}\,\chi_{ij} = 2H'\chi_{ij}\,, \tag{19}$$

where the variable $X_{ij}$ is a back stress, but different from $B_{ij}$, with $H' = H/(1-N)$ as its associated hardening modulus. This transformation enables the flow law (13) to be rewritten in an equivalent form as

$$\dot{\varepsilon}^p_{ij} = \tfrac{1}{2}\,\|\dot{\varepsilon}^p\|\,\frac{S_{ij} - X_{ij}}{\|S - X\|}\,. \tag{20}$$

Likewise, it allows the evolution equation for back stress (18) to be rewritten in the equivalent form

$$\dot{X}_{ij} = 2H'\left(\dot{\varepsilon}^p_{ij} - \frac{X_{ij}}{2L'}\,\|\dot{\varepsilon}^p\|\right), \tag{21}$$

where $L' = L/(1-N)$ is the limiting state for the back stress $X_{ij}$. Upon comparing (20) and (21) with (13) and (16), one observes that they are identical in mathematical structure, but with different values for their constants. These differences lead to differences in the intrinsic dissipation properties of the material [2].

From the perspective of material science [5], the physically correct, internal, state variable has an evolution equation with no $\dot{S}_{ij}$ or $\dot{T}$ dependence. This coincides with the experimental observation that an instantaneous change in either stress or temperature does not produce an instantaneous change in a material's internal structure, *i.e.* its dislocation structure. Consequently, the back stress $X_{ij}$ is the physically correct back stress for Model II, and it is referred to as the physical back stress. (The back stress $B_{ij}$ is the physical back stress of Model I.)

A description of Model II's dissipation response requires knowledge of its material function $\Lambda$, which is found in the expression for the complementary free energy (6). This function is evaluated by combining (12), (14), (15), (19), and (20), and then using $\Lambda$ to cancel out those terms in the dissipation inequality that can become negative valued. This process leads to the differential equation

$$\frac{\partial \Lambda}{\partial S_{ij}} = \frac{N}{1-N}\left(\frac{N}{2H} S_{ij} - \beta_{ij}\right) , \tag{22}$$

which is the simplest of several possible solutions. This equation can be integrated, in conjunction with the constraint given in (10), to produce

$$\Lambda = \frac{N}{1-N}\left(\frac{N}{4H} S_{ij} S_{ij} - \beta_{ij} S_{ij} + \frac{H}{N} \beta_{ij} \beta_{ij}\right) , \tag{23}$$

which when substituted into (6) defines $\Psi$ for Model II; it is the basic constitutive equation of this particular model. It follows then that the intrinsic dissipation (12) for Model II is given by the inequality

$$\left(\| \boldsymbol{S} - \boldsymbol{X} \| + \frac{\| \boldsymbol{X} \|^2}{L'}\right) \| \dot{\epsilon}^p \| \geq 0 , \tag{24}$$

and it is always satisfied. Hence, Model II is thermodynamically admissible. Notice the similarity between (17) and (24).

### Model III

This model considers the irreversible evolution of internal state to be described by

$$\dot{\chi}_{ij} = \dot{\epsilon}^p_{ij} - \frac{H \beta_{ij}}{L} \| \dot{\epsilon}^p \| , \tag{25}$$

which differs from (15) by the exchange of $\chi_{ij}$ with $\beta_{ij}$ in the dynamic recovery term. Like Model II, this model takes $0 < N < 1$, and therefore

$$\dot{B}_{ij} = 2H\left(\dot{\epsilon}^p_{ij} - \frac{B_{ij}}{2L} \| \dot{\epsilon}^p \|\right) + N \dot{S}_{ij} \tag{26}$$

describes the evolution for the back stress $B_{ij}$. Equation 26 was first proposed by Ramaswamy *et al.* [6]. Notice that the only difference between (16) and (26) for Models I and III is the presence of the term $N\dot{S}_{ij}$ found in (26). Model III reduces to Model I when $N = 0$.

Using the same linear transformation that was used in Model II, *i.e.* (19), one determines that Model III has the same transformed flow law (20) as Model II, but it has a different evolution equation for the physical back stress $X_{ij}$, *viz.*

$$\dot{X}_{ij} = \frac{2H}{1-N}\left(\dot{\epsilon}^p_{ij} - \frac{(1-N) X_{ij} + N S_{ij}}{2L} \| \dot{\epsilon}^p \|\right) . \tag{27}$$

To the best of our knowledge, this ia a new expression for the evolution of the physical back stress. Here $N$ proportions the dynamic recovery between the physical back stress $X_{ij}$ and the applied stress $S_{ij}$.

We set out to derive the dissipation response of Model III as we did for Model II. We begin by considering a decomposition

$$\Lambda = \Lambda_1 + \Lambda_2 , \tag{28}$$

where $\Lambda_1$ is taken to be given by (23), *i.e.* $\Lambda_1$ is the $\Lambda$ of Model II. The remaining function is evaluated by combining (12), (14), (19), (20), and (25), and then using $\Lambda_2$ to cancel out those remaining terms in the dissipation inequality that can become negative valued. Because of the constraint equation (10), one obtains two partial differential equations, *viz.*

$$\frac{\partial \Lambda_2}{\partial S_{ij}} = \frac{N}{2L(1-N)} S_{k\ell}\beta_{k\ell} \left( \frac{S_{ij} - 2H\beta_{ij}}{\|S - 2H\beta\|} - \frac{H\beta_{ij}}{L} \right)^{-1} , \tag{29}$$

$$\frac{\partial \Lambda_2}{\partial \beta_{ij}} = \frac{-H}{L(1-N)} S_{k\ell}\beta_{k\ell} \left( \frac{S_{ij} - 2H\beta_{ij}}{\|S - 2H\beta\|} - \frac{H\beta_{ij}}{L} \right)^{-1} , \tag{30}$$

which when integrated will lead to the complementary free energy $\Psi$ for Model III, *i.e.* its fundamental constitutive equation. We have not integrated these equations. For our purpose, it is sufficient to know only that $\Lambda$ exists. The intrinsic dissipation (12) for Model III is therefore given by the inequality

$$\left( \|S - X\| + \frac{\|B\|^2}{L(1-N)} \right) \|\dot{\epsilon}^p\| \geq 0 , \tag{31}$$

and it is always satisfied. Hence, Model III is also thermodynamically admissible.

A preliminary study of these three models [7] indicates that Model III may be able to predict realistic ratchetting behavior; whereas, Model I (and therefore Model II) is known to overpredict ratchetting behavior [8]. Continued research is required to better understand the predictive capabilities of Model III.

## ACKNOWLEDGEMENT

This work is an outgrowth of a personnel exchange program between the agencies of NASA and ONERA.

## REFERENCES

[1] Lubliner, J., On the Structure of the Rate Equations of Materials with Internal Variables, *Acta Mech.*, 1973, **17**, 109–19.

[2] Freed, A.D., Chaboche, J.-L., and Walker, K.P., A Viscoplastic Theory with Thermodynamic Considerations, in press: *Acta Mech.*

[3] Armstrong, P.J. and Frederick, C.O., A Mathematical Representation of the Multiaxial Bauschinger Effect, CEGB, RD/B/N731, Berkeley Nuclear Laboratories, 1966.

[4] Chaboche, J.-L., Viscoplastic Constitutive Equations for the Description of Cyclic and Anisotropic Behavior of Metals, *Bull. Acad. Pol. Sci., Ser. Sci. Tech.*, 1977, **25**, 33[39]–42[48].

[5] Nix, W.D., Gibeling, J.C., and Hughes, D.A., Time-Dependent Deformation of Metals, *Metall. Trans. A*, 1985, **16A**, 2215–26.

[6] Ramaswamy, V.G., Stouffer, D.C., and Laflen, J. H., A Unified Constitutive Model for the Inelastic Response of René 80 at Temperatures Between 538°C and 982°C, *J. Engr. Mater. Tech.*, 1990, **112**, 280–6.

[7] Freed, A.D. and Walker, K.P., Model Development in Viscoplastic Ratchetting, NASA TM 102509, April, 1990.

[8] Inoue, T., Yoshida, F., Ohno, N., Kawai, M., Niitsu, Y., and Imatani, S., Plasticity-Creep Behavior of 2¼Cr-1Mo Steel at 600°C in Multiaxial Stress State, in: *Structural Design for Elevated Temperature Environments—Creep, Ratchet, Fatigue, and Fracture*, ed. C. Becht IV *et al.*, ASME, New York, 1989, PVP **163**, pp. 101–7.

# CONSTITUTIVE EQUATIONS FOR DEFORMATIONS INDUCED BY INTERFACIAL MOTIONS

M. BUISSON, E. PATOOR, M. BERVEILLER
Laboratoire de Physique et Mécanique des Matériaux,UA CNRS 1215,
Institut Supérieur de Génie Mécanique et Productique,
Université de Metz, Ile du Saulcy, 57045 Metz Cedex, France

## ABSTRACT

Interfacial motions in case of reorientation phenomena are taken into account in the modelisation of the global behaviour through a kinematical description of the strain induced by interfacial displacements and through the determination of the thermodynamic driving force.The calculation of such forces is performed using Eshelby's energy momentum formalism ; the interface-movement is taken into account via the concept of an inclusion with moving boundary and the macroscopic flow rule is included in the context of standard generalized media.

## INTRODUCTION

Interfacial motions are at the origin of many physical mechanisms which are responsible for inelastic behaviour. Twinning, phase transformations or some aspects of the classical plasticity like shear banding are related to this phenomena.

In this work, interfacial motions are taken into account in the case of reorientation phenomena between martensitic variants. We propose a general form for the macroscopic behaviour law through a kinematical description of strains and through the determination of the thermodynamic forces which are responsible for the motion of interfaces. The calculation of such forces is performed using Eshelby's energy momentum formalism, we assume that the interfaces between variants are compatible (e.g. no internal stresses are induced, displacements and strain energy density are continuous) and the stresses and strains uniform and constant inside each variant. The macroscopic flow rule is obtained in the context of generalized media ; we observe also the parts that some microstructural parameters are playing.

## TOPOLOGY AND KINEMATIC

We consider a representative macroscopic volume V constituted of N martensitic variants. We assimilate each variant with a moving inclusion (geometric volume $V_I$, I = 1 to N ; boundary $\partial V_I$) embedded in the volume V and bordered by others variants $V_J$ so that we can reconstitute $\partial V_I$ as an union (symbol **U**) of surfaces elements $\partial V^{IJ}$ :

$$\partial V_I = \bigcup_{J=1}^{N} \partial V^{IJ} \tag{1}$$

where $\partial V^{IJ}$ is the surface intersection of $\partial V_I$ with $\partial V_J$ ; we consider in this union that $\partial V^{IJ}$ is zero if $V_I$ and $V_J$ are not adjoining. The mobility of interfaces is taken into account via the concept of the inclusion-problems with moving boundaries [1], [2]; we use the technic of time-derivation $\delta/\delta t$ with respect to the field $\underline{W}$ defined by the eigen-velocities of the interfaces

$$(a)\ \frac{\delta}{\delta t}(\ ) = \frac{\partial}{\partial t}(\ ) + \underline{W}.\underline{\text{grad}}(\ )\ ,\ (b)\ \frac{\delta}{\delta t}\iiint_{V_I(t)}(\ )\,d\underline{x} = \iiint_{V_I(t)}\frac{\partial}{\partial t}(\ )\,d\underline{x} + \iint_{\partial V_I}(\ )\,\underline{W}.\underline{n}\,d\underline{a} \tag{2}$$

From (2b) and the use of the divergence theorem, we deduce the rates $\delta/\delta t$ of the volumic fraction $f^I$ ($f^I = V^I/V$) of variant $V^I$

$$\frac{\delta}{\delta t} f^I = \frac{1}{V}\frac{\delta V_I}{\delta t} = \iiint_{V_I} W_{i,i}(\underline{x},t)\,d\underline{x} = \iint_{\partial V_I} W_i(\underline{a},t)\,n_i(\underline{a},t)\,d\underline{a} \quad \text{with } V_I = \iiint_{V_I} d\underline{x} \tag{3}$$

and the global variation $1/V^2\ \delta V/\delta t$ as negligible.
Assuming that $\underline{W}$ is approximately uniform for each section $\partial V^{IJ}$, we define the velocity $\underline{W}^{IJ}$ and surface $\underline{S}^{IJ}$

$$\underline{W}(\underline{a},t) = \underline{W}^{IJ}(t) \text{ for } \underline{a} \in \partial V^{IJ}\ ,\ S_i^{IJ}(t) = \iint_{\partial V^{IJ}} n_i(\underline{a},t)\,d\underline{a} \tag{4}$$

with $S_i^{IJ} = -S_i^{JI}$ due to the convention of the unit external normal $\underline{n}$ and $S_i^{IJ}=0$ if $V_I$ and $V_J$ are not adjoining or (I)=(J) or $\partial V^{IJ}$ is on the external boundary $\partial V$. Finally (3) becomes

$$\frac{\delta f^I}{\delta t} = \sum_{J=1}^{N} \frac{1}{V}\iint_{\partial V^{IJ}} W_i\, n_i\, d\underline{a} = \sum_{J=1}^{N} W_i^{IJ} S_i^{IJ}/V \tag{5}$$

here we can note that $(S_i^{IJ}\ S_i^{IJ})^{1/2}/V$ is the volumic-fraction of the interface (IJ) which constitutes a notable microstructural parameter. The macroscopic total strain $\mathbf{E}(t)$ is the volumic mean-value of the total local strains $\varepsilon(\underline{x},t)$ and is written as a discrete summation with respect to the N volumes $V_I$;respectively

$$\frac{\delta}{\delta t}E_{ij}(t) = \frac{\delta}{\delta t}\left[\frac{1}{V}\sum_{I=1}^{N}\iiint_{V_I(t)}\varepsilon_{ij}(\underline{x},t)\,d\underline{x}\right] = \frac{1}{V}\sum_{I}\left[\iiint_{V_I}\frac{\partial}{\partial t}\varepsilon_{ij}(\underline{x},t)d\underline{x} + \iint_{\partial V_I}\varepsilon_{ij}(\underline{a},t)\,W_k(\underline{a},t)\,n_k(\underline{a},t)\,d\underline{a}\right] \tag{6}$$

In the specific case of the reorientation of martensitic variants we assume for each geometrical volume $V_I(t)$

$$\varepsilon(\underline{x},t) = \varepsilon^I \text{ and } \partial\varepsilon(\underline{x},t)/\partial t = 0 \quad \text{for } \underline{x} \in V_I(t) \tag{7}$$

due to piece-wize uniformity and time-constancy of physical and mechanical properties of the variants. From (7,3,4,5) and using the antisymmetry of $S_i^{IJ}$, (6) may be written

$$\frac{\delta}{\delta t}E=\sum_{I=1}^{N}\varepsilon^{I}\frac{\delta f^{I}}{\delta t}=\sum_{I=1}^{N-1}\sum_{J>I}^{N}(\varepsilon^{I}-\varepsilon^{J})W_k^{IJ}S_k^{IJ}/V \tag{8}$$

which describes the rate of macroscopic strain with the internal interfaces velocities $W^{IJ}$ and the associated jumps of total strains $(\varepsilon^{I} - \varepsilon^{J})$.

## INTERFACIAL DRIVING FORCE - BEHAVIOUR LAW

The energy momentum tensor and associated driving force are deduced directly from Eshelby [3] and Hill [4] works ; that is we examine the rate of total energy $\delta\phi^{IJ}/\delta t$ when a variant $V_I$ (inclusion) developps to the detriment of an other variant $V_J$ with an associated virtual displacement $W^{IJ}\,\delta t$ and thermodynamical generalized resolved force $F^{IJ}$ via :

$$\frac{\delta}{\delta t}\phi^{IJ}=-F^{IJ}W_k^{IJ}S_k^{IK} \tag{9}$$

In the case of reorientation let us precise that we first assume the compatibility of variants : the Cauchy-stress is uniform ($\sigma_{ij}(\underline{x},t)=\Sigma_{ij}(t)$ where $\underline{\Sigma}$ is the macroscopic external applied stress). Secondly, we assume homogeneous elasticity : the jumps of strain-energy density between variants become zero and the jumps $\varepsilon^{I} - \varepsilon^{J}$ of total strains in (8) may be replaced by $\varepsilon^{tI} - \varepsilon^{tJ}$ where $\varepsilon^{t}$ is the stress-free transformation strain. Finally, the hypothesis of uniformity inside each variant lead us to have zero values for the gradients of inelastic strains. All these hypothesis associated with formula (51) and (60) of [4] give

$$\frac{\delta}{\delta t}\phi^{IJ}=\Sigma_{IJ}(\varepsilon^{tJ}-\varepsilon^{tI})W_k^{IJ}S_k^{IJ}\quad\text{so that from (9)}\quad F^{IJ}=\Sigma_{ij}(\varepsilon_{ij}^{tI}-\varepsilon_{ij}^{tJ}) \tag{10}$$

The description of the macroscopic behaviour law depends first on the intrinsic behaviour of the interfaces. In this work we consider that the interface will move if the resolved force $F^{IJ}$ comes to a certain critical yield force $F_c^{IJ}$ which depends in a quite complicated way from the internal structural parameters and the volumic fraction $f^{K}$. Consequently we define a yield surface $G^{IJ}(\Sigma_{ij},f^{K})=F^{IJ}-F_c^{IJ}(f^{K})$. From (10) we observe that $\partial G^{IJ}/\partial\Sigma_{ij}=\varepsilon_{ij}^{tI}-\varepsilon_{ij}^{tJ}$ so that (8) becomes

$$\frac{\delta}{\delta t}E=\sum_{I=1}^{N-1}\sum_{J>I}^{N}\frac{\partial G^{IJ}}{\partial\Sigma}W_k^{IJ}S_k^{IJ}/V \tag{11}$$

Using the consistency condition $\delta G^{IJ}/\delta t = 0 = \delta F^{IJ}/\delta t - \delta F_c^{IJ}/\delta t$ we write $W_k^{IJ}\,S_k^{IJ}$ in term of $\delta\Sigma/\delta t$. From (10) we deduce $\delta F^{IJ}/\delta t$ and by composition we have successively from (5) :

$$\frac{\delta F_c^{IJ}}{\delta t} = \sum_{K=1}^{N} \frac{\partial F_c^{IJ}}{\partial f^K}\cdot\frac{\delta f^K}{\delta t} = \sum_{K=1}^{N}\sum_{L=1}^{N} \frac{\partial F_c^{IJ}}{\partial f^K} W_i^{KL}\ S_i^{KL}/V$$

then using the antisymmetry of $S_i^{IJ}$ in an equivalent way as for deriving (8) we have

$$\frac{\delta F_c^{IJ}}{\delta t} = \sum_{K=1}^{N-1}\sum_{L>K}^{N}\left(\frac{\partial}{\partial f^K}F_c^{IJ} - \frac{\partial}{\partial f^L}F_c^{IJ}\right) W_i^{KL}\ S_i^{KL}/V$$

so that the consistency condition becomes

$$(\varepsilon_{ij}^{tI} - \varepsilon_{ij}^{tJ})\frac{\delta}{\delta t}\Sigma_{ij} = \sum_{K=1}^{N-1}\sum_{L>K}^{N}\left(\frac{\partial}{\partial f^K}F_c^{IJ} - \frac{\partial}{\partial f^L}F_c^{IJ}\right) W_l^{KL}\ S_l^{KL}/V$$

Let us define the square-matrix **A** with upper-indices(IJ),(KL)by

$$A^{(IJ)(KL)} = \left(\frac{\partial}{\partial f^K}F_c^{IJ} - \frac{\partial}{\partial f^L}F_c^{IJ}\right) \qquad (J>I, L>K)$$

that we assume generally inversible (each pair (IJ) and (KL) stands for an indice-number which identifies the interface (IJ) and (KL)). Eliminating $W^{KL}$, $S^{KL}$ and $\varepsilon^{tI} - \varepsilon^{tJ}$ we obtain a generalized macroscopic flow rule

$$\frac{\delta}{\delta t}E_{ij} = \sum_{I=1}^{N-1}\sum_{J>I}^{N}\sum_{K=1}^{N-1}\sum_{L>K}^{N} \frac{\partial G^{IJ}}{\partial \Sigma_{ij}}\left[A^{-1}\right]^{(IJ)(KL)}\frac{\partial G^{KL}}{\partial \Sigma_{kl}}\frac{\delta}{\delta t}\Sigma_{kl}$$

In this results, the matrix $A^{(IJ)(KL)}$ is analogous to a hardening matrix from which we take into account the influence of the evolution of the substructures on the critical stresses.

The knowledge of this matrix is lying with a precise description of the material structure (topology of interfaces) and with their critical intrinsic resolved stresses.

## REFERENCES


1. SABAR, H.,BERVEILLER, M.and BUISSON, M., Problème d'inclusion à frontière mobile,C.R.Acad Sci Paris,1990, t.310 série II, pp. 447-452.

2. SABAR, H., BUISSON, M.and BERVEILLER, M., The Inhomogeneous and plastic Inclusion Problem with Moving Boundary, Inter. Journal of Plasticity, éd Pergamon, in press.

3. ESHELBY, J.D., Inelastic Behaviour of Solids, éd : M.F. KAN NINEN, W.F. ADLER, A.R. ROSENFIELD and R.I. JOFFEE, p 77, Mac Graw Hill, New York 1970.

4. HILL, R., Energy-Momentum Tensors in Elastostatics : some reflections on the general theory, J. Mech. Phys. Solids, 1986,Vol 34, N° 3, pp. 305-317.

# EXPERIMENTAL ANALYSIS OF THERMOMECHANICAL COUPLING BY INFRA-RED THERMOGRAPHY

A. CHRYSOCHOOS and J.C. DUPRE
Laboratoire de Mécanique Générale des Milieux Continus
Université Montpellier II, Place E. Bataillon 34095 (France)

## ABSTRACT

In the first part we consider the energy balance form in case of thermoelastoplastic material. The relationships between the dissipation, the stored energy of cold work and the state variables are reviewed through the heat equation. Then, an experimental set-up consisting of an infra-red camera placed directly on the crosshead of a testing machine is introduced. Its numerization system allows recording thermal pictures on the surface of the sample during the deformation process. The numerical data analysis allows us to estimate the mechanical heat sources : plastic dissipation and the isentropic term due to thermoelastic effects. In the case of homogeneous mechanical tests performed at room temperature under several uniaxial loading paths, energy balances are shown. Finally, in the case of heterogeneous experiments, the performances of such an approach are tested to observe shear band evolution or to detect localized necking.

## ENERGY BALANCE FOR ELASTOPLASTIC MATERIALS

For homogeneous and uniaxial solicitations when the elastic deformation remains small, most classical kinematical approaches [1] lead to an additive decomposition of the strain rate tensor and of the volumic deformation energy. In the case of a quasi-static deformation process :

$$w_{ext} = w_e + w_a \tag{1}$$

We note $w_e$ is the volumic elastic energy and $w_a$ is the volumic anelastic energy.

Classical results of Thermodynamics of Irreversible Processes are used [2]. A homogeneous volume element of the sample is characterized by a set of n+1 state variables.
Let us take $T = \alpha_0$ as the absolute temperature, $\varepsilon_e = \alpha_1$ as the elastic strain, and $(\alpha_j)_{2\leq j\leq n}$ as the set of n-1 variables describing the hardening state. The volumic Helmholtz free energy is classically decomposed in [3]:

$$\psi ( T, \varepsilon_e, \alpha_j) = \psi_e ( T, \varepsilon_e) + \psi_s ( T, \alpha_j)_{2\leq j\leq n} \qquad (2)$$

The term $\psi_e$ is due to the thermoelastic part of the transformation, the second one $\psi_s$ is due to the strain hardening.

During our quasi-static tests, the temperature variations remain small. In such conditions it has been shown [1] that internal energy variations related to the hardening (stored energy of cold work) correspond to free energy variation. This leads to the assumption that the (small) temperature variations induced by deformation do not modify the hardening state.

Let $w_d$ be the volumic amount of heat associated with the intrinsic dissipation, and let $w_s$ be the stored energy of cold work per unit volume of sample then:

$$w_a = w_d + w_s \qquad (3)$$

Then, if we note $w_{is}$ the isentropic term due to thermoelastic effects, the heat conduction equation can be simplified as:

$$\rho C_\alpha \dot{\theta} - k \Delta\theta = \rho \left(\dot{w}_d - \dot{w}_{is}\right) \qquad (4)$$

The mechanical heat source will be noted $\dot{w}_{ch}$ with:

$$w_{ch} = w_d - w_{is} \qquad (5)$$

The energy balance form can be summed up in eqns (1), (3) and (5).

## EXPERIMENTAL SET-UP

The experimental set-up essentially consists of a 100 kN Screw Machine and an Infra-red Camera. The device is equipped with the numerization system. A micro-computer drives the data storing and processing, and provides the visualization of the "thermal scenes". The camera is placed directly on the crosshead of the Testing Machine . The thermoregulation is carried out carefully. On fig.1 a basic sketch of the experimental set-up can be seen.

The thermal pictures are matrices of 256 lines per 180 columns numerized on 12 bits. The numerization system allows recording at the beginning of each line both signals like "thermal level" and "thermal range" characterizing the state of the camera, and six electrical signals through two-stage amplifiers, like load and deformation signals, room temperature, etc...

A special warming target has been made to calibrate the variations in video signal. The response of the IR detector is strongly non linear even in the vicinity of thermal equilibrium. The calibration law is approximated to a quadratic law.

A numerical treatment of eqn (4), using the thermal data, leads to the determination of mechanical heat sources.

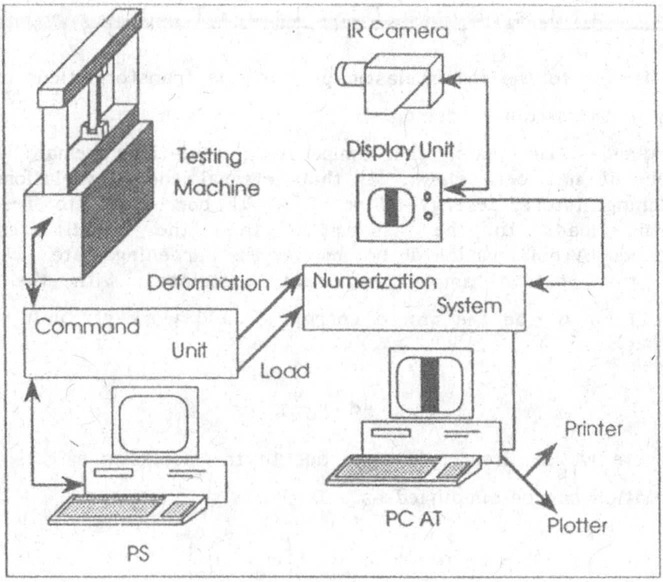


Figure 1. Basic sketch of the experimental set -up.

## EXAMPLES OF APPLICATION

*Energetical Validity Of Thermomechanical Behaviour Laws:*

To describe the mechanical behaviour the 'Generalized Standard Material' formalism can be used. In general, models can be developed which give results in good agreement with experiments. Constituted of a superposition of classical models taking into account the isotropic, linear and nonlinear kinematic hardening rules, these models can predict energy balance evolution, which can be compared with the experimental one. On fig.2, a cyclic test has been made on duralumin. Four stages of deformation have been used (±.5%,±1.%±1.5%,±2.%), each stage being made of four cycles. The experimental evolution of the energy balance is presented. A release of stored energy can be seen after deformation.

*Dissipated Phenomena Associated With Mechanical Instabilities:*

Because of the diffusivity, the thermal response of materials generally gives hazy information on the heat source distribution. Recent progresses made in data storing and processing allow us to associate to "thermal scenes", "energetical charts" on the surface of the sample during its transformation. In the case of heterogeneous phenomena, these technics have been tested to detect instabilities. For instance during a compressive test, shear bands appear, emanating from the hole made in the center of a parallelipipedic and thin plate. On fig.3, the distribution of dissipation (divided by the specific heat), is shown.

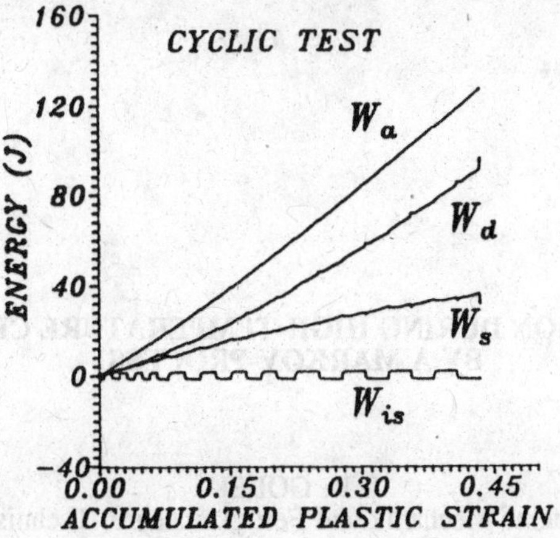


Figure 2. Energy balance during a cyclic test on duralumin; $V_0$ = 614 $mm^3$

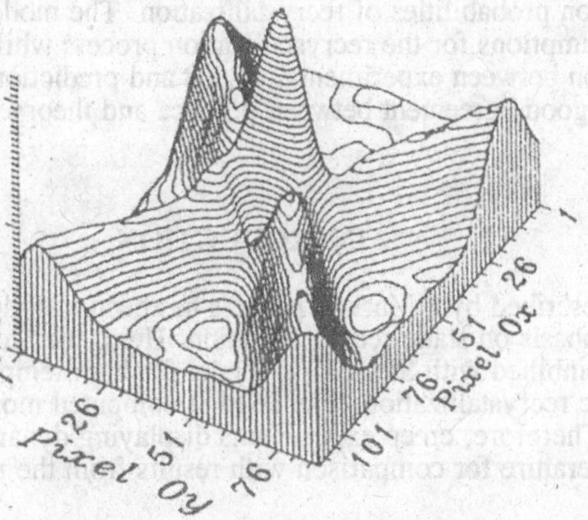


Figure 3. Dissipation chart in shear bands

## REFERENCES


[1] A. CHRYSOCHOOS, *Bilan énergétique en élastoplasticité grandes déformations*, J. de Mec. Théo. et Appl.,1985, 4, n°5, 589-614.

[2] P.GERMAIN, Q.S. NGUYEN, P. SUQUET, *Continuum Thermodynamics*, J. of Appl. Mech., 105, 1983.

[3] J.LEMAITRE, J.L. CHABOCHE, *Mechanics of Solid Materials*, Cambridge University Press, 1990.

# RECRYSTALLIZATION DURING HIGH-TEMPERATURE CREEP DESCRIBED BY A MARKOV-PROCESS

I. R. GÖBEL
Institut für Allgemeine Mechanik und Festigkeitslehre, Technische Universität Braunschweig, Gaußstr. 14, W-3300 Braunschweig, Germany

## ABSTRACT

A Markov-model developed by E. Steck for high-temperature creep is extended by introducing transition probabilities of recrystallization. The model's behavior is studied under different assumptions for the recrystallization process while a constant stress is applied. Comparison between experimental results and predictions from this new model shows that there is good agreement between practice and theory.

## INTRODUCTION

Recrystallization described by a Markov process has been introduced and discussed by Göbel [1] with emphasis on static recrystallization. Here, the Stochastic Recrystallization Model (SRM) is combined with Steck's [2] model for high-temperature plasticity in order to describe dynamic recrystallization. The SRM is computed most easily when the external stress is constant. Therefore, creep experiments displaying dynamic recrystallization have been taken from literature for comparison with results from the model.

## THE STOCHASTIC MODEL FOR DYNAMIC RECRYSTALLIZATION

Recrystallization leads to a total restoration of the structure of an undeformed crystal by the movement of grain boundaries. Dynamic recrystallization means that recrystallization and plastic deformation take place at the same time. The Markov-process for high temperature deformation as introduced by Steck [2] is described by the difference equation

$$\mathbf{z}_{t+\delta t} = \mathbf{SM}\ \mathbf{z}_t \tag{1}$$

where $\mathbf{z}$ denotes the distribution of "flow units" (FUs) at time t, or $t+\delta t$, respectively. **SM** is the stochastic matrix of dimension (N x N) containing the transition probabilities (TPs) of hardening (V) and recovery (E). The overall plastic strain rate is calculated by

$$\dot{\epsilon}_p = \lambda \sum_{i=1}^{N} v_{i-1,i}\ z_i\ , \tag{2}$$

where $\lambda$ is a constant. According to eqn. (2), FUs being moved by TPs of hardening are responsible for plastic deformation. The SRM [1] adds TPs of recrystallization to the stochastic matrix, leading to:

$$
\mathbf{SM} = \left(\begin{array}{ccccccc}
 & & & & \multicolumn{3}{c}{\overbrace{\hphantom{R \quad R \quad R}}^{\texttt{recrystallization range}}} \\
B & E & & & R & R & R \\
V & B & E & & & R & R \\
 & V & B & E & 0 & & R \\
 & & V & B & E & & \\
 & & & V & B & E & \\
 & 0 & & & V & B & E \\
 & & & & & V & B
\end{array}\right)
\begin{array}{l}
\\
\left.\begin{array}{c} \\ \\ \\ \end{array}\right\} \begin{array}{l}\texttt{range of virgin distribution} \\ \texttt{of FUs}\end{array} \\
\\
\\
\\
\\
\end{array}
\qquad (3)
$$

The sum of a column has to be equal one, so that the probability that an FU stays in place, B, is calculated from

$$
B = 1 - V - E - n\,R, \qquad 0 \leq B \leq 1, \qquad (3a)
$$

with all the TPs ranging between zero and one. The SRM moves a group of FUs along an axis of internal stress where the TPs are functions of the internal and external stress and the temperature. The TP of recrystallization depends additionally on the degree of recrystallization. The TPs of hardening and recovery allow FUs to move one step in the forward or backward direction, respectively, whereas the TPs of recrystallization move them back into the initial range where they started out when the crystal was undeformed.

## EXPERIMENTAL

Experimental research of creep with recrystallization has been mostly done in the fifties (e.g. Andrade [3]) and sixties (e.g. Hardwick et al. [4]). Typically, after the beginning of stage-II creep the strain-time curves display one or more inflections caused by acceleration of creep due to dynamic recrystallization followed by a deceleration (cf. Fig. 1). The number of inflections depends on stress and temperature (see e.g. Gilbert and Munson [5]). In strain rate-strain diagrams the onset of dynamic recrystallization leads to a fast increase of the strain rate, followed by a slow decrease.

## COMPUTATION

To investigate the behavior of the SRM during recrystallization and creep, eqn. (1) was programmed with N = 15, where the internal stress increases from 1 to 15 in steps of one. The virgin distribution comprises only one class and the TP of recrystallization is assumed to be independent of the internal stress:

$$
R = R_o + R_c \, z[1] \; ( 1 - z[1] ), \qquad (4)
$$

with $R_o = 1{\cdot}10^{-7}$, where $R_o$ simulates small random fluctuations. $R_c$ is a constant. The number of FUs in the virgin class, z[1], is identical with the degree of recrystallization, because the total number of FUs equals one. Fig. 1 shows the results. When recrystallization begins at low values of the internal stress (i.e. RMIN is small) the creep curve, $\epsilon$ over t, shows several inflections whose number decreases with increasing value of RMIN. The strain rate-strain curves show damped oscillations that go towards a steady state strain rate greater than the steady state strain rate without recrystallization, and this is the more the case the earlier recrystallization begins (i.e. the smaller RMIN). Fig. 2 shows strain rate-time curves with variations of the constant $R_c$ in eqn. (4). The influence of a linear dependence of the TP of recrystallization on the internal stress is also demonstrated. One can conclude from the simulations that

* the number of maxima in a strain rate-strain plot decreases with increasing height of the barrier, RMIN, for the onset of recrystallization. That means, the easier the onset of recrystallization the more strain-rate peaks occur, and the bigger the final strain rate,

* a linear dependence of the TP of recrystallization on the internal stress gives bigger strain-rate peaks than an exponential dependence.

When the number of classes of virgin distribution, VMAX, is changed this results in few strain-rate peaks for large VMAX, and many strain-rate peaks for small VMAX.

## CONCLUSIONS

Comparison of Figs. 1 and 2 with experimental data ([3], [4]) shows that the SRM is very well able to describe experimental creep curves with recrystallization. The present model shows an easy way of simulating dynamic recrystallization, at least during creep experiments and qualitatively, so that it seems to be promising to investigate the model further in a qualitative as well as a quantitative way.

## REFERENCES


1. Göbel, I. R., A Stochastic Recrystallization Model: The Description of Recrystallization by a Markov Process, to be published in Int. J. of Plasticity.

2. Steck, E. A., A Stochastic Model for the High-Temperature Plasticity of Metals, Int. J. of Plasticity, **1**, 1985, 243.

3. Andrade, E. N. da C., The Flow of Metals, J. of the Iron and Steel Institute, July, 1952, 217.

4. Hardwick, D., Sellars, C. M., and Tegart, W. J. McG., The Occurrence of Recrystallization During High-Temperature Creep, J. of the Institute of Metals, **90**, 1960-61, 21.

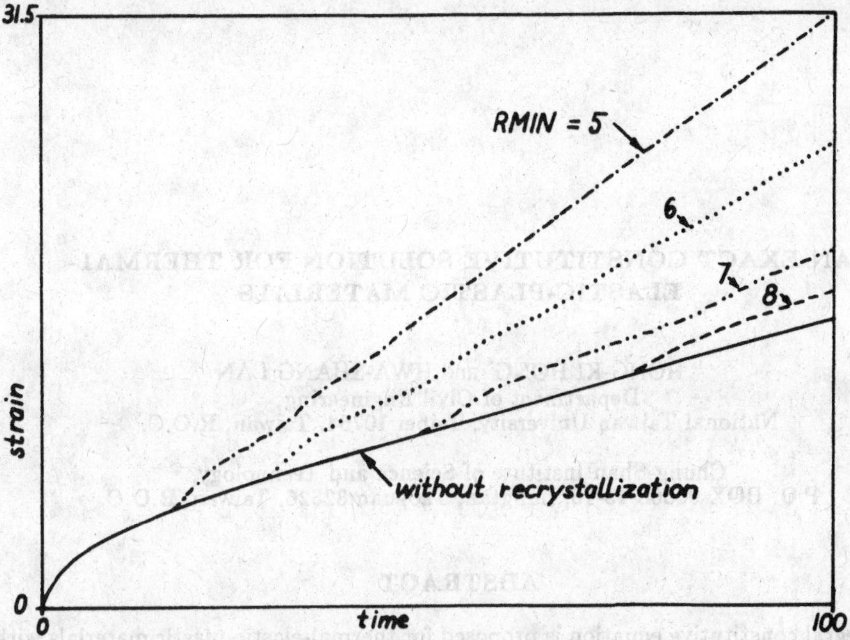


Fig. 1. Creep curves calculated by the SRM for a (15x15)-matrix with decreasing internal-stress barrier, RMIN, for the onset of recrystallization.

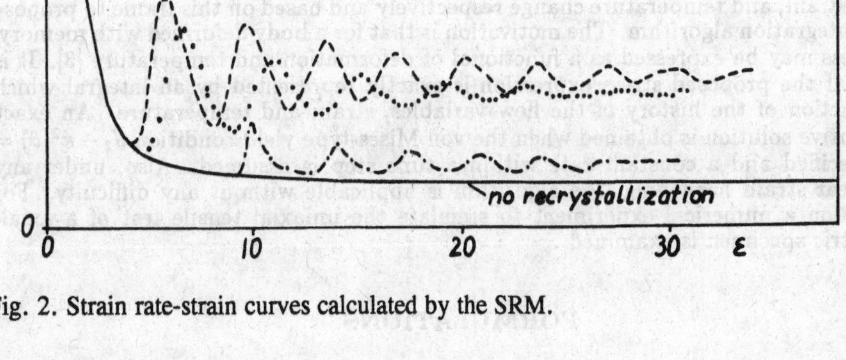

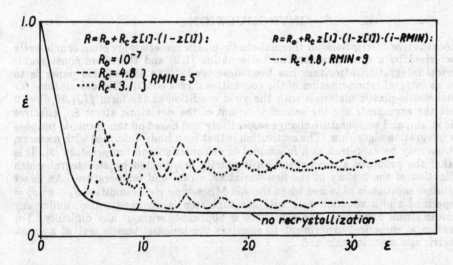

Fig. 2. Strain rate-strain curves calculated by the SRM.

# AN EXACT CONSTITUTIVE SOLUTION FOR THERMAL-ELASTIC-PLASTIC MATERIALS

HONG-KI HONG* and HWA-SHANG LAN
Department of Civil Engineering,
National Taiwan University, Taipei 10764, Taiwan, R.O.C.

Chung-Shan Institute of Science and Technology,
P.O. BOX 90008-15-13, Lungtan, Taoyuan 32526, Taiwan, R.O.C.

## ABSTRACT

An integral constitutive equation is proposed for thermal-elastic-plastic materials with the yield condition of the form $f(J_2, \bar{\epsilon}^p, \theta) = 0$, where stress is completely characterized by strain and temperature histories and which is applicable to mixed, isotropic, and kinematic hardening. When the von Mises type yield condition $J_2 - \kappa^2(\theta) = 0$ is specified and a constant strain rate within a time step is assumed, an exact constitutive solution can be obtained therefrom. For illustration a numerical experiment to simulate the tensile test of a uniaxial coupon is conducted.

## INTRODUCTION

The constitutive descriptions for thermal-elastic-plastic materials are often conclusively characterized by a rate-form constitutive equation [1, 2] and therefrom considerable numerical integration algorithm has been developed. The aim of this paper is to derive an integral representation of the conventional rate constitutive descriptions for thermal-elastic-plastic materials with the yield condition of the form $f(J_2, \bar{\epsilon}^p, \theta) = 0$ , where the arguments are the second invariant of the deviatoric stress $\xi_{ij}$, effective plastic strain, and temperature change respectively and based on this frame to propose a new integration algorithm. The motivation is that for a body deformed with memory, the stress may be expressed as a functional of deformation and temperature [3]. It is seen that the proposed stress expression is exactly represented by an integral which is a function of the history of the flow variables, strain and temperature. An exact constitutive solution is obtained when the von Mises-type yield condition $J_2 - \kappa^2(\theta) = 0$ is specified and a constant rate within a time step is assumed. Also, under any non-linear strain hardening, the algorithm is applicable without any difficulty. For illustration a numerical experiment to simulate the uniaxial tensile test of a axial-symmetric specimen is examined.

## FORMULATIONS

At all stages of the loading history, the total deformation rate tensor is considered to be decomposed into a reversible part, $d_{ij}^{rv}$, and an inelastic (plastic) part, $d_{ij}^{p}$:

$$d_{ij} = d_{ij}^{rv} + d_{ij}^{p} \tag{1}$$

and the rate of stress tensor $\dot{\sigma}_{ij}$ is expressed as (van der Lugt & Huetink [1])

$$\dot{\sigma}_{ij} = E_{ijkl}\, d_{ij}^{rv} + L_{ij}\,\dot{\theta} \tag{2}$$

where $E_{ijkl} = \frac{E}{2(1+\nu)}\left(\delta_{ik}\delta_{jl} + \delta_{il}\delta_{jk} + \frac{2\nu}{1-2\nu}\delta_{ij}\delta_{kl}\right)$, $L_{ij} = -\beta\delta_{ij}$, $\beta = \frac{E\alpha}{1-2\nu}$, and $\theta = T - T_0$, with $E$ as Young's modulus, $\nu$ Poisson's ratio, $\beta$ the thermal modulus, $\alpha$ the coefficient of thermal expansion, $T$ the absolute temperature, and $T_0$ the initial temperature. The elastic moduli $E$ and $\nu$ are assumed to be insensitive to small temperature change $\theta$ [2]. If the von Mises-type yield condition is applied,

$$f(J_2, \bar{e}^p, \theta) = \frac{1}{2}\xi_{ij}\xi_{ij} - \frac{1}{3}\kappa^2(\bar{e}^p, \theta) = 0\ , \quad \text{where } \xi_{ij} = s_{ij} - \alpha_{ij} \tag{3}$$

in which $s_{ij}$ is the stress deviator, $\alpha_{ij}$ deviatoric kinematic hardening tensor (back stress), and $\kappa$ yield stress function of material depending on the effective plastic strain $\bar{e}^p (\equiv \int_0^t (2/3 d_{ij}^p(\tau)\, d_{ij}^p(\tau))^{1/2}\, d\tau)$ and $\theta$. The associative flow rule and kinematic hardening law are respectively

$$d_{ij}^p = \begin{cases} \dfrac{3\dot{\bar{e}}^p}{2\kappa}\xi_{ij} & \text{iff } f = \dot{f} = 0 \\ 0 & \text{otherwise} \end{cases} \quad \text{and } \dot{\alpha}_{ij} = \frac{2}{3}h_{,\bar{e}^p}(\bar{e}^p, \theta)\, d_{ij}^p \ \text{(Prager-Ziegler type)} \tag{4}$$

where $h_{,\bar{e}^p}$, abbreviation for $\partial h/\partial \bar{e}^p$, depends on $\bar{e}^p$ and $\theta$.

Based on eqns (1) through (4), and after performing appropriate manipulations, eqn (2) can be decomposed into a volumetric and a deviatoric part [4]:

- for the volumetric part:

$$\frac{1}{3}\text{trace}(\sigma) = K\int_0^t \text{trace}(\text{d}(\tau))\, d\tau - \beta\theta \tag{5}$$

- for the deviatoric part:

$$\xi_{ij}(t) = \text{e}^{-\Phi(t)}\left(\xi_{ij}(0) + 2G\int_0^t \text{e}^{\Phi(\tau)} d_{ij}'(\tau)\, d\tau\right) \tag{6}$$

where $d_{ij}'$ is the deviator of $d_{ij}$ and the plastic parameter $\Phi$,

$$\Phi(t) = \int_0^t \frac{(3G + h_{,\bar{e}^p}(\bar{e}^p, \theta))\dot{\bar{e}}^p}{\kappa(\bar{e}^p, \theta)}\, d\tau \tag{7}$$

is a monotonically increasing function of the amount of plastic flow.

For purpose of numerical implementation we rewrite eqn (6) in the tensor form

$$\xi = \frac{1}{A}\left(\xi_n + 2G\int_{t_n}^t A(\tau)\text{d}'(\tau)\, d\tau\right)\ , \ t \geq t_n\ , \ \text{where } A(t) = \text{e}^{\Phi(t)-\Phi(t_n)} \tag{8}$$

where $(\cdot)_n$ denotes the converged solutions $(\cdot)$ at time $t = t_n$. To enforce the consistency condition at the end of each time step, we substitute eqn (8) into (3) and obtain

$$\left(\frac{\kappa A}{\kappa_n}\right)^2 = 1 + \omega\xi_n : \int_{t_n}^t A\text{d}'\, d\tau + \eta\int_{t_n}^t A\text{d}'\, d\tau : \int_{t_n}^t A\text{d}'\, d\tau \tag{9}$$

where $\omega = 6G/\kappa_n^2$ and $\eta = 6G^2/\kappa_n^2$. This equation closely relates the effective plastic strain $\bar{e}^p$ with the strain rate deviator $\mathbf{d}'$ and temperature $\theta$.

**For Constant Strain Rate**

Under the consideration of constant strain rate, eqns (8) and (9) can be respectively expressed as

$$\xi = \frac{1}{A}(\xi_n + 2GR\mathbf{d}') \tag{10}$$

and

$$\frac{\kappa A}{\kappa_n} = F(R)^{1/2}, \quad \text{where} \quad F(R) \equiv 1 + \bar{\omega} R + \bar{\eta} R^2 \tag{11}$$

where $R = \int_{t_n}^{t} A(\tau)\, d\tau$, $\bar{\omega} = \omega \xi_n : \mathbf{d}'$, and $\bar{\eta} = \eta \mathbf{d}' : \mathbf{d}'$

**Exact Solution**

Consider that $\kappa$ and $h_{,\bar{e}^p}$ are independent on $\bar{e}^p$, i.e., $\kappa = Y(\theta)$ and $h_{,\bar{e}^p} = D(\theta)$. The exact solution of eqn (11) is

$$R(t) = \frac{1}{\sqrt{\bar{\eta}}} \sinh(\sqrt{\bar{\eta}} Z(t)) + \frac{\bar{\omega}}{2\bar{\eta}}(\cosh(\sqrt{\bar{\eta}} Z(t)) - 1), \text{ where } Z(t) = \int_{t_n}^{t} \frac{Y_n}{Y(\theta(\tau))}\, d\tau \tag{12}$$

and whereupon

$$A(t) = \dot{R}(t) = \frac{Y_n}{Y(\theta)}\left(\cosh(\sqrt{\bar{\eta}} Z(t)) + \frac{\bar{\omega}}{2\sqrt{\bar{\eta}}} \sinh(\sqrt{\bar{\eta}} Z(t))\right) \tag{13}$$

$\xi$ can be found from eqn (10). The $\bar{e}^p$ and therefore $\alpha$ can be easily obtained with the help of eqns (4), (7), and (10).

**Non-linear hardening laws**

In case that $\kappa$ and $h$ are arbitrary functions of $\bar{e}^p$ and $\theta$, the present constitutive model problem is equivalently reduced to the initial value problems given by

$$\dot{\bar{e}}^p = \hat{f}(t, \bar{e}^p, R) \equiv \frac{\kappa_n}{3G + h_{,\bar{e}^p} + \kappa_{,\bar{e}^p}} \left( \frac{\bar{\omega} + 2\bar{\eta} R}{2F(R)^{1/2}} - \frac{\kappa_{,\theta}\, \dot{\theta}}{\kappa_n} \right), \quad \bar{e}^p(t_n) = \bar{e}^p_n \tag{14}$$

$$\dot{R} = \hat{g}(t, \bar{e}^p, R) \equiv \frac{\kappa_n}{\kappa} F(R)^{1/2}, \quad R(t_n) = 0 \tag{15}$$

in which the dependence on time $t$ of functions $\hat{f}$ and $\hat{g}$ results from the dependence on temperature variation $\theta(t)$ of $\kappa$ and $h$.

## ILLUSTRATION

We use the present integration algorithm to calculate the temperature variation of an axial-symmetric specimen which is tensioned by a constant velocity $0.01 mm/sec$, see Figure 1. Consider the energy equation governed by

$$\rho C_v \dot{\theta} - \lambda \nabla^2 \theta = \sigma : \mathbf{d}^p - \beta\, \text{trace}(\mathbf{d}^{rv})(T_0 + \theta) \tag{16}$$

where $\rho$ is the density, $C_v$ specific heat at constant strain, and $\lambda$ coefficient of heat conductivity. Both terms on the r.h.s. of eqn (16) represent the heat generation due to the mechanical work. The results are shown in figures 2 and 3.

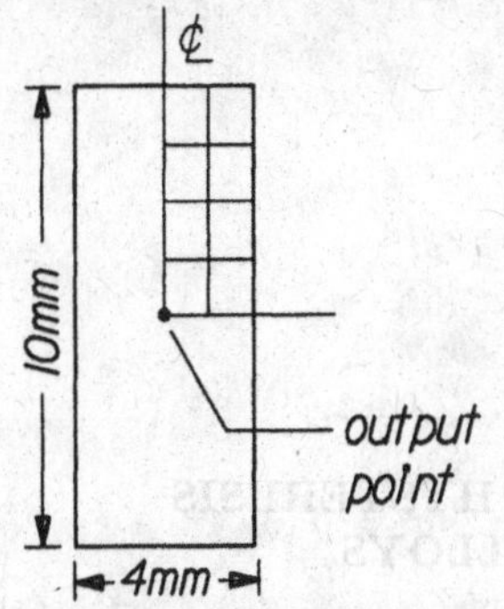


$$E = 2.08 \times 10^5 \text{ MPa}$$
$$\nu = 0.3$$
$$Y_0 = 400 \text{ MPa}$$
$$T_0 = 300\ K$$
$$\rho_0 = 7950\ Kg \cdot M^{-3}$$
$$\alpha = 1.1089 \times 10^{-5}\ K^{-1}$$
$$\lambda = 14.8\ W \cdot M^{-1} \cdot K^{-1}$$
$$C_v = 440\ J \cdot Kg^{-1} \cdot K^{-1}$$
$$h_{,\bar{e}^p} = 0$$
$$\kappa = Q(0.000146 + \bar{e}^p)^n \left(1 - \left(\frac{\theta}{800}\right)^2\right) \text{ MPa}$$

Figure 1. Tensile specimen and its material properties

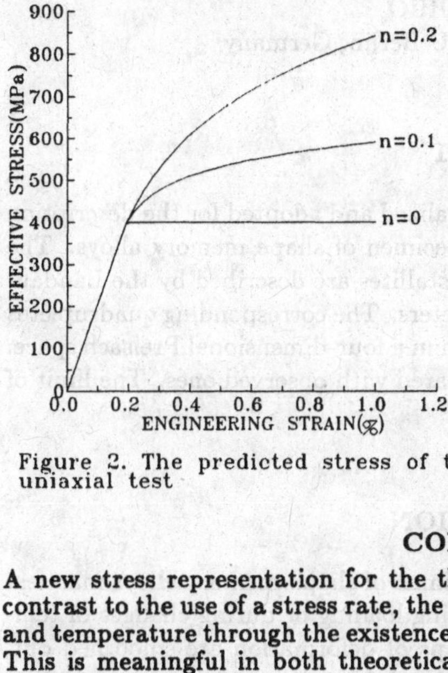


Figure 2. The predicted stress of the uniaxial test

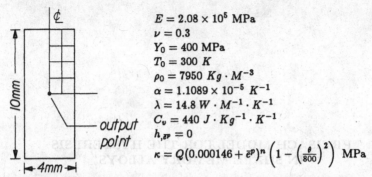

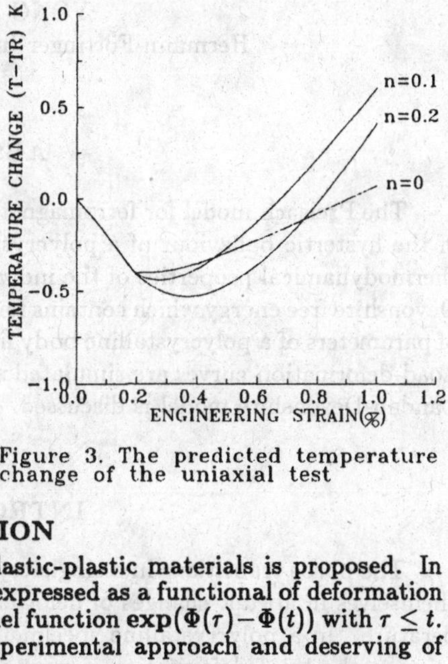


Figure 3. The predicted temperature change of the uniaxial test

## CONCLUSION

**A new stress representation for the thermal-elastic-plastic materials is proposed. In contrast to the use of a stress rate, the stress is expressed as a functional of deformation and temperature through the existence of a kernel function $\exp(\Phi(\tau) - \Phi(t))$ with $\tau \leq t$. This is meaningful in both theoretical and experimental approach and deserving of further research.**

## REFERENCES


1. van der Lugt, J., and Huetink, J., Thermal mechanically coupled finite element analysis in metal-forming process. Comp. Meth. Appl. Mech. Eng., 1986, **54**, 145–160.

2. Argyris, J. H., Doltsinis, J. St., Pimenta, P. M., and Wustenberg, H., Thermomechanical response of solids at high strain-natural approach, Comp. Meth. Appl. Mech. Eng., 1982, **32**, 3–57.

3. Malvern, L. E., Introduction to the Mechanics of A Continuous medium, Prentice-Hall Inc., 1969., pp. 319–324.

4. Hong, H.-K., and Lan, H.-S., A new point-of-view at plastic materials of the prandtl-reuss type, In Advance in Plasticity 1989, ed. A.S. Khan and M. Tokuda, 1989, pp. 95–98.

# PREISACH MODEL FOR THE HYSTERESIS IN SHAPE MEMORY ALLOYS

YONG-ZHONG HUO
Hermann-Föttinger Institut, TU Berlin, Germany

## ABSTRACT

The Preisach model for ferromagnets is generalized and adopted for the description of the hystertic behaviour of a polycrystalline specimen of shape memory alloys. The thermodynamical properties of the individual crystallites are described by the Landau-Devonshire free energy which contains four parameters. The corresponding quadrupletes of parameters of a polycrystalline body fill a region in a four-dimensional Preisach space. Load-deformation curves are simulated and compared with observed ones. The limit of Landau-Devonshire model is discussed.

## INTRODUCTION

The phase transitions in a single-crystal specimen of shape-memory alloys manifest themselves in abrupt changes of deformation during loading or during changes of temperature. In a polycrystalline specimen the jumps of deformation are smoothed out, because the total deformation is now a sum of the deformations of all the crystallite and each crystallite responds differently to changes in load and temperature. The load-deformation diagrams all strongly depend upon temperature and exhibit hysteresis. In order to produce the load- deformation relation of a ploycrystalline body, we need to know the behaviour of each crystallite and then we should be capable to take the sum.

## BEHAVIOUR OF A SINGLE CRYSTAL BODY

For a single crystal of shape memory alloy, the temperature dependent hysteresis in load-deformation diagrams, resulted from martensite transformation, can be qualitatively described by the Landau-Devonshire model. This model proposes a simple analytic form for the free energy as a function of $D$ and $T$, viz.

$$f(D,T) = f_\circ(T) + \frac{1}{2}a(T - T_\circ)D^2 - \frac{1}{4}bD^4 + \frac{1}{6}cD^6, \qquad (1)$$

where $T_\circ, a, b$ and $c$ are positive constants. The load $P$ results from differentiation of $f$ with respect to $D$ and we have

$$P(D,T) = a(T - T_\circ)D - bD^3 + cD^5. \tag{2}$$

Fig.1 shows plots of these functions for different temperatures. The free energies are non-convex and the corresponding load-deformation curves are non-monotone. Such curves imply hysteretic behaviour, because upon loading or unloading there may be yield or recovery along the dotted horizontal lines. The branches of the $(P, D)$-curves on the left and right we interpret as representing the martensitic twins $M^-$ and $M^+$ respectively. The ascending branch through the origin, if any, is interpreted as representing the austenitic phase $A$. Thus the horizontal dotted lines represent phase transitions between martensite and austenite or between the martensitic twins.

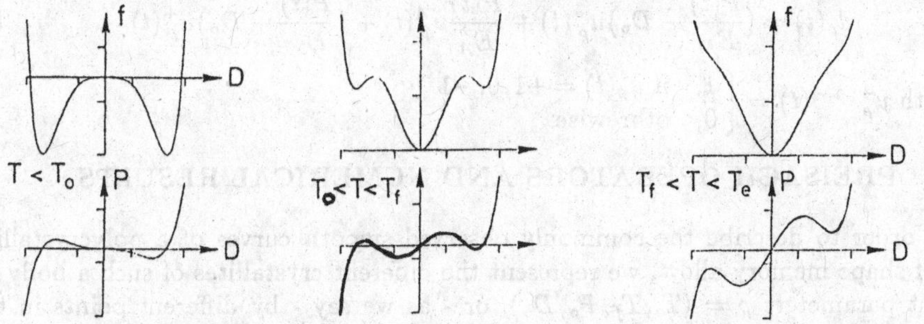


Figure 1. (f,D)-curves and (P,D)-curves of the Landau-Devonshire model.

From their construction it is clear that the horizontal dotted transition lines occur for values of $P$ and $T$ such that $P_i = P(D_i, T)$, where $D_i$ are solutions of $\frac{\partial P(D,T)}{\partial D} = 0$. After some calculations we obtain that

$$P_i = \pm P_\circ c_{1,2}(\Theta) = \pm P_\circ\{\frac{5}{6}\Theta - \frac{1}{2}(1 \pm \sqrt{1 - \frac{5}{9}\Theta})\}\sqrt{\frac{1}{2}(1 \pm \sqrt{1 - \frac{5}{9}\Theta})}, \tag{3}$$

with $T_f = T_\circ + \frac{b^2}{4ac}$, $D_\circ = \sqrt{\frac{b}{c}}$, $P_\circ = \frac{6}{25}\sqrt{\frac{3}{5}}D_\circ\frac{b^2}{c}$, and $\Theta = \frac{T - T_\circ}{T_f - T_\circ}$.

Fig.2 shows the two universal functions $c_{1,2}(\Theta)$ in a $(P/P_\circ, \Theta)$- diagram. The different areas of the figure are marked by $A$ and $M^\pm$ thus identifying the phases that may prevail for pairs $(P, T)$ in these areas. A thermodynamical loading path $L(t) = (P(t), T(t))$ may be represented by a curve in Fig.2. A transition occurs on that path whenever we cross a line on whose other side one phase can no longer exist.

Now let $\rho = (T_\circ, T_f, P_\circ, D_\circ)$ and define a phase-value function as $u_\rho(t) = -1, 0, or +1$, when the body is in phase $M^-, A, or\ M^+$, respectively. Given $L(t) = (P(t), T(t))$ and an initial value $u_\rho(0)$ we may calculate $u_\rho(t)$ by following the path shown in Fig.2.

For each quadruple $\rho$ we introduce the deformation function by assuming linear elasticity for each single phase as

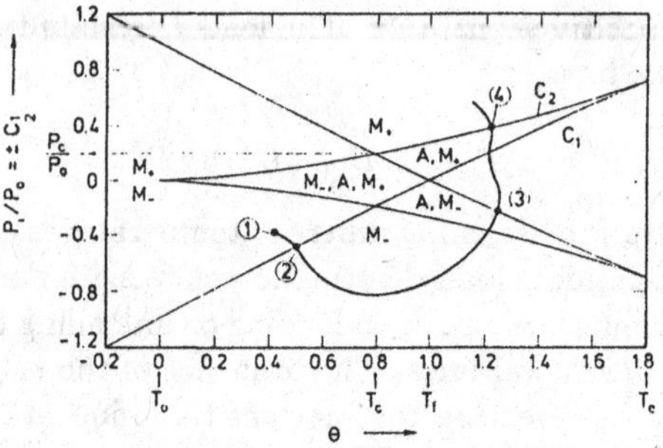


Figure 2. Phase diagram for the L-D model. A typical thermodynamical loading path.

$$d_\rho(t) = (\frac{P(t)}{E_M} - D_\circ)u_\rho^-(t) + \frac{P(t)}{E_A}u_\rho^A(t) + (\frac{P(t)}{E_M} + D_\circ)u_\rho^+(t), \tag{4}$$

with $u_\rho^{+,A,-}(t) = \begin{cases} 1, & \text{if } u_\rho(t) = +1, 0, -1 \\ 0, & \text{othrewise.} \end{cases}$

## PREISACH OPERATORS AND NUMERICAL RESULTS

In order to describe the commonly observed smooth curves of a polycrystalline body of shape memory alloy , we represent the different crystallites of such a body by different parameters $\rho = (T_\circ, T_f, P_\circ, D_\circ)$, or - as we say - by different points in the Preisach space

$$\mathcal{P} = \{\rho = (T_\circ, T_f, P_\circ, D_\circ) | T_f > T_\circ > 0, P_\circ, D_\circ > 0\}. \tag{5}$$

And the phase transition behaviour of each element is characterized by the four critical equations (3). So presisely speaking the Preisach space is a triplet $(\mathcal{P}, c_1, c_2)$ endowed with the critical equations (3).

A polycrystalline body is now a subset $\Omega$ of the Preisach space with a prescribed distribution function $\mu(\rho)$ on $\Omega$. The phase fractions $X^{+,A,-}$ of the martensitic twins and of the austenitic phase may be written as

$$X^{+,A,-}(t) = \frac{1}{V(\Omega)} \int_\Omega u_\rho^{+,A,-}(t)\mu(\rho)d\rho, \tag{6}$$

where $V(\Omega)$ is the volume of $\Omega$ under the measure $\mu$. Accordingly the deformation of the polycrystalline body may be written as

$$D(t) = \int_\Omega d_\rho(t)\mu(\rho)d\rho. \tag{7}$$

Given $L(t) = (P(t), T(t))$ and the set $S_I = \{u_\rho(0) | \rho \epsilon \Omega\}$ of initial conditions within $\Omega$, the phase fractions and deformation at any later time $t$ are uniquely determined by (6-7), which are the so called Preisach operators.

Two appealling characters of Preisach model are: the possible geometrical interpretation in the Preisach space of the phase transition behaviour of the body and the efficiency to make computer calculations. For the first the interested reader is refered to [1], we shall only present here some numerical resutlts. For simplicity we assume:

$$\Omega = \{\rho\epsilon\mathcal{P}|5 < P_\circ < 10, 10 < D_\circ < 20, 20 < X < 30, 355 < Y < 360\}$$

with $X = T_f - T_\circ$ and $Y = T_\circ + \frac{9}{5}X$. $\mu(\rho) = 1, E_M = 0.5, E_A = 1$.
The calculated $(P, D)$ diagrams are prsented in Fig.3.

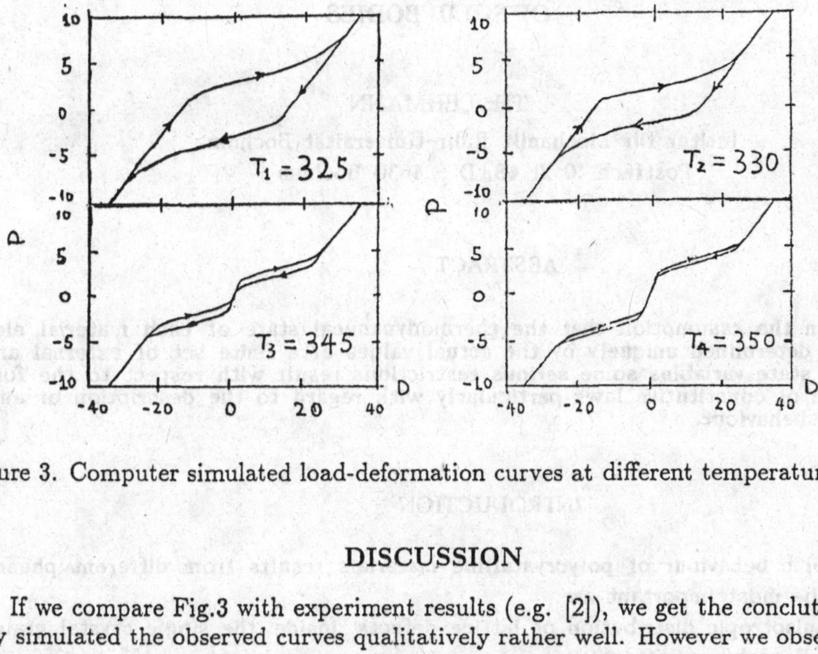


Figure 3. Computer simulated load-deformation curves at different temperatures.

## DISCUSSION

If we compare Fig.3 with experiment results (e.g. [2]), we get the conclution that they simulated the observed curves qualitatively rather well. However we observe also some disadvantages in using Landau-Devonshire model to describe the behaviour of a single crystal. Firstly, the so called Curie temperature $T_c = T_\circ + \frac{9b^2}{20ac}$ is unrealistically too lower, which causes some unexpected features of the Preisach operators. Secondly it predicts a wrong temperature dependence of the area of hysteresis, namely the area decreases for increasing temperature as shown by the last two curves in Fig.3. Better models are availabe, see Müller and Xu [3]. It is also possible to obtain the universal functions $c_{1,2}(\Theta)$ directely by reading from the curves in [2], then we can follow the above procedure to get more realistic deformation-load curves.

## REFERENCES


1. Huo, Y.-Z.: A mathematical model for the hysteresis in shape memory alloys. Continuum Mech.Thermodyn., 1989,1, 283-303.
2. Otsuka, K. and K. Shimizu: Pseudoelasticity and shape memory effects in alloys. Int. Met. Rev., 1986, **31**, 93-114.
3. Müller, I. and H.-B. Xu: On the pseudo-elastic hysteresis. Acta. Metall. Mater., 1991, **39**, 263-271.

# THERMODYNAMICAL CONSIDERATIONS ON ANISOTROPIC DEFORMATIONS OF SOLID BODIES

TH. LEHMANN
Intitut für Mechanik, Ruhr-Universität Bochum
Postfach 10 21 48, D - 4630 Bochum

## ABSTRACT

Based on the assumption that the thermodynamical state of each material element is determined uniquely by the actual values of a finite set of external and internal state variables some serious restrictions result with respect to the formulation of constitutive laws particularly with regard to the description of anisotropic behaviour.

## INTRODUCTION

Anisotropic behaviour of polycrystalline materials results from different phenomena. The most important are:

( a ) anisotropic distribution of lattice defects inside the single crystal grains ( Bauschinger effect ),
( b ) morphological texture due to an oriented distribution of grain shapes,
( c ) crystallographic texture characterized by a non-random orientation of the crystal axes of the grains,
( d ) anisotropic microscopic damage states inside the crystals and at their boundaries.

Phenomena of kind (a) influence the elastic behaviour scarcely wheras the other phenomena affect the inelastic behaviour as well as the elastic behaviour. This is particularly true with respect to phenomena of kinds (c) and (d). We shall focus our considerations to a certain comparison of phenomena of kinds (a) and (c) with respect to some thermodynamical aspects.

## GENERAL FOUNDATIONS

We relate all quantities to a body-fixed coordinate system $\xi^{\alpha}$ in its actual configuration since they are acting in this configuration. The first law of thermodynamics states

$$\dot{u} = \dot{w} - \frac{1}{\rho} q^{\alpha}\big|_{\alpha} + r \qquad (1)$$

where u means the specific internal energy, w denotes the applied specific mechanical work, $\rho$ is the mass density, $q^{\alpha}$ stands for the resulting energy flux, and r represents the sum of specific energy sources. The rate of mechanical work has to be decomposed into the reversible part $\underset{(r)}{w}$, the immediately dissipated part $\underset{(d)}{\dot{w}}$, and into the part $\underset{(h)}{\dot{w}}$ interacting with changes of the internal material structure according to

$$\dot{w} = \underset{(r)}{\dot{w}} + \underset{(d)}{\dot{w}} + \underset{(h)}{\dot{w}} . \qquad (2)$$

The reversible part of work rate can be expressed in the form

$$\underset{(r)}{\dot{w}} = \frac{1}{\overset{o}{\rho}} s^{\alpha}_{\beta} \overset{\nabla}{\underset{(r)}{\varepsilon}}{}^{\beta}_{\alpha} \qquad (3)$$

where $\overset{\nabla}{}$ denotes the objective material (Zaremba-Jaumann) time derivative, $s^{\alpha}_{\beta}$ and $\underset{(r)}{\varepsilon}{}^{\alpha}_{\beta}$ represent a conjugate pair of (weighted) stress and reservible strain, and $\overset{o}{\rho}$ is the mass density in the initial configuration.

According to the fundamental assumption that the thermodynamical state of each material element is determined uniquely by the actual values of a suitably chosen set of external and internal state variables the specific internal energy must be expressible in the form

$$u = u(\underset{(r)}{\varepsilon}{}^{\alpha}_{\beta}, s, a, a^{\alpha}_{\beta}, A^{\alpha\gamma}_{\beta\delta}, \ldots) \qquad (4)$$

where $\underset{(r)}{\varepsilon}{}^{\alpha}_{\beta}$ and the specific entropy s represent the external extensive variables and a, $a^{\alpha}_{\beta}$, $A^{\alpha\gamma}_{\beta\delta}$, etc. stand for a representative set of internal variables. By a double Legendre transformation we replace the external extensive variables by their conjugate intensive variables, i.e. the stress $s^{\alpha}_{\beta}$ and the absolute temperature T. This leads to the definition of the specific free enthalpy

$$\psi = \psi - \frac{1}{\overset{o}{\rho}} s^{\alpha}_{\beta} \underset{(r)}{\varepsilon}{}^{\beta}_{\alpha} - T s = \psi( s^{\alpha}_{\beta}, T, a, a^{\alpha}_{\beta}, A^{\alpha\gamma}_{\beta\delta}, \ldots) . \qquad (5)$$

From (5) we derive the thermic state equation

$$\underset{(r)}{\varepsilon}{}^{\alpha}_{\beta}( s^{\alpha}_{\beta}, T, a, a^{\alpha}_{\beta}, A^{\alpha\gamma}_{\beta\delta}, \ldots) = -\overset{o}{\rho} \frac{\partial \psi}{\partial s^{\beta}_{\alpha}} \qquad (6)$$

and the caloric state equation

$$s( s^{\alpha}_{\beta}, T, a, a^{\alpha}_{\beta}, A^{\alpha\gamma}_{\beta\delta}, \ldots) = -\frac{\partial \psi}{\partial T} . \qquad (7)$$

Concerning the interaction between external energy supply and changes of the internal material structure we obtain the following balance equation [1,2]

$$\underset{(h)}{\dot{w}} - \frac{T}{\rho}\left(\frac{\tilde{q}^{\alpha}}{T}\right)\Big|_{\alpha} + r - T\dot{\eta} = \frac{\partial \psi}{\partial a}\dot{a} + \frac{\partial \psi}{\partial a^{\alpha}_{\beta}}\overset{\nabla}{a}{}^{\alpha}_{\beta} + \frac{\partial \psi}{\partial A^{\alpha\gamma}_{\beta\delta}}\overset{\nabla}{A}{}^{\alpha\gamma}_{\beta\delta} + \ldots \qquad (8)$$

The part $\underset{(h)}{\dot{w}}$ of mechanical work rate, the part $\tilde{q}^{\alpha}$ of energy flux, and the entropy production $\dot{\eta}$ involved in this interaction have to be specified within the constitutive law.

## BAUSCHINGER EFFECT

Since the isotropic and anisotropic distribution of lattice defects inside the crystal grains influences the elastic behaviour scarcely in this case the specific free enthalpy can be decomposed into two parts according to

$$\psi(s^{\alpha}_{\beta}, T, a, a^{\alpha}_{\beta}, A^{\alpha\gamma}_{\beta\delta}, \ldots) = \psi^{*}(s^{\alpha}_{\beta}, T) + \psi^{**}(T, a, a^{\alpha}_{\beta}, A^{\alpha\gamma}_{\beta\delta}, \ldots). \qquad (9)$$

Consequently the integrability conditions for $\psi$ which represent severe restrictions for the formulation of constitutive laws obeying the balance equation (8) divide into two distinct groups

$$\frac{\partial^2 \psi^{*}}{\partial T\, \partial s^{\alpha}_{\beta}} = \frac{\partial^2 \psi^{*}}{\partial s^{\alpha}_{\beta}\, \partial T} \qquad (10\ a)$$

$$\frac{\partial^2 \psi^{**}}{\partial T\, \partial a} = \frac{\partial^2 \psi^{**}}{\partial a\, \partial T}\ ; \quad \frac{\partial^2 \psi^{**}}{\partial a\, \partial a^{\alpha}_{\beta}} = \frac{\partial^2 \psi^{**}}{\partial a^{\alpha}_{\beta}\, \partial a}\ ; \quad \text{etc.} \qquad (10\ b)$$

Some examples how these integrability conditions can be fulfilled in a physically reasonnable manner are discussed in [2]. Some remarks on so-called plastic spin in the formulation of corresponding constitutive laws can be found in [3,4].

## CRYSTALLOGRAPHIC TEXTURE

Since a crystallographic texture influences the inelastic behaviour as well as the elastic behaviour a decomposition of the specific free enthalpy according to equ. (9) is not possible any more. Consequently also the number of integrability conditions increases. This enlarges the difficulties in the formulation of thermodynamically consistent constitutive laws. However, even disregarding the coupling between elastic and inelastic behaviour and the resulting thermodynamical restrictions very few attempts can be found concerning the formulation of phenomenological evolution laws for crystallographic texture during large inelastic deformations (see, for instance, [5,6]). Therefore many questions are still open.

## REFERENCES

[1] Lehmann, Th., Internal Variables in Thermoplasticity. In: Advances in Constitutive Laws for Engineering Materials, eds. Fan Jinghong and S. Murakami, Pergamon Press / Int. Acad. Publ. 1989, 934-50

[2] Lehmann, Th., On the Balance of Energy and Entropy at Inelastic Deformations of Solid Bodies, Europ. J. Mech. A / Solids 1989, 8, 235-51

[3] Lehmann, Th., Some Remarks on So-called Plastic Spin, ZAMM 1991, 71 (in print)

[4] Lehmann, Th., Some Remarks on the Phenomenological Description of Anisotropic Behaviour of Elastic-plastic Bodies (in print)

[5] Boehler, J. P. and Koss, S., Decouplage des effects des textures cristallographiques et morphologiques sur l'anisotropie du comportement macroscopique des metaux en grandes deformations, Inst. Mécanique des Grenoble, Report final (1989)

[6] van der Giessen, E. and van Houte, P., A 2D Analytical Multiple Slip Model for Continuum Texture Development and Plastic Spin, Delft Univ., Rep. No. 927 (1990)

# THE THERMODYNAMICS OF ELASTIC-PLASTIC MATERIALS

MASSIMILIANO LUCCHESI* and MIROSLAV SILHAVY**

* Istituto CNUCE, 36 Via S. Maria, 56100 Pisa, Italy
** Mathematical Institute, Zitna 25, 115 67 Praha 1, Czechoslovakia

## ABSTRACT

The consequences of and the relationship among the second law of thermodynamics and the stability conditions of Il'yushin's type are studied. The construction of the thermodynamic potentials based on these conditions are given and the circumstances under which the potentials are unique are discussed.

## INTRODUCTION

There are several conditions of thermodynamic type describing the dissipation in materials with elastic range. Among them there is first of all the Clausius inequality for cyclic processes expressing the second law of thermodynamics and there are also various "stability postulates" like the Il'yushin condition and the Drucker postulate. The paper discusses these conditions within the framework of the non-isothermal theory of materials with elastic range. It is shown that there are several new issues in comparison with the isothermal theory; for instance, there are several extensions of the isothermal Il'yushin condition to the non-isothermal situations.

## ELASTIC-PLASTIC MATERIAL ELEMENTS

We use the concept of state space to describe the response of the material element. To each material element is attached a set, $\Sigma$, called the state space, and a specified class of processes, $\Pi$. The processes $\pi$ are interpreted as time-evolutions of state. The values of the physical quantities like the deformation, temperature, stress, etc., during a process, are determined by the present state via state functions. The symbols $\pi^i$ and $\pi^f$ denote the initial and the final states of the process $\pi$.

The existence of mappings $F$, $\theta$, $T$, and $\varepsilon$ is postulated which have the following interpretation: for every state $\sigma \in \Sigma$ the values $F(\sigma)$, $\theta(\sigma)$, $T(\sigma)$, and $\varepsilon(\sigma)$ are the values of the deformation gradient, absolute temperature, Cauchy stress tensor, and specific internal energy at the state $\sigma$. In the concept of material element we have not included any state function reflecting the secon law of thermodynamics like the entropy or the free energy.

Now we are going to formulate the properties which express the specific features of leastic-plastic material elements. These are the existence of elastic range and the existence of the plastic reference configuration. The idea underlying the concept of elastic range $E(\sigma)$ corresponding to the state $\sigma$ is that, when restricted to processes starting at $\sigma$ and remaining in the elastic range, the material element behaves as a thermoelastic material in the sense that the state at the end of the process is completely determined by the final value of the deformation gradient and temperature. This allows us to introduce the elastic response functions for the stress and the energy by

$$T^*(\sigma;F(\pi^f),\theta(\sigma^f)) = T(\pi^f)$$

$$\varepsilon^*(\sigma;F(\pi^f),\theta(\sigma^f)) = \varepsilon(\pi^f), \quad (1)$$

for every elastic process $\pi$ starting at $\sigma$. Thus the values $T^*(\sigma;F,\theta)$ and $\varepsilon^*(\sigma;F,\theta)$ are the final values of the stress tensor and the internal energy for any elastic process starting at $\sigma$ and ending at the deformation-temperature pair $(F,\theta)$. The second basic feature of elastic-plastic materials, the existence of the plastic reference configuration, is the assumption that the elastic response functions are expressible in terms of the plastic deformation $P$ corresponding to the state $\sigma$ through the formulas

$$T^*(\sigma;F,\theta) = T_0(FP^{-1},\theta), \qquad \varepsilon^*(\sigma;F,\theta) = \varepsilon_0(FP^{-1},\theta) \tag{2}$$

where $T_0$ and $\varepsilon_0$ are independent of $\sigma$.

## THE SECOND LAW OF THERMODYNAMICS

The material element is said to satisfy the second law of thermodynamics if there exists a state function $\eta$ such that

$$\eta(\pi^f) - \eta(\pi^i) \geq \int_0^d \frac{\dot{\varepsilon} - \rho_0^{-1} J T.L}{\theta}\, dt \tag{3}$$

for every process $\pi$. Here $\rho_0$ is the density in the reference configuration, $J = \det F$, $L = \dot{F} F^{-1}$. Any scalar-valued function $\eta$ obeying (3) will be called a thermodynamical entropy function for the material element.

For every material element obeying the second law there exists a smooth entropy function $\eta_0 = \eta_0(F,\theta)$ such that the thermostatic relations

$$T_0/\theta_0 = -\rho_0(\det E)^{-1}\partial\eta_0/\partial E.E^T,$$

$$1/\theta = \partial\eta_0/\partial\varepsilon, \tag{4}$$

hold; for every thermodynamical entropy function $\eta$ there exists a scalar state function $\eta_1$ such that

$$\eta(\sigma) = \eta_0(F(\sigma)P^*(\sigma)^{-1},\varepsilon(\sigma)) + \eta_1(\sigma), \quad \sigma \in \Sigma \tag{5}$$

and the function $\eta_1$ has the property that it is constant on every elastic range. This means that $\eta_1$ depends only on the inelastic state parameters like the plastic deformation. Every state function $\eta_1$ as in (5) will be called a residual entropy function. The description of all entropy functions is thus reduced to the description of all residual entropy functions.

The material element has a thermodynamical entropy function of the special form

$$\eta(\sigma) = \eta_o(F(\sigma)P^*(\sigma)^{-1}, \varepsilon(\sigma)), \quad \sigma \in \Sigma \tag{6}$$

if and only if there exists a smooth function $\eta_o$ obeying (4) and the plastic power is non-negative in every process.

Consider now a v. Mises ideally plastic material for which the temperature-dependent yield stress $Y(\theta)$ is a decreasing function of temperature. Denote by $c$ the g.l.b. of the ratio $Y(\theta)/\theta$ over the available range of temperatures. Then the function $\eta_1$ is a residual entropy function if an d only if

$$|\eta_1(Re^D P) - \eta_1(P)| \leqq ||D|| \tag{7}$$

for every plastic deformation $P$, every symmetric traceless tensor $D$ and and every rotation $R$. The entropy is unique to within an additive constant if and only if $c = 0$.

## IL'YUSHIN'S CONDITION

A process $\pi$ of the material element is said to be a small $(F,\varepsilon)$-cycle if $(F(\pi^i),\varepsilon(\pi^i)) = (F(\pi^f),\varepsilon(\pi^f))$ an d if the initial deformation-temperature pair belongs to all current elastic ranges during the process. A material element is said to obey the Il'yushin condition for $(F,\varepsilon)$ - cycles if for every small $(F,\varepsilon)$ - cycle for which the temperature during plastic loading never exceeds the initial temperature we have

$$\int_0^d \frac{\dot{\varepsilon} - \rho_o^{-1} JT.L}{\theta} dt \leqq 0. \qquad (8)$$

If the material element satisfies the Il'yushin condition for $(F,\varepsilon)$ - cycles, then there exists a smooth function $\eta_o$ such that relations (4) hold and for every process $\pi$ whose initial deformation temperature pair belongs to all current elastic ranges and for which the temperature during the plastic deformation never exceeds the initial temperature the following inequality holds:

$$\eta_o(F(\pi^f)P_f^{-1},\varepsilon(\pi^f)) - \eta_o(F(\pi^i)P_f^{-1},\varepsilon(\pi^i) \geqq$$

$$\int_0^d \frac{\dot{\varepsilon} - \rho_o^{-1} JT.L}{\theta} dt; \qquad (9)$$

For a material element satisfying Il'yushin's condition the plastic power is non-negative if and only if for every state the elastic region contains an isothermal unloaded point.

If an elastic-plastic material element satisfies Il'yushin's condition for $(F,\varepsilon)$ - cycles an d the plastic power is non-negative in every process, then the material element satisfies the second law of thermodynamics and one of its thermodynamical entropy functions is of the form (6).

## REFERENCES

1. Lucchesi, M., and Šilhavý, M., Il'yushin's conditions in non-isothermal plasticity. Arch. Rational Mech. Anal. 113, 1991, 121 - 163.

# CONSTITUTIVE EQUATIONS FOR METALS COUPLING PLASTICITY AND MICROSTRUCTURAL CHANGES

DIDIER MARQUIS
Laboratoire de Mécanique et Technologie
E.N.S. de Cachan / C.N.R.S. / Université Paris 6
61, Avenue du Président Wilson - 94230 CACHAN - FRANCE

## ABSTRACT

This paper concerns the study of the interaction between aging and plasticity in a 2024 aluminium-copper alloy. The framework of thermodynamics of irreversible processes is used to introduce the effect of aging in the constitutive equations by means of an internal variable.

## INTRODUCTION

This paper presents a mechanical model for metals and alloys where the coupling between aging and plasticity is considered. The definition adopted for aging is the definition used by Krempl in [1]. Aging is the modification of the microstructure of a material such that the same loading process performed in the same environment at different times, gives different mechanical responses. Although all the main ideas are valid for a very general situation, the study will be restricted to a 2024 aluminium-copper alloy . This kind of material was chosen because its aging occurs at room temperature and it is easier to obtain experimental results and to identify the evolution laws.
For the 2024 aluminium-copper alloy, after a dissolution heat treatment followed by a quench into water, diffusion of copper in the aluminium matrix occurs at room temperature. The formation of precipitates (GUINIER-PRESTON's zones), which results from this diffusion has a very strong influence on the yield stress and different stress-strain curves are obtained depending upon the starting time of the tensile test. The plastic strain during a loading process can also change the macroscopic mechanical properties and they are described by the isotropic and kinematic hardenings [2]. A set of cyclic and monotonic uniaxial tests on the material during the process of aging can be found in [3]. They show the different types of couplings between the mechanisms involved. In this case the isotropic hardening and the aging are strongly coupled but the kinematic hardening and aging are almost independent. Considering those global results, this paper introduces an elasto-plastic model, developed in the framework of thermodynamics of irreversible processes, which takes in account the coupling between the different phenomena.

## MODELING

To describe plasticity and aging in metallic materials, internal variables theory in the framework of thermodynamics of irreversible processes is used. For each microscopic mechanism an internal variable is introduced. In this theory, the Helmholtz free energy and a dissipation potential are sufficient to define a complete set of constitutive equations. The coupling between phenomena is described by the state laws, obtained from a free energy potential in which appears the product of function of the associated internal variables. A non-direct coupling [6] between mechanisms is described by the evolution laws obtained from a dissipation potential where the internal variables appear as parameters. This induces an influence of a phenomenon on the rate of another one.

## INTERNAL VARIABLES

The local state method [4] states that the thermodynamical state of a continuum medium at a given material point and at a given time is completely defined by a set of variables. Those variables will be introduced "a priori" to describe some physical mechanisms, and their evolution laws must satisfy the second law of thermodynamics. For a solid material submitted to small transformation there are usually [2] the observable variables, which are the absolute temperature (T) and the total strain ($\varepsilon$), and the internal variables. For the 2024 aluminium alloy four internal variables, noted $\varepsilon^p$, p, $\alpha$ and a are introduced. $\varepsilon^p$ is the classical plastic strain tensor which globally represents the deformation induced by the dislocation movements ; p is a scalar associated to the isotropic hardening and is linked to the density of dislocations and their interactions ; $\alpha$ is a second order tensor associated to the kinematic hardening and takes into account the microstresses due to the incompatibilities between the different slip systems ; $a$ is a scalar associated to the aging and is related to the size and volume fraction of hardening precipitates.

## FREE ENERGY POTENTIAL

The free energy potential $\Psi$ is a differentiable scalar function of all the state variables and contains the first type of coupling, i.e. the direct coupling between the state variables.

$$\Psi\ (\varepsilon\text{-}\varepsilon^p,\ T,\ \alpha,\ p,a) \tag{1}$$

In this work we will suppose that the temperature is constant (hypothesis of isothermal transformations) and for the 2024 aluminium alloy we will propose the following expression:

$$\rho\psi(\varepsilon,T,\varepsilon^p,p,\alpha,a) = W\ (\varepsilon\text{-}\varepsilon^p) + W'\ (\alpha) + W''\ (p,a) \tag{2}$$

where

* $\rho$ is the mass density
* $W(\varepsilon^e) = 1/2\ C_{ijkl}\ \varepsilon^e_{ij}\ \varepsilon^e_{kl}$ , with $\varepsilon^e = \varepsilon - \varepsilon^p$, is the elastic strain energy, C is the elasticity tensor.
* $W'\ (\alpha) = 1/2\ c_1\ c_0\ \alpha_{ij}\ \alpha_{ij}$ is the energy density related to the kinematic hardening.

$c_1$ and $c_0$ are material dependent parameters.

* $W''\ (p,a) = b[p + \frac{1}{\gamma}\ e^{-\gamma p}] + a\ (cp - L) + L$ is the energy density related to the isotropic hardening. The first part of the expression is the classical term giving the isotropic hardening in plasticity and the second part is a coupled term of p and $a$ which gives the isotropic hardening due to the aging. b, $\gamma$, c and L are material dependent parameters.

The associated variables are defined from this potential by partial derivatives. The relations between state variables and associated variables are the so called state laws:

$$\begin{aligned}
\sigma_{ij} &= \rho \frac{\partial \Psi}{\partial \varepsilon^e_{ij}} = C_{ijlk}\ \varepsilon^e_{kl} \Rightarrow \dot{\sigma}_{ij} = C_{ijkl}\ \dot{\varepsilon}^e_{kl} \\
X_{ij} &= \rho \frac{\partial \Psi}{\partial \alpha_{ij}} = c_1\ c_0\ \alpha_{ij} \Rightarrow \dot{X}_{ij} = c_1\ c_0\ \dot{\alpha}_{ij} \\
R &= \rho \frac{\partial \Psi}{\partial p} = b\ (1 - e^{-\gamma p}) + ca \Rightarrow \dot{R} = b\gamma\ e^{-\gamma p}\dot{p} + c\dot{a} \\
Z &= \rho \frac{\partial \Psi}{\partial a} = cp + L \Rightarrow \dot{Z} = c\dot{p}
\end{aligned} \tag{3}$$

From the definition of $W'(\alpha)$ and $W''(p,a)$ and from (3) it is easy to verify that the isotropic hardening is coupled with the aging while the kinematic hardening is not. This is in agreement with the experimental observations in [3].

**POTENTIAL OF DISSIPATION**

To complete the constitutive equations, evolution laws are needed for the internal variables. There are obtained by the introduction of a potential of dissipation $\Phi^*$ depending upon the associated variables.

In the potential $\Phi^*$ appears the second kind of coupling : the non direct coupling obtained by considering the state variables as parameters in the potential of dissipation

$$\Phi^*(\sigma,R,X,Z,\ \text{grad}\ T\ ;\ T,\varepsilon^a,p,\alpha,a) \tag{4}$$

For the 2024 aluminium alloy, we will still make use of the hypothesis of isothermal transformations and propose the following expression for $\Phi^*$

$$\Phi^*(\sigma,R,X,Z,\ \text{grad}\ T\ ;\ T,\varepsilon^p,p,\alpha,a) = \Phi^*_p(\sigma,R,X) + \Phi^*_a(Z;\varepsilon^a,p,a) \tag{5}$$

where

* $\Phi^*_p(\sigma,R,X) = I_C(\sigma,R,X)$ is the part of $\Phi^*$ associated to plasticity.

  It was chosen as the indicator function of a convex set C defined of the following form :
  $C = \{(\sigma, R, X) \mid f(\sigma,R,X) = J_2(\sigma - X) - R - \sigma_y \leq 0\}$
  with $J_2(\sigma - X) = [\,3/2\,(S - X) : (S - X)]^{1/2} = [\,3/2\,(S_{ij} - X_{ij})(S_{ij} - X_{ij})]^{1/2}$
  and $S$ is the deviator of $\sigma$. $\sigma_y$ is the yield stress

* $\Phi^*_a(Z;\varepsilon^a,p,a) = [\alpha - \beta \exp(-\gamma p)] < a_\infty - a >< -Z >$ with $a_\infty = A + (1 - A) \exp[-k'(\varepsilon^p_{II})_{max}]$

  and $(\varepsilon^p_{II})_{max} = \text{Max}_{0 \to t}\left((2/3\ \varepsilon^p : \varepsilon^p)^{1/2}\right)$
  $\alpha$, B, $\gamma$, A, k' are positive material parameters such that $\alpha > \beta$ and $0 \leq A \leq 1$
  $< a_\infty - a >$ is defined by $< a_\infty - a > = \text{Max}\,[(a_\infty - a), 0]$

The evolution laws are derived from the potential of dissipation by the normality rule :

$$\dot{\varepsilon}^p = \frac{\partial f}{\partial \sigma}\dot{\Lambda} = = \frac{3}{2}\,\frac{S - X}{J_2(\sigma - X)}\,\dot{\Lambda}$$

$$\dot{p} = -\frac{\partial f}{\partial R}\dot{\Lambda} = \dot{\Lambda} \tag{6}$$

$$\dot{\alpha} = -\frac{\partial f}{\partial X}\,\dot{\Lambda} = \frac{3}{2}\,\frac{S - X}{J_2(\sigma - X)}$$

$$\dot{a} = [\alpha - \beta\, e^{-\gamma p}] < a_\infty - a > H(cp - L) \quad \text{with H the Heaviside function.}$$

$\dot{\Lambda}$ is the plastic multiplier which may be calculated from the consistency conditions, (3) and (5) make a complete set of constitutive equations for the 2024 aluminium alloy. This model can be improved by considering a term of perturbation in the evolution law of $\alpha$ in order to introduce a non linearity in the kinematic hardening evolution law [5] :

$$\dot{\alpha} = \dot{\varepsilon}^p - \varphi(p)\, X \dot{p} \tag{7}$$

with $\varphi(p) = \frac{1}{c0}\,[\,\varphi_\infty + (1 - \varphi_\infty) \exp(-\eta p)]$ where $\varphi_\infty$ is a material dependant parameter

If (7) is considered we have the following expression for $\dot{\Lambda}$ :

$$\dot{\Lambda} = 0 \quad \text{if } f < 0 \text{ or if } f = 0 \text{ and } \dot{f} < 0$$

$$\dot{\Lambda} = \left\langle \frac{2/3\,(S-X):\dot{\sigma}}{J_2(\sigma-X)} - c\dot{a} \right\rangle \Big/ c_1\left(c_0 - \frac{2}{3}\,\varphi(p)\,\frac{(S-X):X}{J_2(\sigma-X)} + b\gamma e^{-\gamma p}\right) \quad \text{if } f = 0 \text{ and } \dot{f} = 0$$

In the evolution law for $a$ , $a$ (t = 0) is supposed to be zero and the asymptotic value $a_\infty$ is supposed to be equal to one in a process without plastic strain. This expression is based on experimental results obtained by mechanical measures in the following form : from (3) we can see that, if p is constant, the variable $a$ can be obtained by the expression

$$c\,a(t) - c\,a(0) = R(t) - R(0)$$

and so $\quad a(t) = [R(t) - R(0)]/c \qquad (8)$

because $\quad a(0) = 0$

As $\lim_{t \to +\infty} a\,(t, p = 0) = 1$, the constant c is equal to $\lim_{t \to +\infty} R\,(t, p = 0) - R(0, p = 0)$

where $\lim_{t \to +\infty} R\,(t, p = 0)$ is the asymptotic value of R without previous hardening and R(0, p = 0) is the initial value of R without previous hardening.

In this case, the internal variable $a$ represents the relative change of the variable R, whose evolution can be easily measured [2]. Experimental results show that the rate of aging is increased by the plastic strain and that the saturation value of $a$ is reduced by the plastic strain, which is also taken in account by the theory.

## REFERENCES

[1] KREMPL E.
Viscoplasticity based on total strain. The modelling of creep with special considerations of initial strain and aging, J. of Eng. Mat. and Tech., vol. 101,1979.

[2] LEMAITRE J., CHABOCHE J.L.
Mécanique des Matériaux Solides. Dunod, 2nd Ed. 1988.

[3] MARQUIS D.
Thermodynamique et Phénoménologie. Thèse d'état Université de Paris 1989. English translation NASA TT 20828.

[4] GERMAIN P.
Mécanique des milieux continus - Tome 1, Dunod, 1973.

[5] MARQUIS D.
Modélisation et identification de l'écrouissage anisotrope des métaux, Thèse de Troisième Cycle, Paris 6, E.N.S. de Cachan.

[6] MARQUIS D., LEMAITRE J.
Constitutive equations for the coupling between elasto-plasticity, damage and aging, Revue de Physique Appliquée, 1988, 23, 615-624

# BOUNDING SURFACE THERMOPLASTICITY AND THERMOVISCOPLASTICITY

DAVID L. MCDOWELL
Georgia Institute of Technology
Atlanta, Georgia 30332-0405

## ABSTRACT

The multiple backstress decomposition theory proposed by Moosbrugger and McDowell [1] is extended to thermomechanical cyclic loading. Both thermoplastic and thermoviscoplastic forms are presented. Predictions of the thermoviscoplastic model are compared with experiments for in-phase temperature-strain cycling of OFHC copper.

## MODEL STRUCTURE

We adopt the yield condition $f = (3/2)||\underset{\sim}{s}-\underset{\sim}{\alpha}||^2 - R^2$ where $\underset{\sim}{s} = \underset{\sim}{\sigma} - (\sigma_{kk}/3)\underset{\sim}{I}$ is the deviatoric stress and $\underset{\sim}{\alpha}$ is the deviatoric backstress. We decompose $\underset{\sim}{\alpha}$ into multiple components [1-2], i.e. $\underset{\sim}{\alpha} = \Sigma\underset{\sim}{\alpha}_i$ for $i = 1,2,...,N$. The yield surface radius R is composed of a temperature dependent component $R^o$ and a component $R^{iso}$ which depends on the degree of isotropic hardening, i.e. $R = R^o(T) + R^{iso}$. We define $p = ||\dot{\underset{\sim}{\epsilon}}^p|| = (\dot{\epsilon}_{ij}^p\dot{\epsilon}_{ji}^p)^{1/2}$ and T as the absolute temperature. The $i^{th}$ backstress amplitude or saturation level $b_i$ is decomposed according to $b_i = b_i^o(T) + b_i^{iso}$, where $b_i^o(T)$ depends only on temperature and $b_i^{iso}$ depends on the degree of isotropic hardening. Isotropic hardening is partitioned according to [1-2]

$$\dot{R}^{iso} = \sqrt{\frac{3}{2}}\,(1 - \sum_{i=1}^{N} \omega_i)\,\dot{\chi} , \qquad \dot{b}_i^{iso} = \omega_i\dot{\chi} \tag{1}$$

where $\dot{\chi} = \mu[\bar{\chi}(\phi_j,T)-\chi]\dot{p} + (\partial\chi/\partial T)\dot{T}$ is the evolution rate of isotropic hardening variable $\chi$ representative of the increase of dislocation density, dislocation substructure development, etc. Parameter set $\phi_j$ denotes dependence on nonproportionality of loading and plastic strain amplitude, for example. The factor $\omega_i$ ($0 \le \Sigma\omega_i \le 1$) partitions the isotropic hardening between the $b_i^{iso}$ and $R^{iso}$. Parameter $\bar{\chi}$ represents the saturation level of $\chi$ for isotropic hardening. Components $R^o(T)$ and $b_i^o(T)$ are independent of isotropic hardening.

The incompressible flow rule is given by

$$\dot{\underset{\sim}{\epsilon}}^p = \frac{1}{H} \langle \dot{\underset{\sim}{s}}:\underset{\sim}{n} - \Lambda \dot{T} \rangle \underset{\sim}{n} = \frac{1}{H} \langle \psi \rangle \underset{\sim}{n} \quad \text{(thermoplasticity)} \tag{2}$$

$$\dot{\underset{\sim}{\epsilon}}^p = \sqrt{\frac{3}{2}}\, \xi^n \exp(B_0 \xi^{n+1})\ \theta\ \underset{\sim}{n} \quad \text{(thermoviscoplasticity)} \tag{3}$$

where $< >$ denote Macauley brackets and $\underset{\sim}{n}$ is the unit vector in the plastic strain rate direction, i.e. $\underset{\sim}{n} = (\underset{\sim}{s}-\underset{\sim}{\alpha})/||\underset{\sim}{s}-\underset{\sim}{\alpha}||$. Note that a temperature rate term enters into the argument $\psi$ in addition to the projection of the stress rate onto the direction of plastic flow. The temperature rate terms contribute to evolution of $\underset{\sim}{\alpha}$ and $\chi$ even for stresses within the yield surface. The $\theta$ function introduces explicit temperature dependence in the viscoplastic flow rule [2]. Parameter $\xi$ is defined by

$$\xi = \langle \sqrt{\frac{3}{2}}\ ||\underset{\sim}{s} - \underset{\sim}{\alpha}|| - R \rangle / D = \Sigma_v / D \tag{4}$$

where $\Sigma_v$ is the viscous stress and D is the drag stress. To complete the stress-strain relations, we assume linearized, isotropic, decoupled thermohypoelasticity.

A convenient assumption is that of temperature history independent evolution of internal variables. We neglect static thermal recovery. One may assume invariance of the ratios $\underset{\sim}{m}_i = \underset{\sim}{\alpha}_i/(C_i b_i)$ and $M = \chi/(\mu\bar{\chi})$ with a temperature change, leading to the requirements that C and $\mu$ must be temperature independent and that each component of $b_i$ and $\bar{\chi}$ must be multiplicatively separable in T and p, i.e. [3]

$$\Sigma(T)\ \hat{b}_i(p) = b_i^o + b_i^{iso} \tag{5}$$

$$\bar{\chi} = \Sigma(T)\ \hat{\chi}(p) \tag{6}$$

Hence, we may select $b_i^o$ to be of the form $b_i^o = B_i\Sigma(T)$ where $B_i$ is at most a function of p. The complete theory for temperature path history independent evolution of internal variables is summarized as follows, assuming $\omega_i$ are temperature independent:

$$\dot{\underset{\sim}{\alpha}}_i = C_i \left[ b_i \underset{\sim}{n} - \underset{\sim}{\alpha}_i \right] \dot{p} + \left( \omega_i \frac{\underset{\sim}{\alpha}_i}{b_i} \frac{\chi}{\bar{\chi}} \frac{\partial \bar{\chi}}{\partial T} + \frac{\underset{\sim}{\alpha}_i}{b_i} \frac{\partial b_i^o}{\partial T} \right) \dot{T} \tag{7}$$

$$\dot{\chi} = \mu \left[ \bar{\chi} - \chi \right] \dot{p} + \frac{\chi}{\bar{\chi}} \frac{\partial \bar{\chi}}{\partial T} \dot{T} \tag{8}$$

For thermoplasticity, H and $\Lambda$ in (2) are determined by consistency. The initial condition $\chi(0)$ uniquely determines the initial values of $R^{iso}$ and $b_i^{iso}$. The form of the temperature rate terms introduced in this section has adopted elsewhere (c.f. [4] and [5]).

## SOME EXPERIMENTAL CORRELATIONS

It is useful to investigate the performance of the model outlined in the last section, with emphasis on the thermoviscoplastic theory. The experiments to be considered herein were conducted on initially annealed OFHC Copper as reported by Freed [5]. The temperature dependent elastic moduli and thermal expansion coefficient are given in [5].

The backstress $\underset{\sim}{\alpha}$ is decomposed into two components, with one assuming a linear form [1-2]. Material constants and parameters for the temperature path history independent thermoviscoplasticity model include $R^o = 0$, $b^o = 92.6\Sigma(T)$ MPa, $k_1 = 500$, $k_2 = 5000$, $k_3 = 1$, $k_4 = 3$, $\omega = 0.7$, $\mu = 30$, $\bar{\chi} = 200$ MPa, $H^* = 138 + 2000\exp(-0.0105T)$ MPa, $\Sigma(T) = 0.24 - 1.9\text{x}10^{-4}T + 0.45\exp(-0.008T)$, $B_o = 2.2\text{x}10^{-14}$, $n = 5$, $T_m = 1356°K$, $A_o = 0.024$ ($D = A_oR$), $Q = 2\text{x}10^5$J/mol, $k = 8.314$ J/mol-°K and the initial condition $\chi(0) = 3$ MPa. The model was fit to isothermal, cyclically stable hysteresis loops at several temperatures [2]. Figure 1 compares the predictions of the model with the in-phase TMF experimental results. The agreement is qualitatively reasonable. Figure 1 also compares the prediction based on elimination of temperature rate terms with experimental results. The viscoplastic model more accurately captures TMF response when the temperature rate terms are neglected, probably due to the temperature dependence of the viscous flow rule. This is not the case for thermoplasticity theory, as reported elsewhere [6]. Also, as shown by Ohno and Wang [7], temperature rate terms should be retained for weakly hardening kinematic hardening variables (e.g. linear term) to avoid unrealistic cycle-by-cycle shifting of TMF hysteresis loops due to temperature path history dependence.

## CONCLUSIONS & ACKNOWLEDGEMENTS

Nonlinear kinematic-isotropic hardening thermoplastic and thermoviscoplastic theories have been presented. A thermomechanical experiment on OFHC Copper was predicted with the thermoviscoplastic theory. Inclusion of temperature rate terms is apparently not essential, for the case considered, to match nonisothermal response.

The author is grateful to Texas Instruments, Inc. and the National Science Foundation (ENG MSM 860-1889) for support of this and related research.

## REFERENCES

1. Moosbrugger, J.C. and McDowell, D.L., "A Rate-Dependent Bounding Surface Model with a Generalized Image Point for Cyclic Nonproportional Viscoplasticity," Journal of the Mechanics and Physics of Solids, Vol. 38, No. 5, 1990, pp. 627-656.
2. McDowell, D.L., "Bounding Surface Thermoplasticity and Thermoviscoplasticity, MOMRG Report 105, Mechanical Engineering, Georgia Institute of Technology, Atlanta, GA, December 1990.

3. Ohno, N. and Wang, J., "Nonisothermal Constitutive Modeling of Inelasticity Based on Bounding Surface," Proc. Seventh International Seminar on Inelastic Analysis, Fracture and Life Prediction, SMiRT 10 Post-Conference, University of California, Santa Barbara, August 21-22, 1989, pp. B.1/1-B.1/24.
4. Chaboche, J.L., "Constitutive Equations for Cyclic Plasticity and Cyclic Viscoplasticity," Int. J. Plasticity, Vol. 5, No. 3, 1989, pp. 247-302.
5. Freed, A.D., "Thermoviscoplastic Model with Application to Copper," NASA Tech. Paper 2845, December 1988.
6. McDowell, D.L., "Thermoplastic and Thermoviscoplastic Forms of Bounding Surface Theories," Constitutive Laws for Engineering Materials, Eds. Desai, Krempl, Frantziskonis and Saadatmanesh, ASME Press, 1991, pp. 350-362.
7. Ohno, N. and Wang, J.-D., "Multisurface and Multicomponent Forms of Nonlinear Kinematic Hardening: Application to Nonisothermal Plasticity," Constitutive Laws for Engineering Materials, Eds. Desai, Krempl, Frantziskonis and Saadatmanesh, ASME Press, 1991, pp. 219-222.

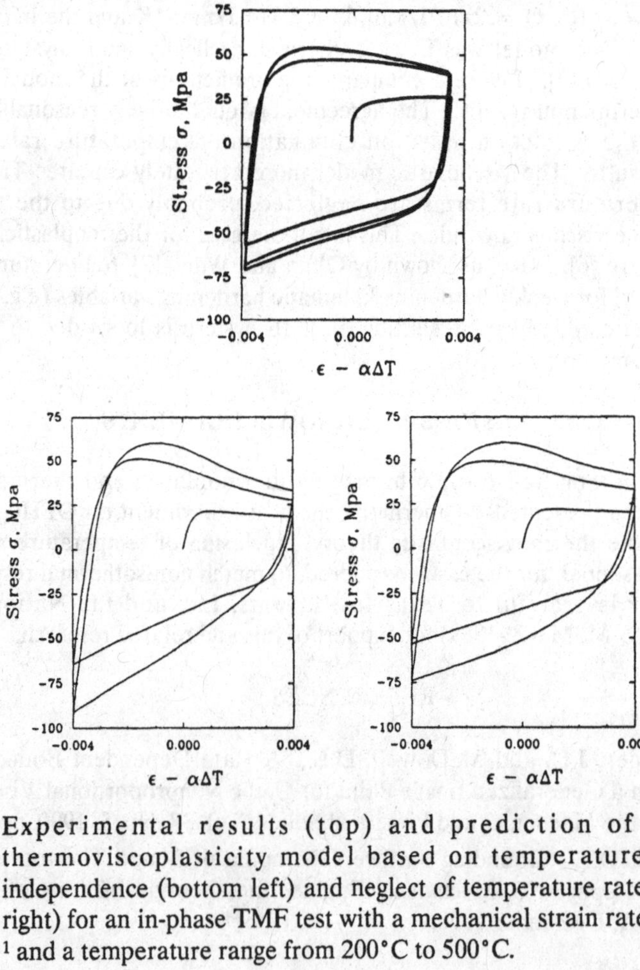


Figure 1. Experimental results (top) and prediction of two surface thermoviscoplasticity model based on temperature path history independence (bottom left) and neglect of temperature rate terms (bottom right) for an in-phase TMF test with a mechanical strain rate of $1.5 \times 10^{-5}$ sec$^{-1}$ and a temperature range from 200°C to 500°C.

# EQUILIBRIUM BETWEEN COHERENT PHASES

INGO MÜLLER
Physikalische Ingenieurwissenschaft
TU Berlin

## ABSTRACT

**If two solid phases are coherent, the free energy of the phase mixture will contain a term representing the coherency energy. That energy will stabilize the pure phases. Thus there will be a hysteresis loop in the load-deformation diagram and phase equilibria will all be unstable.**

## PHASE EQUILIBRIA

The mathematical description of phase equilibrium requires a non-convex free energy $\Psi$ as a function of length $D$ so that the load

$$P = \frac{\partial \Psi}{\partial D} \tag{1}$$

may be equal in the two phases at different values of $D$. If we denote the two phases by $'$ and $''$ the free energy and the length of the phase mixture may be written in the form

$$\Psi = \Psi(D') + \Psi(D'') + Am'\frac{m''}{m} \quad \text{and} \quad D = D' + D''. \tag{2}$$

Here the term $Am'\frac{m''}{m}$ represents a coherency energy along interfacial boundaries whose number we take to be proportional to $m'\frac{m''}{m}$ where $m$ denotes the mass. That assumption may be satisfactory for a random distribution of interfaces. We introduce specific values

$$\psi = \frac{\Psi}{m}, \psi(d') = \frac{\Psi(D')}{m'}, \psi(d'') = \frac{\Psi(D'')}{m''}, d = \frac{D}{m}, d' = \frac{D'}{m'}, d'' = \frac{D''}{m''} \tag{3}$$

and the phase fraction $z = \frac{m''}{m}$ and obtain from (2)

$$\psi = (1-z)\psi(d') + z\psi(d'') + A(1-z)z \qquad \text{and} \qquad d = (1-z)d' + zd''. \tag{4}$$

$\psi$ is a function of three variables, viz. $d', d'', z$ and for equilibrium to prevail this function must have a minimum, if $d$ is fixed. The mathematical procedure of finding a minimum subject to a constraint is well-known and we obtain three equations for the determination of the three equilibrium values $d'_E, d''_E, z_E$, viz.

$$\frac{\partial\psi}{\partial d}\Big|_{d'_E} = \frac{\partial\psi}{\partial d}\Big|_{d''_E} = \frac{\psi(d''_E) - \psi(d'_E) + A(1-2z_E)}{d''_E - d'_E} \quad \text{and} \quad d = (1-z_E)d'_E + z_E d''_E \tag{5}$$

The exploitation of these conditions requires the knowledge of the form of the function $\psi(d)$ which is not given here explicitly. All we know is that $\psi(d)$ is non-convex so that $P = \frac{\partial\psi}{\partial d}$ is a non-monotone function of $d$. That, however, is enough to understand the principal results following from (5). Let us consider:

Equation $(5)_1$ states that the loads on the two phases must be equal. Therefore the equations $(5)_{1,2}$ may be written in the form

$$P_E(d''_E - d'_E) - \int_{d'_E}^{d''_E} P(\alpha)d\alpha = A(1-2z_E), \tag{6}$$

where $P_E = \frac{\partial\psi}{\partial d}|_{d'_E} = \frac{\partial\psi}{\partial d}|_{d''_E}$ is the common load on the phases in equilibrium. Equation (6) lends itself for an easy graphical solution which is illustrated in Figure 1: The area of the rectangle formed by $P_E$ and $d''_E - d'_E$ must differ from the area under the $(P,d)$-curve between $d'_E$ and $d''_E$ by the amount $A(1-2z_E)$. It is obvious that for any given $z_E$ this construction provides unique values $P_E$ and $d'_E, d''_E$ and hence follows, by $(5)_3$, the corresponding value of $d$, see Figure 1. We conclude that $P_E, d'_E, d''_E$ all depend on the phase fraction and that the curve $P_E(d)$ has a negative slope so that all phase-equilibria are unstable.

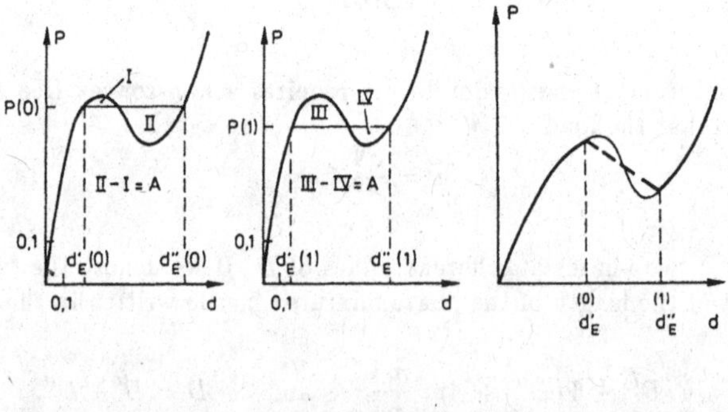


Figure 1: Phase equilibria for different values of the phase fraction.

If $A$ were zero the graphical construction of Figure 1 would be the same for all $z$ and it would amount to the "Maxwell rule of equal areas". In that case the phase equilibria would be indifferent.

## HYSTERESIS

As we start loading, beginning in the origin with $P = 0, d = 0$ we have the body in phase $'$, i.e. with $z = 0$. Equilibrium is reached with phase $''$ at the load $P_E(0)$, see Figure 1a. This is therefore the "yield load", where the phase transition $' \rightarrow ''$ occurs. On the other hand, if we start with phase $''$ at a large load and a large deformation and unload, we reach phase equilibrium with phase $'$ at the load $P_E(1)$ and that determines the "recovery load". Thus we conclude that in a loading-unloading process the states run through a hysteresis loop and the area inside that loop is given by 2A, see Figure 2.

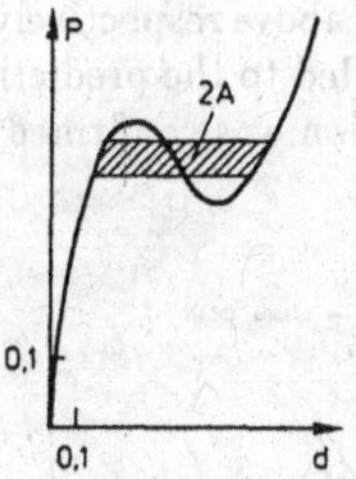


Figure 2: The hysteresis loop in a phase transition between coherent phases.

## METASTABLE STATES

When the yield is interrupted before the phase transition is complete and the load is reduced, the state of a tensile specimen moves steeply into the hysteresis loop and, at some interior point, it bends in what we call an "internal recovery". The load of that internal recovery is higher the earlier the yield was interrupted. This phenomenon is illustrated by Figure 3a which represents a plot taken from a tensile testing machine. A similar phenomenon arises when the recovery is interrupted and the load is increased: In that case there is "internal yield" as illustrated in Figure 3b.

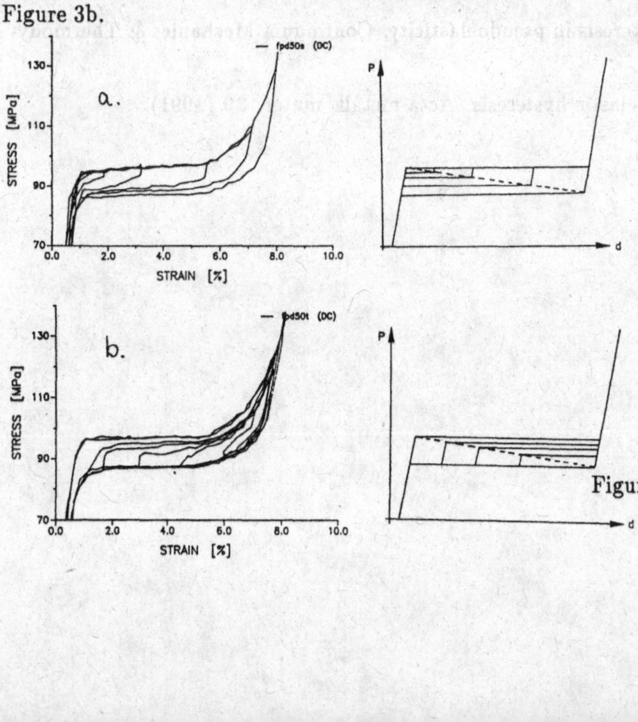


Figure 3: Internal recovery and yield. Left: experiment. Right: theory.

From the theory presented in Sections 1 and 2 the following conjecture seems inescapable for an interpretation of the experimental curves of Figure 3: The steep lines into the hysteresis loop must be metastable states of constant phase fraction. We call them metastable, because along these lines the specimen is obviously not in phase equilibrium. But as soon, as the state along those metastable lines reaches the line of unstable phase equilibrium, the metastability is lost and internal yield or recovery sets in depending on whether we were loading or unloading This conjecture is illustrated on the right hand side of Figure 3.

We may thus say that the line of unstable phase equilibrium triggers internal yield or recovery when it is reached from below or above respectively. This expectation was first expressed by an extrapolation of the theory and it led to the prediction of internal spirals as shown on the right hand side of Figure 4. The prediction was confirmed by experiment, see the left hand side of Figure 4.

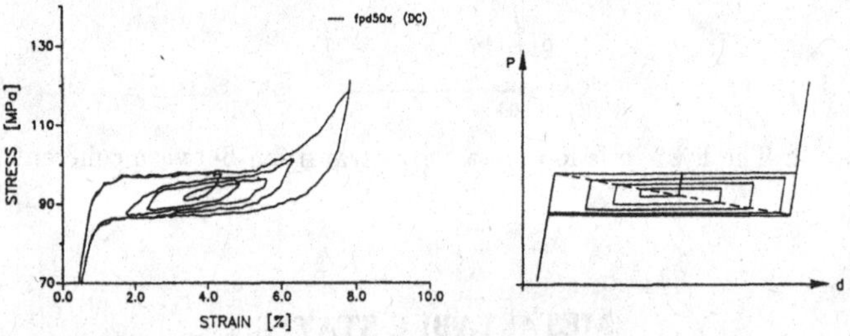


Figure 4: Internal hysteresis loop

## REFERENCES

Müller, I. On the size of the hysteresis in pseudoelasticity. Continuum Mechanics & Thermodynamics **1** (1989)

Müller, I. Xu, H. On the pseudoelastic hysteresis. Acta metall. mater. **39** (1991).

# STABILITY ON THE YIELD SURFACE

A. PAGLIETTI
Istituto di Meccanica Teorica ed Applicata,
Università di Udine, 33100–Udine, Italy

ABSTRACT

It is proved that, if plastic deformation is unrestrained, the yield surface has to be equipotential for free energy. On the other hand, if plastic deformation is isochoric, then the intersections of the yield surface with the planes at constant volume must be equipotential.

## 1.INTRODUCTION

Plastic deformation dramatically increases the degrees of freedom of an elastic–plastic material. Under purely elastic conditions, the state of deformation of a material element is fully determined by the six independent components of the elastic strain tensor $\epsilon^e$. By properly choosing the reference configuration, the latter can be made to coincide with the total strain $\epsilon$. As the material reaches the yield limit, however, the whole range of plastic deformations becomes accessible. The independent deformation variables increase from six to twelve (or eleven), since six components of the plastic strain tensor $\epsilon^p$ (or only five independent components, if plastic deformation is assumed to be isochoric) must now join the elastic strain to fully describe the deformation of the material.

Such a change in deformation kinematics makes it possible for a material element to undergo changes in its elastic and plastic deformation, still leaving its total strain unaltered. However, the forces (stress) that act on the element surface cannot perform any work if the total strain is kept constant. This suggests that the values of free energy at the yield surface must somehow be restricted. If not, the element could move to a more energetically convenient state of elastic deformation on the yield surface, still keeping its total deformation unaltered and, therefore, without absorbing any work from the surroundings. This would clearly mean instability of the deformation at yield. It would also mean that once a material element reaches the yield surface, its stress could no more be controlled by controlling its total deformation. Both elastic and plastic strain would start to change at constant total strain as the material seeks the equilibrium state. In the process, the point representing the state of the material would clearly move along the yield surface, as this is the only place where changes in plastic deformation can occur.

An instability of this kind has not been observed experimentally. It has not, at least, for the elastic–plastic materials usually exploited in practice. We shall presently prove that, if this instability is to be ruled out, then the yield surface must be equipotential for the free energy of the material, provided that there is no constraint to plastic deformation. On the other hand, if plastic deformation occurs at constant volume, then it is its intersections with the constant volume planes – and not the yield surface as a whole– that must be equipotential.

The stability considered in this paper is, of course, quite another thing from Drucker's celebrated stability of elastic–plastic materials. The latter concerns the way in which the yield limit is affected by plastic straining. On the contrary, the present kind of stability concerns the state of deformation at yield, irrespectively of whether the material becomes harder or softer when plastically strained. As is true for Drucker's stability, the present stability too entails restrictions on the form of the yield surface. It may be worth noting, for instance, that for plastically incompressible materials the von Mises yield condition is stable in the sense of the present paper, while the Tresca condition is not.

## 2. UNCONSTRAINED PLASTIC DEFORMATION: EQUIPOTENTIAL YIELD SURFACE

Isothermal processes will be considered throughout. The explicit dependence of temperature can accordingly be omitted from the constitutive equations. In strain space, the yield surface of an elastic–plastic material is therefore represented by the equation

$$f(\epsilon^e, \xi) = 0, \tag{2.1}$$

where $\xi$ denotes the set of variables that help to define the state of inelastic deformation of the material – plastic strain may be included. Once the relation between $\epsilon$, $\epsilon^e$ and $\epsilon^p$ is specified (for instance $\epsilon = \epsilon^e + \epsilon^p$ for small deformation plasticity), then eq.(2.1) can be represented equally well in $\epsilon^e$–space or in $\epsilon$–space.

In the present section the six components of the plastic strain tensor are supposed to be free from any kinematic constraint. The case of isochoric plastic deformation will be dealt with in the next section. Under this hypothesis and for fixed values of $\xi$ every point of surface (2.1) must be at the same free energy. Should this not be so, the states of deformation at the yield surface could not be stable. This can be proved in the following way. Let us refer to an infinitesimal element of material, whose state of deformation lies on the yield surface. Starting from this, let us consider a virtual variation in strain, say $\delta\epsilon$, $\delta\epsilon^e$ and $\delta\epsilon^p$, meeting the following conditions:

$$\sigma_{ij}\,\delta\epsilon_{ij} = 0 \tag{2.2}$$

$$\delta\epsilon^e_{ij} = \mu \frac{\partial f}{\partial \epsilon^e_{ij}} \tag{2.3}$$

$$\delta\epsilon^p_{ij} = \delta\epsilon_{ij} - \delta\epsilon^e_{ij} \tag{2.4}$$

and

$$\sigma_{ij}\,\delta\epsilon^p_{ij} \geq 0, \tag{2.5}$$

the quantity $\mu$ appearing in (2.3) being an infinitesimal constant. Condition (2.2) can always be met since the six components of tensor $\delta\epsilon$ can be assigned arbitrarily, while eq. (2.2) sets just one condition among them. A possible choice could be $\delta\epsilon_{ij} = 0$ , but this would be a much stronger assumption than what eq.(2.2) would strictly demand. Condition (2.3) is needed in order to guarantee that the considered virtual strain does not take the material out of the yield surface, i.e. back into the elastic region. This is an essential condition, as no plastic deformation –and hence no virtual plastic strain – could otherwise occur. Condition (2.4) follows simply from the fact that virtual strains are infinitesimal, which means that additive composition applies to them. Condition (2.5) is finally needed in order to make the considered virtual plastic strain consistent with the well–known fact that plastic deformation dissipates work. Of course, this condition implies the following restriction on $\delta\epsilon^e_{ij}$

$$\sigma_{ij}\,\delta\epsilon^p_{ij} = \sigma_{ij}\,\delta\epsilon_{ij} - \sigma_{ij}\,\delta\epsilon^e_{ij} = -\,\sigma_{ij}\,\delta\epsilon^e_{ij} \geq 0\,, \tag{2.6}$$

as immediately follows from eqs. (2.4) and (2.5). This can always be met though,as we can always switch from $\delta\epsilon^e_{ij}$ to $-\,\delta\epsilon^e_{ij}$, in case $\sigma_{ij}$ should turn up to be negative.

When expressed in terms of virtual work, equilibrium requires that external virtual work $\delta L_{ext}$ and internal virtual work $\delta L_{int}$ should meet the relation $\delta L_{ext} + \delta L_{int} \leq 0$, the inequality sign holding true for non–invertible virtual strains. For the considered material element, $\delta L_{ext}$ is the work of the forces (stress) acting on the element surface. In view of eq. (2.2) we have, therefore, $\delta L_{ext}=0$. This means that equilibrium of the element requires that

$$\delta L_{int} \leq 0 \tag{2.7}$$

But, $\delta L_{int}$ can be expressed as

$$\delta L_{int} = \rho\,\delta\psi + \delta w^p\,, \tag{2.8}$$

where $\rho$ is mass density, $\psi=\psi(\epsilon^e,\xi)$ is the specific free energy of the material (a single–valued function of $\epsilon^e$ for constant $\xi$), while $w^p$ is plastic work per unit volume.

The latter is given by

$$\delta w^p = \sigma_{ij}\, \delta\epsilon^p_{ij} \geq 0 \,, \tag{2.9}$$

the inequality following from condition (2.5).

The combine of (2,7), (2.8) and (2.9) implies that

$$\delta\, \psi \leq 0. \tag{2.10}$$

This result must be met at every point of the yield surface. It means that the free energy of the material cannot increase along the yield surface. For constant $\xi$, however, the free energy $\psi$ is a single–valued function of $\epsilon^e$, while there is no constraint whatsoever on the direction of the purely elastic processes connecting any two points on the yield surface. This means that relation (2.10) can only be met if

$$\delta\psi = 0 \tag{2.11}$$

at the yield surface, which in turn implies that the latter is an equipotential surface for free energy.

## 3. PLASTIC INCOMPRESSIBILITY: EQUIPOTENTIAL SECTIONS OF THE YIELD SURFACE

Plastic incompressibility adds the further condition

$$\delta\epsilon^p_{ii} = 0 \tag{3.1}$$

to restrictions (2.2)–(2.5). In order to meet it, let us limit our attention to elastic virtual strains that occur in the [$I_{\epsilon^e}$ = const.]–sections of the yield surface; $I_{\epsilon^e}$ denoting the first strain invariant of $\epsilon^e$. For such elastic virtual strains we have that

$$\delta\epsilon^e_{ii} = 0 \,, \tag{3.2}$$

which in view of eqs (3.1) and (2.4) also implies that

$$\delta\epsilon_{ii} = 0. \tag{3.3}$$

The same analysis of the previous section can now be repeated by considering virtual strains $\delta\epsilon_{ij}$ and $\delta\epsilon^e_{ij}$ that meet restrictions (3.2) and (3.3). It leads to the conclusion that $\psi$ must be constant on the considered sections of the yield surface.

It may be interesting to observe that, being a limitation to the distortion free energy the material can store, the von Mises yield surface can be expressed as

$$\psi(\epsilon^e, \xi) = \psi_v(I_{\epsilon^e}) + k \,, \tag{3.4}$$

where $\psi_v(I_{\epsilon^e})$ is the part of $\psi$ that is due to volume changes, while k is the maximum energy of distortion. This clearly means that the sections $I_{\epsilon^e}$ = const. of this surface are equipotential. Thus, the von Mises yield condition is stable in the sense of this paper, while an otherwise reasonable surface, as that following from the Tresca condition, is not.

# INELASTIC BEHAVIOR OF POLYCRYSTALLINE METAL UNDER VARYING TEMPERATURE CONDITIONS

M.TOKUDA, Y.INAGAKI, F.HAVLICEK
Department of Mechanical Engineering , Mie University
Kamihama 1515, Tsu 514 JAPAN

## ABSTRACT

The inelastic behaviors of polycrystalline metals were investigated experimentally by using thin-walled tubular specimens of aluminium alloy subjected to combined loads of axial force and torque under varying temperature conditions. The temperature range was selected to be between the room temperature and 250 ℃. Also, a set of constitutive equations were formulated on the basis of crystal plasticity and the equations were confirmed to reproduce the complex behaviors of polycrystalline metals observed in the experiments within a reasonable accuracy.

## INTRODUCTION

The numerical analysis may be quite powerful, for example, Finite Element Method, for the assessment of reliability and safety for newly designed machines and structures. In the analysis, however, reasonably compact and reliable constitutive equations are indispenesable. We investigated the constitutive equations valid for the varying temperature/complex loading condition. In this investigation, first, fundamental experiments were systematically performed by using thin-walled tubular specimens subjected to complex loadings of axial force and torque under the varying temperature. Several typical features of inelastic behaviors related with changes of temperature were observed in these experiments. Concerning the formulation of constitutive equations, almost all proposed constitutive equations [1-4] were formulated on the basis of the phenomenological approach. However the inelastic behaviors of polycrystalline metals observed in the experiments seem to be too complex for us to formulate the constitutive equations by any phenomenological approach. In authors' opinion, such complex behaviors may be reproduced by using the combination of a few simple (micro-) mechanisms well-investigated in the crystal plasticity. In this paper, an example of such (a set of) constitutive equations was presented and confirmed to reproduce the observed complex behaviors in a reasonable accuracy.

## EXPERIMENTS AND RESULTS

The specimens are thin-walled tubes made of aluminium alloy (A5056, JIS) . The experimental apparatus is the SHIMADZU AUTOGRAPH AG-10TCS which can control automatically combined loads of axial force and torque as well as the temperature (max.1100 ℃). In this paper, the experimental results are described by using the parameters related with stress and strain vectors $\underline{s}$ and $\underline{e}$ proposed by Ilyushin. In the present case , the following two-dimensional forms of these vectors are convenient as

$$\underline{e} = \varepsilon \underline{n}_1 + (\gamma / \sqrt{3})\underline{n}_3, \qquad \underline{s} = \sigma \underline{n}_1 + (\sqrt{3}\ \tau)\underline{n}_3 \tag{1}$$

where $\varepsilon$ and $\gamma$ are axial and shear strains of tubular specimen, respectively and $\sigma$ and $\tau$ are axial and shear stresses, respectively and, $\underline{n}_1$ and $\underline{n}_3$ are a set of orthonormal base vectors. The strain path is given in the vector plane of strain ($\varepsilon$, $\gamma / \sqrt{3}$). In the every experiment, the strain rate is controlled to be constant as $|\dot{\underline{e}}| = |d\underline{e}/dt| = 3x10^{-5}$/sec, where $|*|$ denotes the magnitude of vector *. Figure 1 shows the stress - strain relations obtained by the strain path along the $\varepsilon$-axis (uniaxial tension test), under constant temperature conditions. Figure 2 shows the $|\underline{s}|$ - L relations obtained by the bi-linear strain path (a typical example of complex loading paths) with right angle corner as shown in the inserted figure, under constant temperature conditions, where L is an accumulated strain defined as $L = \int |\dot{\underline{e}}|dt$. Figures 3 and 4 shows the results obtained by by the uniaxial tension tests with a change of temperature at $|\underline{e}|$=3% (150℃ → 200 ℃, 200 ℃ → 150 ℃, 150 ℃ → 250 ℃, 250 ℃ → 150 ℃). Figure 5 shows the results for the bi-linear strain path in which the temperature changed at $|\underline{e}|$ = 1.5% and the right angle corner is at $|\underline{e}|$ = 3%. These figures reveal several typical features of inelastic behaviors of polycrystalline metal (in this case , alluminium alloy) under complex loading/varying temperature conditions.

## CONSTITUTIVE EQUATIONS OF POLYCRYSTALLINE METALS UNDER VARYING TEMPERATURE CONDITIONS

**We consider the following three important semi-micro mechanisms (1)-(3) which may control the inelastic behavior of polycrystalline metal observed in the above experiments. (1) a slip mechanism in a slip system (the driving mechanism of dislocations in the slip system closely related with the interaction mechanisms among dislocations); In order to incorporate this mechanism, the following thermo-activated shear slip and shear stress relation is employed:**

$$\dot{\gamma}^{(n)} = \dot{\gamma}_0 \exp[-F_0/(kT)\{1-\tau^{(n)}/\tau_y^{(n)}\}] \qquad (2)$$

where $\dot{\gamma}$ is slip strain rate $\dot{\gamma}_0$ is a material constant, $F_0$ is the activation energy, k is the Boltzmann constant, T is the absolute temperature, $\tau$ is the resolved shear stresse , $\tau_y$ is the critical resolved shear stress by which the dislocations can move without any aid of thermal activation, and $*^{(n)}$ denotes that the parameter * belongs to the n-th slip system (n = 1,2,.., N;N is the number of slip systems in a single crystal), (2) hardening of single crystal component (related closely with the interaction mechanisms between slip systems in a grain) ; in order to incorporate this mechanism in proper way we use the following hardening rule( including the softening mechanisms caused by the anihilations among dislocations with opposite sign).

$$\dot{\tau}^{(n)} = H\{\sum_{j=1}^{N}\dot{\gamma}^{(j)}\}^{p-1}\{\sum_{j=1}^{N}\dot{\gamma}^{(j)}\} - B(\tau_y^{(n)} - \tau_0)^m \exp[-Q_D/(kT)] \qquad \ldots\ldots\ldots\ldots (3)$$

where H,B,and P are the material constants, $Q_D$ is the activation energy for diffusion, (3) strain constant idea is used for incorporating the interaction mechanisms among crystal components in the polycrystal, that is,

$$e_{ij(k)} = e_{ij} \quad (i,j=1,2,3) \qquad (4)$$

where $e_{ij}$ and $e_{ij(k)}$ are strain deviators of the polycrystal model and of the k-th grain component (k = 1,2,.., M; M is the number of single crystal components in the polycrystal model, which is enough large to give the polycrytal model an initial isotropy), respectively. The concrete form of constitutive equations based on the above three mechanisms has been already described in the previous paper [5], in a form of internal variable theory.

The results reproduced by the constitutive equations are shown by using the solid and dashed curves in Figs. 1-5. As found from the figures, the presented constitutive equations can reproduce the typical features observed in the experiments on aluminium alloy under fundamental loading conditions with the change of temperature. The used material parameters are as follows.

$\dot{\gamma}_0 = 1.3\times10^5$/sec, $F_0/k = 2.3\times10^4$K, H = $2.4\times10^2$ MPa, P = 0.8,
B = $4.0\text{MPa}^{-2}$/sec, m = 3.0, $Q_D/k = 1.0\times10^4$K

## CONCLUSIONS

Several typical features of inelastic behaviors of polycrystalline metal under varying temperature conditions were observed in the systematic and fundamental experiments. Also it was confirmed that the semi-microscopic approach for the formulation of constitutive equations might be useful for polycrystalline metals under varying temperature/complex loading conditions.

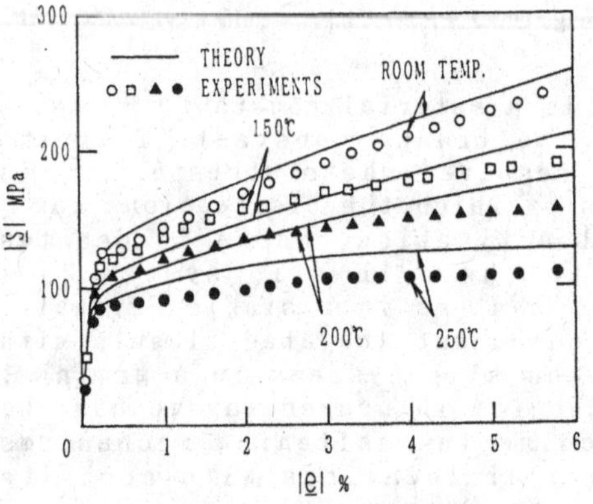


Fig. 1 Stress-strain relations for uniaxial tension under constant temperature conditions

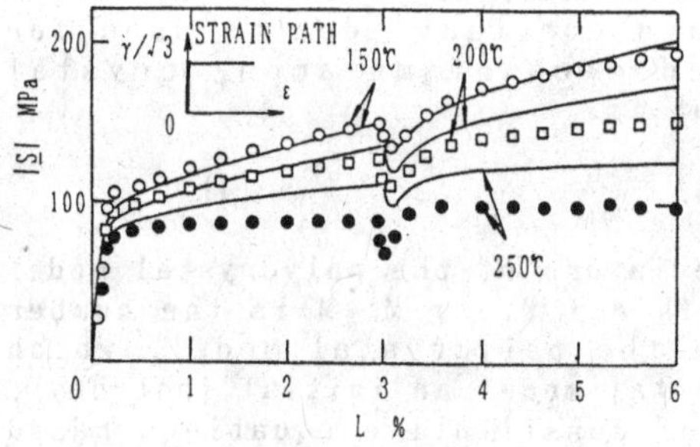


Fig. 2 Stress-strain relations for bilinear strain path under const.temperature conditions

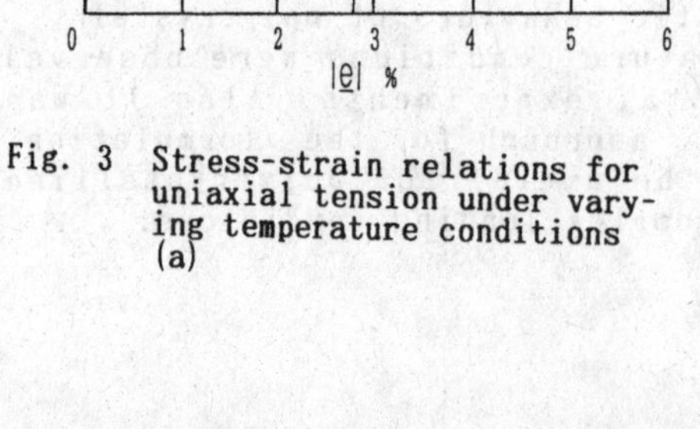

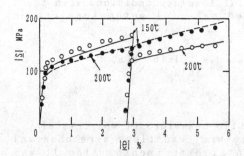

Fig. 3 Stress-strain relations for uniaxial tension under varying temperature conditions (a)

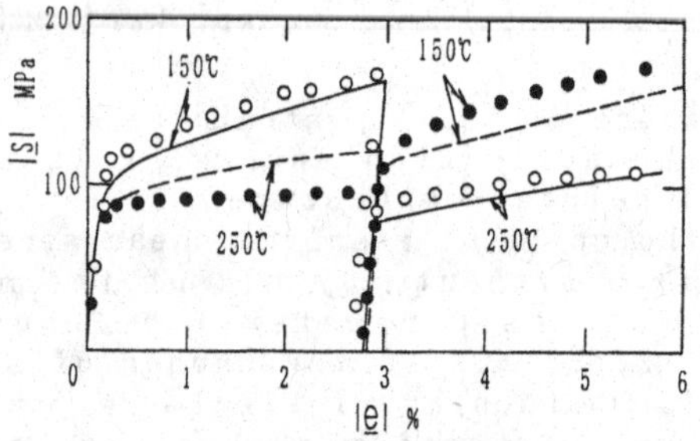


Fig. 4 Stress-strain relations for uniaxial tension under varying temperature conditions (b)

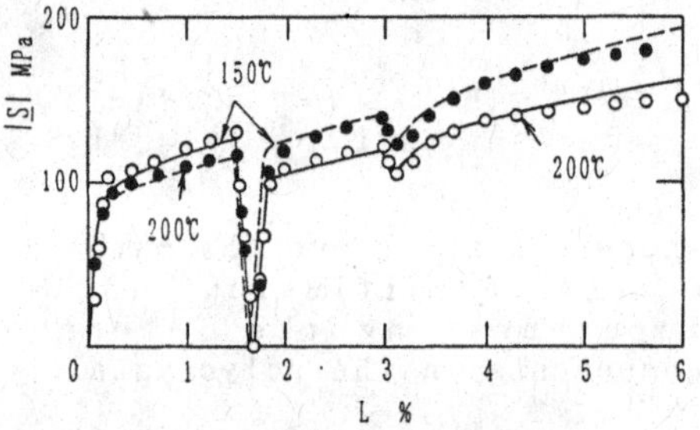


Fig. 5 Stress-strain relations for bilinear strain path under varying temperature conditions

## REFERENCES

(1) A.Horiguchi, Y.Niitsu, K.Ikegami, Effect of Temperature on Plastic Behavior of Mild Steel, Trans.of the JSME, 54-499A, p.565,1988

(2) Y.Niitu, K.Ikegami, Effect of Tempe-ature Variation on Cyclic Plastic Behavior of SUS 304 Stainless Steel, Trans. of the JSME, p.1621,52-478A, 1621, 52-478A, 1986

(3) S.Murakami, M.Kawai, Y.Ohmi, Effect of Strain Amplitude and Temperature History on Multi-axial Cyclic Hardening of Type 316 Stainless Steel, Trans. of the JSME,54-501A, p.1131, 1989

(4) Y.Iwasaki, T.Hiroe, T.Igari, An Application of the Viscoplasticity Theory to the Inelastic Analysis at Elevated Temperature, Trans.of the JSME, 53-493A, p.1838, 1987

(5) M.Tokuda,N.Ohno, J.Kratochvil, "Unified Constitutive Equations for Inelastic Behavior of Polycrystalline Metals Based on a Semi-micro Approach", Proc.Int.Conf.of Creep, Tokyo, p.411, 1986

# METAL FORMING , STRUCTURAL ANALYSES AND COMPUTATIONAL ASPECTS

# A VARIATIONALLY CONSISTENT TIME MODELLING OF ELASTIC-PLASTIC CONSTITUTIVE EQUATIONS

G. BORINO and C. POLIZZOTTO
Università di Palermo, DISEG
Dipartimento di Ingegneria Strutturale & Geotecnica
Viale delle Scienze, I-90128 Palermo, Italy

## ABSTRACT

A general energy-based time discretization method for evolutive analysis is presented. Most known time integration procedures (mid-point rule, backward difference, etc.) are shown to be particular cases of it. For space continuous systems, a sequence of weighted boundary value problems of deformation-theory plasticity are obtained, each characterizable by a number of variational principles useful for finite element discretization.

## INTRODUCTION

Elastic-plastic structural analysis requires discretizations both in space and in time. Space discretization methods are well-known as being variationally consistent, that is derivable from suitable variational principles [1], whereas time discretization makes use of *ad-hoc* rules and related algorithms aimed at the time integration of the nonlinear constitutive equations and thus at the forward advancement of the numerical solution through the time steps [2]. This kind of time discretization is deprived of a firm rational basis. Martin and Ponter [3] proposed the concept of "minimum energy strain path", which is equivalent to a constitutive model with plastic strain rates following a linear strain path in strain space, that is a strain path which minimizes the plastic dissipation work with respect to any other curved path joining the same strain points [4,5]. This view has enabled the well-known backward difference method to possess a variational consistency [6,7,8].

On considering that any *ad-hoc* rule always produces some approximation into the actual evolution of the system within every step, in which the infinite plastic evolution modes are replaced by a discrete number of modes, and in which therefore the rate-form plastic yielding laws take on the shape of holonomic, or deformation-theory, plasticity yielding laws, here a more general view-point is adopted for the time discretization of the elastic-plastic evolutive problem [9]. This method, referred to as the Finite Interval Method (FIM), is based on the concept of *material evolutive model*, and uses the maximum plastic work theorem to obtain the holonomic-type plastic yielding laws appropriate to a material evolving in time following the specified modes. In this way, the original initial/boundary value problem of incremental-type plasticity theory is transformed into a sequence of boundary value problems of holonomic-type plasticity theory, one for every step.

## FINITE INTERVAL MODELLING

A rate-independent associative material model is assumed, described by

$$\dot{\boldsymbol{\varepsilon}}^p = \dot{\lambda}\frac{\partial\phi}{\partial\boldsymbol{\sigma}}, \qquad \dot{\boldsymbol{\xi}} = -\dot{\lambda}\frac{\partial\phi}{\partial\boldsymbol{\chi}} \tag{1a}$$

$$\phi(\boldsymbol{\sigma},\boldsymbol{\chi}) \leq 0, \quad \dot{\lambda} \geq 0, \quad \dot{\lambda}\phi(\boldsymbol{\sigma},\boldsymbol{\chi}) = 0 \tag{1b}$$

where $\boldsymbol{\sigma}$ denotes the stress tensor; $\boldsymbol{\varepsilon}^p$ the plastic strain tensor; $\boldsymbol{\chi}$ and $\boldsymbol{\xi}$ dual (stress-like and strain-like) internal variable tensors; $\phi$ the (convex, smooth) yield function; $\dot{\lambda}$ the plastic parameter. Let $(0,h)$ be a closed time interval and let the histories $\dot{\boldsymbol{\varepsilon}}^p(t)$ and $\dot{\boldsymbol{\xi}}(t)$ be assumed in the form (*evolutive mode*)

$$\dot{\boldsymbol{\varepsilon}}^p(t) = \frac{1}{h}g(t)\,\mathring{\boldsymbol{\varepsilon}}^p, \qquad \dot{\boldsymbol{\xi}}(t) = \frac{1}{h}g(t)\,\mathring{\boldsymbol{\xi}}, \quad \text{in } (0,h) \tag{2}$$

where $\mathring{\boldsymbol{\varepsilon}}^p$ and $\mathring{\boldsymbol{\xi}}$ are time-independent tensor parameters and $g(t)$ is a specified nonnegative scalar function. The total intrinsic dissipation in $(0,h)$ is

$$W = \int_0^h \left(\boldsymbol{\sigma} : \dot{\boldsymbol{\varepsilon}}^p - \boldsymbol{\chi} : \dot{\boldsymbol{\xi}}\right) dt = \bar{\boldsymbol{\sigma}} : \mathring{\boldsymbol{\varepsilon}}^p - \bar{\boldsymbol{\chi}} : \mathring{\boldsymbol{\xi}} \tag{3}$$

where $\bar{\boldsymbol{\sigma}}$ and $\bar{\boldsymbol{\chi}}$ are *weighted* values of $\boldsymbol{\sigma}(t)$ and $\boldsymbol{\chi}(t)$, i.e.

$$\bar{\boldsymbol{\sigma}} = \frac{1}{h}\int_0^h g(t)\,\boldsymbol{\sigma}(t)\,dt, \qquad \bar{\boldsymbol{\chi}} = \frac{1}{h}\int_0^h g(t)\,\boldsymbol{\chi}(t)\,dt \tag{4}$$

dual of $\mathring{\boldsymbol{\varepsilon}}^p$ and $\mathring{\boldsymbol{\xi}}$, respectively, in the sense that their scalar product gives the actual total intrinsic dissipation. The use of the maximum intrinsic dissipation theorem enables one to derive a total dissipation function $W = W(\mathring{\boldsymbol{\varepsilon}}^p, \mathring{\boldsymbol{\xi}})$ such that

$$\frac{\partial W}{\partial \mathring{\boldsymbol{\varepsilon}}^p} = \bar{\boldsymbol{\sigma}}, \qquad \frac{\partial W}{\partial \mathring{\boldsymbol{\xi}}} = -\bar{\boldsymbol{\chi}}, \tag{5}$$

as well as the related (generalized) plastic flow laws, namely

$$\mathring{\boldsymbol{\varepsilon}}^p = \mathring{\lambda}\frac{\partial\bar{\phi}}{\partial\bar{\boldsymbol{\sigma}}}, \qquad \mathring{\boldsymbol{\xi}} = -\mathring{\lambda}\frac{\partial\bar{\phi}}{\partial\bar{\boldsymbol{\chi}}} \tag{6a}$$

$$\bar{\phi}(\bar{\boldsymbol{\sigma}},\bar{\boldsymbol{\chi}}) \leq 0, \quad \mathring{\lambda} \geq 0, \quad \mathring{\lambda}\bar{\phi}(\bar{\boldsymbol{\sigma}},\bar{\boldsymbol{\chi}}) = 0. \tag{6b}$$

where

$$\bar{\phi}(\bar{\sigma}, \bar{\chi}) = \frac{1}{h}\int_0^h g(t)\,\phi(\tilde{g}(t)\,\bar{\sigma},\, \tilde{g}(t)\,\bar{\chi})\,dt \tag{7}$$

is the generalized yield function and $\tilde{g}(t)$ is a complementary shape function such that

$$\frac{1}{h}\int_0^h g(t)\,\tilde{g}(t)\,dt = 1. \tag{8}$$

Specializations of the above FI modelling are obtained by making particular choices for $g(t)$.

## FINITE INTERVAL DISCRETIZATION OF THE EVOLUTIVE PROBLEM

An elastic-plastic body of volume $V$ and boundary surface $S = S_t \cup S_u$ is loaded by quasi-static external actions. Equilibrium, compatibility, Hooke's law and the thermodynamic relationship between the internal variables read in case of small displacements $\boldsymbol{u}$:

$$\boldsymbol{C}^T\boldsymbol{\sigma} + \boldsymbol{F} = \boldsymbol{0} \text{ in } V, \qquad \boldsymbol{\sigma}\cdot\boldsymbol{n} = \boldsymbol{T} \text{ on } S_t \tag{9}$$

$$\boldsymbol{\varepsilon} = \boldsymbol{C}\boldsymbol{u} \text{ in } V, \qquad \boldsymbol{u} = \boldsymbol{c} \text{ on } S_u \tag{10}$$

$$\boldsymbol{\sigma} = \boldsymbol{E} : (\boldsymbol{\varepsilon} - \boldsymbol{\varepsilon}^\theta - \boldsymbol{\varepsilon}^p) \text{ in } V, \qquad \boldsymbol{\chi} = \boldsymbol{H} : \boldsymbol{\xi} \text{ in } V. \tag{11}$$

Here, $\boldsymbol{C}$ is the compatibility differential operator, $\boldsymbol{C}^T$ its adjoint, $\boldsymbol{n}$ the unit normal vector to $S$, $\boldsymbol{E}$ and $\boldsymbol{H}$ are the elastic and the hardening moduli (both symmetric end positive definite) tensors. $\boldsymbol{F}, \boldsymbol{T}, \boldsymbol{c}$, and $\boldsymbol{\varepsilon}^\theta$ denote body forces, tractions, imposed displacements and imposed strains (e.g. thermal strains), all variable in the interval $(0, t_f)$. The evolutive analysis problem is governed by eqs. (9)–(11) and (1) written for all $t$ in $(0, t_f)$ and supplemented by initial conditions at $t = 0$. A FI discretization of this problem is achieved by subdividing the interval $(0, t_f)$ in small steps of equal length $h$, with subdivision instants $t_0 = 0, t_1 = h, t_2 = 2h, \ldots, t_n = nh, \ldots$. The yielding laws (1) are modelled within the $n$-th step as

$$\mathring{\boldsymbol{\varepsilon}}_n^p = \mathring{\lambda}_n \frac{\partial\bar{\phi}}{\partial\bar{\boldsymbol{\sigma}}_n}, \qquad \mathring{\boldsymbol{\xi}}_n = -\mathring{\lambda}_n \frac{\partial\bar{\phi}}{\partial\bar{\boldsymbol{\chi}}_n}, \tag{12a}$$

$$\bar{\phi}(\bar{\boldsymbol{\sigma}}_n, \bar{\boldsymbol{\chi}}_n) \le 0, \quad \mathring{\lambda}_n \ge 0, \quad \mathring{\lambda}_n \bar{\phi}(\bar{\boldsymbol{\sigma}}_n, \bar{\boldsymbol{\chi}}_n) = 0. \tag{12b}$$

Applying the weighting operation in (4) upon eqs. (9)-(11) for the $n$-th step gives

$$\boldsymbol{C}^T\bar{\boldsymbol{\sigma}}_n + \bar{\boldsymbol{F}}_n = \boldsymbol{0} \text{ in } V, \qquad \bar{\boldsymbol{\sigma}}_n\cdot\boldsymbol{n} = \bar{\boldsymbol{T}}_n \text{ on } S_t \tag{13}$$

$$\bar{\boldsymbol{\varepsilon}}_n = \boldsymbol{C}\,\bar{\boldsymbol{u}}_n \text{ in } V, \qquad \bar{\boldsymbol{u}}_n = \bar{\boldsymbol{c}}_n \text{ on } S_u \tag{14}$$

$$\bar{\boldsymbol{\sigma}}_n = \boldsymbol{E} : (\bar{\boldsymbol{\varepsilon}}_n - \boldsymbol{\varepsilon}_n^* - \mathring{\boldsymbol{\varepsilon}}_n^p) \text{ in } V, \qquad \bar{\boldsymbol{\chi}}_n = \boldsymbol{H} : \left(\boldsymbol{\xi}_n^* + \mathring{\boldsymbol{\xi}}_n\right) \text{ in } V. \tag{15}$$

where $\bar{\boldsymbol{\sigma}}_n$, $\bar{\boldsymbol{\chi}}_n$, etc. are the step weighted quantities, and $\boldsymbol{\varepsilon}_n^*$, $\boldsymbol{\xi}_n^*$ denote known strains at the $n$-th step.

Equations (12)–(15) enable one to solve the $n$-th step problem once the previous one has been solved. This weighted b.v. problem admits a number of variational principles, suitable for space discretizations.

## CONCLUSIONS

The Finite Interval Method presented is an energy-method based procedure for time discretization of evolutive elastic-plastic analysis problems and thus for transforming the original rate-type plasticity problem into a sequence of holonomic plasticity step problems. Most of the known ad-hoc rules for time integration can be viewed as particular cases of it. Variational principles of holonomic plasticity offer a firm basis for space discretizations. The solution of every step problem provides mean values for the unknown displacements, stresses and plastic strain increments. An appropriate interpolation of these mean values may provide continuous time histories.

*Acknowledgements* — This paper has been completed with the financial support of the Ministero dell'Università e della Ricerca Scientifica e Tecnologica (MURST). G. Borino thanks the IEREN-CNR Institute of Palermo for the financial support given under the C.N.R. Fellowship 201.07.49.

## REFERENCES

1. Washizu, K., *Variational Methods in Elasticity & Plasticity*, Pergamon Press, 1982.
2. Ortiz, M. and Popov, E.P., "Accuracy and Stability of Integration Algorithms for Elastoplastic Constitutive Relations", *Int. J. Num. Meth. in Engng.*, 1985 **21**, pp. 1561–1576.
3. Ponter, A.R.S. and Martin J.B., "Some Extremal Properties and Energy Theorems for Inelastic Materials and Their Relationship to the Deformation Theory of Plasticity", *Jour. Mech. Phys. Solids*, 1972, **20**, pp. 281–300.
4. Martin, J.B., *Plasticity: Fundamentals and General Results*, MIT Press, Cambridge (MA), 1975.
5. Reddy, B.D., Martin, J.B. and Griffin, T.B., "Extremal Paths and Holonomic Constitutive Laws in Elastoplasticity", UCT/CSIR Appl. Mech. Res. Unit, *Technical Report*, **82**, University of Cape Town, 1986.
6. Ortiz, M. and Martin, J.B., "Simmetry Preserving Return Mapping Algorithms and Incrementally Extremal Paths: A Unification of Concepts", *Int. J. Num. Meth. in Engng.*, 1989 **28**, pp. 1839–1853.
7. Martin, J.B., "Integration along the Path of Loading in Elastic-Plastic Problems", in D.R.J. Owen, E. Hinton and E. Oñate (eds.), *Computational Plasticity II*, Pineridge Press, Swansea, U.K., Vol. I, 1989, pp. 1–15.
8. Maier, G. and Novati, G., "Extremum Theorems for Finite-Step Backward-Difference Analysis of Elastic-Plastic Nonlinearly Hardening Solids", *Int. J. Plasticity*, 1990 **6**, pp. 1–10.
9. Borino, G., Fuschi, P, and Polizzotto, C., "Consistent Time Modelling for the Evolutive Analysis of Elastic-Plastic Solids", in D.R.J. Owen, E. Hinton and E. Oñate (eds.), *Computational Plasticity II*, Pineridge Press, Swansea, U.K., Vol. I, 1989, pp. 85–98.

# DETERMINATION OF DUCTILE CRACK GROWTH USING GURSON'S MODEL WITH DIFFERENT VOID EVOLUTION LAWS

WOLFGANG BROCKS and DIETMAR KLINGBEIL

Bundesanstalt für Materialforschung und -prüfung (BAM),
D-1000 Berlin 45, Germany

## ABSTRACT

A new approach to fracture phenomena basing on micro-mechanical damage models becomes of increasing interest. The failure mechanism of ductile tearing is dominated by the initiation, growth and coalescence of voids. This material behaviour can be described conveniently by the Gurson model as its constitutive equations include a parameter describing the volume fraction of voids. The model has been implemented into the FE program ABAQUS. Two different evolution laws for the nucleation of voids are tested with respect to their ability to predict the crack growth inside of a smooth tensile bar made of a ductile steel.

## INTRODUCTION

The influence of stress triaxiality on ductile fracture has been emphasized recently in explaining the geometry dependent resistance of specimens and structures to ductile tearing [1]. A high hydrostatic stress state inhibits plastic flow and, hence, promotes fracture, because the external work will to a lesser part be dissipated by plastic deformation but be available to enhance material degradation and damage. Stress triaxiality is also known for promoting void growth on the micromechanical level [2,3] and thus causing "damage" in the process zone. Constitutive equations who account for damage as e.g. Gurson's model [4] will therfore be able to describe ductile fracture phenomena without using any additional global "tearing resistance curves" and without introducing any "initial" crack in the structure. This approach will be studied by a finite element (FE) analysis of a smooth tensile bar made of the American steel A 710. the material data are courtesy of FhIWM Freiburg [5].

## THE MICROMECHANICAL MODEL

The constitutive equations for modeling ductile failure base on a yield condition by Gurson [4] which has been modified by Needleman and Tvergaard [6-8]. According to this model, the plastic flow in porous media is influenced by a single "damage" parameter, $0 \leq f < 1$, which represents the void volume fraction. The plastic potential or yield function, $\Phi$, is given by

$$\Phi(\underline{T}', \mathrm{tr}\underline{T}, f, \sigma_m) = \frac{3}{2}\frac{T' \cdot\cdot T'}{\sigma_m^2} + 2qf^* \cosh\left(\frac{\mathrm{tr}T}{2\sigma_m}\right) - \left(1 + (qf^*)^2\right) \quad (1)$$

$\underline{T}$ and $\underline{T}'$ are the stress tensor and its deviator, $\sigma_m$ is the actual flow stress of the matrix material. The additional parameter q was introduced by Tvergaard [8] to improve the numerical results for small values of f, and the modified damage parameter $f^*(f)$ by Tvergard and Needleman [6]

$$f^* = \begin{cases} f & \text{for } f \leq f_c \\ f_c + K(f - f_c) & \text{for } f > f_c \end{cases} \quad (2)$$

accounts for the coalescence of two neighbouring voids due to slip planes or necking. The ultimate value of $f^*$, at which the macroscopic stress carrying capacity is lost, equals $1/q$, and for this state K can be calculated from eq. (2) if the void volume fraction at final fracture is known from experiments.

f and $\sigma_m$ are internal variables which follow the evolution equations

$$\dot{\sigma}_m = H\,\dot{\varepsilon}^p_m \quad \text{with} \quad H = \frac{E\,E_T}{E - E_T}, \quad (2)$$

$$\dot{f} = \dot{f}_{growth} + \dot{f}_{nucl} \quad (3)$$

$$\text{with} \quad \dot{f}_{growth} = (1-f)\,\mathrm{tr}\underline{D}^p \quad (3a)$$

$$\text{and} \quad \dot{f}_{nucl} = A\,\dot{\sigma}_m + - B/_3\,\mathrm{tr}\dot{\underline{T}} \quad (3b)$$

The nucleation of voids occurs by a decohesion process of the ductile matrix material from rigid inclusions. It may be controlled by the accumulated plastic strain in the matrix material, $\varepsilon_m{}^p$, or by the hydrostatic stress. The factors A and B in eq. (3b) are

$$A = \frac{1}{H}\frac{f_n}{s_n\sqrt{2\pi}}\exp\left[-\frac{1}{2}\left(\frac{\varepsilon_m{}^p - \varepsilon_n}{s_n}\right)^2\right] \quad \text{and} \quad B = 0 \quad (4a)$$

for strain control and $\varepsilon_m{}^p > \varepsilon_n$ and

$$A = B = \frac{f_n}{s_n\,\sigma_y\sqrt{2\pi}}\exp\left[-\frac{1}{2}\left(\frac{\sigma_m + \frac{1}{3}\mathrm{tr}\underline{T} - \sigma_n}{s_n\,\sigma_y}\right)^2\right] \quad (4b)$$

for stress control and $\sigma_m + {}^1/_3\,\mathrm{tr}\underline{T} \geq \sigma_n$, respectively. If $\varepsilon_m{}^p \leq \varepsilon_n$ or $\sigma_m + {}^1/_3\,\mathrm{tr}T \leq \sigma_n$ no nucleation occurs, i.e. A=B=0. $\varepsilon_n$ and $\sigma_n$ are the respective mean critical values for decohesion between matrix and inclusions, and $s_n$ is the standard deviation. $\sigma_y$ denotes the initial yield stress of the matrix material.

## FE CALCULATIONS

The described modified Gurson model has been implemented in the FE program ABAQUS. The calculations have started with a rather simple structure, namely a smooth tensile bar, see Fig. 1, allowing for large deformations and necking of the mid section. One of the basic problems of applying the constitutive equations to real materials is the identification and determination of the various material parameters which exceed the data obtained by a simple tensile test. For the present numerical studies data of a steel ASTM A710 will be used which have been determined by the FhIWM in Freiburg [5]. One of the future perspectives of the numerical studies will be to propose clearly defined identifying experiments for the model. The first step, which is reported in this presentation, is to investigate and compare the two different possibilities of void nucleation, i.e. stress control and strain control.

## REFERENCES


1. Brocks, W. and Schmitt, W.: The role of crack tip constraint for ductile tearing. In Fracture Behaviour and Design of Materials and Structures, ed. D. Firrao, Preprints of the 8th European Conference on Fracture, Vol. II, EMAS, Warley, 1990, pp. 1023-1032.

2. McClintock, F.A., A criterion for ductile fracture by the growth of holes, Trans. ASME, J. Appl. Mech., 1968, **35**, pp. 363-371.

3. Rice, J.R. and Tracey, D.M., On the ductile enlargement of voids in triaxial stress fields, J. Mech. Phys. Solids, 1969, **17** pp. 201-217.

4. Gurson, A.L., Continuum theory of ductile rupture by void nucleation and growth: Part I - Yield criteria and flolow rules for porous ductile media. J. Engng. Mat. Tech. 1977, **99**, pp. 2-15.

5. Sun, D.-Z., Voss, B. and Schmitt, W., Numerical prediction of ductile fracture resistance behaviour based on micromechanical models. In Elements of Defect Assessment, ed. H.Blauel and K.-H. Schwalbe, Proceedings of Symposium on Elastic-Plastic Fracture Mechanics, Freiburg, 1989, to be published.

6. Needleman, A. and Tvergaard, V., An analysis of ductile rupture in notched bars, J. Mech. Phys. Solids, 1984, **32**, pp. 461-490.

7. Needleman, A. and Tvergaard, V., An analysis of ductile rupture at a crack tip, J. Mech. Phys. Solids, 1987, **35**, pp. 151-183.

8. Tvergaard, V., On localization in ductile materials containing spherical voids, Int. J. Fract., 1982, **18**, pp. 237-252.

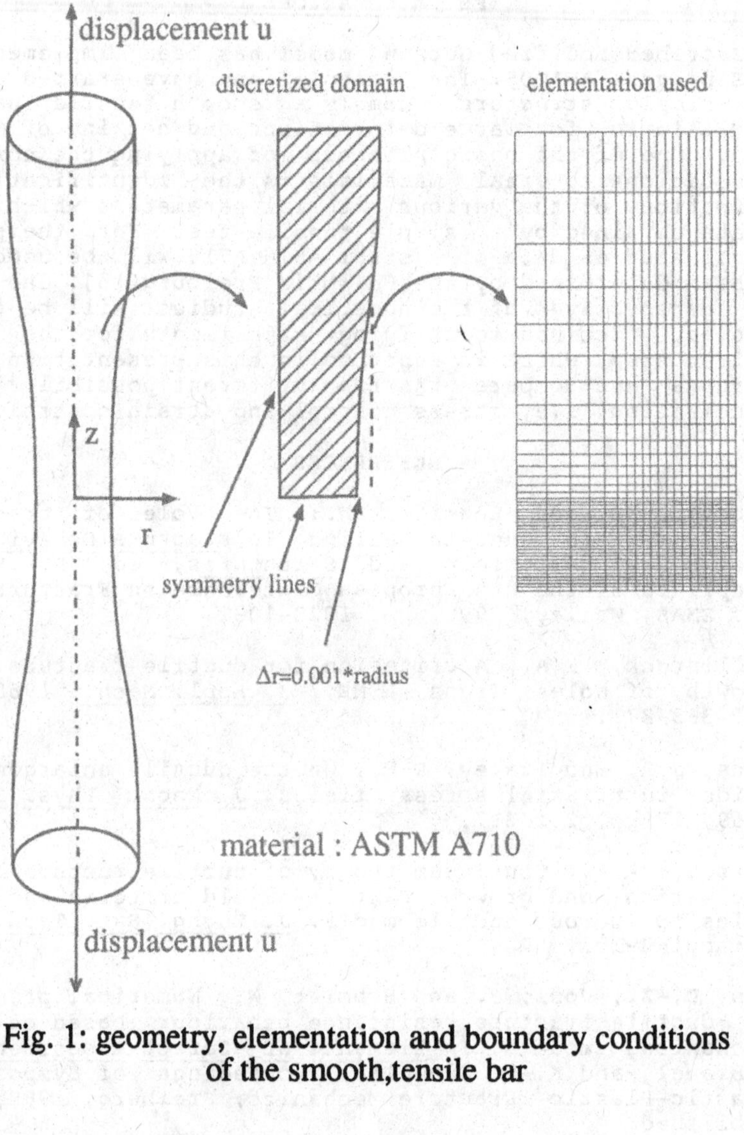


**Fig. 1: geometry, elementation and boundary conditions of the smooth,tensile bar**

# THE MID-POINT RULE AND THE TRAPEZOIDAL RULE FOR THE INTEGRATION OF THE VON MISES ELASTIC-PLASTIC CONSTITUTIVE LAWS

S. CADDEMI

Centre for Research in Computational and Applied Mechanics
University of Cape Town
Rondebosch 7700, South Africa

## ABSTRACT

A spatially continuous, time discrete formulation of the loading of an elastic, perfectly plastic body governed by a von Mises yield condition is considered. The incremental problem is formulated as a convex nonlinear programming problem by making use of a generalised mid-point rule and a suitable trapezoidal rule. Explicit expressions of the functional to be minimised are provided. The iterative solution procedure for this problem is based on the classical Newton-Raphson algorithm. In particular, a recent interpretation of the trapezoidal rule, which leads to a symmetric consistent tangent modulus, is adopted and discussed.

## INTRODUCTION

The governing equations of the elastic-plastic evolutive problem require a time discretisation in order to be integrated. Assumptions concerning the constitutive behaviour of the material within each time interval have to be adopted and lead to algorithms for the integration of the nonlinear constitutive equations (return algorithms). In recent work [1,2] the piecewise holonomic approximation has been adopted. It is based on the concept that the strain follows a minimum work path between the discrete instants [3] and provides a one-to-one relationship between stress and strain. The well known fully implicit or Euler backward difference return algorithm has been shown to be fully equivalent to the piecewise holonomic approximation [4]. The backward difference assumption leads to a symmetric consistent tangent modulus indeed, but the integration algorithm which follows is only first order accurate. Recently the question concerning a more appropriate choice of the return algorithm has been addressed. Ortiz and Popov [5] proposed a mid-point and a trapezoidal rule which are both second order accurate but the related consistent tangent moduli dot not preserve the desirable feature of the symmetry. A suitable generalised mid-point rule has been then introduced by Simo and Govindjee [6] which is second order accurate and also leads to a symmetric tangent modulus.

In this work we consider a continuous elastic, perfectly plastic body governed by a von Mises yield condition and subjected to some history of loading. The incremental problem is formulated as a convex nonlinear programming problem, by making use, first, of a generalised mid-point rule, then, of a suitable trapezoidal rule. The functionals to be minimised are presented in explicit form. An iterative solution algorithm based on the Newton-Raphson technique is adopted. The symmetry of the consistent tangent modulus is preserved both for the generalised mid-point and trapezoidal rule. The trapezoidal rule here adopted has been presented in [7], it is different from that one of Ortiz and Popov [5] and is further discussed in this paper.

## THE MID-POINT RULE AND THE TRAPEZOIDAL RULE

We consider a continuous elastic,perfectly plastic body governed by a von Mises yield condition $\phi = \frac{1}{2}s_{ij}s_{ij} - k^2 \leq 0$, where $s_{ij}$ are the deviatoric stress components and $k$ is a constant. The body is subjected to a loading history of quasi-static external tractions over the time interval [0,T]. A time discretisation is introduced by dividing the interval [0,T] into subintervals. If we consider the n-th interval $\Delta t_n = [t_{n-1}, t_n]$, the aim is to compute increments of displacements $\Delta u_i^n(x_i)$, stresses $\Delta\sigma_{ij}^n(x_i)$, total strains $\Delta\varepsilon_{ij}^n(x_i)$ and deviatoric plastic strains $\Delta e_{ij}^{pn}$, given the external tractions at $t_n$ and the values $u_i^{n-1}(x_i)$, $\sigma_{ij}^{n-1}(x_i)$, $\varepsilon_{ij}^{n-1}(x_i)$, $e_{ij}^{p(n-1)}$ at $t_{n-1}$.

The first assumption concerning the constitutive behaviour we adopt over the time interval is the generalised mid-point rule. Deviatoric plastic strain increments $\Delta e_{ij}^{pn}$ are assumed to have a constant direction over the time interval given by the normal to the yield surface at the point $s_{ij}^{n\alpha} = (1-\alpha)s_{ij}^{n-1} + \alpha s_{ij}^n$, which is the stress point at time $t_{n\alpha} = t_{n-1} + \alpha\Delta t_n$, for chosen value $\alpha$ such that $0 \leq \alpha \leq 1$. For $\alpha = 0$ we recover the fully explicit forward difference assumption, for $\alpha = 1$ the fully implicit backward difference assumption. According to [6], the consistency condition is imposed at $t_{n\alpha}$, i.e. $\phi(s_{ij}^{n\alpha}) = 0$. By following the same reasoning path presented in [8] it is possible to write the deviatoric constitutive relations through a convex nonquadratic deviatoric potential function $W_n^d$ as follows

$$s_{ij}^{n\alpha} = \left.\frac{\partial W_n^d}{\partial \Delta e_{ij}}\right|_{\Delta e_{ij}^{n\alpha}} \tag{1}$$

where

$$W_n^d = \frac{1}{4G}s_{ij}^E s_{ij}^E \qquad \text{if } \frac{1}{2}s_{ij}^E s_{ij}^E \leq k^2 \tag{2a}$$

$$W_n^d = \frac{k}{G}\left\{\frac{1}{2}s_{ij}^E s_{ij}^E\right\}^{\frac{1}{2}} - \frac{k^2}{2G} \qquad \text{if } \frac{1}{2}s_{ij}^E s_{ij}^E \geq k^2 \tag{2b}$$

and

$$s_{ij}^E = (s_{ij}^{n-1} + 2G\Delta e_{ij}) \quad . \tag{3}$$

Eqs. (1)–(3) relate deviatoric stresses $s_{ij}^{n\alpha}$ and deviatoric strain increments $\Delta e_{ij}^{n\alpha}$ at $t_{n\alpha}$.

Stresses $s_{ij}^n$ and plastic strain increments $\Delta e_{ij}^{pn}$ at $t_n$ are obtained as functions of $s_{ij}^{n\alpha}$ and $\Delta e_{ij}^{pn\alpha}$, respectively, by means of a forward linear projection as follows

$$s_{ij}^{n} = \frac{1}{\alpha} s_{ij}^{n\alpha} + \frac{1-\alpha}{\alpha} s_{ij}^{n-1} \quad , \qquad \Delta e_{ij}^{pn} = \frac{1}{\alpha} \Delta e_{ij}^{pn\alpha} \quad . \tag{4a, b}$$

The second assumption we consider is that the deviatoric plastic strain increment $\Delta e_{ij}^{pn}$, instead of following a straight path, is divided into two parts over the n-th time interval

$$\Delta e_{ij}^{pn} = \Delta \bar{e}_{ij}^{pn} + \Delta e_{ij}^{pon} \qquad \text{where} \qquad \Delta e_{ij}^{pon} = \lambda \left. \frac{\partial \phi}{\partial s_{ij}} \right|_{s_{ij}^{n}} \tag{5a, b}$$

where $\lambda$ is the plastic multiplier, and the component $\Delta \bar{e}_{ij}^{pn}$ is obtained by projecting forward $\Delta e_{ij}^{po(n-1)}$ of the previous time interval as follows

$$\Delta \bar{e}_{ij}^{pn} = \frac{1-\alpha}{\alpha} \Delta e_{ij}^{po(n-1)} \tag{6}$$

and is, hence, known. The assumption given by eq. (5a) is an interpretation of the trapezoidal rule different from that one given by Ortiz and Popov [5]. For this assumption the deviatoric constitutive relations can also be written by making use of a convex nonquadratic deviatoric potential function $W_n^d$

$$s_{ij}^{n} = \left. \frac{\partial W_n^d}{\partial \Delta e_{ij}} \right|_{\Delta e_{ij}^{n}} \tag{7}$$

where

$$W_n^d = \frac{1}{4G} s_{ij}^{E*} s_{ij}^{E*} \qquad \text{if} \quad \frac{1}{2} s_{ij}^{E*} s_{ij}^{E*} \leq k^2 \tag{8a}$$

$$W_n^d = \frac{k}{G} \left\{ \frac{1}{2} s_{ij}^{E*} s_{ij}^{E*} \right\}^{\frac{1}{2}} - \frac{k^2}{2G} \qquad \text{if} \quad \frac{1}{2} s_{ij}^{E*} s_{ij}^{E*} \geq k^2 \tag{8b}$$

and

$$s_{ij}^{E*} = s_{ij}^{n-1} - 2G \Delta \bar{e}_{ij}^{pn} + 2G \Delta e_{ij} \quad . \tag{9}$$

Eqs. (7)–(9) relate deviatoric stresses $s_{ij}^{n}$ and deviatoric strain increments $\Delta e_{ij}^{n}$ at $t_n$.

## THE MINIMUM PRINCIPLE AND THE ITERATIVE SOLUTION ALGORITHM

The incremental elastic-plastic analysis of the body described in the previous section can be cast in the form of a minimisation of a nonquadratic convex functional $U_n(\Delta \varepsilon_{ij}, \Delta u_i)$ (nonlinear programming problem) for each time interval. The functional $U_n$ depends on the deviatoric potential function $W_n^d$, and its minimisation is based on a predictor-corrector iterative procedure according to the classical Newton-Raphson procedure. The predictor step in the k-th iteration provides new estimates $\Delta \varepsilon_{ij}^{nk}$ and $\Delta u_i^{nk}$ of the increment in strain and displacement by means of the minimisation of a quadratic approximation of $U_n$ which can be obtained by replacing $W_n^d$ with its second order Taylor expansion. The second derivative of $W_n^d$ about the deviatoric strain increment at the end of the previous iteration $\Delta e_{ij}^{n(k-1)}$ represents the consistent tangent modulus

$$D_{ijkl}^{C} = \left. \frac{\partial^2 W_n^d}{\partial \Delta e_{ij} \partial \Delta e_{kl}} \right|_{\Delta e_{ij}^{n(k-1)}} \tag{9}$$

where $W_n^d$ is given by eqs. (2) for the mid-point rule and by eqs. (8) for the trapezoidal rule. $D_{ijkl}^{C}$ is a symmetric and positive definite forth order tensor for both the assumptions presented in the previous section.

## CONCLUSIONS

Explicit incremental constitutive relations for a von Mises elastic, perfectly plastic material with a generalised mid-point and trapezoidal rule have been presented. It has to be noted that the trapezoidal rule adopted leads to a symmetric consistent tangent modulus as well as the generalised mid-point rule. The same formulation adopted in [8] for the backward difference approximation has been utilised in this paper which proves that the backward difference, the generalised mid-point rule and the trapezoidal rule are related to each other. Extension of the present formulation to the case of kinematic and isotropic hardening will be reported in the full paper together with the explicit expressions of the consistent tangent moduli.

*Acknowledgements* — The support of the Foundation for ResearchDevelopment is gratefully acknowledged.

## REFERENCES

1. Martin, J.B., Reddy, B.D., Griffin T.B. and Bird, W.W., "Application of Mathematical Programming Concepts to Incremental Elastic-Plastic Analysis ", *Eng. Struct.*, 1987 **9**, pp. 171–176.
2. Reddy, B.D., Martin, J.B. and Griffin T.B., "Extremal Paths and Holonomic Constitutive Laws in Elastoplasticity ", *Quart. Appl. Math.*, 1987 **45**, pp. 487–502.
3. Ponter, A.R.S. and Martin J.B., "Some Extremal Properties and Energy Theorems for Inelastic Materials and their Relationship to the Deformation Theory of Plasticity", *Jour. Mech. Phys. Solids*, 1972, **20**, pp. 281–300.
4. Ortiz, M. and Martin, J.B., "Simmetry Preserving Return Mapping Algorithms and Incrementally Extremal Paths: A Unification of Concepts", *Int. J. Num. Meth. in Engng.*, 1989 **28**, pp. 1839–1853.
5. Ortiz, M. and Popov, E.P., "Accuracy and Stability of Integration Algorithms for Elastoplastic Constitutive Relations", *Int. J. Num. Meth. in Engng.*, 1985 **21**, pp. 1561–1576.
6. Simo, J.C. and Govindjee, S., "Nonlinear B-Stability and Symmetry Preserving Return Mapping Algorithms for PLasticity and Viscoplasticity", *Int. J. Num. Meth. in Engng.*, 1991 **31**, pp. 151–176.
7. Recontre, L.J., Caddemi, S. and Martin, J.B., "The Relationship between the Generalised Mid-Point and Trapezoidal Rule in Incremental Elastoplasticity", UCT/CSIR CERECAM, *Technical Report*, **152**, University of Cape Town, 1990.
8. Caddemi, S. and Martin, J.B., "Convergence of the Newton-Raphson Algorithm in Elastic-Plastic Incremental Analysis", *Int. J. Num. Meth. in Engng.*, 1991 **31**, pp. 177–191.

# DETERMINATION OF PLASTIC ZONE AT CRACK TIP USING LASER SPECKLE DECORRELATION

F.P. CHIANG & Y.Z. DAI
Laboratory for Experimental Mechanics Research
State University of New York at Stony Brook
Stony Brook, N.Y. 11794-2300, USA

## ABSTRACT

Speckle patterns from a metallic surface vary as a function of plastic deformation the material experiences and hence may be used to detect the plastic zone size. The speckle patterns are recorded and compared using cross correlation analysis based on an image processing system. This technique, which has the merits of being non-contact, remote sensing, and high sensitivity, is applied to measuring the size of plastic zone around a hole in an aluminum plate with finite width. A comparison with theoretical and finite element results is presented.

## INTRODUCTION

The size and shape of plastic zone is of importance to the understanding of ductile fracture of metals. Most plastic zones are calculated numerically. Experimental determination usually requires quantitative values of plastic strains – a time consuming process. A non-contact, remote sensing technique for measuring plastic zone size is presented based on plasticity induced surface roughness and speckle pattern decorrelation.

The surface roughness of a metallic material changes with increase in surface strain [1]. As a result, the speckle pattern from the material surface changes correspondingly. This phenomenon suggests methods to relate plastic strain quantitatively with the variation in speckle pattern. Some techniques based on this principle have been developed [2,3]. Yet, the sensitivity and reliability of these techniques need improvement.

The present non-contact, remote method for determining plastic zone is based on an image processing system. A cross correlation technique is employed to analyze the digitized speckle patterns. In particular the technique is applied to the determination of the plastic zone around a central hole in a plate with finite width. The experimental result is in good agreement with that by finite element analysis.

## SPECKLE PATTERN CORRELATION

The proposed technique makes use of a set-up as schematically shown in Fig.1. The point of interest on a test object was illuminated with a $20mw$ He-Ne laser. To increase the lateral resolution of the technique, the diameter of the laser beam was reduced through a system consisting of a spatial filter, an aperture, a collimating lens and a converging lens. A piece of ground glass was placed at a distance of $0.8m$ from the test object. A digital camera was connected to a monitor and supported by a computer which provided correlation analysis capabilities.

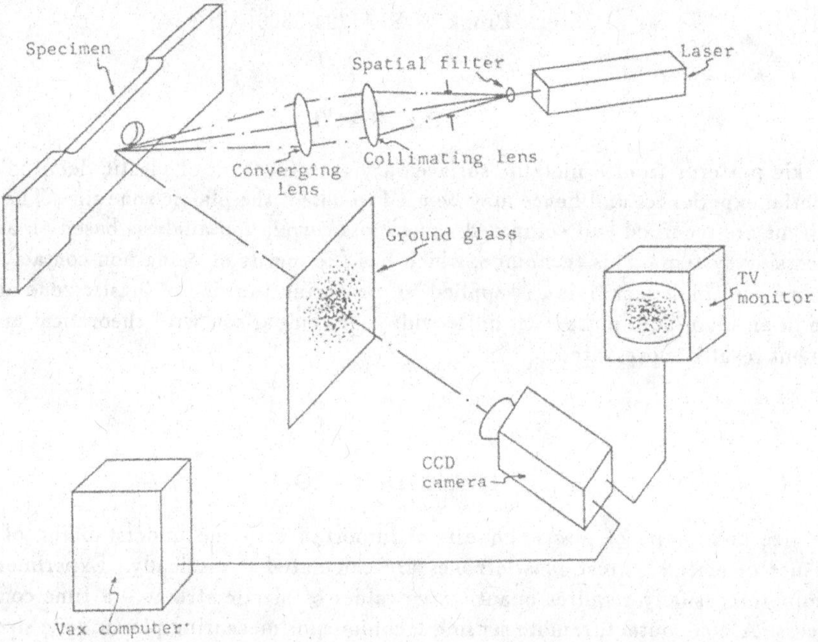


Figure 1 Schematic illustration of the experimental set-up

Cross correlation [3] is used to characterize the digitized speckle patterns such as those shown in Fig.2. For two speckle patterns described by $g(i,j)$ and $f(i,j)$, the normalized zero shift cross correlation $C_c(g,f)$, which is a measure of the resemblance between the two functions, is defined as

$$C_c(g,f) = \frac{\sum_{i=1}^{M}\sum_{j=1}^{N} g(i,j) \times f(i,j)}{\sqrt{\sum_{i=1}^{M}\sum_{j=1}^{N} g^2(i,j) \times \sum_{i=1}^{M}\sum_{j=1}^{N} f^2(i,j)}} \tag{1}$$

where $M, N$ are the dimensions of $g(i,j)$ and $f(i,j)$, respectively. The cross correlation coefficient of two speckle patterns, one from elastic zone and the other from plastic zone, is smaller than that all from elastic zone and hence may be used to detect plastic zone.

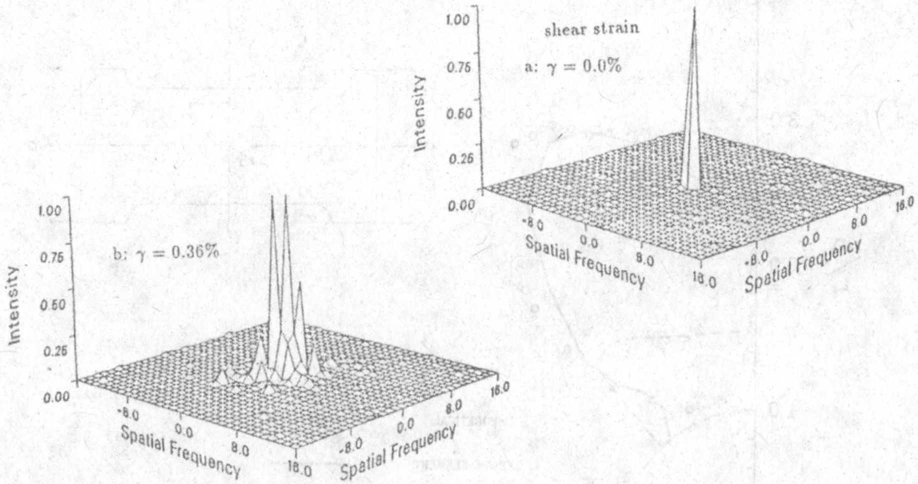


Figure 2 Speckle patterns from elastic and plastic zones

## EXPERIMENTAL RESULT

An aluminum plate of 3.2$mm$ thick with a central hole was used in the experiment. The geometry of the specimen is shown in Fig.1 where the width of the specimen is 76$mm$ and the radius of the hole is 6.4$mm$. One side of the specimen surface was polished until a RMS roughness value of approximately 0.05$\mu m$ was obtained. The specimen was loaded axially at $\sigma = 0.765\sigma_o$, where $\sigma$ is the remote normal stress along the loading direction, and $\sigma_o$ is the yield strength of the material. The specimen was then mounted on a stage which is capable of translating the specimen in two directions in 0.006$mm$ increments. Speckle patterns at different points on the specimen surface was digitized and processed to yield cross correlation coefficient.

Figure 3 shows the plastic zone determined using the method described above. The solid line represents a computed effective strain of 0.01% by a finite element analysis. As can be seen both the experimental and computed results agree reasonably well. The theoretical solution for plastic zone around a circular hole in an infinite flat plate is also included (broken line). As can be seen that it underestimated the size of plastic zone.

It has been demonstrated that speckle pattern cross correlation technique can be applied to the detection of the incipience of plasticity with a sensitivity of 0.01%. With further automation, the technique has the potential of being developed into a practical tool which may be applied to determining plastic zone at crack tip.

The authors wish to thank the Army Research Office, Engineering Science Division for providing the financial support through contract No. DAAL0388K0033.

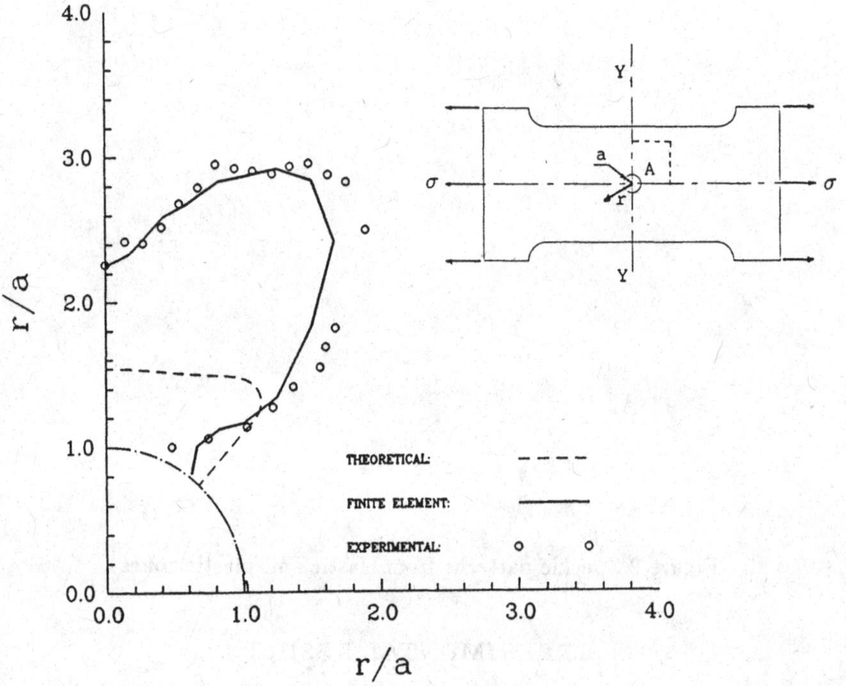


Figure 3 Plastic zone around a hole in a finite width plate ($\sigma = 0.765\sigma_o$)

## REFERENCES


1. P.F. Thomson, B.V. Shafer, The roughening of free surface during plastic working, Int. J. of Mach. Tool Des. Res., **22**,(4), pp.261-264, 1982

2. M. Miyagawa, A. Azushima, Measurement of plastic deformation by means of a laser beam, Proceedings of the 9th World Conference on Nondestructive Testing, Session 4a-4, pp.1-7, Melbourne, Australia, Nov.,1979

3. C. Lee, Y.J. Chao, M.A. Sutton, et al., Determination of plastic strain at notches by image-processing methods, Experimental Mechanics, **29**,(2), pp.214-220, 1989

4. A. Rosenfeld, C.A. Kak, Digital Picture Processing, Academic Press, pp.19, 1982

# A LARGE TIME INCREMENT APPROACH FOR CYCLIC LOADINGS

J.-Y. COGNARD, P. LADEVEZE
Laboratoire de Mécanique et Technologie
(E.N.S. de Cachan / C.N.R.S. / Université Paris 6)
61, avenue du Président Wilson - 94235 CACHAN CEDEX (France)
G.R.E.C.O.- G.I.S. "Calcul des Structures"

## ABSTRACT

The principles of a new approach for cyclic viscoplastic problems are described in this paper. A computational method to deal with non-linear mechanical behaviours, described by internal variables, is presented. This approach contrasts with the classical step by step method since it is an iterative procedure which accounts for the whole loading process in a single time increment. Special tools have been developed to simulate, within the large time increment method, several hundred cycles in a single increment. This method also allows a quick calculation of the limit response of the structure, as long as the whole stress and strain histories are not sought. In this case very few iterations of the method are necessary.

## INTRODUCTION

The mechanical material models for viscoplasticity or plasticity are numerous and have become more and more accurate. Numerous internal variables are necessary to describe correctly cyclic behaviour including multiaxial and out-of-phase loading conditions, ratcheting effects, etc ([2], [8]). The costs and complexity of the numerical simulations of these non-linear evolution problems are still dissuasive [5], which explains why, today, there are few complex non-linear computations carried out in industry.

The classical method of simulating such problems consists of splitting the loading path into small increments (a step by step method). The LArge Time INcrement method (LATIN method) introduced by P. Ladevèze [6] uses an approach which differs from the step by step methods, and allows a substantial reduction of the numerical cost [1], [3]. It is an iterative procedure which accounts for the whole loading process in a single time increment, which is not a priori limited.

A new version of the LATIN method adapted to behaviours described by internal variables [7], where convergence has been proved, is described. The performances on a simple example, with a large number of cycles in case of viscoplastic behaviour, are presented.

## THE LARGE TIME INCREMENT METHOD

Let us consider a structure which occupies a domain $\Omega$. At each time $t \in [0, T]$, the displacement $U_d$ is given on $\partial_1\Omega$, a partition of the boundary of $\Omega$. The surface force $F_d$ is given on the complementary part of the boundary $\partial_2\Omega$. A body force $f_d$ is applied on $\Omega$. For quasi-static response the structure evolution is described by the problem :

Find $(U(M,t))$ , $\sigma(M,t)$, $M\in\Omega$, $t\in[0,T]$, verifying $\forall\, t \in [0,T]$ :

- the kinematic equation (U kinematically admissible, KA)
- the equilibrium equation ($\sigma$ statically admissible, SA)
- the constitutive relations $\dot{\varepsilon}(t) = \mathcal{A}\,(\sigma(\zeta)\,,\, \zeta \le t)$ $\quad\mathcal{A}$ material characteristic.

According to the properties of the mechanical equations, the difficulties can be split into two groups : - (1) linear equations, which are possibly global in space,
- (2) equations which are local in space and possibly non linear.

Classically we note $\mathbf{A_d}$ (resp $\Gamma$) the subspace of elements s verifying the condition (1) (resp (2)). The LATIN method starts from an initial $s_0$, an element of $\mathbf{A_d}$ (elastic response over [0,T]), and then successive elements belonging to $\Gamma$ and to $\mathbf{A_d}$, are built up to converge to the solution denoted by $s_{ex} = \mathbf{A_d} \cap \Gamma$ (Fig. 1). An iteration, i.e. the construction of a new admissible element $s_{n+1}$ from a given admissible element $s_n$, is composed of two stages :

- local stage : find $\hat{s} \in \Gamma$ such that $\hat{s} = (\Gamma) \cap (s_n + \mathbf{E^+})$
- global stage : find $s_{n+1} \in \mathbf{A_d}$ such that $s_{n+1} = (\mathbf{A_d}) \cap (\hat{s} + \mathbf{E^-})$

$\mathbf{E^+}$ is a given direction, linear and local in space, and $\mathbf{E^-}$ is a given linear direction.

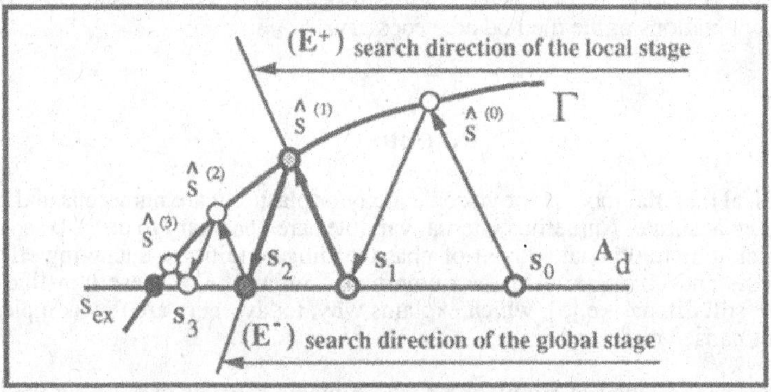


Fig. 1. Presentation of the LATIN method

## APPLICATION TO CYCLIC LOADINGS

Several versions of the LATIN method have been studied [1], [3], [4]. For cyclic loadings new search directions $\mathbf{E^+}$ and $\mathbf{E^-}$ are introduced. The classical behaviour modelling described by internal variables is changed in order to introduce the concept of "normal formulation" which is characterized by a quadratic free energy. It has been shown that most of the present plasticity and visco-plasticity models allow such a formulation. This new approach is presented for

Chaboche's model which is well adapted to cyclic processes [2]. We note p the isotropic hardening parameter and $\alpha$ the kinematic one. The associated forces are noted R and **x**. $\dot{\varepsilon}^e$ and $\dot{\varepsilon}^p$ are the elastic and the inelastic strain rate. A change of variables (isotropic hardening parameters) is introduced so that the state equations become linear :

$$\tilde{R} = A\,\tilde{p} \qquad (A > 0)\,; \qquad \tilde{p} = \int_0^p \left(\frac{\partial R}{\partial p}\frac{1}{A}\right)^{1/2} dp$$

The only non-linear relations are the evolution laws which are local in space. The partition of the equations proposed uses the set of variables **s** :

$$\mathbf{s} = (\dot{\varepsilon}^p, \dot{\underline{X}}\,;\ \sigma, \underline{Y}) \qquad M \in \Omega \text{ and } t \in [0,T]$$

$$\underline{X} = (\alpha, \tilde{p})^T \text{ (internal variables)} \qquad \underline{Y} = (\mathbf{x}, \tilde{R})^T \text{ (associated thermodynamic forces)}$$

We note $\mathbf{A_d}$ the set of elements **s** such that :

$\sigma$ SA ; $\dot{\varepsilon} = (\mathbf{K_e}^{-1}\dot{\sigma} + \dot{\varepsilon}^p)$ KA — admissibility conditions ($\mathbf{K_e}$ elastic operator)

$\underline{X} = \Lambda\,\underline{Y}$ — equations of state ( $\Lambda$ linear operator)

and we note $\Gamma$ the set of elements **s** such that :

$$\frac{d}{dt}\begin{bmatrix}\varepsilon^p \\ -\underline{X}\end{bmatrix} = \mathbf{B}\begin{bmatrix}\sigma^D \\ \underline{Y}\end{bmatrix} \quad \text{and} \quad \begin{bmatrix}\dot{\varepsilon}\,|_{t=0} \\ \underline{X}\,|_{t=0}\end{bmatrix} = 0$$

evolution laws (**B** non-linear operator)

The general framework of the LATIN method can still be used, only the search directions have to be defined. The real tangent subspace to $\Gamma$ is introduced, for $\mathbf{s} \in \Gamma$ we note **L** the operator verifying the relation :

$$\begin{bmatrix}\Delta\dot{\varepsilon}^p \\ -\Delta\dot{\underline{X}}\end{bmatrix} = \mathbf{L}\begin{bmatrix}\Delta\sigma^D \\ \Delta\underline{Y}\end{bmatrix}$$

where $\Delta U$ defines the derivative of U.

At each iteration, the local stage requires the resolution of a simple non-linear equation where time t is a parameter over [0, T]. In practice another search direction is used.

Similar techniques to those used in the previous versions of the LATIN method ([1], [3]) can be applied to solve the global stage at a relatively low cost. The sum of the products of space fields by scalar time functions gives very good results for quasi-static problems :

$$\Delta v_n = v_{n+1}(t, M) - v_n(t, M) = \sum_{i=1}^{m} \alpha_i(t)\, A_i(M) \qquad (v = \dot{\varepsilon}^p, \dot{\underline{X}}, \sigma)$$

Introducing the different corrections, the variational form of the global stage is such that :

$$\int_0^T\!\!\int_\Omega \left\{ \mathrm{Tr}\,[\,\mathbf{K_e}^{-1}\Delta\dot{\sigma}_n . \sigma^*\,] + \Delta\underline{X}_n . \Lambda\underline{X}^* + \mathbf{L}\begin{bmatrix}\Delta\sigma_n^D \\ \Delta\underline{X}_n\end{bmatrix} . \begin{bmatrix}\sigma^* \\ \Lambda\underline{X}^*\end{bmatrix} \right\} d\Omega\, dt =$$

$$\int_0^T\!\!\int_\Omega \left\{ \mathrm{Tr}\,[\,(\varepsilon_n^p - \hat{\dot{\varepsilon}}^p). \sigma^*\,] + (\underline{X}_n - \hat{\underline{X}}). \Lambda\underline{X}^* + \mathbf{L}\begin{bmatrix}\hat{\dot{\sigma}}^D - \sigma_n^D \\ \hat{\underline{Y}} - \Lambda\underline{X}_n\end{bmatrix} . \begin{bmatrix}\sigma^* \\ \Lambda\underline{X}^*\end{bmatrix} \right\} d\Omega\, dt$$

$\forall\,\sigma^*$ kinematically admissible for homogeneous equations, $\forall\,\underline{X}^*$

This proposed representation allows this problem to be split into two simple ones. An iterative procedure (time-space) is proposed to find the best couples of the functions [3]. If the time functions are assumed to be known, the best associated space fields are solution of a classical linear global problem. A small system of differential equations gives the best time functions if the space fields are assumed to be known, these equations take the following form :

$$\mathbb{M}\,\frac{d}{dt}\mathbb{G} + \mathbb{L}(t)\,\mathbb{G} = \mathbb{F}(t) \qquad \forall\, t \in [0,T] \qquad \text{and} \quad \mathbb{G}(t=0) = 0$$

For cyclic loadings the functions are described using two different time-scales. For the "slow" representation a finite element discretisation is used, for a linear "slow" time interpolation between $T_i$ and $T_{i+1}$ any time-function is defined by :

$$t \in [T_k, T_{k+1}] \qquad \alpha(t) = \alpha^k(\tau)\,\frac{T_{k+1} - t}{T_{k+1} - T_k} + \alpha^{k+1}(\tau)\,\frac{t - T_k}{T_{k+1} - T_k} \;; \quad \text{with } \tau = \frac{t}{\Delta T}$$

where $\alpha^k(\tau)$ are periodic functions defined over the cycle length $\Delta T$. Special tools have been developed in order to obtain an efficient computational method for computing $\alpha^k(\tau)$. Only a very small number of iterations of the method are necessary [4]. For simple examples (uniaxial problem) with Chaboche's viscoplastic model, and for 1000 cycles of a sine shaped loading, an excellent accuracy is reached after only six iterations. The whole loading path is computed in single increment with the LATIN method. More complex examples will be shown. Let us note that a variant of this approach also allows a quick calculation of the limit response of the structure, as long as the whole of the stress and strain histories are not sought.

## CONCLUSION

The feasibility of the LATIN method has been shown using various examples. Convergence has always been obtained after a very small number of iterations. The saving in numerical cost with respect to the classical step by step methods increases when the loading path becomes more complex. Moreover parallel computers can be used very efficiently without much difficulty. Therefore, it can be concluded that this new computational method for cyclic problems could prove to be very interesting in the case of industrial simulations.

## REFERENCES


1. Boisse Ph., Bussy P., Ladevèze P., "A new approach in non linear mechanics : the large time increment method", Int. J. Numer. Methods Eng.,Vol. 29, pp. 647-663, 1990.
2. Chaboche J.L., "Constitutive equations for cyclic plasticity and cyclic viscoplasticity", Int. J. of Plasticity, n°5, pp. 245-302, 1989.
3. Cognard J. Y., "Le traitement des problèmes non linéaires à grand nombre de degrés de liberté par la méthode à grand incrément de temps", Calcul des structures et intelligence artificielle, Vol 4, J.M. Fouet, P. Ladevèze, R. Ohayon, Pluralis 1990.
4. Cognard J. Y., Ladeveze P., "The large time increment method applied to cyclic loadings", Proceedings IUTAM "Creep in Structures IV", Cracow 1990.
5. Horneberger K., Stamm H., "An implicit integration algorithm with a projection method for viscoplastic constitutive equations", Int. J. Numer. Methods Eng., vol 28, p. 2397-2421, 1989.
6. Ladevèze P., "Sur une famille d'algorithmes en mécanique des structures", Comptes-rendus, Académie des Sciences - Paris, 300, série II, n° 2, pp. 41-44, 1985.
7. Ladevèze P., "La méthode à grand incrément de temps pour l'analyse de structures à comportement non linéaire décrit par variables internes", Comptes-rendus, Académie des Sciences - Paris, 309, série II, n°11, pp. 1095-1099, 1989.
8. Watanabe O., Atluri S. N., "Internal time, general internal variable, and multi-yield-surface theories of plasticity and creep : a unification of concepts", Int. J. of Plasticity, vol. 2, p. 37-57, 1986

# PREDICTION OF FRACTURE INITIATION USING A FINITE ELEMENT METHOD AND VARIOUS DAMAGE MODELS

N. L. DUNG
Technical University of Hamburg-Harburg
MT2-Structural Mechanics Division
Eißendorferstr. 42, 2100 Hamburg 90, Germany

## ABSTRACT

In this paper the failure analysis is introduced to predict the initiation of the flow localization and the ductile fracture of plastic media used in bulk forming processes. Those examined are axisymmetric upsetting and extrusion and plane-strain side-pressing of some metals (steel, aluminium alloys, brass,...). The damage models are evaluated for these processes by analyzing the stress and strain fields obtained step-by-step from a rigid-plastic finite element simulation for each forming process. The theoretically predicted results (fracture strain and site) are compared with experimental observations.

## INTRODUCTION

In the last 20 years, there are many fracture criteria developed for predicting the fracture initiation in metal forming processes [1 - 12]. According to the model type, these criteria are subdivided into macroscopic fracture criteria and models of void growth. Fig. 1 illustrates a few reprensentative fracture criteria for both groups which were often used to analyze the fracture initiation in bulk and sheet metal forming. Except of *Cockcroft & Latham* criterion [3, 4], they are almost path-dependent [1].

The theoretical background of the fracture criteria is based either on the macroscopic mechanics (limited local plasticity) or on the microscopic mechanisms (void growth and coalescence). Only few of them can take into account both macroscopic and microscopic factors, such as the attempt made by *Gurson* or *Lemaitre* [4, 9, 10]. Generally, the fracture criteria analyze the onset of cracking in metal forming with the same concept. They propose that the ductile fracture will appear in the body of the workpiece if the value of accumulated damage reaches its critical value. The critical value at fracture is constant for a material. The value of accumulated damage is, for instance, the tensile plastic work introduced by *Cockcroft & Latham*, the growing volume fraction of voids introduced by *Oyane* or the critical void growth and coalescence introduced by *McClintock* and *Thomason* [3-5, 12].

To calculate the accumulated damage, it is neccessary to know about the material behavior throughout the forming process. The stress and strain fields can

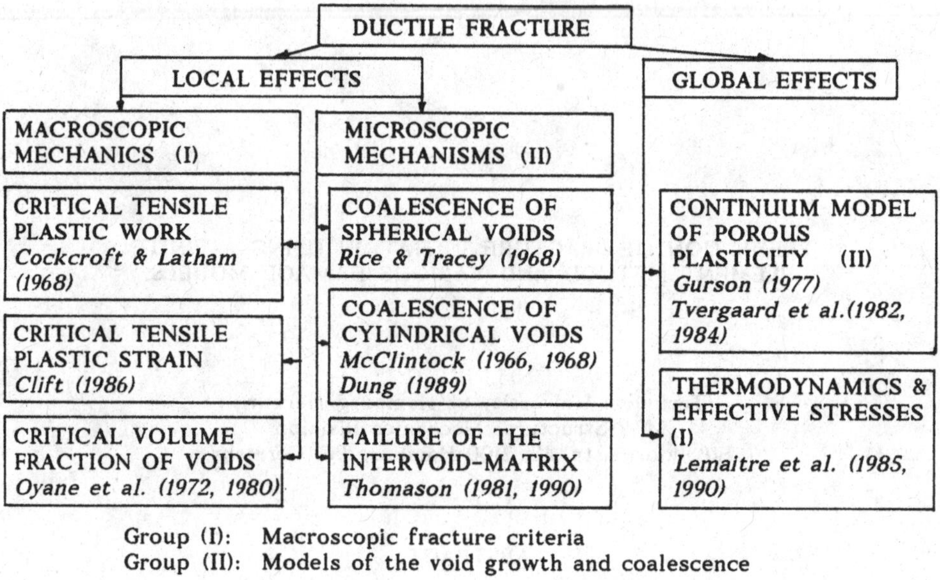


Figure 1. The representative damage models in the metal forming.

be adequately obtained stepwise during the unsteady process by means of a finite element method. Therefore, the fracture criteria are often implemented into a finite element procedure in order to predict the fracture initiation. In this case, the macroscopic fracture criteria seem to be simple and economic for problems of bulk forming. The criteria of *Gurson* and *Tvergaard* analyze the ductile fracture in sheet forming precisely, but not for bulk forming with a triaxial stress system [9].

The paper deals with the theoretical analysis of the fracture initiation in bulk forming by means of a rigid-plastic finite element method together with some fracture criteria under consideration of the local effects (Fig. 1).

## METHOD OF ANALYSIS

### Finite Element Method

The finite element method used is based on a modified *Markov's* principle for rigid-plastic materials, in which the incompressibility condition is included by means of a *Lagrange* multiplier (the hydrostatic stress) and the friction work is regarded as internal energy dissipation. The results of the variational problem are the admissible field of velocities and the distribution of the hydrostatic stress. Fig. 2 shows the flow chart of the finite element simulation [5]. An instationary forming process is divided in many stationary deformation steps in order to analyze stepwise the material behavior with the stress and strain fields. After each deformation step, the contact problem is checked at the interfacial surface between tool and workpiece, and the FE mesh is updated accordingly or remeshed if neccessary. The rigid-plastic finite element method provides the results in an adequate accuracy and in an economic manner.

### Fracture Criteria

The fracture criteria are implemented into the finite element method as shown in the flow chart, Fig. 2 (in the window covered with dots). In case of the empirical criteria of *Cockcroft & Latham*, *Clift* [2,3,8], the tensile plastic work is calculated in each stationary deformation step and then compared with the critical value for the tested material. If the models of void growth and coalescence (*McClintock* [11], *Dung* [5, 6], *Thomason* [12],...) are subjected, one calculates the value of the accumulated damage accordingly. For the criterion of *Lemaitre* [10], for example, the material behavior must also be corrected due to the growth of void volume and the changing of the intervoid-matrix.

**Solution**

The fracture initiation is investigated during bulk forming process as upsetting, side-pressing and extrusion using the above described method of analysis. As an example for this analysis, *McClintock's* model of void growth and coalescence [5, 6] is now used for predicting the fracture site and strain. Then, the growth of void in the direction a is given for the system of transverse stresses $\sigma_a$ and $\sigma_b$ as:

$$d(\ln F_{ca}) = \left( \frac{\sqrt{3}}{2(1-n)} \sinh\left\{ \frac{\sqrt{3}(1-n)}{2} \frac{\sigma_a + \sigma_b}{\bar{\sigma}} \right\} + \frac{3}{4} \frac{\sigma_a - \sigma_b}{\bar{\sigma}} \right) d\bar{\varepsilon} = f_{ca}(\sigma_a, \sigma_b)\, d\bar{\varepsilon} . \quad (1)$$

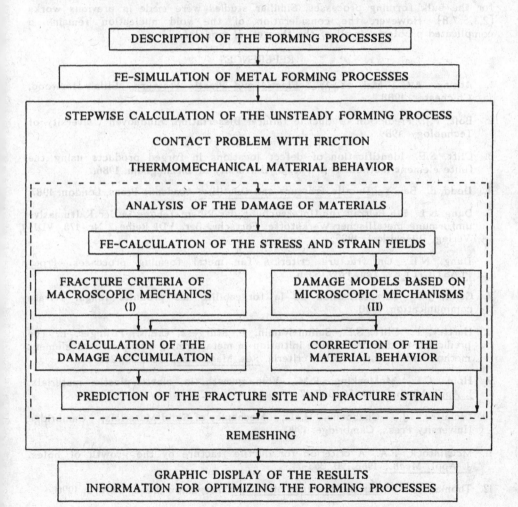

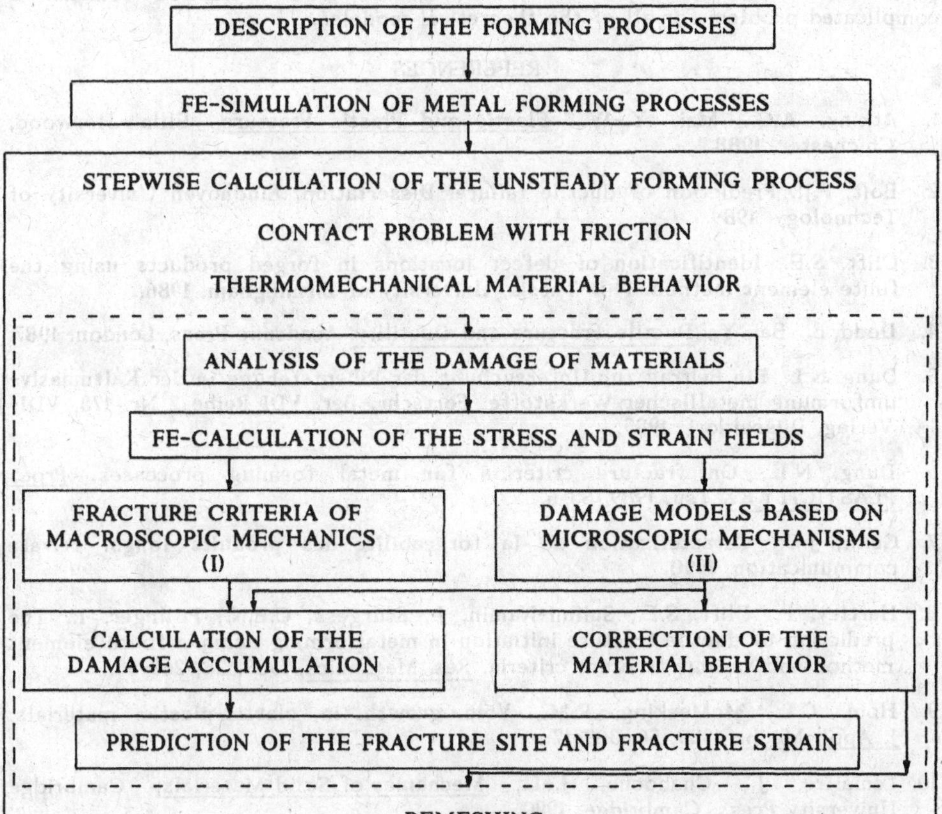


Figure 2. FE-Prediction of the ductile fracture initiation during forming processes.

It follows that the damage is accumulated during the process as

$$A_{ca} = \int_0^{\bar{\varepsilon}} f_{ca}(\sigma_a, \sigma_b)\, d\bar{\varepsilon} \;; \text{ and at fracture: } A_{ca} = \int_0^{\bar{\varepsilon}_f} f_{ca}(\sigma_a, \sigma_b)\, d\bar{\varepsilon} = A^f_{ca} \quad (2)$$

The 6 values $A_{ca}$, $A_{cb}$, $A_{ab}$, $A_{ac}$, $A_{ba}$ and $A_{bc}$ are obtained at every node of the FE mesh in each deformation step. One of them will dominate the fracture initiation. If $A_{ca}$ is assumed to be always positive and larger than the other values, the crack will appear at this particular node at which $A_{ca}$ is the largest and will reach its critical value $A^f_{ca}$ at first. $A^f_{ca}$ of a material is experimentally determined from the relation between spacing and shape of voids at the moment of coalescence. The modified version [5, 6] of *McClintock's* model [11] is satisfactory for a series of bulk forming processes (upsetting, side-pressing and extrusion).

The predictions of the fracture initiation are carried out in this paper in order to introduce the representative fracture criteria, whose results are satisfactory for the bulk forming processes. Similiar studies were made in previous works [2,3,5,7,8]. However, the consideration of the void nucleation remains a complicated problem for all of the theoretical models.

## REFERENCES


1. Atkins, A.G., Mai, Y.-W., Elastic and Plastic Fracture, Ellis Horwood, Chichester, 1988.

2. Bolt, P.J., Prediction of ductile failure. Dissertation, Eindhoven University of Technology, 1989.

3. Clift, S.E., Identification of defect locations in forged products using the finite element method. PhD-Thesis, University of Birmingham, 1986.

4. Dodd, B., Bai, Y., Ductile Fracture and Ductility, Academic Press, London, 1987.

5. Dung, N.L., Ein Beitrag zur Untersuchung der Rißentstehung in der Kaltmassivumformung metallischer Werkstoffe. Fortschr.-Ber. VDI Reihe 2 Nr. 175, VDI-Verlag, Düsseldorf, 1989.

6. Dung, N.L., On fracture criterion for metal forming processes. Proc. PLASTICITY'89, Tsu, 1989, 53-6.

7. Gelin, J.C., Caractèrisation de la forgeabilitè des produits longs. Private communication, 1990.

8. Hartley, P., Clift, S.E., Salimi-Namin, J., Sturgess, C.E.N., Pillinger, I., The prediction of ductile fracture initiation in metalforming using a finite element method and various fracture criteria. Res Mechanica, 1989, 28, 269-93.

9. Hom, C.L., McMeeking, R.M., Void growth in elastic-plastic materials. J. Appl. Mech., 1989, 56, 309-17.

10. Lemaitre, J., Chaboche, J.-L., Mechanics of Solid Materials, Cambridge University Press, Cambridge, 1990.

11. McClintock, F.A., A criterion for ductile fracture by the growth of holes. J. Appl. Mech., 1968, 90, 363-71.

12. Thomason, P.F., Ductile Fracture of Metals, Pergamon Press, Oxford, 1990.


**Acknowledgement**
This work is carried out under the grant No. Du 178/2-1 of the DFG (Germany).

# GENERALIZED YIELD STRENGTH CRITERIA FOR BAR STRUCTURES

by B. GUESSAB and S. TURGEMAN
Institut de Mécanique de Grenoble
BP 53 X
38041 GRENOBLE CEDEX

## ABSTRACT

The plastic calculation of bar structures is based on a transformation of the three-dimensional continuum into a one-dimensional continuum which obeys generalized yield criterion. A new method to obtain this criterion is presented here. This method gives limit loads for the one-dimensional continuum, which are upper bounds of the limits loads for the three-dimensional continuum.

## INTRODUCTION

The dimensional design of bar structures is based on the schematic representation of the two- or three-dimensional continuum by a one-dimensional continuum in which the state of stress is defined by the stress tensor acting on a bar cross-section. The force components (normal and shear forces, bending and torsion moments) are called generalized stresses. Using the expression for the plasticity criterion of the continuum, so-called "local" yield criteria are obtained by a classical approach. The arguments of these criteria are the generalized stresses [1]. These local criteria are used to obtain values for the limit loads of the structure, considered to be a generalized continuum. These values cannot be interpreted as either the upper or lower bounds of the values of the same loads calculated within the formalism of a two- or three-dimensional continuum.

This paper presents a method for determining the plasticity criteria in generalized stress conditions, based on the study of limit loads for an elementary bar subjected to special boundary conditions. While remaining within the generalized continum, these conditions can be used to calculate plastic deformation velocity fields that are admissible for the two- or three-dimensional continuum. It is then possible to define the mechanical sense of the limit load values obtained within the theoretical framework of a generalized continuum.

In order to simplify the explanation of this method, the structures considered here will be assumed to consist of straight bars, with an embedded plane, loaded in this plane, and with planar, homogeneous and weightless stress conditions. Moreover, the standard, perfectly plastic rigid material is assumed to obey Von Mises yield criterion.

## 1 . Presentation of the method

Let (e) be an elementary bar of length $l_o$ , called generic element (fig. 1). The right and left sections of this element (respectively $S_d$ and $S_g$) are subjected to a rigid body movement defined by the vectors $X_d$ for $S_d$, and $X_g$ for $S_g$. The components of $X_g$ are $U_x^g$ , $U_y^g$ (translation speed components) and $\omega^g$ (rotation speed). Identical notations exist for $X_d$. The boundary AB and CD of the generic element is not loaded. For these boundary conditions, the principle of virtual powers is written as follows:

$$\int_{(e)} \sigma_{ij}\, \vartheta_{ij}\, dV = \Sigma_i\, (X_g \setminus X_d)_i \,. \qquad (1)$$

where: $\sigma$ = stress field statically permissible in (e)
$\vartheta$ = deformation velocity field kinematically permissible in (e)
$\Sigma$ = section loading vector, of generalized stress components N (normal force), T (shear force), M (bending moment) on $S_g$
$X_g \setminus X_d$ = generalized deformation velocity vector of component
$[U_x] = U_x^g - U_x^d ;\; [U_y] = U_y^g - U_y^d + \omega^d\, l_o \,; [\omega] = \omega^d - \omega^g \,.$ (2)

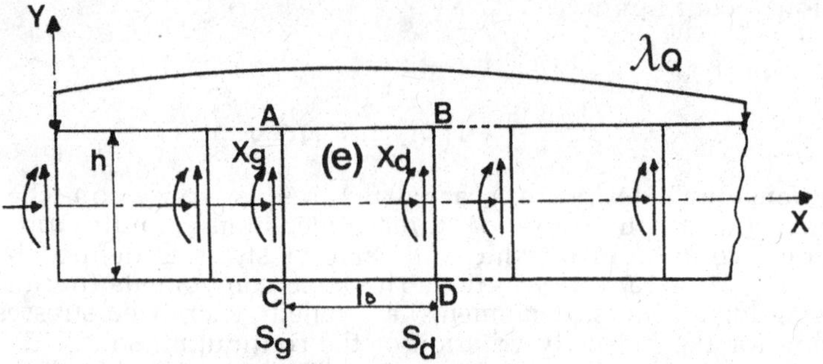


Figure 1. generic element .

The loadings $\Sigma$ that can be supported by (e) belong to a convex $K_{l_o} = (\Sigma \,/\, \phi_{l_o}(N,T,M) \leq 0)$ whose origin is the centre of symmetry and whose the plane N=0 is the plane of symmetry, and which decreases as a function of $l_o$ $(l > l' \rightarrow K_{l'} \subset K_l)$. By definition, $\phi_{l_o}$ (N,T,M) is the plasticity criterion under the generalized stress conditions required.
If $\phi_{l_o}$ (N,T,M) is known (or its external approximation, obtained using the kinematic limit analysis method), a admissible displacement velocity field $\tilde{u}\,(X_g, X_d, l_o)$ can be associated with any rigid movement pair $(X_g\,,\, X_d)$, the plastically dissipated power in this field, $\pi\,(X_g \setminus X_d)$ being minimal
Consider now a bar structure subjected to an external loading Q, of limit value denoted by $\lambda_l Q$ $(\lambda_l > 0)$. Let this structure then be divided into elements $(e_i)$ of length $l_o$, either contiguous or separated by rigid elements of any length, and give each of the various sections $S_i$ a rigid body movement $X_i$.

The loading $\lambda^*Q$ where:

$$\lambda^* = \text{Min} \sum_{i=1}^{N} \pi (X_i \setminus X_{i+1}, l_o) \,., \quad X = \{X_i, i = 1, N\} \text{ admissible} \,. \quad P_{ext}(Q, X) = 1 \,. \tag{3}$$

thus constitutes an upper bound of the limit load $\lambda_l Q$. In (3), $P_{ext}$ (Q, X) designates the power of the external loading Q in the displacement velocity field defined explicitly by X (note that, in each ($e_i$), a velocity field u ($X_i$, $X_{i+1}$, $l_o$) can be associated with the $X_i$, $X_{i+1}$ pair).

In order for $\lambda^*$ to be a good approximation of $\lambda_l$, the length $l_o$ of the generic element must satisfy the two following criteria:

- criterion 1: it must be as small as possible (as the convex $K_{lo}$ was obtained on the assumption that (e) was not loaded on AB and CD).
- criterion 2: it must correspond to the smallest possible convex $K_{lo}$.

## 2 . Choice of optimum length

The two criteria involved in the choice of length are conflicting, given that $K_{lo}$ decreases as a function of $l_o$. In order to choose a length (if any) that satisfies the two criteria correctly, the variation in $K_{lo}$ must be studied as a function of $l_o$. Since this problem is extremely heavy to handle, it will be restricted simply to the variations in the projections $N_{max}$, $T_{max}$ and $M_{max}$ of $K_{lo}$ on the N, M and T axes (fig. 2). Note that, when $l_o$ is small (of the order of h, height of the bar, and small reference dimension), $N_{max}$ and $M_{max}$ reach a value close to their minimum, contrary to $T_{max}$. The analysis of the velocity fields $\tilde{u}$ ($X_1$, $X_4$, $l \geq 1.5$ h) (where $X_1$, $X_4$ are such that $[u_x] = [\omega] = 0$) (fig. 3), shows that these fields are in fact made up of two fields $\tilde{u}$ ($X_1$, $X_2$, $l^*$) and $\tilde{u}$ ($X_3$, $X_4$, $l^*$) (where $l^* = 3h/4$) separated by a rigid zone. It follows that a knowledge of $K_{lo}$ (with $l_o = 3h/4$) enables the $T_{max}$ (l) curve to be obtained for $l \geq 1.5$ h.from this study,a value of 3h/4 was obtained which correctly satisfies the two choice criteria .

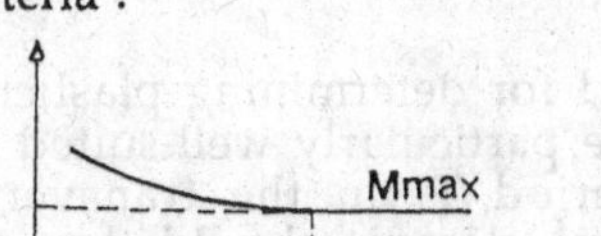


Figure 2 . Nmax,Tmax & Mmax variations as fonctions of l .

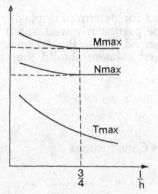

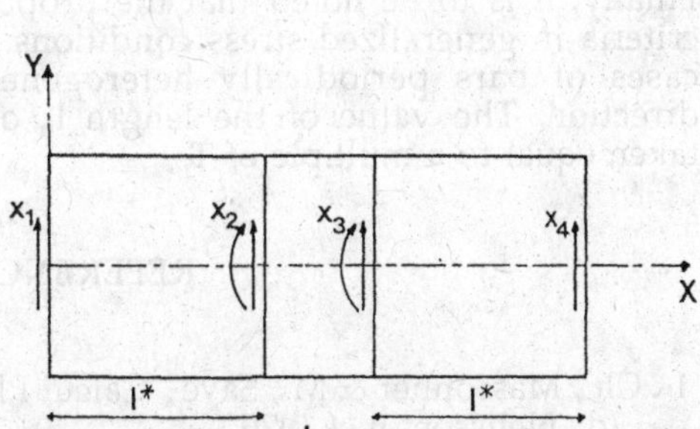


Figure 3 . velocity field in the element of length L>1.5 h and subjected to shear force .

## 3 . Conclusion

When the ratio $l_o/L$ (L = bar length) is very small, it may be advantageous to use the limit equation $\wp_o$ instead of equation (3). The limit equation $\wp_o$ is obtained by assuming that $l_o/L$ tends to 0. In this problem, the velocity fields in the bar are described by means of a function $X(x) = (U_x(x), U_y(x), \Omega(x))$ that is derivable element by element, with the macroscopic variable x marking the section considered (fig. 4). In the calculation of external force level, consideration should be given to the fact that the displacement velocity to be taken into account also depends on a microscopic variable s ($0 \leq s \leq l_o$). The limit equation $\wp_o$ can then be written:

$$\lambda^* = \text{Min} \int_0^L \pi\,(X'(x))dx\,.$$

$$X(x) \text{ admissible}\,.$$

$$P_{ext}(Q, X) = \int_0^L (Q_x U_x + Q_y U_y)(x)\,dx + T_c(Q, X'(x)) = 1\,. \quad (4)$$

where $X'(x) = \left(-U_x'(x), -U_y'(x) + \Omega(x), \Omega'(x)\right)$, and $T_c$ is a complementary term due to the displacement velocity variation as a function of s.

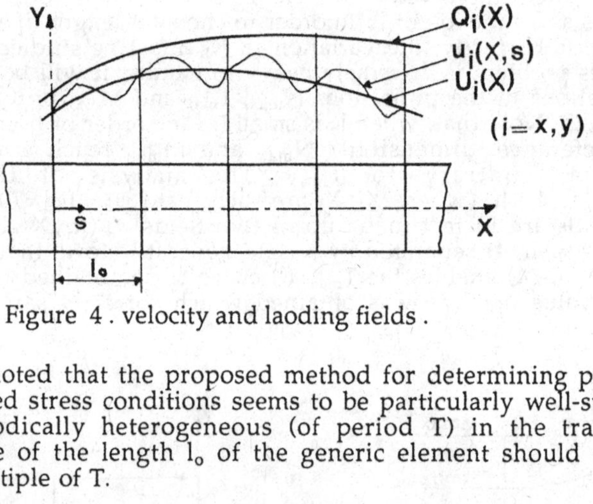


Figure 4 . velocity and laoding fields .

Finally, it is to be noted that the proposed method for determining plasticity criteria in generalized stress conditions seems to be particularly well-suited to cases of bars periodically heterogeneous (of period T) in the transverse direction. The value of the length $l_o$ of the generic element should then be taken equal to a multiple of T.

## REFERENCES

1 . Ch , Massonnet & M , Save , Calcul Plastique des Constructions . Ed . Nelisson B . ( 1976 ) .

2 . B , Halphen & J , Salençon , Elasto-plasticité . Press . Ponts & Chaussées . ( 1987 ) .

3 . J , Salençon , Calcul à la Rupture et analyse limite . Press . Ponts & Chaussées . ( 1983 ) .

4 . Y , Lescouarch , Calcul Plastique des Structures . COTECO ( 1983 ) .

# PREDICTING OF THE RESIDUAL STRESS FIELD CREATED BY INTERNAL PEENING OF NICKEL-BASED ALLOY TUBES

N. HAMDANE* , G. INGLEBERT** and L. CASTEX***

* Ecole Nationale Supérieure d'Arts et Métiers, Laboratoire de microstructure et mécanique des matériaux, 151, boulevard de l'Hôpital, 75013 Paris, France.

** Institut Supérieure des Matériaux et de Construction Mécanique, 3, rue Fernand-Hainault, 93407 Saint-Ouen, France et E.N.S.A.M Paris.

*** Ecole Nationale Supérieure d'Arts et Métiers, Centre d'enseignement et de recherche d'Aix-en-Provence, 2, cours des Arts et Métiers, F-13617 Aix-en-Provence Cedex 1.

## ABSTRACT

Modelling the residual stresses induced by peening requires the applied stress field to be known as a function of the impact characteristics: the impact energy, the nature of the peening medium and the mechanical behavior of the processed material. Shot size, material, and velocity and in some instances the impingement angle and the coefficient of friction of the contacting bodies are the main factors which have been taken into account.
A realistic approximation of the constitutive law of the investigated material has to be developed. A four-parameter representation leads us to fairly good predicted results as well for materials having a well-defined behaviour [ferritic steels, titanium alloys] as for those showing a much more complex one (austenitic, aluminum, nickel-based alloys). In our opinion it constitutes an acceptable trade-off between simplicity of identification and accuracy of description of the material behavior.

## 1- INTRODUCTION

Surface treatments are used to improve the life of workpiece subjected to wear and fatigue loading. Shot peening is one of the most employed techniques: the treated surface is orthogonally impacted by free spherical media.
We focus this paper on roto-peening designed by EDF for use in tubes of steam generators in nuclear power plant. For this technique, the spherical media are fixed on an elastic flag guided in translation and rotation. So, all shots can be evacuated from the tube after processing, [and risks of contamination are supressed]. So a constant impingement angle is observed, generating a great deal of shear stresses on the inner face of the tube due to frictional effects.

## 2- THEORETICAL MODEL

The present work [1] is based on previous studies on residual stresses created by shot-peening. The following approach was used [2], [3], [4]: applied elastic stresses generated by peening are estimated, and associated stabilised residual stresses and plastic strains are deduced assuming a semi-infinite body ; then these residual stresses are redistributed to account for the actual geometry. For shot-peening with normal impact, uniaxial modelling was sufficient. Guechichi and al. [5] use an elasto-plastic constitutive law with linear kinematic hardening and some complementary isotropic hardening. To describe more complex hardening such as those which exist in Aluminium or Nickel based alloy, Khabou and al. [6] define an improved constitutive law with two yield coupled plastic mechanisms. So a non linear continuous description of the stress-strain curve is achieved.
In roto-peening, the impingement angle generates strong frictional effects. A three dimensional modelling of the elastic applied stresses and so of the elasto-plastic constitutive law is thus needed. Our particular roto-peened material is INCONEL 600, from the austenitic family, showing complex hardening; we had to derive a three dimensional generalization of Khabou's constitutive law to obtain residual stresses through the above approach.

### 2-1 ELASTIC APPLIED STRESSES ON SEMI-INFINITE BODY

In the roto-peening process, the impacting spherical media are fixed side by side on the flag. This bar was modelised by a rigid cylinder sliding on a semi-infinite body. The applied stress field $\mathbf{S}^{el}$ corresponds to a bidimensional contact with plane strains assumption [7], [8]. Physical parameters for the contact are obtained by an elastic equivalence and are given as functions of the parameters of the roto-peening process. Besides the roto-peening parameters, the coefficient of friction of the contacting bodies is a key-parameter to define roto-peening applied stresses.

### 2-2 FUNDAMENTAL EQUATIONS FOR THE SEMI-INFINITE BODY

In a cylindrical basis [r, θ, z] where z axis is parallel to the axis of the tube, the inelastic displacement field is assumed as only a function of r. In this part, the radius of the tube is infinite compared to the radius of the impacting media. Stress equilibrium and boundary conditions imply the following residual stresses ρ, total **E** and plastic $\mathbf{E}^p$ strain tensors:

$$\rho = \begin{bmatrix} 0 & 0 & 0 \\ 0 & \rho_{\theta\theta} & 0 \\ 0 & 0 & \rho_{zz} \end{bmatrix} \; ; \; \mathbf{E}^P = \begin{bmatrix} E^P_{rr} & E^P_{r\theta} & 0 \\ E^P_{r\theta} & E^P_{\theta\theta} & 0 \\ 0 & 0 & E^P_{zz} \end{bmatrix} \; ; \; \mathbf{E} = \begin{bmatrix} E_{rr} & E_{r\theta} & 0 \\ E_{r\theta} & E_{\theta\theta} & 0 \\ 0 & 0 & 0 \end{bmatrix}$$

Then Hooke's law allows to define the $\mathbf{M_p}$ matrix which links residual stresses to plastic strains:

$$\text{dev}\, \rho_{ij} = \mathbf{M_p}\, E^P_{ij}$$

### 2-3 CONSTITUTIVE LAW AND PREDICTION

A three dimensional constitutive law is used following basic models by Zarka and Casier [2] and Inglebert and al [3], [4]. Each volume element is considered as a fictitious microstructure with two perfectly plastic mechanisms [k = 1;2]. Local stresses σ on these mechanisms, internal parameters α [local strains], and local **y** or global **Y** transformed parameters are written:

$$\sigma^k_{ij} = A_k\, S_{ij} - y^k_{ij} = A_k\, S^{el}_{ij} - [y^k_{ij} - A_k\, \text{dev}\, \rho_{ij}] = A_k\, S^{el}_{ij} - Y^k_{ij} \qquad i, j = r, \theta, z$$

Where $A_k$ is the elastic localisation tensor in the volume element governing the plastic strains too. **y** and **Y** are introduced simplify the local writing the yield condition. The influence of the actual **S** or applied loading $\mathbf{S^{el}}$ on the volume element appears clearly; **y** and **Y** are only functions of plastic strains; **y** and $\alpha$ are linked through a hardening matrix **B**:

$$|| A_k S_{ij} - y^k_{ij} || = || A_k S^{el}_{ij} - Y^k_{ij} || \leq S_0$$

$$y^k_{ij} = B_{kl} \alpha^l_{ij} \; ; \; E^p_{ij} = A_k \alpha^l_{ij}$$

A, B and $S_0$ introduce four material constants. This identification is performed on a uniaxial cyclic test. The stabilised residual stresses under cyclic or repetitive loading is predicted from the applied elastic stress, and the elasto-plastic response $\mathbf{Y_i}$ to the first half cycle. $\mathbf{Y_f}$ and the associated $\mathbf{Ep_f}$ and $\rho_f$ are obtained by projecting $\mathbf{Y_i}$ on the final yield convex centered on zero elastic stresses.

2-4 ACCOUNTING FOR THE ACTUAL GEOMETRY

Plastic strains calculated in 2-3 for a semi-infinite body are introduced as an initial strain field on the tube. Actual axisymetric equilibrium equations and boundary conditions on the tube are analytically solved as $\rho$ and $\mathbf{E^p}$ are function of r, only. New $\rho$ and **Y** parameters are obtained. An iterative procedure is used to correct $\mathbf{E^p}$ if some **Y** do not obey the yield condition.

## 3- EXPERIMENTAL STUDY AND RESULTS

Residual stresses have been measured by X ray diffractometry on a tube submitted to known roto-peening conditions. They have been predicted through the above model for precise roto-peening conditions. Data from a uniaxial cyclic test were used to define A, B, $S_0$ and the coefficient of friction has been experimentally measured. Experimental and predicted results are presented on figure. 1:

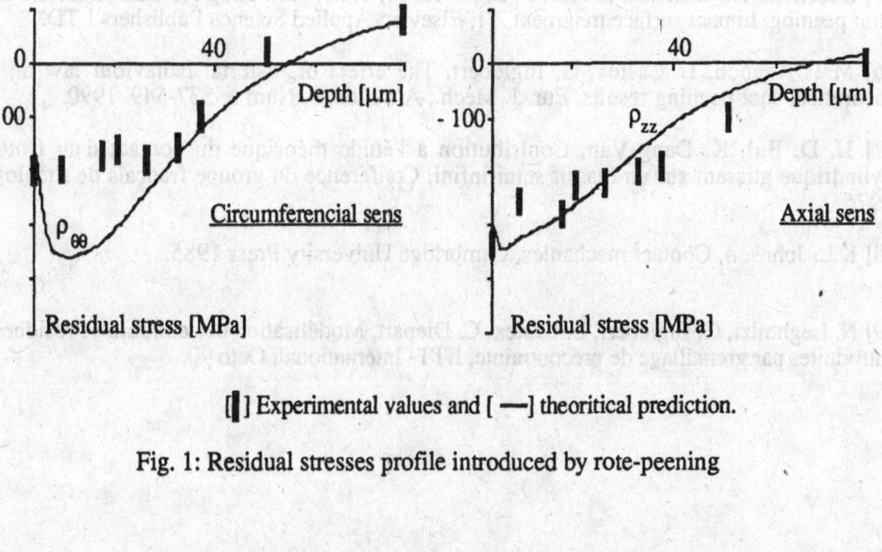


[▮] Experimental values and [ —] theoritical prediction.

Fig. 1: Residual stresses profile introduced by rote-peening

## 4- CONCLUSION

The very good agreement observed on fig. 1 shows that the four parameters three dimensional plastic constitutive law used for prediction is sufficient to describe the plastic behaviour of this material. Axial experimental values are in good agreement with the predicted values. The gap observed for circumferencial values can be explained by the necessary opening of the tubes for X-ray mesurement. An optimisation of the roto-peening treatment can be performed and the influence on the residual stresses of the various parameters of the process can be explored theoretically as it was done in [9] for shot-peening. Experiments on other materials should be done to explore the application field of this model.

## 5- ACKNOWLEDGMENT

We hereby express our deepest appreciation and thanks to "Direction des Etudes et Recherches" of Electricité de France and particularly to Mr P. Vidal "chef de groupe", for the continous support they extended to us for the development of this study.

## REFERENCES

[1] N. Hamdane, Modélisation des contraintes résiduelles introduites par martelage dans les parois d'un tube de générateur de vapeur, Thèse de doctorat, ENSAM Paris, 90 ENAM0019.

[2] Zarka J., Casier J., 1979, Elastic-plastic reponse of a structure to cyclic loading Mechanics today, 6, Ed. Nemat-Nasser, Pergamon press.

[3] Inglebert G., Frelat J., Proix J. M., 1985, Structures under cyclic loading. Arch. Mech., Warszawa, 37, 4-5, 365.

[4] Inglebert G., Frelat J., 1989, Quick analysis of inelastic structures using a simplified method, to be published in Num. Engng. Design, NED116, 1989, pp 281-291.

[5] Guechichi H., Castex L., Frelat J., Inglebert G., 1986, Predicting residual stresses due to shot peening. Impact surface treatment, 11, Elsevier, Applied Science Publishers LTD.

[6] M.T. Khabou, L. Castex, G. Inglebert, The effect of material behaviour law on the theoretical shot peening results. Eur. J. Mech., A/Solids, 9, Num 6, 537-549, 1990.

[7] H. D. Bui, K. Dang-Van, Contribution à l'étude théorique du contact d'un frotteur cylindrique glissant sur un massif semi-infini, Conférence du groupe français de rhéologie, 1975.

[8] K.L. Johnson, Contact mechanics, Cambridge University Press 1985.

[9] N. Leghmizi, G. Inglebert, L. Castex, C. Diepart, Modélisation des contraintes résiduelles introduites par grenaillage de précontrainte, IITT- International, Octo 90.

# HYDROFORMING OF MONOLITHIC PARTS TO PRODUCE RF CAVITIES FOR PARTICLE ACCELERATORS

C. HAUVILLER
CERN
1211 Geneva 23, Switzerland

## ABSTRACT

Superconducting accelerating structures made of niobium coated copper cavities are now installed in various new particle accelerators. The standard manufacturing technique for the copper shells associates spinning and welding. But a parallel way, hydroforming of monolithic parts, leads to better dimensional and surface quality. Parts are obtained by plastic deformation of copper tubes up to 200%. Hydroforming is performed at room temperature with intermediate annealings in a vacuum oven. Pieces have been produced in a large range of dimensions : tube diameters from 60 to 300 mm and lengths up to 3 m. Manufacturing procedures, in particular optimisation of the forming and heat treatment sequence, are presented together with the characteristics of the final products. Comparisons are done with theoretical computations. The advantages of the present solution are discussed from the technical and financial points of view.

## INTRODUCTION

The world's largest scientific machine - the LEP electron-positron collider - is used to probe the innermost constituents of matter to find how the whole of the Universe works. Particle physics needs very large and powerful accelerators to understand the smallest parts of matter. Particle and anti-particle beams are accelerated to higher and higher energies in order to study smaller and smaller objects. This acceleration is produced by radio-frequency power transferred by cavities.

The radio-frequency structures proposed for existing and future large particle accelerators will be superconducting, operating at cryogenic temperatures down to 2°K. These superconducting cavities have the tremendous advantage that RF losses are reduced enormously in comparison with the conventional copper versions.

For two decades, niobium has been used both as superconductor and as structural material. Recently a version made from copper with a thin internal layer of sputter-deposited niobium has been developed. It permits not only a reduction of the price owing to the lower cost of the raw material, but also a decrease in power losses. This latter version implies the production of copper parts with a good surface quality, suitable for coating.

Hydroforming of monolithic cavities from copper tubes presents several advantages compared to the standard manufacturing technique of spinning and welding: better geometrical accuracy and surface quality, lower cost and shorter manufacturing time. However this innovative technique demands a thorough study of the mechanical and structural behaviour of copper submitted to successive heat treatments and forming phases.

The successful application of this technique to the forming of circular / elliptical cavities in the decimetric range (300 MHz to 3 GHz) is illustrated by figure 1: 2.1 GHz as a first demonstration model, 1.5 GHz for CEA Saclay and 352 MHz for the LEP 200 project [1] [2].

## HYDROFORMING

### Principle

Hydroforming is a common manufacturing procedure, used for example in the production of bellows. In this way, large volume changes can be achieved. The principle is to push the part against a rigid die by applying a large pressure through a liquid or polymer. This method has the advantage of producing monolithic pieces and therefore suppressing the need for welding, particularly in critical regions. Although the tooling is expensive, reproducible parts can be obtained in a straightforward way.

Cavities are produced from a copper tube, which has to withstand very large deformations, typically up to about 200%. Since the ultimate elongation of annealed copper is only of the order of 50%, the process must be multistage, including preliminary swaging and several expansions with intermediate annealing. Figure 2 shows the successive manufacturing steps of a 352 MHz cavity. The main dimensions of the different types of cavities are given in table 1.

TABLE 1

| Frequency | Diameters (external value) (mm) | | | D/d | Thickness (mm) | |
|---|---|---|---|---|---|---|
| | Tube | Maximum (D) | Minimum (d) | | Tube | Minimum |
| 2.1 GHz | 59.3 | 126.1 | 39.6 | 3.18 | 2.15 | 1.04 |
| 1.5 GHz | 86 | 184.1 | 75 | 2.45 | 3.0 | 1.36 |
| 352 MHz | 304 | 759.9 | 259 | 2.93 | 9.0 | 3.67 |

### Swaging

The principle of swaging is shown in figure 3 : an internal steel core is introduced into the initial annealed tube. The tube is then slid inside a high resistance polyurethane membrane embedded in a steel support. The oil-pressurised membrane pushes the annealed copper tube onto the internal core, thus creating a toroidal groove.

The problem is to prevent plastic buckling due to high compressive stresses, creating irreparable ripples. This phenomenom has been modelled using the BOSOR5 software and checked experimentally. A good correlation was found for the critical buckling values in the case of uniform thickness but divergence appears in the case of a local thinning, probably because, in this case, local defects greatly influence the buckling mode (20). After optimisation of the process, a reduction up to 35% of the diameter has been achieved in only one stage with a pressure up to 650 bars.

### Expansion

Figure 4 shows the expansion principle: the tube is put into a multi-part die, which is initially open and will close progressively during expansion, while the length of the tube decreases. This process allows virtually no axial elongation while keeping a reasonable thickness. The closed die (all parts in precise contact and located with pins) has the exact external shape of the final cavity. A progressive internal hydraulic pressure up to 200 bars pushes the tube against the die. The expansion of each cell is simply monitored by dial gauges, which measure the increase of the maximum diameter. After swaging, the tube is only locally hardened, and it can be directly submitted to the first expansion stage. The total number of expansion steps depends on the total radial deformation and on the behaviour of the annealed copper.

As for the swaging, the expansion phase has been computer modelled. The CASTEM software has been used to study the complex phenomena of hydroforming: plasticity, large deformations and displacements, variable boundary conditions and contacts. The mesh is based on axisymmetric shell elements and the non-linearities are taken into account by an updated Lagrangian formulation. This modelling precisely predicts the behaviour during all stages of the process, the measured values agreeing to better than 5% with the measured ones.

## OFE COPPER AND HEAT TREATMENT

OFE (Oxygen Free Electrolytic) copper is preferred for its purity and high conductivity at cryogenic temperatures. It is annealed under vacuum but the temperature and duration of this operation have a drastic influence on its metallurgical and mechanical behaviour. Heat treatment must be adapted to the degree of strain-hardening of the material to allow a maximum elongation but a too long or too hot heat treatment will rapidly enlarge grain size. Several tensile tests have been performed with samples annealed from $300^0$C to $600^0$C for times between 30 to 90 mn. The influence of time duration is found only marginal. Heat treatment starts to influence material properties only at $400^0$C, particularly ultimate elongation (figure 5) and hardness, but also surface roughness which is of prime importance for coating quality. However, no direct relation has been found between roughness and grain size. Increase of roughness during hydroforming process is very marked but surface quality is well improved by chemical polishing and/or tumbling.

In order to optimise the process, the hydroforming phases have been simulated on test samples and extensive measurements conducted after each step: initial (as received + annealed), phase 1 (elongation (37%) + annealed), phase 2 (elongation (30%) + annealed),... Ultimate elongations obtained for two different annealing temperatures are given on figure 6. It shows that it was impossible to recover entirely from damage generated by the previous steps, especially from the first one where the strain corresponds to the copper ultimate load. Moreover, in these uniaxial tests, it was impossible to go further than phase 3 (total deformation of 122%). That has been explained not only by the behaviour of copper but also by the poor geometry of the test samples after two successive elongations. The biaxial loading generated by hydroforming is less damaging and therefore applicable on more steps.

## CONCLUSION

It has been demonstrated that hydroforming coupled with a series of heat treatments is applicable to a wide range of axisymmetrical parts like superconducting radio-frequency cavities. The main advantages of this process are: the suppression of the critical welds and a better geometrical accuracy together with a cheaper series production.

## REFERENCES

1. Hauviller C., Fully hydroformed RF cavities, IEEE 1989 Accelerator Conference, Chicago, March 1989.

2. Dujardin S., Genest J., Hauviller C., Jaggi R., Jean-Prost B., Hydroforming monolithic cavities in the 300 MHz range, European Particle Accelerator Conference 90, Nice, Editions Frontieres, June 1990.

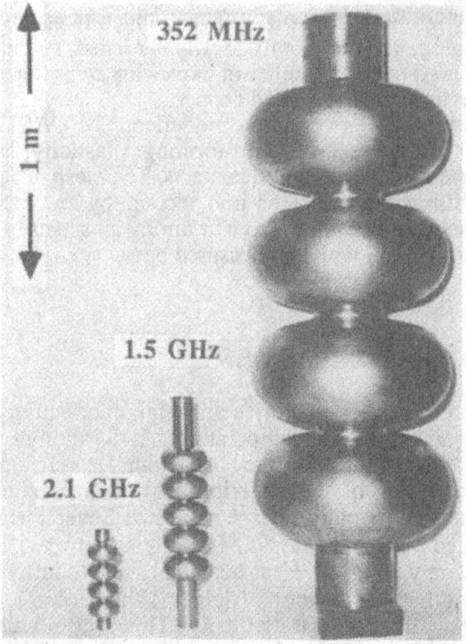


Figure 1. Hydroformed cavities

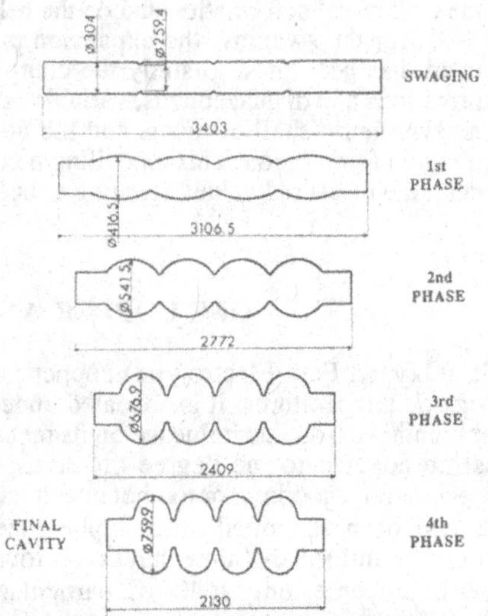


Figure 2. Manufacturing steps (352 MHz)

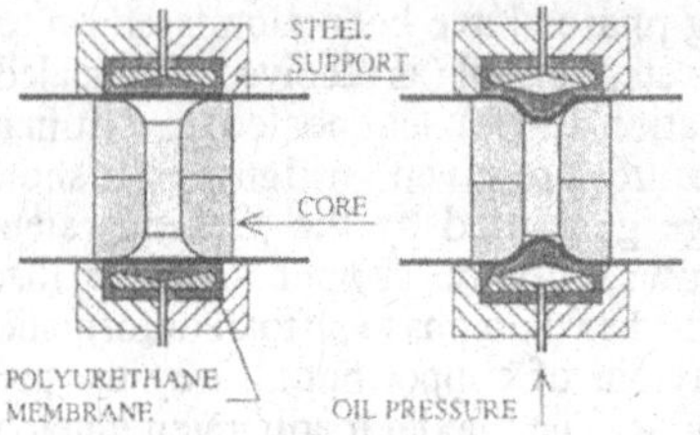


Figure 3. Swaging principle

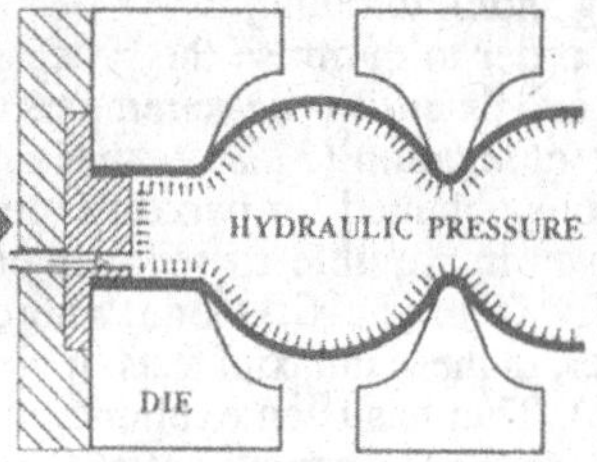


Figure 4. Expansion principle

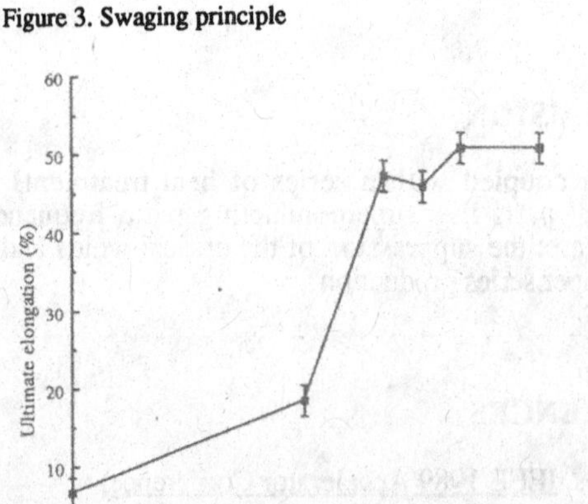


Figure 5. Ultimate elongation after 30 mn annealing

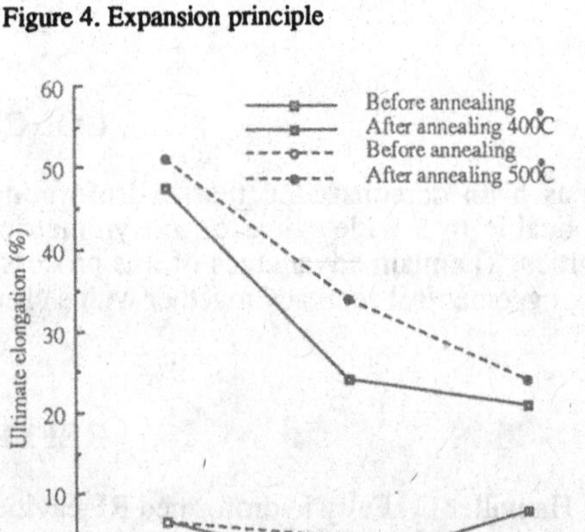


Figure 6. Ultimate elongation (process simulation)

# DIRECT LIMIT ANALYSIS OF ELBOWS BY FINITE ELEMENT AND MATHEMATICAL PROGRAMMING

REINALDO JACQUES JOSPIN*, NGUYEN DANG HUNG** and GERY DE SAXCE***
* Instituto de Engenheria Nuclear-CHEN, Rio de Janeiro-BRASIL, ** Universite de Liege-LTAS, Liege-BELGIQUE, *** Universite de Mons, Mons-BELGIQUE

## ABSTRACT

**This work is concerned with the kinematical admissible calculation of plastic limit loads of constant or variable loads for thin straight or curved tubes using a displacement elbow finite element. The difficulties of convergence in the mathematical programming due to the non-differentiability of the plastic dissipation is avoided by replacing it by the strain energy of a fictitious elastic-perfectly plastic material with a nearly infinite Young modulus. The numerical results are compared with simplified methods of limit loads or sophisticated analysis using shell elements and realistic materials. An important reduction of computer time is observed so this constitutes an effective alternative practical tools for pipe assemblage analysis.**

## INTRODUCTION

**Mechanical engineering structures like power plants, reactors, pressure vessels are exposed to constant and variable loads. In the situation of constant loads, classical limit analysis with proportional loads can be applied. For variable loads structural failure can occurs before a mechanism appearing, either by low cycle fatigue,neither by incremental collapse (ratchetting) . The most efficient way to handle with this problem is to apply the shakedown theory. The use of Von-Mises yielding criterion leads to a non-differentiable and nonlinear mathematical programming (NMP). Unfortunately the majority of NMP solvers concerns problems where the objectif function is differentiable. A tool is proposed in this paper to permit the use of this solvers.**

## 1. STATEMENT OF THE PROBLEM:

**A body of volume V is considered. Its boundary $\Gamma$ is subjected to imposed surface traction $\bar{t}$ on the part $\Gamma_t$ and to imposed displacement rates $\dot{\bar{u}}$ on the part $\Gamma_u$ . The body loads $\bar{b}$ are defined in the volume V.**

**LIMIT ANALYSIS: If the material is supposed rigid-plastic, the forces are proportional and the kinematic boundary condition are homogeneous:**

$$\dot{\mathbf{E}} = \dot{\mathbf{E}}^p,\ \bar{\mathbf{b}} = \alpha \mathbf{b}^0,\ \bar{\mathbf{t}} = \alpha \mathbf{t}^0,\ \dot{\bar{\mathbf{u}}} = \mathbf{0}\ on\ \Gamma_u$$

**then the kinematical theorem of upper bound load multipliers can be stated as following:**

***The exact load multiplier $\alpha$ is the minimum among the load multipliers $\alpha^+$ corresponding to the incompressible and compatible displacement rate fields $\dot{\mathbf{u}}$***

$$\alpha = \min\left\{\alpha^+ = \frac{W_p}{W_e}\right\} \quad where \quad W_p(\dot{\mathbf{u}}) = \int_V D(\dot{\mathbf{E}}(\dot{\mathbf{u}}))\,dV \ \text{ and } \ W_e(\dot{\mathbf{u}}) = \int_V \mathbf{b}^0.\dot{\mathbf{u}}\,dV + \int_{\Gamma_t} \mathbf{t}^0.\dot{\mathbf{u}}\,d\Gamma$$

For the Von-Mises criterion with the yield stress $\sigma_y$, the plastic dissipation has the following form:

$$D(\dot{\mathbf{E}}) = \sigma_y \sqrt{\frac{2}{3} tr(dev\dot{\mathbf{E}})^2}$$

which is not differentiable at the points where the strain rates vanish (i.e. in the rigid parts of the body). In order to avoid this difficulty this function is regularized by an elasto-plastic energy density $\Omega(\mathbf{E})$ with a Young modulus that tends to infinity [1].

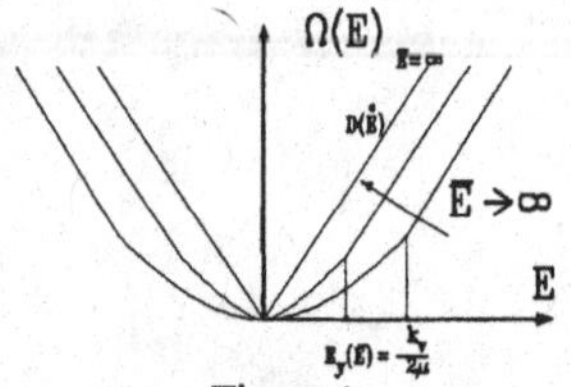


Figure 1
Strain Elasto-Plastic Energy

The modified kinematic theorem can be stated as follow:

*The exact load multiplier $\alpha$ is the minimum among the load multipliers $\alpha^+$ corresponding to the incompressible and compatible displacement field **u** when the elastic modulus of the regularized material tends to infinity.*

$$\alpha = \min\left\{\alpha^+ = \frac{\Omega(\mathbf{E})}{W_e(\mathbf{u})}\right\} \quad where \quad \Omega(\mathbf{E}) = \int_V U(\mathbf{E})dV \quad \text{and}$$

$$U(\mathbf{E}) = \frac{k_c}{6}(tr\mathbf{E})^2 + \mu tr(dev\mathbf{E})^2 - \beta\left[\sqrt{tr(dev\mathbf{E})^2} - \frac{k_v}{2\mu}\right]^2 \quad \text{with } k_c = \frac{\bar{E}}{3(1-2\nu)},\ k_v = \frac{\sigma_y}{\sqrt{3}},\ \mu = \frac{\bar{E}}{2(1+\nu)}$$

$$\beta = 0 \ \text{ if } \sqrt{tr(dev\mathbf{E})^2} - \frac{k_v}{2\mu} \le 0 \qquad \bar{E}, \nu : \textit{material properties}$$

$$= 1 \ \text{ if } \sqrt{tr(dev\mathbf{E})^2} - \frac{k_v}{2\mu} > 0$$

SHAKEDOWN ANALYSIS: If the material is now supposed elasto-plastic and the loads are variables;

$$\mathbf{E} = \mathbf{E}^e + \mathbf{E}^p, \quad \gamma^- \mathbf{b}^0 \le \bar{\mathbf{b}} \le \gamma^+ \mathbf{b}^0, \quad \omega^- \mathbf{t}^0 \le \bar{\mathbf{t}} \le \omega^+ \mathbf{t}^0$$

then the shakedown theorem is presented as follow:

*The exact load multiplier $\alpha$ is the minimum among the load multipliers $\alpha^+$ corresponding to the incompressible and compatible strain increment fields $\Delta\mathbf{E}^p$ when the elastic modulus of the regularized material tends to infinity.*

$$\alpha = \min \alpha^+ = \int dt \int_V U(\dot{\mathbf{E}}^p)\, dV$$

$$\textit{subjected to: } \int dt \int \mathbf{T}^{E0}.\dot{\mathbf{E}}^p\, dV = 1, \quad \Delta\mathbf{E}^p = \int \dot{\mathbf{E}}^p\, dt, \quad \Delta\mathbf{E}^p = \nabla^s(\Delta\mathbf{u}) \qquad \mathbf{T}^{E0} : \textit{purely elastic stresses}$$

The time integration is practical achieved by considering the most severe cycle loading for which plastic yielding occurs at the vertices of the loading domain [2].

$$\alpha = \min \alpha^+ = \sum_n \Omega(\mathbf{E}^{pn})$$

$$\textit{subjected to: } \sum_n \mathbf{T}^{E0n}.\mathbf{E}^{pn}\, dV = 1, \quad \Delta\mathbf{E}^p = \sum_n \mathbf{E}^{pn}, \quad \Delta\mathbf{E}^p = \nabla^s(\Delta\mathbf{u})$$

## 2. DOMAIN DISCRETISATION:

The elbow finite element used is the one develloped by Bathe[3] and Militiello[4]. Its field displacement **u** is splitted off in two parts; the beam field displacement $\mathbf{u}^b$ that satisfy the Bernoulli hypothesis and the shell displacement field $\mathbf{u}^c$ that incorporates the cross section elbow distortion and ovalization

$$\mathbf{u} = \mathbf{u}^b + \mathbf{u}^c \quad \text{where} \quad \mathbf{u}^c = (u^c, v^c, w^c)$$

$$\mathbf{u}^b = \sum_{k=1}^{4}\left\{ L_k(r)\mathbf{U}^k + t\, a_k L_k(r)[\Phi^k \times \mathbf{V}_t^k] + s\, a_k L_k(r)[\Phi^k \times \mathbf{V}_s^k] \right\}$$

$$u^c = \sum_{k=1}^{4} \left\{ \sum_{m=2}^{N_p} (u^s_m)^k L_k(r) \cos m\varphi + \sum_{m=2}^{N_q} (u^a_m)^k L_k(r) \sin m\varphi \right\}$$

$$v^c = \sum_{k=1}^{2} \left\{ \sum_{m=1}^{N_p} \left[ (v^s_n)^k H_k(r) + ((v_{,r})^s_n)^k h_k(r) \right] \sin n\varphi + \sum_{m=1}^{N_q} \left[ (v^a_n)^k H_k(r) + ((v_{,r})^a_n)^k h_k(r) \right] \cos n\varphi \right\}$$

$$w^c = -\frac{\partial v^c}{\partial \varphi} + w_0$$

where $L_k$ is the cubic Lagrange interpolation functions of classe $C^0$; $H_k$ , $h_k$ are the Hermite interpolation functions of classe $C^1$; $U^k_i$ , $\Phi^k_i$ are the generalized displacements and rotations of the beam axis in the global direction i at node k, and $u^l_n$ , $v^l_n$ , $(v_{,r})^l_n$ are the axial and circunferential generalized displacements and the circunferential displacement derivative for symmetric (l = s) and antisymmetric (l = a) harmonics n ( = 2m). In order to represent the ring concentrated forces, the internal pressure and the temperature loads, the radial symmetric displacement $w_0$ is defined for the harmonic n = 0.

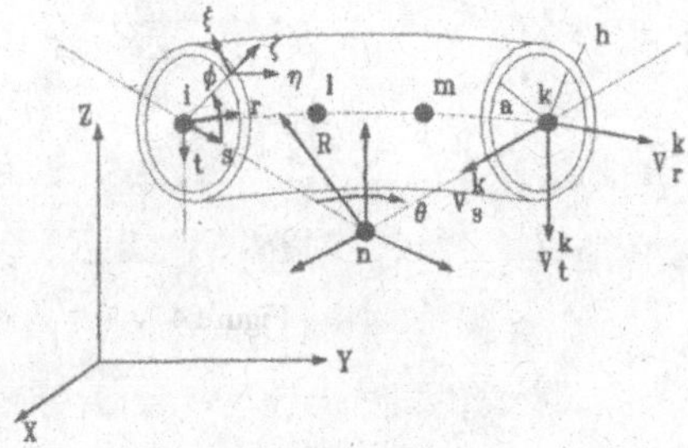


Figure 2
Elbow geometry

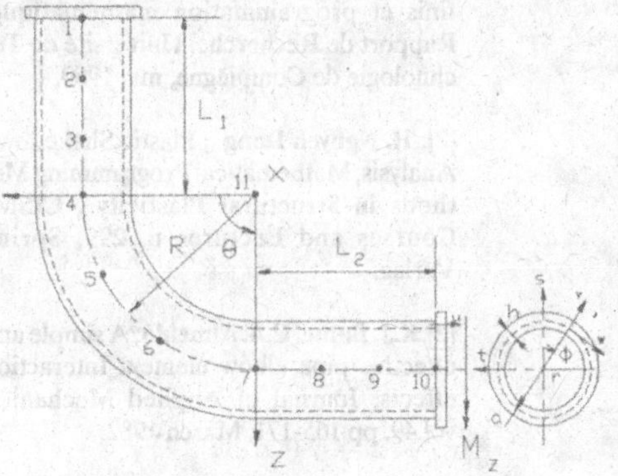


Figure 3
Pipe Assemblage Geometrie

## 3. NUMERICAL RESULTS:

The first numerical example presents a comparison of limit loads and cpu time solving of the pipe assemblage of figure 3 subjected to an out-of-plane moment. The limit moment $M^+_p$ obtained by the present formulation and the $M^{++}_p$ obtained by Save[6] using an step by step method and a three dimensional shell analysis are presented in the table 1.

The structure is discretized in the present formulation analysis by 3 elbow finite elements. The first three antisymmetrical axial and circunferential shell displacements are activated. The numerical integration is performed by (3,12,3) integration points in the $(\eta,\xi,\zeta)$ directions respectively.

| Cross section type | 2a x h (mm x mm) | R (mm) | $L_1 = L_2$ (mm) | $M^+_p \times 10^{-7}$ (N x mm) | Cpu (min) | $M^{++}_p \times 10^{-7}$ (N x mm) | Cpu (min) |
|---|---|---|---|---|---|---|---|
| DN-250 | 273 x 6.3 | 381 | 600 | 7.1 | 66 | 6.16 | 840 |
| DN-400 | 406.4 x 8.8 | 609.5 | 600 | 22.0 | 66 | 18.2 | 900 |

Table 1

The axial, circunferential and shear stresses are presented in figure 4 and indicate the structure reach the limit loads due to the plastic yielding developed near the load end and at the connection elbow-straight pipe. The main shell mode of failure concerns the first harmonic of ovalisation and distortion displacements of the pipe structure.

A second example (figure 5) presents the limit and shakedown domain of a infinite cylindrical pipe subjected to an internal pressure in the lenght l and a radial ring concentrated force in the symmetric cross section. The results presented in the figure 6 are compared with the ones obtained by Cocks and Leckie [7].

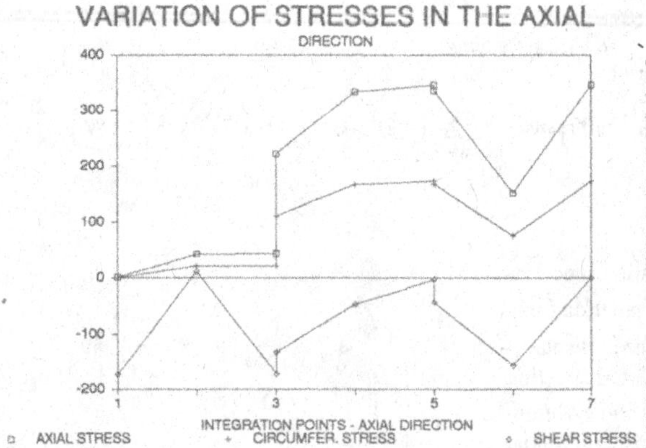


Figure 4

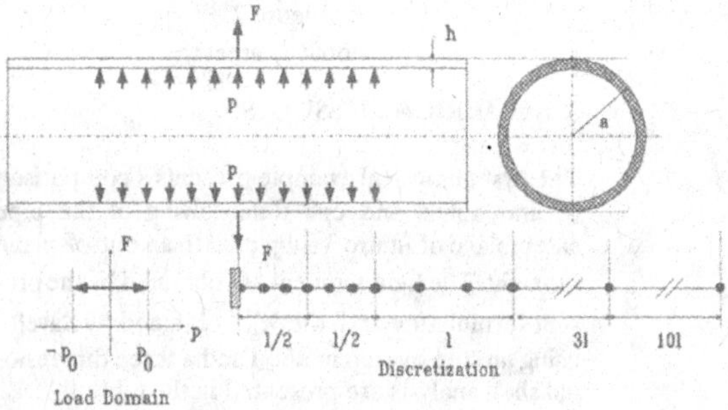


Figure 5
Straight Pipe

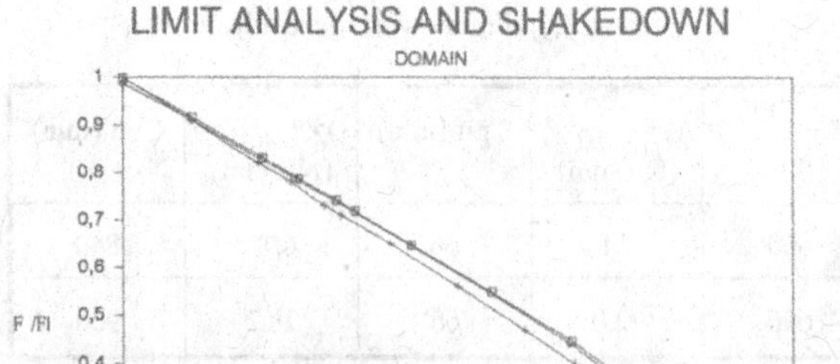

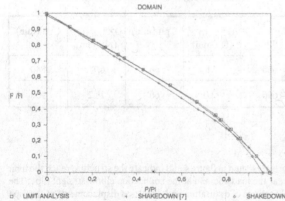

Figure 6

## 4.CONCLUSIONS:

The results presented above agree very well with the literature at a low cpu cost because of the low number of degrees of freedom of the finite element. This regularization method proved to be a reliable tool and represent a good compromise for people working in the projects of pipe assemblage.

## 5. ACKNOWLEDGEMENTS:

The authors are grateful to Brazilian Organisations CNPq and CNEN/IEN for the financial support of the first author of this paper

## 6. REFERENCES:

[1]. **G. De Saxce**; Resolution du problème elasto-plastique incremental par éléments finis et programmation mathématique, Rapport de Recherche, Université de Technologie de Compiègne, mai 1989.

[2]. **H. Nguyen Dang** ; Plastic Shakedown Analysis, Mathematical Programming Methods in Structural Plasticity , CISM-Courses and Lectures n. 299, Spring Verlag.

[3].**K.J. Bathe, C.A. Almeida**; A simple and effective pipe elbow element-Interaction effects, Journal of Applied Mechanics, vol.49, pp 165-171, March 1982.

[4]. **C. A. Militiello, A.E.Huespe**; Displacement Based Pipe Elbow Element, Computers and Structures, Vol. 29 n. 2, pp 339-343, 1988

[5]. **R.J.Jospin, H. Nguyen Dang**; Analyse Limite des Coudes par la Méthode des Eléments Finis et la Programmation Mathématique, Rapport de Recherche LTAS MR-01, Université de Liège , 1990

[6]. **M. Save**; Limit Loads of Pipe Elbows, Comission of European Communities, Contract RAI-0134-B, Faculté Polytechnique de Mons, September 1990.

[7].**A.C.F. Cocks, F.A. Leckie**;Deformation Bounds for Cyclically Loaded Shell Structures Operating Under Creep Condition, Journal of Applied Mechanics, Vol. 55, pp. 509-516, September 1988.

# THE EFFECT OF A TEMPERATURE CYCLE ON THE STRESS DISTRIBUTION IN A SHRINK FIT

ADAM KOVACS[1] and HORST LIPPMANN[2]
[1]Dept. of Technical Mechanics, Technical University of Budapest, HUNGARY
[2]Lehrstuhl A für Mechanik, Technische Universität München, FRG

## ABSTRACT

Semi-analytical and finite element methods were carried out for the calculation of the stress distribution in a shrink fit made from different materials under a low range temperature cycle.

## INTRODUCTION

Two, completely different methods were developed and used for the calculation of the stress distribution due to a temperature cycle. In the first method, using the Tresca yield condition and the associated flow rule, and assuming the yield limits as well as the elastic constants of the two members to depend only on the temperature - which should be constant in space, - an almost analytical solution has been found. This solution, based on the isothermal formulation of KOLLMANN [1], permits the calculation of the initial stress state from the known interference and then, the investigation of the thermoelastic-plastic state. We assume plane stress conditions and an elastic-perfectly plastic material for the inner and the outer ring, respectively.

In the second method the stress state has been determined using finite elements. The Tresca, as well as the von Mises yield criteria can be taken into account with linear isotropic work hardening. In both formulation the deformations are assumed to be small enough for linear thermoelasticity to hold. Rate sensitivity of the yield limits were disregarded.

## SEMI-ANALYTICAL METHOD

The shrink fit is constructed from two rings. The inner and outer radii are a and $b_{Io}$ for the inner, $b_{Ao}$ and c for the outer ring. The interference $i_o$ is defined as $i_o = b_{Io} - b_{Ao}$. After the assemblage the common radius is $b \approx b_{Io} \approx b_{Ao}$.

The temperature cycle consists of two monotonous branches: from $T_o$=293 [K] to $T_1$ and from $T_1$ to $T_o$. If the first part of the cycle is heating, the uniaxial yield limit decreases and therefore, plastification starts from the inner boundary of the rings. The plastic zone extends to a certain radius x in the inner and y in the outer ring. The equations which give the unknown radial and hoop stresses from the known joint pressure $p_b$ and the radii x and y, have been published in [1]. The calculation of $p_b$, x and y is possible using the following continuity conditions:

$$\sigma_r\Big|_{r\to -x} = \sigma_r\Big|_{r\to +x}, \quad \sigma_r\Big|_{r\to -b} = \sigma_r\Big|_{r\to +b}, \quad \left|\varepsilon_t\Big|_{r\to -b} - \varepsilon_t\Big|_{r\to +b}\right| = \frac{i_o}{b} \quad (1)$$

r and t refer to the radial and circumferential direction, respectively. $\varepsilon_t$ consists of a thermoelastic and a plastic part.

Since two equations of (1) are transcendent, they can be solved in a semi- analytical way. The unloading process is assumed to be completely elastic. The material parameters E, $\nu$, $\alpha$ can change during this process, this is treated as if the rings were first mechanically and thermally unloaded under the old parameters and then reloaded under the new ones. Using this unloading-reloading procedure, the unknown final joint pressure can be determined.

## FINITE ELEMENT METHOD

The change in the temperature means an initial thermal loading and is imposed incrementally on the structure. The equilibrium equations are derived from the principle of the virtual work. (See e.g. OWEN and HINTON [2]). The balancing internal forces due to the stress state can be then calculated with a Newton-Raphson type iteration process.

## NUMERICAL EXAMPLE

As an example, we consider the inner ring to be produced from aluminium and the outer ring from copper. Let a/b=0.25 and b/c= 0.8. The material data collected in [3] allow the linear approach of the uniaxial yield limit:

$$Y = Y_o - m\cdot(T - T_o), \qquad (2)$$

provided $T - T_o < 150$ [K].

The material parameters are the following ([3], [4]):

$Y_{oAl} = 50$ [MPa], $m_{Al}/Y_{oAl} = 5\cdot10^{-3}$[1/K], $E_{Al} = 68.67$ [GPa]

$\nu_{Al} = 0.3$, $\alpha_{Al} = 2.38\cdot10^{-5}$[1/K]

$Y_{oCu} = 130$ [MPa], $m_{Cu}/Y_{oCu} = 4.23\cdot10^{-3}$[1/K], $E_{Cu} = 113.8$ [GPa]

$\nu_{Cu} = 0.35$, $\alpha_{Cu} = 1.698\cdot10^{-5}$[1/K]

Fig.1 and Fig.2 show the intermediate and final stress state during one temperature cycle. The maximum temperature difference is $\Delta T = 55$ [K]. The initial joint pressure is $p_{bo}$=17.5 [MPa] due to an initial interference of $i_o/b = 9.667\cdot10^{-4}$. The good agreement between the analytical and the numerical results is obvious.

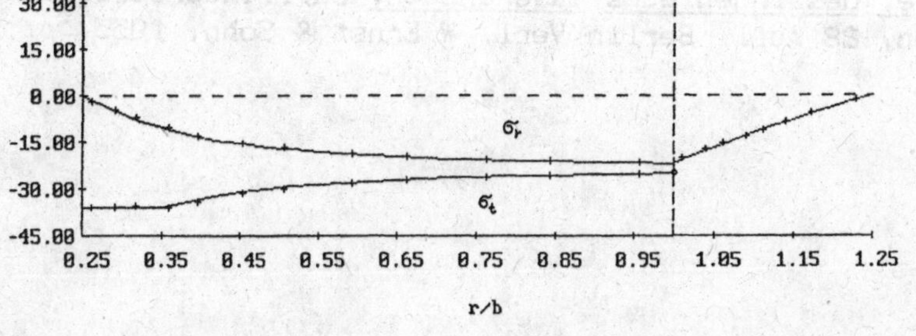

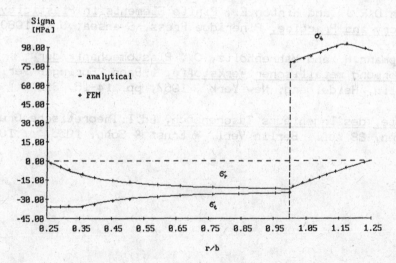

Figure 1. Stress distribution after temperature elevation $\Delta T$=55 [K]

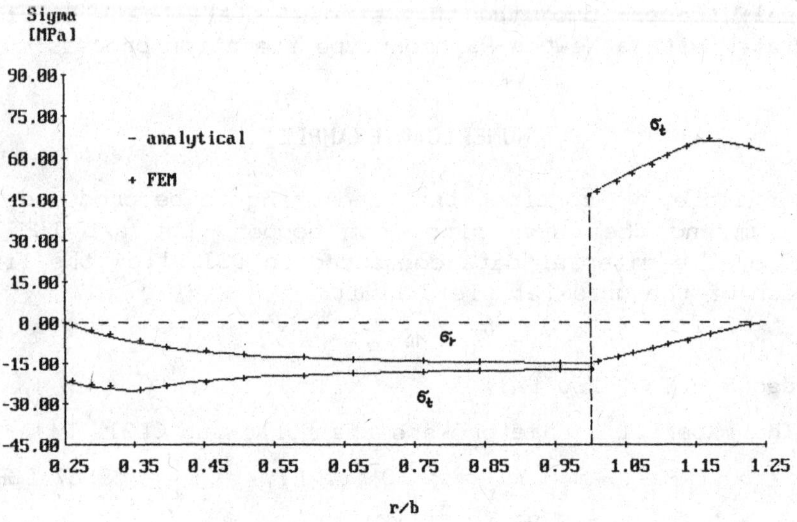


Figure 2. Residual stress distribution after a complete cycle

## REFERENCES


1. Kollmann, F. G., Die Auslegung elastisch-plastisch beanspruchter Querpreßverbände. Forsch. Ing.-Wes., 1978, **44**, 1-11.

2. Owen, D. R. J. and Hinton, E., Finite Elements in Plasticity, Theory and Practice, Pineridge Press, Swansea, U.K. 1980.

3. Lippmann, H. and Mahrenholtz, O., Plastomechanik der Umformung metallischer Werkstoffe, 1. Band, Springer Verlag Berlin, Heidelberg, New York , 1967, pp. 14-15, 313.

4. Hutte, des Ingenieurs Taschenbuch, Bd. I: Theoretische Grundlagen, 28. Aufl. Berlin: Verl. W. Ernst & Sohn, 1955, p. 1049.

# ANALYSIS OF A TYPICAL LMFBR-STRUCTURE BY MEANS OF A COMPONENT TEST

KARL KUSSMAUL, KARL MAILE
Staatliche Materialprüfungsanstalt
Pfaffenwaldring 32, D-7000 Stuttgart
HARTWIG LAUE, UWE LOHSE
Interatom GmbH
Friedrich-Ebert-Straße, D-5060 Bergisch Gladbach

## ABSTRACT

The paper describes the verification of simplified inelastic methods for the design of the reactor vessel wall by means of a representative testing facility "TAKEV". The concept of the test rig and the operating conditions were described to show the transferability of the test results to the real tank wall.

## INTRODUCTION

Changes of the operational state of the German LMFBR SNR 300 (load change, startup and shutdown processes) lead to a moving level of the coolant in the reactor tank. Connected herewith are cyclic variations of the temperature gradients along the tank wall. In connection with external loads (dead weight) these changes in thermal loads lead to fatigue and creep in the affected zone. The structural behaviour must be assessed in relation to high-temperature and time-dependent characteristic material data. This leads to additional requirements concerning analysis and evaluation procedures.

### Design methods and criteria

A simplified inelastic approach close to the ORNL-model has been used in the creep fatigue assessment of the tank wall. Fatigue damage and creep damage ist determined separately and combined in an interaction diagram according to ASME CC N47. Based on elastic calculated stress results the applied method allows the estimation of the strain range by the use of a plastic strain concentration factor [1] and the residual stress state at the end of the cycle ist determined by a simplified construction of the elastic-plastic stress-strain hysteresis. Stressrelaxation is considered when calculating the creep damage using a time faction rule. An application example of the procedure ist given in [2].

**Testing facility for verification (TAKEV)**

In order to verify the computation resp. the creep-fatigue evaluation in the reactor tank wall a representative test is being carried out carry out. The following major generic requirements were taken into account when planning the test rig with regard to the transferability of the test results to the real component:

1 component like loading situation
2 component like dimensions of the specimen
3 service like damage mechanisms
4 monitoring of local deformation which enables a comparison with calculated results
5 monitoring of initiation and time dependent development of damage (operation surveillance)

**Matching the initial requirements**

**Loading situation and specimen geometry:** In the test a model plate of AISI 304 SS type, Fig. 1, (original wall thickness 40 mm, dimensions 750 x 460 mm) with SAW T-shape seams is subjected to loadings, which simulate mechanically the calculated loadings of the tank wall. Two plates of which one has sparkeroded flaws to simulate corrosion cracks are tested.

The plate is kept under a constant primary tensile stress of 10 MPa to simulate the dead weight load of the tank. In addition a displacement controlled bending moment in order to create a strain amplitude of +0.2 % at the surface is superimposed, which is reversed after 40 h holdtime. In compression the same strain amplitude is applied. No hold time is included in the compression phase. This alternating bending strain simulates the cyclic loading resulting from temperature changes along tank wall. A schematic view of the test rig is given in Fig. 2.

**Damage mechanism:** The 100000 h design life of the real tank is reduced to about 12000 h duration by means of a time-lapse process: an increase of temperature and bending strain will accelerate the creep deformation rate, but also the damage development. The test temperature of 570 °C, which is raised by 25 °C in comparison to the maximum service temperature will not induce a change of the damage mechanism.

**Local deformation and damage development:** Another aim of the test is not only to check the inelastic analyses but also to screen the applicability of nondestructive test methods for damage identification. As shown in Fig. 1 sufficient instrumentation by means of capacitive strain gages with high long-term stability provide the boundary conditions for the inelastic analyses which are performed in parallel. Nondestructive test methods such as Penetrant Testing, Eddy Current Testing,

Replica Technique, Optical Methods (Microscop) are used. All methods were applied before beginning of the test to determine the as-received state of the material.

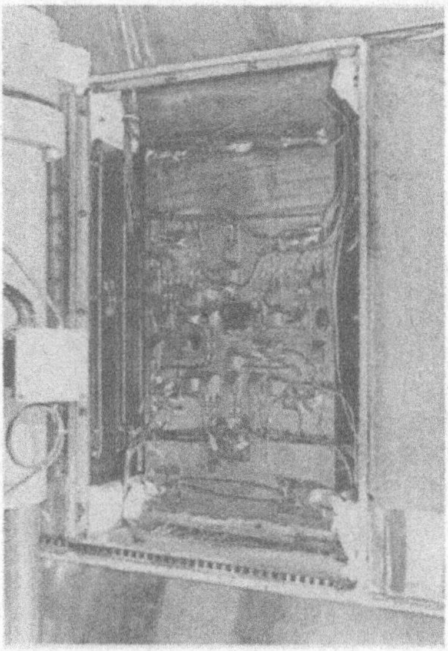

Figure 1. Test plate

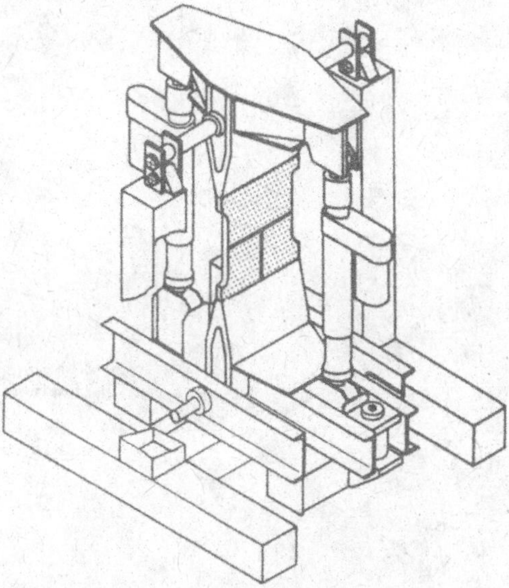

Figure 2. Test rig

**Results**

**Experimental results:** The experiment started in March 1991. The first revision is planned after approx. 3000 h. No problems controlling temperature and load were observed.

**Analytical results:** Inelastic Finite Element (FE) calculations are performed in parallel to predict the lifetime of the test plate including the time-dependent deformation behaviour and damage history under such complex loading conditions. The FE-calculations are carried out by Interatom and MPA Stuttgart using the FE-code ABAQUS [2] on the CRAY 2 computer at the University of Stuttgart. The creep behaviour of the material was characterized by Blackburn's uniaxial creep law for primary and secondary creep. The total creep strain hardening rule incorporated in the ORNL model was used [3]. In the further development it is planned to use other creep-laws, which also describe the tertiary creep, to incorporate other hardening rules which descibe the real material hardening more realistic and to install a damage hypothesis. It is also planned to carry out inelastic Finite Element calculations with the real material data of the specific cast of the plate. This data will be gained from a complementary small scale specimen program.

An example of the results of the calculation already carried out is given in Fig. 3. The axial strain in the middle resp. the edge of the specimen is depicted versus holdtime. The calculated strains have to compared with experimental results in the further progress of the experiment.

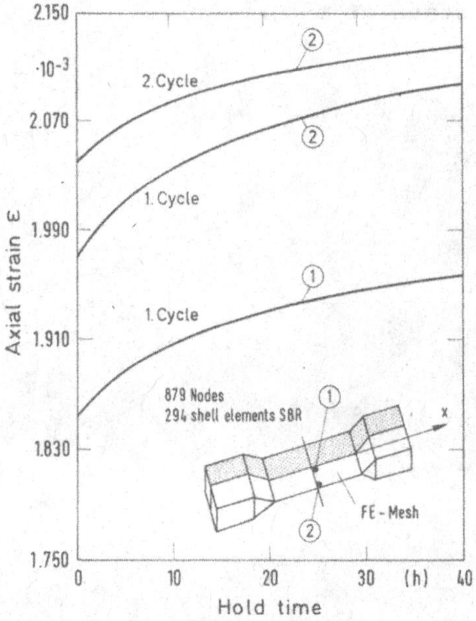


Figure 3. Calculated axial strain in the test plate.

## Conclusions

The component test programme TAKEV is a project to proof the specified design life of the reactor tank of German LMFBR. The analytical determination of accumulated creep-fatigue damage in the wall is verified by experiments with component like specimens under service like loading conditions. The applicability of NDE methods for damage identification are checked.

## References

1. Hübel, H., Plastische Dehnungsüberhöhungsfaktoren in Regelwerken und Vorschlag zur Etablierung angemessener Faktoren. Dissertation, Universität Kassel, 1985.

2. Angerbauer, A., Laue, H., Meyer-Reumers, H., Sapci, A., Application of modified elastic creep fatigue damage evaluation methode to SNR 300. 2nd International Seminar on Standards and Structural Analysis in Elevated Temperature Application for Reactor Technology, Venice 1986

3. Hibbit, Karlson and Sorensen, Inc., Abaqus User Manual, Version 4.6. Providence, Rhode Island, U.S.A..

4. Corum, J. M. et al, Interim Guidelines for Detailed Inelastic Analysis of High Temperature Reactor System Components. ORNL-Report 5014 (1974).

# SOME PROBLEMS WITH CONTACT FORCES OF THREEDIMENSIONAL METAL FORMING

OSKAR MAHRENHOLTZ, JOSEF APPELTAUER
Technical University Hamburg-Harburg
D 2100 Hamburg 90, Germany

## FRICTIONAL CONTACT PROBLEMS

Friction exercises essential influence on numerical simulation of mainly threedimensional forming processes. Nevertheless, it cannot be mechanically defined in a simple, well determined manner. There are too many structural, technological and mechanical parameters influencing friction forces between workpiece and forming tool.

The main mechanical parameter is the normal pressure between contact surfaces. Classical pointwise friction law of Amonton-Coulomb only allows introduction of full contact area into the linear relation

$$\boldsymbol{\tau} = -\mu\sigma\frac{\Delta \mathbf{v}}{|\Delta \mathbf{v}|} \qquad (1) \qquad ; \qquad \sigma = \frac{N}{A_f}, \qquad (2)$$

where $\boldsymbol{\tau}$ represents the friction stress vector at start of sliding, directed in concordance with the vector $\Delta\mathbf{v}$ of relative velocity between workpiece and tool; $\sigma$ is the normal pressure, with $N$ the normal force and $A_f$ the full contact area; finally $\mu$ synthesizes structural and technological parameters into the concept of frictional coefficient. In nonlocal theory of friction, reduced contact area $A_r$ only is considered, dependent on local deformation of so-called junctions between contact surfaces, according to oxide or impurity layer also. Then, frictional coefficient $\mu$ can be written as

$$\mu = \frac{t}{p_y}, \qquad (3)$$

where $t$ means the real shear stress of junctions, and $p_y$ their local plastic yield pressure. The nonclassical friction law has obviously nonlinear character:

$$\boldsymbol{\tau} = -k_\tau \Phi_\epsilon(\Delta \mathbf{v})\frac{\Delta \mathbf{v}}{|\Delta \mathbf{v}|}, \qquad (4)$$

where $k_\tau$ represents some critical value for tangential stress and $\Phi_\epsilon(\cdot)$ a monotonous function, depending on a parameter $\epsilon > 0$. Thus, $\boldsymbol{\tau}$ becomes a function of weighted measure of normal stresses in a neighbourhood of the considered point.

It will be shown that Amonton-Coulomb friction law and its nonlocal generalization are really incompatible with numerically obligated variational form of the mathematical simulation model everytime if physical model concerns with rigid-plastic material behaviour. Thus, Siebel friction law

$$\boldsymbol{\tau} = -mk\frac{\Delta \mathbf{v}}{|\Delta \mathbf{v}|} \tag{5}$$

is introduced, wherein $m \in [0;1]$ is a frictional coefficient again, and $k$ means shear yield stress of the material.

# EXTENDED BOUNDARY VALUE PROBLEMS

Neglecting mass and inertia forces, in the control volume hold the equilibrium conditions

$$\nabla \cdot \boldsymbol{T} = 0 \tag{6}$$

and the first order kinematical relations

$$\boldsymbol{D} = \frac{1}{2}(\boldsymbol{L} + \boldsymbol{L}^T) \quad (7) \quad ; \quad \boldsymbol{L} = \nabla \mathbf{v}\,, \tag{8}$$

where $\mathbf{T}$ represents the Cauchy stress tensor ($\boldsymbol{T}'$ will be its deviator), $\boldsymbol{D}$ the rate of deformation tensor, $\boldsymbol{L}$ the spatial gradient of velocity tensor, $\mathbf{v}$ the velocity vector, and $\nabla$ the Nabla operator. Flow rule

$$\boldsymbol{T}' = \frac{k}{\sqrt{II_D}}\boldsymbol{D}\,, \tag{9}$$

coupled with yield condition and isotropic hardening condition,

$$II_{T'} = k^2 \quad (10) \quad ; \quad k = k(\boldsymbol{B}^{-1}, \theta)\,, \tag{11}$$

transforms (6), (7) in field equations for velocity and stress determination. Herein $II_D$ and $II_{T'}$ are second invariants, $\boldsymbol{B}^{-1}$ is the left Cauchy-Green tensor, and $\theta$ the constant temperature of isothermal forming. This constitutive definition is valid for homogeneous, isotropic, rigid-plastic material behaviour, by simultaneous neglection of plastic spin tensor, and implies with

$$trD = 0 \tag{12}$$

plastic incompressibility.

The boundary conditions expressed in friction stresses have special importance, since amounts only can be prescribed, for example in free forging, while directions are initially unknown velocity directions. Thus, incremental Euler description (6), ... (12) does not constitute a normal boundary value problem, but rather an extended one [1].

## EXTENDED VARIATIONAL PROBLEM

Supplemental restriction by the incompressibility condition (12) requires the replacement of extended boundary value problem by a likewise extended variational problem. Therefore, the weak (Galerkin) form of the principle of virtual velocities, largely applied in this field, must be replaced by the variation of an extended Markov potential [2],

$$\delta_v \Pi = \delta_v [\int_V (\nabla \cdot \boldsymbol{T}' + \sigma_m \nabla \cdot \mathbf{v}) dV - \int_S \mathbf{n}\boldsymbol{T}\mathbf{v} dS] = 0 \;, \tag{13}$$

where $\sigma_m$ denotes the mean normal stress, interpreted as a penalty function, and $\mathbf{n}$ the normal unit vector at a surface point. Any competitional velocity field requires $\partial\Pi/\partial v = 0$, and differentiation of the last term in (13) must be carried out with respect to $(\mathbf{n}\boldsymbol{T})$ and $\mathbf{v}$, both. In profile rolling for instance derivative of $(\mathbf{n}\boldsymbol{T})$ refers to amount and direction simultaneously.

## FRICTIONAL COMPATIBILITY

In [3], the third term of (13) is replaced by

$$\int_{S_\sigma} (\lambda \mathbf{n}\mathbf{v} + \tau_0 |\mathbf{v}|) dS \;, \tag{14}$$

where $\lambda$ represents a penalty factor, and $\tau_0$ a constant friction stress value, like $mk$ in (5). Then, the natural boundary condition leads to

$$\mathbf{n}\mathbf{v} = 0 \quad (15) \qquad ; \qquad \mathbf{n}\boldsymbol{T} + \lambda\mathbf{n} + \tau_0 \frac{\mathbf{v}}{|\mathbf{v}|} = 0 \,. \tag{16}$$

Condition (15) is clearly satisfied. By multiplying with $\mathbf{n}$ (16) gives

$$\lambda = -\mathbf{n}\boldsymbol{T}\mathbf{n} \equiv p_n \;, \tag{17}$$

the normal pressure component, so that correctly

$$\mathbf{n}\boldsymbol{T} = -\tau_0 \frac{\mathbf{v}}{|\mathbf{v}|} \ . \tag{18}$$

In the Amonton-Coulomb friction case

$$\tau_0 = \mu\lambda = \mu p_n \ , \tag{19}$$

and the natural boundary condition is

$$(\mathbf{n}\boldsymbol{T} + p_n\mathbf{n} + \mu p_n \frac{\mathbf{v}}{|\mathbf{v}|})\delta\mathbf{v} + (\mathbf{n}\mathbf{v} + \mu|\mathbf{v}|)\delta p_n = 0 \ . \tag{20}$$

Now,

$$\mathbf{n}\mathbf{v} + \mu|\mathbf{v}| \neq 0 \ , \tag{21}$$

which shows the mentioned incompatibility.

## FINITE ELEMENT CONSEQUENCES

If instead of Markov procedure the Galerkin form is used, iterative solution of nonlinear system of equations must begin with a very well estimated velocity field, to avoid possible false convergencies. The numerical advantage of Markov process consists mainly in excellent iteration convergence even by beginning with a simple estimate without friction.

## REFERENCES


1. Appeltauer, J., Dung, N.L. and Mahrenholtz, O., Untersuchung des dreidimensionalen Freiformschmiedens mit der Finite-Element-Methode. In Präzisionsumformtechnik, ed. K. Lange and H.G. Dohmen, Springer-Verlag, Berlin etc., 1990, pp. 274–285.

2. Lung, M. and Mahrenholtz, O., A finite element procedure for analysis of metal forming processes. Trans. CSME, 1973/74, **2**, 31–36.

3. Besdo, D., Numerik. In Plastomechanische und metallkundliche Grundlagen der Umformtechnik, ed. R. Kopp, RWTH Aachen, Verein Deutscher Eisenhüttenleute, Düsseldorf, to be published.

# AN ENERGETIC CONTROL ON NUMERICAL INSTABILITY FOR A RATE-TYPE VISCOPLASTIC OSCILLATOR

M. MIHAILESCU-SULICIU and I. SULICIU
Institute of Mathematics
Str. Academiei 14, Bucharest, ROMANIA

## ABSTRACT

The constitutive equations we deal with can describe combined kinematic and isotropic work hardening. In the case of free oscillations of such a viscoplastic oscillator the total energy of the exact solutions is a non-increasing function of time. It is concluded that the numerical schemes which preserve better this property give more accurate and stable numerical solutions for both free and forced oscillations.

## INTRODUCTION

We consider an oscillator consisting of a bar of initial length L and cross section area A. One end of the bar is fixed and to the other end a rigid body of mass M is attached. We use a rate-type viscoplastic model with combined kinematic and work hardening [1] to describe the behaviour of the bar.

Our model leads to a Cauchy problem for a system of four differential equations when stress, strain and work hardening parameter are assumed to be uniform along the bar. For the validity of such an assumption in the elastic case see [4]. The exact solution of our problem must satisfy certain energy inequality. Following [5] we require to the numerical schemes we use, to verify as close as possible the same inequality. This requirement turns out to be a numerical stability requirement. We find an upper bound for time integration step h, which is here $2/k$ and $k>0$ is the Maxwell type viscosity coefficient. If $h \leq 2/k$ the numerical solution is stable and if $h>2/k$ it is unstable.

Our model reduces to an elastic one when $k=0$. When $k>0$ and a certain yield function $Y(\kappa)=0$ our model reduces to the linear standard model of viscoelasticity. In both elastic and viscoelastic cases we can construct exact solutions and we compare them with the computed ones in order to check the accuracy of the discussed numerical schemes. We conclude that the numerical schemes which preserve better the energy inequality give more accurate numerical solutions.

## ANALYSIS

The rate-dependent problem we deal with is

$$M\dot{v}= f- A\sigma \quad , \quad L\dot{\varepsilon}= v \quad , \quad \dot{\sigma}-E\dot{\varepsilon}=-kF(\sigma-\tau)$$
$$\dot{\kappa}= \frac{k}{E-E_0}(\sigma-E_0\varepsilon)F(\sigma-\tau) \quad , \quad (v,\varepsilon,\sigma,\kappa)(0)=(v_0,\varepsilon_0,\sigma_0,\kappa_0) \ , \tag{1}$$

with the unknowns $(v,\varepsilon,\sigma,\kappa)$=(mass velocity,strain,stress,work hardening parameter). $f(t)$ is a given exciting force, $E>E_0\geq 0$ are the dynamic and static Young's moduli, respectively. $F(r)$, $r\in R$ is a relaxation function such that $rF(r)\geq 0$. In the overstress $\sigma-\tau$ , $\tau$ is a continuous function of $\varepsilon,\sigma,\kappa$ defined as

$$\tau= \sigma \text{ if } |\sigma-E_0\varepsilon|\leq Y(\kappa) \quad \text{and} \quad |\tau-E_0\varepsilon|=Y(\kappa) \text{ otherwise} \tag{2}$$

where $\kappa\geq 0$ and $Y(\kappa)\geq 0$. We use $Y(\kappa)= \sqrt{2\alpha\kappa}$ , $\alpha$=const.$\geq 0$.

When k tends to infinity the problem (1) tends to a rate-independent problem [1],i.e. $\sigma-\tau \to 0$. We used the rate-independent model and the slow rate experiments of [2,3] to identify the material constants $E,E_0$ and $\alpha$ with $E$=60.9 GPa,$E_0$=45.5 GPa, $\alpha$=6.3 GPa. We obtain a good overall agreement with the experimental data except a small neighborhood of the origin ($\varepsilon=0,\sigma=0,\kappa=0$).

We define the total energies as

$$e= \frac{Mv^2}{2} +AL[\frac{\sigma^2}{2E}+ \frac{E_0(\sigma-E\varepsilon)^2}{2E(E-E_0)} ] \ , \quad e_1= e+ AL\kappa \ . \tag{3}$$

The solutions of the problem (1) must satisfy certain energy identities which for f=0 reduce to

$$\dot{e}= - \frac{ALk}{E-E_0}(\sigma-E_0\varepsilon)F\leq - \frac{ALk}{E-E_0}(\sigma-\tau)F(\sigma-\tau) \quad , \quad \dot{e}_1=0 \ . \tag{4}$$

From $(4)_2$ we obtain bounds in energy for $(v,\varepsilon,\sigma,\kappa)(t)$. Since $rF(r)\geq 0$, from $(4)_1$ we get

$$\dot{e}(t)=0 \text{ if } k=0 \quad \text{or} \quad \dot{e}(t)\leq 0 \text{ if } k>0. \tag{5}$$

The numerical schemes we discuss will be analysed in connection with their ability to preserve the energy property (5)

In what follows we take $F(r)=r$, $f(t)=0$ and $h>0$. We denote $v_n = v(nh)$, etc., $n=0,1,\ldots$

If the problem (1) is written as $\dot{x}= g(x)$ then an Euler's scheme-(E) is $x_{n+1} = x_n + hg_n$, $g_n = g(x_n)$. We denote $\overline{x}_n = x_{n+1}$ where $x_{n+1}$ is obtained by using scheme-(E). A second order Runge-Kutta scheme-(RK-2) is $x_{n+1} = x_n + h(g_n + \overline{g}_n)/2$, $\overline{g}_n = g(\overline{x}_n)$. An implicite scheme is $x_{n+1} = x_n + h(g_n + g_{n+1})/2$. We construct an implicite-explicite scheme-(IE-1) by substituting $g_n$ with $(g_n + g_{n+1})/2$ in scheme-(E) only for those components of $g(x)$ which are linear in x. With the notations

$$\omega^2 = \frac{AE}{ML}, \quad \omega_0^2 = \frac{AE_0}{ML}, \quad D = 1 + \frac{\omega^2 h^2}{4}, \quad D_0 = 1 + \frac{\omega_0^2 h}{4}, \quad D_1 = 1 - \frac{\omega^2 h^2}{4} \tag{6}$$

we can write the (IE-1)-scheme under the explicite form

$$v_{n+1} = [D_1 v_n - \frac{Ah}{M}(\sigma_n - \frac{hk}{2}F_n)]/D, \quad \sigma_{n+1} = [D_1 \sigma_n + h(\frac{Ev_n}{L} - kF_n)]/D$$
$$\sigma_{n+1} - E\varepsilon_{n+1} = \sigma_n - E\varepsilon_n - hkF_n, \quad \kappa_{n+1} = \kappa_n + hk(\sigma_n - E_0\varepsilon_n)F_n/(E-E_0). \tag{7}$$

A second order implicite-explicite scheme-(IE-2) can be written as a (RK-2)-scheme with $\overline{v}_n = v_{n+1}$, etc., where $v_{n+1}$, etc. are obtained from (7).

We note that $\omega$ and $\omega_0$ are frequency oscillations due to the instantaneous and equilibrium elasticities.

The energy $e_{n+1}$ of the numerical solution at time $t_{n+1} = (n+1)h$ is obtained from (3) if we substitute $(v_{n+1}, \varepsilon_{n+1}, \sigma_{n+1})$ instead of $(v, \varepsilon, \sigma)$. We have the following relations between $e_{n+1}$ and $e_n$ for the (E)-and (EI-1)-schemes

$$e_{n+1} \leq (1+\omega^2 h^2)e_n - Ah^2 kv_n F_n \quad \text{if} \quad h \leq 2/k \tag{8}$$

$$e_{n+1} \leq e_n - h^2 kF_n v_n/(2D) \quad \text{if} \quad h \leq 2/k \quad \& \quad |\varepsilon_n| \leq 4D_0 Y_n/[E_0(D-D_0)] \tag{9}$$

respectively. The (E)-scheme (as well as (RK-2)-scheme) has an amplifying factor of the energy as in the elastic case [4]. From (9) we conclude that the (EI-1)-scheme preserves the property (4), up to $O(h^2)$ if $h \leq 2/k$ and $|\varepsilon_n| \leq 4D_0 Y_n/E_0(D-D_0)$ . The (EI-2)-scheme is more accurate but otherwise it has similar properties with (EI-1).

We selected different input data as $(v_0, \varepsilon_0, \sigma_0, \kappa_0) = (1m/s, 0,0,0)$, $(A,L,M) = (1m^2, 10m, 1000Kg)$ or $(0.25m^2, 1m, 10000Kg)$. We get from (7) $\omega$=2467.8/sec or 1230.85/sec. Thus for accuracy reasons we have to use small h as $h=1\times10^{-4}$, $1\times10^{-5}$, $1\times10^{-6}$sec. The four numerical schemes have the accuracy in the following order: (E), (RK-2), (IE-1), (IE-2).

Since stability requires $h\leq2/k$, we use $k=1\times10^{5}$, $1\times10^{6}$/sec. All four schemes are stable for $h\leq2/k$ and unstable for $h>2/k$. For instance if $h=2.2/k$ the obtained numerical solutions have large oscillations with respect to the exact viscoelastic solutions after several time integration steps.

Finally, it is also important to know the behaviour of the solutions of problem (1) in the viscoelastic case ($Y(\kappa)=0$) in order to identify the value of k. The eigenvalues of this problem are the roots of $r^2(r+k)+a(rE+kE_0)=0$, $a=A/(ML)$. For all $k>0$ and any fixed $E,E_0$ and a such that $E\neq 9E_0$ this equation has one real root $r_1=-\lambda<0$ and two complex roots $r_{2,3}=-\nu\pm i\omega_v$ $(i=\sqrt{-1})$, $\nu>0$. In addition $0<\lambda<k$, $\nu=(k+\lambda)/2$, $\omega_v\in(\omega_0,\omega)$ and there is a $k_1>0$ such that $0<\nu(k)\leq\nu(k_1)$ for all $k\in(0,\infty)$. For our data $k_1$ is of order $10^3$/sec and $\nu(k)$ decreases sharply outside the interval $(10^3,10^4)$/sec.

Acknowledgement. This work was partly done at Hermann-Föttinger Institut, Technische Universität Berlin. The support from the Leibnitz Program is acknowledged.

REFERENCES

1. Mihailescu-Suliciu,M.,Suliciu,I. and Williams,W., On viscoplastic and elastic-plastic oscillators, Quart.Appl.Math., 1989,47,105-116

2. Sano,O.,Ito,I. and Terada,T., Influence of strain rate on dilatancy and strength of Oshima granite under uniaxial compression, J.Geoph.Res.,1981,86,9299-9311

3. Zobak,M.D. and Byerlee,J.D., The effect of cyclic differential stress on dilatancy in westerly granite under uniaxial and triaxial conditions, J.Geoph.Res.,1975,80,1526-1530

4. Liu,I-Shih and Suliciu,I.,Energy control of the numerical solution of an elastic oscillator, to be published

5. Mihailescu-Suliciu,M. and Suliciu,I., On the method of characteristics in rate-type viscoelasticity, ZAMM,1985,65, 479-486

# ENERGY CONSIDERATION FOR NOTCHES IN SMALL SCALE YIELDING

A. MOFTAKHAR and G. GLINKA
Department of Mechanical Engineering
University of Waterloo
Waterloo, Ontario, Canada N2L 3G1

## ABSTRACT

**A relation for energy density in the notch tip is derived. It enables the notch tip strains and stresses to be estimated in the presence of the localized plastic yielding. This relation is validated against experimental results.**

## INTRODUCTION

Notches in machine parts cause stress concentration resulting in localized plastic yielding. It is generally accepted that, these local inelastic strains and stresses determine the fatigue crack initiation life, and as a consequence, the life of the components as a whole.

The Neuber relation [1] is widely used to calculate the non-linear stress-strain behaviour at notches. A new relation based on energy considerations, has been developed recently [2]. This paper aims to investigate a more general approach to elastic-plastic notch stress-strain analysis, and then to evaluate the 'Neuber' and 'Energy' relations under the assumption of small scale yielding.

**Analysis of a notched body**

Consider a body with a notch subjected to traction $T_i$, applied on the portion $S_t$ of the boundary. Denote the stress field by $\sigma_{ij}$ and the corresponding displacement by $u_i$. The stress components $\sigma_{ij}$ are in equilibrium within the body and satisfy boundary conditions on $S_t$ and on the notch surface, thus:

$$\sigma_{ij,j} = 0; \quad \sigma_{ij} = \sigma_{ji} \tag{1}$$

$$\sigma_{ij}\, n_j = T_i \qquad on \quad s_t \tag{2}$$

$$\sigma_{ij}\, n_j = 0 \qquad on\ notch\ surface \tag{3}$$

The infinitesimal strain tensor is defined by,

$$\varepsilon_{ij} = \frac{1}{2}\left(u_{i,j} + u_{j,i}\right) \tag{4}$$

Using (1), (4) and theorem of Gauss, we can write,

$$\int_V \sigma_{ij}\, \varepsilon_{ij}\, dV = \int_V (\sigma_{ij}\, u_i)_{,j}\, dV = \int_{S_t} T_i\, u_i\, ds \tag{5}$$

Note that eq. (5) holds independently of the stress-strain relation. It is assumed that the plastic zone at the notch tip is small compared to the elastic surrounding field. In the case of such a small scale yielding, stress state at a distance $R \gg R_p$ is not perturbed much by the stress relaxation in the plastic zone $R_p$ and the tractions and displacements vectors on the contour R are essentially given by the linear elastic solution. Therefore in the case of small scale yielding, we can write,

$$\left.\begin{aligned} \bar{T}_i = \sigma_{ij}^a\, \bar{n}_j \approx \sigma_{ij}^e\, \bar{n}_j \\ u_i^a \approx u_i^e \end{aligned}\right] \quad on\ R \tag{6}$$

Now by using eq. (5) we have,

$$\int_{V_R} \sigma_{ij}^a\, \varepsilon_{ij}^a\, dV = \int_{V_R} \sigma_{ij}^e\, \varepsilon_{ij}^e\, dV \tag{7}$$

Where $V_R$ is the volume enclosed by R, and $\sigma_{ij}^a$, $\varepsilon_{ij}^a$, $u_i^a$ denote the actual stress, strain and displacement field respectively, and $\sigma_{ij}^e$, $\varepsilon_{ij}^e$, $u_i^e$ are analogous values obtained from linear elastic solution of the problem. Eq. (7) is a statement of equality of the total strain energy within the region enclosed by contour R in the case of small scale yielding. The energy term can be split into the strain energy W (ε) and the complementary energy W (σ),

$$\int_V \sigma_{ij}\, \varepsilon_{ij}\, dV = \int_V [W(\varepsilon) + W(\sigma)]\, dV \tag{8}$$

The representation of the elastic-plastic material behaviour is assumed to be of the type suggested by Ramberg and Osgood which in the case of proportional loading may be written in the general form (9):

$$\varepsilon_{ij} = (1 + \nu)\, S_{ij} + \frac{1-2\nu}{3}\, \sigma_{kk}\, \delta_{ij} + \frac{3}{2}\, \alpha\, \sigma_{eq}^{n-1}\, S_{ij} \tag{9}$$

where: $S_{ij} = \sigma_{ij} - \frac{1}{3}\, \sigma_{kk}\, \delta_{ij}$ and $\sigma_{eq}^2 = \frac{3}{2}\, S_{ij}\, S_{ij}$.

By using eq. (9), the complementary strain energy density W (σ) and strain energy density W (ε) can be written as,

$$W(\sigma) = \int_0^{\sigma_{ij}} \varepsilon_{ij}\, d\sigma_{ij} = \frac{1}{3}\,(1 + \nu)\, \sigma_{eq}^2 + \frac{1-2\nu}{6}\, \sigma_{kk}^2 + \frac{\alpha}{n+1}\, \sigma_{eq}^{n+1} \tag{10}$$

$$W(\varepsilon) = \int_0^{\varepsilon_{ij}} \sigma_{ij}\, d\varepsilon_{ij} = \frac{1}{3}\,(1+\nu)\, \sigma_{eq}^2 + \frac{1-2\nu}{6}\, \sigma_{kk}^2 + \frac{\alpha n}{n+1}\, \sigma_{eq}^{n+1} \tag{11}$$

Eqns. (10) and (11), enable the ratio of W (σ) to W (ε) to be calculated, as follows,

$$m = \frac{W(\sigma)}{W(\varepsilon)} = \frac{\frac{1}{3}(1+\nu)\,\sigma_{eq}^2 + \frac{1-2\nu}{6}\,\sigma_{kk}^2 + \frac{\alpha}{n+1}\,\sigma_{eq}^{n+1}}{\frac{1}{3}(1+\nu)\,\sigma_{eq}^2 + \frac{1-2\nu}{6}\,\sigma_{kk}^2 + \frac{\alpha n}{n+1}\,\sigma_{eq}^{n+1}} \tag{12}$$

Therefore, using (12),(8),(7) can be written in the following form,

$$\int_{V_R} \frac{m+1}{2}\, W^a(\varepsilon)\, dV = \int_{V_R} W^e(\varepsilon)\, dV \tag{13}$$

Note that the value of the parameter, m, is such that, m = 1, in the elastic domain and, m ≥ o, in the plastic zone (i.e. it is bounded by o ≤ m ≤ 1). It follows that,

$$\int_{VR_p} \frac{m+1}{2}\, W^a(\varepsilon)\, dV + \int_{VR_e} W^a(\varepsilon)\, dV = \int_{VR_p} W^e(\varepsilon) + \int_{VR_e} W^e(\varepsilon)\, dV \tag{14}$$

The contour enclosing the region of the first integration, $VR_p$ is the plastic zone boundary and for the second integral, it is the elastic region $VR_e$. The presence of the plastic zone results in the increase of the stresses in the remaining elastic domain in comparison with the purely elastic solution. This is due to the stress redistribution caused by plastic yielding. Therefore, the following relations can be formulated.

$$\left.\begin{array}{c} \sigma_{ij}^a\,\varepsilon_{ij}^a > \sigma_{ij}^e\,\varepsilon_{ij}^e \\ or \\ W(\varepsilon)^a > W^e(\varepsilon) \end{array}\right] \quad \begin{array}{l}\textbf{in the elastic region}\\ \textbf{include the boundary of}\\ \textbf{plastic zone}\end{array} \tag{15}$$

Substitution of eq. (15) in (14) gives,

$$\int_{VR_p} \varepsilon_{ij}^a\,\sigma_{ij}^a\, dV < \int_{VR_p} \varepsilon_{ij}^e\,\sigma_{ij}^e\, dV \tag{16}$$

By considering eqns. (15) and (16) it may be concluded that, there is a region $A_1$ inside the plastic zone, where on its boundary, eq. (17a), is valid, while inside the region $A_1$ relation (17b) is true,

$$\sigma_{ij}^a\,\varepsilon_{ij}^a = \sigma_{ij}^e\,\varepsilon_{ij}^e \quad \text{at the boundary of } A_1 \tag{17a}$$

$$\sigma_{ij}^a\,\varepsilon_{ij}^a < \sigma_{ij}^e\,\varepsilon_{ij}^e \quad \text{inside } A_1 \tag{17b}$$

Relation (17a) is equivalent of the Neuber's rule and it shows that it is true on a contour $A_1$ inside the plastic zone and not at the notch tip as indicated originally by Neuber. Consequently, Neuber's rule in the form of eq. (17a) will result in overestimation of strains and stresses in the notch tip. Having known that, m ≤ 1 the following inequalities can be derived from eqn. (13),

$$\int_{V_R} \sigma_{ij}^a \varepsilon_{ij}^a \, dV > \int_{V_R} \frac{m+1}{2} \varepsilon_{ij}^e \sigma_{ij}^e \, dV \tag{18a}$$

or

$$\int_{V_R} W^a(\varepsilon) \, dV > \int_{V_R} W^e(\varepsilon) \, dV \tag{18b}$$

Subsequently relation (19) can be derived on the basis of eqns. (15) and (18a),

$$\varepsilon_{ij}^a \sigma_{ij}^a \geq \frac{m+1}{2} \varepsilon_{ij}^e \sigma_{ij}^e \tag{19}$$

By comparing relation (19), with the hypothesis of equivalent strain energy density it may be found that, the hypothesis underestimates the stresses and strains at the notch tip.
Eqns. (19) and (18a) made it possible to establish a band within which the true total strain energy density at the elastic-plastic notch tip lies,

$$\frac{1+m}{2} \varepsilon_{ij}^e \sigma_{ij}^e \leq \varepsilon_{ij}^a \sigma_{ij}^a \leq \varepsilon_{ij}^e \sigma_{ij}^e \tag{20a}$$

or

$$\varepsilon_{ij}^E \sigma_{ij}^E \leq \varepsilon_{ij}^a \sigma_{ij}^a \leq \varepsilon_{ij}^N \sigma_{ij}^N \tag{20b}$$

where $\varepsilon_{ij}^E$ and $\sigma_{ij}^E$ are the notch tip stresses and strains predicted using the hypothesis of equivalent strain energy density [2] while $\varepsilon_{ij}^N$ and $\sigma_{ij}^N$ are those values obtained utilizing the Neuber's rule [1].

## Comparison with experimental data

Available experimental results [3,4,5] for stress and strains at the notch tip lie inside the bond predicted by eq. (20b). A formula based on an average of the two methods may be suggested as an approximate solution for the notch tip stresses and strains. A detailed analysis and discussion may be presented later.

## REFERENCES


1. Neuber, H., Theory of stress concentration for shear strained prismatical bodies with arbitrary non-linear stress-strain law, ASME J. Appl. Mech. 28 (1961) 544-50.

2. Molski, K. and Glinka, G., A Method of elastic-plastic stress and strain calculation at notches, Proc. 8th Int. Conf. Structural Mechanics in reactor Technology, Brussels, August 1985, Paper L 4/3.

3. Umeda, H. and Sakane, M. and Ohnami, M., - Comparison of local strain at notch root between FEM analysis and experimental strain measurement under creep-fatigue condition, JMSE Int. J., vol. 30, No. 268, 1987, P. 1543.

4. Polak, J., Stress and Strain Concentration factor evaluation using the equivalent energy concept, Material Science and Engineering, 61 (1983) 195-200.

5. Sharpe, W. and Yong, C. and Tregoning, E., An evaluation of the Neuber and Glinka relations for monotonic loading (yet to be published).

# FREE SURFACE DUCTILITY IN UPSETTING

HITOSHI MORITOKI
Department of Mechanical Engineering for Production
Mining College, Akita University
1-1 Gakuen-cho Tegata Akita 010 Japan

## ABSTRACT

Surface cracking in upsetting is investigated with the analytical method based on the modified criterion for the collapse of unique solution, and the modes of cracking are determined by comparing the stabilities of deformation from the forming limit to the two modes able to permit strain rate discontinuity on the plane of localized necking. The forming limit and the mode predicted are in very good agreement with the experimental results.

## THE CRITERION FOR SURFACE DUCTILITY

Multiple solutions can be realized, when the following conditions are satisfied:

$$\left.\begin{array}{lll} (i)\ \Delta\dot{s}_1=0, & \Delta\dot{s}_2=0, & \Delta\dot{s}_3=0 \\ (ii)\ \Delta\dot{s}_1=0, & \Delta\dot{\varepsilon}_2=0, & \Delta\dot{s}_3=0 \\ (iii)\ \Delta\dot{s}_1=0, & \Delta\dot{s}_2=0, & \Delta\dot{\varepsilon}_3=0, \end{array}\right\} \quad (1)$$

where the first is statical collapse for uniqueness, and the second and the third are kinematical collapse. $s_i$ is nominal stress and $\varepsilon_i$ strain. A dot represents the time derivative, and $\Delta$ the difference between any two multiple solutions. From the definition of nominal stress

$$\left.\begin{array}{l} \dot{s}_i=\dot{\sigma}_i-\sigma_i\dot{\varepsilon}_1 \\ \Delta\dot{s}_i=\Delta\dot{\sigma}_i-\sigma_i\Delta\dot{\varepsilon}_1, \end{array}\right\} \quad (2)$$

where $\sigma_i$ is true stress. Subscript 1 means to be referred to the direction of maximum principal stress. In the present analysis the Mises' yield condition and Levy-Mises constitutive relations are used. The material flow curve is represented with the power law where n denotes work hardening exponent. Stress ratios, and strain rate ratio are defined as

$$\alpha = \frac{\sigma_2}{\sigma_1}, \quad \beta = \frac{\sigma_3}{\sigma_1}, \quad \gamma = \frac{\dot{\varepsilon}_2}{\dot{\varepsilon}_1}. \tag{3}$$

Limit strains based on the conditions in eqn (1)

$$\left.\begin{array}{ll} \text{(i)} & \dfrac{\varepsilon_{gD}}{n} = \dfrac{4}{3\sqrt{3}} \dfrac{(2-\alpha-\beta)\left(\sqrt{1+\gamma+\gamma^2}\right)^3}{1+\alpha\gamma^2+\beta(1+\gamma)^2} \\ \text{(ii)} & \dfrac{\varepsilon_{gL}}{n} = \dfrac{2}{\sqrt{3}} \dfrac{1-\beta}{1+\beta}, \quad \gamma = 0, \ \alpha = \dfrac{1}{2}(1+\beta) \\ \text{(iii)} & \dfrac{\varepsilon_{gL}}{n} = \dfrac{2}{\sqrt{3}} \dfrac{1-\alpha}{1+\alpha}, \quad \gamma = -1, \ \beta = \dfrac{1}{2}(1+\alpha) \end{array}\right\} \tag{4}$$

are obtained [1], where $\varepsilon_g$ is equivalent strain, and subscripts D and L denote the limit strain based on statical and kinematical collapse for uniqueness respectively.

What we can control in deforming process is force (nominal stress) or displacement. So, in substance, the path should be prescribed as the relation between the components of nominal stress, and stress or strain path results from the corresponding loading path and the constitutive relation of material. When the first principal nominal stress $s_1$ larger than the value material can endure is prescribed on loading path, it can not be reached. Therefore, the terminal point on loading path is the point where $s_1$ is stationary, depending upon material strength. Then, now, we assume the stationary condition of first principal nominal stress as the statical criterion for multiplicity:

$$\dot{s}_1 = 0, \tag{5}$$

which is proved fulfilling the condition, eqn (1-i) of statical criterion for multiplicity.

On the above discussion, only force balance was considered, but the velocity field does not always have the conformability of strain rates for multiplicity. In order to satisfy the strain rate conformability, either mode P or O must be realized on the boundary next to rigid region. And the strain rates of these modes are $\gamma$ =0 and $\gamma$ =-1 respectively. But, in general, these modes are different from the mode which material has taken just before the state shown in eqn (5). However, it is difficult to determine quantitatively when the transition begins from stationary state to the state with mode P or O. Then, it is assumed here that the deforming process can go on the assigned path exactly to the stationary state which instantaneously changes to the state with mode P or O.

Eqn (5) gives following criterion

$$\frac{\varepsilon_{gS}}{n} = \frac{4}{\sqrt{3}} \frac{\left(\sqrt{1+\gamma+\gamma^2}\right)^3}{(2+\gamma)(1+\gamma\xi)}, \tag{6}$$

where $\xi = \dot{\sigma}_2/\dot{\sigma}_1$. Subscript S indicates evaluation for the limit based on statical multiplicity with $s_1$ stationary.

## CRACKING MODES

Which mode is realized, P or O ? It depends upon the stability between these modes and the deformation mode finally achieved just a little before the collapse of uniqueness. It might be naturally accepted that the mode showing lower stability of P or O can be realized. Here, we discuss the stability.

The stability of the state for deforming body is represented with

$$\int_V \dot{\bar{\sigma}} \cdot \dot{\varepsilon}\, dV - \int_S \dot{F} \cdot \dot{u}\, dS = \int_V (\dot{\bar{\sigma}} - \dot{\sigma}) \cdot \dot{\varepsilon}\, dV, \tag{7}$$

which is the difference between the external work done and the energy absorbed into material by plastic deformation and then dissipated as heat [2]. In the above equation, $\sigma$ and $\varepsilon$ are stress and strain tensors vectorically represented in stress space. A bar indicates the value at the cracking mode. As the multiplicity considered here is local, if the integration is done in the infinitesimal region where plastic deformation continues, the stability is compared with the integrand in eqn (7), that is,

$$\dot{E} = (\dot{\bar{\sigma}} - \dot{\sigma}) \cdot \dot{\varepsilon}. \tag{8}$$

The index of stability is as follows.
For mode P:

$$\dot{E}_P = \frac{\sqrt{3}}{2} \dot{\bar{\varepsilon}}_g \{2(1-\bar{\beta}) g_1 - (1-\beta) g_2\}. \tag{9}$$

For mode O:

$$\dot{E}_O = \frac{\sqrt{3}}{2} \dot{\bar{\varepsilon}}_g \{2(1-\bar{\beta}) g_1 - (1-\alpha) g_2\}, \tag{10}$$

where

$$g_1 = \frac{1}{\sqrt{3}} \frac{\dot{\bar{\sigma}}_g}{1-\bar{\beta}}, \quad g_2 = \frac{\sqrt{3}\,\dot{\sigma}_g}{(2-\alpha-\beta)\sqrt{1+\gamma+\gamma^2}} \tag{11}$$

and subscript P or O indicates evaluation at mode P or O respectively. On the assumption of instantaneous change from stationary state to the state with either of the modes, eqns (9) and (10) turn into

$$\left.\begin{aligned} \dot{E}_P &= \frac{1}{2} \dot{\bar{\varepsilon}}_g \dot{\sigma}_g e_P, \quad e_P \equiv 2 - \frac{2+\gamma}{\sqrt{1+\gamma+\gamma^2}} \\ \dot{E}_O &= \frac{1}{2} \dot{\bar{\varepsilon}}_g \dot{\sigma}_g e_O, \quad e_O \equiv 2 - \frac{1-\gamma}{\sqrt{1+\gamma+\gamma^2}}. \end{aligned}\right\} \tag{12}$$

Since $\dot{\bar{\varepsilon}}_g$ and $\dot{\sigma}_g$ are common on the comparison between $\dot{E}_P$ and $\dot{E}_O$, they vary in the same way as $e_P$ and $e_O$, which intersect at $\gamma = 1/2$. Within the usual range of $\gamma$ encountered in forming operation, $e_P$ is smaller than $e_O$ at $\gamma > -1/2$ and then, the critical state takes mode P. When $\gamma < -1/2$, it takes mode O. When $\gamma = -1/2$, it is possible that both modes P and O occur simultaneously.

## THE LIMIT OF DUCTILITY AND ITS MODE

Figure 1 shows the experimental results in upsetting by Kudo & Aoi [3]. The limits of ductility (the strain where cracking begins) are represented with the symbols ● or ○ , where ● means to be mode P, and ○ mode O. Experimental No.4 has the symbol ◐ , which means that both mode P and O

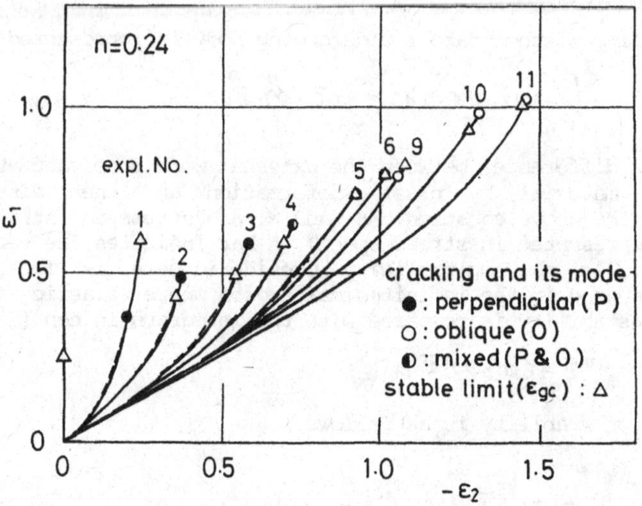


Figure 1. Comparison between strain in cracking [3] and stable limit.

were observed.

If the equivalent strain $\varepsilon_g$ on the path is smaller than $\varepsilon_{gs}$, the process can be considered stable. As deformation proceeds, $\varepsilon_g$ linearly increases, but $\varepsilon_{gs}$ generally decreases, though it has relatively large value at the beginning of deformation. Therefore, the deformation reaches the state before long where $\varepsilon_g = \varepsilon_{gs}$, which denotes the limit of stable deformation.

The analytical results referred to the experiment by KUDO et al. are also shown in Figure 1. The strain paths approximated are shown with broken lines, and the stable limits are denoted with the symbol $\Delta$. On the whole, good agreement is found between cracking and stable limits. The value of $\gamma$ at stable limit (denoted with subscript c) determines the type of cracking mode P or O. In the cases of the experiments with smaller number than 4, $\gamma_c > -1/2$, and for the larger number experiments, $\gamma_c < -1/2$. At No.4 it is very close to -1/2, that suggests the possibility of the simultaneous occurrence of P and O modes. In reality, both modes are observed in the experiment. The prediction for the mode exactly agrees with the experimental observation, though No.4 should take mode O strictly speaking. Good agreement with respect to forming limit and cracking mode confirms the validity of this analysis for predicting free surface ductility.

## REFERENCES

1. Moritoki,H., Central bursting in drawing and extrusion under plane strain, Advanced Technology of Plasticity 1990, Japan Society for Technology of Plasticity, Tokyo, 1990, 1, pp.441-446.
2. Hill,R., Stability of rigid-plastic solids, J. Mech. Phys. Solids, 1975, 6, 1-8.
3. Kudo,H. and Aoi,K., Effect of compression test condition upon fracturing of a medium carbon steel, J. Japan Soc. Tech. Plasticity, 1967, 8, 17-27.

# INTEGRATION ALGORITHM FOR FRICTIONAL MATERIALS INCLUDING PLASTICITY, DAMAGE AND RATE EFFECTS

V. P. PANOSKALTSIS and J. LUBLINER
Department of Civil Engineering
University of California at Berkeley
Berkeley, CA 94720, USA

## ABSTRACT

A new unified model for the description of frictional materials was developed [1], whose basic idea is to fully describe the viscoelastic-plastic-damage behavior of frictional materials, with a representation of the rate effects on the entire stress-strain curve. This paper presents a time-integration algorithm for the model in the case of biaxial compression.

## INTRODUCTION

This note presents the foundations for the numerical implementation of a recently developed unified model for the description of frictional materials. The model is based on the strain decompositions $\boldsymbol{\varepsilon} = \boldsymbol{\varepsilon}^e + \boldsymbol{\varepsilon}^v + \boldsymbol{\varepsilon}^p$ and $\boldsymbol{\varepsilon}^v = \sum_{m=0}^{N} \mathbf{q}_m$, where $\boldsymbol{\varepsilon}$, $\boldsymbol{\varepsilon}^e$, $\boldsymbol{\varepsilon}^v$, $\boldsymbol{\varepsilon}^p$ are the total, elastic, viscous and plastic strains, respectively, and $\mathbf{q}_m$ is the three-dimensional generalization of a partial strain due to the $m$th element of a series representing the discrete modified Kuhn model [1]. The rate equations for the viscous and plastic internal variables are

$$\begin{gathered} \dot{\mathbf{q}}_m + \frac{C}{r^m}\mathbf{q}_m = \frac{C}{r^m} B \ln r\, \mathbf{s} \\ \dot{\boldsymbol{\varepsilon}}^p = \dot{\lambda}\frac{\partial f}{\partial \boldsymbol{\sigma}}, \quad \dot{\kappa} = \mathbf{h}^T(\boldsymbol{\sigma}, c, \kappa)\dot{\boldsymbol{\varepsilon}}^p, \quad \dot{c} = k(\boldsymbol{\sigma}, c, \kappa)\dot{\kappa}, \end{gathered} \tag{1}$$

where $\mathbf{s}$ is the stress deviator and $J_2$ its second invariant, $\kappa$ is a plastic-damage variable, $c$ is the cohesion, and $\mathbf{h}$ and $k$ are respectively matrix- and scalar-valued functions [2]. It is assumed here that the viscous strain is purely deviatoric. This assumption is not essential.

The yield function is $f(\boldsymbol{\sigma}, c, \kappa) = F(\boldsymbol{\sigma}) - c$, with

$$F(\boldsymbol{\sigma}) = \frac{1}{1-\alpha}[\sqrt{3J_2} + \alpha I_1 + \beta\sigma_{\max} - \gamma < -\sigma_{\max} >],$$

where $I_1$ is the first invariant of stress, $\sigma_{\max}$ the algebraically largest principal stress, and $\alpha$, $\beta$, $\gamma$ are dimensionless constants (for more details see [1, 2]).

## BIAXIAL COMPRESSION

In the case of biaxial compression ($\sigma_3 = 0$, $\sigma_2 \leq \sigma_1 \leq 0$), the evolution of $\kappa$ may take the form $\dot{\kappa} = c|\dot{\varepsilon}_2^p|/g_c$, where $\dot{\varepsilon}_2^p$ is the algebraically smallest eigenvalue of $\dot{\boldsymbol{\varepsilon}}^p$ and $g_c$ can be defined from the one-dimensional case as $G_c/l$, where $G_c$ is an assumed material property with dimension of energy/area and $l$ is a characteristic length related to mesh sensitivity in the softening range [2]. Furthermore, in this case a direct relation may be assumed between $c$ and $\kappa$. A suitable form for concrete, for example, has been found to be $c(\kappa) = (f_0/a)[(1+a)\sqrt{\phi(\kappa)} - \phi(\kappa)]$, where $f_0 = c(0)$, $a$ is a dimensionless constant, and $\phi(\kappa) = 1 + a(2+a)\kappa$.

With the definitions $\boldsymbol{\varepsilon} = [\varepsilon_{11}\ \varepsilon_{22}\ 2\varepsilon_{12}]^T$, $\boldsymbol{\sigma} = [\sigma_{11}\ \sigma_{22}\ \sigma_{12}]^T$ and $\mathbf{q} = [q_{11}\ q_{22}\ 2q_{12}]^T$, and with $\boldsymbol{\varepsilon}^v = \sum_m \mathbf{q}_m$ as before, the rate equations in the reduced 1-2 space become

$$\dot{\mathbf{q}}_m + \frac{C}{r^m}\mathbf{q}_m = \frac{C}{r^m}B\ln r\,\mathbf{P}\boldsymbol{\sigma}$$

$$\dot{\boldsymbol{\varepsilon}}^p = \dot{\lambda}\frac{1}{1-\alpha}\left[\frac{3}{2\sqrt{\frac{3}{2}\boldsymbol{\sigma}^T\mathbf{P}\boldsymbol{\sigma}}}\mathbf{P}\boldsymbol{\sigma} + \alpha\mathbf{U}\right], \quad \dot{\kappa} = -\frac{c}{g_c}\left[\frac{1}{2}\mathbf{U}^T\dot{\boldsymbol{\varepsilon}}^p - \frac{1}{2}\sqrt{\dot{\boldsymbol{\varepsilon}}^{pT}\mathbf{R}\dot{\boldsymbol{\varepsilon}}^p}\right], \tag{2}$$

where

$$\mathbf{P} = \frac{1}{3}\begin{bmatrix} 2 & -1 & 0 \\ -1 & 2 & 0 \\ 0 & 0 & 6 \end{bmatrix}, \quad \mathbf{R} = \begin{bmatrix} 1 & -1 & 0 \\ -1 & 1 & 0 \\ 0 & 0 & 1 \end{bmatrix}$$

and $\mathbf{U}^T = [1\ 1\ 0]$. Since $I_1 = \mathbf{U}^T\boldsymbol{\sigma}$ and $J_2 = \mathbf{s}^T\mathbf{s}/2 = \boldsymbol{\sigma}^T\mathbf{P}\boldsymbol{\sigma}/2$, the yield condition is

$$f = \frac{1}{1-\alpha}\left[\sqrt{\frac{3}{2}\boldsymbol{\sigma}^T\mathbf{P}\boldsymbol{\sigma}} + \alpha\mathbf{U}^T\boldsymbol{\sigma}\right] - c \leq 0.$$

## TIME-INTEGRATION ALGORITHM

It is assumed that the solution has been obtained at time $t_n$ and that at time $t_{n+1}$ the solution is sought. The implicit backward Euler integration scheme, which is unconditionally stable and first-order accurate, is applied to Equations (2), resulting in

$$(\mathbf{q}_m)_{n+1} = \frac{1}{1+C\Delta t/r^m}(\mathbf{q}_m)_n + \frac{C\Delta t/r^m}{1+C\Delta t/r^m}B\ln r\,\mathbf{P}\boldsymbol{\sigma}_{n+1}$$

$$\boldsymbol{\varepsilon}^p_{n+1} = \boldsymbol{\varepsilon}^p_n + \zeta_{n+1}\mathbf{r}_{n+1} \tag{3}$$

$$\kappa_{n+1} = \kappa_n - \zeta_{n+1}\frac{c(\kappa_{n+1})}{g_c}\left[\frac{1}{2}\mathbf{U}^T\mathbf{r}_{n+1} - \frac{1}{2}\sqrt{\mathbf{r}^T_{n+1}\mathbf{R}\mathbf{r}_{n+1}}\right]$$

where $\zeta_{n+1} = \lambda_{n+1}\Delta t$ and

$$\mathbf{r}_{n+1} = \frac{1}{1-\alpha}\left[\frac{\mathbf{P}\boldsymbol{\sigma}_{n+1}}{\sqrt{\frac{2}{3}\boldsymbol{\sigma}^T_{n+1}\mathbf{P}\boldsymbol{\sigma}_{n+1}}} + \alpha\mathbf{U}\right].$$

The discrete consistency condition is

$$f_{n+1} = \frac{1}{1-\alpha}\left[\sqrt{\frac{3}{2}\boldsymbol{\sigma}_{n+1}^T \mathbf{P}\boldsymbol{\sigma}_{n+1}} + \alpha \mathbf{U}^T\boldsymbol{\sigma}_{n+1}\right] - c(\kappa_{n+1}) = 0 \tag{4}$$

and the elastic stress-strain relation is

$$\boldsymbol{\sigma}_{n+1} = \mathbf{D}(\boldsymbol{\varepsilon}_{n+1} - \boldsymbol{\varepsilon}_{n+1}^p - \boldsymbol{\varepsilon}_{n+1}^v), \tag{5}$$

where $\mathbf{D}$ is the plane-stress elastic stiffness matrix and $\boldsymbol{\varepsilon}_{n+1}^v = \sum_m (\mathbf{q}_m)_{n+1}$.

In order to solve the equations (3)-(5) the predictor-corrector algorithm is used [4]. Unlike the elastoplastic case, in the present model the predictor phase is not purely elastic, since some of the internal variables (namely the viscous strain) change. During the corrector phase the total strain $\boldsymbol{\varepsilon}_{n+1}$ remains constant, while the plastic variables as well as the viscous strain evolve. For the implementation of the solution algorithm in a finite-element framework in such a way that the quadratic rate of convergence associated with Newton's method at the global level of iteration is preserved, it is necessary to find, for both the predictor and corrector phases, the algorithmic ("consistent") tangent moduli [3, 5]; these are defined here as the Gateaux derivative of $\boldsymbol{\sigma}$ with respect to $\boldsymbol{\varepsilon}$ at $t_{n+1}$ after convergence has been achieved, that is, after the yield surface has been found at $t_{n+1}$. In the predictor phase the algorithmic moduli are found to be

$$\left.\frac{d\boldsymbol{\sigma}}{d\boldsymbol{\varepsilon}}\right|_{n+1} = [\mathbf{D}^{-1} + \theta_1 \mathbf{P}]^{-1},$$

where

$$\theta_1 = \left(\sum_m \frac{C\Delta t/r^m}{1 + C\Delta t/r^m}\right) B \ln r.$$

It is noted that $d\boldsymbol{\varepsilon}^v = \theta_1 \mathbf{P} d\boldsymbol{\sigma}$ and also that in the corrector phase

$$d\mathbf{r} = \eta_{n+1}(\mathbf{P} - \boldsymbol{\xi}_{n+1}\boldsymbol{\xi}_{n+1}^T) d\boldsymbol{\sigma}$$

where

$$\eta_{n+1} = \frac{1}{1-\alpha}\sqrt{\frac{3}{2}}\frac{1}{\sqrt{\boldsymbol{\sigma}_{n+1}^T \mathbf{P}\boldsymbol{\sigma}_{n+1}}}, \quad \boldsymbol{\xi}_{n+1} = \frac{\mathbf{P}\boldsymbol{\sigma}_{n+1}}{||\mathbf{s}_{n+1}||} = \frac{\mathbf{P}\boldsymbol{\sigma}_{n+1}}{\sqrt{\boldsymbol{\sigma}_{n+1}^T \mathbf{P}\boldsymbol{\sigma}_{n+1}}}$$

The systematic linearization of the corrector equations in the Gateaux sense yields finally

$$\left.\frac{d\boldsymbol{\sigma}}{d\boldsymbol{\varepsilon}}\right|_{n+1} = \boldsymbol{\Xi} - \frac{\boldsymbol{\Xi}\mathbf{r}_{n+1}\mathbf{r}_{n+1}^T\boldsymbol{\Xi} - \theta_3 c'(\kappa_{n+1})\boldsymbol{\Xi}\mathbf{r}_{n+1}\mathbf{t}_{n+1}^T\boldsymbol{\Xi}}{\mathbf{r}_{n+1}^T\boldsymbol{\Xi}\mathbf{r}_{n+1} + c'(\kappa_{n+1})\theta_4},$$

where

$$\boldsymbol{\Xi} = [\mathbf{D}^{-1} + \zeta_{n+1}\eta_{n+1}(\mathbf{P} - \boldsymbol{\xi}_{n+1}\boldsymbol{\xi}_{n+1}^T) + \theta_1\mathbf{P}]^{-1},$$

which is of the form $[\mathbf{B} - \zeta_{n+1}\eta_{n+1}\boldsymbol{\xi}_{n+1}\boldsymbol{\xi}_{n+1}^T]^{-1}$, where $\mathbf{B} = \mathbf{D}^{-1} + \zeta_{n+1}\eta_{n+1}\theta_1\mathbf{P}$, and therefore can be evaluated in closed form by means of the rank-reduction method [6, p.127], and

$$\theta_2 = -\frac{c(\kappa_{n+1})}{g_c} m_{n+1} \left(1 + \frac{c'(\kappa_{n+1})}{g_c} \zeta_{n+1} m_{n+1}\right)^{-1},$$

$$m_{n+1} = \frac{1}{2}\mathbf{U}^T \mathbf{r}_{n+1} - \frac{1}{2}\sqrt{\mathbf{r}_{n+1}^T \mathbf{R} \mathbf{r}_{n+1}},$$

$$\theta_3 = -\zeta_{n+1} \frac{c(\kappa_{n+1})}{2g_c} \eta_{n+1} \left(1 + \frac{c'(\kappa_{n+1})}{g_c} \zeta_{n+1} m_{n+1}\right)^{-1}, \quad \theta_4 = \theta_2 - \theta_3 \mathbf{t}_{n+1}^T \boldsymbol{\Xi} \mathbf{r}_{n+1},$$

and

$$\mathbf{t}_{n+1}^T = \left(\mathbf{U}^T - \frac{\mathbf{r}_{n+1}^T \mathbf{R}}{\sqrt{\mathbf{r}_{n+1}^T \mathbf{R} \mathbf{r}_{n+1}}}\right) (\mathbf{P} - \boldsymbol{\xi}_{n+1} \boldsymbol{\xi}_{n+1}^T).$$

The algorithmic moduli are not symmetric. When $\Delta t \to 0$ ($\zeta_{n+1} \to 0$) the algorithmic moduli have as limit the continuum elastoplastic moduli of the model with the viscous part not included.

Damage in the sense of stiffness degradation as well as nonlinearity in the viscous response have not been included in the present algorithm but can be incorporated without difficulty.

## REFERENCES


1. Panoskaltsis, V.P., Lubliner, J. and Monteiro, P.J.M., A viscoelastic-plastic-damage model for concrete. In *Constitutive Laws for Engineering Materials*, ed. C.S. Desai et al., ASME Press, New York, 1991, pp. 317–320.

2. Lubliner, J., Oliver, J., Oller, S. and Oñate, E., A plastic-damage model for concrete. *Int. J. Solids Structures*, 1989, **25**, 299-326.

3. Nagtegaal, J.C., On the implementation of inelastic constitutive equations with special reference to large deformation problems, *Computer Meth. Appl. Mech. Engrg.*, 1982, **33**, 469-484.

4. Simo, J.C. and Taylor, R.L., A return mapping algorithm for plane stress elastoplasticity. *Int. J. Num. Meth. Eng.*, 1986, **22**, 649-670.

5. Simo, J.C. and Hughes, T.J.R., *Elastoplasticity and Viscoplasticity, Computational Aspects*, Dept. of Appl. Mech., Stanford Univ., 1988.

6. Ciarlet, P.G., *Introduction to Numerical Linear Algebra and Optimisation*, Cambridge Univ. Press, Cambridge (England), 1989.

# A THEORY FOR PLANAR ANISOTROPY IN METALLIC SHEETS

SHYAM K. SAMANTA and N. R. SENTHILNATHAN
Plasticity Laboratory, Department of Mechanical Engineering
University of Nevada-Reno, Reno, NV 89557-0030, U.S.A.

## ABSTRACT

A new planar orthotropic plasticity theory is proposed for metallic sheets. Attractive features of the present formulation against those of Hill's (1990) theory are discussed.

KEYWORDS: sheet metal, planar orthotropy, quadratic yield function, associated flow rule, incremental constitutive equations, plasticity.

## INTRODUCTION

With increasing cold working (such as rolling, drawing and extrusion) an originally isotropic metallic material becomes anisotropic in respect of many mechanical, electrical and magnetic properties. Constitutive relations for the plastic yielding and deformation of anisotropic metals were proposed long ago by Hill (1948), and is classically known as Hill's quadratic yield function or Hill's 'old' criterion. In 1979 Hill re-appraised the classical theory of 1948, and proposed "a new type of yield function to account for the so called 'anomalous' behavior of some materials". Hill's both quadratic and non-quadratic theories, as often referred to in the literature, were devised for sheet metals with in-plane isotropy i.e., so-called normal anisotropy (Hill 1990). By contrast, in the literature on plasticity there is a large volume of experimental evidence which strongly indicates that pre-strained metallic materials i.e., sheet metals with orthotropic textures, exhibit in-plane anisotropy i.e., so-called planar anisotropy, if afterwards subjected to uniaxial tension at arbitrary in-plane orientations. In that case, *planar anisotropy is rather a rule than an exception* ('anomaly'). To our knowledge, only the simplest form of experiments have been attempted, namely through uniaxial or biaxial loading, to map the yield locus of prestrained metallic materials. This has been our position that the yield surfaces should be determined directly and should be free from any *a priori* hypothesis. This implication has never been *adequately* explored. The needed experiments are admittedly taxing, but not prohibitively so (Hill 1979). Based on Hill's (1950) hypothesis that the orthotropy axes coincides with the principal directions of stretch, the change in orthotropy directions has been studied theoretically by Kumar, Samanta and Mallick (1991). Novel experiments were then conducted by Mallick, Samanta and Kumar (1991) on the basis that a sequence of yield loci can be mapped by testing thin-walled tubular specimens subjected to simultaneous triaxial loading (tension, torsion and internal pressure). These carefully conducted experiments clearly demonstrate the nature of in-plane anisotropy present in a prestrained SAE 1020 steel, a material widely used in technological design. Hence, the motivation for the present work derives from the need for an *in-plane anisotropy theory* consistent with the phenomenological model of the material and the observations from such non-trivial experiments as reported in the Ref. of Mallick, Samanta and Kumar (1991). In a subsequent paper (Senthilnathan and Samanta, 1991), the new proposal has been fully derived and adopted to study the flange wrinkling in Swift Cup test.

## PROBLEM FORMULATION

To give the readers a perspective , we first recall some relevant features of Hill's (1990) new orthotropic plasticity theory and then we present our new theory of anisotropic plasticity.

The significant departure of the 'new' orthotropic plasticity theory by Hill (1990) from Hill's (1948) orthotropic criterion is the non-quadratic nature of the yield function. The non-quadratic criterion requires an arbitrary material constant *m*, which is to be determined by fitting the yield criterion to the experimentally obtained yield points. This process involves extensive experimental data and iterative computations. The criterion also requires experimentally determined values for the yield stress in equibiaxial tension ($\sigma$), the yield stress in pure shear ($\tau$), and the yield stresses in uniaxial tension along the three directions ($\sigma_0$, $\sigma_{45}$ and $\sigma_{90}$) to the orthotropic axes in the plane of the sheet. For a work-hardening material, Hill assumes that the ratios of the yield stresses $\sigma/\sigma_0$, $\sigma/\sigma_{45}$ and $\sigma/\sigma_{90}$ are constants (Hill 1990, p. 412). This assumption, although may be true for some materials may not be true for all metallic materials with strong planar anisotropy. In the following, a new orthotropic plasticity theory, which does not need an empirical constant, *m*, and which does not need the yield stress from an equibiaxial test, $\sigma$, and as a consequence does not require to assume constant values for the stress ratios, is now being presented.

## A NEW PLANAR ORTHOTROPIC YIELD CRITERION

The new criterion is obtained by assuming that it is quadratic in the stresses and that Bauschinger effect can be neglected. The yield criterion is expressed (the full derivation is presented in Senthilnathan and Samanta, 1991) in the form

$$f = a_1\sigma_x^2+2a_2\sigma_x\sigma_y+a_3\sigma_y^2+a_4\tau_{xy}^2 = 1 \tag{1}$$

where $x$ is along 0° and $y$ is along 90° of orthotropy in the plane of the sheet. $a_1$, $a_2$, $a_3$ and $a_4$, are material parameters which are defined by the yield stresses in uniaxial tension and pure shear as follows:

$$a_1 = \frac{1}{\sigma_0^2},\quad a_3 = \frac{1}{\sigma_{90}^2},\quad a_4 = \frac{1}{\tau^2},$$
$$a_2 = \left(\frac{2}{\sigma_{45}^2}-\frac{1}{2\sigma_0^2}-\frac{1}{2\sigma_{90}^2}-\frac{1}{2\tau^2}\right) \tag{2}$$

where $a_2$ is obtained by a uniaxial test at 45° to the orthotropic axes.

Albeit the form of the yield criterion (1) is the same as that in Hill's (1948) 'old' orthotropic yield criterion, the coefficients $a_1$, $a_2$, $a_3$ and $a_4$ in the present case, however, are assumed variables for a work-hardening material. In other words, the yield stresses are assumed functions of the respective plastic strains and are given by

$$d\sigma_0 = H_0 d\varepsilon_0^p, \quad d\sigma_{45} = H_{45} d\varepsilon_{45}^p, \qquad (3)$$
$$d\sigma_{90} = H_{90} d\varepsilon_{90}^p, \quad d\tau = H_\tau d\gamma^p$$

where $H$ is the plastic modulus for the corresponding stress-strain curve.

The incremental plastic strains are defined in terms of the total stresses by the usual associated flow rule and are given by

$$d\varepsilon_0^p = d\lambda \frac{\partial f}{\partial \sigma_x} = d\lambda (2a_1\sigma_x + 2a_2\sigma_y) = d\lambda . \sigma'_x$$

$$d\varepsilon_{90}^p = d\lambda \frac{\partial f}{\partial \sigma_y} = d\lambda (2a_2\sigma_x + 2a_3\sigma_y) = d\lambda . \sigma'_y$$

$$d\gamma^p = d\lambda \frac{\partial f}{\partial \tau_{xy}} = d\lambda (2a_4\tau_{xy}) = d\lambda . \tau'_{xy} \qquad (4)$$

$$d\varepsilon_{45}^p = d\lambda (\sigma'_x \cos^2 45^0 + \sigma'_y \sin^2 45^0 + 2\tau'_{xy} \sin 45^0 \cos 45^0)$$
$$= \frac{d\lambda}{2} (\sigma'_x + \sigma'_y + 2\tau'_{xy})$$

where

$$\sigma'_x = (2a_1\sigma_x + 2a_2\sigma_y), \quad \sigma'_y = (2a_2\sigma_x + 2a_3\sigma_y) \text{ and } \tau'_{xy} = 2a_4\tau_{xy} \qquad (5)$$

The consistency condition is given by

$$df(\sigma_x, \sigma_y, \tau_{xy}, \sigma_0, \sigma_{45}, \sigma_{90}, \tau) = 0 \qquad (6)$$

which gives

$$\frac{\partial f}{\partial \sigma_x} d\sigma_x + \frac{\partial f}{\partial \sigma_y} d\sigma_y + \frac{\partial f}{\partial \tau_{xy}} d\tau_{xy} + \frac{\partial f}{\partial \sigma_0} d\sigma_0 + \frac{\partial f}{\partial \sigma_{45}} d\sigma_{45} \qquad (7)$$
$$+ \frac{\partial f}{\partial \sigma_{90}} d\sigma_{90} + \frac{\partial f}{\partial \tau} d\tau = 0$$

The incremental orthotropic stress-strain relations are assumed as

$$(d\sigma) = [C^e](d\varepsilon - d\varepsilon^p) \qquad (8)$$
$$= [C^e](d\varepsilon) - [C^e](\sigma') . d\lambda$$

where the plastic strains given by the equations (4) are made use of and $[C^e]$ is the elastic constitutive matrix in plane-stress condition.

Use of the equations (8) in the consistency condition (7) in conjunction with the definitions of the plastic modulii (3) and the flow rule relations (4) gives

$$d\lambda = (\phi(\sigma_x, \sigma_y, \tau_{xy}, \sigma_0, \sigma_{45}, \sigma_{90}, \tau, H_0, H_{45}, H_{90}, H_\tau))^T (d\varepsilon) \qquad (9)$$

where $\phi$ is a known vector function of the current stresses and the material parameters.

Substitution of the scalar value $d\lambda$ into equations (8) finally yields the incremental elastoplastic stress-strain relations given by

$$(d\sigma) = [C^{e}](d\varepsilon) - [C^{e}](\sigma')(\phi)^{T}(d\varepsilon) \qquad (10)$$
$$= [C^{ep}](d\varepsilon)$$

where

$$[C^{ep}] = [C^{e}] - [C^{e}](\sigma')(\phi)^{T} \qquad (11)$$

## REMARKS ON THE FINITE CONSTITUTIVE EQUATIONS

It is interesting to note that the present formulation does not require an equibiaxial stress-strain relationship, an equivalent stress- equivalent strain relationship and the empirical material parameter *m*. The evolution of the yield function is non-isotropic but non-kinematic. The material parameters required in the proposed criterion can be determined by *simple experiments such as uniaxial tension tests along* 0°, 45° and 90° *to the orthotropic axes and a pure shear test.* Since the new formulation uses the just mentioned stress-strain curves without invoking the definition of equivalent stress- equivalent strain, it should be more appealing than Hill's new 1990 criterion. The quadratic nature of the present yield function results in algebraically simpler elements in the elasto-plastic matrix than those given by Hill's non-quadratic yield function and therefore should lead to faster numerical computation than the latter one.

## ACKNOWLEDGEMENT

The authors wish to acknowledge the support of US National Science Foundation under the grant DDM 9011060.

## REFERENCES

Hill, R. (1948) : Proc. Roy. Soc. London Ser. A 193, 281-297.
Hill, R. (1950) : *Mathematical Theory of Plasticity*, Clarendon Press, Oxford.
Hill, R. (1979) : Math. Proc. Cambridge Phil. Soc. 85, 179-191.
Hill, R. (1990) : J. Mech. Phys. Solids, 38, 405-417.
Hill, R. (1991) : J. Mech. Phys. Solids, 39, 295-307.
Kumar, A., Samanta, S.K. and Mallick, K. (1991) : J. Eng. Mat. Tech., 113, 187-191.
Mallick, K., Samanta, S.K. and Kumar, A. (1991) : J. Eng. Mat. Tech., 113, 192-198.
Senthilnathan, N.R. and Samanta, S.K. (1991) : An In-plane Anisotropy Theory and its application to flange wrinkling, J. Mech. Phy. Solids (Submitted).

# FINITE ELEMENTS MODELLING OF ELASTOPLASTIC STRUCTURES USING SELF-CONSISTENT CONSTITUTIVE RELATION

E. SCACCIATELLA, J. M. GENEVAUX, P. LIPINSKI, M. BERVEILLER
J. M. DETRAUX*, F. HORKAY*
Laboratoire de Physique et Mécanique des Materiaux
(URA CNRS 1215), ENIM, Ile du Saulcy, 57000 - METZ
(*) RNUR, Boulogne Billancourt

## ABSTRACT

In order to describe very complicated and highly anisotropic evolution of the internal structure of materials during complex loading paths, the classical phenomenological constitutive laws use numerous adjustable parameters rather difficult to identify.

On the other hand, the recent developpment of the self-consistent modelling based on a few (three in our case) physically well-defined parameters has proved its advantages [1], [2].

In this work, we propose to couple the finite element code and the self-consistent modelling of the elastoplastic constitutive laws. To this end, macro-homogeneous polycrystalline volumes are associated with each Gauss' point of the finite elements mesh. These macro-homogeneous volumes are defined by N grains described by their crystallographic and morphological orientations, internal stress states, and critical shear stresses on the glide systems.

In this paper, the practicability of such an approach is shown and some examples are presented.

## INTRODUCTION

The determination of the stiffness matrix $[k_e]$ of a given element is one of the most important operation of the finite element method. This matrix is defined by the following expression:

$$[k_e] = \int_V [B]^T [L][B]\, dV$$

where V is the element' s volume,
[B] is the matrix relating strains to nodal displacements,
[L] defines the elastoplastic properties of the material

The analytical expression of $[k_e]$ exists only for a few simple elements, for instance a linear triangular element. In the general case the stiffness matrix is obtained using the numerical integration methods, among which the Gauss' quadrature rule is the most popular one.

In order to determine the approached value of the $[k_e]$ matrix, we introduce the set of sampling points called "Gauss' points", and we determine the value of the [B] and [L] matrices for this position inside the element. The [B] matrix is exactly defined by the geometry of the element and kinematic relations (definition of the strain tensor).
Until now, the [L] matrix has been approached using phenomenological descriptions. To evaluate this matrix at each Gauss' point, we use the micro-macro transition operations and exactly speaking, the elastoplastic self-consistent scheme [1]. In the same way, we evaluate the effective stresses which are necessary to determine the residual forces.

## Modelling at the Gauss' point

In order to determine the elastoplastic properties of the material, the polycrystal at a given Gauss' point is represented by a set of N grains (fig 1). Each grain is characterized by:

- the crystallographic orientation of the lattice
- the critical shear stresses on all glide systems
- the local stress state
- the grain shape which is assumed to be ellipsoïdal

We suppose that the crystallographic glide is the unique mechanism of plastic deformation. This crystallographic glide is governed by Schmid's law and we assume the intra-crystalline hardening may be described by the hardening matrix [H], linking the critical shear stress rate and the rate of the plastic slip on the slip systems [3].

This approach permits to follow the evolution of the crystallographic texture which may be considered as the most important parameter responsible for the anisotropy of the material.

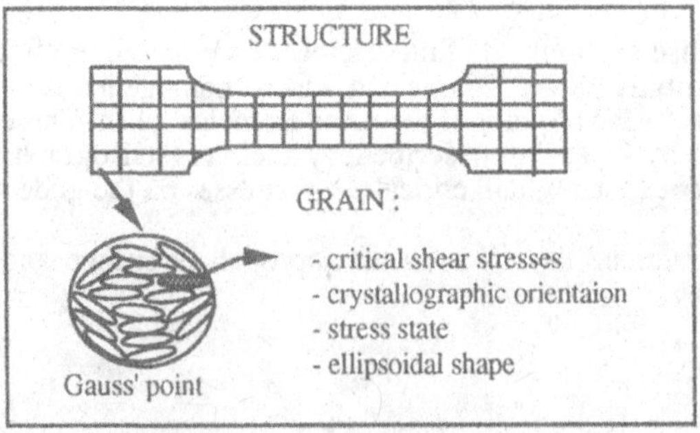


*fig. 1 Modelling at the Gauss' point*

## Implementation on the finite element code

To show the practicability of this coupling, we have used a simple 2D finite element code in which we have changed the part concerning the phenomenological constitutive relation by the appropriate sub program executing micro-macro transitions.

The Newton Raphson method is used to solve the system of non linear equations of elastoplasticity. For each iteration the [L] matrix, the macroscopic stress state and the macroscopic plastic strains are obtained by the self-consistent approximation.

The convergence is achived after a few iterations (less than three) because a small increment of the external forces is imposed by the self-consistent modelling. Figure 2 shows the convergence of the solution versus size of the increment of external loading, in the case of a cantilever beam under tension .

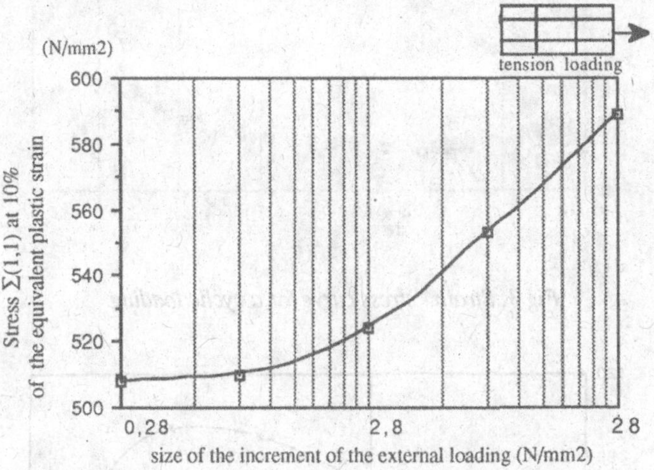


size of the increment of the external loading (N/mm2)

*Fig 2: Influence of the increment size on the convergence of the solution*

## Applications and results

In this paragraph, an example concerning a cantilever beam under cyclic loading (tension, compresion, unloading) is presented. We use eight node quadratic elements for the discretization of the beam. In this case one hundred FCC grains of a spherical shape and random initial crystallographic orientations have been chosen to represent the polycrystal at each Gauss' point.

The elastic behavior of the single crystal is supposed to be isotropic and defined by Lamé's constant $\mu = 30000$ N/mm$^2$ and Poisson's ratio $\nu = 0{,}3$. The initial critical shear stresses are the same for all grains and glide systems and equal to 140 N/mm$^2$.

The interactions between glide systems which are described by the hardening matrix, correspond to the aluminium alloy.

Figure 3 presented bellow exhibits, at a given Gauss' point, the stress-strain curve obtained during the cyclic loading mentionned above. This curve shows a very important macroscopic Bauschinger effect.

We emphasise that locally (at single crystal level) this effect is absent and it results only from the internal stresses due to plastic incompatibilities between the grains.

Figure 4 shows the yield surface as a function of the prestraining level. A very complex evolution is visible, manifesting an anisotropic and a kinematic hardening of the material.

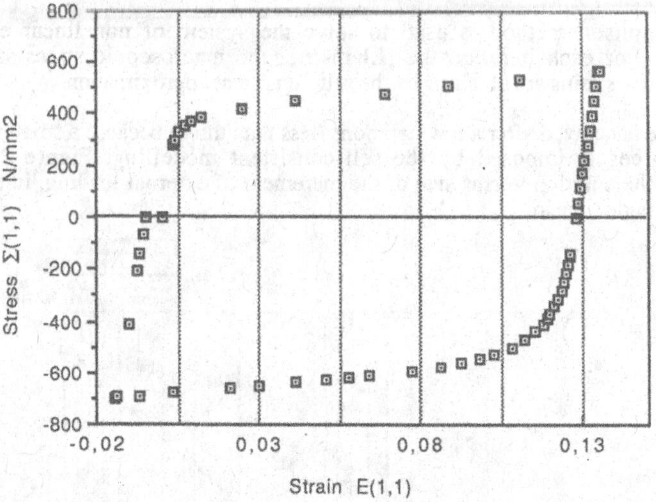


*Fig 3: Strain - stress curve for a cyclic loading*

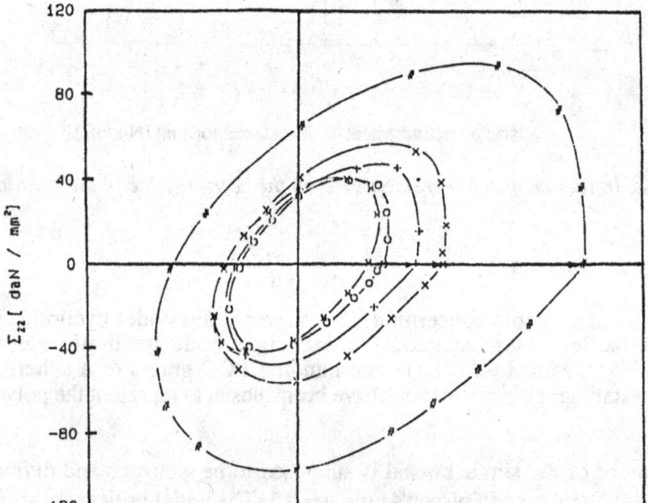


*Fig 4: Evolution of the yield surface (prestraining up to 0 ,10 ,20, 60% of equivalent plastic strain)*

## REFERENCES


1. Lipinski, P., Berveiller, M., Int. J. of Plasticity , 1989, 5 , p 149-172.
2. Krier, J., Lipinski, P., Berveiller, M., Rev. Phys. Appl., 1990, 25, p 361-388.
3. Franciosi, P., Berveiller, M., Zaoui, A., Acta Metall, 1980, 28, p 273-283.

# RECENT DEVELOPMENTS IN THE FORMULATION AND NUMERICAL ANALYSIS OF THERMOPLASTICITY

J. C. SIMO[1], C. MIEHE[2] & F. ARMERO[1]

Division of Applied Mechanics, Stanford University

## ABSTRACT

**This paper reviews the current mathematical status of thermoplasticity, presents and generalizes a new formulation of coupled associative thermoplasticity at finite strains recently introduced in [10], and addresses in detail fundamental numerical analysis aspects relevant to large-scale numerical simulations.**

## 1. INTRODUCTION

This work addresses thermomechanical, mathematical and numerical analysis aspects involved in a formulation of thermoplasticity at finite strains. On the mechanical side, two distinctive features characterize the proposed thermomechanical formulation of metal plasticity. First, the part of the (configurational) entropy arising at the micro-level as a result of dislocation and lattice defect motion (see e.g., [3]) is characterized in the present theory by means of an additional independent internal variable referred to as the *plastic entropy*. Second, the plastic flow evolution equations compatible a multiplicative decomposition of the deformation gradient, including the evolution equation for the plastic configurational entropy, are derived by an extended thermomechanical version of the classical principle of maximum dissipation. This postulate is the cornerstone of the modern mathematical theory of plasticity, completed in the late 70's with the work of Johnson [6] and others, see the recent review in [4].

On the numerical analysis side, the coupled nonlinear initial boundary value problem (IBVP) is treated by means of a new global product formula algorithm first suggested in [10], which leads to a *two-step*, *unconditionally stable* solution procedure. This approach circumvents the conditional stability property inherent to classical staggered coupled algorithms, as in [1], while maintaining their computational advantages. The key idea underlying this new methodology is to view the coupled field equations of thermoplasticity as the generator of a nonlinear quasi-contractive semigroup (see [9]) and exploit a natural *operator split* of this generator into a mechanical phase, followed by a heat conduction phase at fixed configuration. These two problems are strongly coupled via the structural heating and the plastic dissipation. A two-step unconditionally stable algorithm is then constructed by exploiting the algorithmic counterpart of the classical *Trotter-Lie-Kato* formula; see e.g., [4]. The integration of the plastic flow evolution equations is accomplished via a new class of local return mapping algorithms which preserve *exactly* the incompressibility constraint on the plastic flow. Finally, the spatial finite element discretization incorporates a new class of mixed finite element methods recently introduced in [11], which circumvents well-known difficulties associated with the incompressibility constraint on the plastic flow, provide a sound mathematical justification for ad-hoc incompatible mode techniques, and are ideally suited for localization analysis.

---

[1] Supported by NSF under Grant 2-DJA-792 with Stanford University

[2] Supported by the Deutche Forschungsgemeinschaft (DFG)

## 2. THE LOCAL IBVP FOR THERMOPLASTICITY AT FINITE STRAINS

Following [10], we give a brief outline of a formulation of thermoplasticity at finite strains in which the thermoelastic domain, denoted by $\mathsf{E}$, is specified in a classical format by means of a yield criterion expressed in terms of *true stresses* defined in the current configuration of the body; i.e.,

$$\mathsf{E} := \{ (\boldsymbol{\tau}, \boldsymbol{\beta}, \Theta) : \quad \phi(\boldsymbol{\tau}, \boldsymbol{\beta}, \Theta) \leq 0 \}. \tag{1}$$

Here, $\boldsymbol{\tau}$ is the Kirchhoff stress tensor, $\Theta$ is the absolute temperature field and $\boldsymbol{\beta}$ is a suitable vector of internal variables characterizing the state of strain hardening in the material. We remark that the principle of ***objectivity*** restricts the possible forms of $\phi$ in (1) to ***isotropic functions***.

**2.1. Constitutive Equations. Dissipation Inequality.** Motivated by the structure of single crystal plasticity models, see e.g. [2] and references therein we assume the local multiplicative decomposition $\boldsymbol{F}(X,t) = \boldsymbol{F}^e(X,t)\boldsymbol{F}^p(X,t)$, $\forall (X,t) \in \mathcal{B} \times \mathsf{I}$. From a micromechanical point of view, $\boldsymbol{F}^p$ is an internal variable related to the amount of dislocation flow through the crystal lattice. From a phenomenological standpoint, $\boldsymbol{F}^{e-1}$ defines the local, stress-free, unloaded configurations. Let $\boldsymbol{b}^e := \boldsymbol{F}^e \boldsymbol{F}^{e\,T}$ be the elastic left Cauchy-Green tensor. Consistent with the restriction to isotropy implied by the thermoelastic domain (1), we restrict our attention to an internal energy function of the form

$$e = \bar{e}(\boldsymbol{b}^e, \alpha, \eta^e), \quad \text{with} \quad \eta^e := \eta - \eta^p. \tag{2}$$

Here $\eta(X,t)$ is the total entropy, while $\eta^p(X,t)$ denotes that part of the total entropy related to dissipative plastic structural changes which does not affect the internal energy. For the model the model problem at hand, the internal energy is simply the stored energy associated with the elastic lattice deformation. The free energy function $\bar{\psi}$ is obtained from the internal energy (2) via the standard Legendre transformation as

$$\bar{\psi}(\boldsymbol{b}^e, \alpha, \Theta) := \bar{e}(\boldsymbol{b}^e, \alpha, \eta^e) - \eta^e \Theta. \tag{3}$$

Constitutive equations consistent with this assumed form for the free energy function are derived by systematically exploiting the second law of thermodynamics. Here, we shall adopt the Clausius-Plank form of the second law (see [12, p.259]):

$$\mathcal{D} := \Theta \gamma_{\text{loc}} := \Theta \dot{\eta} + \boldsymbol{\tau} \cdot \boldsymbol{d} - \dot{e} \geq 0, \tag{4}$$

where $\gamma_{\text{loc}}$ is the local entropy production, $\boldsymbol{d} = \text{sym}[\boldsymbol{l}]$ is the rate of deformation tensor, $\boldsymbol{l} := \dot{\boldsymbol{F}}\boldsymbol{F}^{-1}$ is the spatial velocity gradient and $\boldsymbol{\tau}$ is the Kirchhoff stress tensor. Time differentiation of the free energy function (3) and use of the identity

$$\dot{\boldsymbol{b}}^e = \boldsymbol{l}\boldsymbol{b}^e + \boldsymbol{b}^e \boldsymbol{l}^T + \boldsymbol{\pounds}_{\boldsymbol{v}} \boldsymbol{b}^e \qquad \text{with} \qquad \boldsymbol{\pounds}_{\boldsymbol{v}} \boldsymbol{b}^e := \boldsymbol{F} \frac{\partial}{\partial t} [\boldsymbol{C}^{p-1}] \boldsymbol{F}^T, \tag{5}$$

where $\boldsymbol{C}^p := [\boldsymbol{F}^{p\,T} \boldsymbol{F}^p]$ is the plastic right Cauchy-Green tensor and $\boldsymbol{\pounds}_{\boldsymbol{v}} \boldsymbol{b}^e$ denotes the ***Lie derivative*** of $\boldsymbol{b}^e$, yields after some manipulation the following expression for the local dissipation:

$$\begin{aligned} \mathcal{D} = & \left[ -(\eta - \eta^p) - \partial_\Theta \bar{\psi} \right] \dot{\Theta} + \left[ \boldsymbol{\tau} - 2\partial_{\boldsymbol{b}^e} \bar{\psi}\, \boldsymbol{b}^e \right] \cdot \boldsymbol{d} \\ & + \left[ 2\partial_{\boldsymbol{b}^e} \bar{\psi}\, \boldsymbol{b}^e \right] \cdot \left[ -\tfrac{1}{2} (\boldsymbol{\pounds}_{\boldsymbol{v}} \boldsymbol{b}^e) \boldsymbol{b}^{e-1} \right] + \left[ -\partial_\alpha \bar{\psi} \right] \dot{\alpha} + \Theta\, \dot{\eta}^p \geq 0. \end{aligned} \tag{6}$$

By demanding that (6) holds for all admissible processes a standard argument then yields the constitutive equations:

$$\boldsymbol{\tau} = 2\,\partial_{\boldsymbol{b}^e}\bar{\psi}\,\boldsymbol{b}^e \quad \text{and} \quad \eta = \eta^p - \partial_\Theta \bar{\psi}. \tag{7}$$

Finally, upon setting $\beta := -\partial_\alpha \bar{\psi}$ the the dissipation inequality (6) takes the reduced form

$$\mathcal{D} = \underbrace{\boldsymbol{\tau}\cdot\left[-\tfrac{1}{2}(\pounds_{\boldsymbol{v}}\boldsymbol{b}^e)\boldsymbol{b}^{e-1}\right] + \beta\,\dot{\alpha}}_{\mathcal{D}_{\text{mech}}} + \underbrace{\Theta\,\dot{\eta}^p}_{\mathcal{D}_{\text{ther}}} \geq 0. \tag{8}$$

In view of (8) we conclude that the introduction of the internal variable $\eta^p$ leads to an additional purely thermal contribution to the mechanical dissipation $\mathcal{D}_{\text{mech}}$, given by $\mathcal{D}_{\text{ther}} := \Theta\dot{\eta}^p$ and associated with the entropy production $\dot{\eta}^p$. We remark that in the present context, $\boldsymbol{\tau}$, $\boldsymbol{b}^e$ and $\pounds_{\boldsymbol{v}}\boldsymbol{b}^e$ commute as a result of the restriction to isotropy.

**2.2. Associative Evolution Equations. Maximum Dissipation.** To define suitable evolution equations for the internal variables in the model we shall adopt the simplest assumption corresponding to ***associative plasticity*** and characterized by the property that the ***actual dissipation in the material attains a maximum.*** Accordingly, the actual state $(\boldsymbol{\tau}, \beta, \Theta) \in \mathsf{E}$ in a plastic deformed body, corresponding to a given (fixed) configuration with prescribed intermediate configuration and prescribed rates $\{\pounds_{\boldsymbol{v}}\boldsymbol{b}^e, \dot{\alpha}, \dot{\eta}^p\}$, renders a maximum of the dissipation function (8); i.e.,

$$[\boldsymbol{\tau} - \bar{\boldsymbol{\tau}}]\cdot\left[-\tfrac{1}{2}(\pounds_{\boldsymbol{v}}\boldsymbol{b}^e)\boldsymbol{b}^{e-1}\right] + [\beta - \bar{\beta}]\,\dot{\alpha} + [\Theta - \bar{\Theta}]\,\dot{\eta}^p \geq 0, \tag{9}$$

for all *admissible* $(\bar{\boldsymbol{\tau}}, \bar{\beta}, \bar{\Theta}) \in \mathsf{E}$. A well-known result in convex analysis shows that inequality (9) holds if and only if the coefficients $\{-\frac{1}{2}(\pounds_{\boldsymbol{v}}\boldsymbol{b}^e)\boldsymbol{b}^{e-1}, \dot{\alpha}, \dot{\eta}^p\}$ lie in the cone normal to $\partial\mathsf{E}$ at the point $(\boldsymbol{\tau}, \beta, \Theta)$. In particular, if $\partial\mathsf{E}$ is defined by a smooth yield function as in (1), the evolution equations read

$$-\tfrac{1}{2}\,\pounds_{\boldsymbol{v}}\boldsymbol{b}^e = \lambda\,[\partial_{\boldsymbol{\tau}}\bar{\phi}]\,\boldsymbol{b}^e, \quad \dot{\alpha} = \lambda\,\partial_\beta\bar{\phi}, \quad \dot{\eta}^p = \lambda\,\partial_\Theta\bar{\phi}. \tag{10}$$

where the consistency parameter $\lambda \geq 0$ is the Lagrange multiplier satisfying the Kuhn–Tucker conditions $\lambda \geq 0$, $\phi(\boldsymbol{\tau}, \beta, \Theta) \leq 0$ and $\lambda\,\phi(\boldsymbol{\tau}, \beta, \Theta) = 0$. We remark that for the Mises yield criterion $(10)_1$ coincides with the expression derived in [7,8]. Furthermore, the change in plastic flow predicted by $(10)_1$ is given by

$$d(\log J^p)/dt = \lambda\,\mathrm{tr}\left[\partial_{\boldsymbol{\tau}}\phi\right]. \tag{11}$$

This expression shows that for a pressure–insensitive yield criterion the flow rule (10) predicts the well-known property of isochoric plastic flow.

**2.3. Balance of Energy: The Temperature Evolution Equation.** The thermomechanical response of the solid undergoing plastic deformation is governed by the balance of linear momentum equation and the energy equation, supplemented by the constitutive equations along with appropriate boundary conditions. Balance of mass is enforced at the outset merely by requiring $\partial\rho_0/\partial t = 0$, and balance of angular momentum reduces to the symmetry condition $\boldsymbol{\tau} = \boldsymbol{\tau}^T$ also enforced at the outset. In the present context, it can be shown that the local balance of energy yields the following evolution equation for the absolute temperature:

$$c\,\dot{\Theta} = \left[\mathcal{D}_{\mathrm{mech}} - \mathcal{H}\right] + \left[-J\mathrm{div}[\boldsymbol{q}/J] + \mathcal{R}\right] \quad \text{in} \quad \Omega \times \mathsf{I}, \tag{12}$$

where $\mathcal{R}$ the heat source, $\boldsymbol{q}$ is the ***Kirchhoff heat flux*** and div[·] denotes the divergence in the current configuration. The factor $J$ and a composition implicit in (12) are the result of the Piola–transformation. In equation (12) $c$ is the ***specific heat*** at constant deformation and internal variables per unit of the reference volume and $\mathcal{H}$ is the ***elastic-plastic structural heating***, respectively defined by the general relations:

$$c := -\Theta\partial^2_{\Theta\Theta}\bar{\psi}, \qquad \mathcal{H} := -\Theta\partial_\Theta[\,\boldsymbol{\tau}\cdot\boldsymbol{d} - \mathcal{D}_{\mathrm{mech}}\,]. \tag{13}$$

Finally, consistent with the restriction to isotropy, we complete the model of associative thermoplasticity at finite strains by assuming the classical Fourier's law $\boldsymbol{q} = -k\nabla\Theta$ in $\Omega \times \mathsf{I}$ which ensures satisfaction of the Clausius-Duhem inequality for $k \geq 0$.

The foregoing theory can be immediately specialized to obtain a thermomechanical model of $J_2$–flow theory at finite strains. The elastic domain is given by (1) with the function $\phi$ defined by the Mises yield criterion and the thermomechanical flow rule is given by (10). A particularly simple form of the free energy can be obtained under the assumption of constant heat capacity. We refer to [10] for further details.

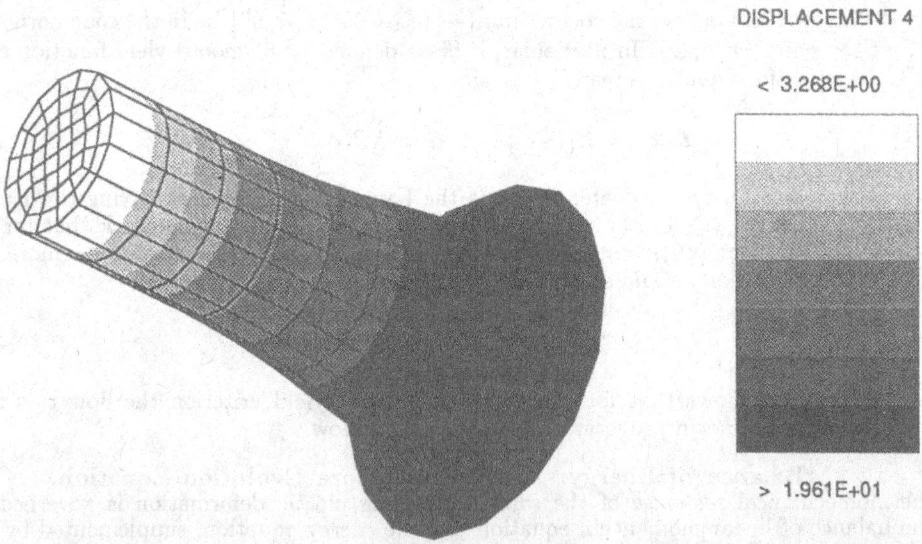


**FIGURE 1.** Dynamic impact of a circular bar on a wall: Temperatures contours. 576 mixed elements, 40 time steps. Isotropic hardening$J_2$ flow theory.

## 3. GLOBAL PRODUCT FORMULA ALGORITHMS

The proposed class of global time stepping algorithm for the IBVP of thermoplasticity is a new ***product formula algorithm*** arising via a systematic exploitation of an ***operator split*** of the nonlinear evolution equations. The method provides a rigorous justification for the solution of the coupled problem in terms of a nonlinear plasticity problem at fixed (initial) temperature or fixed (initial) entropy [Phase 1], and a heat conduction problem at fixed (final) configuration [Phase 2]. In sharp contrast with the lack of symmetry in the original IBVP when formulated in configuration/temperature variables, these two problems are ***symmetric***. Furthermore, it is possible to prove ***unconditional nonlinear stability*** of the resulting scheme. The representative numerical simulation shown in Figure 1 illustrates the performance of both the couple solution strategy and the new class of return mapping algorithms.

## 4. REFERENCES


[1] Argyris, J.H.; Doltsinis, J.St.; Pimenta, P.M.; Wüstenberg, H. [1982], "Thermomechanical Response of Solids at High Strains — Natural Approach", ***Computer Methods in Applied Mechanics and Engineering*** **32**, 3–57.

[2] Asaro, R. [1983],: "Micromechanics of Crystals and Polycrystals" In: ***Advances in Applied Mechanics*** (Ed.: T.Y. Wu, J.W. Hutchinson), **23**, 1–115.

[3] Cottrell, A.H. [1967], "Dislocations and Plastic Flow in Crystals", Oxford University Press, London.

[4] Demengel, F., [1989], "Compactness Theorems for Spaces of Functions with Bounded Derivatives and Applications to Limit Analysis Problems in Plasticity," ***Archive for Rational Mechanics and Analysis,*** **105**, No.2, pp.123-161.

[5] Kato, T. [1974], "On the Trotter-Lie Product Formula," ***Prococeedings Japon Academy,*** **50**, 694-698.

[6] Johnson, C., [1978], "On Plasticity with Hardening," ***Journal of Applied Mathematical Analysis,*** Volume **62**, 325-336.

[7] Simo, J.C. [1988a], "A Framework for Finite Strain Elastoplasticity Based on Maximum Plastic Dissipation and Multiplikative Decomposition: Part I. Continuum Formulation", ***Computer Methods in Applied Mechanics and Engineering*** **66**, 199–219.

[8] Simo, J.C. [1988b], "A Framework for Finite Strain Elastoplasticity Based on Maximum Plastic Dissipation and Multiplicative Decomposition: Part II. Computational Aspects", ***Computer Methods in Applied Mechanics and Engineering*** **68**, 1–31.

[9] Simo, J.C. [1990], "Nonlinear Stability of the Time Discrete Variational Problem in Nonlinear Heat Conduction and Elastoplasticity," ***Computer Methods in Applied Mechanics and Engineering,*** (In press)

[10] Simo, J.C. & Miehe, C. [1991], "Associative Coupled Thermoplasticity at Finite Strains: Formulation and Numerical Analysis, ***Computer Methods in Applied Mechanics and Engineering,*** in press

[11] Simo, J.C.; Armero, F. [1991], "Nonlinear Enhanced Mixed Finite Element Methods and the Method of Incompatible Modes, ***I. J. Numerical Methods in Engineering,*** Submitted for publication.

[12] Truesdell, C.; Noll, W. [1965], "The Nonlinear Field Theories of Mechanics", In: ***Handbuch der Physik Bd. III/3*** (Ed.: S. Fluegge), Springer–Verlag, Berlin

# SHAKEDOWN OF STRUCTURES UNDERGOING LARGE ELASTIC-PLASTIC DEFORMATION

HELMUT STUMPF
Lehrstuhl für Allgemeine Mechanik,
Ruhr-Universität Bochum
W-4630 Bochum, Germany

## ABSTRACT

A fully nonlinear shakedown analysis is considered including physical as well as geometrical nonlinearities. The underlying kinematics of finite elastoplasticity are based on the multiplicative decomposition of the deformation gradient using a local, current relaxed intermediate configuration. It is shown that the notion of self-equilibrated residual stress field in Melan's linear shakedown theorem has to be replaced by the notion of a self-equilibrated residual state. Theorems for path-dependent and path-independent shakedown are presented, which can be applied by an incremental step-by-step procedure.

## INTRODUCTION

In many engineering problems structures are subjected to repeated cyclical loads varying between certain prescribed limits. The statical shakedown theorem of Melan [1] allows to predict for every possible loading history whether or not a linear-elastic-perfectly plastic structure will fail according to unlimited accumulation of irreversible plastic strains or alternating plasticity. To prove shakedown or adaptation of a structure we have to find a fictitious, self-equilibrated, time-independent residual stress field $\sigma_R$ and the stress field $\sigma^e(t)$ of an "elastic comparison body". If the sum of these two stress fields satisfies the yield condition everywhere, then shakedown occurs.

There are only a few papers taking into account geometrical nonlinearities in a shakedown analysis. Here we want to mention the contributions of Weichert [2] and Groß-Weege [3], who investigated small time-dependent deformations superposed upon time-independent moderate deformations.

In this paper we want to consider the aspects, which have to be taken into account in a general nonlinear shakedown analysis. For simplicity thermal effects are neglected.

## CHANGE OF MELAN'S PARADIGMA

In order to formulate the constitutive relations in finite elastoplasticity we have to decompose the total deformation into elastic and plastic parts. One generally accepted concept is to introduce a local, current, relaxed intermediate configuration $\bar{\mathcal{B}}$ allowing to decompose the deformation gradient $\mathbf{F}$ according to

$$\mathbf{F} = \mathbf{F}^e \mathbf{F}^p , \tag{1}$$

where $\mathbf{F}^p$ describes the instanteneous irreversible plastic deformation and $\mathbf{F}^e$ an incompatible elastic deformation.

With (1) an additive decomposition of the Green strain tensor $\mathbf{E}$ and of the Almansi strain tensor $\mathbf{e}$ into purely elastic and purely plastic parts is not possible. Only in the intermediate configuration an additive decomposition can be obtained (see Stumpf [4]):

$$\bar{\mathbf{E}} = \bar{\mathbf{E}}^e + \bar{\mathbf{e}}^p . \tag{2}$$

For an incremental shakedown analysis we have to use objective strain rates and decompose them into elastic and plastic parts. It can be shown that neither in the Eulerian nor in the Lagrangian or intermediate discription an additive decomposition of objective strain rates is possible:

$$\mathbf{d} = \mathbf{d}^e_* + \mathbf{d}^p_* , \quad \bar{\mathbf{d}} = \bar{\mathbf{d}}^e_* + \bar{\mathbf{d}}^p , \quad \mathbf{D} = \mathbf{D}^e_* + \mathbf{D}^p , \tag{3}$$

where an asterisk indicates that the corresponding quantity does not consist of a purely elastic and purely plastic part, respectively.

Following [4] we formulate the elastic response, the flow rule and the hardening law in the intermediate configuration, which enables us to determine the corresponding relations in Eulerian or Lagrangean description by push-forward or pull-back operation.

From these considerations we can conclude that for a general nonlinear shakedown analysis we have to change Melan's paradigma: The notion of self-equilibrated residual stress field must be replaced by the notion of self-equilibrated residual state, for which we have to determine the residual second Piola-Kirchhoff stress field $S_R$ together with the "elastic"and "plastic" parts of the residual Green strain fields $\bar{E}^e_R$ , $E^p_R$ and the residual field of internal plastic variables $Q_R$. It follows immediately that for large deformations in general it will be not possible to find a fictitious residual state. Therefore we have to determine real residual states by an incremental step-by-step procedure.

## PATH-DEPENDENT AND PATH-INDEPENDENT RESIDUAL STATES

For elastic-plastic structures undergoing large deformations the residual state obtained after loading and unloading is in general not only depending on the loading path, but also on the unloading path. Therefore we formulate a path-dependent and a path-independent shakedown theorem.

Let us assume that a structure is subjected to a loading history corresponding to a specific load path in a given load domain: Then we can define path-dependent adaptation as follows:

**THEOREM:** Path-dependent adaptation is obtained for an arbitrary fixed point on a given load path, if there exists one common residual state for all various extremal unloading paths.

With extremal loading/unloading paths we denote those paths, which enclose the load domain with its vertices.

Theorem I does not define adaptation in Melan's sense, because it assures adaptation only for one specific load path. To prove path-independent adaptation we have to consider repeated load cycles enclosing the load domain. Then we can formulate the following theorem:

**THEOREM II:** If for the extremal load cycles enclosing the load domain and if for their inverse cycles the same residual state is obtained the structure shakes down for arbitrary loading paths within the load domain.

## STABILITY BOUNDARIES AND SHAKEDOWN DOMAIN

To determine the shakedown domain of structures undergoing large elastic-plastic deformations we have to take into account that the shakedown domain can be restricted also by the stability boundaries of the structure, where snap-through or bifurcation buckling can occur. In an incremental structural analysis those instability points on the loading/unloading paths can be determined in the usual way.

Theory and application of the nonlinear shakedown analysis for elastic-plastic structures undergoing finite deformations is considered in detail in [4].

## REFERENCES


1. Melan, E., Theorie statisch bestimmter Systeme aus ideal plastischem Baustoff, Sitzber. Akad. Wiss., Wien, IIa, 1936, 145, 195-218.

2. Weichert, D., On the influence of geometrical nonlinearities on the shakedown of elasto-plastic structures, Int. J. Plasticity, 1986, 2, 135-148.

3. Groß-Weege, J., A unified formulation of statical shakedown criteria for geometrically nonlinear problems, Int. J. Plasticity, 1990, 6, 433-447.

4. Stumpf, H., Theoretical and computational aspects in the shakedown analysis of finite elastoplasticity, to be published.

# APPLICATION OF AN ELASTIC-PLASTIC FINITE ELEMENT MODEL FOR THE SIMULATION OF FORMING PROCESSES

A. VAN BAEL*, P. VAN HOUTTE*, E. AERNOUDT*,
I. PILLINGER**, P. HARTLEY**, C.E.N. STURGESS**
*Department of Metallurgy and Materials Engineering,
K.U. Leuven, de Croylaan 2, B-3001 Leuven, Belgium
**School of Manufacturing and Mechanical Engineering, University of Birmingham,
Edgbaston, Birmingham B15 2TT, U.K.

## ABSTRACT

A model for the anisotropic material behaviour of textured materials is currently being incorporated into an elastic-plastic finite-element (FE) program for the simulation of forming processes of such materials. In this paper the developed anisotropic software is tested by simple FE simulations of tensile tests on steel sheet specimens for the prediction of r-values.

## MATERIALS AND METHODS

A cold rolled low carbon steel sheet suited for deep drawing applications is used. R-values have been measured in tensile tests at 0, 45 and 90 degrees to the rolling direction. For an anisotropic FE program based on Hill's yield description these r-values would be needed for the calculation of the yield locus used in FE simulations. In the present approach however it is assumed that the anisotropic plastic behaviour of the steel sheet depends completely on the crystallographic texture of the sheet. This texture is characterised by the C.O.D.F. (crystallographic orientation distribution function) which has been obtained from pole figures measured by X-ray diffraction. In [1] it is explained how polycrystal deformation models such as the Taylor-Bishop-Hill (TBH) theory can be used to predict the anisotropic yield locus of a material with a known C.O.D.F., and the analytical yield locus formulation used here is the series expansion method described in [1]. The principles of the incorporation of this formulation into the elastic-plastic FE program are outlined in [2] and [3].

In the FE simulations only the homogeneously deforming part of a tensile specimen has been considered, and a uniform axial elongation is imposed on the specimen leaving transverse and thickness contractions unconstrained. The width and thickness dimensions of the specimen after each FE increment have been treated as if they were experimental results in order to calculate the plastic strains $\varepsilon_{ij}^{pl}$ and from these the r-value.

## RESULTS

The $\phi_2 = 45$ degrees section in Euler space of the C.O.D.F. of the steel sheet is given in figure 1. This type of texture is commonly observed in rolled steel sheets, and it consists of a strong $\gamma$-fibre with a maximum at the (111) <110> component [4]. Figure 2 shows the $\pi$-plane section of the yield locus derived from this C.O.D.F. with the series expansion method described in [1] using a 6th order expansion. The principal stress directions I, II, III in figure 2 correspond to the rolling, transverse and thickness directions of the sheet respectively. Point A in figure 2 corresponds to the stress state in a tensile test in the rolling direction. Using the series expansion method the plastic strain rate direction $\dot{\varepsilon}^{pl}$ in A is also known [1], and the r-value can then be calculated by

$$r = \frac{\dot{\varepsilon}^{pl}_{22}}{\dot{\varepsilon}^{pl}_{33}} \tag{1}$$

The solid line in figure 3 shows the variation of the r-value for tensile tests at angles between 0 and 90 degrees to the rolling direction derived in this way from the 6th order series expansion. Furthermore r-values can be calculated directly from the C.O.D.F. using a TBH calculation, and these results are given in figure 3 for 7 angles. The difference between TBH and series expansion predictions is small, and probably due to the smoothing effect introduced by using a 6th order expansion for the yield locus. Finally figure 3 also shows the experimental r-values at 0, 45 and 90 degrees. Only at 90 degrees the experimental result deviates from the theoretical predictions.

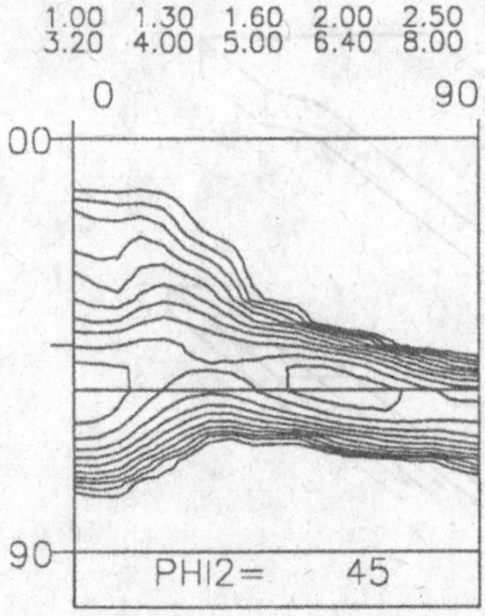


Figure 1.

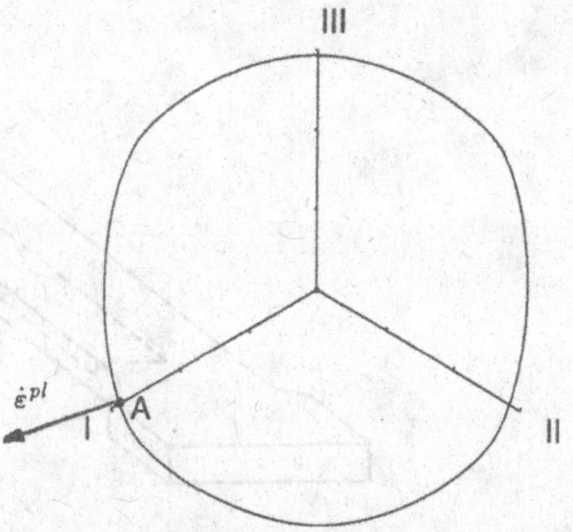


Figure 2.

Figure 1. $\phi_2$-section of the C.O.D.F. of a cold rolled steel sheet.

Figure 2. $\pi$-plane section of the yield locus for the cold rolled steel sheet of figure 1.

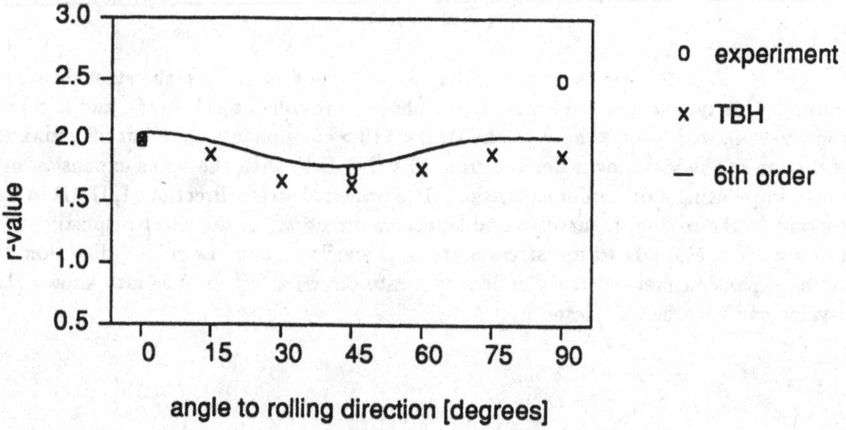


Figure 3. Experimental and predicted r-values.

FE simulations at 0 and at 45 degrees to the rolling direction have been performed using the FE mesh shown in figure 4, and imposing a total of 11 increments of 0.25 % engineering strain each. The derived plastic strains and r-values are shown in figures 5 (a) and (b). It can be seen that the r-values remain practically constant, which is expected because no texture variation is considered during the FE simulation. Furthermore the FE predictions in figure 5 (b) are in good agreement with the r-values obtained directly from the series expansion method in figure 3.

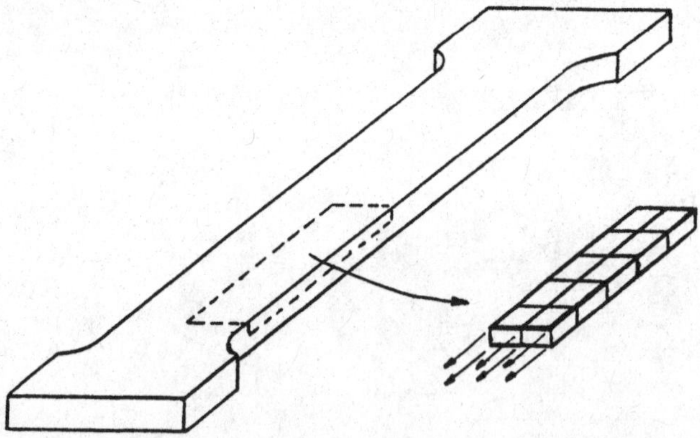

Figure 4. FE mesh consisting of brick-type elements.

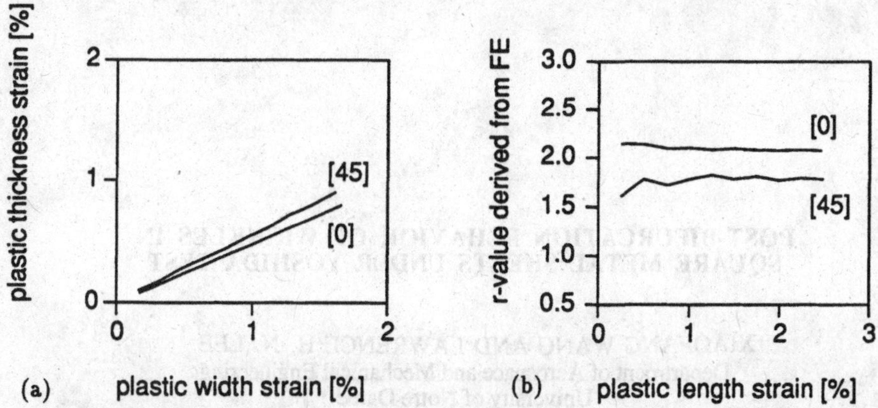


Figure 5. (a) Plastic strains and (b) r-values derived from FE simulations of tension tests at 0 and 45 degrees to the rolling direction.

## CONCLUSION

The simple FE simulations of tensile tests show that the anisotropic behaviour derived from the C.O.D.F. and TBH models can be incorporated into a FE program for the prediction of anisotropic material flow. At present other analytical yield locus representations are being developed which are expected to have numerical advantages compared to the series expansion method [5].

## ACKNOWLEDGEMENT

This work has been carried out as part of the Brite/Euram project BREU⋆ 0107-c funded by the Commission of the European Communities.

## REFERENCES

1. Van Houtte, P., Mols, K., Van Bael, A. and Aernoudt, E., Application of Yield Loci Calculated from Texture Data. Textures and Microstructures, 1989, 11, 23-29
2. Pillinger, I., Hartley, P., Sturgess, C.E.N., Van Bael, A., Van Houtte, P. and Aernoudt, E., Finite-element analysis of anisotropic material deformation. In Suppl. Proc. 3rd Int. Conf. on Numerical Methods in Industrial Forming Processes, Fort Collins, Colorado, 26-30 June 1989, Thompson, E.G., Wood, R.D., Zienkiewicz, O.C. and Samuelsson, A., eds., A.A. Balkema, Rotterdam, 1989, 6 pages
3. Van Bael, A., Van Houtte, P., Aernoudt, E., Hall, F.R., Pillinger, I., Hartley. P., and Sturgess, C.E.N., Anisotropic finite-element analysis of plastic metalforming processes. In Proc. ICOTOM9, Avignon (in print)
4. Van Houtte, P., Deformation textures. In Directional Properties of Materials, Bunge, H.J., ed., DGM Informationsgesellschaft mbH, Overursel, Germany, 1988, 65-76
5. Toth, L.S., Van Houtte, P. and Van Bael, A. Analytical representation of polycrystal yield surfaces. In this proceedings.

# POST-BIFURCATION BEHAVIOR OF WRINKLES IN SQUARE METAL SHEETS UNDER YOSHIDA TEST

XIAOFANG WANG AND LAWRENCE H. N. LEE
Department of Aerospace and Mechanical Engineering
University of Notre Dame
Notre Dame, Indiana 46556, USA

## ABSTRACT

The stretching of a square sheet along one of its diagonals, called "Yoshida Test", has been developed to simulate the wrinkling behavior in press forming of steel sheets into autobody panels. The finite deformation, onset of wrinkling and growth of wrinkles in such a specimen are investigated. Hill's yield criterion for sheet materials have the normal anisotropy and Hill's quasi-static bifurcation criterion are employed. The growth of wrinkles in the finite deformation process is incrementally and numerically determined by a thin shell finite element in a convected coordinate system. The Lagrangian formulation of the thin shell finite element is based on Hill's variational principle for elastic-plastic solids, a modification of Love-Kirchhoff postulates and a quasi-conforming element technique. The shell element fulfills the inter-element $C^1$ continuity condition in a variational sense. Reasonable agreements between the present numerical results and available analytical and experimental results are shown.

## INTRODUCTION

In sheet metal forming, wrinkling is a common mode of failure which is unacceptable in the finished product for functional or aesthetic reasons. Wrinkling may occur where a sheet metal is unevenly stretched. The behavior of wrinkles after their initiation is called here the post-bifurcation behavior. As a simulation test of wrinkling occurred in forming autobody panels, a simple test called Yoshida test has been developed by Yoshida [1]. The test involves the stretching of a square sheet along one of its diagonals. A standard test piece of 100mm square is specified with a gripping width, at the corners, of 40mm, a gauge of 75mm and with the wrinkle height at the center measured across a span of 25mm as shown in Figure 1. The Yoshida test is to provide a reference of the wrinkling-resistant properties of various sheet metals. Several numerical and/or experimental investigations have been conducted to correlate Yoshida test results with material properties. However, experimental difficulties concerning the design of grip forces and gripping influence the test results [2].

The nonlinear behavior of wrinkles in a Yoshida test piece has been numerically investigated by Tomita and Shindo [3]. They employed a corner type constitutive relationship and a finite element method with an isoparametric shell element based on an updated Lagrangian formulation to trace the growth of wrinkles. However, a survey of yield surface experimentations [4] suggests that, while a yield surface with relatively high curvature at the loading point is often observed, sharp corners are seldom seen. It has been shown by Mellor [5] that Hill's yield criterion for sheet materials of normal anisotropy lead to a reasonable

* To whom all correspondence should be addressed.

description of material response. Therefore, Hill's yield criterion is employed in this investigation.

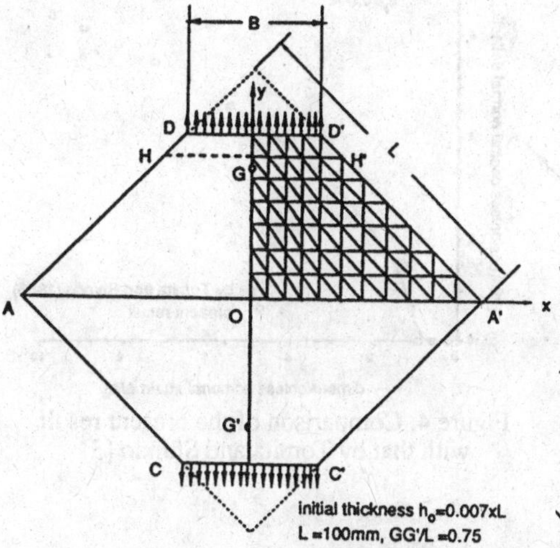


Figure 1. Thin square sheet subject to diagonal tension.

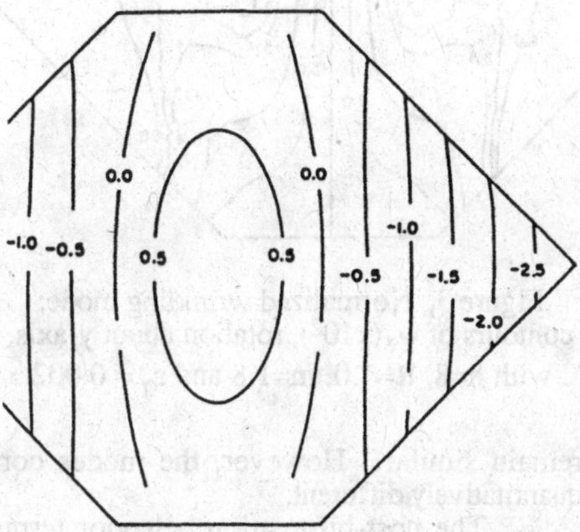


Figure 2. Normalized wrinkling mode: contours of normal deflection w (mm) with n=8, R=1.0, m=1.8 and $\varepsilon\gamma$=0.002.

In this paper, the post-bifurcation behavior of wrinkles in a square sheet under Yoshida test is incrementally and numerically determined by a thin shell finite element technique. It is based on a nonlinear shell theory in a convected Lagrangian formulation [6] which treats the finite deformation of an elastic-plastic sheet metal including finite stretching, rotation, bending and change in thickness. The nonlinear shell theory is based on the principle of virtual work, Hill's variational principle [7] for elastic-plastic solids, and a modification of Love-Kirchhoff postulates. In formulating the present thin shell finite element [8], a quasi-conforming technique [9] of a multivariate concept consistent with Hu-Washizu principle is employed. The shell element fulfills the inter-element $C^1$ continuity condition in a variational sense. Numerical results and their comparison with available numerical and experimental results are presented.

## NUMERICAL RESULTS

Figure 1 shows a square sheet stretched along a diagonal. The truncated edges D-D' and C-C' of width B, where stretching forces are applied, are considered as clamped without lateral displacement and rotations along x and y axes. In the following description e represents the nominal strain defined in Figure 1. It is evaluated in the span of 75mm along the line of symmetry G-G'. $e_w$ denotes the value of e at the onset of wrinkling and is referred as critical nominal strain. Throughout the investigation described below, constant values of E=206 Gpa, ν=0.333, L=100mm and $h_o$=0.7mm are used.

The normalized wrinkling modes at $e=e_w$, with the maximum value of normal displacement of 1.0 at the center are displayed in terms of contours of normal displacement w, and slopes $w_x$ and $w_y$ in Figures 2 and 3. These contours are very similar to those presented by Tomita and Shindo [3]. The effects of material properties on the wrinkling mode have been examined. It appears that the general patterns of the modes of various material constants

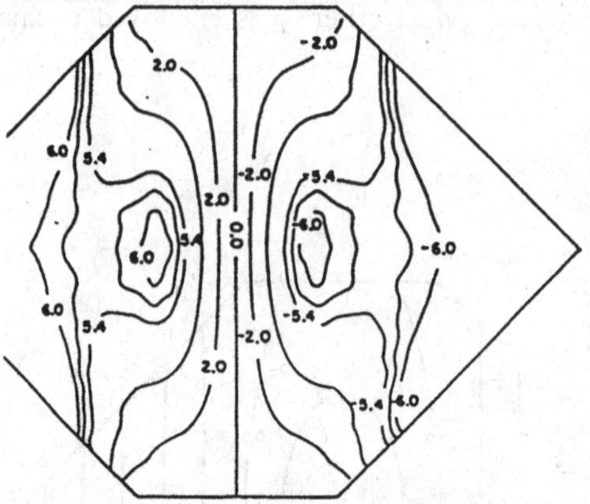


Figure 3. Normalized wrinkling mode: contours of $w_x(x10^2)$, rotation about y-axis, with n=8, R=1.0, m=1.8 and $\varepsilon_\gamma = 0.002$

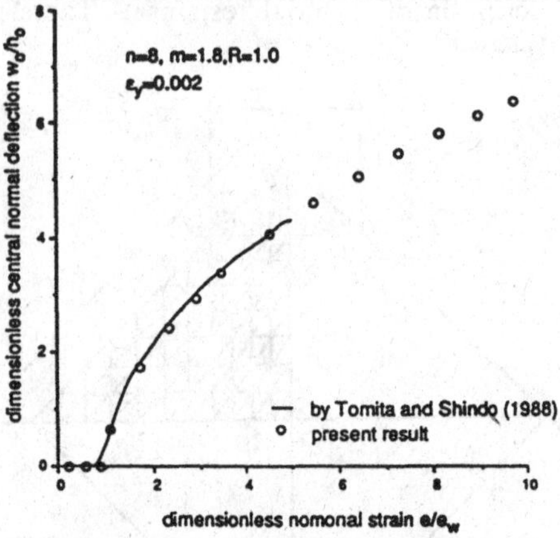


Figure 4. Comparison of the present result with that by Tomita and Shindo [3]

remain similar. However, the modes corresponding to different material constants are quantitatively different.

The post-bifurcation behavior terms of the histories of displacements, strains and stresses at each nodal point, has been investigated by the present approach. However, for a clear view, only the relationship between the central normal deflection $w_o$ and the normal strain e is illustrated in Figure 4. Here, the material constants are n=8, m=1.8, R=1.0 and $\varepsilon_y$=0.002.

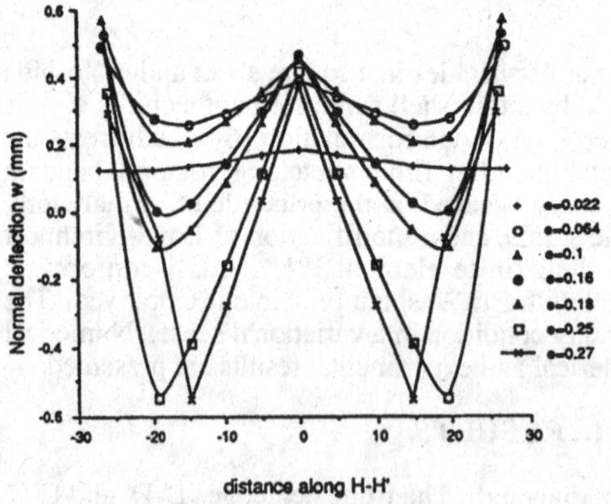


Figure 5. Development of wrinkling.

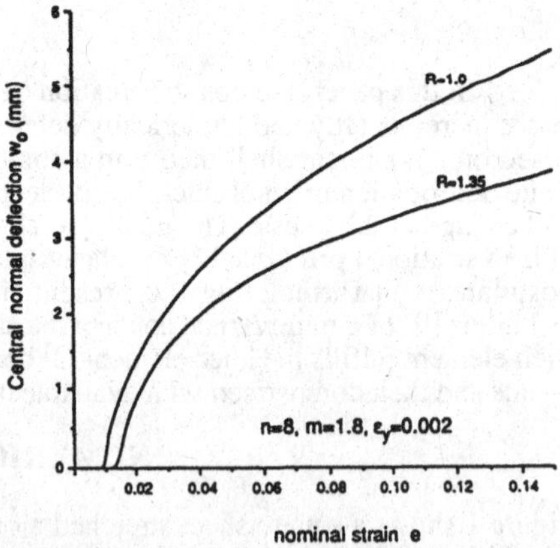


Figure 6. Effect of anisotropy parameter R on the post-bifurcation deformation

A comparison of the present result with that by Tomita and Shindo [3] is shown. A good agreement is found. It should be pointed out that the material planar orthotropy is considered by Tomita and Shindo while it is not in the present study.

The available experimental result by Lake [2] and others employed commercial aluminum-stabilized steels with $\varepsilon_y$ in the range of 0.0009 to 0.0014, thickness $h_o$ of 0.6 to 1.0mm, R of 1.0 to 2.0 and n of 3 to 6. At the deformation stage of e=0.04, an average central wrinkle height from the present numerical output is 2.7mm, while the experimental wrinkle height varies from 3.0mm to 4.2mm. The discrepancies may be attributed to the possible slippage of material from the outer regions of the grip due to uneven plastic deformation.

The post-bifurcation behavior of the case (with R=1.0, m=1.8, n=8 and $\varepsilon_y$=0.002) is further illustrated by profiles of normal deflection w along line H-H' (marked in Figure 1) as shown in Figure 5. It appears that while the center of the sheet is raised by the stretching action, some other regions are restrained by the nearby clamped edge. Figure 6 depicts the effects of anisotropy parameter R on $w_o$ vs. e curve. The curve with R=1.35 is below the curve of the isotropic case of R=1.0. It seems that the large value of R tends to lessen the wrinkling process as stretching progresses.

## CONCLUDING REMARKS

An incremental finite element procedure is developed for the determination of the wrinkling behavior of an initially flat sheet metal subject to finite deformation. The procedure is illustrated by an investigation on the post-bifurcation behavior of wrinkles in square metal sheets under Yoshida test. The finite deformation, onset of wrinkling and growth of wrinkles in such a specimen are investigated. Hill's yield criterion for sheet metals having the normal anisotropy is employed. The effects of material properties on the wrinkling behavior are studied. Reasonable agreements between the present numerical results and available analytical and experimental results are shown. However, a conclusive comparison is yet to be reached by further research on the constitutive modelling of anisotropic plasticity and more precise experimental verifications of sheet metals subject to finite deformation.

**Acknowledgment**
The support of this work by the National Science Foundation under Materials Engineering and Processing Program through Grant MSM-8805748 to the University of Notre Dame is gratefully acknowledged.

## REFERENCES

1. Yoshida, K., Aim and character of Yoshida Buckling Test (YBT). J. Jap. Soc. Techn. Plasticity, 1983, **24**, 901.

2. Lake, J.S.H., The Yoshida Test - A critical evaluation and correlation with low-strain tensile parameters. Proc. of 13th Congr. IDDRG, Melbourne, Australia, 1984, pp.554-564.

3. Tomita, Y. and Shindo, A., Onset and growth of wrinkles in thin square plates subjected to diagonal tension. Int. J. of Mech. Sci., 1988, **30**, pp. 921-931.

4. Hecker, S.S., Yield surfaces in prestrained aluminum and copper. Metallurgical Transactions Vol. 2, 1971, pp.2077-2086.

5. Mellor, P.B., Experimental studies of plastic anisotropy in sheet metal. "Mechanics of Solids" - The Rodney Hill 60th Anniversary Volume. Eds. H.G. Hopkins and M.J. Sewell, Pergamon Press, New York, 1982, pp. 383-415.

6. Lee, L.H.N., Wrinkling in sheet metal forming. Interdisciplinary Issues in Materials Processing and Manufacturing Vol. 2, Eds. Samanta, S.K., Komanduri, R., McMeeking, R., Chen, M.M., Tseng, A., ASME, New York, 1987, pp. 419-436.

7. Hill, R., A general theory of uniqueness and stability of elastic-plastic solids. J. Mech. Phys. Solids, 1958, **6**, pp.236-249.

8. Wang, X., Jiang, H.Y., and Lee, L.H.N., Finite deformation formulation of a shell element for problems of sheet metal forming. To appear in J. Computational Mechanics.

9. Tang, L.M., Chen, J, and Liu, Y.S., Formulation of quasi-conforming element and Hu-Washizu principle. Comput. Struct., 1984, **19**, pp.247-250.

# INELASTIC ANISOTROPIC SHELLS AT MODERATE ROTATIONS

D. WEICHERT and R. SCHMIDT*
Université de Lille 1-LML (EUDIL)
F-59655 Villeneuve d'Ascq Cedex (France)
*Bergische Universität Wuppertal, FG Baumechanik
Pauluskirchstr.7, D-5600 Wuppertal 2 (F.R.G.)

## ABSTRACT

In this paper, a rate theory for shells is presented allowing for elastic-plastic material behaviour including initial elastic and plastic anisotropy, kinematic and isotropic hardening and a simple model of material damage. The problem is treated in the framework of the first-order shear approximation theory and geometrical nonlinearity is taken into account by using strain- and rotation based order estimates assuming that the shell material elements can undergo moderate rotations.

## INTRODUCTION

The use of strongly anisotropic materials in structural engineering requires more adequate methods of analysis in the elastic as well as in the inelastic range of the material response. So it is well known that even for thin elastic shells strong anisotropy may render Kirchhoff-Love-type theories inadequate for the prediction of the structural behaviour. As a very popular material composite laminates are of particular interest; if finite deflections in shell-like structures are admitted, strong anisotropic effects may be met in the elastic as well in the plastic range of deformation, accompanied eventually by material damage effects.
The question how to take geometrical nonlinearities correctly into account, has been discussed by many authors, in particular as far as purely elastic processes are concerned. Also the problem of elastic plastic shells has been studied, mainly in connection with the problem of ultimate loads, but also rate formulations related to the initial boundary value problem have been given. For extended reviews we refer to (1-4).

In this paper a rate-type theory is for shells is presented which admits anisotropic elastic-plastic material behaviour for combined kinematical and isotropic hardening. Furthermore, a simple model for material damage is adapted to the presented theory. The theory does not require that the thickness h of the shell is small. The problem is treated in the framework of the so-called first order shear approximation theory and geometrical non-linearity is taken into account by using consistent strain- and rotation-based order estimates. We assume that the strains remain small everywhere in the shell, but that the shell material elements can undergo moderate rotations.

## NOTATIONS AND BASIC RELATIONS

Starting point of the theory is the representation of the Green's strain tensor E by linearized strains $\eta$ and linearized rotations $\Omega$ (see (1)) with

$$\eta_{ij} = 1/2\,(V_{i;j} + V_{j;i}),\quad \Omega_{ij} = 1/2\,(V_{i;j} - V_{j;i}),$$
$$E_{ij} = \eta_{ij} + \Omega_{ri}\Omega^{r}_{.j}\;\; 1/2(\eta_{ri}\Omega^{r}_{.j} + \eta_{rj}\Omega^{r}_{.i}) + \eta_{ri}\,\eta^{r}_{.j}\,, \tag{1}$$

with V as displacement vector of an arbitrary point $\mathcal{P}$ of the shell body. Covariant differentiation with respect to the metric of the undeformed shell space is defined by $(.)_{;}$ with Latin indices running from 1 to 3. Summation convention over repeted indices is adopted.

As physical basis of the approximation we assume that (a) strains remain small everywhere in the shell body ($E_{ij} \ll \mathcal{O}(\theta^2)$, where $\theta^2 \ll 1$) (b) in-surface rotations of the shell elements are small and (c) that rotations of the normals to the midsurface $M$ of the shell are moderately large i.e. that they are larger compared to the strains but of magnitude such that their squares and products are small as compared to unity (1). Projection of all quantities on the midsurface $M$ with the help of the shifter tensor $\mu$ (1) and omission of all terms of order $\theta^4$ and higher then leads to the following 2D strain-displacement relations

$$\overset{(0)}{E}_{\alpha\beta} = \overset{(0)}{\theta}_{\alpha\beta} + 1/2\,\overset{(0)}{\varphi}_{\alpha}\overset{(0)}{\varphi}_{\beta} + \mathcal{O}(\theta^4)$$

$$\overset{(1)}{E}_{\alpha\beta} = \overset{(1)}{\theta}_{\alpha\beta} - 1/2(b^{\lambda}_{\alpha}\overset{(0)}{\varphi}_{\lambda\beta} + b^{\lambda}_{\beta}\overset{(0)}{\varphi}_{\lambda\alpha}) + 1/2(\overset{(0)}{\varphi}_{\alpha}\overset{(1)}{\varphi}_{\beta} + \overset{(1)}{\varphi}_{\alpha}\overset{(0)}{\varphi}_{\beta}) + \mathcal{O}(\theta^4/h)$$

$$\overset{(2)}{E}_{\alpha\beta} = -\,1/2(b^{\lambda}_{\alpha}\overset{(1)}{\varphi}_{\lambda\beta} + b^{\lambda}_{\beta}\overset{(1)}{\varphi}_{\lambda\alpha}) + 1/2\,\overset{(1)}{\varphi}_{\alpha}\overset{(1)}{\varphi}_{\beta} + \mathcal{O}(\theta^4/h^2)$$

$$\overset{(0)}{E}_{\alpha 3} = 1/2(\overset{(0)}{\varphi}_{\alpha} + \overset{(1)}{v}_{\alpha}) + 1/2\,v^{\lambda}\,\overset{(0)(1)}{\varphi}_{\lambda\alpha} + 1/4\,\overset{(1)}{v}_{3}(\overset{(0)}{\varphi}_{\alpha} - \overset{(1)}{v}_{\alpha}) + \mathcal{O}(\theta^4/h)$$

$$\overset{(1)}{E}_{\alpha 3} = 1/2\ \overset{(1)}{v}_{3,\alpha} + 1/2\ \overset{(1)}{v}{}^{\lambda}\ \overset{(1)}{v}_{\lambda|\alpha} - 1/4\ \overset{(1)}{v}_{3,\lambda}\ \overset{(0)}{\varphi}{}^{\lambda}_{.\alpha} + O(\theta^4/h)$$

$$\overset{(0)}{E}_{\alpha 3} = \overset{(1)}{v}_{3} + 1/2\ \overset{(1)}{v}{}^{\lambda}\ \overset{(1)}{v}_{\lambda} + O(\theta^4/h^2)\ , \tag{2}$$

with $v_i$ as displacement vector of points on $M$, $b_{\alpha\beta}$ as curvature, $\theta_{\alpha\beta}$, $\varphi_{\alpha}$ and $\varphi_{\alpha\beta}$ represent membrane strain, rotations about tangents and about normals to $M$, respectively. Vertical stroke denotes covariant differentiation with respect to $M$. Superposed numbers define the order of the considered term and greek indices run from 1 to 2. Details on these definitions and the approximation method are found in (1).

## INCREMENTAL VARIATIONAL FORMULATION

Starting point is the 3D rate variational functional

$$I(\dot{V}) = \int_V (\tfrac{1}{2}\dot{s}^{ij}(\dot{V})\dot{E}_{ij}(\dot{V}) + \tfrac{1}{2}s^{ij}\ddot{E}_{ij}(\dot{V}) - \rho\dot{F}^i\dot{V}_i)dV - \int_A {}^*\dot{t}^k\dot{V}_k dA\ , \tag{3}$$

given in (5), where the velocities $\dot{V}$ are the independent variables subject to variation. Here $\ddot{E}$ denotes the second time derivative for quasi-static processes (e.g. for $\ddot{V}=0$) of E and $\rho F$, ${}^*t$ are the body force vector and prescribed tractions on the boundary $A$ of the body, respectively. Using the time derivatives (denoted by superposed dot) of the strain tensor components of eqns. (2) and appropriately defined representatives of the stress tensor, the traction vector and traction couple on $M$ and the shell boundary $C$, respectively (ref. (1)), one arrives at the 2D-variational functional of the rate problem for the presented shell theory

$$I(\dot{v}) = \int_M (\tfrac{1}{2}(\sum_{n=0}^{2} \overset{n}{\dot{L}}{}^{\alpha\beta}\ \overset{n}{\dot{E}}_{\alpha\beta} + 2\sum_{n=0}^{1} \overset{n}{\dot{L}}{}^{\alpha 3}\ \overset{n}{\dot{E}}_{\alpha 3} + \overset{0}{\dot{L}}{}^{33}\ \overset{0}{\dot{E}}_{33}) +$$

$$\tfrac{1}{2}(\sum_{n=0}^{2} \overset{n}{L}{}^{\alpha\beta}\ \overset{n}{\ddot{E}}_{\alpha\beta} + 2\sum_{n=0}^{1} \overset{n}{L}{}^{\alpha 3}\ \overset{n}{\ddot{E}}_{\alpha 3} + \overset{0}{L}{}^{33}\ \overset{0}{\ddot{E}}_{33}) +$$

$$-\sum_{n=0}^{1} ((\overset{n}{\dot{B}}{}^{\alpha} + \overset{n}{\dot{p}}{}^{\alpha})\overset{n}{\dot{v}}_{\alpha} + (\overset{n}{\dot{B}}{}^{3} + \overset{n}{\dot{p}}{}^{3})\overset{n}{\dot{v}}_{3})\ dM -$$

$$\int_{C_s} (\sum_{n=0}^{1} ({}^*\overset{n}{\dot{L}}{}^{\alpha\beta}\nu_{\alpha}\nu_{\beta}\overset{n}{\dot{v}}_{\nu} + {}^*\overset{n}{\dot{L}}{}^{\alpha\beta}t_{\alpha}\nu_{\beta}\overset{n}{\dot{v}}_{t} + {}^*\overset{n}{\dot{L}}{}^{3\beta}\nu_{\beta}\overset{n}{\dot{v}}_{3}))\ ds,$$

where L, p, and B are stress resultants, surface couples and body couples, resp., whilst $\nu$ and t denote normal and tangent unit vectors on the shell boundary $C$ composed of $C_s$ and $C_k$, where statical and kinematical conditions are prescribed, respecively. The velocity rates $\dot{v}$ have to satisfy the kinematic boudary conditions on $C$ . Prescribed quantities are indicated by asterix.

## CONSTITUTIVE RELATIONS

In principle the presented theory is valid for any rate-type material allowing to express the rates of the stress tensor components as explicit functions of the rates of the components of the Green's strain tensor. Here, the shell is considered to be composed of a finite number N of layers of constant thickness and the state of stress in each layer is assumed to be constant across its thickness. This approach is chosen because inelastic behaviour of shell-like bodies is essentially related to their three-dimensional nature although in an approximate manner 2D-constitutive laws can be applied sucessfully. As example, orthotropic elastic behaviour is considered. Anisotropic damage is taken into account according to the suggestions of Lemaitre and Chaboche (6). As far as the plastic behaviour is concerned, Hill's yield criterion with kinematic and isotropic hardening and associated flow rule is considered as extension of the approaches presented in (1,2).

## REFERENCES

1. Schmidt, R. and Weichert, D., A refined theory of elastic plastic shells at moderate rotations. ZAMM, 1989, 69, 1, 11-21.

2. Basar, Y. and Weichert, D., A finite-rotation theory for elastic-plastic shells under consideration of shear deformations. ZAMM, (in print), 70, 7.

3. Pietraszkiewicz, W., Geometrically nonlinear theories of thin elastic shells. Advances in Mechanics, 1989, 12, 1, 51-130.

4. Sawzuk, A., On plastic shell theories at large strains and displacements. Int.J.Mech.Sci., 1982, 24, 231-44.

5. Neale, K.W., A general variational theorem for the rate problem in elasto-plasticity. Int.J.Solids Struct., 1972, 8, 865-76.

6. Lemaitre, J. and Chaboche, J.L., Mécanique des matériaux solides, Dunod, Paris, 1988.

# SIMPLIFIED ANALYSIS OF ELASTOPLASTIC STRUCTURES

JOSEPH ŻARKA, JOEL FRELAT, SAHBI BRAHAM
Laboratoire de Mécanique des Solides
Ecole Polytechnique, PALAISEAU, France

## ABSTRACT

Some numerical analyses of elastoplastic structures requires enormous computational efforts when dealing with classical methods.
To adress these limitations a simplified analysis of these structures was developed by J. Zarka and coll.. Approximated evaluations of limit state or stabilized state of such structures under statical or dynamical, monotoneous or cyclic loadings, are obtained from a few elastic computations.
This approch enables a simple modelization of critical problems such as elastoplastic contact. We illustate this abablity by a presentation of modelizations of shot-peening, rolling and roller-burnishing.

## I. INTRODUCTION

Many elastoplastic calculations of structures need a large amount of memory and computer time. New generations of computers allow very complicated calculations, but cost remains too expensive.
In order to calculate elastoplastic structures under complicated loadings, J. Zarka has proposed a new simplified method. One of the best results was obtained for calculation of some classical mechanical surface treatments: shot-peennig, rolling, and roller-burnishing. The loading path of such treatments are very complicated, but they are modelized as cyclic loadings, field for which this new approach can be successfully applied.
In the first part, we shortly recall the principles of the simplified analysis of inelastic structures, for the linear kinematic hardening material.
In the second part, we present the principles of the modelization of shot-peening and rolling.

## II. SIMPLIFIED ANALYSIS OF ELASTOPLASTIC STRUCTURES

In order to present shortly the new approach, we consider only structures made of elastoplastic materials with linear kinematic hardening.

### II.1 Elastic response

We consider a structure represented by a finite volume V with its boundary $\partial V$. It is subjected to: surface forces $F_i^d$ (t) on $\partial_{F_i}$ V, body forces $\underline{X}^d$ (t) in V, initial strain $\underline{\underline{E}}^I$ (t) in V, displacements $U_j^d$ (t) on $\partial_{U_j}$ $V^i$.

We assume the structure is made of a purely linear elastic material. We denote: $\underline{\Sigma}^{el}$ (t) the stress field, $\overrightarrow{\underline{U}}^{el}$ (t) the displacement field, $\underline{E}^{el}$ (t) the strain field.

**II.2 Actual response and inelastic fields**

In the real structure we have:

$\overrightarrow{\underline{U}}$ , $\underline{E}$ (t) , $\underline{E}^{P}$ (t) real displacement and total strain and plastic strain fields,

$\underline{\Sigma}$ (t) real stress field, $\underline{E}$ (t) = $\underline{M}$ $\underline{\Sigma}$ (t) + $\underline{E}^{I}$ (t) + $\underline{E}^{P}$ (t)

We denote $\underline{R}$ (t) the residual stress field. We may decompose the fields into:

$\underline{R}$ (t) = $\underline{\Sigma}$ (t) -$\underline{\Sigma}^{el}$ (t) statically admissible with $\underline{0}$ in V and 0 on $\partial_{F_i}$ V.

$\overrightarrow{\underline{U}}^{ine}$ (t) = $\overrightarrow{\underline{U}}$ (t) -$\overrightarrow{\underline{U}}^{el}$ (t), $\underline{E}^{ine}$ (t) = $\underline{E}$ (t) -$\underline{E}^{el}$ (t) kinematically admissible with 0 on $\partial_{U_j}$ V.

We can compute easily these inelastic fields using elasticity, with body forces $\underline{0}$ in V , surface forces 0 on $\partial_{F_i}$ V, displacement 0 on $\partial_{U_j}$ V, and initial strain $\underline{E}^{P}$ (t) in V.

**II.3 Linear kinematically hardening material and transformed parameters for the structure**

Very often a kinematically hardening material is associated with the Mises criterion, represented by a sphere in the deviatoric stress space :

$$f ( \underline{S} , \underline{E}^{P} ) = 1/2 ( \underline{S} - C \underline{E}^{P} )^{T} (\underline{S} - C \underline{E}^{P}) - \sigma_y^2 \le 0$$

with $\sigma_y$ , radius of the sphere, is the elastic limit in shear, C is the hardening modulus, $\underline{S}$ = dev $\underline{\Sigma}$ is the stress deviator

The evolution low, ie plastic flow, obeys to the rule of normality.

We denote x=α=$E^{P}$ the internal parameters, we define the transformed parameters in the local stress space y = C$E^{P}$.

We introduce the transformed parameters for the structure, $\underline{Y}$ = C $\underline{E}^{P}$ - dev$\underline{R}$

The Mises criterion can be written, using Y: 1/2 ($\underline{S}^{el}$ - $\underline{Y}$)$^{T}$ ($\underline{S}^{el}$ - $\underline{Y}$) ≤ $\sigma_y^2$ .

These transformed parameters for the structure are dependent on the geometry of the structure and its particular surfaces $\partial_{F_i}$ V and $\partial_{U_j}$ V, and must be "plastically" admissible for each time of the loading path. When all the convex set is locally constructed, using the purely elastic response, we are able to consider each point of the structure independenly and to construct the transformed parameters Y.

**II.4 Inverse operation**

Now, it is necessary to come back to the classical internal parameters ie, for this material, the plastic strain.

The inelastic strain field can be written:

$\underline{E}^{ine}$ = ($\underline{M}$ + $C^{-1}$) R + $C^{-1}$ $\underline{Y}$ or $\underline{E}^{ine}$ = $\underline{M}'$ R + $C^{-1}$ $\underline{Y}$

where $\underline{M}'$ is the elastic matrix : $\underline{M}'$ = ($\underline{M}$ + dev $C^{-1}$)

If we assume that the material is elastically isotropic with E Young modulus and ν Poisson ratio, the new elastic matix is isotropic too with the new coefficients E′ and ν′:

ν′/E′ = (ν/E + 1/3C) (1 + ν′)/E′ = ((1 + ν))/E + 1/C

If the field $\underline{Y}$ is determined, we shall obtain, using the elastic analysis with the modified coefficients,

$\overrightarrow{\underline{U}}^{ine}$, $\underline{E}^{ine}$, then, $\underline{R}$ = $\underline{L}'$ ($\underline{E}^{ine}$ - $C^{-1}$ $\underline{Y}$), and, at last, $\underline{E}^{P}$ = $C^{-1}$ [$\underline{M}$ + dev $\underline{R}$]

## III MODELIZATION OF SUCCESSIVE CONTACT PROBLEMS

Various technics are used to improve the fatigue life of a structure, introducing "favourable residual stresses". Among them, those using contact, such rolling, shot-peening, or roller-burnishing, are not easily modelized using classical elastoplastic analysis. Two main problems must be solved by a double numerical iteration: the determination of the area of contact, and the plastic strain field. This would involve expensive calculations.

The three important industrial problems can be modelized using the present framework. The following hypothesis are necessary:
- we consider several repetitive loadings,
- the loading is applied either a great number of times along a fixed surface (rolling or roller-burnishing), or at all points of the external surface (shot-peening),
- a limiting state is obtained.

We shall limit this paper to the case of a very simple geometry of the structure , the semi-infinite body bounded by a plane, and the linear kinematically hardening material.

### III.1 Hypothesis

i) The semi-infinite body occupies the halfspace $z>=0$. The elastic constants are denoted $Y_G$ (Young modulus) and $\nu$ (Poisson ratio), the plastic constants are $\sigma_y$ (shear elastic limit) and C (hardening modulus).

ii) The elastic loading is given by the Hertz theory, with particular case of cylindrical contact for rolling, spherical contact for shot-peening and ellipsoidal contact for roller-burnishing.

iii) The residual state is obtained.

### III.2 Form of the residual fields

***III.2.1 Rolling***. We denotes Oxyz the cartesian coordinates. The cylinder $x^2+(z-R)^2<=0$ is infinite along y, pressed upon the body by a normal force F per unit of length. The residual state obeys to plane strain(Oyz). For this particular geometry, considering that the components of the residual stress and plastic strain field depends only on z, and integrating the equilibrium equations, we obtain $\underline{U}^{ine} = (0,0,W(z))$

The non-zero components of $\underline{E}^{ine}$ (z) are $E_{xz}$, $E_{yz}$, $E_{zz}$. The non-zero components of $\underline{R}$ (z) are $R_{xx}$, $R_{yy}$, $R_{xy}$ .

These equations lead to the following explicit relations beetween the main components of $\underline{R}$, $\underline{E}^p$ and $\underline{E}^{ine}$ :

$$( R_{xx} -\nu R_{yy} ) / Y_G + E^p_{xx} = 0 \; ; \; ( R_{yy} -\nu R_{xx} ) / Y_G + E^p_{yy} = 0$$

$$\nu( R_{xx} + R_{yy} ) / Y_G + E^p_{zz} = E^{ine}_{zz} \; ; \; (1+\nu) E^p_{xy} / Y_G + E^p_{xy}$$

Consequently, the non-zero components of the transformed parameters for the structure, which depends only on z, must be: $Y_{xx}$ , $Y_{yy}$ , $Y_{zz}$ , $Y_{xy}$ .

Using the definition of Y, we can write:

$$Y_{xx} = C\, E^p_{xx} - (2R_{xx}-R_{yy}) / 3 \; ; \; Y_{yy} = C\, E^p_{yy} - (2R_{yy}-R_{xx}) / 3$$

$$Y_{zz} = - (Y_{xx}+Y_{yy}) \; ; \; Y_{xz} = C\, E^p_{xz} \; ; \; Y_{xy} = - R_{xy}$$

Using the previous relations, we have also the relations:

$$( 1/Y_G + 2C/3) R_{xx} - (\nu/Y_G +C/3) R_{yy} + C^{-1} Y_{xx} = 0$$

$$( 1/Y_G + 2C/3) R_{yy} - (\nu/Y_G +C/3) R_{xx} + C^{-1} Y_{yy} = 0$$

***III.2.2 Shot-peening***. This technics consists by making normal projections of sphere which radius $R_a$ , with an impact speed V all over the plane Oxy.

The two coordinates x and y are equivalent. The residual state depends only on z, the non-zero components of $\underline{E}^p$ are $E^p_{xx} = E^p_{yy} = -0.5E^p_{zz} = E^p$,

the non-zero components of $\underline{R}$ are $R_{xx} = R_{yy} = R = -Y_G\, E^p /(1-\nu)$,

and the non-zero components of $\underline{Y}$ are

$$Y_{xx} = Y_{yy} = -0.5Y_{zz} = Y = R/3 +CE^p = (C+Y_G /3(1-\nu))E^p .$$

***III.2.3 Roller-burnishing***. For this case, the problem cannot be solve so easily. The technic consists to roll linearly with the direction Ox an ellipsoidal body pressed upon the semi-infinite body with a normal force F, and then the residual state depends on y and z. Explicit relations cannot be written, we have to solve all the equations numerically.

### III.3 Loading path: characteristic points

Either elastic or plastic shakedown may occur. The loading path is not radial. However, we are able to simplify it in order to build the transformed parameters for the residual state.

i) Case of elastic shakedown

Any new loading do not create new plastic strain: the response is purely elastic, then the stress field is given analytically using the Hertz theory of contact, and for shot-peening, the elastic impact formulaes.

For the rolling (or roller-burnishing) the residual state is independent of x. We suppose that the loading path is sufficiently characterized by the following two positions of the cylinder (or the roller) :

- the minimum loading is $\underline{\Sigma}^{el} = \underline{\Sigma}^{min} = \underline{0}$
- the maximum loading is obtained for y=0, ie, $\underline{\Sigma}^{max} = \Sigma(x,y=0,z)$

For the shot-peening, the contact area varies during the impact: the part $\partial_{F_i}$ V is inot fixed. However we adopt for the maximum loading the one reached for the maximum penetration of the bullet.

ii) Case of plastic shakedown

Now, for each new loading, the plastic strain field has a cyclic evolution: the loading path is not elastic. However we suppose that the previous characterization can be adopted.

### III.4 Local determination of the transformed parameters

We are now able to determine $\underline{Y}$ in the following way. Consider a particular depth z ( or, for roller-burnishing, a particular point y,z ). In the Y-space, we observe the positions of the two convex of elasticity $C^{min}$ and $C^{max}$ centered at $S^{min}$ and $S^{max}$. We suppose that $Y_0$, the initial state is equal to zero. Three cases may occur:

i) $C^{min}$ and $C^{max}$ have a common intersection, and $Y_0$ is inside it; then , at this point $E^p$ is equal to zero;

ii) $C^{min}$ and $C^{max}$ have a common intersection, and $Y_0$ is not inside it; we project $Y_0$ on the intersection with respect to the particular form of Y to obtain the value of the local transformed parameter Y;

iii) $C^{min}$ and $C^{max}$ do not have a common intersection; Y is choosen on the boundary of $C^{min}$ (centered at $S^{min} = 0$) with respect to the form of Y, in the direction of $S^{max}$.

### III.5 Global plastic strains and residual stresses

To obtain the residual state for the classical fields, ie $\underline{E}^p$ and $\underline{R}$ , we just have to solve an elastic boundary values problem with initial strains, and variable elastic constants. Where $E^p = 0$, the elastic constants are $Y_G$ and $\nu$, where Y is not equal to zero, the initial strain is Y/C and the alstic constants are $Y'_G$ and $\nu'$ .

For rolling and shot-peening, the previous explicit relations lead to the fields $\underline{E}^p$ and $\underline{R}$ , in the whole domain V. For roller-burnishing, the limit state obeys to plane strain in the plane Oyz, but we have to solve the problem numerically.

## IV CONCLUSION

These important industrial examples, for which numerical results will be showed, demonstrate the how this new approach can be applied succesfully.

## REFERENCES

[1] J. FRELAT, G. INGLEBERT, *"Modélisation du roulement et du grenaillage"* - chap. 15 - dans *"Nouveaux matériaux métalliques et nouveaux procédés de fabrication"*. Conseil National de Recherches - CANADA (1987).

[2] K.L. JOHNSON. *"Contact mechanics"*. Cambridge University Press (1985).

[3] J. MANDEL, J. ZARKA et B. HALPHEN, *"Adaptation d'une structure élastoplastique à écrouissage cinématique"*, Mech. Res. Comm. 4, 309, (1977).

[4] J. ZARKA, J. FRELAT, G. INGLEBERT and P. KASMAI-NAVIDI, *"A new approach in inelastic analysis of structures"*. Martinus Nijhoff Publishers (1988).

# USE OF ARBITRARY YIELD FUNCTIONS IN FEM

D. ZHOU and R. H. WAGONER
Department of Materials Science and Engineering
116 West 19th Avenue
The Ohio State University
Columbus, OH 43210-1179, U. S. A.

## ABSTRACT

Many standard yield functions allow derivation of closed-form equations relating the effective strain rate to principal strain rates. Such yield functions are easily implemented in finite element modeling (FEM) programs, which require explicit equations and directional derivatives. Newer anisotropic yield functions (for example, those introduced by Hosford, Bassani and Barlat) do not allow this simple approach. A numerical method is proposed to solve this problem. The method is found to be efficient, accurate and robust. As a demonstration of the capability, Hosford's yield function was introduced into a rigid-visco-plastic FEM program, **SHEET-3**, for calculation of forming limit diagrams (FLD) and simulation of punch stretching.

## INTRODUCTION

FEM Analysis of sheet forming operations relies on accurate knowledge of the plastic response and flow behavior of a material under a range of stress states. Continuum plasticity attempts to describe the stress-strain behavior of aggregate polycrystals on the basis of yield criteria without regard for internal structure. The yield criteria are subject to modifications to represent more closely the experimental results.

After Hill's 1948 yield criterion[1], many different anisotropic yield criteria have been proposed. So far, only a few of them, Hill's 1979 yield function case (IV)[2], for example, allow derivation of a closed-form equation relating the effective strain rate to principal strain rates. Newer anisotropic yield functions (Hosford, Bassani, Barlat, [3-5] e.g.) do not allow such a derivation. The absence of such a relationship makes introduction of such yield function into incremental FEM programs difficult because there is no closed-form integration of the constitutive equations available.

In this study, we present a general method for introducing an arbitrary yield function into FEM by using the effective stress expression alone, without knowing its associated effective strain rate expression explicitly. As a demonstration, Hosford's yield function is introduced into a 3-D rigid-visco-plastic finite element code, **SHEET-3**, previously developed at Ohio State University [6]. Sample results from the program are presented in this paper.

## PLASTICITY CONCEPTS

In a typical FEM program, the nodal velocity vector (v) found by using the virtual work principle is converted to an elemental strain rate vector ($\dot{\varepsilon}$ or D) based on the shape function. The only variables are the effective strain rate and its first derivative $\partial\dot{\bar{\varepsilon}}/\partial\dot{\varepsilon}_{ij}$. When the Newton-Raphson method is used for solving this nonlinear system, the second derivatives $\partial^2\dot{\bar{\varepsilon}}/\partial\dot{\varepsilon}_{ij}\partial\dot{\varepsilon}_{kl}$ are also required.

A numerical procedure is proposed to solve these problem. It consists of constructing a "normalized" effective strain rate surface from the yield function at a series of discrete stress ratios and then interpolating to obtain the required terms.

We define $\bar{\sigma}^N = 1$ and $\dot{\bar{\varepsilon}}^N = 1$ to be the "normalized" effective stress and effective strain rate, using the superscript 'N' to represent the normalized term. From plastic normality we have

$$\dot{\varepsilon}^N_{ij} = d\lambda\frac{\partial\bar{\sigma}^N}{\partial\sigma_{ij}} = \frac{\partial\bar{\sigma}^N}{\partial\sigma_{ij}} \qquad (1)$$

where $d\lambda$ is unity because of the normalization. The $\dot{\bar{\varepsilon}}^N_{ij}$ will be used to build the normalized effective strain rate surface in the strain rate space. By geometric mapping, the relation between the practical strain rate, $\dot{\varepsilon}$, in FEM and the normalized strain rate, $\dot{\varepsilon}^N$, in the same direction of strain rate space is given as:

$$\frac{\dot{\bar{\varepsilon}}}{\dot{\bar{\varepsilon}}^N} = \frac{\dot{\bar{\varepsilon}}_{ij}}{\dot{\bar{\varepsilon}}^N_{ij}} = K_\varepsilon = \dot{\bar{\varepsilon}} \qquad (2)$$

The proportional coefficient, $K_\varepsilon$, has the same value as the effective strain rate in the normalized condition. Another two interesting properties of the normalized yield locus are presented as follows:

$$\frac{\partial\dot{\bar{\varepsilon}}}{\partial\dot{\varepsilon}_{ij}} = \frac{\partial\dot{\bar{\varepsilon}}^N}{\partial\dot{\varepsilon}^N_{ij}} = \sigma^N_{ij} \quad \text{and} \quad \frac{\partial^2\dot{\bar{\varepsilon}}}{\partial\dot{\varepsilon}_{ij}\partial\dot{\varepsilon}_{kl}} = \frac{1}{k_\varepsilon}\frac{\partial^2\dot{\bar{\varepsilon}}^N}{\partial\dot{\varepsilon}^N_{ij}\partial\dot{\varepsilon}^N_{kl}} \qquad (3)$$

All the above-mentioned properties show that it is necessary to store only one set of data for the FEM calculations. Comparing the memory space used for run an FEM program with analytic derivatives, the memory space occupied for storing the yield function data is very little. For any point on the locus of normalized effective strain rate, one can obtain the first and second derivatives numerically by the equidistant differential and interpolation scheme.

## APPLICATION OF THE NUMERICAL METHOD

We first use the numerical and analytical methods to introduce Hill's new yield function, keeping all the input data (tool geometry, materials parameters, mesh,..) the same. Our testing shows that the numerical precision of the numerical method is equivalent to the analytical one and all the output results are identical within the digital accuracy of the computer. The computer time difference is less than 4% when identical problem with identical material laws are solved. Two examples are given in the following sections. Because the yield functions used are different in these examples, the cpu time difference is larger than that for the same yield function.

**1. Calculation of Forming Limit Diagrams:**
Figure 1 shows two FLDs calculated by FEM using Hosford's and Hill's yield functions, and one FLD determined by Marciniak and Kuszynski (M-K) model [7] using Hosford's yield function for different R values (R=0.7-1.8). A remarkable characteristic of Hosford's yield function is that the R value has a negligible effect on FLD compared to Hill's new yield function. A detailed discussion about the influence of R value on FLD can be found elsewhere [8]. Cpu times shown refers to CRAY Y-MP8/864 computer.

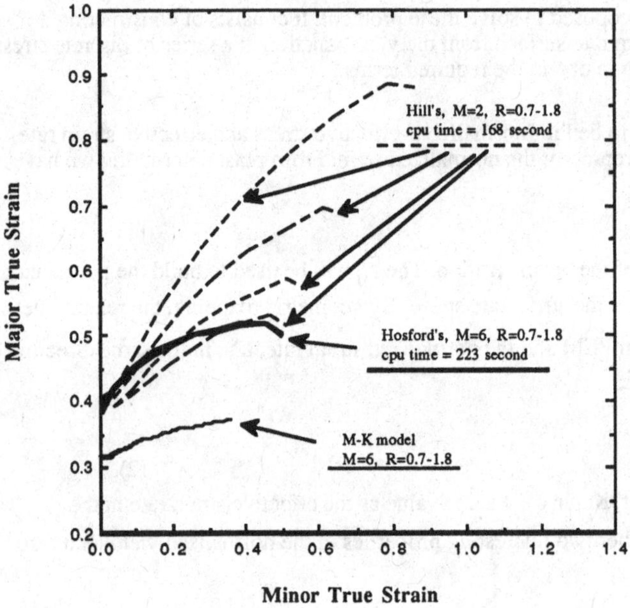


Figure 1, Simulated FLD's using the M-K model [7] and FEM based on Hill's and Hosford's yield functions. The Hosford calculations used the numerical technique.

**2. Strain Distribution of Hemispherical Punch Stretching**

Figure 2 compares strain distributions for hemispherical punch stretching by the numerical technique (Hosford) and closed-form technique (Hill) using **SHEET-3**. Cpu times on a DEC VAX 8550 Computer are also presented.

## CONCLUSIONS

A general method has been developed to introduce an arbitrary yield function into a rigid-visco-plastic FEM. Only the effective stress expression is needed. The effective strain and its first and second partial derivatives can be calculated by the new numerical method. Comparison of this numerical method with the analytical method in conjunction with FEM shows that the methods produce identical results at nearly the same cpu time. As examples of application of this general method, Hosford's and Hill's non-quadratic yield functions were introduced in calculate FLDs and simulate a punch-stretching problem.

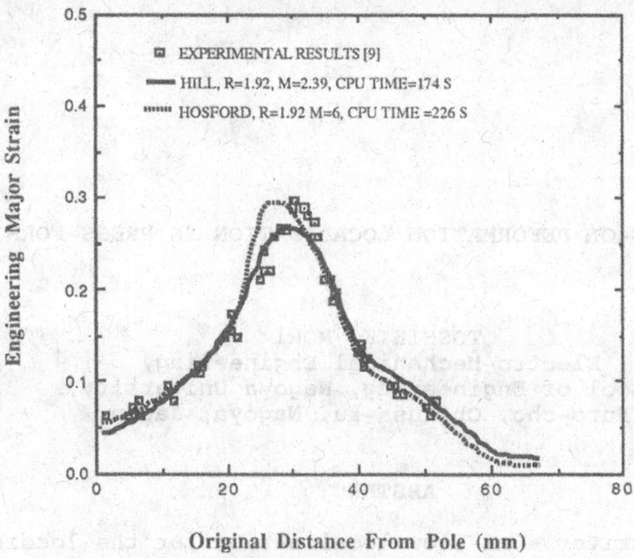


Figure 2, Comparison of the FEM results with experimental results for punch stretching by using Hosford's and Hill's yield function.

## REFERENCES


1. R. Hill, A Theory of the Yield Flow of Anisotropic Metals. Proc. R. Soc. London. 1948, **193A**, pp.281.
2. R. Hill, Theoretical Plasticity of Textured Aggregates. Math. Proc. Camb. Phil. Soc. 1979, **85**, pp.179.
3. W. F. HOSFORD and R. M. CADDELL, Metal Forming: Mechanics and Metallurgy, Prentice-Hill, Englewood Cliff, N. J., 1983. p.44M.
4. J. L. Bassani, Yield Characterization of Metals with Transversely Isotropic Plastic Properties, Int. J. Mech. Sci. 1977, **19**, pp. 651.
5. F. Barlat, A Six-Component Yield Function For Anisotropic Materials. Part I: Theoretical Derivation and Part II: Application To Aluminum Alloys, Submitted for publication in Int. J. Plasticity. (1990).
6. Y.Germain, K. Chung and R. H. Wagoner, A rigid-viscoplastic Finite Element Program For Sheet Metal Forming Analysis. Int. J. Mech. Sci. 1989, **31**, pp. 1.
7. Z. Marciniak and K. Kuszynski, Limit Strains in the Processes of Stretch-forming Sheet Metal. Int. J. Mech. Sci. 1967, **9**, pp.609.
8. K. Narasimhan, D.A. Burford and R. H. Wagoner, FEM Simulation of In-plane Forming Limit Diagram of Sheets Containing Realistic Finite Defects. Submitted to Metall. Trans.
9. J. R. Knibloe and R.H.Wagoner, Experimental Investigation and Finite Element Modeling of Hemispherically Stretched Steel Sheet, Metall. Trans., 1989, **20A**, pp.1509.

# SOME VIEWS ON DEFORMATION LOCALIZATION IN PRESS-FORMING

TOSHIHIKO MORI
Electro-Mechanical Engineering,
School of Engineering, Nagoya University,
Furo-cho, Chikusa-ku, Nagoya, Japan

## ABSTRACT

The definite criteria of flow localization for the loading path of positive strain ratio in sheet metal forming is not yet clarified theoretically and experimentally. This study deals with the onset of localized necking in such cases. It is pointed out that the onset occurs by the softening due to the gradient along the maximum stress direction and is promoted by the plane strain condition.

## INTRODUCTION

The formability of sheet metals is often controlled by flow localization, so it is important to clarify the instability criteria followed by localization. Although the classical Hill instability criteria for diffuse and local necking (1) are applicable when one of the two principal in-plane strains describing the deformation is negative and the other is positive, reliable criterion has not yet been proposed when both strains are positive. In various proposed criteria of local necking, the Marciniak-Kuczynski (2) analysis has been investigated by many researchers. This analysis assumes a uniform and proportional loading for simplicity, and postulates the presence of a material imperfection at which a shift in strain state from both positive strains to plane strain occurs and thence permits a localized necking to form. This study considers the gradient of the maximum stress along the direction which is usually encountered in practical forming. The softening grows at the peak point of maximum stress and becomes the localized necking. Moreover, if the plane strain state is close to the peak stress point from the beginning of deformation, the onset of localized necking becomes earlier.

## LOCALIZED NECKING DUE TO STRESS GRADIENT

In the M-K model, the gradient of the maximum stress arises by the imperfection whereby the cause may be material related, such as roughness. This leads to higher stress gradient which consequently grows the imperfection. Finally the localized necking occurs when the deformation outside the imperfection stops. However, if the stress gradient precedes, the onset of localized necking can occur without existence of the imperfections necessary to the M-K model.

This paper deals with the sheet metal forming by the hemispherical punch.

## STRESS STATE IN SHEET METAL FORMING BY HEMISPHERICAL PUNCH

According to Fig. 1, the following equation is derived by equilibrium of the forces acting in the meridian and thickness directions in a small element of sheet metal stretched by the hemispherical punch,

$$\frac{d\,\sigma_\phi}{d\,r} = \frac{\mu\,(\sigma_\phi + \sigma_\theta)}{r_p \cdot \cos\phi} - \frac{\sigma_\phi - \sigma_\theta}{r} - \frac{r \cdot \sigma_\phi}{t} \cdot \frac{d\,r}{d\,t} \qquad (1),$$

or

$$\frac{d\,\sigma_\phi}{d\,\phi} = \mu\,(\sigma_\phi + \sigma_\theta) - (\sigma_\phi - \sigma_\theta)\cot\phi - \frac{\sigma_\phi}{t} \cdot \frac{d\,t}{d\,\phi} \qquad (2),$$

Figure 2 shows schematic of the stress gradient along the meridian direction given by equation 2. The broken line shows first term of the right side due to friction between punch and metal, the dot-dash line shows second term due to the axisymmetrical flow, the double-dots-dash line shows third term due to the thickness distribution. The summation is the meridian stress gradient shown by the solid line. The stress gradient changes from positive to negative and the peak stress is at the end of the contact zone. The thickness of the contact surface decreases prior to the surface without contact (the thickness of the element spreading between the end of the punch and die contact zone) resulting in a higher peak stress. Figure 3 shows the meridian and circumferential stress distributions which are calculated by the calculus of finite differences of equation 2, using stress-strain relation of n

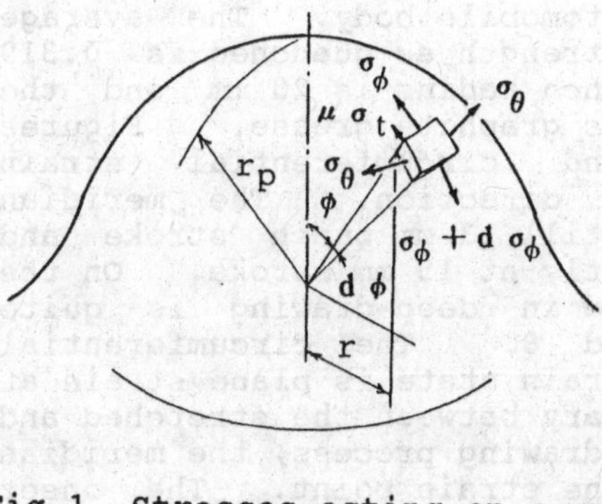


Fig.1 Stresses acting on a small element of sheet metal stretched hemispherical punch

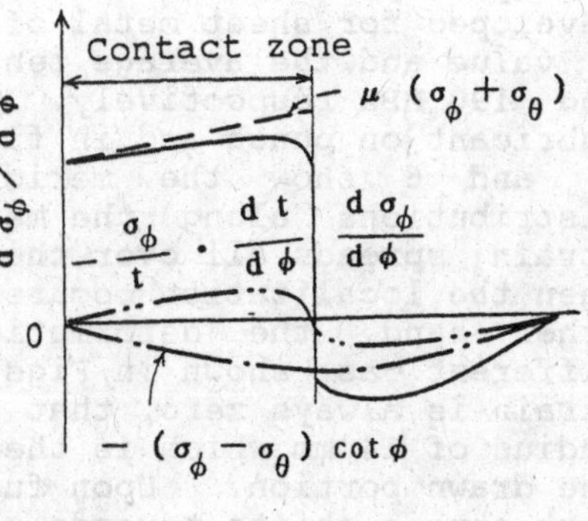


Fig. 2 Schematic of stress gradient along meridian direction

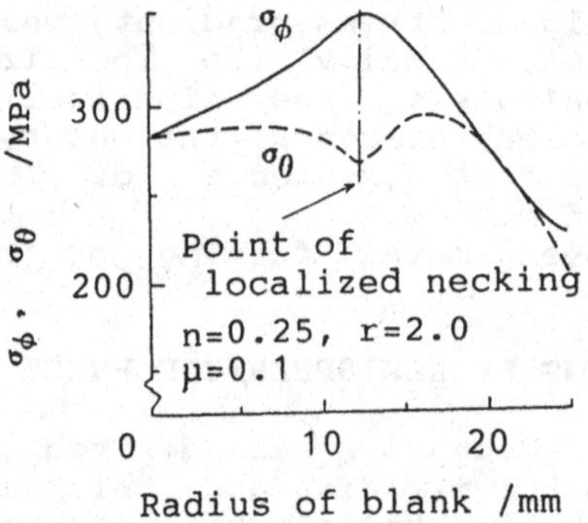


Fig.3 Stress distributions (Stretching)

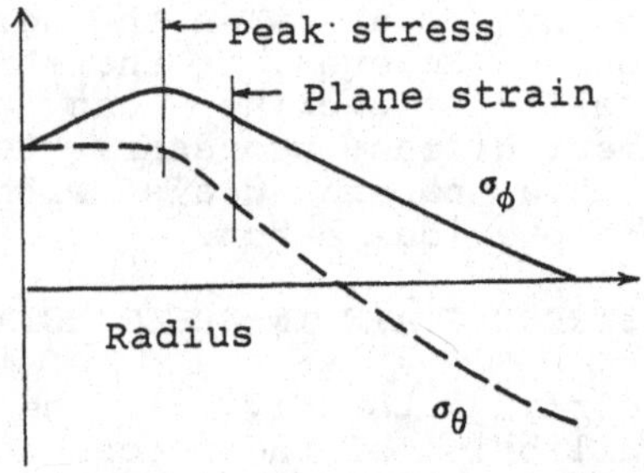


Fig.4 Stress distributions (Deep drawing)

power law and flow rule of strain increment. As the localized necking which occurs at the peak point of the meridian stress grows, the circumferential stress decreases locally and therefore the strain state at that point shifts towards the plane strain. Moreover, in the case of deep-drawing, the plane strain state occurs from the beginning of the process. Figure 4 shows schematically the meridian and circumferential stress distributions along the meridian direction. The point of the plane strain exists on the surface without contact. Upon further drawing process, the point of the meridian peak stress approaches the plane strain point and hence the onset of localized necking is promoted.

## OBSERVED STRAIN DISTRIBUTIONS ALONG MERIDIAN DIRECTION IN STRETCHING AND DEEP-DRAWING BY HEMISPHERICAL PUNCH

The process of localized necking in sheet metal forming by hemispherical punch was studied by the measurement of strain distributions. The test material is Al-Mg-Si series alloy, A6061 sheet (thickness is 0.8 mm). This material is one of the precipitation hardening type aluminum alloys which is developed for sheet metal of an automobile body. The average n value and the average tensile strength as quenched is 0.319 and 190 MPa respectively. The punch radius is 20 mm and the lubricant on punch and in flange is graphite grease. Figures 5 and 6 show the meridian and circumferential strain distributions along the meridian direction. The meridian strain spreads all over the area till 13 mm punch stroke and then the localization occurs slightly at 15 mm stroke. On the other hand, the deformation mode in deep-drawing is quite different as shown in Figs. 7 and 8. The circumferential strain is always zero, that is, strain state is plane strain at radius of 14 mm which is the boundary between the stretched and the drawn portion. Upon further drawing process, the meridian peak strain shifts towards the plane strain point. The onset of localized necking occurs abruptly when both points coincide.

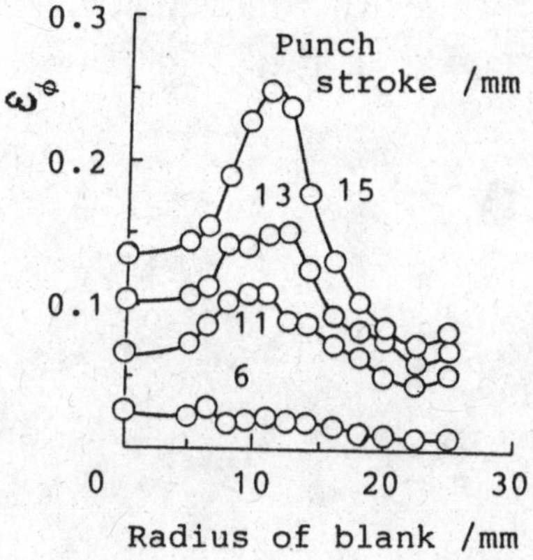


Fig.5 Meridian strain distribution (Stretching)

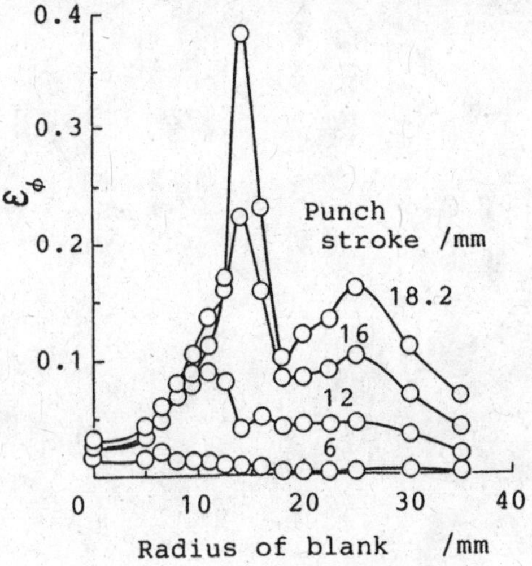


Fig.7 Meridian strain distribution (Deep drawing)

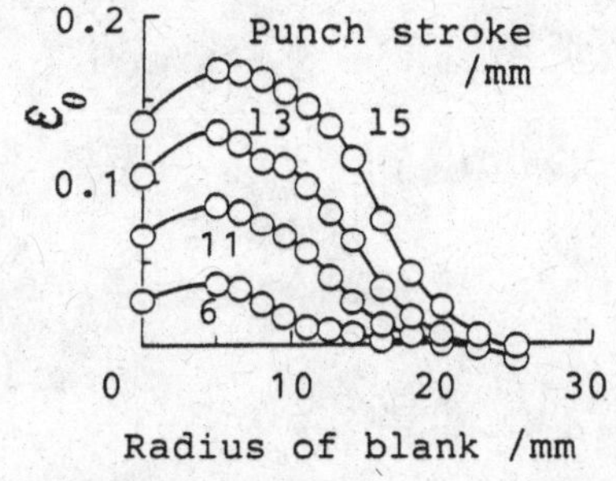


Fig.6 Circumferential strain distribution (Stretching)

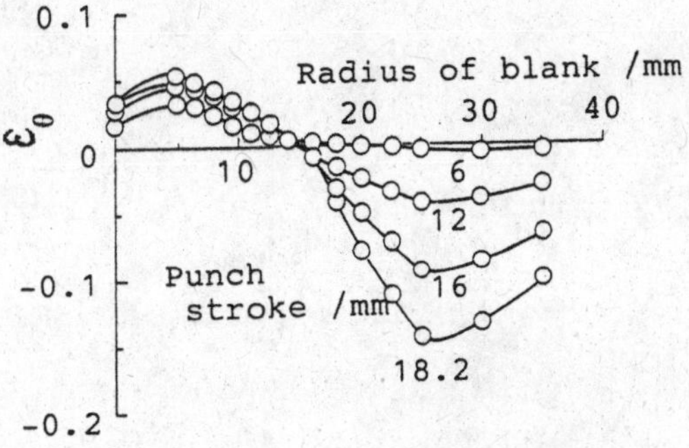


Fig.8 Circumferential strain distribution (Deep drawing)

## REFERENCES


1. Hill, R., J. Mech. Phys. Solids, 1952, 1, pp1-6.
2. Marciniak, Z. and Kuczynski, K., Int. J. Mech. Sci., 1967, 9, pp609-614.

# AUTHOR INDEX

Printed in the United States
By Bookmasters